Nanodiamond

RSC Nanoscience & Nanotechnology

Editor-in-Chief:
Paul O'Brien FRS, *University of Manchester, UK*

Series Editors:
Ralph Nuzzo, *University of Illinois at Urbana-Champaign, USA*
Joao Rocha, *University of Aveiro, Portugal*
Xiaogang Liu, *National University of Singapore, Singapore*

Honorary Series Editor:
Sir Harry Kroto FRS, *University of Sussex, UK*

Titles in the Series:
  1: Nanotubes and Nanowires
  2: Fullerenes: Principles and Applications
  3: Nanocharacterisation
  4: Atom Resolved Surface Reactions: Nanocatalysis
  5: Biomimetic Nanoceramics in Clinical Use: From Materials to Applications
  6: Nanofluidics: Nanoscience and Nanotechnology
  7: Bionanodesign: Following Nature's Touch
  8: Nano-Society: Pushing the Boundaries of Technology
  9: Polymer-based Nanostructures: Medical Applications
  10: Metallic and Molecular Interactions in Nanometer Layers, Pores and Particles: New Findings at the Yoctolitre Level
  11: Nanocasting: A Versatile Strategy for Creating Nanostructured Porous Materials
  12: Titanate and Titania Nanotubes: Synthesis, Properties and Applications
  13: Raman Spectroscopy, Fullerenes and Nanotechnology
  14: Nanotechnologies in Food
  15: Unravelling Single Cell Genomics: Micro and Nanotools
  16: Polymer Nanocomposites by Emulsion and Suspension
  17: Phage Nanobiotechnology
  18: Nanotubes and Nanowires: 2nd Edition
  19: Nanostructured Catalysts: Transition Metal Oxides
  20: Fullerenes: Principles and Applications, 2nd Edition
  21: Biological Interactions with Surface Charge Biomaterials
  22: Nanoporous Gold: From an Ancient Technology to a High-Tech Material
  23: Nanoparticles in Anti-Microbial Materials: Use and Characterisation
  24: Manipulation of Nanoscale Materials: An Introduction to Nanoarchitectonics

25: Towards Efficient Designing of Safe Nanomaterials: Innovative Merge of Computational Approaches and Experimental Techniques
26: Polymer–Graphene Nanocomposites
27: Carbon Nanotube-Polymer Composites
28: Nanoscience for the Conservation of Works of Art
29: Polymer Nanofibers: Building Blocks for Nanotechnology
30: Artificial Cilia
31: Nanodiamond

How to obtain future titles on publication:
A standing order plan is available for this series. A standing order will bring delivery of each new volume immediately on publication.

For further information please contact:
Book Sales Department, Royal Society of Chemistry, Thomas Graham House, Science Park, Milton Road, Cambridge, CB4 0WF, UK
Telephone: +44 (0)1223 420066, Fax: +44 (0)1223 420247
Email: booksales@rsc.org
Visit our website at www.rsc.org/books

Nanodiamond

Edited by

Oliver Williams
Cardiff School of Physics and Astronomy, Queen's Buildings, Cardiff
Email: Oliver.Williams@astro.cf.ac.uk

RSC Nanoscience & Nanotechnology No. 31

ISBN: 978-1-84973-639-8
ISSN: 1757-7136

A catalogue record for this book is available from the British Library

© The Royal Society of Chemistry 2014

All rights reserved

Apart from fair dealing for the purposes of research for non-commercial purposes or for private study, criticism or review, as permitted under the Copyright, Designs and Patents Act 1988 and the Copyright and Related Rights Regulations 2003, this publication may not be reproduced, stored or transmitted, in any form or by any means, without the prior permission in writing of The Royal Society of Chemistry or the copyright owner, or in the case of reproduction in accordance with the terms of licences issued by the Copyright Licensing Agency in the UK, or in accordance with the terms of the licences issued by the appropriate Reproduction Rights Organization outside the UK. Enquiries concerning reproduction outside the terms stated here should be sent to The Royal Society of Chemistry at the address printed on this page.

The RSC is not responsible for individual opinions expressed in this work.

Published by The Royal Society of Chemistry,
Thomas Graham House, Science Park, Milton Road,
Cambridge CB4 0WF, UK

Registered Charity Number 207890

For further information see our web site at www.rsc.org

Preface

Nanodiamond has a rich diversity of forms ranging from particles of a few nanometres to films constituting crystallites of hundreds of nanometres and microns thick. The allotropic behaviour of carbon leads to complicated structures at the nanoscale comprised of both sp^3 and sp^2 bonding. Unfortunately, the variety of such forms results in an inevitable layer of jargon, which can be unhelpful for the uninitiated. One of the easiest ways to diffuse such confusion is to divide the field with clear definitions, something that has been especially opaque with nanodiamond films. The images below are an effort in this direction, subdividing the fields of nanodiamond films and particles.

Diamonoids	Ultra-dispersed diamond (UDD)	Sub micron diamond grit
< 1 nm	< 10 nm	> 20 nm
Purification of petroleum	Detonation synthesis	Mechanical size reduction of bulk diamond

Ultra-nanocrystalline diamond	Nanocrystalline diamond
< 10 nm	< 100nm
Re-nucleation results in no evolution of surface roughness	Little or no re-nucleation, films roughens with thickness

The smallest known diamond-structured particles are in fact not strictly speaking diamonds, but diamondoids. These particles are generally isolated from petroleum, being a source of many problems in natural gas, gas condensates and light crude oil flow systems, where they can act as flocculation sources, blocking flow paths. A single adamantane molecule weighs 10^{-21} carats, i.e., 2×10^{-22} g. Adamantane is not truly a diamond as every carbon atom is at the surface and thus is bonded to at least one hydrogen atom. To date, there are few applications of adamantane and the higher diamondoids and, as they are not true diamonds, they will not be discussed in detail in this book.

The next biggest diamond particles, and thus the smallest true diamonds, are made from detonation synthesis, such as with trinitrotoluene (TNT) and hexogen (see Chapters 1 to 5). During the detonation shockwave, the pressure and temperature reach the stability region of diamond in the phase diagram. Thus, for a brief moment of typically around 1 µs, diamond particles can be grown. These particles are usually called ultra-dispersed diamond (UDD) and their mean size is quoted as around 4 nm. This size results in a specific surface area greater than 400 m^2 g^{-1}, with more than 15% of UDD particle carbon atoms located at the surface. This has profound implications on the surface chemistry and stability of such particles. It has been shown that diamond is actually energetically favoured over polycyclic aromatics for diameters of less than 3 nm with hydrogen termination. Thus, at this length scale, diamond cannot truly be said to be meta-stable. The reactivity of such fine particles can also differ substantially from bulk diamond surfaces.

Finally, as one gets to sizes of greater than 20 nm, nanodiamond particles behave like bulk diamond. This is predominantly due to the far reduced concentration of atoms at the surface with regards to the bulk. These diamonds are usually produced from top down methods, such as jet milling, or the abrasion of larger diamonds, which is an expensive and relatively time-consuming process. However, their quality generally exceeds that of smaller diamonds due to their reduced surface-to-volume fraction. Their Raman and X-ray diffraction spectra are far more reminiscent of bulk diamond than there smaller counterparts. These types of nanodiamond have been exploited in the abrasives industry for decades.

For films, the difference is very much complicated by mostly historical terminology. In the earlier days of diamond growth, when chemical vapour deposition (CVD) reactor design was in its infancy and nucleation densities were very low, nanocrystalline diamond was a name given to thin diamond films, which generally had low quality. Diamond growth evolved and high-quality single crystal and microcrystalline films dominated research, leaving nanocrystalline diamond very much in the background. Recently, nanocrystalline diamond has developed into a sophisticated material with a wide variety of applications and associated terminology (see Chapters 10 to 20). Figure 2 attempts to clarify some of this terminology.

The smallest grain size diamond films are called ultrananocrystalline diamond (UNCD), a term originating from Argonne National Laboratory. These films have grain sizes around 5 nm, with a considerable amount of amorphous grain boundaries, which are very similar to diamond-like carbon (DLC). DLC is, strictly speaking, not diamond but is included in the table for clarity, and because it has many similarities with UNCD. DLC has no crystalline structure, as seen in the TEM image (Figure 2), whereas UNCD clearly exhibits the lattice planes of diamond, with amorphous DLC-like regions between grains. UNCD is very much a special case, with all other types of nanodiamond film being termed nanocrystalline diamond (NCD). NCD films have grain sizes generally below 100 nm, but sometimes films with grains up to 500 nm are also labelled NCD. Generally speaking, NCD films contain less sp^2 hybridisation and are thus more transparent than UNCD films; this is particularly acute when the films are grown without re-nucleation, *i.e.*, with a low methane concentration and high power density.

Perhaps the most convincing definition between the various forms of diamond is in their resulting properties. In this way it is easy to distinguish between nano-carbons, as measurements such as Young's modulus, optical transparency, thermal conductivity *etc.* are objective real world properties that can be quantified and exploited in real world applications. Ultimately the real world application arena is the true measure of a useful material and relegates all ambiguity and argument about material classification to semantics.

This book aims to clarify some of the idiosyncrasies of the field of nanodiamond, and highlight the wide application space within which it is exploited. The majority of this field is covered within these pages, with self-contained chapters authored by known experts. These chapters overlap in places, due to the encouraging merging of fundamental and applied science. The interdisciplinary nature of materials science blurs this boundary even further, and for this reason chapters are not grouped into distinct sub sections.

The first few chapters focus on the production, purification, and fundamental properties of nanodiamond particles, with routes to applications (see Chapters 1 to 6). The chapters that follow detail the broad horizon space of nanodiamonds from electrochemistry to cell labelling and drug delivery (see Chapters 6 to 10). The diversity of the applications space of nanodiamond particles is startling, and impossible to cover exhaustively, not least as new ideas surface frequently. However, only a detailed understanding of the surface properties can lead to the full exploitation of this unique material, which is still work in progress.

The nanodiamond film section begins with nucleation, a section related to both films and particles and a key area where diamond nanoparticles have made a major impact. The chapters that then follow deal with the growth and doping of films, with some attention to new low-temperature growth capabilities (Chapters 10 to 14). The applications of nanodiamond films are probably even more diverse than nanodiamond particles and certainly well established. The chapters in this book deal almost entirely with doped films with electrochemistry, superconductivity, and field emission (Chapters 15 to 20).

It is hoped that the chapters within this book, coupled with their citations, will help in some way to clarify the field of nanodiamond. At the very least, it is envisioned that this text will serve as a reliable entry point to the various disciplines of the field.

Oliver A. Williams, Cardiff

Contents

Chapter 1	**Distribution, Diffusion and Concentration of Defects in Colloidal Diamond** *Amanda S. Barnard*	**1**
	1.1 Introduction	1
	1.2 Defect-free Diamond Nanoparticles	3
	1.3 Modelling Defects in Diamond Nanoparticles	7
	1.3.1 Mechanical and Thermodynamic Stability	8
	1.3.2 Kinetic Stability and Probability of Observation	9
	1.4 Point Defects	10
	1.4.1 Intrinsic Defects	10
	1.4.2 Incidental Impurities	13
	1.4.3 Photoactive N–V Centres	14
	1.4.4 H3 Centres, N3 Centres and V–N–V Defects	17
	1.5 Comparison and Concentration	20
	1.6 Conclusions	22
	Acknowledgments	23
	References	23
Chapter 2	**Detonation Nanodiamonds: Synthesis, Properties and Applications** *A. Ya. Vul', A. T. Dideikin, A. E. Aleksenskii and M. V. Baidakova*	**27**
	2.1 Introduction	27
	2.2 Main Features of DND Technology	28
	2.2.1 Synthesis of Detonation Carbon	28
	2.2.2 Isolation of Nanodiamonds from Detonation Carbon	30

	2.3	Structure of Detonation Soot Particles and Nanodiamonds	33
	2.4	DND Suspensions	35
	2.5	DND Applications	36
	2.6	Unsolved Problems	41
	2.7	Conclusion	43
	Acknowledgements		43
	References		44

Chapter 3 The Chemistry of Nanodiamond — 49
Anke Krueger

3.1	The Surface Structure of Nanodiamond		49
	3.1.1	The Initial Surface Structure	49
	3.1.2	Agglomeration and Deagglomeration of Nanodiamond	51
3.2	Modifying the Surface Termination of Nanodiamond Particles		53
	3.2.1	Direct Termination of the Nanodiamond Surface	54
	3.2.2	Using Linkers for the Surface Modification of Nanodiamond	68
3.3	Grafting of Complex Moieties onto Pre-functionalized Nanodiamond		76
	3.3.1	Peptide Coupling onto Nanodiamond	77
	3.3.2	Click Chemistry on Nanodiamond	78
	3.3.3	Miscellaneous Reactions for the Surface Modification of Nanodiamond	80
3.4	Summary and Outlook		81
References			82

Chapter 4 Nanodiamond Purification — 89
Sebastian Osswald

4.1	Purification of Nanodiamond Powders		89
	4.1.1	Introduction	89
	4.1.2	Nanodiamond Structure and Composition	90
	4.1.3	Assessment of Nanodiamond Purity	92
	4.1.4	Thermal Stability and Oxidation Behavior of Nanodiamond	94
4.2	Chemical Purification of Nanodiamond		96
	4.2.1	Liquid-phase Purification	98
	4.2.2	Gas-phase Purification	99
	4.2.3	Effect of Metal Impurities on Oxidation Behavior	100

4.3	Properties of Purified Nanodiamond		102
	4.3.1	Structure, Composition, and Surface Chemistry	102
	4.3.2	Optical and Electrical Properties	105
4.4	Summary and Outlook		107
	4.4.1	Purity Assessment	107
	4.4.2	Purification Treatment	108
References			109

Chapter 5 Pure Nanodiamonds Produced by Laser-assisted Technique 112
Boris Zousman and Olga Levinson

5.1	Introduction, Nanodiamond Market, Problems and Prospects		112
5.2	Synthesis Nanodiamonds by Laser Ablation in Liquid		114
5.3	Light Hydro-dynamic Pulse for Nanodiamond Fabrication		116
	5.3.1	Technological Process	116
	5.3.2	Physical Mechanism	117
5.4	Characterization of Nanodiamonds Obtained by Laser Synthesis		117
	5.4.1	X-Ray Diffraction (XRD) Analysis	117
	5.4.2	Scanning Electron Microscopy	119
	5.4.3	Transmission Electron Microscopy	120
	5.4.4	Raman Spectroscopy	122
	5.4.5	Inductively Coupled Plasma Mass Spectrometry	122
5.5	Controlled Nanodiamond Synthesis		123
5.6	LHDP *vs.* Existing Technology for Detonation Nanodiamond Synthesis		124
5.7	Conclusion and Prospects in the Development of the Nanodiamond Industry		125
Acknowledgments			125
References			125

Chapter 6 Electrochemistry of Nanodiamond Particles 128
Katherine B. Holt

6.1	Introduction	128
6.2	Diamond in Electrochemistry	129
6.3	Structure and Surface Chemistry of Nanodiamond	129

	6.4	Electrochemical Behaviour of Diamond Nanoparticles	130
		6.4.1 Electrode Preparation Methods for the Study of Nanodiamond Particles	130
		6.4.2 Electrochemical Response of Electrode-immobilised Nanodiamond Particles	131
	6.5	Electronic Structure of the Nanodiamond Surface	135
	6.6	Interaction of Nanodiamond Particles with Solution Redox Species	136
		6.6.1 Interaction with $Fe(CN)_6^{4-/3-}$	136
		6.6.2 Interaction of Nanodiamond Particles with Other Solution Redox Species	140
	6.7	*In Situ* Spectroscopy Studies of the Nanodiamond Surface in Redox Solutions	141
	6.8	Electrochemical Impedance Spectroscopy Studies of Nanodiamond Particles	142
	6.9	Applications of Nanodiamond in Electrochemical Technologies	143
		6.9.1 Nanodiamond and Electrochemical Sensors	143
		6.9.2 Nanodiamond in Biosensors	143
		6.9.3 Nanodiamond as Fuel Cell Catalyst Electrode Supports	144
		6.9.4 Nanodiamond as a Supercapacitor Electrode Material	145
		6.9.5 Nanodiamond in Electrode Coatings and Composites	146
		6.9.6 Nanodiamond in Photocatalysis	146
	6.10	Electrochemistry of Other Diamond Nanoparticles	146
		6.10.1 Boron-doped Nanodiamond	146
		6.10.2 Electrochemistry with 100 nm Diamond Particles	147
		6.10.3 Electrochemical Studies with HTHP Type 1b Diamond	147
	6.11	Conclusions and Outlook	148
		References	148
Chapter 7	**Nanodiamonds for Drug Delivery and Diagnostics**		**151**
	Han Man, Joshua Sasine, Edward K. Chow and Dean Ho		
	7.1	Introduction	151
	7.2	Nanodiamond Systemic Drug Delivery	152
		7.2.1 Nanodiamond–Doxorubicin Complexes	152
		7.2.2 Targeted Nanodiamond Drug Delivery	153

Contents   xv

7.3	Nanodiamond-based Implantable Devices	154
	7.3.1 Nanodiamond–Chemotherapeutic Tablets	155
	7.3.2 Nanodiamond–PLLA Complexes for Tissue Engineering	156
7.4	Nanodiamonds and Imaging	156
	7.4.1 Fluorescent Nanodiamonds for Bio-imaging Applications	157
	7.4.2 Nanodiamonds for Magnetic Resonance Imaging	158
7.5	Nanodiamond Safety and Biocompatibility Studies	158
	7.5.1 Pre-clinical Evaluation of Nanodiamond Safety	159
	7.5.2 *In Vitro* Validation of Nanodiamond Safety	159
7.6	Nanodiamonds and Diagnostics	160
	7.6.1 Novel Approaches for Nanodiamond-based Detection	160
7.7	Modeling of Nanodiamond Vehicles for Optimized Applications	161
	7.7.1 Nanodiamond-based Gene Delivery	161
7.8	Concluding Remarks	163
Acknowledgements		163
References		164

Chapter 8 Biophysical Interaction of Nanodiamond with Biological Entities *In Vivo* 170
J. Mona, E. Perevedentseva and C.-L. Cheng

8.1	Introduction	170
8.2	Nanodiamond Surface Functionalization/Conjugation with Biomolecules	171
8.3	Essential Characterizations for the Biophysical Interactions of Nanodiamond	173
8.4	Interactions of Nano-carbon with Unicellular Microorganisms	177
	8.4.1 Interactions with Bacteria	178
	8.4.2 Interactions with Protozoa	184
8.5	Summary	187
References		188

Chapter 9 Neuron Growth on Nanodiamond 195
Robert Edgington and Richard B. Jackman

9.1	Introduction: Neuronal Biomaterials	195

9.2	\multicolumn{3}{l	}{The Use of Nanodiamond Monolayer Coatings to Promote the Formation of Functional Neuronal Networks}
			197
	9.2.1	Introduction	197
	9.2.2	Neuronal Cell Attachment on ND Layers	198
	9.2.3	Substrate Dependence of Surface Roughness Following ND Layering	199
	9.2.4	Neuronal Attachment and Outgrowth on Different Materials	201
	9.2.5	Formation of Neuronal Networks	201
	9.2.6	Intrinsic Electric Excitability of ND-grown Neurons	203
	9.2.7	Synaptic Connectivity	204
	9.2.8	Electrochemical Network: Calcium Oscillations	205
	9.2.9	Discussion	206
9.3	\multicolumn{3}{l	}{The Effect of Varying Nanodiamond Properties on Neuronal Adhesion and Outgrowth}
			207
	9.3.1	Introduction	207
	9.3.2	Coating of Different Nanodiamond Coatings and Characterisation	208
	9.3.3	Neuronal Cell Attachment on ND Coatings	211
	9.3.4	Mechanism of Neuronal Adhesion with Respect to ND Surface Properties	213
9.4	Patterned Neuronal Networks Using Nanodiamonds		214
9.5	Conclusions		216
References			217

Chapter 10 Diamond Nucleation and Seeding Techniques: Two Complementary Strategies for the Growth of Ultra-thin Diamond Films 221
J. C. Arnault and H. A. Girard

10.1	Introduction		221
10.2	Diamond Nucleation		222
	10.2.1	What is Nucleation?	222
	10.2.2	Diamond Homogeneous Nucleation	223
	10.2.3	Diamond Heterogeneous Nucleation	223
	10.2.4	Different Approaches for Enhanced Diamond Nucleation	227
10.3	Nanodiamond Seeding		233
	10.3.1	Abrasion Techniques Involving Particles	233
	10.3.2	Controlled Deposition of Nanodiamonds on Substrates	234
10.4	Conclusion		244
References			245

Contents xvii

Chapter 11 The Microstructures of Polycrystalline Diamond, Ballas and Nanocrystalline Diamond   253
Roland Haubner

    11.1 Introduction   253
    11.2 Polycrystalline Diamond   254
    11.3 Low-pressure Diamond   255
          11.3.1 Characterisation of the Diamond Deposits   255
          11.3.2 Growth Mechanism for the Different Diamond Morphologies   258
    11.4 Correlation of Growth Conditions with Diamond Morphology   261
          11.4.1 Gas Activation (Atomic Hydrogen and Carbon Species)   261
          11.4.2 Gas-phase Composition   262
          11.4.3 Influence of Impurities   262
    11.5 Summary and Conclusions   263
    References   265

Chapter 12 Low-temperature Growth of Nanocrystalline Diamond Films in Surface-wave Plasma   268
Kazuo Tsugawa and Masataka Hasegawa

    12.1 Introduction   268
    12.2 Surface-wave Plasma CVD   270
    12.3 Nanocrystalline Diamond Grown in Surface-wave Plasmas   273
    12.4 Growth Mechanism   277
    12.5 Tribological Properties   282
    12.6 Summary   287
    References   288

Chapter 13 Low Temperature Diamond Growth   290
Tibor Izak, Oleg Babchenko, Stepan Potocky, Zdenek Remes, Halyna Kozak, Elisseos Verveniotis, Bohuslav Rezek and Alexander Kromka

    13.1 Introduction   290
    13.2 Synthesis of Diamond Thin Films   292
          13.2.1 CVD Diamond Process Temperature: Limiting Factor for Substrates   292
    13.3 Low Temperature Diamond Growth   297
          13.3.1 Strategy 1: Modification of CVD Deposition Systems   297
          13.3.2 Strategy 2: Influence of Gas Chemistry   302

13.4	Determination of Substrate Temperature and Calculation of Activation Energy		307
	13.4.1	Substrate Temperature	307
	13.4.2	Activation Energy	309
13.5	LTDG in Pulsed Linear Antenna Microwave Plasma		312
13.6	Diamond Growth on Amorphous Silicon		317
	13.6.1	Nucleation and Growth of Diamond on a-Si	317
	13.6.2	Selective Growth of Diamond Nanocrystals	320
13.7	Diamond Growth on Glass		322
	13.7.1	Diamond Growth Using Focused MW Plasma	323
	13.7.2	Diamond Growth Using PLAMWP	326
13.8	Diamond Growth on Optical Elements for IR Spectroscopy		327
	13.8.1	Diamond Growth on Metallic Optical Mirrors for GAR-FTIR	329
	13.8.2	Diamond Growth on Si Prism for ATR-FTIR Applications	331
	13.8.3	Diamond Growth on Ge Substrates	333
13.9	Conclusions		335
Acknowledgements			336
References			337

Chapter 14 P-type and N-type Conductivity in Nanodiamond Films   343
Oliver A. Williams

14.1	Introduction	343
14.2	Film Structure and sp^2 Content	344
14.3	Film Conductivity	346
14.4	Carrier Transport in NCD Films	348
14.5	Optical Properties of NCD Films	351
14.6	Conclusions	352
References		352

Chapter 15 Electrochemistry of Nanocrystalline and Microcrystalline Diamond   354
Inga V. Shpilevaya and John S. Foord

15.1	Introduction		354
15.2	Electrochemical Measurements		355
15.3	Diamond Films in Electrochemistry		358
15.4	Electrochemistry of Nanodiamond Powder Films		362
15.5	Functionalised Diamond Electrodes		369
	15.5.1	Types of Diamond Surface Termination	371
	15.5.2	Photochemical, Electrochemical and Chemical Functionalisation	375

	15.6	Adsorption Properties of Diamond Films	378
	15.7	Concluding Remarks	380
	References		381

Chapter 16 Superconductivity in Nanostructured Boron-doped Diamond and its Application to Device Fabrication — 385
Soumen Mandal, Tobias Bautze and Christopher Bäuerle

16.1	Introduction	385
16.2	Nanofabrication of Polycrystalline Boron-doped Diamond	387
16.3	Low-temperature Characterization of Nanofabricated Samples	389
16.4	Superconducting Quantum Interference Device Made From Superconducting Diamond	393
	16.4.1 Low Temperature Studies on Non-shunted Devices	395
	16.4.2 First Trials on Shunted Devices	398
16.5	Nanomechanical Systems	402
16.6	Summary and Conclusions	407
Acknowledgement	407
References	407

Chapter 17 Diamond Nano-electromechanical Systems — 411
Pritiraj Mohanty and Matthias Imboden

17.1	Introduction	411
	17.1.1 What NEMS Do	412
	17.1.2 Why They Work	412
	17.1.3 Diamond: The Ideal NEMS Material	412
17.2	NEMS Dynamics	414
	17.2.1 Elasticity Theory and the Euler–Bernoulli Equations	414
	17.2.2 Dynamic Solutions	415
	17.2.3 Beyond the Standard Euler–Bernoulli Equation	419
	17.2.4 The Simple Harmonic Oscillator and Nonlinear Terms	419
17.3	Methods of Fabrication, Actuation, and Detection	422
	17.3.1 Fabrication	422
	17.3.2 Drive and Detection	425
17.4	Dissipation in Diamond NEMS	432
	17.4.1 Clamping Losses (Q_{CL}^{-1})	433
	17.4.2 Thermoelastic Dissipation (Q_{TE}^{-1})	434

		17.4.3	Circuit Loading (Q_{CL}^{-1}) and Multiple Materials (Q_{MM}^{-1})	435
		17.4.4	Mechanical Defects, Surfaces (Q_{SL}^{-1}) and Bulk (Q_{MD}^{-1})	437
		17.4.5	Viscous Damping (Q_{VD}^{-1})	438
		17.4.6	Quantum Dissipation at Ultra-low Temperatures	439
	17.5	Novel Diamond NEMS Devices and Future Capabilities		441
		17.5.1	Field Emission in Diamond	442
		17.5.2	Superconductivity and Diamond NEMS	443
		17.5.3	The Nitrogen Vacancy Defect in Diamond	444
	17.6	Conclusions		444
	References			445

Chapter 18 Diamond-based Resonators for Chemical Detection — 448
Emmanuel Scorsone and Adeline Trouvé

	18.1	Introduction		448
	18.2	Diamond Materials: Some Remarkable Properties for the Development of High Performance Chemical Sensors		450
		18.2.1	Physical Properties	450
		18.2.2	Chemical Properties	453
	18.3	Diamond Cantilevers		457
		18.3.1	Transduction Principles	457
		18.3.2	Fabrication Methods	461
		18.3.3	Diamond-based Resonant Cantilever Chemical Sensors	461
	18.4	Surface Acoustic Wave Resonators		465
		18.4.1	Generality	465
		18.4.2	Diamond Nanoparticles-coated SAW Chemical Sensors	467
		18.4.3	Toward Artificial Olfaction Using Diamond-based SAW Sensors	470
	18.5	Conclusions		472
	References			472

Chapter 19 All-diamond Electrochemical Devices: Fabrication, Properties, and Applications — 476
Nianjun Yang, Waldemar Smirnov and Jakob Hees

	19.1	Introduction	476

19.2	Fabrication		478
	19.2.1	Microelectrode Arrays and Ultramicroelectrode Arrays	478
	19.2.2	Nanoelectrode Ensembles and Arrays	480
	19.2.3	Atomic Force Microscope–Scanning Electrochemical Microscope (AFM–SECM) Tip	482
19.3	Properties		484
	19.3.1	Microelectrode Arrays and Ultramicroelectrode Arrays	484
	19.3.2	Nanoelectrode Ensembles and Arrays	487
	19.3.3	Atomic Force Microscope–Scanning Electrochemical Microscope (AFM–SECM) Tip	488
19.4	Applications		491
19.5	Summary and Outlook		491
Acknowledgements			492
References			492

Chapter 20 Electron Field Emission from Diamond — 499
Travis C. Wade

20.1	Mechanisms of Electron Emission		499
20.2	Field Emission		500
20.3	Fowler–Nordheim Field Emission Theory		500
20.4	Improvements upon the F–N Form		500
20.5	Factors Relevant to Electron Emission from Diamond		502
	20.5.1	Topology/Geometric Enhancement	502
	20.5.2	Temperature	504
	20.5.3	Surface States and Electron Affinity	504
	20.5.4	Carrier Transport	508
References			511

Subject Index — 516

CHAPTER 1

Distribution, Diffusion and Concentration of Defects in Colloidal Diamond

AMANDA S. BARNARD

CSIRO Materials Science and Engineering, 343 Royal Parade, Parkville, Victoria, 3052, Australia
Email: amanda.barnard@csiro.au

1.1 Introduction

It is often convenient to think of nanodiamond as pure, and free of defects, but this is not necessarily realistic. Nanodiamonds can (and do) contain a variety of defects, whether we want them there or not. These include intrinsic point defects, such as lattice vacancies, and incidental impurities, such as nitrogen, which are a result of the synthesis and/or purification processes. In general, defects are always thermodynamically unstable, but the relative (in)stability of these defects, and hence the probability that they can be removed from the particle, can depend on the location of the defect within the particle. This is quite different to the case of bulk diamond, where all lattice sites are geometrically (and, therefore, energetically) equivalent.

There are of course, types of defects that are very useful, and are therefore introduced deliberately. Well known examples are the p-type or n-type dopants used in electronic applications, but there are other types of useful point defects that are not dopants. Collectively these are often referred to as "functional defects" (as they provide some functionality), and include the range of optically active defects and colour centres.[1-3] The most simple defect in diamond is a single, neutral lattice vacancy, which is commonly referred to as a GR1 defect (where GR stands for general radiation). Vacancies are omnipresent in diamond, and so this defect has been extensively studied in various states.[4]

Since nitrogen is also widespread in diamond, numerous studies have also focused on characterising and understanding the properties of different types of N-related defects, including the single substitutional nitrogen impurity, known as the C-centre. However, arguably, the most widely studied defect in nanocrystalline diamond is the paramagnetic nitrogen–vacancy complex (N–V),[5,6] which forms when a vacancy (GR1) migrates to bind with a C-centre.[7-9] The energy-level structure of the negatively charged N–V defect results in emissions characterised by a narrow zero-phonon line (ZPL) at 637 nm (the neutral N–V centre has a zero-phonon line at 575 nm)[10,11] accompanied by a wide-structured side band of lower energy due to transition from the same excited state, but with formation of phonons localised on the defect. The optical emission from N–V centres in diamond nanocrystals has been shown to strongly depend on the crystal size,[6,7,12-14] and the charge state is related to the temperature.[15]

Defects, such as GR1 centres and N–V centres, are mobile within diamond, and may migrate if a driving force is sufficient to overcome the kinetic energy barriers associated with diffusion. The diffusion of an N–V centre is vacancy assisted, and the rate-limiting step is the C–N exchange energy. During this migration, if an N–V centre interacts with another single nitrogen atom (or a migrating vacancy interacts with a nitrogen dimer, known as an A-centre), then an H3 centre is formed. The H3 centre consists of two N atoms surrounding a vacancy. It is one of the most studied in diamond,[1,16] and may be formed abundantly by irradiation with 1 to 2 MeV electrons to doses of 10^{18}–10^{20} electrons cm^{-2}, and annealing at 1200 K for 20 h in a vacuum.[17] If we continue this logical progression, then if an N–V centre migrates to an A-centre, or alternatively an H3 centre migrates to a C-centre (which is far less likely), then an N3 centre may be formed. The N3 centre consists of three nitrogen atoms surrounding a vacancy (whereas a vacancy completely surrounded by four N atoms is known as a B-centre). Both the H3 and N3 provide photoluminescence and cathodoluminescence, are known to be thermally stable, and exhibit high quantum efficiency up to temperatures in excess of 500 K.[18]

A summary of the point group symmetry, zero phonon line (ZPL), central wavelength (λ_0) and photoacoustic imaging quantum yield for the GR1, N–V, H3 and N3 defects is provided in Table 1.1, where we can see that the quantum yield for each of these defects is quite different. The quantum yield is defined as the number of photons emitted *via* photoluminescence *versus* the number of photons absorbed during excitation.[19] The N–V and H3 defects are the most efficient, which somewhat explains their popularity in the scientific community. In addition to the quantum yield the wavelength itself is very important as it determines the penetration depth of the irradiation.

Table 1.1 Point group symmetry, zero phonon line (ZPL), central wavelength (λ_0), and photoacoustic imaging quantum yield (Q) for optically active defects consisting of combinations of nitrogen and lattice vacancies in diamond.[20]

Defect	Point group	ZPL (nm)	λ_o (nm)	Q
GR1	T_d	741	898	0.014
N–V°	C_{Ih}	575	600	–
N–V¯	C_{3v}	638	685	0.99
H3	C_{3v}	504	531	0.95
N3	C_{2v}	415	445	0.29

The penetration depth for 532 nm light is 0.6 mm, for 670 nm is 2.4 mm, and for 750 nm is 2.6 mm. This is insufficient to penetrate human skin, irrespective of the quantum yield.

To be able to effectively exploit any of these functional defects, we need to draw upon a reliable understanding of the thermochemical stability of the defect within the host particle, in addition to the photostability. The optical properties of each of these defects are intrinsically linked to the physical structure of the defect within the lattice as this determines the energy levels of the excited states. In conventional experiments this can be difficult to probe as information on the physical (or structural) stability must be extracted indirectly by measuring the optical spectra, and if the emission decays, blinks or disappears any underlying physiochemical explanations will be obscured. In contrast, by using computer simulations the stability and properties of different configurations can be accessed directly. It is possible to establish the structure of a defect definitively, irrespective of the location within the particle, and the thermochemical stability can be determined unambiguously.

In this chapter a collection of computational studies examining the stability of a range of different defects will be briefly reviewed, with a focus on how the stability of the host nanodiamond affects the stability of the defect.

1.2 Defect-free Diamond Nanoparticles

Before beginning an exploration of defective diamond nanoparticles, it is firstly important to select the right host structures to use, and develop a general understanding of issues related to other "defects" intrinsic to diamond nanoparticles: the surfaces, edges, and corners.

To determine the lowest energy shape for nanodiamonds enclosed by low-index surfaces, Barnard and Sternberg used the density functional-based tight-binding method with self-consistent charges (SCC-DFTB) to simulate a set of nineteen different diamond nanoparticle structures ranging from 1 nm to 3.3 nm in diameter (142 to 1798 atoms).[21] This method was selected as it had previously been shown to provide good agreement with higher level quantum chemical methods for all-carbon systems, and is capable of accommodating sizes much larger than those accessible to the purely first principles methods (mentioned above). Within this structure set there were four subsets consisting of octahedral, truncated octahedral, cuboctahedral, and cuboid shapes, respectively. The complete octahedral subset contains C_{286}, C_{455}, C_{680}, C_{969}, C_{1330}, and C_{1771} structures enclosed entirely (100%) with {111} surfaces. The truncated octahedral subset contains C_{268}, C_{548}, C_{837}, C_{1198}, and C_{1639} structures enclosed with ~76% {111} surfaces and ~24% {100} surfaces. The cuboctahedral subset contains C_{142}, C_{323}, C_{660}, and C_{1276} structures enclosed with ~36% {111} surfaces and ~64% {100} surfaces. The final subset of cuboid structures contains C_{259}, C_{712}, C_{881}, and C_{1798} with ~34% {100} surfaces and ~66% {110} surfaces. All of the structures were fully relaxed using the conjugate gradient scheme to minimise the total energy.

In this article the authors systematically modelled the evolution of the core–shell structure for octahedral, truncated octahedral, cuboctahedral, and cuboid shapes over this size range, including explicit examination of the fraction of sp^3, sp^{2+x} and sp^2-bonded atoms, and their location. This can be seen in the figure plate provided in Figure 1.1, which is based on a

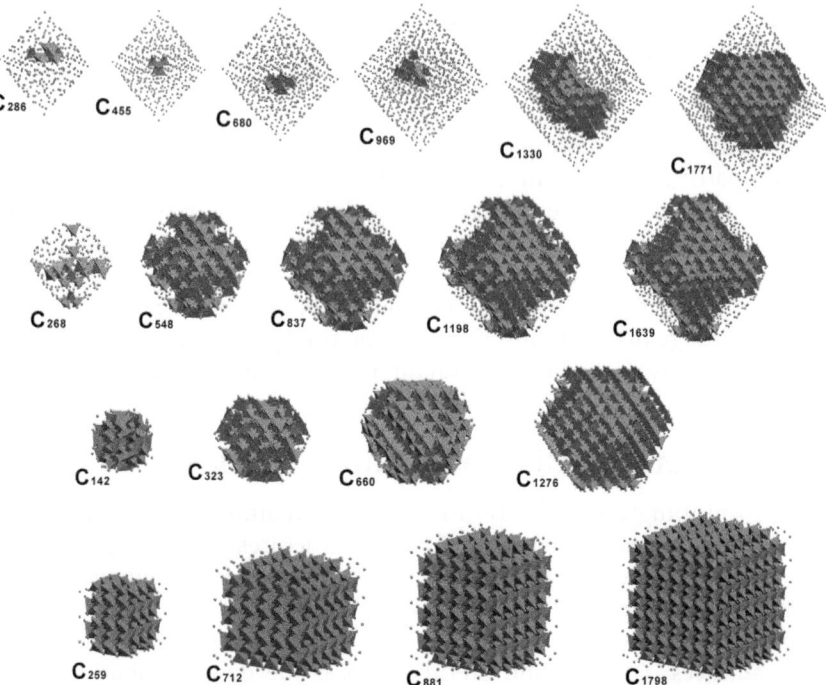

Figure 1.1 Optimised diamond nanoparticles reported in ref. 21. (Top row) the octahedral subset, (second row) the truncated octahedral subset enclosed with ~76% {111} surfaces, (third row) the cuboctahedral subset enclosed by ~36% with {111} surfaces, and (bottom row) the relaxed structures of the cuboid subset, 0% with {111} surfaces.

combination of visualisation modes, employing a simple ball method for the sp^2 and sp^{2+x} hybridised atoms (where $0<x<1$), and the polyhedron method for the tetrahedrally coordinated sp^3 hybridised atoms (each of which is surrounded by a coordination tetrahedron spanned by the four neighbours of the central atom). In this figure the diamond-like regions appear as collections of interpenetrating tetrahedra, and sp^2 and sp^{2+x} atoms participating in the fullerenic (or graphitic) regions appear as simple spheres decorating the outer surface of the diamond-like regions (connecting bonds not shown). This was designed to make the shape and extent of the diamond-like cores easily discernible,[21] at the expense of detail in the shell region (which is more prevalently displayed in other works).[22–27]

Based on these results it was determined that there is a relationship between the size of the particle and the fraction of diamond-like and/or fullerenic carbon, but that it depended significantly on the overall shape. In shapes when there is greater than 76% {111} surface area, nanodiamonds are likely to prefer a core–shell (bucky-diamond) structure, and the core/shell ratio depends on the overall size. The authors noted a distinct cross-over between predominately sp^2 and predominantly sp^3 structures at ~ 1100 atoms. If there is less than 76% {111} surface area, particles are likely to be stable in the diamond structure with a thin (either single or double layer) shell down to approximately 600 atoms (which is discussed below). It was presumed that a type of confinement by multiple layers is responsible for inhibiting relaxation of sp^3-bonded atoms into a sp^2-bonded shell, and promoting the stability of diamond-like cores at the centre of the structures with a high fraction of {111} surface area.

The reconstruction of the {111} surfaces on smaller (<2.5 nm) nanodiamonds lowers the surface energy (as the anti-bonding electrons become part of the aromatic character of the fullerenic shell), but introduces considerable surface stress. This was explicitly investigated by Barnard et al.[28] using a simple thermodynamic theory to compare nanodiamonds and fullerenes directly. By treating only dehydrogenated nanodiamonds (i.e., nanodiamond structures consisting of mostly sp^3-bonded atoms as opposed to bucky-diamond), a direct comparison with fullerenes was made.[28] The method was based on the enthalpy of formation as a function of size, expressed in terms of the bond energies for diamond-like and fullerenic particles, the surface dangling bond energy, the number of carbon atoms, the number of dangling bonds on the surface of the particle, and the standard heat of formation of carbon at $T = 298.15$ K.

In the case of fullerenes the closed shell eliminates the dependence on the effective surface-to-volume ratio and, therefore, the size dependence. Thus, a term for the strain energy that vanishes in the graphene limit was added by first making the assumption that a fullerene may be approximated as a homogeneous and isotropic elastic sphere. This was derived by considering the bending and stretching of a suitable elastic sheet in terms of the bending energy per unit area, the bending modulus of the sheet, and the mean radius of curvature. A spherical model was assumed and an expression for the strain energy per carbon atom for fullerenes that is proportional to the inverse of the square of the radius of curvature was derived. Using this model the cross-over in the enthalpy of formation of dehydrogenated (stable) nanodiamond crystals and fullerenes was found to be at ~1100 atoms, which is approximately equivalent to cubic nanodiamond crystals of 1.9 nm in diameter. An important point in this work was the selection of the chemical reservoir and the frame of reference. The model used a reservoir of free (isolated) C atoms, and included the formation enthalpy of a dangling bond so that the nanoparticles were assumed to be in mutual equilibrium with a continuous diamond or graphitic surface, not the bulk.[28]

To investigate the cases where carbon nanoparticles may contain both sp^2 and sp^3 bonding simultaneously, Barnard et al.[29] addressed the stability of multi-shell carbon nanoparticles using the same model described above for comparing the stability of nanodiamonds and fullerenes, and applied it to bucky-diamond and carbon onions. The onions were treated as nested fullerenes by adding a term for the van der Waals attraction (0.056 eV) to the expression used to describe fullerenes.[30] The bucky-diamonds were treated in the same manner as nanodiamonds, although, obviously, the dangling-bond-to-carbon-atom ratio is different for nanodiamonds and bucky-diamonds (of similar diameter) due to the formation of the graphitised fullerenic outer shells.[29]

The enthalpy of formation (as a function of particle size) for bucky-diamonds and carbon onions was calculated, and extrapolated along with the nanodiamond and fullerene results mentioned above. From this comparison three main conclusions were drawn. First, the sp^2-bonded onion and fullerene results were indistinguishable (within uncertainties) below approximately 2000 atoms. Second, the enthalpy of formation of a bucky-diamond is more akin to carbon onions than to nanodiamonds. Finally, in the region from ~500 to ~1850 atoms the results predicted that a thermodynamic coexistence region is formed, within which bucky-diamonds coexist (within uncertainties) with the other carbon nanoparticles.[29] This region was then further broken into three sub-regions. From ~1.4 nm to 1.7 nm the enthalpy of formation of bucky-diamonds was found to be indistinguishable from that of fullerenes (within uncertainties), although carbon onions represent the most stable form of nanocarbon. Between ~1.7 nm and 2.0 nm bucky-diamonds and carbon onions coexist (within uncertainties), and bucky-diamond was found to coexist with nanodiamond (within uncertainties) between ~2.0 nm and 2.2 nm. Further, the intersection of the bucky-diamonds and carbon onions stability was found to be very close to the intersection for nanodiamonds and fullerenes at ~1100 atoms, suggesting that at approximately 1100 atoms an sp^3-bonded core becomes more favourable than an sp^2-bonded core, irrespective of surface structure.[29] Once again, the model used in this study assumed a reservoir of free (isolated) C atoms, and the nanoparticles were assumed to be in mutual equilibrium with a continuous diamond or graphitic surface.[28]

As we can see from these examples, advances have been made in understanding the relative stability of sp^2- and sp^3-bonded particles at the nanoscale, and the basic structure of diamond nanoparticles has been established. These studies have clearly identified the two important size regimes, where (depending upon the phases under consideration) sp^2-to-sp^3 or sp^3-to-sp^2 phase transitions may be readily expected. In the case of larger particles the cross-over in stability between nanodiamond and nanographite may be expected at around 5 nm to 10 nm in diameter, and for smaller particles, the crossover between nanodiamond and fullerenic particles may be expected at 1.5 nm to 2 nm. They have also identified the lowest energy morphology (the truncated octahedron) in this size regime.

Based on these results, it is acceptable to use a single, model nanodiamond for exploring the stability of different point defects, provided it is a truncated octahedral (and, therefore, provides a range of both diamond-like and graphitised facets, if unpassivated), and sufficiently large to transcend the quantum confinement regime and (at least) occupy the coexistence regime from $\sim$500 to $\sim$1850 atoms. It is important that the model structure meets these criteria so as to ensure that all of the possible local bonding environments are included because the stability of a given defect will be sensitive to these issues.

1.3 Modelling Defects in Diamond Nanoparticles

In the following sections the thermodynamic stability (potential energy surface) and kinetic stability (probability of observation) will be reviewed for the GR1, N–V^0, H3, and N3 defects in model nanodiamonds. The particles used in the studies reviewed in this chapter are a C$_{837}$ truncated octahedral bucky-diamond and a hydrogenated C$_{837}$H$_{252}$ truncated octahedral nanodiamond, each displaying six {100} facets and eight {111} facets.

All of the calculations were originally performed using SCC-DFTB,[31,32] which is a two-centre approach to density functional theory (DFT), where the Kohn–Sham density functional is expanded to second order around a reference electron density. In this approach the reference density is obtained from self-consistent density functional calculations of weakly confined neutral atoms, and the confinement potential is optimised to anticipate the charge density and effective potential in molecules and solids. A minimal valence basis is established and one- and two-centre tight-binding matrix elements are explicitly calculated within DFT. A universal short-range repulsive potential accounts for double counting terms in the Coulomb and exchange-correlation contributions, as well as the internuclear repulsion, and self-consistency is included at the level of Mulliken charges.[32] This method was selected in these studies as it is more computationally efficient than DFT when such a large number of individual calculations are required.

1.3.1 Mechanical and Thermodynamic Stability

Since there is currently no way of determining experimentally where a defect is likely to be located in a given particle, a study of the stability of defects in diamond nanoparticles must include a range of possible substitution sites in the bucky-diamond and passivated nanodiamond structures in order to develop a reliable statistical description.

In bulk materials many (if not all) lattice sites can be considered as equivalent, so a single site (calculation) is representative. However, this is not the case in nanoparticles because the properties of a defect at one location within the nanoparticle may be very different to the properties at another location. Since the structure is finite, the location of any lattice site may be uniquely defined relative to the position of any collection of surfaces, edges, and corners. Therefore, by definition, all lattice sites in the nanoparticle are unique. To build a robust statistical description of point defects in a nanostructure, individual defects must be introduced at a number of different lattice sites, so as to sample the full range of crystallographically and geometrically unique lattice sites within the particles.

As mentioned above, in the present chapter fully relaxed C_{837} and $C_{837}H_{252}$ nanodiamonds were used as initial configurations, and the various point defects were substituted for carbon atoms located along specific (albeit zig-zagged) substitution paths within the lattice. The "paths" extend from the centro-symmetric atom to different points on the surfaces, edges, and corners. The directions of these substitution paths are shown in Figure 1.2, for the substitution paths terminating at the centre of the {100} surface, {111} surface, {100}/{111} edge, {111}/{111} edge, and the {111}/{111}/{100} corner, respectively. If we consider the nanoparticle morphology to be analogous to the shape of the diamond Brillouin zone, the substitution paths begin at the Γ-point and extend along the X, L, U, K, and W directions, respectively. In all, the point defects were introduced at over 50 geometrically unique sites

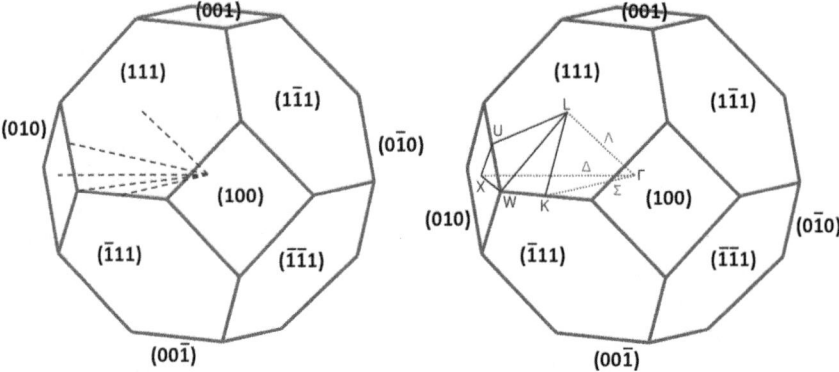

Figure 1.2 Substitution paths for the inclusion of a geometrically and crystallographically diverse range of point defects in a truncated octahedral diamond nanoparticle.

within the diamond nanoparticles to effectively sample configuration space. Following inclusion of the defect, the entire structure was re-relaxed using the same method as described above.

1.3.2 Kinetic Stability and Probability of Observation

The probability of observation ($P_{obs}(R,E_K)$) of a point defect in a diamond nanoparticle of radius (R) is a function of the kinetic energy imparted during probing E_K, the probability that the defect will diffuse to the surface and escape ($P_{esc}(R,E_K)$), and the probability that a defect will be initially created during synthesis ($P_{form}(R)$), such that:

$$P_{obs}(R, E_K) = P_{form}(R)[1 - P_{esc}(R, E_K)] \quad (1.1)$$

The probability of the formation of a specific defect will be proportional to the *limiting concentration* available during synthesis (c), the kinetic energy during growth ($E_{K,growth}$), and will be a function of the characteristic energy of the defect E_d at a position r. This may be approximated by a Boltzmann function, so that:

$$P_{form}(R) = c \sum_{r=0}^{R} P(r) \exp\left(\frac{-E_d(r)}{E_{K,growth}}\right), \quad (1.2)$$

where $P(r)$ is the probability of the defect being at r, when $0 < r < R$. As briefly outlined above, there are two distinct structural environments that may surround a point defect. It may be in a sp^3-bonded environment, in the bulk-like *core* region, or in a sp^2-bonded environment, such as in the *shell*. We may therefore simplify this to:

$$P_{form}(R) = c\left[P_{core}(R_{core}) \exp\left(\frac{-E_{d,core}}{E_{K,growth}}\right) + P_{shell}(R - R_{core}) \exp\left(\frac{-E_{d,shell}}{E_{K,growth}}\right)\right], \quad (1.3)$$

where $E_{d,core}$ is the characteristic energy of the defect in the *sp^3*-bonded core region, $E_{d,shell}$ is the characteristic energy of the defect in the sp^2-bonded shell region, R_{core} is the radius of the core, and P_{core} and P_{shell} are the probability of the defect being located in the core and shell, respectively. The extent of the shell region has been shown to be related to the excitonic radius of the donor or acceptor, and is important in determining the fraction of atoms occupying each region. To define this quantity we may use $P_{core} = N_{core}/N$ and $P_{shell} = 1 - N_{core}/N$. Shenderova et al.[33] determined that the total number of atoms (N) in a facetted diamond particle with n atoms along the (111)/(111) edge is given by:

$$N = \begin{cases} \frac{1}{12}n(2n+1)(5n+2) & \forall n \in (2,4,6,8...) \\ \frac{1}{12}(10n^3 + 9n^2 + 2n - 9) & \forall n \in (1,3,5,7...) \end{cases} \quad (1.4)$$

As we will see in the subsequent sections, the shell region consists of anywhere from four to eight atomic layers for H-terminated nanodiamond and unpassivated bucky-diamond, respectively, so this formula may be used to determine N_{core} by simply calculating the number of atoms in a particle that is the size of the core. Note that if the nanoparticle has less than five atoms along the (111)/(111) edge then it is effectively "all shell".

Similarly, the total probability of escape will be a combination of contributions from the core and shell regions. Each probability of escape will be a function of the input kinetic energy (E_K) and the escape energies. These escape energies are denoted by $E_{esc,core}(E_K)$ and $E_{esc,shell}(E_K)$ for the core and shell, respectively, and are once again described using the Boltzmann function. If E_K is significantly lower than the escape energy for each region then $P_{esc}(E_K)$ will be negligible, whereas, when $E_K = E_{esc}(E_K)$, the probability for diffusion approaches unity in that region. Hence, the total probability of escape, $P_{esc}(R,E_K)$, is:

$$P_{esc}(R, E_K) = [P_{core}(R) P_{esc,core}(E_K) + P_{shell}(R)] P_{esc,shell}(E_K)$$
$$= \left[\frac{N_{core}}{N} \exp\left(\frac{-E_{esc,core}}{E_K} \right) + \frac{N - N_{core}}{N} \right] \exp\left(\frac{-E_{esc,shell}}{E_K} \right), \quad (1.5)$$

where $E_{esc,core} = |E_{diff,core} - E_{d,core}|$ and $E_{esc,shell} = |E_{diff,shell} - E_{d,shell}|$ are the differences in the kinetic barrier to diffusion E_{diff} and the energy of the static defect E_d in the core and shell, respectively. Therefore, the calculation of $P_{obs}(R,E_K)$ requires only c, $E_{d,core}$, $E_{d,shell}$, $E_{diff,core}$, and $E_{diff,shell}$.

If we combine this $P_{obs}(R,E_K)$ with the total number of atoms (which is the total number of possible defect sites) then we may estimate the concentration of defects per particle (C) that may be expected in a nanodiamond of radius R. In the following sections the formation temperature was assumed to be 3000 K, which is consistent with the formation temperatures of detonation nanodiamond,[33] and the kinetic energy is obtained at a temperature of 300 K. In each case the limiting concentration (c) is assumed to be 1%.

1.4 Point Defects

1.4.1 Intrinsic Defects

In the first of the studies we will review, Barnard and Sternberg used SCC-DFTB computer simulations to investigate the structural and energetic stability of vacancies in the 837 atom model diamond nanoparticles with clean (reconstructed) and hydrogen-passivated surfaces. The concentration of vacancies (or GR1 defects[34]) in bulk synthetic diamond has been estimated to be of the order of ~26 ppm,[35,36] and the concentration of monovacancies in polycrystalline diamond has been measured at ≤7 ppm.[37] There is also evidence to suggest that the majority of these vacancies will be located in the vicinity of the diamond surface.[38,39] Previous studies have also examined the diffusion barrier for individual vacancy defects in bulk diamond, both theoretically[40–42] and experimentally,[43,44] but this was the first study to consider the preferred location, concentration, and stability of these point defects in isolated nanodiamonds.[45]

Figure 1.3 (left) reproduces the radial distribution in the vacancy defect energies for the hydrogen passivated nanodiamond (red symbol and line), and bucky-diamond (blue symbol and line). In this study the relative defect energy, $E(r)-E(0)$, was used, defined as the total energy of the nanoparticle with a given vacancy site relative to the energy of the nanoparticle with the vacancy in the centrosymmetric position. The x-axis represents a scaled (dimensionless) nanoparticle radius defined by dividing the distance from the centre to the average vacancy site (r_{defect}) by the total distance from the centre to the extremum (R_{total}), and averaging over each path. Hence $r_{defect}/R_{total}=0$ is the centre, and $r_{defect}/R_{total}=1$ is the outermost vacancy site located on a surface, edge, or corner. In Figure 1.3 (left) the uncertainties in the x-axis are related to the geometric differences between paths, and the uncertainties in the y-axis are the statistical variance in the results, which provide a measure of instability. In general the energetic uncertainties are comparable to (or larger than) the diffusion barrier for neutral vacancies in diamond.[39–44]

The results of this study showed that the (in)stability of the GR1 defect does depend on the location of the defect within the nanoparticle, and on

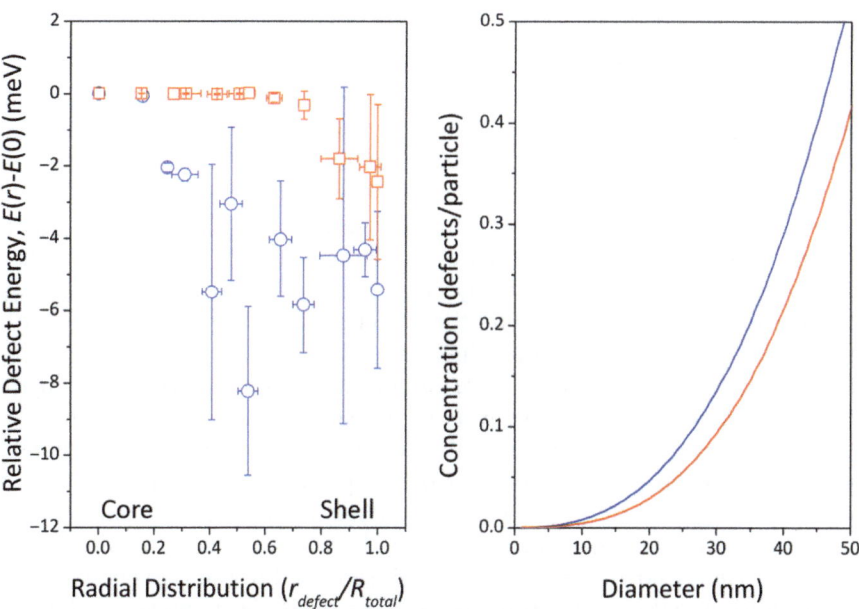

Figure 1.3 (Left) Stability of vacancy point defects in a $C_{837}H_{252}$ nanodiamond (red, □), and C_{837} bucky-diamond (blue, ○), and (right) the predicted concentration in each type of particle as a function of diameter (D).

the type of surface structure. Even when the surfaces are stabilised by hydrogen, the vacancy is thermodynamically unstable when a substitution site is within six atomic layers from the surface/edge/corner. At this point there is a thermodynamic driving force for diffusion that increases the closer the vacancy is to the surface/edge/corner. The study also showed that the stability of the nanodiamond itself is also affected by the presence of vacancies. This was particularly significant in the case of the bucky-diamond, where dramatic changes in energy were reported, due to sub-surface graphitisation of the "inner surface" of the bucky-diamond core. In these cases the defect did not need to reside in (or near) the "shell" of the bucky-diamond as the sub-surface graphitisation could be activated when the defect was up to as many as eight atomic layers away from the extrema. This resulted in structural asymmetry, where (111) facets in the vicinity of the defect exhibit a dual-shell (onion-like) structure, while the remaining (111) facets retain the single-shell surface structure.[45]

These results suggested that diffusion is likely to occur spontaneously at temperatures used during synthesis, or possibly during irradiation, as there is a strong thermodynamic driving force for diffusion of vacancies in diamond nanoparticles towards the surface (escape) even when the surfaces are stabilised with hydrogen. By combining these energetic results with the diffusion barriers in the core and in the shell, we can see from Figure 1.3 (right) that the probability of observing a GR1 defect is different for hydrogenated nanodiamond and a sp^2/sp^3 bucky-diamond.

The defects are more likely to be observed in a bucky-diamond compared to surface-passivated nanodiamond, which is counterintuitive when we consider that one of the primary benefits of surface passivation with hydrogen is to stabilise the particle and preserve the sp^3 hybridisation. This is because, in the case of the H-terminated nanodiamond, while the defect energy is higher in the "core" region (less favourable than in the shell region), the kinetic energy barrier to diffusion is also higher. In the "shell" region the defect energy is lower (more favourable), but the kinetic barrier to diffusion is lower (than in the "core"), making diffusion of the defect out of the particle (escape) more likely: $E_{d,core} > E_{d,shell}$ and $E_{diff,core} > E_{diff,shell}$. In the case of the bucky-diamond the defect energy is significantly higher in the "core" region than in the "shell" region, with a kinetic energy barrier to diffusion the same as the core of the nanodiamond. However, in the "shell" region the defect energy is significantly lower (being much more favourable), but the barrier to diffusion is higher (than in the "core"), making diffusion of the defect out of the particle more unlikely: $E_{d,core} > E_{d,shell}$ but $E_{diff,core} < E_{diff,shell}$. The graphitised shell effectively traps the defect and prevents escape as the energetic barrier associated with the interlayer diffusion (moving from one graphitised shell to the next) is too high.

In addition to this (as the probability of escape is always present) when the limiting concentration is low the majority of the defects are likely to escape. Under the conditions used in this example ($c = 1\%$) only half of the defects are retained when the particles are ~ 50 nm in size.

1.4.2 Incidental Impurities

Nitrogen is ubiquitous in diamond nanoparticles. This is due, in part, to the fact that it is a primary constituent of the source explosive. Detonation-induced transformations of powerful explosives and their mixtures with the composition $C_aH_bN_cO_d$, with a negative oxygen balance in a non-oxidising medium, yield a number of condensed carbon phases, including diamond nanoparticles.[46] Therefore, commercial diamond nanoparticles contain a small fraction of nitrogen, usually between 1% and 4%,[46,47] but it can sometimes be as high as 7% to 8%.[48] In this case a single nitrogen atom substitutes for a carbon lattice atom, and the paramagnetic electron is located in an anti-bonding orbital between the nitrogen and one of the nearest-neighbour carbon atoms. The defect consequently has <111> (trigonal) symmetry, and is known as the C-centre, as mentioned before.

Using the same approach described above for GR1 point defects, and the same model nanodiamond structures, Barnard and Sternberg also investigated the stability of substitutional nitrogen defects using the DFTB method. The study used the same sampling of substitution paths (and sites), and produced results that are directly comparable with the relative stability of vacancies, as shown in Figure 1.4 (left).[49] In the case of the relaxed hydrogenated nanodiamond the results showed an interesting relation between the relative defect energy and the location of the substitutional site. It was

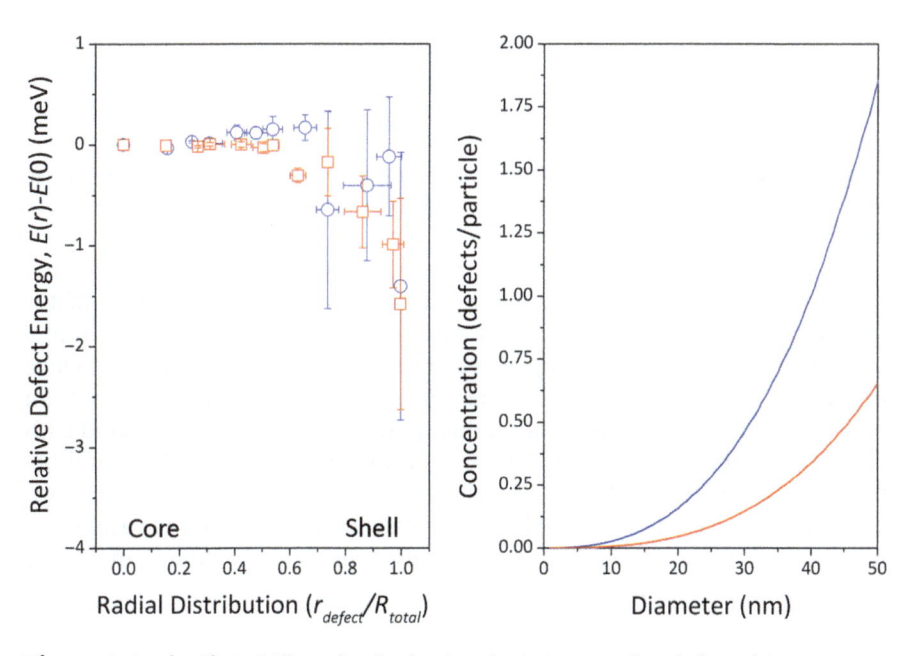

Figure 1.4 (Left) Stability of substitutional nitrogen point defects in a $C_{837}H_{252}$ nanodiamond (red, □), and C_{837} bucky-diamond (blue, ○), and (right) the predicted concentration in each type of particle as a function of diameter (D).

found that, while there is little energetic difference between the substitution paths in the core region of the particle (between 0% to ~50% of the distance from the centre), beyond this distance the general trend is toward lower energies, indicating that it is energetically preferable for nitrogen to be located near the surface of the nanodiamond.[49]

In the case of the relaxed bucky-diamond, with the exception of the interior of the core, the results showed a gradual increase in energy for nitrogen substitution sites approaching the "inner-surface" of the bucky-diamond core, which is also shown in Figure 1.4 (left). This was followed by a sharp decrease in energy for sites located on the "inner-surface" (approximately 75% of the distance from the centre). These sites represent a "trap" if the nitrogen atoms were driven to diffuse. One path did not follow this trend (and did not include a trap), since there is no surface graphitisation (and therefore no inner-surface) in that direction. In addition to this the *co-ordination* of N in the bucky-diamond was found to depend upon the distance from the centre. Within the core the nitrogen atoms were found to be four-fold coordinated, and three-fold coordinated in the shell. This correlates with the general "flattening" of the atomic layers on the inner surface of the bucky-diamond core and all subsequent outer layers comprising the shell.[49]

More detailed results for a variety of sizes and particles shapes were also reported, and it was confirmed that the stability of the defects were actually independent of particle size, and related only to the location of the defect with respect to the extrema, regardless of shape.[50] In general these results revealed that nitrogen atoms prefer to reside near the surfaces/edge/corners of diamond nanoparticles, and not within the core; and there was an obvious thermodynamic incentive for diffusion. This fuelled speculation that nitrogen, and hence other more functional nitrogen-complexes (discussed in the following sections), would not be stable with respect to diffusion, particularly in small particles.

However, when we calculate the probability of observation, we find that this is not necessarily the case. By inserting a limiting concentration of 1% for the nitrogen during formation, we can see from Figure 1.4 (right), that a significant proportion of nitrogen is retained under these conditions. Once again, the bucky-diamond is more efficient at retaining nitrogen than hydrogen-passivated nanodiamond due to the energy barrier associated with the diffusion of substitutional nitrogen from the core to the shell, and between layers in the shell. This barrier is moderated by the C–N exchange energy, which is extremely high in the absence of an adjacent vacancy to relieve the strain. Nitrogen is actually more likely to be found in bucky-diamonds than was originally assumed.

1.4.3 Photoactive N–V Centres

Among the various types of optically active defects, the nitrogen–vacancy (N–V) complex is the most widely studied,[51,52] and forms when a GR1 defect migrates to bind with a substitutional N impurity (a combination of the defects described in the last two sections).[40,53] The energy level structure of the negatively charged N–V⁻ defect results in emissions characterised by a narrow zero-phonon line (ZPL) at 637 nm, while the neutral N–V⁰ centre has a zero-phonon line at 575 nm, both accompanied by a wide-structured side band of lower energy due to transition from the same excited state, but with formation of phonons localised on the defect.[51] For many years, all available data pointed to a strong dependence on crystal size and the surface-to-volume ratio, and the optical emission from such defects was rarely seen in small diamond nanoparticles (<40 nm in diameter).[52,54] Photo-physical characteristics for 25 nm particles were later reported,[54] and most recently N–V emission from 5 nm detonation nanodiamond agglomerates[55] and isolated 8 nm diamonds[56] has been shown.

In order to realise any of the diverse applications for N–V centres in nanodiamonds,[3] a clearer understanding of this dependence is imperative. To these ends a series of simulations has also been reported on the relationship between the location of the defect, the stability of the nanodiamond, and the probability of observation (akin to those described above). Once again, the computational work used DFTB and employed the 837 atom truncated octahedral model particle, sampling the configuration space of N–V⁰ and N–V⁻ defects by substituting individual defects at the familiar set of 54 geometrically unique sites along specific lattice directions.[13] The advantage of using the same method, model particle, and configurational sampling is that these results are directly comparable with the results of the intrinsic and incidental defects described above. The relative energy results for N–V⁰ and N–V⁻ centres were found to be thermodynamically degenerate, and will be termed N–V from this point on.

Presented in Figure 1.5 (left) are the site-dependent defect energies for the hydrogen-passivated nanodiamond and the reconstructed bucky-diamond. In the case of the passivated $C_{837}H_{252}$ structure the N–V defects were found to be relatively stable within the particle until the substitution site fell within three atomic layers from the surface/edge/corner. Although the defect is thermodynamically unfavourable within the core (with respect to the "shell"), the energetic barrier for a transformation to a lower energy configuration is still high. At $r/R > 0.7$ there is a ~1.5 eV to 4.5 eV thermodynamic driving force for diffusion that increases the closer the N–V defect is to the surface/edge/corner.[13]

In the C_{837} bucky-diamond structure the defect was found to be highly unstable, and with a substantial thermodynamic (up to ~6 eV to ~9 eV) driving force for diffusion within the particle core. In bucky-diamond nanoparticles, where the sp^2 shell exists, the lattice parameter is different from the bulk, and a significant amount of strain already exists in the particle. This means that the energy barrier for distortion of the N–V centres is lowered, and depending on the position of the defect, the structure of the defective region changes to reduce the total stress and the total energy. This was shown to manifest as sub-surface graphitisation since the defect energy

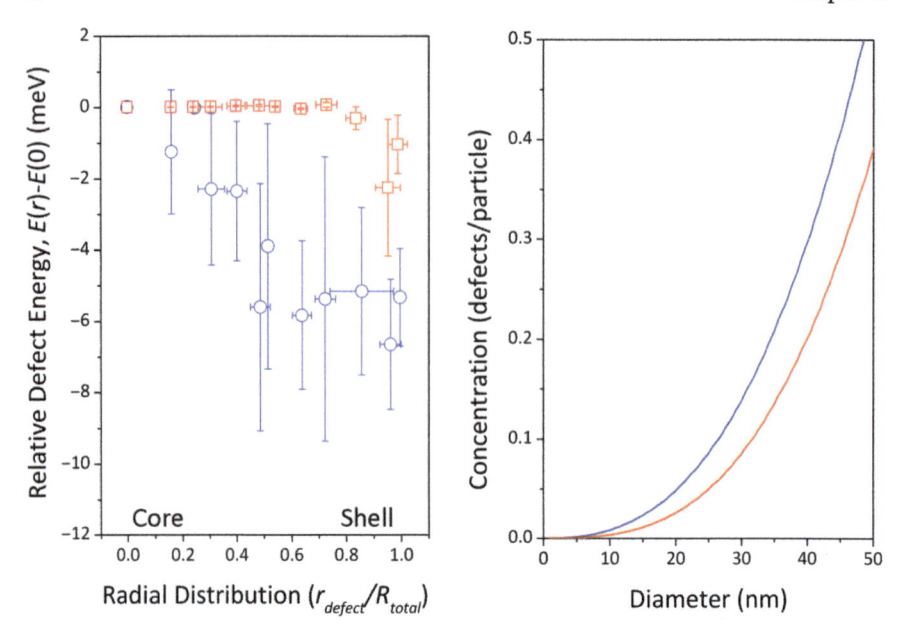

Figure 1.5 (Left) Stability of neutral N–V point defects in a $C_{837}H_{252}$ nanodiamond (red, □), and C_{837} bucky-diamond (blue, ○), and (right) the predicted concentration in each type of particle as a function of diameter (D).

for N–V was lower in sp²-bonded regions, and this provides a significant reduction in the site-dependent defect energy. These transformations are common near the surface of nanodiamonds, even when hydrogen terminated, and can be identified in the cores of bucky-diamonds by the large uncertainties in Figure 1.5 (left), which also provide a local minima.[13]

Using these radially averaged values of $E_{d,core}$ and $E_{d,shell}$ (see Figure 1.5, right), we find that the vacancy assisted diffusion of N–V defects is more likely than a simple substitutional nitrogen defect (C-centre). The diffusion barrier is still dominated by the N–C exchange energy, but this is (once again) higher in the graphitised shell of the bucky-diamond. A defect diffusing out of the stable core region will be required to penetrate the shell in order to escape, and is more likely to be trapped in a bucky-diamond than the H-terminated counterpart. Using a synthesis temperature of 3000 K and nitrogen concentration of $c = 1\%$, an estimate of the probabilities of observing an N–V defect in H-terminated nanodiamond and unterminated bucky-diamond are shown in Figure 1.5 (right).

Based on this model and technique, complementary predications for diamond nanoparticles produced with different synthesis techniques have previously been reported, including for detonation or ultra-dispersed diamond (UDD), chemical vapour deposition (CVD) nanodiamond, and high pressure and high temperature (HPHT) diamond nanoparticles.[13] These results were rigorously validated by explicitly measuring the emission for

3690 individual nanodiamonds, and correlating the size (determined using atomic force microscopy) with single photon fluorescence. By measuring the second order correlation function it was possible to determine if the detected fluorescence was due to single or multiple N–V centres. In this case the probability of detecting two simultaneous photons was normalised by the probability of detecting two photons at once for a random photon source, where an "anti-bunching" dip in the second order correlation function indicated sub-Poissonian statistics of the emitted photons. This revealed the presence of a single quantum system, which cannot simultaneously emit two photons, and was sufficient to confirm the predictions, and provided additional identification of the critical dimension for which the probability of finding a single N–V defect is optimal (under the conditions employed).[13]

Subsequent work by the same researchers and their colleagues has further confirmed the existence and behaviour of N–V centres in isolated 5 nm nanodiamond, and reported on the direct, room-temperature observation of profound surface-controllable luminescence intermittency (blinking), which offered a fresh insight into colour centre behaviour in isolated nanodiamonds.[57] Based on the DFTB simulations (described and shown above) it was determined that the blinking was related to the defects located in or near the shell of the particles, where the structural instability can alter the electronic states (such as the lowest unoccupied molecular orbital, LUMO), and extinguish the luminescence.

1.4.4 H3 Centres, N3 Centres and V–N–V Defects

Since the diffusion of an N–V centre is vacancy-assisted, then it is reasonable to assume that an N–V centre may migrate and potentially interact with other point defects. If the N–V centre interacts with another single nitrogen atom (or a GR1 migrates and interacts with a nitrogen dimer, known as an A-centre) then an H3 centre is formed. The H3 centre has an N–V–N structure, and can be generated abundantly by the radiation damage of diamond with 1 to 2 MeV electron beams, followed by heat treatment at temperatures above 800 °C in a vacuum.[58] As indicated in Table 1.1, this defect emits in the green region, and the radiative lifetime is stable up to 500 °C.

When the sampling and modelling procedure describe above is repeated for the H3 defect, the stability of the defect is found to be similar to the N–V defect (comparing Figure 1.5, left, and Figure 1.6, left), with less uncertainty (more like the C-centre). The defect energy of the H3 in the bucky-diamond is not as low as the N–V defect, indicating that the defect is not as easily accommodated by the strained lattice of the bucky-diamond shell. However, the probability of observation of the H3 defect in the two model particles shows an interesting difference. As shown in Figure 1.6 (right), the probability of observation of an H3 centre in the hydrogenated nanodiamond is similar to the case of the N–V, but the probability of observation of

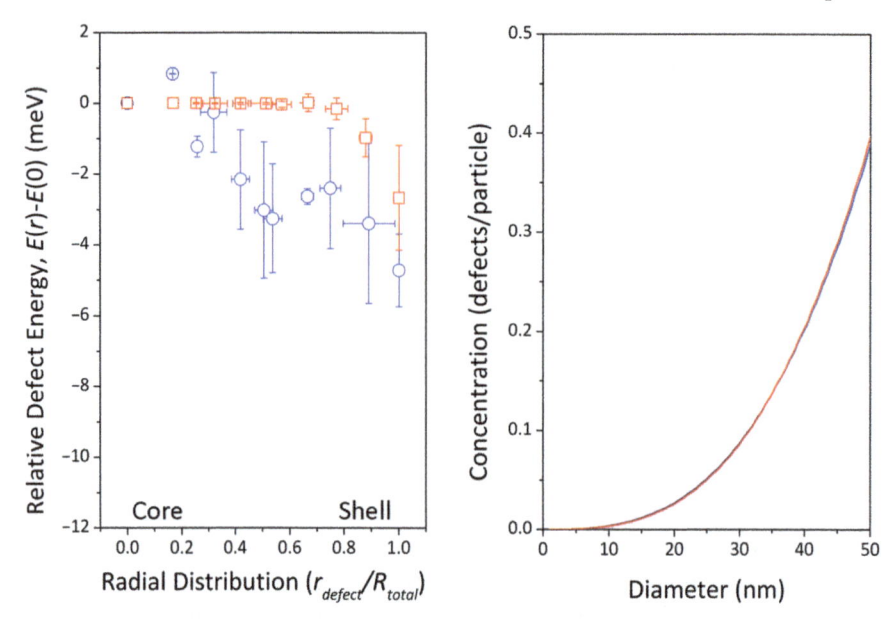

Figure 1.6 (Left) Stability of H3 point defects in a $C_{837}H_{252}$ nanodiamond (red, □), and C_{837} bucky-diamond (blue, ○), and (right) the predicted concentration in each type of particle as a function of diameter (D).

an H3 centre in a bucky-diamond is not. The kinetic stability of the H3 centre is independent of the surface chemistry, which indicates that the fullerenic shell does not effectively trap this defect as it did for the GR1 and the N–V.

Returning again to the issue of N–V migration, there are some other defects that can be formed. If the N–V centre migrates to a nitrogen dimer (A-centre), or alternatively an H3 centre migrates to a single nitrogen atom, then an N3 centre is formed. As mentioned above, the N3 centre consists of three nitrogen atoms surrounding a vacancy (whereas a vacancy completely surrounded by four N atoms is known as a B-centre). The N3 centre is also paramagnetic.

When the sampling and modelling procedure described above is repeated for the N3 defect, the stability of the defect is once again found to be similar to the N–V defect (comparing Figure 1.5, left, and Figure 1.7, left), and the kinetic stability (as characterised by the probability of observation) is similar to the N–V defect too. As shown in Figure 1.7 (right), the probability of an N3 centre is lower than N–V (Figure 1.5, right), which is consistent with experimental observations in bulk diamond.[18]

In each of the cases described above, the N–V centre is assumed to be mobile, and interacting with a (presumably stationary) nitrogen impurity or complex. However, as the diffusion barrier for a GR1 defect is lower than that of an N–V centre, it is also possible that a vacancy may combine with an N–V, and a V–N–V defect may be formed. This is far more likely than the

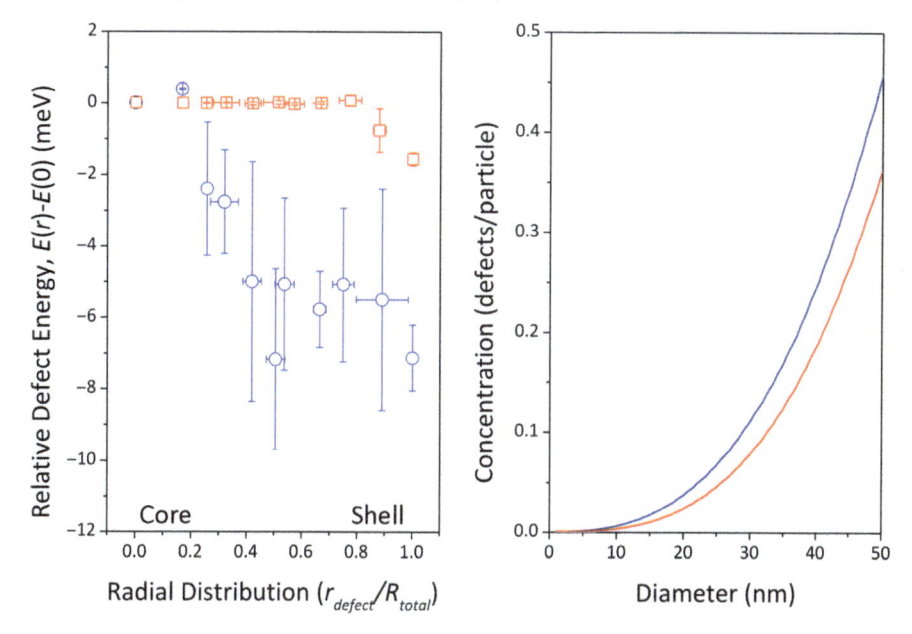

Distribution, Diffusion and Concentration of Defects in Colloidal Diamond 19

Figure 1.7 (Left) Stability of N3 point defects in a $C_{837}H_{252}$ nanodiamond (red, □), and C_{837} bucky-diamond (blue, ○), and (right) the predicted concentration in each type of particle as a function of diameter (D).

conditions required to form an H3 or an N3, and is significant since a V–N–V defect is not known to be luminescent.

When the sampling and modelling procedure described above is repeated for a V–N–V defect, we can see that the stability of this defect is different to the other complexes described in this section. As we can see from Figure 1.8 (left), the defect energies for a V–N–V defect within hydrogenated nanodiamond and a bucky-diamond are entirely different. This defect is accommodated much more readily by the bucky-diamond, with up to a 10 meV difference when the defect is located in proximity to a surface, edge, or corner. The defect energies in the bucky-diamond are also much lower than the simple GR1 or N–V, so it is reasonable to assume that if this defect forms it will not separate again into a GR1 and N–V. If this defect were to form, the fluorescence from the participating N–V centre would be permanently quenched.

The kinetic stability of this defect is interesting too (see Figure 1.8, right). The probability of observation for a V–N–V defect is much greater in a bucky-diamond than in an H-terminated nanodiamond, except at small sizes where the situation is reversed. Below ~12 nm the V–N–V defect is more likely to be found in an H-terminated nanodiamond (unlike all other defects discussed here) as the probability of trapping this defect in a bucky-diamond of this size is negative (due to the exceedingly low defect energy in the shell). Although not shown in Figure 1.8 (right), this disparity increases when we increase the limiting concentrations.

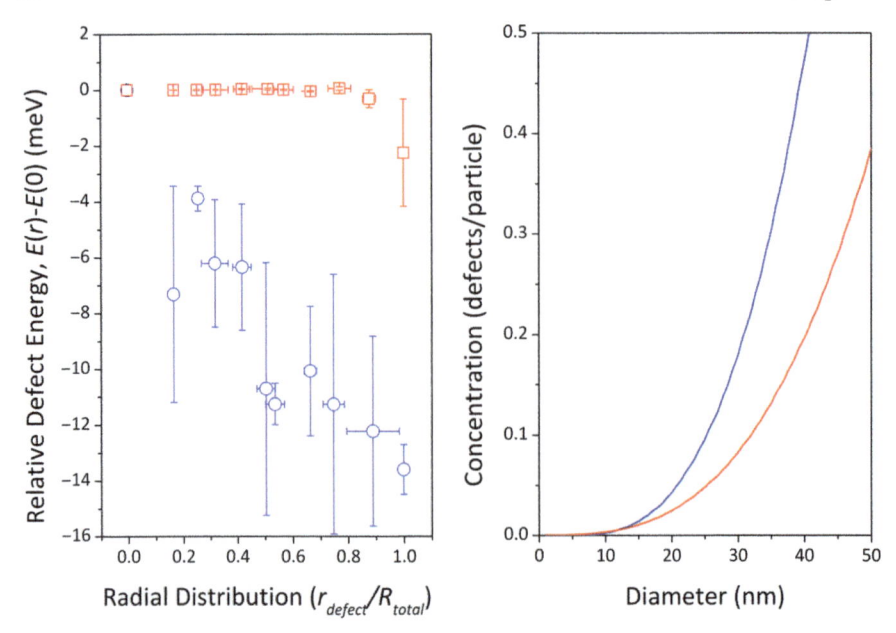

Figure 1.8 (Left) Stability of V–N–V point defects in a $C_{837}H_{252}$ nanodiamond (red, □), and C_{837} bucky-diamond (blue, ○), and (right) the predicted concentration in each type of particle as a function of diameter (D).

1.5 Comparison and Concentration

When we compare the stability of all of these defects, some overarching trends become apparent. Figure 1.9 (left) shows the relative stability of all of the N/V defects in the model bucky-diamond, and Figure 1.9 (right) shows the same results for the H-terminated nanodiamond. In the case of the bucky-diamond we can see that, in general, the energy of the defect decreases as the ratio of vacancies to nitrogen increases. When no vacancies are present at all (the nitrogen C-centre) the defect energy is highest. In contrast, when the particle is H-terminated the results for the different defects are practically indistinguishable (within the uncertainties associated with the different possible distributions), irrespective of the composition of the defect.

The relationship between the defect energy and the ratio of nitrogen and vacancies in the defect does not extend to the relative probabilities of observation (see Figure 1.10). As the probabilities are governed by the diffusion barriers, the defects with the greatest C–N exchange energies are the ones that are most likely to be trapped. For this reason the probability of observing the nitrogen C-centre is far greater than the other defects discussed here. The mobility of this defect is effectively zero under ambient conditions, and the observed concentration of substitutional nitrogen impurities is ultimately determined by the formation conditions.

By altering the synthesis temperature, $E_{K,growth}$, and c, the probability of observing all of these defects may be improved. In general, increasing the

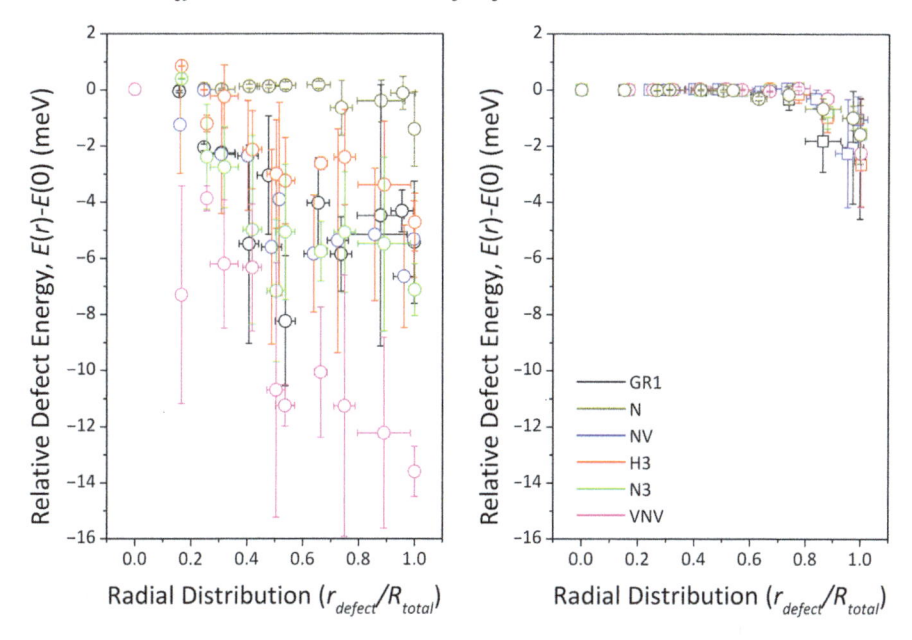

Figure 1.9 (Left) Stability of the GR1, N, N–V, H3, N3, and V–N–V point defects in a C_{837} bucky-diamond, and (right) $C_{837}H_{252}$ nanodiamond. The legend in the right-hand-side panel applies to both the left and right graphs.

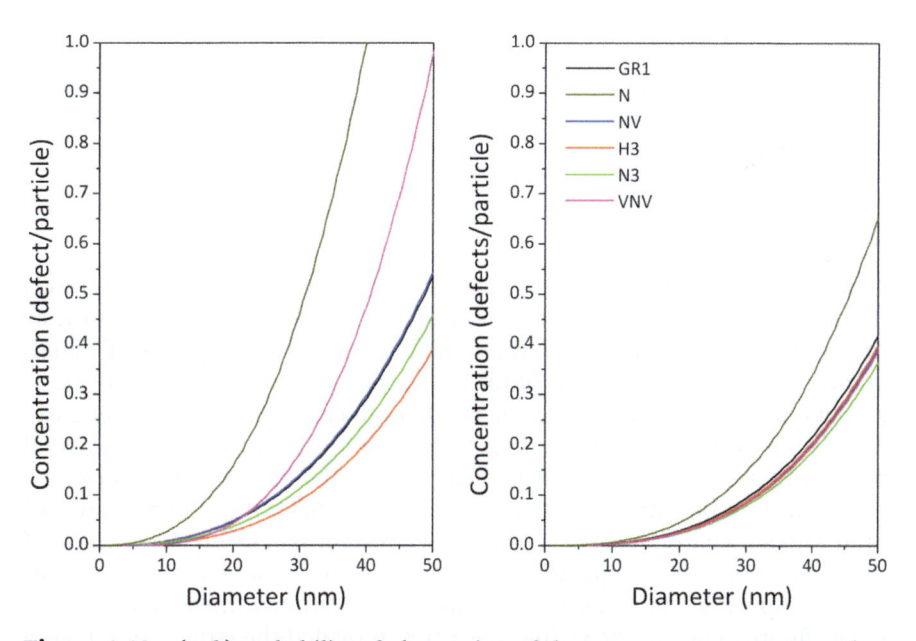

Figure 1.10 (Left) Probability of observation of the GR1, N, N–V, H3, N3, and V–N–V point defects in a C_{837} bucky-diamond, and (right) $C_{837}H_{252}$ nanodiamond, with a limiting concentration, c, of 1%. The legend in the right-hand-side panel applies to both graphs.

Table 1.2 Expected defect concentration per particle for an H-passivated diamond particle with a limiting concentration of 4%.

Size (nm)	GR1	N–V	H3	N3
3.7	0.0009	0.0006	0.0007	0.0005
4.8	0.0022	0.0017	0.018	0.014
10	0.015	0.013	0.014	0.011
35	0.58	0.54	0.55	0.50
50	1.67	1.56	1.58	1.45
1000	12640	12316	12379	11616

synthesis temperature and quantity of N in the precursors increases the probability that luminescent N–V and H3 defects will be present in the lattice, assuming that diffusion occurs on the same timescale as observed in bulk diamond and graphite. We can also see that the probability of observation is size-dependent, and so by increasing the concentration of N in larger HPHT nanodiamonds a reasonable concentration of functional defects is virtually assured. This is summaries in Table 1.2, which provides a comparison of the luminescent defects present in diamond particles of different characteristic sizes.

The study of other impurities and defects within isolated diamond nanomaterials[46,59] and nanocrystalline diamond films[60–63] with grain sizes in the order of ~5 nm to 100 nm is also receiving some attention. Some results have also been reported on other types of optically active defects, such as the analogous silicon–vacancy complex.[64] This defect has a much higher probability of observation,[65] but due to difficulties associated with Si introduction during synthesis, the N–V defect remains the firm favourite for the majority of relevant applications.

1.6 Conclusions

Diamond nanoparticles are never perfect. Given that nitrogen is ubiquitous during the synthesis of diamond nanoparticles (irrespective of the synthesis method), the inclusion of N in the nanodiamond lattice is practically guaranteed. Although it is preferable for N to reside near the surfaces, edges, or corners, diffusion of a C-centre is highly unlikely (due to the barrier associated with the C–N exchange energy), and so under normal conditions it will be trapped at or near the site of its inclusion. On the other hand, GR1 centres are highly mobile, and by introducing lattice vacancies that can migrate and bind with a C-centre, it is possible to generate a measurable concentration of "functional" defects. Depending on the relative concentration of vacancies and N impurities (and the different N configurations), these defects emit stable fluorescence in the visible spectrum, provided the structural integrity of the host particle and the defect can be preserved.

In the case of the host particle the crystallinity is sensitive to the shape of the particle (which determines the crystallographic orientation of the surface facets), and the types of surface reconstructions. In the absence of stable surface passivation, the low energy {111} facets of nanodiamond reconstruct to form a fullerenic shell around the diamond-like core. The resultant "bucky-diamond" has reduced crystallinity, but the formation of this sp^2/sp^3 core–shell structure has its advantages. The stability of defects in colloidal diamond particles is sensitive to the structure of the exterior shell, and deliberate removal of surface groups to encourage the bucky-diamond reconstructions can be useful to trap defects and prevent diffusion.

Another way of increasing the probability of observation of stable functional defects in diamond nanoparticles is to simply increase the concentration of nitrogen and/or vacancies either during formation or during post-synthesis treatment. This will be challenging in smaller nanodiamonds, less than ∼40 nm, where the expected concentration of functional defects (per particle) is less than one, even with 4% nitrogen in the precursors. In cases such as these a better strategy may be to simply add more nanodiamonds to the sample as then we can be confident that at least some of them will be sufficiently defective.

Acknowledgments

This project was supported by the Australian Research Council under grant number DP0986752.

References

1. J. Walker, *Rep. Prog. Phys.*, 1979, **42**, 1606.
2. Yu. D. Glinka, K.-W. Lin, H.-C. Chang and S. H. Lin, *J. Phys. Chem. B*, 1999, **103**, 4251.
3. A. S. Barnard, *Analyst*, 2009, **134**, 1751.
4. F. Jelezko and J. Wrachtrup, *Phys. Stat. Sol. (A)*, 2006, **203**, 3207.
5. A. T. Collins, G. Davies, H. Kanda and G. S. Woods, *J. Phys. C*, 1988, **21**, 1363.
6. J. R. Rabeau, A. Stacey, A. Rabeau, S. Prawer, F. Jelezko, I. Mirza and J. Wrachtrup, *Nano Lett.*, 2007, **7**, 3433.
7. Y.-R. Chang, H.-Y. Lee, K. Chen, C.-C. Chang, D.-S. Tsai, C.-C. Fu, T.-S. Lim, Y.-K. Tzeng, C.-Y. Fang, C.-C. Han, H.-C. Chang and W. Fann, *Nat. Nanotechnol.*, 2008, **3**, 284.
8. A. Mainwood, *Phys. Rev. B*, 1994, **49**, 7934.
9. K. Iakoubovskii and G. J. Adriaenssens, *J. Phys.: Condens. Matter*, 2001, **13**, 6015.
10. G. Davies and M. F. Hamer, *Proc. R. Soc. Lond. A*, 1976, **348**, 285.
11. G. Davies, *J. Phys. C: Solid State Phys.*, 1979, **12**, 2551.
12. B. R. Smith, D. Inglis, B. Sandnes, J. R. Rabeau, A. V. Zvyagin, D. Gruber, C. Noble, R. Vogel, E. Ōsawa and T. Plakhotnik, *Small*, 2009, **5**, 1649.
13. C. Bradac, T. Gaebel, N. Naidoo, J. R. Rabeau and A. S. Barnard, *Nano Lett.*, 2009, **9**, 3555.

14. Y. Shen, T. M. Sweeney and H. Wang, *Phys. Rev. B*, 2008, **77**, 033201.
15. B. T. Webber, M. C. Per, D. W. Drumm, L. C. L. Hollenberg and S. P. Russo, *Phys. Rev. B*, 2012, **85**, 014102.
16. M. D. Crossfield, G. Davies, A. T. Collins and E. C. Lightowlers, *J. Phys. C*, 1974, **7**, 1909.
17. *The Properties of Diamond*, ed. J. E. Field, Academic, New York, 1979, pp. 23–77.
18. S. C. Rand and L. G. DeShazer, *Opt. Lett.*, 1985, **10**, 481.
19. T. Plakhotnik and R. Chapman, *New J. Phys.*, 2011, **13**, 045001.
20. H.-C. Chang, in *Nanodiamonds, First Applications in Biology and Nanoscale Medicine*, ed. D. Ho, Springer Science + Business Media, New York, 2009, pp. 127–150.
21. A. S. Barnard and M. Sternberg, *J. Mater. Chem.*, 2007, **17**, 4811.
22. A. S. Barnard, S. Russo and I. K. Snook, in *Handbook of Theoretical and Computational Nanotechnology*, ed. M. Rieth and W. Schommers, American Scientific Publishers, Stevenson Ranch, CA, 2005, Chapter 36.
23. J.-Y. Raty and G. Galli, *Nat. Mater.*, 2003, **2**, 792.
24. A. S. Barnard, S. P. Russo and I. K. Snook, *Diamond Relat. Mater.*, 2003, **12**, 1867.
25. A. S. Barnard, S. P. Russo and I. K. Snook, *Int. J. Mod. Phys. B*, 2005, **17**, 3865.
26. J. Y. Raty, G. Galli, C. Bostedt, T. W. van Buuren and L. J. Terminello, *Phys. Rev. Lett.*, 2003, **90**, 37402.
27. C. Wang, B. Zheng, W. T. Zheng and Q. Jiang, *Diamond Relat. Mater.*, 2008, **17**, 204.
28. A. S. Barnard, S. P. Russo and I. K. Snook, *J. Chem. Phys.*, 2003, **118**, 5094.
29. A. S. Barnard, S. P. Russo and I. K. Snook, *Phys. Rev. B*, 2003, **68**, 73406.
30. Y. Guo, *I. Molecular simulations of buckyball fullerenes. II. Quantum chemistry studies on high-T_c superconductors*, Ph.D. Thesis, California Institute of Technology, 1992.
31. D. Porezag, Th. Frauenheim, Th. Köhler, G. Seifert and R. Kaschner, *Phys. Rev. B*, 1995, **51**, 12947.
32. Th. Frauenheim, G. Seifert, M. Elstner, T. Niehaus, C. Köhler, M. Amkreutz, M. Sternberg, Z. Hajnal, A. Di Carlo and S. Suhai, *J. Phys: Condens. Matter*, 2002, **14**, 3015.
33. O. A. Shenderova, Z. Hu and D. Brenner in *Synthesis, Properties and Applications of Ultrananocrystalline Diamond*, Proceedings of the NATO Advanced Research Workshop on Ultrananocrystalline Diamond, ed. D. Gruen, A. Ya. Vul', and O. Shenderova, NATO Science, St. Petersburg, Russia, 2004, pp. 5–6.
34. G. Davies, M. F. Thomaz, M. H. Nazare, M. M. Martin and D. Shaw, *J. Phys. C: Solid State Phys.*, 1987, **20**, L13.
35. K. Iakoubovskii, I. Kiflawi, K. Johnston, A. Collins, G. Davies and A. Stesmans, *Physica B*, 2003, **340**, 67.
36. A. T. Collins and A. Dahwich, *J. Phys.: Condens. Matter*, 2003, **15**, L591.

37. S. Dannefaer, W. Zhu, T. Bretagnon and D. Kerr, *Phys. Rev. B*, 1996, **53**, 1979.
38. L. Allers and A. Mainwood, *Diamond Relat. Mater.*, 1998, 7, 261.
39. J. Orwa, K. Ganesan, J. Newnham, C. Santori, P. Barclay, K. Fu, R. Beausoleil, I. Aharonovich, B. Fairchild, P. Olivero, A. Greentree and S. Prawer, *Diamond Relat. Mater*, 2012, **24**, 6.
40. A. Mainwood, *Phys. Rev. B*, 1994, **49**, 7934.
41. X. J. Hu, Y. B. Dai, R. B. Li, H. S. Shen and X. C. He, *Solid State Comm.*, 2002, **122**, 45.
42. R. Q. Hood, P. R. C. Kent, R. J. Needs and P. R. Briddon, *Phys. Rev. Lett.*, 2003, **91**, 076403.
43. G. Davies, S. C. Lawson, A. T. Collins, A. Mainwood and S. J. Sharp, *Phys. Rev. B*, 1992, **46**, 13157.
44. D. C. Hunt, D. J. Twitchen, M. E. Newton, J. M. Baker, J. K. Kirui, J. A. van Wyk, T. R. Anthony and W. F. Banholzer, *Phys. Rev. B*, 2000, **62**, 6587.
45. A. S. Barnard and M. Sternberg, *J. Comput. Theo. Nanosci*, 2008, 5, 2089.
46. V. Y. Dolmatov, *Russ. Chem. Rev.*, 2001, **70**, 607.
47. G. Post, V. Yu. Dolmatov, V. A. Marchukov, V. G. Sushchev, M. V. Veretennikova and A. E. Sal'ko, *Rus. J. Appl. Chem.*, 2002, **75**, 755.
48. A. E. Aleksenskiĭ, V. Yu. Osipov, A. Ya. Vul', B. Ya. Ber, A. B. Smirnov, V. G. Melekhin, G. J. Adriaenssens and K. Iakoubovskii, *Phys. Sol. State*, 2001, **3**, 145.
49. A. S. Barnard and M. Sternberg, *J. Phys. Chem. B*, 2005, **109**, 17107.
50. A. S. Barnard and M. Sternberg, *Nanotech.*, 2007, **18**, 025702.
51. A. T. Collins, G. Davies, H. Kanda and G. S. Woods, *J. Phys. C*, 1988, **21**, 1363.
52. J. R. Rabeau, A. Stacey, A. Rabeau, S. Prawer, F. Jelezko, I. Mirza and J. Wrachtrup, *Nano Lett.*, 2007, 7, 3433.
53. A. Lenef, S. W. Brown, D. A. Redman, S. C. Rand, J. Shigley and E. Fritsch, *Phys. Rev. B*, 1996, **53**, 13427.
54. Y.-R. Chang, H.-Y. Lee, K. Chen, C.-C. Chang, D.-S. Tsai, C.-C. Fu, T.-S. Lim, Y.-K. Tzeng, C.-Y. Fang, C.-C. Han, H.-C. Chang and W. Fann, *Nat. Nanotechnol.*, 2008, **3**, 284.
55. I. I. Vlasov, O. Shenderova, S. Turner, O. I. Lebedev, A. A. Basov, I. Sildos, M. Rähn, A. A. Shiryaev and G. Van Tendeloo, *Small*, 2010, **8**, 687.
56. J. Tisler, G. Balasubramanian, B. Naydenov, R. Kolesov, B. Grotz, R. Reuter, J.-P. Boudou, P. A. Curmi, M. Sennour, A. Thorel, M. Borsch, K. Aulenbacher, R. Erdmann, P. R. Hemmer, F. Jelezko and J. Wrachtrup, *ACS Nano*, 2009, **3**, 1959.
57. C. Bradac, T. Gaebel, N. N. Naidoo, M. J. Sellars, J. Twamley, L. Brown, A. S. Barnard, T. Plakhotnik, A. V. Zvyagin and J. R. Rabeau, *Nat. Nanotechnol.*, 2010, **5**, 345.
58. T.-L. Wee, Y.-W. Mau, C.-Y. Fang, H.-L. Hsu, C.-C. Han and H.-C. Chang, *Diamond Relat. Mater.*, 2009, **18**, 567.
59. V. V. Danilenko, *Phys. Solid State*, 2004, **6**, 595.

60. S. Sattel, J. Robertson, Z. Tass, M. Scheib, D. Wiescher and H. Ehrhardt, *Diamond Relat. Mater.*, 1997, **6**, 255.
61. D. M. Gruen, *Annu. Rev. Mater.. Sci.*, 1999, **29**, 211.
62. T. Sharda, T. Soga, T. Jimbo and M. Umeno, *Diamond Relat. Mater.*, 2001, **10**, 1592.
63. T. Wang, H. W. Xin, Z. M. Zhang, Y. B. Dai and H. S. Shen, *Diamond Relat. Mater*, 2004, **13**, 6.
64. I. I. Vlasov, A. S. Barnard, V. G. Ralchenko, O. I. Lebedev, M. V. Kanzuba, A. V. Saveliev, V. I. Konov and E. Goovaerts, *Adv. Mater.*, 2008, **21**, 808.
65. A. S. Barnard, I. I. Vlasov and V. G. Ralchenko, *J. Mater. Chem.*, 2009, **19**, 360.

CHAPTER 2

Detonation Nanodiamonds: Synthesis, Properties and Applications

A. YA. VUL',* A. T. DIDEIKIN, A. E. ALEKSENSKII AND
M. V. BAIDAKOVA

Ioffe Physical-Technical Institute, 26 Polytechnicheskaya, St. Petersburg, 194021, Russia
*Email: Alexandervul@mail.ioffe.ru

2.1 Introduction

This chapter addresses present-day views on the synthesis, properties and applications of nanodiamonds fabricated by the detonation of strong explosives, which form the so-called detonation nanodiamonds (DNDs).

Studies of DNDs have recently become a kind of fashion, which accounts for the appearance in the literature of a number of monographs and reviews on the subject.[1-7] This increasing interest in DNDs can be traced, on the one hand, to the extensive possible applications of these materials, and on the other, to their commercial availability because DNDs are presently produced in a number of countries on a commercial scale. In this regard, it is noteworthy that the structure of DND particles has already been largely established and the properties of DND powders and suspensions are not as strongly dependent on the supplier, as was the case in the comparatively recent past. One more factor which, in our opinion, has prompted increased interest in DNDs was the preparation of stable DND suspensions consisting

RSC Nanoscience & Nanotechnology No. 31
Nanodiamond
Edited by Oliver Williams
© The Royal Society of Chemistry 2014
Published by the Royal Society of Chemistry, www.rsc.org

predominantly of 4 nm diamond particles,[8-11] which considerably broadened the areas of their application, primarily in biology and medicine and as an element of composite materials.

In this chapter we focus primarily on the results of DND studies, which have been reliably established and universally accepted, while concurrently stressing the points that, in our opinion, still remain controversial. We intend to also acquaint the reader with results published only in Russian-language editions, which are largely inaccessible to the broader audience.

Being physicists by profession, we will focus the discussion primarily on the physical aspects of the subject and we recommend that readers refer to the other chapters for the more "chemical" aspects related to DNDs.

2.2 Main Features of DND Technology

The technology employed in the preparation of detonation nanodiamonds can be separated into two main stages: the preparation of detonation carbon, which is sometimes called detonation soot, and the isolation of nanodiamonds from detonation carbon. We consider these stages in more detail in the following sections.

2.2.1 Synthesis of Detonation Carbon

It is well known that, under normal conditions, the thermodynamically stable form of bulk carbon is graphite. In the industrial-scale synthesis of microcrystalline diamond from graphite, the graphite–diamond phase transition occurs, as a rule, at temperatures in the range of 1500–1800 °C and static pressures above 5 GPa.[12]

The detonation synthesis of nanodiamonds is actually based on two original ideas:

- use of the shock wave that forms in the detonation of strong explosives to generate high pressures and temperatures;[†]
- formation of the diamond crystal by the self-assembly of the carbon atoms of the explosives themselves.

The chemical reactions that drive an explosion are similar to those involved in burning. Just as in the burning of organic materials in air, an explosion produces carbon dioxide from carbon, water from hydrogen and elementary nitrogen from complex chemical compounds. However, in

---

[†]Recall that, in considering a shock wave, one understands the process of propagation of perturbation in a medium with supersonic speed. The source of energy for the shock wave in an explosion is the energy of the chemical transformation of the starting explosive into the reaction products. The shock wave driven through the medium of the explosive by the energy released in the chemical transformation is called the detonation wave. A comprehensive description of the physics underlying the processes involved in detonation waves can be found in refs 13–15.

contrast to conventional burning that occurs in an excess of the oxidizer—oxygen in the air—the starting-combustible-to-oxidizer ratio in the chemical reaction of explosive transformations is governed by the composition of the explosive itself.

Sticking to a simplified model,[14] an explosive may be considered as a mixture of fuel (hydrogen and carbon), an oxidizer (oxygen) and, on rare occasions, chlorine, as well as bound nitrogen. Explosive transformations produce, in the first instance, substances that provide the largest contribution to the energy of the detonation process, more specifically, elementary nitrogen and water. The remaining oxygen is distributed in accordance with the equilibrium distribution for the reaction: $2CO \rightarrow C + CO_2$. It is the ratio of carbon to oxygen present in the starting explosives and the products of detonation—the so-called oxygen balance—that determines the applicability of an explosive to diamond synthesis.

Obviously, if the amount of oxygen is not large enough to ensure complete transformation of carbon to CO, *i.e.*, under a negative oxygen balance (C/O > 1), some of the carbon will segregate. If the pressures and temperatures reached in this case at detonation lie in the region of the thermodynamic stability of diamond, the detonation will culminate in the synthesis of diamond crystals. Obviously, the short duration of the synthesis, which takes up to, as a rule, a few microseconds, should dictate a small size of the crystallites obtained. Indeed, they vary in size from 3 to 5 nm.[‡] Oxidation of the segregated carbon by oxygen in the air makes the synthesis of diamond possible only in special explosion chambers that ensure the proper conservation of the explosion products.[16,17]

Consider now the main features of industrial detonation synthesis.[3,17] As a rule, starting explosives are usually a mixture of trinitrotoluene (TNT) and hexogen (the TNT/hexogen ratio varies from 40/60 to 70/30) with a negative oxygen balance. The governing factor in diamond crystallization is the kinetics of the process. The reason for this lies in that, during the course of the synthesis, the *P–T* parameters pass through the region of the kinetic instability of diamond. Indeed, after detonation of the explosives, the temperature and pressure drop sharply, with the *P–T* parameters falling into the region where diamond is thermodynamically unstable. Now, if the temperature is still high enough to maintain a high mobility of the carbon atoms (the Debye temperature, $T_D \approx 1800$ K, may be chosen here as a criterion), the reverse diamond–graphite transition will be more probable than the transition to the region of the thermodynamic stability of graphite, which occurs at $T < T_D$. The higher the cooling rate, the shorter the time that the product of the detonation synthesis resides in the region of the kinetic instability of diamond will be and, accordingly, the smaller the probability of the reverse diamond–graphite phase transition in the course of the synthesis. Thus, it is the ratio of the rates of cooling and the pressure drop (d*T*/d*t* and d*P*/d*t*) that

---

[‡]A detailed discussion of the size distribution of synthesized diamond crystallites is deferred to Section 2.4.

defines the formation of the sp^2 hybridized carbon, *i.e.*, of the nano- and micro-particles of graphite contained in the detonation carbon powder that forms after the explosion.[3] Note that this point was first stressed in ref. 18.

Depending on what non-oxidizing medium the synthesis was conducted in, the various forms of this process are subdivided into "dry", if the explosion chamber was filled with an inert gas, "water" or "ice" versions, the latter subdivisions being the case if the explosives were blown up in water or dry ice, respectively. The fraction of the diamond phase in detonation carbon before its chemical purification depends, naturally, on the heat capacity of the medium in which the detonation occurred, which increases in the order of gas < water < ice, reaching about 40 wt %, 63 wt % and 75 wt %, accordingly.[19] This is why the "water" synthesis, being more efficient than the gas method, is employed on a commercial scale. The "ice" synthesis, while being the most efficient, is economically less attractive than the other two and, thus, is not appealing to the industry.

2.2.2 Isolation of Nanodiamonds from Detonation Carbon

X-Ray diffraction analysis (see, for example, refs 20–22) has provided copious evidence for detonation carbon containing, besides crystalline grains of nanodiamond measuring 4–5 nm, micro- and nano-particles of graphite as well. In detonation carbon one also finds iron oxides and silicon dioxide produced by erosion of the walls of the explosion chamber that came into contact with the shock wave.

Nanodiamond is isolated from detonation carbon by chemical methods based on the differences in the chemical properties of the nanodiamonds and the other components of detonation carbon. The method involving the oxidation of the non-diamond part of detonation carbon with water-diluted nitric acid (concentration 50–67%) has gained wide attention. The process is conducted under a moderate pressure (80–100 atm) in equipment made from titanium alloys.[17] The level of contamination by inorganic compounds of the nanodiamonds isolated in this way is low because nitric acid forms soluble salts with practically all known metals. The main disadvantage of this technology is that it involves additional contamination with titanium dioxide (TiO_2), which is a product of corrosion of the reaction equipment. The content of residual TiO_2 in nanodiamond may be as high as 0.3–0.5 wt %. However, since TiO_2 is an inert and nonmagnetic compound, its presence in nanodiamond in small amounts may be considered acceptable for most applications

Note that commercially produced nanodiamonds isolated from detonation carbon by chemical means contain, in small amounts, non-diamond forms of carbon, as well as metallic salts and inert impurities (*e.g.*, SiO_2, TiO_2) in concentrations of up to 1 wt %.[6,23] The presence of metallic salts may result in other artifacts, in particular, "magnetic carbon", because the thermal treatment initiates the reduction of the metallic salts with the formation of metal carbides, which exhibit magnetic properties. This has stimulated the development of additional methods of purification for detonation nanodiamond following their isolation from detonation carbon using strong acids (for instance, hydrochloric or hydrobromic acid). For details of these additional purification methods, which are aimed at producing "magnetically pure" DND, the reader is referred to ref. 24. Incidentally, for the complete removal of metallic impurities, simple washing of the nanodiamonds with acids, even under vigorous stirring, is insufficient. The reason for this is because nanodiamonds have a tendency to agglomerate (*i.e.*, the self-assembly of porous diamond microparticles 500–1000 nm in size; deagglomeration is discussed below). The rate of diffusion of an etchant inside an agglomerate is low and, to increase it, intense ultrasonic irradiation is required.[24,25]

Inert impurities, such as SiO_2 or TiO_2, are among the most difficult to remove, although it is possible to obtain nanodiamond samples with ash contents below the sensitivity threshold with hydrofluoric acid treatment; however, needless to say, this requires preliminary purification from magnetic impurities.[26]

It appears reasonable that the last stage in DND technology (see the scheme in Figure 2.1) is the removal of acidic residues, *i.e.*, through washing. Both DND suspensions and powders, which are readily obtained by drying the suspensions, are currently commercially available.[§] The properties of suspensions are largely determined by the size of the DND particles and the processes of their agglomeration. This aspect will be considered in Section 2.3 of the present chapter.

It appears only natural that the decisive factor in producing high-purity DNDs is the development of appropriate methods for controlling the content of impurities. The diversity of possible impurities makes offering a universal method capable of determining, in one measurement, the precise composition of a nanodiamond impossible. For an overall estimation of the composition and degree of purity of nanodiamond one resorts to X-ray diffraction analysis, which permits the identification of any kind of impurity that has a crystal lattice and allows an estimate of the content of amorphous carbon to be made. X-Ray diffraction also permits one to establish the presence of impurities; however, this is only possible on a semi-quantitative level with accuracy no more than 20%. The threshold of determination with this technique depends on the actual impurity and is a few percent. The total content of non-carbon impurities is evaluated most frequently from the ash content of the DND powder, which is determined by burning a powder sample under standard conditions (1000 °C). Under these conditions, all forms of carbon, as well as carbides, become oxidized. The residue of calcination in air (ash) contains nonvolatile oxides only. It is their mass in relation to the total mass of the sample that gives the result of the analysis. The accuracy and the threshold of determination provided by this method is

---

[§]Incidentally, it is the use of detonation carbon rather than DND that is the most reasonable approach for a number of applications, primarily for polymer–nanodiamond blends.[17]

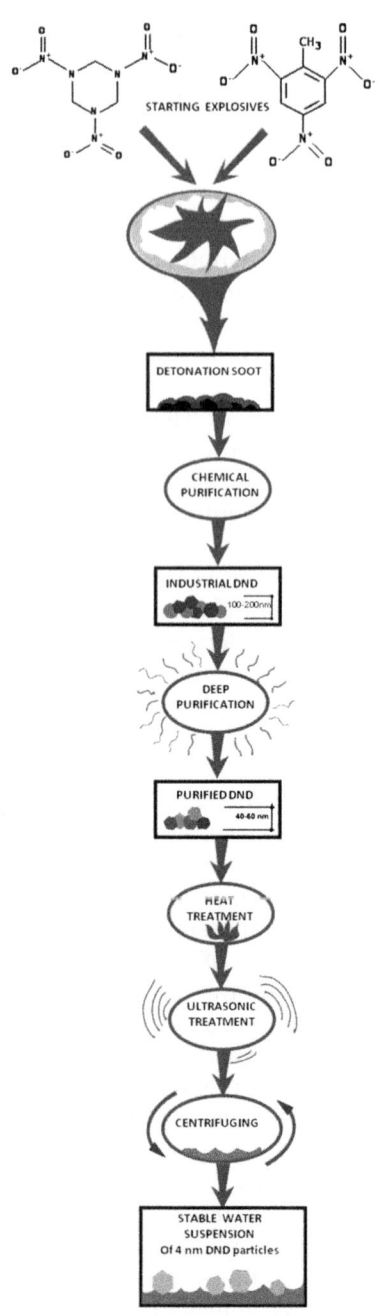

Figure 2.1 Scheme of the main technological stages in the production of detonation nanodiamond.

usually 0.1%. More sophisticated methods, such as X-ray fluorescence analysis,[15] can also be employed to determine the elemental composition; regrettably, only a few communications reporting its application have thus far appeared in the literature. The most sensitive and well-tested approaches to identify transition metal ions impurities in DNDs are electron paramagnetic (EPR) and nuclear magnetic resonance (NMR) methods. These methods, as applied to nanodiamonds, are discussed in considerable detail in refs 23, 27 and 28.

By properly combining the above methods, it is possible to evaluate the degree of DND purity with sufficiently high accuracy.

2.3 Structure of Detonation Soot Particles and Nanodiamonds

The first model concepts treating the structure of a single particle of detonation soot and detonation diamond are believed to have been formulated in the late 1990s. It was at that time that the model visualizing a particle of detonation soot as a diamond core encapsulated by a carbon shell in an amorphized or partially graphitized state containing also submicron-sized particles of crystalline graphite and metal won recognition. The isolation of detonation diamond from detonation soot results in the removal of the shell, with the emergence of DND particles in the form of perfect diamond-lattice crystallites, which have an average size of about 4 nm. Thus, despite the complex structure of detonation soot, a comprehensive analysis of the results obtained from experiments, which utilized different chemical purification regimes, culminated in the isolation of DND powder of acceptable purity, making possible the investigation of structure variations for single particles of detonation soot under thermal annealing[18,29–32] and permitted the development of a core–shell model for the structure of detonation soot particles and detonation nanodiamond.[31,33] This model of the structure of detonation soot particles[33] and of nanodiamonds in the bucky-diamond form[34] has won well-earned recognition as it provides an explanation for many experimental findings (see ref. 3 and references therein for additional details).¶

It is currently universally accepted that a detonation nanodiamond particle is cuboctahedron in shape, with the surface partially graphitized to the extent determined by the conditions chosen for the DND isolation from the detonation soot (see Figure 2.2). The existence of an sp^2-bonded fullerenic shell has been demonstrated in more than one experiment (see ref. 6 and references therein), including a quite recent study,[35] which employed an elegant chemical approach to demonstrate that the Bingel–Hirsch reaction,

---

¶Recall that, by definition, bucky-diamond is an all-carbon core–shell particle that is characterized by a crystalline diamond sp^3-bonded core encapsulated by a single- or multi-layer sp^2-bonded fullerenic shell that either partially or fully covers the particle surface.

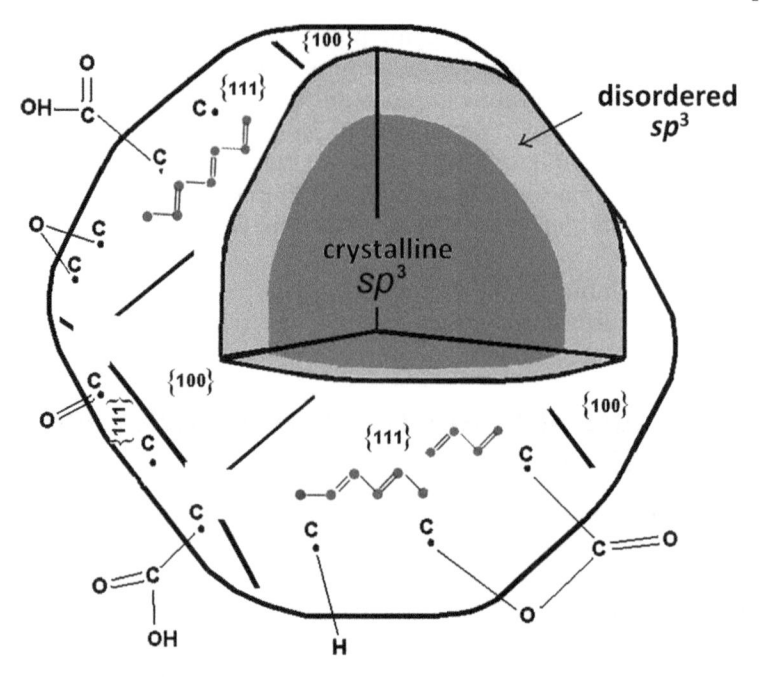

Figure 2.2 Schematic model for the DND structure in the shape of a truncated octahedron.

characteristic of sp² carbon materials, is applicable to nanodiamonds as well.

Such graphitization (the formation of a layer of sp²-hybridized carbon atoms) can be initiated not only by a heat-stimulated sp³–sp² phase transition[36] but by a natural reconstruction of the surface of the diamond cuboctahedron driven by the minimization of surface energy.[37] It is apparently the latter factor that has thus far impeded the production of detonation nanodiamonds with a surface free of traces of the sp² phase, which accounts for the strong optical absorption in the visible part of the spectrum and is responsible for the unusual dark-brown color of suspensions of 4 nm DND particles.[11,38–40]

Among the interesting and certainly unusual features revealed by single DND particles is the conclusion drawn from first-principles computer modeling, which indicated the presence of charges of opposite signs on different facets of the cuboctahedron.[37,41,42] It is the electrostatic interaction caused by these charges that is believed to drive aggregation of the DND particles into submicron-size agglomerates.[43] Some authors have suggested the existence of an intermediate layer with mixed sp²–sp³ bonding between the core and the shell of DND particles.[44–46] The appearance of this layer could be linked to the strong drop of pressure and temperature realized in the concluding stage of the formation of DND particles in the shock wave.

Dangling bonds on the surface of the DND particles are always saturated by a variety of functional groups, whose composition depends on the actual conditions of the interaction with the surrounding medium. The composition of the functional groups can be varied by chemical methods. This problem has been a subject of interest in recent years[47] and it is also treated in considerable detail in other chapters of the present book.

2.4 DND Suspensions

Commercially available detonation nanodiamond appears in the form of powders and suspensions. Suspensions are regarded, in most cases, as adequate starting materials for biomedical applications. However, unexpected problems arose during their initial use in practice as the DND particles formed strong agglomerates with sizes substantially in excess of 4 nm, i.e., the size of a single crystalline particle.[48] The formation of aggregates is known to be a typical property of colloidal systems but, in the case of nanodiamonds, the strength of the aggregates with varying dimensions around 100 nm turned out to be so high that the problem of disaggregation required intense, dedicated studies and it took a number of years to reach a solution.[3,7,11,49] The first success came from the application of mechanical crushing, which utilized cubical-modified zirconium dioxide microspheres to crush the aggregates in a water medium.[9,49] Attempts to find more readily available materials for milling has not resulted in noticeable progress[50,51] and using other liquid media brought about only a small decrease in the aggregates' dimensions, down to 20–40 nm.[52,53] Although thermal oxidation in air, which resulted in partial graphitization of the DND particle surfaces, did not initially resolve the issue of the aggregates,[54–58] the approach did eventually culminate in the production of 4 nm DND suspensions. Thus, the problem was finally solved through the thermal processing of dry, high-purity DND powder in an atmosphere of molecular hydrogen or oxygen, followed by intense ultrasonic treatment in water and separation by centrifuging.[10,11]

Incidentally, there is still no unambiguous opinion concerning the origin of the high strength exhibited by 100 nm DND agglomerates. One model has proposed[41,42] that this strength is related to the electrostatic interaction between oppositely charged faces of individual DND particles (see above),[59,60] while another model has assumed the existence of covalent C–C bonds between neighboring particles in the aggregates and suggested that these may form as early as the closing stage of the detonation synthesis. The latter viewpoint seems more credible as, after disaggregation, strong aggregates are not formed again.

It is an established fact that a water suspension of 4 nm DND particles maintains its stability through a double electric layer, which forms during the dissociation of carboxyl groups in the water medium.[61] These carboxyl groups appear during thermal annealing in an oxygen atmosphere, resulting in negatively charged particles. Thermal annealing in a hydrogen atmosphere reverses the sign of 4 nm particles to positive in a stable water suspension; however, the manner in which the double electric layer is generated remains unknown in this case. Interestingly, the zeta potentials in both cases are close in absolute magnitude and are in the range of 30–40 mV.

It is noteworthy that 4 nm particles have not been successfully achieved in media other than water and the DND aggregates that are produced in stable suspensions in other liquid media varying in size (20–40 nm).

2.5 DND Applications

The unique properties of DNDs have sparked promising applications for them in many areas of industry and technology. These areas are conventionally divided into two large groups, depending on the properties (Figure 2.3). The first group of applications are based primarily on the traditional attractive bulk properties of diamond (hardness, thermal conductivity, mechanical strength *etc.*). In these areas DNDs provide an economically feasible alternative to small fractions of natural and artificial diamonds synthesized from graphite at high pressures and temperatures. The presence of hard DND aggregates and, quite frequently, foreign impurities produce only a small adverse effect and the applications based on the bulk properties of DNDs are responsible for the initial interest that was paid to DND production.

The area where the bulk properties of detonation nanodiamonds can be advantageously used is the abrasive treatment of superhard materials. The application of nanodiamonds in abrasive compositions for the hyperfine mechanical polishing of the surfaces of hard materials, including optical elements, semiconductor plates and hard magnetic discs for data information storage, allowed the production of surfaces with an average roughness of less than 0.3 nm. This is apparently the limit for mechanical polishing.[62] Another noticeable use of nanodiamonds in this area is diamond-based cutting tools. Significant improvement of these tools was achieved by adding DND to the binder, which holds the large diamond grains together. In addition, nanodiamonds have been utilized in abrasive ceramics for the treatment of medium hard materials.[63] In this regard DND has also been used as a superfine abrasive in lubricating oils, which improve the wear resistance of friction surfaces by favoring their alignment, and reducing friction and fuel consumption in internal combustion engines.[64]

DND has also enjoyed use in the biochemical industry, where it is employed in nanoporous ceramic mixtures based on silicon carbide, which are produced by sintering at a high pressure and temperature, to produce highly stable and elastic filters.[65] DND composites, to which this ceramic belongs, offer a fertile field of industrial applications and a major part of DND production is focused on them. The introduction of DND particles into various materials upgrades the hardness and elasticity, imparts wear resistance and immunity to atmospheric air and corrosive media, and improves other operating characteristics as well. Varied applications have been discovered for

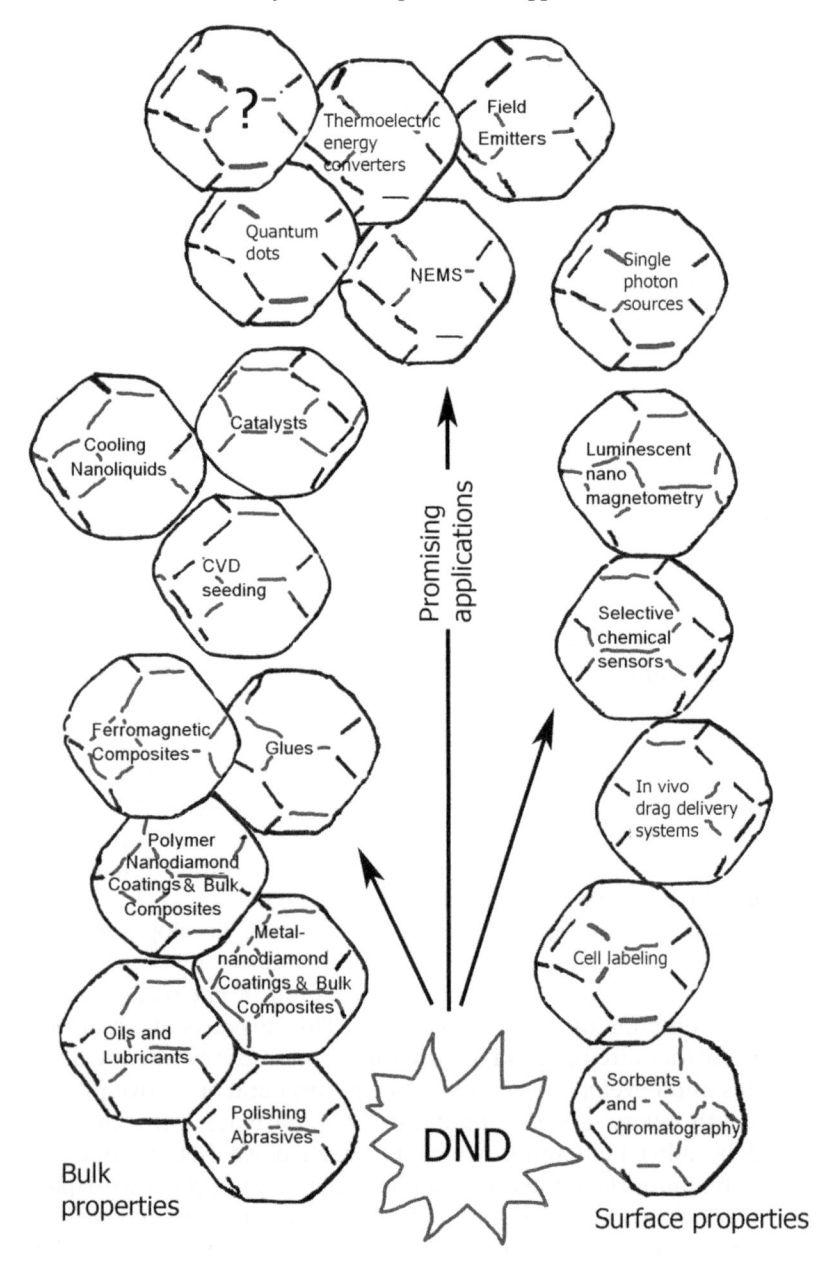

Figure 2.3 Current and future applications of detonation nanodiamonds.

composites based on Cu, Zn, Sn, Au, Ag, Cr and Ni metals, which have been used to produce platings prepared by galvanic deposition. Such platings are characterized also by their reduced sliding friction and are employed in the engineering industry for the most demanding units that must operate in severe service conditions. Falling into this category are, for instance, aircraft engines and oil pumps operating in oil wells. Due to their high thermal conductivity, DND composites based on copper and aluminum alloys[66] are also used in cooling systems for electronic devices.

No less significant among the DND composites are the materials based on organic polymers. Examples include rubber for car tyres, which can be prepared through the addition of detonation soot and possess a stronger road grip,[67] thermoplastic elastomers with improved thermal stability, which are based on ethylene-propylene-dien rubber[68] and polyvinyl alcohol.[69] Polymer–nanodiamond composites based on polyimide and polyurethane with DND contents of up to 1–3% have also been applied in microelectronics. These composites began with the use of insulating layers with low dielectric permittivity and high thermal conductivity[70] and opened up the possibility of using such composites in wiring insulation materials in transformers and high-power compact electric motors for ecologically clean vehicles.

Of particular importance among DND composites are thin-film fluorine-elastomer-based and polysiloxane-based coatings. Such coatings are widely employed in engineering and the chemical industry to reduce friction and protect metals and other materials from exposure to the atmosphere and corrosive media, and have been used in sea, land and air transport, and the chemical and mining industries. The introduction of nanodiamonds into such coatings reduces the penetrability to polar solvents of the coatings by up to 50 times. The highest chemical resistance to attack by salt, and acidic or alkaline media has been demonstrated by polymer–nanodiamond coatings based on ethylene copolymers with perfluoroalkylvinyl ethers.

A series of cosmetics based on novel organic-plastic composites with nanodiamonds has been developed for skin care and protection against UV radiation, which has been patented in recent years.[71]

Significantly, in most of the studies aimed at the development of composite materials, the DNDs were not disaggregated or subjected to any surface modification process aimed at the formation of chemical bonds with the organic components of the composites. This suggests an emerging potential for further attempts to increase the strength of the already designed composites and develop new materials with novel and attractive properties. High-strength epoxy glues containing DND particles with the surface modified by ethylenediamine, which is attached *via* NH_2 amino groups, serves as an appropriate illustration in this regard. Such glues are capable of polymerization without the addition of a hardening component because that part is played by the ethylenediamine immobilized on the surface of the nanodiamond particles.[72]

Also of note with regards to composites is the combination of DNDs with ferromagnetic materials, in which the DND particles either perform an auxiliary function of reducing friction and increasing the wear resistance of the ferromagnetic platings in data storage devices[63] or are used as abrasives in modern systems for the selective treatment of high-precision devices (Magnetic Abrasive Lapping Materials).[73]

The growing recognition of the possibilities inherent to the use of diamond and diamond-like coatings prepared by chemical vapor deposition (CVD)[74] in engineering has initiated the emergence of one of the most promising areas of DND applications. It is known that, in order for a CVD film to grow on a surface, one should disperse on the surface centers of crystallization (precursors) in the form of diamond particles. Significantly, the quality of the film thus grown will depend directly on the number of such centers per unit surface area. Using DNDs has allowed the attainment of precursor concentrations that have been unachievable for disperse diamonds of other kinds.[3] Besides a significant increase of diamond film quality, the preparation of suspensions of 4 nm DND particles has offered radically new perspectives for the CVD method by making it possible to transform polycrystalline diamond into bulk structures, including inverted photonic crystals—periodic structures with elements about 500 nm in size—which have unique optical properties.[75,76]

Recent years have witnessed intense progress in a new area of DND applications, more specifically, the development of novel catalytic systems for fine chemical synthesis, fuel elements and industrial chemistry. In such systems DND particles are used as a neutral, chemically stable phase with a layer of catalyst, for instance, platinum metal, which is deposited on the surfaces of the DNDs. This approach has provided a way to increase the catalysis efficiency, with a concomitant sharp drop in the consumption of precious and rare metals.[77,78]

Another promising approach in the development of catalysts is based on the use of the intrinsic catalytic activity of DND particles and bucky-onions obtained from the thermal annealing of DND in a vacuum or a neutral medium.[79] It is possible that these applications of DNDs will open up ways in the future to develop selective catalytic systems.

An interesting direction in the use of DND particles that shows considerable promise has been the introduction of DNDs in heat-transfer media for use in electrical plants and cooling systems. It has been demonstrated that a small (not more than a few tenths of a percent) amount of DND particles can be added to a liquid coolant and that this sharply increases heat transfer. This effect is beyond the standard concepts of heat conduction. To account for it, a mechanism of resonant phonon propagation in a heterogenic system has been proposed.[80] The first results of practical significance were obtained by adding DND particles to transformer oil. In this way it was possible to eliminate the breakdowns initiated by local overheating and lower the overall operating temperature.[81]

Despite the rich diversity of the lines along which, as we have just seen, DND can be used to advantage, current research interest is focused primarily on the second group of applications that involve the surface properties of these particles. Analogous to other nanoparticles, the surface of DNDs accommodates about one fifth of the total number of atoms making up each particle and this is the main feature that differentiates DNDs from other modified natural and artificial diamonds. Realization of these applications in practice has become possible only after the problem of disaggregation and the resulting preparation of suspensions of 4 nm DND particles was solved, thus opening the way to the effective functionalization of DND surfaces aimed at imparting the desired properties to the particles.

Liquid chromatography became a major field making use of the remarkable surface properties of DNDs. In this regard compacted (by hot pressing) DND particles have been used as a fixed phase for ion chromatography to separate alkali-earth metal ions, transition metals ions, polyaromatic hydrocarbons, monoalkyl benzenes, polymethyl benzenes *etc.* with an efficiency that is much higher than all other known materials.[82] DNDs also exhibit high efficiency in the separation of biologically active substances, including proteins. An essential factor making the application of DNDs in this area possible is its lack of toxicity.[83] This remarkable feature, combined with the broad potential provided by surface functionalization, permits one to employ DNDs as highly efficient absorbents, which can be used to fix toxins directly inside living organisms.[84] Furthermore, the sorptive properties exhibited by modified surfaces of DND particles have served as a basis for the development of unique sensors for the detection of specific chemical and biological molecules in the environment. The operation of such sensors is based on the mechanical resonance of sub-micron sized silicon plates; incidentally, their sensitivity is high enough to permit the measurement in some cases of the mass of individual adsorbed molecules.[85]

The absence of toxicity, combined with the small size of the particles and the possibility of surface functionalization, has made DNDs a practically ideal base for systems intended for the targeted delivery of biologically active substances (drugs, genes, vaccines) inside living organisms,[86] including cancer therapy.[87] Studies have been performed on living cells, which have demonstrated that the small size of DND particles makes them capable of penetrating cell membranes to enter the cell protoplasm without interfering with normal and vital cell functions.[88] To develop such delivery systems, the surface of DND particles is first modified by immobilization of carboxyl groups or hydrogen through thermal treatment in oxygen or hydrogen atmospheres, or in plasma.[7] Next, depending on the application that is being pursued, one performs chlorination,[89] fluorination, or saturation of the surface by hydroxyls or amino groups,[90] after which the active molecules or proteins are functionalized through the use of specific reactions. The successful intracellular delivery of large DNA plasmids by means of nanodiamond particles has also been reported.[91]

While the problems associated with preventing coagulation of DND particles in liquid media inside living organisms and their excretion are still awaiting a solution, it is this direction of DND applications that appears to hold forth the most promise in the nearest future.

The capability of DND particles to traffick inside living cells is increasingly used in the development of biomarkers, such as optical or luminescent markers that permit investigation of the vital processes occurring within living organisms. The position of the nanodiamond particles can be determined by confocal Raman spectroscopy *via* the characteristic diamond line[92] at 1332 cm^{-1} and the use of luminescent molecules attached to the DNDs.[93] The most expedient method to determine the position of DNDs within living organisms is to monitor the intrinsic strong diamond luminescence originating from the presence of charged NV$^-$ centers associated with nitrogen impurities.[94] It is pertinent to note, however, that despite the observation of luminescent NV$^-$ centers in DND particles,[95] the luminescence is yet to be made strong enough; therefore, crushed natural or high pressure high temperature (HPHT) artificial diamonds are currently used as luminescent markers, despite the limitations originating from their relatively large size.

Besides the biological applications, the luminescence of the NV$^-$ centers in nanodiamond crystals has been employed in the development of scanning nanomagnetometers, which have been shown to be sensitive to single electron and even nuclear spins at room temperature.[96] Another approach regarding the use of NV$^-$ center luminescence in nanodiamonds is the development of single-photon light sources for quantum cryptography, quantum computers and communication systems based on the use of entangled photon states.[97]

Among the more remote possibilities of nanodiamond applications is the development of thermoelectric energy converters, which make use of the anomalous thermoelectric effect derived from the ballistic phonon drag of electrons in nanoparticles.[98] No less potentially promising is the development of an efficient plane field electron emitter,[99] a problem that still remains unsolved despite the colossal efforts expended in trying to cope with it. Its solution will probably materialize in the form of a structure representing a smooth conducting surface with single 4 nm diamond particles playing the part of emission centers or a combination of DND particles on a graphene plane.[100] The main difficulty impeding solutions to these problems is the total absence of aggregates among free 4 nm particles, which is required for the active layers of such devices to be assembled.

The capability of DND particles to form onion-like shells under an electron beam or thermal annealing will eventually permit the design of quantum dot electronic devices, which may permit the generation and conversion of electromagnetic radiation and the switching of electric signals. Combined with carbon nanotubes and graphene planes, DND particles, and the onion-like structures incorporating them, will become components of future nanoelectromechanical systems (NEMS).

Our analysis of present-day and the possible future applications of DNDs cannot naturally claim completeness due to the continued emergence of new areas of their use, which shows the truly enormous potential of this unique material.

2.6 Unsolved Problems

In our opinion one of the key problems regarding DND technology still remaining unsolved is the preparation of stable, monodisperse suspensions in various media, where the particles equal in size that of the diamond crystal core, *i.e.*, 3–4 nm. In the experiments reported thus far, none have been able to obtain monodisperse distributions. At the same time, while the solution of this problem is of considerable current interest for an array of applications, it is impossible to reach without first understanding the mechanism governing the hardness of the large-size aggregates.

Indeed, the electrostatic interaction model,[41,42] first, assumes an ideal, perfect crystalline surface with faces of a diamond polyhedron. However, this means that the model disregards both the various functional chemical groups, whose presence on the surface of DND particles has been experimentally verified repeatedly[3,47] and possible surface reconstruction with a concomitant decrease in the number of dangling bonds, for instance, the formation of Pandey chains.[38] Moreover, the assumption of the electrostatic character of the interactions fails to explain why, after completion of the disaggregation procedure by thermal treatment,[10,11] solid aggregates do not form again in the solid phase (after removal of water in the drying stage). It could be assumed that aggregation is actually a dynamic process, *i.e.*, that aggregates of weakly coupled particles that can be separated into 4 nm particles by intense centrifugation. It would seem that the experiments[100] in which centrifugation (with accelerations of up to $9 \times 10^4 g$) produced particles about 4 nm in size may be considered as supportive evidence for this hypothesis. The authors of ref. 100 did not, however, specify the relative magnitude of this fraction, thus leaving room for the possibility that a certain amount of 4 nm particles could be present in the suspension initially, *i.e.*, before the centrifugation.

The mechanism underlying disaggregation is intimately linked to the nature of the zeta potential of the particles in an aqueous suspension obtained after disaggregation through annealing in various atmospheres. While the nature of the negative zeta potential that occurs after thermal treatment in oxygen finds a ready explanation by considering the dissociation of carboxyl groups in water,[61] the positive zeta potential and, all the more so, the high conductivity of the suspensions[101] obtained after thermal treatment in a hydrogen atmosphere remain unclear. Their explanation could possibly be approached by assuming the density of electronic states near the surface of DND particles changes under hydrogenation, similar to the effect observed in CVD diamond films.[102] These comments extend to aqueous DND suspensions only; more complicated questions naturally arise when suspensions in other liquid media are considered.

The model accounting for the strong light absorption in the visible region,[38,40] which provides explanation for the absorption of suspensions composed of 4 nm particles and, as a consequence, for its experimentally observed dark color, has not yet received significant attention. This model assumes essentially instantaneous reconstruction of the surface of a 4 nm DND particle, involving the formation of Pandey chains immediately after disaggregation. Direct and compelling evidence in support of this model would be gained if removal of the Pandey chains and the concomitant functionalization of the surface resulted in a change of the suspension color, an experiment that has not yet been performed. The iridescent color of water suspensions, which has been observed in one experiment only,[40] is most probably the result of light scattering from particles of different dimensions.[103]

Two more problems that effect potential applications and await solutions should be noted in connection. One of them is related to the nature of the efficient field emission of electrons observed in carbon nanostructures. The presently accepted model associates this emission with the sp^2/sp^3 hybridized electronic shells of carbon atoms in an optimum ratio.[98] The possibility of easily controlling this ratio by properly varying the time or temperature of DND heat treatment makes DNDs a convenient model to test this hypothesis. The second problem in urgent need of attention is the production of strongly photoluminescent nitrogen-vacancy defects that can be found in high concentrations in nanodiamonds used as biomarkers.[95] No definitive answer has thus far been obtained regarding the possibility of suppressing surface recombination to obtain efficient luminescence in diamond crystallites less than 10 nm in size.

Note that the above problems have been discussed, in one form or another, in review papers published in recent years.[3,5–7] It is noteworthy that the lists of unsolved problems discussed in these reviews and in this chapter are similar and this stresses the significance of the current issues.

2.7 Conclusion

This chapter has presented our opinion on the most significant results obtained in the investigation of detonation nanodiamonds. We naturally focused our attention on the problems that have direct bearing on our studies regarding the physics and chemistry of DNDs, which started in the early 1990s.

In our opinion we are currently witnessing a significant expansion in the scope of activities in the field of nanodiamond applications in various areas, and it is this trend that accounts for the sharp increase in the number of papers and reviews discussing the associated problems. We hope, in this regard, that our chapter makes a sizable contribution to the growing literature.

Acknowledgements

Grant support from the Program of the Presidium of RAS and RFBR is gratefully acknowledged.

References

1. D. Gruen, O. Shenderova, A. Ya. Vul' (eds), *Synthesis, Properties and Applications of Ultrananocrystalline Diamond*, NATO Science Series II: Mathematics, Physics and Chemistry, Springer, Dordrecht, The Netherlands, 2005, p. 192.
2. O. Shenderova and D. Gruen (eds), *Ultra-Nanocrystalline Diamond: Syntheses, Properties and Applications*, William Andrew Publishing, New York, USA, 2006.
3. M. Baidakova and A. Vul', *J. Phys. D: Appl. Phys.*, 2007, **40**, 6300.
4. A. Schrand, S. A. Hens and O. A. Shenderova, *Crit. Rev. Solid State Mater. Sci.*, 2009, **34**, 18.
5. A. Krueger, *Chem. Eur. J.*, 2008, **14**, 1382.
6. A. Aleksenskiy, M. Baidakova, V. Osipov, A. Vul', in *Nanodiamonds: Applications in Biology and Nanoscale Medicine*, ed. D. Ho, Spinger, New York, Dordrecht, Heidelberg, London, 2009, p. 55.
7. V. Mochalin, O. Shenderova, D. Ho and Y. Gogotsi, *Nat. Nanotechnol.*, 2012, **7**, 11.
8. E. Ōsawa, *Pure Appl. Chem.*, 2008, **80**, 365.
9. A. Krüger, F. Kataoka, M. Ozawa, T. Fujino, Y. Suzuki, A. E. Aleksenskii, A. Ya. Vul' and E. Ōsawa, *Carbon*, 2005, **43**, 1722.
10. O. A. Williams, J. Hees, C. Dieker, W. Jager, L. Kirste and C. E. Nebel, *ACS Nano*, 2010, **4**, 4824.
11. A. E. Aleksenskiy, E. D. Eydelman and A. Ya. Vul', *Nanosci. Nanotechnol. Lett.*, 2011, **3**, 68.
12. T. D. Burchell (ed.), *Carbon Materials for Advanced Technologies*, Elsevier Science Ltd, Amsterdam, 1999.
13. W. Fickett, W. Davis in *Detonation: Theory and Experimen*, 2nd edition, Dover Publications, Mineola, N.Y., USA, 2000.
14. A. N. Dremin in *Toward Detonation Theory, Shock Wave and High Pressure Phenomena*, Springer, Heidelberg, 1999; Ch. Mader in *Numerical Modeling of Explosives and Propellants*, 3rd edition, CRC Press, Boca Raton, FL, USA, 2007.
15. *Advanced Energetic Materials*, Committee on Advanced Energetic Materials and Manufacturing Technologies, Board on Manufacturing and Engineering Design, Division of Engineering and Physical Sciences, National Research Council of the National Academy, The National Academic Press, Washington, DC, 2004.
16. V. Danilenko, in *Ultra Nanocrystalline Diamond*, ed. O. Shenderova and D. Gruen, William Andrew Publishing, Norwich, N.Y., USA, 2006, p. 335.
17. V. Dolmatov, in *Ultra Nanocrystalline Diamond*, ed. O. Shenderova and D. Gruen, William Andrew Publishing, Norwich, N.Y., USA, 2006, p. 347.
18. A. Aleksenskii, M. Baidakova, A. Vul', V. Davydov and Yu. Pevtsova, *Phys. Sol. State*, 1997, **39**, 1007.
19. V. Yu. Dolmatov, M. V. Veretennikova, V. A. Marchukov and V. G. Sushchev, *Phys. Sol. State, Phys. Sol. State*, 2004, **46**, 611.

20. A. I. Lymkin, E. A. Petrov, A. P. Ershov, G. V. Sakovitch, A. M. Staver and V. M. Titov, *Dokl. Akad. Nauk USSR*, 1988, **302**, 611.
21. N. R. Greiner, D. S. Philips, J. D. Johnson and F. Volk, *Nature*, 1988, **333**, 440.
22. M. V. Baidakova, V. I. Siklitsky and A. Ya. Vul', *Chaos*, 1999, **10**, 2153.
23. A. E. Alexenskii, M. V. Baidakova, W. Kempi ski, V. Yu., E. Osipov, M. Ozawa Ōsawa, V. I. Siklitski, A. M. Panich, A. I. Shames and A. Ya. Vul', *J. Phys. Chem. Solids*, 2002, **63**, 1993.
24. A. Ya. Vul', E. D. Eydelman, L. V. Sharonova, A. E. Aleksenskiy and S. V. Konyakhin, *Diam. Relat. Mater.*, 2011, **20**, 279.
25. H. Huang, L. Dai, D. H. Wang, L.-S. Tan and E. Ōsawa, *J. Mat. Chem.*, 2008, **18**, 1347.
26. V. Pichot, M. Comet, E. Fousson, C. Baras, A. Senger, F. Le Normand and D. Spitzer, *Diam. Relat. Mater.*, 2007, **16**, 2098.
27. A. I. Shames, A. M. Panich, S. Porro, M. Rovere, S. Musso, A. Tagliaferro, M. V. Baidakova, V. Yu. Osipov, A. Ya. Vul', T. Enoki, M. Takahashi, E. Ōsawa, O. A. Williams, P. Bruno and D. Gruen, *Diam. Relat. Mater.*, 2007, **16**, 1806.
28. A. M. Panich, *Crit. Rev. Solid State Mater. Sci.*, 2012, **37**, 276.
29. V. L. Kuznetsov, M. N. Aleksandrov, I. V. Zagoruiko, A. L. Chuvilin, E. M. Moroz, V. N. Kolomiichuk, V. A. Likholobov, P. M. Brylyakov and G. V. Sakovitch, *Carbon*, 1991, **29**, 665.
30. V. L. Kuznetsov, A. L. Chuvilin, E. M. Moroz, V. N. Kolomiichuk, Sh. K. Shaikhutdinov and Yu. V. Butenko, *Carbon*, 1994, **32**, 873.
31. V. L. Kuznetsov, A. L. Chuvilin, Yu. V. Butenko, A. K. Gutakovskii, S. V. Stankus and R. A. Khairulin, *Chem. Phys. Lett.*, 1998, **289**, 353.
32. B. L. V. Prasad, H. Sato, T. Enoki, Y. Hishiyama, Y. Kaburagi, A. M. Rao, P. C. Eklund, K. Oshida and M. Endo, *Phys. Rev. B*, 2000, **62**, 11209.
33. A. E. Aleksenskii, M. V. Badakova, A. Ya. Vul' and V. I. Siklitskii, *Phys. Solid State*, 1999, **41**, 668.
34. J.-Y. Raty, G. Galli, C. Bostedt, T. W. van Buuren and L. J. Terminello, *Phys. Rev. Lett.*, 2003, **90**, 037401.
35. P. Betz and A. Krueger, *Chem. Phys. Chem.*, 2012, **13**, 2578.
36. T. Petit, J.-C. Arnault, H. A. Girard, M. Sennour and P. Bergonzo, *Phys. Rev. B*, 2011, **84**, 233407.
37. Q. Xu, T. Yang, S.-T. Li and X. Zhao, *J. Chem. Phys.*, 2012, **137**, 154702.
38. K. V. Reich, *JETP Lett*, 2011, **94**, 23.
39. C.-L. Park, A. Y. Jee, M. Lee and S.-g. Lee, *Chem. Commun*, 2009, **37**, 5567.
40. A. E. Aleksenskii, A. Y. Vul', S. V. Konyakhin, K. V. Reich, L. V. Sharonova and E. D. Eidel'man, *Phys. Solid State*, 2012, **54**, 578.
41. A. S. Barnard and M. Sternberg, *J. Mater. Chem.*, 2007, **17**, 4811.
42. A. S. Barnard, *J. Mater. Chem.*, 2008, **18**, 4038.
43. H. Huang, L. Dai, D. H. Wang, L.-S. Tanc and E. Osawa, *J. Mater. Chem.*, 2008, **18**, 1347.

44. O. O. Mykhaylyka, Y. M. Solonin, D. N. Batchelder and R. J. Brydson, *J. Appl. Phys.*, 2005, **97**, 074302.
45. B. Palosz, C. Pantea, E. Grzanka, S. Stelmakh, Th. Proffen, T. W. Zerda and W. Palosz, *Diam. Rel. Mater.*, 2006, **15**, 1813.
46. L. Hawelek, A. Brodka, J. C. Dore, V. Honkimaki, S. Tomita and A. Burian, *Diam. Rel. Mater.*, 2008, **17**, 1186.
47. A. Krueger and D. Lang, *Adv. Funct. Mater.*, 2012, **22**, 890.
48. A. E. Aleksenskii, V. Y. Osipov, A. T. Dideykin, A. Y. Vul', G. J. Adreaenssens and V. V. Afanasev, *Tech. Phys. Lett.*, 2000, **26**, 819.
49. E. Osawa, *Pure. Appl. Chem.*, 2008, **80**, 1365.
50. A. Krueger, M. Ozawa, G. Jarre, Y. Liang, J. Stegk and L. Lu, *Phys. Stat. Sol. (A)*, 2007, **204**, 2881.
51. S. Osswald, G. Yushin, V. Mochalin, S. O. Kucheyev and Yu. Gogotsi, *J. Am. Chem. Soc.*, 2006, **128**, 11635.
52. C.-C. Li and C.-L. Huang, *Colloids Surf., A*, 2010, **353**, 52.
53. O. Shenderova, S. Hens and G. McGuire, *Diam. Rel. Mater.*, 2010, **19**, 260.
54. K. Xu and Q. Xue, *Phys. Sol. State*, 2004, **46**, 649.
55. K. Xu and Q. Xue, *Diam. Rel. Mater.*, 2007, **16**, 277.
56. O. Shenderova, I. Petrov, J. Walsh, V. Grichko, V. Grishko, T. Tyler and G. Cunningham, *Diam. Rel. Mater.*, 2006, **15**, 1799.
57. V. Pichot, M. Comet, E. Fousson, C. Baras, A. Senger, F. Le Normand and D. Spitzer, *Diam. Rel. Mater.*, 2008, **17**, 13.
58. Y. Liang, T. Meinhardt, G. Jarre, M. Ozawa, P. Vrdoljak, A. Schöll, F. Reinert and A. Krueger, *J. Colloid Interface Sci.*, 2011, **354**, 23.
59. L.-Y. Chang, E. Osawa and A. S. Barnard, *Nanoscale*, 2011, **3**, 958.
60. J. Hees, A. Kriele and O. A. Williams, *Chem. Phys. Lett.*, 2011, **509**, 12.
61. A. M. Panich and A. E. Aleksenskii, *Diam. Rel. Mater.*, 2012, **27–28**, 45.
62. Y. Zhu, Z. Feng, B. Wang and X. Xu, *China Particuology*, 2004, **2**(4), 153.
63. V. Y. Dolmatov, *Russ. Chem. Rev.*, 2001, **70**, 607.
64. C. C. Chou and S. H. Lee, *J. Mater. Proc. Technol.*, 2008, **201**, 542.
65. J. B. Wachtman, Jr, P. Darrell Ownby and J. Liu, *Ceram. Eng. Sci. Proc.*, 2008, **12**(7–8).
66. K. Mizuuchi, K. Inoue, Y. Agari, Y. Morisada, M. Sugioka, M. Tanaka, T. Takeuchi, J.-I. Tani, M. Kawahara and Y. Makino, *Composites Part B*, 2011, **42**, 825.
67. Y.-X. Wang, Y.-P. Wu, W.-J. Li and L.-Q. Zhang, *Appl. Surf. Sci.*, 2011, **257**, 2058.
68. F. H. Jahromi and A. A. Katbab, *J. Appl. Polym. Sci.*, 2012, **125**, 1942.
69. U. Maitra, K. E. Prasad, U. Ramamurty and C. N. R. Rao, *Sol. State Comm.*, 2009, **149**, 1693.
70. O. Shenderova, T. Tyler, G. Cunningham, M. Ray, J. Walsh, M. Casulli, S. Hens, G. McGuire, V. Kuznetsov and S. Lipa, *Diam. Relat. Mater.*, 2007, **16**, 1213.
71. C.-M. Sung, M. Sung and E. Sung, US Pat. 7294340, Issues Nanodiamonds for Cosmetics, November 13, 2007.

72. V. N. Mochalin, I. Neitzel, B. J. M. Etzold, A. Peterson, G. Palmese and Y. Gogotsi, *ACS Nano*, 2011, **5**, 7494.
73. Z. X. Zheng, R. Wang and C. M. Wang, *Curr. Appl Phys.*, 2011, **11**, 227.
74. D. M. Gruen and I. Buckley-Golder, *MRS Bull.*, 1998, **23**, 16.
75. J. F. Galisteo-López, M. Ibisate, R. Sapienza, L. S. Froufe-Pérez, A. Blanco and C. López, *Adv. Mater.*, 2011, **23**, 30.
76. D. A. Kurdyukov, N. A. Feoktistov, A. V. Nashchekin, Yu. M. Zadiranov, A. E. Aleksenskii, A. Ya. Vul' and V. G. Golubev, *Nanotechnology*, 2012, **23**, 015601.
77. O. V. Turova, E. V. Starodubtseva, M. G. Vinogradov, V. I. Sokolov, N. V. Abramova, A. Ya. Vul' and A. E. Alexenskiy, *Catal. Commun.*, 2011, **12**, 577.
78. N. N. Vershinin, O. N. Efimov, V. A. Bakaev, A. E. Aleksenskii, M. V. Baidakova, A. A. Sitnikova and A. Ya. Vul', *Fullerenes Nanotubes Carbon Nanostruct.*, 2011, **19**, 63.
79. D. Su, N. I. Maksimova, G. Mestl, V. L. Kuznetsov, V. Keller, R. Schlo and N. Keller, *Carbon*, 2007, **45**, 2145.
80. X. W. Wang, X. Xu and S. U. S. Choi, *J. Thermophys Heat Transfer*, 1999, **13**(4), 474.
81. J. L. Davidson, W. P. Kang in *Synthesis, Properties and Applications of Ultrananocrystalline Diamond*, NATO Science Series, Series II: Mathematics, Physics and Chemistry, Springer, Netherlands, 2005, vol. 122, pp 357.
82. P. N. Nesterenko, O. N. Fedyanina, Y. V. Volgin and P. Jones, *J. Chromatogr. A*, 2007, **1155**, 2.
83. K. V. Purtov, A. P. Puzyr' and V. S. Bondar', *Dokl. Biochem. Biophys.*, 2008, **419**(1), 72.
84. N. Gibson, Z. Fitzgerald, T.-J. Luo, O. Shenderova, V. Grichko, V. Bondar, A. Puzur and D. Brenner, *Technical Proceedings of the 2007 NSTI Nanotechnology Conference and Trade Show*, 2007, **2**, 838.
85. R. Kaur and I. Badea, *Int. J. Nanomed.*, 2013, **8**, 203.
86. P. Kulha, A. Kromka, O. Babchenko, M. Vanecek, M. Husak, O. A. Williams and K. Haenen, *Vacuum*, 2010, **84**, 53.
87. R. Sinha, G. J. Kim, S. Nie and D. M. Shin, *Mol Cancer Ther.*, 2006, **5**(8), 1909.
88. K. K. Liu, C. C. Wang, C. L. Cheng and J. I. Chao, *Biomaterials*, 2009, **30**(26), 4249.
89. T. Ando, M. Nishitani-Gamo, R. E. Rawles, K. Yamamoto, M. Kamo and Y. Sato, *Diam Relat. Mater.*, 1996, **5**, 1136.
90. Y. Liu, Z. Gu, J. L. Margrave and V. N. Khabashesku, *Chem Mater.*, 2004, **16**(20), 3924.
91. C. M. Wiethoff and C. R. Middaugh, *J. Pharm. Sci.*, 2003, **92**, 203.
92. E. Perevedentseva, C. Y. Cheng, P. H. Chung, J. S. Tu, Y. H. Hsieh and C. L. Cheng, *Nanotechnology*, 2007, **18**, 315102.
93. M. Chen, X. Q. Zhang, H. B. Man, R. Lam, E. K. Chow and D. Ho, *J. Phys. Chem. Lett.*, 2010, **1**, 3167.

94. M. F. Weng, S. Y. Chiang, N. S. Wang and H. Niu, *Diam. Relat. Mater.*, 2009, **18**, 587.
95. I. I. Vlasov, O. Shenderova, S. Turner, O. I. Lebedev, A. A. Basov, I. Sildos, M. Rähn, A. A. Shiryaev and G. Van Tendeloo, *Small*, 2010, **6**, 687.
96. R. Maze, P. L. Stanwix, J. S. Hodges, S. Hong, J. M. Taylor, P. Cappellaro, L. Jiang, M. V. Gurudev Dutt, E. Togan, A. S. Zibrov, A. Yacoby, R. L. Walsworth and M. D. Lukin, *Nature*, 2008, **455**, 644.
97. I. Aharonovich, S. Castelletto, D. A. Simpson, A. Stacey, J. McCallum, A. D. Greentree and S. Prawer, *Nano Lett.*, 2009, **9**(9), 3191.
98. E. D. Eidelman and A. Ya. Vul', *J. Phys. Condens. Matter.*, 2007, **19**, 266210.
99. Y. Wang, M. Jaiswal, M. Lin, S. Saha, B. Ozyilmaz and K. P. Loh, *ACS Nano*, 2012, **6**, 1018.
100. Y. Morita, T. Takimoto, H. Yamanaka, K. Kumekawa, S. Morino, S. Aonuma, T. Kimura and N. Komatsu, *Small*, 2008, **4**, 2154.
101. A. N. Zhukov, F. R. Gareeva and A. E. Aleksenskii, *Colloid J.*, 2012, **74**, 463.
102. A. N. Andriotis, G. Mpourmpakis, E. Richter and M. Menonx, *Phys. Rev. Lett.*, 2008, **100**, 106801.
103. V. Grichko, T. Tyler, V. I. Grishko and O. Shenderova, *Nanotechnology*, 2008, **19**, 225201.

CHAPTER 3

The Chemistry of Nanodiamond

ANKE KRUEGER

Julius Maximilians Universität Würzburg, Institut für Organische Chemie, Am Hubland, 97074 Würzburg, Germany
Email: anke.krueger@uni-wuerzburg.de

3.1 The Surface Structure of Nanodiamond

3.1.1 The Initial Surface Structure

Diamond nanoparticles can be produced in many ways and hence their inherent properties can vary to a great extent. Typical procedures that yield larger amounts of the material include the detonation and shock wave syntheses, which were invented in the 1960s,[1-5] as well as the crushing of microdiamond originating from the HTHP (high temperature high pressure) synthesis.[6,7] Chemical vapor deposition (CVD) diamond films can also be used as a starting material for the production of diamond nanocrystallites.[8] Laser ablation of carbonaceous targets is another way to obtain nanoparticles with a diamond structure.[9-11]

Obviously, the production method influences the resulting product to a great extent. Not only do the particles vary in morphology (see Figure 3.1) but, in particular, their surface reflects the history of the diamond material from its original production through the many processing steps.[5,12,13] However, the lattice structure itself typically does not differ too much, except for the densities of dislocations or twins and other defects that can result from mechanical treatment (*e.g.*, crushing), thermal stress or pressure

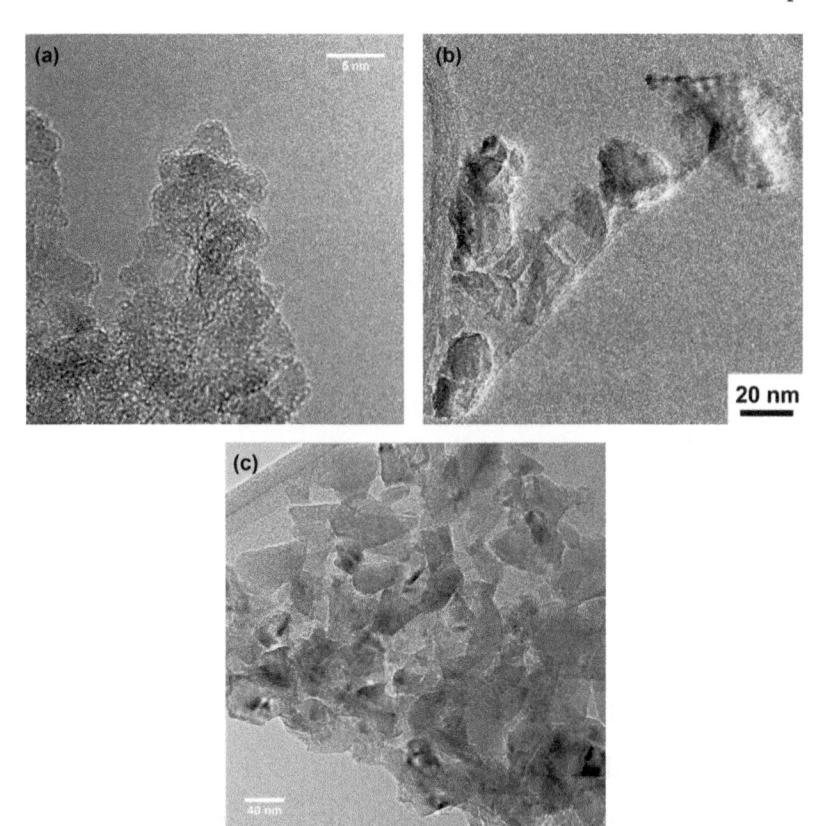

Figure 3.1 Morphologies of nanodiamond particles produced by different methods vary to a great extent. a) Detonation nanodiamond occurs preferentially in the form of roundish particles with a strong tendency for agglomeration (scale bar 5 nm); b) HTHP diamond derived nanodiamond produced by mechanochemical treatment (scale bar 20 nm), as well as nanodiamond obtained by the crushing of polycrystalline diamond films (c, scale bar 40 nm) show a faceted morphology and are much less agglomerated in solution.

gradients. On the other hand, due to an oxidative environment during production or purification, most nanodiamond (ND) materials initially have an oxidized surface. Conditions promoting the formation of oxygen-containing surface groups include the supercritical aqueous environments induced by the addition of ice for cooling purposes during "wet" detonation synthesis. This leads to the formation of highly reactive hydroxyl species. The use of oxidizing mineral acids during soot purification plays an important role, too.[14–18] Such process steps lead to the formation of carbonyl and carboxyl groups (and related structures, such as bridging –O– atoms). The structure of keto groups depends on the location of the group on the different surface orientations with bridging ethers being more likely on [100] surfaces than the isolated keto functions.[19,20] Carboxylic acids, on the other hand, extend from the surface by the bonding of single sp^2 hybridized carbon atoms (*i.e.*, the carbonyl C) to the actual lattice. Bridging positions are not possible for this functional group.

Diamonds that were produced by CVD and did not undergo oxidative surface treatment after production represent an exception. In this case the surface will be hydrogenated regardless of its orientation due to the reactive hydrogen plasma that is formed in the deposition reactor.[21–23] However, any kind of mechanical treatment (such as milling in water) or acid purification will lead to an at least partial oxidation of such surfaces. Therefore, the surface of diamond nanoparticles produced by crushing CVD films will inevitably be oxygenated unless the milling step is followed by hydrogenation.[24]

Besides carbonyl and carboxylic groups, different types of alcohols can be bound to the diamond surface. Depending on the orientation of the surface and the direct surface atom, primary as well as secondary and tertiary alcohols can be found.[25–30] On unpurified nanodiamond, especially from detonation syntheses, different types of carbonaceous structures can be observed. These include not only isolated double bonds, but also conjugated π-systems up to graphene-like or graphitic structures.[31,32]

Another aspect of the initial particle structure is the shape of the individual particles. In the case of nanodiamond it also depends on the production method. All nanodiamond particles produced by crushing with direct impact on the lattice of the starting material will be faceted, as crushing preferentially occurs along lattice planes.[8] Therefore, an extended flat surface will be formed. On the other hand, detonation nanodiamond (DND) exists in the form of more or less spherical primary particles due to the isotropic growth in a kinetically controlled environment.[4,5] Although these roundish objects, in principle, show the same chemistry as their faceted counterparts, the large number of steps and edges on the surface result in a significantly increased reactivity when compared to faceted diamond even of a very similar size. This leads to longer reaction times and harsher conditions for faceted ND when comparable surface loadings with functional moieties are desired (often the surface loading on faceted ND does not reach the values for DND even under such adapted conditions).

3.1.2 Agglomeration and Deagglomeration of Nanodiamond

Due to the presence of different surface groups, varying particles sizes, particle shapes *etc.*, the term "nanodiamond" comprises many different materials. One of the most obvious differences in their properties is the agglomeration behaviour. Nanoparticles of all kinds tend to form larger clusters consisting of anything from a few to a huge number of particles. They are typically held together by non-covalent interactions, such as electrostatic (Coulomb) interactions, hydrogen bonds, van der Waals forces, π-stacking and other weak interactions.[33] A single interaction might contribute only a tiny amount to the exchange energy, but the large number of these binding events can add up to a strong bonding between the particles of an agglomerate. There are several ways to avoid agglomeration of nanoparticles in solution. Typically, electrostatic repulsion, steric hindrance or a combination of the two—electrosteric stabilization—can be achieved by proper surface termination.[33–35] The stabilization of nanoparticles in colloidal solution by surface functionalization will be discussed in the respective chapters on surface chemistry. The second relevant parameter for stabilization of colloidal systems is the surface charge. Highly charged particles do not agglomerate as much as those with low surface charges due to Coulomb repulsion.[35] Therefore, the knowledge of surface charge is of special importance for the understanding or even prediction of the agglomeration behaviour of nanoparticle colloids. This is of course also true for nanodiamond particles. A useful physical value to explain colloid stability is the so-called zetapotential. It does not exactly represent the actual surface charge but the electric potential between the bulk dispersion medium and the static (solvent) layer around the nanoparticle.[36] Hence, it is influenced by the nature and thickness of the solvate shell around the particles. Nevertheless, a negative zetapotential typically relates to a negatively charged surface (as the solvate shells' overall potential is mainly influenced by the charge of the surface groups). The zetapotential is obviously dependant on the pH value of the solution as the charge of the surface groups will be strongly influenced by different pH (*e.g.*, carboxylic acids are deprotonated at high pH and uncharged (protonated) at sufficiently acidic pH). Further influence on the zetapotential can be attributed to phenomena such as surface hole doping *etc.*[37] Another factor contributing to the dispersion stability of a given material is the size of the individual primary particles: the smaller the particles are, the higher the probability is to find conditions where a stable colloid can be obtained. Furthermore, the shape of the particles influences colloidal stability: more or less spherical particles tend to be more dispersible than similarly sized objects with strongly faceted shapes (which is, in part, due to irregular charge distribution on such faceted objects).

Depending on the surface structure, the nanodiamond particles exhibit very different colloidal properties. The values for zetapotentials range from about −50 to +50 mV (higher values are occasionally observed). They depend on the actual surface termination, as well as the pH of the solution, which is the reason for the distinct agglomeration properties of different types of nanodiamond under various conditions. It is therefore very important to know the origin and pre-processing of the nanodiamond before using it, *e.g.*, in biomedical applications. Depending on the production and purification method (see 3.1.1), the samples can differ tremendously. Generally, nanodiamonds originating from HTHP or CVD syntheses show a fairly weak tendency towards strong agglomeration. On the other hand, detonation nanodiamond exhibits unusually strong agglomeration, which involves not only the above-mentioned non-covalent interactions between surface groups and charges but also covalent bonding between the individual particles.[38–42] This is due to the high density of reactive surface

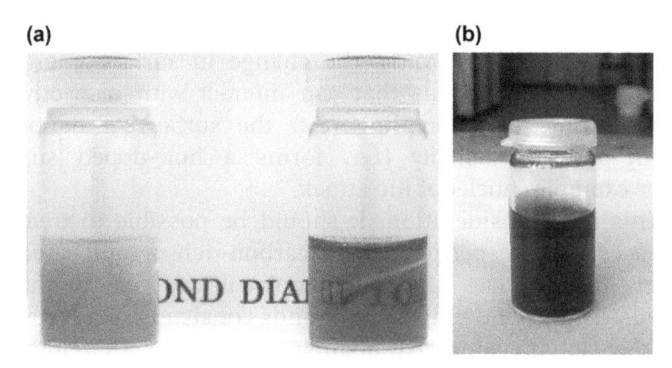

Figure 3.2 Colloidal solutions of nanodiamond. a) Detonation nanodiamond in water before (left) and after deagglomeration by attrition milling using zirconia beads (right)[43] © 2007 Wiley, with kind permission; b) nanodiamond produced from CVD diamond in dimethyl sulfoxide (DMSO) after crushing the film and purifying the resulting suspension from non-carbon and graphitic impurities.

groups that undergo bond formation once the aqueous environment is either dried away or exchanged for another solvent. Even in solution bond formation is possible because of the high local concentration of reactive surface groups and the confined environment in loosely bound agglomerates. Heating or partial evaporation further promote such condensations.

Several techniques have been developed to overcome this obstacle that impedes many DND applications. These include mechanic, *i.e.*, milling or sonication, and mechanochemical techniques, as well as chemical methods, such as the hydroxylation of the surface.[43–48] In the end all approaches are aimed at the establishment of sufficiently charged surface groups and a highly negative or positive zetapotential, thus yielding colloidal solutions of nanodiamond in different solvents depending on the surface termination, size and morphology of the particles (see Figure 3.2).

3.2 Modifying the Surface Termination of Nanodiamond Particles

As a bulk material, diamond is not really reactive except under very harsh conditions. Fluorination, for example, is possible by the treatment of diamond with elemental fluorine.[49] Combustion using air requires temperatures of more than 1000 °C for bulk crystal as kinetic hindrance makes it really difficult to start the reaction (once burning, though, diamond behaves like any other carbon modification and yields a combustion enthalpy of −94.50 kcal mol^{-1} as compared to −94.05 kcal mol^{-1} for graphite).[50] In pure oxygen (O_2 torch) the required temperature is significantly reduced. Depending on the grain size, oxidation sets in at temperatures as low as ∼600 °C.[51] Diamond is inert against most other harsh reactants, such as mineral acids or strong bases. However, when the particle size approaches the nanometre range, reactivity is highly increased. This is due to the much higher number of surface atoms, the change in surface energy and the occurrence of electronic levels that can interact with respective reaction partners. Electronic band bending near the surface of nanodiamond[52] changes the electron density (*i.e.*, forms a hole-doped surface) and enables, for example, nucleophilic attack.

Taking this into consideration, it should be possible to treat nanodiamond surfaces like the periphery of a carbon-rich organic molecule. The most closely related class of organic molecules is that of the so-called diamondoids. These are defined compounds consisting of a certain number of three-dimensionally annulated adamantane units.[53] Therefore, the outer atoms of these molecules show binding situations similar to those found on the different surface facets of larger diamond objects. There are tertiary "bridgehead" or apical positions and secondary "bridge" or medial positions with distinct reactivity. In the parent diamondoids all these positions are terminated with hydrogen atoms. In the meantime a large number of functionalized derivatives have been reported. Typical reactions include, among others, thiolation, hydroxylation and amination.[54] Not all of these reactions can be directly applied to nanodiamond as the surface does not display, for example, the required bridgehead positions in large numbers and in an easily accessible way. Another phenomenon distinguishing the surface chemistry of diamond from adamantane and diamondoid chemistry is the reconstruction of larger areas of the surface. Such transformations are the result of surface energy minimization and they do not play a role in small molecules, like diamondoids. Reconstruction will be discussed in the chapter on surface annealing as these processes are strongly related to each other. On different types of nanodiamond, a large variety of organic reactions are viable. In the end, nanodiamond is nothing but a large, carbon-rich organic compound that can be derivatized in many ways. The extent of functionalization is governed by the amount of available surface positions with suitable reactivity. Hence, smaller particles typically show higher reactivity as the ratio of surface over bulk atoms is higher.

3.2.1 Direct Termination of the Nanodiamond Surface

Although the model compounds of nanodiamond—the diamondoids—are simple hydrocarbons (and hence their "surface" is hydrogenated), the surface atoms of larger diamonds are, in most cases, already terminated with different oxygen-containing groups due to the production and purification processes (see section 3.1.1). In addition, the surface is usually not really homogeneous (see Figure 3.3). Therefore, in a first step, the nanodiamond surface needs to be homogenized. In the following chapters several ways for the homogenization of the nanodiamond surface will be discussed. All these methods represent ways to change the direct termination (the groups directly bound to carbon atoms that are part of the diamond lattice) of

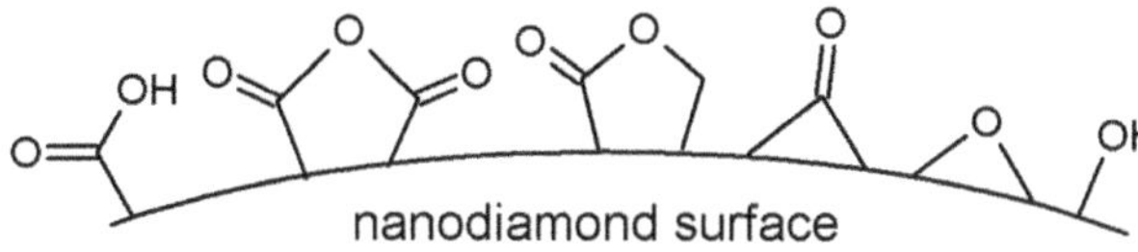

Figure 3.3 The initial surface termination of nanodiamond particles strongly depends on the production and purification processes. In most cases a large variety of oxygen-containing groups, such as acids, ketones and alcohols, are present due to the harsh conditions during purification with oxidizing mineral acids.

nanodiamond. The grafting of larger moieties will be discussed at the end of this chapter.

3.2.1.1 Hydrogenated Nanodiamond

Hydrogenating an organic compound nowadays is a well-explored operation. Transition metal catalyzed addition of molecular hydrogen to double bonds and reduction of carbonyl and hydroxyl compounds, as well as radical reactions of haloalkanes, are well established and give high yields.[55] However, most of these processes are not viable for the hydrogen termination of diamond surfaces. Especially the addition of hydrogen to π-bonds and the defunctionalization of oxygen-terminated diamond are rather difficult. There are several reasons for this: some of the most efficient hydrogenation reactions rely on the use of heterogeneous catalysts (based on Pd, Pt or Ni) or metal-organic compounds as homogeneous catalysts due to the high activation barrier, which prohibits the direct thermal hydrogenation (*e.g.*, addition to double bonds). In the case of nanodiamond this poses a serious problem. First, the removal of the catalyst after the reaction is virtually impossible. Furthermore, for heterogeneous catalysts, a powder mixture with the nanodiamond is obtained, and for homogeneous systems the diamond surface behaves in a similar way to any other carbonaceous nanomaterial—it simply adsorbs such species. On the other hand, with most of the available, non-catalytic reducing agents (such as complex hydrides), the complete reduction fails and alcohols are formed instead. Therefore, it is necessary to find different ways to terminate the diamond surface with hydrogen. The easiest way, of course, is to choose the right production method, where hydrogen termination represents the natural status of the surface, like in CVD production of diamond. However, for the production of truly nanoscale particles with hydrogen termination, this method has not been explored so far. There are very few reports of substrate-free diamond CVD, like the one by Howard *et al.*[56] This process yields rather large crystallites with a broad size distribution. However, this gives a hint toward possible techniques: the treatment of nanodiamond with molecular hydrogen at elevated temperatures or in H_2 plasma should lead to the desired surfaces. In early accounts on the hydrogenation of detonation diamond at high temperatures or in plasma it was reported that the main reactions are decarbonylation and decarboxylation processes.[57] They do, however, not always lead to purely hydrogenated surfaces. Only elemental (not molecular) hydrogen is able to form C–H bonds in an uncatalyzed way accompanied by the complete removal of the oxygen-containing groups as mentioned before. One way to make hydrogen more reactive is to increase the temperature. For normal organic compounds this procedure is limited by the thermal stability of the individual bonds (resulting in cleavage and subsequent decomposition at temperatures above 500 °C). For diamond, however, one can go far beyond that stability limit. The material, even as nanoscale particles, resists thermal treatment up to much higher temperatures. Spitsyn and coworkers investigated the hydrogen treatment of detonation diamond at 850 °C.[58] However, they found not only C–H bonds but also a large amount of OH groups, indicating the reaction of molecular hydrogen with existing carbonyl groups on the DND surface. Unfortunately, going to even higher temperatures is not easily possible with single-digit nanometre-size crystallites. Surface graphitization or decomposition induced by sublimation of larger hydrocarbon fragments or carbon clusters result in an etching process that eventually consumes most of the diamond. Still, for somewhat larger diamond particles (from ~100 nm), the reaction with molecular hydrogen at >900 °C is a promising method. Tsubota *et al.* successfully hydrogenated sub-micron diamond using hydrogen at 900 °C, leading to a hydrophobic material with spectroscopically proven (FTIR) C–H bonds.[59] For smaller particles, only lower temperatures are permitted, hence molecular hydrogen alone will not suffice. Doing the reaction at a lower temperature leads to surfaces that carry not only C–H bonds but also a large number of OH groups. Williams and coworkers reported on a hydrogenation procedure that yielded nanodiamond with a highly positive zetapotential over a rather broad pH range.[60] Most likely, a large number of OH groups is formed during the process as indicated by the IR data. Several other groups reported on similar techniques leading to partially hydrogenated surfaces as well.[61,62]

If thermal hydrogenation is not sufficient, one has to think of ways to make the hydrogen more reactive at acceptable temperatures. One way of doing so is the application of hydrogen plasma. In this technique H_2 is broken apart and highly reactive species, as well as a reducing atmosphere, are present. In such an environment the addition of hydrogen to many kinds of π-bonds is possible.

Obviously, the hydrogen plasma treatment is a well-known technique for diamond films.[63–65] As they are produced in a CVD reactor, this termination occurs inevitably. However, for particulate samples an additional issue arises: as nanodiamond powders cluster into large aggregates, the surface is not necessarily accessible in a homogeneous way. The simplest method for a more homogeneous hydrogenation is to repeat the reaction twice or several times, as reported by Loh and coworkers.[66] They used hydrogen plasma at 800 °C for 60 min twice on oxidized detonation diamond to receive a quite homogeneous surface. Another way to increase reactivity is the formation of atomic hydrogen using a hot filament, followed by the hydrogenation of graphitized detonation diamond.[67] A technically more demanding procedure was reported by Girard *et al.*, whereby, in order to achieve a more homogeneous reaction of the powder, the quartz tubes were rotated in a microwave reactor while hydrogen was passed through (see Figure 3.4). This technique leads to a rather homogeneous exposure of the nanodiamond to the hydrogen plasma.[68–70]

Hydrogenated diamond is a very useful variant of homogenized diamond particles. A variety of grafting techniques leads to stable bonding of organic moieties to the diamond core, typically by C–C bonds of different types. Besides conventional wet-chemical reactions, photochemistry and electrochemistry are valuable means for the modification of hydrogenated diamond. Loh and Zhong have published a review on the different ways of functionalizing *H*-terminated diamond.[71]

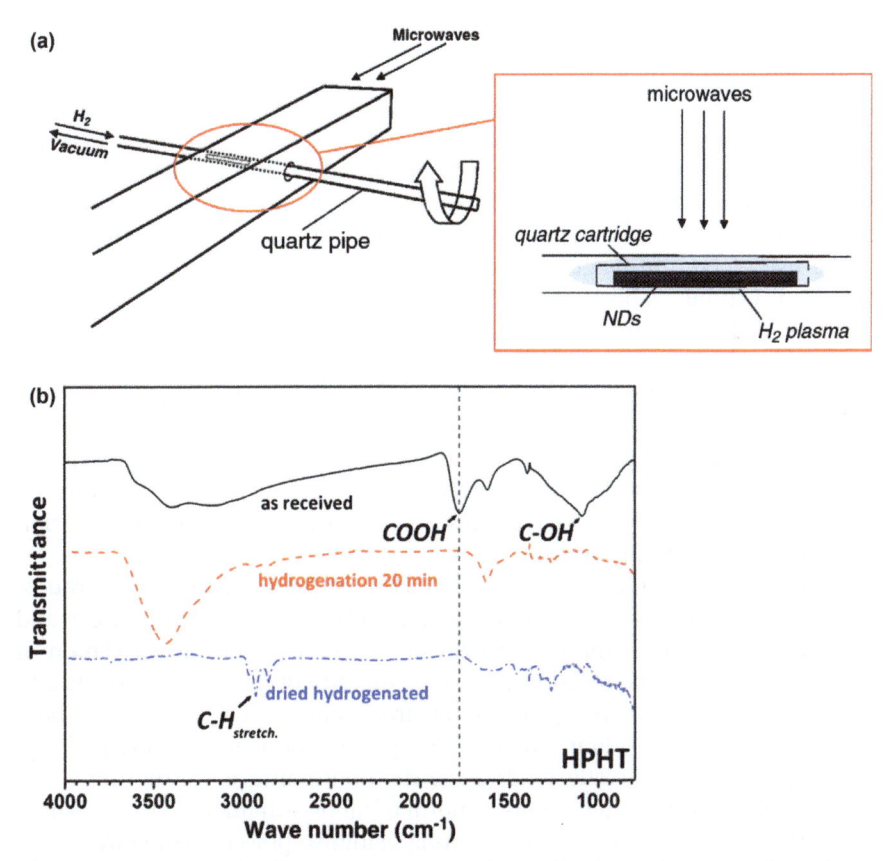

Figure 3.4 Hydrogenation of NDs using plasma leads to the formation of C–H bonds and can be monitored using IR spectroscopy. a) Apparatus for the homogeneous treatment of nanodiamond powders in hydrogen plasma; b) FTIR spectra of HTHP diamond before and after hydrogenation.[69]
© Elsevier 2010, with kind permission.

3.2.1.2 Hydroxylated Nanodiamond

For many applications a hydrophilic but not acidic surface of diamond is desirable. The easiest way to fulfil these requirements is to establish a fully hydroxylated diamond surface—a diamond polyol. However, the typical procedures for the synthesis of low-molecular alcohols,[55] such as the acid-catalyzed addition of water to double bonds, the formation and opening of epoxides *etc.*, have not proved their viability. Again, adapted methods respecting the special situation on the diamond surface have to be found. In section 3.2.1.1, one possibility has already been described: in an attempt to hydrogenate nanodiamond at rather low temperatures using molecular hydrogen, at least a partial hydroxylation (accompanied by C–H bond formation) or even a homogeneous OH termination can be achieved. This works well for detonation diamond and somewhat larger particles alike.

Another option is the transformation of already existing surface groups, namely, all kinds of carbonyl moieties. Depending on the oxidation level achieved during the purification of the nanodiamond, ketones and carboxylic acids (and in some cases derivatives thereof) are present on the surface. Reducing agents include complex hydrides (*i.e.*, LiAlH$_4$), as reported by several groups.[72–74] Utmost care must be taken to remove residual aluminium hydroxide as it is strongly adsorbed on the particle surface or within agglomerates. To this end, complexing agents (like citric acid) are best suited. Conventional acidic workup using HCl is not sufficient.

The use of borane reagents, such as BH$_3$·THF, avoids this issue as the hydrolysis product is boric acid, which can be easily washed away with water. It allows for the hydroxylation of different types of nanodiamond with up to 0.5 mmol g^{-1} of OH groups.[75,76] However, depending on the reducing agent, only part of the carbonyl groups are accessible. Borane, for instance, cannot reduce esters or lactones, whereas lithium aluminium hydride is also able to reduce all kinds of carboxylic acid derivatives. On the other hand, existing double bonds (from small sp^2 structures on the surface) can be transformed into sp^3 carbon carrying OH groups using borane in a hydroborination reaction and a suitable workup.

Besides hydrogen addition and the reduction of pre-existing surface groups, the direct grafting of OH moieties is an attractive technique. García and coworkers reported on the use of Fenton's reagent for the hydroxylation of nanodiamond.[77,78] The reaction employs a mixture of hydrogen peroxide and iron(II) sulphate in a strongly acidic environment, which is known to be a strong and versatile oxidant. The reactive species in this process are hydroxyl radicals, which can directly bind to the diamond surface. Furthermore, the reagent enables the simultaneous removal of all non-diamond carbon (which is "burned" to CO$_2$) due to its very strong oxidative power (see Figure 3.5).

Hydroxyl radicals can be formed by mechanical methods too. Using the so-called beads-assisted sonication (BASD) or attrition milling in water, the surface of the resulting deagglomerated nanodiamond is strongly hydroxylated.[38,43] The obtained particles are highly hydrophilic (zetapotential

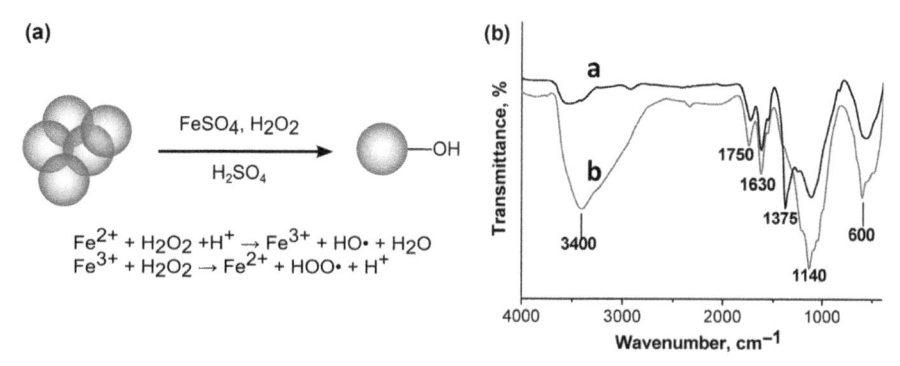

Figure 3.5 Hydroxylation and purification of nanodiamond using Fenton's reagent. The hydroxyl radicals generated from the hydrogen peroxide act as a strong oxidant. a) Reaction scheme, b) IR spectrum of pristine (a) and hydroxylated (b) ND.[77]
© 2010 ACS, with kind permission.

~50 mV at pH 7) and form stable colloidal solutions or even gels at higher concentrations. The mechanism of this hydroxylation process is not yet fully clear; surface radicals on the diamond surface might play an important role.

Photochemical hydroxylation of hydrogen-terminated nanodiamond is another way to introduce OH groups in a controlled manner. When submitted to water vapor under UV irradiation, hydrogenated detonation diamond can be transformed into hydrophilic DND-OH.[68] The reactive species are again OH radicals generated *in situ*.

One has to be aware of the fact that all the methods mentioned above do not produce a purely hydroxylated surface when pristine detonation diamond, with its inhomogeneous surface termination, is used. None of the methods is able to transform all kinds of existing structures into sp^3 carbon carrying OH groups. Therefore, in these cases it is recommended to first oxidize the nanodiamond with a really strong oxidizing agent and do the reduction in a second step using a scheme involving OH radicals. Of course, hydrogenation will also produce suitable starting materials for a subsequent hydroxylation.

3.2.1.3 Carboxylated Nanodiamond

On the upper end of the oxidation scale of carbon stand the carbonyl and carboxyl species. Carboxylic acids and their derivatives represent the most oxidized state of the carbon atom in an organic compound (only inorganic compounds, such as CO_2, possess +IV) and obviously on diamond surfaces too. The attachment of a COOH group to the diamond particle is necessarily given by one single C–C bond as all other valences of the oxidized carbon are occupied by the functional group. In the case of ketones, an "end on" and a "bridging" binding mode can occur depending on the structure of the respective surface. So far, the chemistry of the ketone-functionalized surface has not been explored in detail, and hence will not be discussed here. Highly oxidized, *i.e.*, carboxylated nanodiamond, on the other hand, is a highly versatile ND derivative and is commonly used for a variety of applications.

The carboxylation of nanodiamond requires very strong oxidizing agents to ensure the homogeneous establishment of carboxylic groups and to remove all lower oxidation states at the surface. Due to these harsh conditions, part of the nanodiamond is typically lost and the particle size can be reduced—in some cases a desired side effect when aiming, for instance, at small, fluorescent ND. The released carbon is removed in the form of carbon dioxide and sometimes carbon monoxide.[79] Typical reagents include mixtures of strong oxidizing mineral acids, such as concentrated sulphuric, nitric (and sometimes perchloric) acid, or a mixture of sulphuric, nitric and hydrochloric acid.[80–85] As these reaction can, at times, be rather violent, utmost care has to be taken when treating nanodiamond (or any other carbon material) with these mixtures. Another option is the use of "piranha water", a mixture of sulphuric acid and hydrogen peroxide.[86] This reagent is equally dangerous as it is able to detonate when in contact with easily oxidizable matter. Hydrogen peroxide alone is also a potent oxidant for nanodiamond.[83] However, sometimes it turns out that the surface does not only carry COOH groups but also ketones and ether moieties after oxidation with H_2O_2 alone. A frequently applied method for the surface oxidation of ND is treatment with a 3:1 mixture of concentrated sulphuric and nitric acid.[30,87] This reagent not only establishes the desired COOH groups, but it also removes residual non-diamond carbon. The latter has a higher reactivity towards oxidation and therefore reacts with an increased rate to the fully oxidized CO_2.[16] Another suitable, although technically highly demanding, technique is treatment with supercritical water.[88,89] Water molecules in this state actually represent a strong oxidant. Non-diamond carbon is easily removed using this kind of treatment. But attention: if the temperature and pressure are too high, the diamond will be oxidized as well! The nature of the resulting surface groups on the remaining nanodiamond has to be investigated in more detail. So far, it is clear that the surface is oxygenated. However, the presence of OH groups in addition to higher oxidation states seems quite likely. For purposes other than purification, treatment with supercritical water is, therefore, not yet applicable.

Another useful aspect of an oxidative treatment is the removal of non-carbon impurities, such as metals and metal oxides. These originate from, for example, reactor linings in the case of detonation diamond, milling debris from jet or attrition milling, or metal impurities from industrial grade nitric acid, which is used for the first cleaning of the raw nanodiamond powder.

Besides wet-chemical oxidation, the use of oxygen or ozone in a heterogeneous reaction is yet another way to oxidize the surface atoms of nanodiamond. However, very careful control of the treatment conditions is required in order to keep diamond losses as small as possible and still oxidize the entire surface and completely remove the non-diamond carbon.

The Chemistry of Nanodiamond

Gogotsi and coworkers have published a detailed study on the control of the sp^3/sp^2 ratio using oxidation in air.[18] The temperature of the reaction zone needs to be chosen with respect to the size and morphology of the diamond to be oxidized (see Figure 3.6). A minimum of 400 °C is required even for the smallest ND; above 450 °C such nanodiamonds undergo oxidation together with all non-diamond carbon. For detonation diamond a temperature of 425 °C has turned out to be suitable. However, for each type of nanodiamond the optimal temperature should be determined, for example, by oxidizing a small amount of the material in a thermogravimetric analysis (TGA) apparatus to monitor the mass loss with temperature.

Air oxidation not only removes sp^2 carbon but it also establishes oxygenated surface groups. The surface becomes quite homogeneous and also transparent to light from luminescent lattice defects as reported by Smith

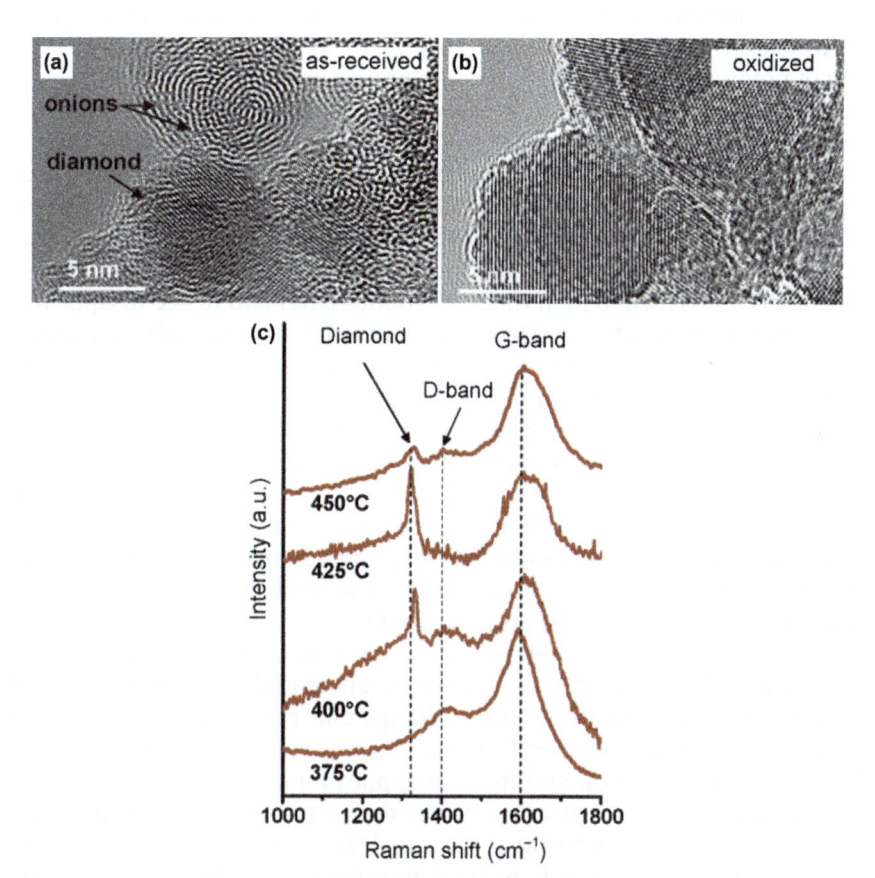

Figure 3.6 Air oxidation of nanodiamond. HRTEM images of pristine (a) and air-oxidized (b) material of detonation diamond (UD50); c) Raman spectra before and after oxidation at different temperatures of the same sample. The optimal oxidation temperature is found to be 425 °C.[18] © ACS 2006, with kind permission.

and colleagues.[90] They reported on the improved luminescent properties of nanodiamond after oxidation in air. In some cases oxidation at much higher temperatures has been reported.[81] In these cases material loss is more substantial; however, the homogeneity of the obtained samples is often quite high.

After air oxidation the surface typically exhibits a mixture of COOH groups with varying amounts of ketones, aldehydes, lactones, ethers and esters. Depending on the envisaged application, a characterization of the existing groups by IR spectroscopy, X-ray photoelectron spectroscopy (XPS) and/or Boehm titration (see below), as well as an optimization of the air oxidation have to be carried out.

Air oxidation has been used to reduce the size of nanodiamonds by partial oxidation of the diamond core. In 2004 a study reported on the size dependence of the reactivity and the resulting size reduction during the treatment of different nanodiamond samples.[91] In 2012 Gaebel *et al.* applied air oxidation for the size reduction of luminescent nanodiamond with the aim of producing the smallest possible luminescent ND.[92] In their study the smallest ND with a stable luminescence from nitrogen–vacancy (NV) centres had a diameter of 8 nm. However, the oxidative size reduction seems to have a lower limit: below ~ 4 nm ND particles undergo graphitization in the air stream and are then rapidly oxidized to CO_2. Obviously, the low energetic difference between sp^2 and sp^3 carbon at this size plays a major role (together with the elevated temperature). Such small objects of sp^2 carbon are much more reactive than their sp^3 counterparts and hence disappear in the course of the reaction.[93] Therefore, the oxidation of nanodiamond quite often results in an apparent increase in particle size, which is due to the disappearance of the smaller diamond particles by oxidation and hence an increasing average diameter. Nevertheless, air oxidation is a suitable method for the homogenization and purification of nanodiamond materials.[94]

If particularly strong oxidation is required, one can also turn to ozone treatment. It has been reported that ozonized ND exhibits not only a high acidity (indicating a large number of COOH groups at its surface) but also an improved colloidal stability in dispersion. The latter is most likely an effect of the large number of polar surface groups.[95,96]

Independent of the chosen oxidation technique, the resulting oxidized nanodiamond is hydrophilic, exhibits zetapotentials of ¬30 to ¬50 mV depending on the density of COOH groups and can be dispersed in polar solvents, including cell culture media. This renders it highly attractive for a large variety of biological and medical applications.

However, in order to use the established carboxylic groups in further reactions or for the adsorption of polar molecules, the number of groups on a certain surface area needs to be determined. One option is to use the reaction with a standard compound. Such a technique has been described by Cheng *et al.* with the adsorption of dodecylamine onto carboxylated diamond surfaces.[87] A zwitterion is formed and the gravimetric determination of the mass increase gives the number of available COOH groups. For a very dense loading with COOH groups this method, however, is not fully reliable. Here, the classical Boehm titration (originally developed for the quantification of oxidized surface groups on soot) is best suited.[28] Care must be taken to remove all adsorbed carbon dioxide from the surface just before the titration as it would lead to the overestimation of the actual number of COOH groups.[97] Another advantage of the Boehm titration is the ability to distinguish between different oxygenated groups when a stepwise titration of the different groups is carried out.

3.2.1.4 Annealing the Surface of Nanodiamond

As already mentioned, the diamond surface does not necessarily carry surface groups. The removal of all foreign elements, including hydrogen, would result in the formation of a highly unsaturated all-carbon surface that exhibits dangling bonds at either the apices or the bridge atoms of the surface adamantane units, depending on the orientation of the crystal's facets. These dangling bonds are then saturated by a phenomenon called surface reconstruction. For diamond, it has been studied in detail. The process leads to highly ordered structures of π bonds on the diamond surface.[98,99] Of course, this process occurs not only on continuous films but on diamond particles as well, provided sufficient energy is supplied and the surface is free of hindering groups.[100,101] The transformation is not limited to the direct surface of ND particles. There have been several reports on the complete or partial transformation of NDs into bucky diamonds (ND core with a multilayer graphitic shell) and carbon onions (complete transformation into multi-shell fullerenes). This change of modification can be induced either thermally[102-104] or by irradiation with high energy particles, such as electrons in a transmission electron microscope (TEM).[105-107] So far, only treatment of nanodiamond at elevated temperatures has been used for the production of macroscopic amounts of graphitized ND materials as electrons with sufficient energy are not easily available on a large scale.

As with air oxidation, the parameters of the thermal treatment for surface annealing have to be chosen with care. In particular, when NDs with a very thin (preferentially monoatomic) layer of sp^2 carbon is desired, process control is highly important. Not only the temperature but also the duration and the vacuum in the chamber (or quartz tube) play decisive roles in the formation of sp^2 carbon. Furthermore, the size and morphology of the ND particles have an influence as the surface reactivity differs significantly between different types of ND. Typically, roundish particles are more prone to thermal annealing than faceted ones. Already existing sp^2 carbon at the particle surface promotes further graphitization.

Larger particles require temperatures well above 1000 °C for significant graphitization. For instance, Hwang and colleagues annealed air-oxidized diamond with sizes around 250 nm at 1200 °C for 1–3 h and observed the formation of sp^2 carbon, which could subsequently be modified in radical reactions.[108] The sp² carbon was found to be well ordered and resembled the structures in carbon nanotubes.

Small (single-digit nm) diamond would at least be transformed to heavily graphitized bucky diamond at this temperature. It is hence necessary to work under less harsh conditions in order to obtain a very thin layer of sp² carbon in this case.

A typical protocol is thermal annealing at 700–750 °C under a moderate vacuum of $\sim 10^{-3}$ mbar for a duration of 2 h (see Figure 3.7). These conditions are sufficient to remove the oxygenated surface groups and to form of a thin layer (albeit not continuous) of sp² carbon.[109] If higher temperatures are chosen, multilayer graphitic structures are inevitably formed. Usually, the formation of such a multilayer shell is not desired as the particle properties are then governed by this shell and not by the diamond core

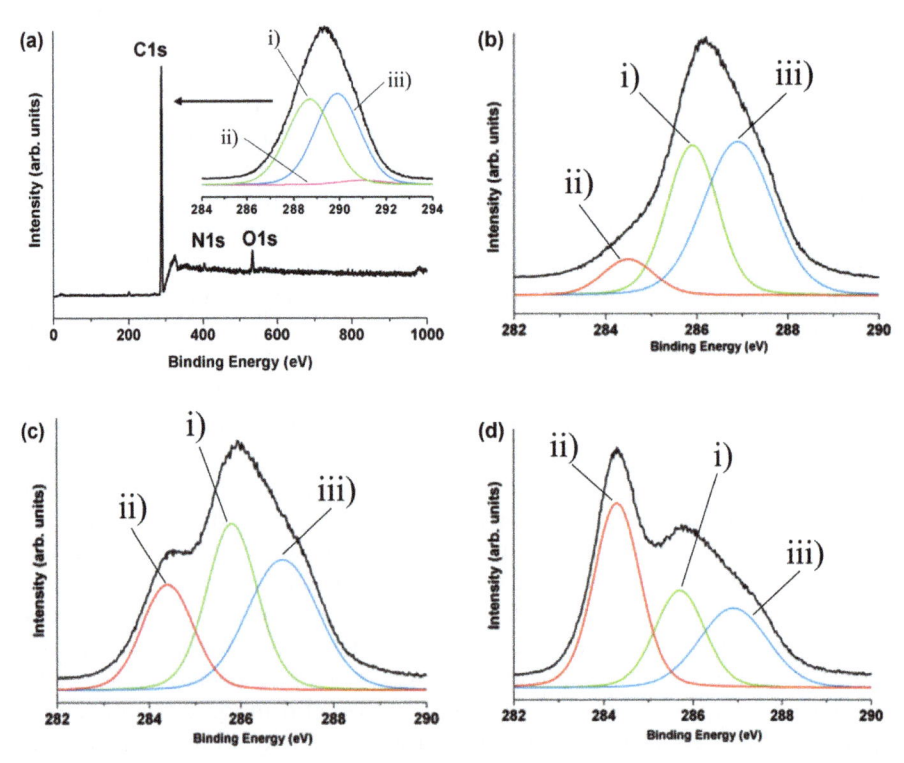

Figure 3.7 The initial stages of thermal annealing can be followed in detail by XPS. The peak deconvolution shows sp³ carbon (i), sp² carbon (ii) and defective structures (iii). Initially, curved graphene-like caps are formed on the surface, which coalesce to closed graphitic shells. The latter reduces the number of defective structures as well. XPS data of a) pure nanodiamond, b) ND annealed at 700 °C, c) ND annealed at 900 °C, d) ND annealed at 1100 °C.[32]
© 2011 American Physical Society, with kind permission.

anymore. It was found that a better vacuum does not mean better quality of the annealed ND. We observed that, in these cases, residual carbonyl and OH groups are still present. One possible explanation for this surprising finding is the absence of residual oxygen and humidity in the case of a "good" vacuum, which might be necessary for an initial etching step.

It was found that detonation nanodiamond withstands higher temperatures when deposited onto a substrate, such as silicon nitride-covered silicon. This could be due to the stabilizing effect of the nitride layer or the low surface reactivity of the ND in the complete absence of humidity and oxygen in the reactor chamber.[110]

Besides actual thermal annealing, the energy that is required for the surface graphitization can also be introduced using mechanical means. Intense milling and beads-assisted sonication (BASD) lead to the formation of sp^2 carbon at the particle surface as well. This can be explained by the recombination of dangling surface bonds that are generated by the break-up of larger particles or in the deagglomeration process.[44]

Some of the available nanodiamond varieties already contain graphitic carbon. However, due to the harsh production conditions, for example, in detonation synthesis, the amount and degree of graphitization cannot easily be controlled and the materials' properties vary from one batch to the next.[111–113] It is therefore recommended to first remove all existing non-diamond carbon by oxidation or hydrogenation and then to start from this homogeneous material for the formation of a reproducibly thin layer of sp^2 carbon. Suitable methods include ozone treatment,[95,96] mixtures of HF/HNO$_3$ or HNO$_3$/H$_2$O$_2$[16], supercritical water[88] or boric anhydride.[114] As most of the purification protocols include at least one oxidation step, the removal of sp^2 carbon can be achieved simultaneously with the removal of non-carbon impurities. It is highly recommended to use such homogenized material for the production of surface-annealed nanodiamond.

3.2.1.5 Further Terminations of the Surface of Nanodiamond

Besides the most common surface terminations of NDs described in the sections 3.2.1.1–3.2.1.4, there are a few other possibilities for the establishment of useful surface groups; one of them is the amination of NDS. So far, a complete and homogeneous direct amination of the ND surface has not been achieved. However, several promising methods on the way to this ultimate goal have been reported. The reason for the appeal of NH$_2$ groups attached directly to the surface is their suitability for the coupling of, for example, biomolecules or polymers by peptide coupling or reductive amination and several other well-known organic transformations involving amines. The latter have a highly nucleophilic character and could, for example, react with electrophilic species in a rather straightforward manner.

For diamond films, several treatments leading to the formation of amino groups on their surface have been described. These include plasma treatment (using gas feed containing ammonia)[115] and photochemical amination.[116] For slightly larger diamond particles (ca. 500 nm) a thermal amination using gaseous ammonia has been reported.[117] As these techniques did not lead to fully aminated ND particles, other approaches have to be taken to establish NH_2 on the ND surface. Usually, this includes the grafting of a short spacer molecule that acts as a bridge between the diamond surface and the amino group (see below and section 3.2.2.1 on alkylation.

The shortest linker reported so far is a methylene bridge. Shenderova and coworkers transformed initial OH groups into nitriles (*via* tosylation and subsequent nucleophilic substitution), which then underwent reduction to $-CH_2-NH_2$ moieties. However, due to the existence of $-CH_2OH$ groups at the ND surface, originating from the reduction of acids or lactones, ethylene bridges should be present as well. The intensity of the respective alkyl signals in the IR spectra support this hypothesis.[74] Other linkers include silanes[75,76] and aromatic rings.[66]

Similar to amino groups, thiols are another sought after termination for nanoparticles in general and of NDs as well. Thiols enable the grafting of biomolecules by the formation of disulfide bridges or by thiol-ene reactions. Furthermore, thiols could be used to attach NDs onto noble metal surfaces and structures, *e.g.*, gold. For instance, the placement of NDs in plasmonic nanostructures relies on a stable positioning of the nano-object, *e.g.*, by the formation of a bond between surface thiols and the gold.[118] A very simple scheme for the introduction of sulphur-containing surface groups was reported by Nakamura and colleagues (see Figure 3.8). They used the

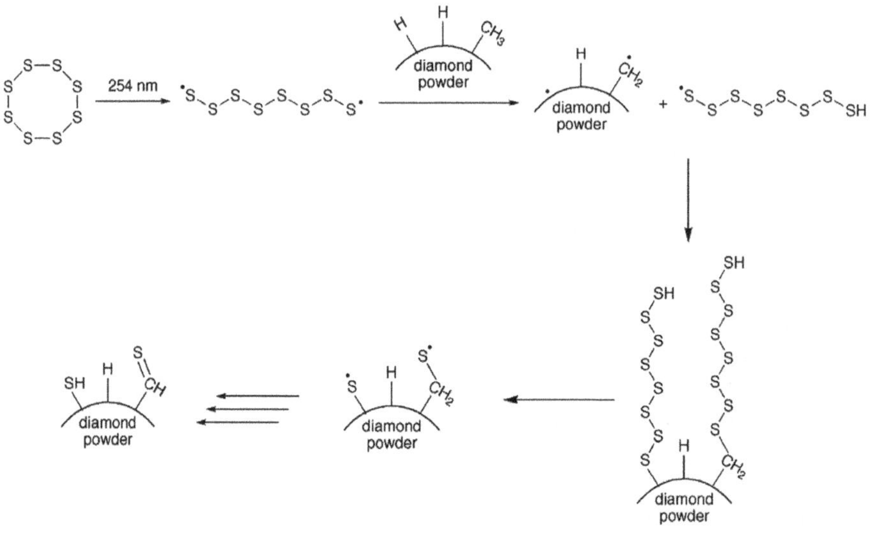

Figure 3.8 The photochemical reaction of nanodiamond with elemental sulphur and the formation of thiol groups at the surface.[119]
© 2009 Royal Society of Chemistry, with kind permission.

photochemical generation of reactive species (polysulfide radicals) starting from elemental sulphur S_8.[119] The structure of the resulting groups is not fully clear (possibly thiols, but moieties with more than one sulphur atom in a row might be involved too). The interaction with gold, however, indicates a successful sulphur termination of the ND surface. As with amines, another way for the establishment of SH groups is the use of a short linker. In this case the functional group can be generated after the actual grafting step, *e.g.*, by the reduction of sulfonic acids to thiols.[109]

Another highly versatile group in organic chemistry are alkyl halides as they can easily be substituted in a variety of ways. For diamond this holds true as well. Unfortunately, the portfolio of available halogenation protocols is very limited. Basically, fluorination is the only halogenation that can be used on a preparative scale.

A well-established route to fluorinated NDs is the treatment of the pristine powder in a tubular oven at ~500 °C with a mixture of F_2 and hydrogen.[120,121] This has been applied to generate highly hydrophobic NDs. The fluorine atoms are then easily replaced by nucleophiles, most likely in an elimination–addition reaction. However, this can only hold true when, in the elimination step, the intermediate carbanion-like transition state is not tertiary but less alkylated instead. Nucleophiles that can be used in the second step (the nucleophilic addition) include amines and metal-organic compounds, such as lithium alkyls or Grignard reagents. Furthermore, the reaction with CF_4 in atmospheric plasma affords fluorinated diamond.[122] However, the nature of the surface is not fully clear. It is highly possible that a mixture of fluorine atoms and CF_3 groups attach to the surface.

Although fluorination of NDs has been reasonably well explored in recent years, halogenation with the heavier representatives of this group has not. This is mainly because the resulting materials show a pronounced reactivity and are not stable under usual ambient conditions. Chlorination is still the most feasible possibility and photochemical chlorination of hydrogenated diamond leads to at least partially Cl-covered nanodiamond.[117] However, thermal chlorination using elemental chlorine or tetrachloromethane can also be used to produce such products.[123] The published examples of attempted bromination of nanodiamond are very scarce. There has been a report on the partial bromination of hydroxylated nanodiamond using *N*-bromo succinimide (NBS).[78]

As already mentioned above, the stability of the halogenated ND samples is an issue. Typically, these samples are quite susceptible to hydrolysis leading to mixed surfaces containing halogens and OH groups. The reason for this different behavior as compared to fluorinated ND is the fact that chlorine and bromine atoms are good leaving groups and lead to the formation of carbenium ions when bound to tertiary or secondary carbon atoms in protic environments. These carbenium ions are then easily attacked by nucleophiles, such as water, which results in the partial substitution mentioned above.

3.2.2 Using Linkers for the Surface Modification of Nanodiamond

The direct establishment of suitable surface groups on the diamond surface is not always possible. In such cases the use of linker molecules is recommended. Several requirements apply to these moieties: a) they should be readily available, b) they should be stable against, for example, hydrolysis when the envisaged application requires so, and c) the grafting reaction should be specific and efficient. Furthermore, a versatile linker enables the presentation of a variety of different terminal groups. Several systems have been used so far for the functionalization of nanodiamond. This chapter presents the most versatile and frequently used linkers for the modification of nanodiamond.

As a starting point, one might consider the nature of nanodiamond as a huge organic molecule with a core of quaternary carbon atoms and a rather small number of peripheral atoms. In that sense one could expect to see a surface reactivity very similar to adamantane, which is the smallest member of the diamondoid family.[53] However, this is not fully the case. Several additional parameters, such as steric hindrance, electronic effects of the diamond core and, of course, pre-existing surface groups, influence the surface behaviour.

3.2.2.1 Alkylation

The formation of bonds between the ND surface and alkyl chains is a versatile tool to control the hydrophobicity of the ND particles. The longer the lipophilic chains and the higher their density on the ND surface, the less polar the respective functionalized nanodiamond. Depending on the termination of the surface before the alkylation, these groups can be either attached by the formation of C–C bonds, ethers, esters or amides. Furthermore, alkyl chains with an additional terminal functional group enable the further grafting of larger moieties with a flexible linker between the particle surface and the functional unit.

Starting from halogenated NDs, one can use metal-organic compounds to substitute the halogen by a C–C bound alkyl residue. So far, this kind of reaction has been reported for fluorinated nanodiamond.[120] As fluorine is not a good leaving group, it is rather unlikely that a conventional S_N2 reaction mechanism is followed. It seems more likely that an elimination–addition process is taking place, where the carbanionic species act both as the base (in the elimination step, which could be considered to be of the E1cB type)[124] and a nucleophile (in the subsequent nucleophilic addition to the electron-poor, rather strained double bonds). This goes along with the observations for diamond films and C_{60}.[125] Reagents that have been successfully used to alkylate NDs are butyl-lithium and a variety of Grignard reagents (alkyl magnesium halides). Besides these carbanionic reagents, amines can also be used resulting in the formation of C–N bonds between the ND surface and the residue.

The formation of ethers between the hydroxylated ND and alkyl chlorides using a base (i.e., NaH), like in a classical Williamson ether synthesis, has been reported by Cheng and coworkers.[73,126] The acylation of ND-OH with acid chlorides with different alkyl chain lengths has been used to increase the solubility of the increasingly hydrophobic ND materials in rather non-polar organic solvents, such as tetrahydrofurane.[127] In protic solvents these acylated ND species are not stable, which is due to the fact that the ester bonds are typically formed between tertiary or neopentyl alcohols (on the ND surface) and the carboxylic acid chloride. Such esters are known to be rather sensitive to hydrolysis (and cleavage by primary alcohols).

When the diamond surface is initially covered with COOH groups, the formation of esters or amides can be used for the grafting of alkyl chains. That is how carboxylated NDs were first reacted with thionyl chloride to form the respective acid chlorides and, subsequently, dodecylamine, which was added to produce a highly hydrophobic nanodiamond.[128,129] However, also in this case, the alkylation does not occur directly on the diamond surface and the hydrophobic shell is prone to hydrolytic removal.

The most stable bonding would be established if the alkyl groups were bound directly to the diamond surface by C–C bonds. Such reactions can be carried out using fluorinated nanodiamond (see above), but also hydrogenated NDs could be functionalized in such a manner. For this purpose the well-known photochemical grafting of terminal alkenes[130] has been employed by Girard et al.[68] They used UV light to generate surface radicals, which attack the double bond. In a similar way Nakamura and colleagues immobilized optically active amides on the surface of sub-micron diamond particles.[131] Besides the photochemical grafting, it is possible to use radical species that are generated chemically. This has been reported by Tsubota and coworkers for the grafting of lauryl residues, as well as aromatic species.[132] Komatsu and coworkers immobilized ω-amino carboxylic acids using benzoyl peroxide as the radical initiator. Although not fully clarified, it seems that, in this case, the immobilization was achieved by the formation of ester bonds.[133] Perfluorinated residues were bound on different types of diamond particles using the respective azo precursors.[134]

3.2.2.2 Silanization

Hydroxylated nanodiamond can also be used to establish siloxane shells around the ND particle. Such shells can contain terminal surface groups that enable further functionalization of the resulting hybrid material (see Figure 3.9). Depending on the silanes used and the nature of the nanodiamond, the silanization can either yield individual nanoparticles with individual siloxane linkers attached to them or coated particles with a fully condensed shell. In the latter case this can be used to round the outer shape of faceted ND particles.[135]

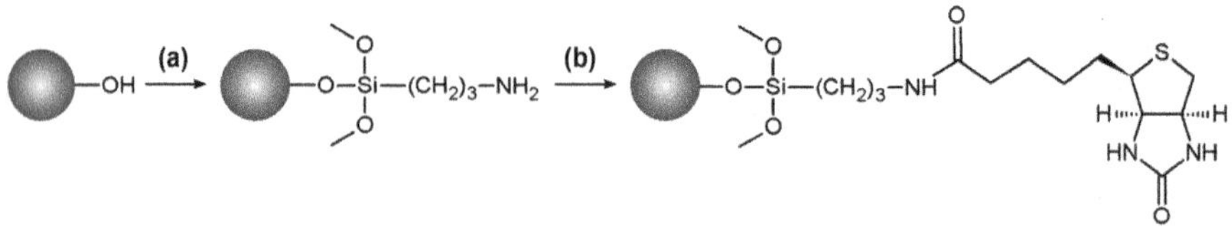

Figure 3.9 Silanization of nanodiamond can be used to establish a large variety of functional groups on the ND. Either trialkoxysilanes or dialkoxysilanes are used in order to achieve a stable binding. In this example biotin has been immobilized *via* silanization using the silane (MeO)$_3$Si–(CH$_2$)$_3$–NH$_2$) in step a) and amide formation using ethyl-dimethylaminopropyl carbodiimide (EDC) and a biotin conjugate with valerian acid in step b). The ND–biotin conjugate exhibited unaltered activity in a streptavidin assay.[136]

The first account of silanized NDs was reported in 2006, when 3-aminopropyltrimethoxy silane was used to functionalize the surface of detonation diamond.[75] In that case the subsequent grafting of amino acids by conventional peptide coupling was demonstrated. Further reports on the same aminosilanized ND have included the grafting of biomolecules,[136] dyes[137] and catalysts for carbon nanotube growth.[138] In addition, phenylboronic acids have been immobilized on ND using the silanization technique.[139] These have been subsequently used to selectively capture glycoproteins and detect them in a MALDI setup. When the SiO$_2$ shell is not fully closed (and hence active alkoxy groups are still available), the silanized diamond surface is not fully inert, especially in aqueous media at slightly acidic pH. Therefore, a variety of applications requires either a fully condensed silica shell (by using, for example, tetraethyl orthosilicate (TEOS) and suitable reaction conditions) or the use of other grafting methods.

A typical side effect in silanization reactions is the formation of interparticle bonds by the condensation of remaining unbound alkoxy groups. As, usually, a maximum of two alkoxy groups of one moiety are grafted onto the particle surface due to steric reasons, such reactions can take place rather easily.[140] The same process is also responsible for the formation of larger siloxane shells around the ND particles as several layers of trialkoxy silanes will be grafted in such condensations. To avoid these undesirable processes, one can use mono- or di-alkoxy silanes, which do not show such a strong tendency towards condensation after grafting (usually all alkoxy groups react with the surface in this case). However, to some extent, the condensation can actually help to stabilize the siloxane shell when the condensation reactions occur between neighboring groups on the same particle. This can, for example, be achieved by sufficient dilution of the reaction mixture or the simultaneous application of mechanical force either by milling or beads-assisted sonication (BASD).[44] By doing so, the surface of each particle is functionalized in a very homogeneous way as the initially resulting agglomerates are constantly destroyed by shear force (see Figure 3.10). The freshly generated surface does then react immediately

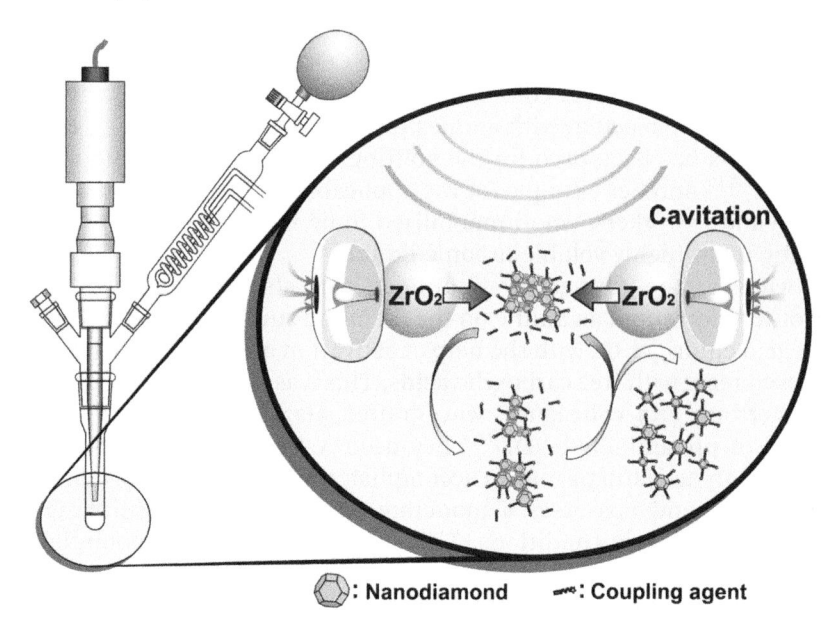

Figure 3.10 Apparatus for the mechanochemical treatment of nanodiamonds and other nanoparticles. Simultaneous beads-assisted ultrasound deagglomeration (BASD) and surface functionalization lead to fully dispersed and homogeneously functionalized ND particles. The freshly generated surface is immediately saturated by reagent molecules, thus preventing re-agglomeration and gradually producing functionalized primary particles.[44]
© American Chemical Society 2009, with kind permission.

with the available reagent molecules and, hence, by running the process for a sufficient period of time, isolated primary particles with a (mono)layer of siloxanes are produced.

3.2.2.3 Ester and Amide Formation

The formation of esters and amides is a well-established reaction in many areas of chemistry. It is therefore highly desirable to use this reaction for the modification of carboxylated nanodiamond.

It can be used to couple peptides onto carboxylated nanodiamond using amino groups from side chains in the peptide. Of course, the terminal amino group of the peptide strand could be used as well. In principle, any molecule containing a suitable amino group (preferentially a primary amine) can be coupled using techniques developed for protein chemistry. These include the use of so-called coupling agents and, in some cases, also the activation of reactants prior to the coupling. Reagents include the carbodiimides, such as EDC, but also compounds like EEDQ (*N*-ethoxycarbonyl-2-ethoxy-1,2-dihydroquinoline).[141] In cases where the ND species involved and the functional molecules are not sensitive to acid chlorides and the stereochemistry in the linkage is not an issue, the COOH group on diamond can be transformed into an acid chloride using oxalyl chloride, thionyl chloride or alike. Then, the reaction with amines proceeds without the use of coupling agents. This has been used for the grafting of dodecylamine onto nanodiamond.[128,129] Another example for the applicability of this method was given by Lee and coworkers, who immobilized ionic moieties on the ND surface making them highly soluble in ionic liquids.[142]

Obviously, the formation of amides can also be achieved starting the other way round: an amine on the diamond surface should readily react with acid chlorides or, with the help of activation and/or coupling agents, it can even react with free carboxylic acids. This was, for instance, reported by Boukherroub and colleagues, who grafted 4-azidobenzoic acid onto the surface of pre-aminated ND.[143] They used N,N'-dicyclohexylcarbodiimide (DCC) as an activating agent to accomplish the coupling. They found, that the amide bond possesses a moderate stability against hydrolysis even under slightly basic conditions. This fact is important for potential uses in biomedical applications, where water stability is an absolute requirement. However, even with this criterion fulfilled, the ND material is not necessarily applicable under physiological conditions: Enzymatic decay, for example, by proteases, could cleave the linkage between the ND particle and the functional molecules on its surface and hence hamper the proper function of the conjugate.[144] In such a case non-natural linker systems using direct C–C coupling between the ND and the functional moiety, or cycloadditions and related transformations, should be used (see below).

Besides the formation of amides, the respective esters can be generated using carboxylic acids and suitable alcohols. As with amides, the carboxylic acid can be either established on the ND surface or react with its counterpart (here an alcohol) on the ND particle. For both variations one can find examples in the literature. Komatsu and coworkers reported on the immobilization of polyglycidol on the ND surface, which reacted, at least in part, with surface-bound COOH groups.[145] The formation of esters using hydroxylated ND has been reported as well. Acid chlorides of different fatty acids have been coupled to ND-OH produced by the borane reduction method (see ref. 127). Due to the long alkyl chains grafted onto the diamond surface, these derivatives become soluble in organic solvents. However, esters are more sensitive towards hydrolysis than amides both under acidic and basic conditions. In particular the acylation of ND-OH leads to derivatives with enhanced hydrolytic sensitivity. This is due to the tertiary or neopentyl esters that are formed during the esterification. Furthermore, a large variety of enzymes is able to cleave ester bonds (different types of esterases). Therefore, the coupling of functional moieties for bioapplications should rather be carried out with either amide linkages or hydrolytically stable bonds, such as C–C bonds.

3.2.2.4 Arylation Using Diazonium Salts

Under certain conditions, groups attached to the surface by heteroatom bonds are prone to cleavage. In order to reduce this unwanted decomposition, the connection of the functional moiety with the ND can be achieved using non-polar C–C bonds. In order to do so, the reagents have to be attached to the carbon atoms immediately on the surface. If these are hydrogenated, the reaction needs either radical initiators or photochemical activation of the diamond surface (see section 3.2.2.1). Surface radicals can then undergo bond formation by attacking terminal alkenes. The generation of surface radicals is also discussed for the formation of C–C bonds between the diamond surface and aromatic moieties. In this case the bond is formed between an sp^3 hybridized surface atom on the ND and the sp^2 hybridized carbon of the aromatic linker.

The precursor for this type of arylation is typically an aromatic diazonium salt (see Figure 3.11). It can be generated *in situ* or applied as an isolated salt (mostly tetrafluoroborates are used). The arylation can take place on hydrogenated ND,[68,69] but also sp^2 carbon can be reacted using this technique.[146] In the latter case the reaction of thermally treated ND is related to the processes observed for the arylation of carbon nanotubes with the same reagents. In particular aryldiazonium salts carrying electron-withdrawing groups in the *para* position of the aromatic ring show a high reactivity towards annealed ND.[109] During the addition of the aryl moieties the previously sp^2 hybridized carbon returns to sp^3, which improves the diamond character of the resulting derivative by significantly reducing the amount of surface sp^2 carbon. That does not mean, however, that the transformation of the partially graphitized surface is complete. Usually, a certain amount of sp^2 carbon will remain, as not all surface atoms can be reacted with aryl moieties due to steric constraints. In the case of hydrogenated ND the surface carries residual surface hydrogen atoms. Therefore, the starting material for arylation should be chosen according to the respective application. Conductivity, photophysical and other properties are influenced by the residual surface termination originating from the starting material.

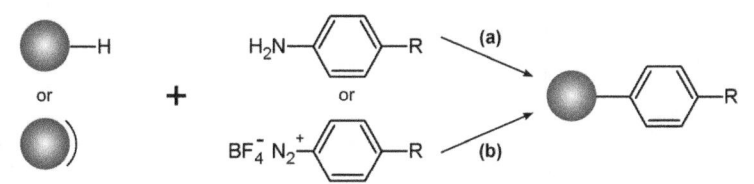

Figure 3.11 Arylation of ND using diazonium salts enables the stable grafting of aromatic moieties with a broad range of additional functional groups. These groups can be used to further ligate the ND particles to even more complex organic structures (R = COOH, NO$_2$, CN, alkyl, *etc.*). In the reaction with annealed or hydrogenated diamond, diazonium salts can be used directly (b) or the respective anilines are transformed *in situ* using amyl nitrite (a) before the formation of the C–C bonds.

3.2.2.5 Arylation by Diels–Alder Reactions

In the case of arylation using diazonium salts the functional moiety is attached to the diamond particle by a single C–C bond. An even stronger connection can be achieved by the formation of more than one bond between the surface and the linker molecule. This can be achieved, for example, by addition to the double bonds on the surface of annealed ND. In this case, also the amount of residual sp^2 carbon atoms is reduced (as compared to the reaction with diazonium salts) because neighboring atoms of a π-system are reacted in the same event.

The most prominent reaction for the simultaneous formation of two C–C bonds in one reaction step without using any further reagents is the Diels–Alder reaction (see Figure 3.12). However, it requires the existence of suitable alkene or diene structures on the surface of the diamond. It has been shown by computations that thermally annealed nanodiamond carries both isolated π-bonds and extended, fullerene-like aromatic structures on its surface.[147] These are suitable for different types of cycloadditions. From fullerene chemistry it is known that curved condensed aromatic structures are usually electron poor and act as the dienophile in [4 + 2] cycloadditions. It is therefore reasonable to choose electron-rich dienes. It has been reported that NDs undergo Diels–Alder reactions with *ortho*-quinodimethanes that have been generated *in situ* from dibromo precursors.[148] The diene was chosen for its irreversible addition to the surface alkenes, which is due to the formation of an aromatic ring in the product. "Classical" dienes, such as cyclopentadiene or anthracene, yield only poor surface loadings as the Diels–Alder reaction is reversible in these cases and the amount of grafted dienes is rather low in the equilibrium state. The reaction was also used to demonstrate the minimum temperature for the formation of sufficient amounts of sp^2 carbon as the reaction exclusively takes place on such structures (see section 3.2.1.4). For detonation nanodiamond, an annealing temperature of 750 °C was found to be necessary to observe significant surface loadings after the reaction with suitable dienes. The same temperature was found

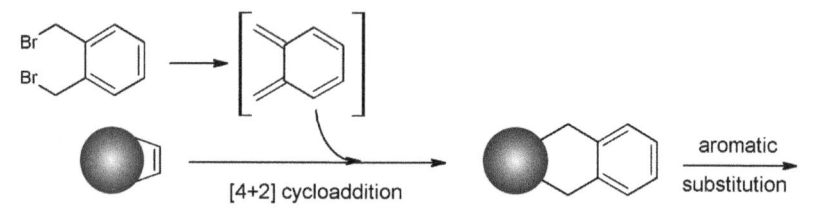

Figure 3.12 Diels–Alder reaction on the surface of annealed nanodiamond. In the case of this reaction a sufficient amount of sp^2 carbon is required for the successful application of this reaction. Each moiety is bound by two C–C bonds, making the conjugate very stable. The driving force of this reaction is the formation of the aromatic ring. Subsequent aromatic substitution enables the production of a large variety of ND materials with different moieties.[148]

with independent X-ray photoelectron spectroscopy (XPS) investigations by T. Petit *et al.*[32]

3.2.2.6 Prato Reaction

As shown in section 3.2.2.5, surface π-bonds can undergo [4 + 2] cycloaddition reactions. Another versatile cycloaddition is the reaction of alkenes with azomethine ylides, a special class of 1,3-dipoles. It has been known for a long time that this transformation can be used for the efficient functionalization of fullerenes and carbon nanotubes. It is named the Prato reaction.[149] For nanodiamond, the first report on the use of azomethine ylides was published only in 2011; nevertheless, the reaction is equally useful for the functionalization of annealed diamond surfaces (see Figure 3.13).[150]

The reaction principle is typically based on the *in situ* formation of the ylide using an aldehyde and an amino acid under basic conditions. However, other precursors, such as pyridinium, chinolinium and isochinolinium salts, can be used too. This opens the way for the establishment of a multitude of additional functional groups to which further moieties can be grafted in a selective fashion.

3.2.2.7 Bingel–Hirsch Reaction

Another useful reaction in the field of fullerene and nanotube chemistry is the Bingel–Hirsch reaction. It consists of the addition of a malonate onto the π-bond of the carbon materials *via* the *in situ* generation of the respective bromomalonate and the formation of a carbanion in the double α-position to the carbonyl moieties. Other electron-withdrawing groups can be chosen instead of the carbonyl moieties too (see Figure 3.14).[151–153]

In the case of C_{60}, the reaction has been used to produce a multitude of functionalized derivatives with highly symmetric grafting patterns and a large variety of further functional groups at either end of the malonate. The same is valid for nanodiamond. As soon as the diamond carries π-bonds, they can react with a large variety of malonates using CBr_4 and a base, such

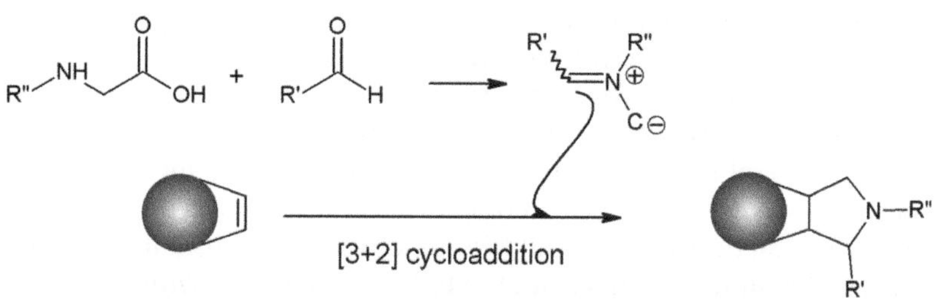

Figure 3.13 The Prato reaction employs azomethine ylides that are generated *in situ* from an aldehyde and an amino acid. By carefully choosing the reaction components, a large variety of conjugates is possible.

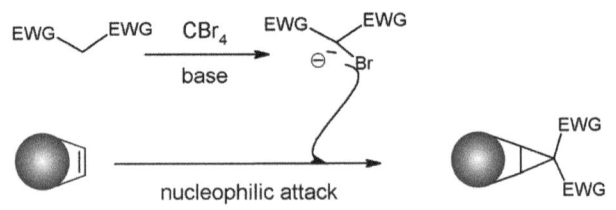

Figure 3.14 The reaction of a deprotonated malonate (EWG = electron-withdrawing group, *e.g.*, COOR, COOH, CN) with the curved double bonds of a carbon nanomaterial is related to the analogous reaction on C_{60}. Not only alkene-like structures can react, but carbonyl functions will undergo a similar reaction as well. The difference lies in the formation of a cyclic structure (reaction on the double bond) or open structures with additional OH groups.

as diazabicycloundec-7-ene (DBU). Depending on the steric demand of the side chains, the surface loadings vary between ~0.15 and 0.7 mmol g^{-1}.[154] Besides simple esters, cyanides or oligoethylene glycols can also be immobilized in this way. Using asymmetric malonates, *e.g.*, those carrying a carbonyl and an ester group, two independently addressable groups can be attached in one step.

However, it was found that carbonyl groups at the diamond surface can undergo a reaction with bromomalonates as well. The electrophilic carbonyl carbon can be attacked by the nucleophile leading to ring-opened structures, where the malonate is attached to the carbonyl carbon. As hydroxyl functions are formed in the process as well, a mixed surface termination is the result of this transformation. For several biological applications such mixed terminations can be useful as they enable the colloidal stability of the material to be tuned in different environments.

3.3 Grafting of Complex Moieties onto Pre-functionalized Nanodiamond

The grafting of complex moieties, such as biomolecules, is typically executed on pre-functionalized surfaces carrying linkers with suitable terminal groups. This strategy enables not only higher surface loadings (due to larger spacing between the reactive surface groups), but also avoids the already mentioned sensitivity of the conjugation bond against nucleophilic attack (see section 3.2.2). However, the conjugation method needs to be carefully selected as several requirements apply: first, the grafting should be possible in suitable solvents, *i.e.*, for biomolecules, water is typically required. Second, the reaction itself should be as efficient as possible to avoid the formation of residual groups from the linker (either the terminal group itself or a hydrolysis product). Third, the reaction should preferentially neither require reagents nor form by-products that are difficult to remove, *e.g.*, due to poor solubility or strong adsorption onto the diamond surface. Several techniques have already proven their value and will be described in the following paragraphs.

In some cases it can be useful to simply adsorb the functional moiety onto the diamond surface. This is the case especially when cleavage is desired at some point of the application of the conjugate. Ho and coworkers have used non-covalent conjugates of anti-tumor therapeutics with detonation diamond for the efficient delivery of the drug to its site of action.[155] Several other reports on the successful application of non-covalent ND conjugates have been published as well.[156,157] However, when the conjugate is supposed to act as such, the non-covalent adsorption of biomolecules can hamper their proper function. The accessibility of the active site can be hindered by the multiple interactions with the ND surface. Additionally, the three-dimensional shape of the functional molecule can be altered by the adsorption. Here, it is recommended to rely on defined geometries induced by pre-grafted linker structures that put the biomolecule at a certain distance from the diamond surface and that can be attached far away from the important functional sites. Such conjugates are formed in a defined manner more easily by covalent chemistry.

In each case the linker chemistry has to be chosen very carefully to avoid the introduction of additional sites for non-specific interactions or the non-specific attachment of the functional molecule.

3.3.1 Peptide Coupling onto Nanodiamond

One of the most frequently used coupling techniques is the formation of amide bonds between the nanoparticle surface and the functional moiety. On one hand this can be done right at the diamond surface using either carboxyl or amino moieties directly attached to the diamond core (see section 3.2.2.3). On the other hand the existence of a terminal amino or acid function on a linker moiety enables the formation of the respective amide further away from the ND surface. As explained above, this might help to conserve the function of sensitive molecules and enhance the stability of the conjugate by establishing C–C bonds at the ND surface and keep the heteroatom bonds in the periphery, where they are less prone to cleavage.

Many examples for peptide coupling on nanodiamond using different methods have been published. In the frame of this chapter only a few can be mentioned. There have been the formation of small peptides and biotinylation on aminosilanized detonation diamond,[75,136] the immobilization of ethylene diamine onto carboxylated ND (see Figure 3.15),[158] and the grafting of dyes by an amide bond.[137] In some cases the amino groups were part of a non-covalently bound organic shell around the diamond nanoparticle, *e.g.*, a poly-lysine layer. Here, the polymeric nature of the shell ensures the conjugate stability. The functional moiety is then attached in a covalent manner.[159]

The peptide coupling itself can be carried out using conventional protein chemistry. Typically, a coupling agent or an activated ester (like the

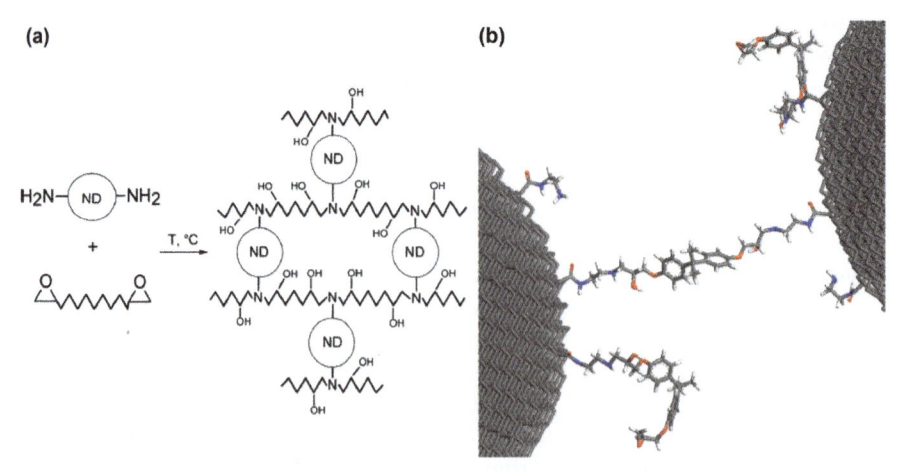

Figure 3.15 Immobilization of ethylene diamine on carboxylated ND and the subsequent covalent incorporation of aminated ND (the NH$_2$ groups are linked to the ND *via* an ethylene amide linker) into an epoxy resin. a) Reaction scheme; b) model of the covalently bonded composite.[158] © 2011 ACS, with kind permission.

N-hydroxysuccinimide (NHS) ester) have to be used as the simple mixing of the carboxylic acid and the amine leads to salt formation, but not to amides. A frequently used coupling agent is EDC because the resulting urea by-product is more soluble than the one of DCC (the usual coupling agent in conventional organic chemistry, see also section 3.2.2.3).

3.3.2 Click Chemistry on Nanodiamond

As stated in the seminal article by Sharpless *et al.* in 2001, coupling with high efficiency, *i.e.*, yield, in water and with no or very little side products and/or waste is the most desirable type of reaction.[160] This is even truer when looking at the surface chemistry of nanoparticles. Side products can be adsorbed so strongly that the purification of the conjugates is difficult, and non-aqueous solvents make it difficult to work with complex biomolecules in many cases. It was therefore only a question of time until the concept of click chemistry was adapted to the surface functionalization of nanoscale diamond particles.[161,162]

The prototype "click reaction" is the coupling of terminal alkynes with organic azides in a copper-catalyzed reaction. Related to the Huisgen [3 + 2] cycloaddition of a π-system with a 1,3-dipole (here, the azide), this reaction leads to the formation of triazole rings in, usually, a quantitative yield with no side products besides the catalyst, which can be removed by complexation and washing. Catalyst-free versions of the reaction have been shown to be efficient for the grafting of organic molecules onto quantum dots (using, for example, strained cycloalkynes),[163] but have not been used so far for

The Chemistry of Nanodiamond 79

grafting onto nanodiamond. The reaction of azides immobilized on the surface of different types of nanodiamond (detonation or HTHP) was shown to proceed smoothly, as well as the inverse case of reacting of an immobilized alkyne with an organic azide. The reactions have been used for the immobilization of various molecules, such as dyes,[164] saccharides,[165] and therapeutic molecules, such as CO-releasing complexes (see Figure 3.16).[166] The reaction is quantitative, specific and compatible with most biomolecules. Under physiological conditions, triazoles are stable, which makes this linker structure highly attractive for the conjugation of NDs with targeting moieties, *etc.*

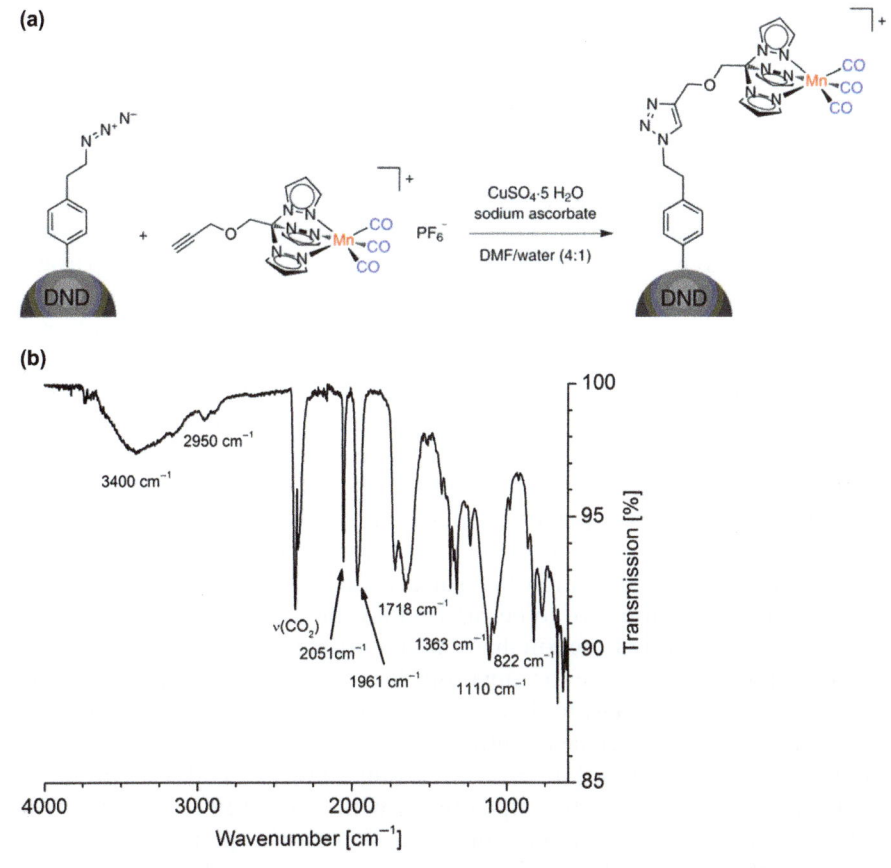

Figure 3.16 The immobilization of therapeutic molecules is an important application field for nanodiamond. Here, a molecule that releases CO photochemically (photo-CORM) is grafted onto ND using an alkyne–azide coupling (a). The carbonyl ligands are clearly identifiable in the IR spectrum (b). The efficiency of the photo-CORM is not hampered by the immobilization.[166]
© Royal Society of Chemistry 2012.

Besides the reaction of azides and alkynes, the so-called thiol-ene reaction is often considered to be another click reaction. The formation of a thioether, starting from a thiol and a terminal alkene, proceeds smoothly and selectively in most cases. This reaction has been used, for example, for the immobilization of several saccharides.[167]

3.3.3 Miscellaneous Reactions for the Surface Modification of Nanodiamond

Besides the classic click reactions, the formation of oximes and imines can be considered as one of the most economic ways to immobilize sensitive structures on nanoparticles. These reactions do not require activation agents and are usually compatible with buffered, aqueous environments, making them especially useful for the grafting of peptides and other biomolecules. There have been some examples for the formation of these structures. The formation of an imine was used to immobilize NADH-dependant alcohol dehydrogenase onto detonation diamond[168] and for the functionalization of submicron diamond with 4-(trifluoromethyl) benzylamine.[169]

Depending on the application, for example, targeted drug delivery, transport of hydrophobic cargo in aqueous environments, *etc.*, it can be useful to not rely exclusively on one type of surface group on the nanodiamond particle. In such a case a mixed termination can help to achieve the desired performance. For instance, the partial termination of the above mentioned saccharide-modified ND[167] with sulfonate groups helps to enhance the colloidal stability of the samples in water and cell culture medium. It keeps the zetapotential in a negative range and therefore promotes the electrostatic stabilization of the colloid. Thiol groups alone would not be able to keep the material in buffered aqueous solution as such surfaces typically exhibit only moderately positive zetapotential values and, furthermore, they promote agglomeration through the formation of hydrogen bonds.

On the other hand a multiple surface functionalization is necessary when several functions are to be realized on the same particle. This is the case when covalently functionalized nanoparticles are used for targeting, for example, tumor tissue (or any other location in the organism) and for the delivery of a drug, *etc.* to this respective location. Here, an independent reactivity of the two surface moieties should be obtained with the targeting moiety selectively bound in a very strong manner (and not prone to hydrolysis or enzymatic decomposition), while the drug is attached in a labile way for a rapid availability of the unaltered therapeutic molecule. In doing so, the cargo could be released by an external stimulus, like a change in pH or the presence of a specific enzyme, while the targeting unit would inevitably direct the cargo to its site of action without being cleaved from the surface of the diamond. Such systems can be realized by binding the targeting unit, for example, by azide–alkyne click chemistry (forming a very stable triazole ring) and the drug *via* peptide coupling (prone to enzymatic cleavage). Other systems are conceivable using different types of linkers. A model system has been reported using aromatic diazonium salts carrying a benzoic acid or an alkyne. Two dyes with the suitable counterpart terminal groups were grafted and could be addressed independently.[164]

3.4 Summary and Outlook

As can be seen from all the chemistry in this chapter, nanodiamond is neither inert nor "difficult" in its chemistry. It behaves more or less like a very large organic molecule with a core of quaternary carbon atoms. Hence, only the imminent surface is accessible for chemical modification. Reactions so far include radical reactions (*e.g.*, the arylation using diazonium salts or the grafting or perfluoralkyl radicals) and ionic reactions (like the esterification of surface hydroxyl groups or the Bingel–Hirsch reaction), as well as pericyclic reactions, like the Diels–Alder reaction of *ortho*-quinodimethanes with π-bonds on the diamond surface. Depending on the chosen bonding between the diamond surface and the functional moieties, the conjugation withstands different conditions. Where it is intended to obtain conjugates that are inert against hydrolysis, enzymatic attack or other nucleophilic decomposition, it is preferable to opt for nonpolar carbon–carbon single bonds, which usually exhibit a rather low reactivity in the scenarios mentioned above. The required anchor groups can be established at the terminus of inert linker molecules, where problems due to the binding situation are easily avoided by the choice of suitable linker architecture. For some of the surface modifications, the full reaction mechanism has not yet been elucidated. Thus, in such cases, it may be difficult to predict the reactive behavior of different types of nanodiamond. Further efforts are necessary to continue "filling the toolbox" for the surface functionalization of nanodiamond. Another issue that has not been solved to date is the defined 1 : 1 conjugation of an individual nanoparticle and a large functional molecule, such as an enzyme. So far, statistical distributions are typically obtained, where one diamond on average carries one functional moiety, but non-conjugated and multiply grafted ND particles are also found in the product mixture. The isolation of the 1 : 1 conjugate could be, in principle, achieved by chromatography or other size or charge selective methods. However, the solubility of the ND conjugates is often not sufficient for the successful separation of different conjugates.

In summary the surface functionalization of nanodiamond is a valuable tool for the adjustment of the surface properties of the nanoparticles, as well as for the conjugation of NDs with functional moieties for a broad variety of applications.

References

1. V. Yu. Dolmatov, *Russ. Chem. Rev.*, 2001, **70**, 607.
2. E. Osawa, *Diamond Relat. Mater.*, 2007, **16**, 2018.
3. A. Krueger, *J. Mater. Chem.*, 2008, **18**, 1485.
4. J.-B. Donnet, C. Lemoigne, T. K. Wang, C.-M. Peng, M. Samirant and A. Eckhardt, *Bull. Soc. Chim. Fr.*, 1997, **134**, 875.
5. O. A. Shenderova, V. V. Zhirnov and D. W. Brenner, *Crit. Rev. Solid State Mater. Sci.*, 2002, **27**, 227.
6. L.-J. Su, C.-Y. Fang, Y.-T. Chang, K.-M. Chen, Y.-C. Yu, J.-H. Hsu and H.-C. Chang, *Nanotechnol.*, 2013, **24**, 315702.
7. J.-P. Boudou, P. A. Curmi, F. Jelezko, J. Wrachtrup, P. Aubert, M. Sennour, G. Balasubramanian, R. Reuter, A. Thorel and E. Gaffet, *Nanotechnol.*, 2009, **20**, 235602.
8. E. Neu, C. Arend, E. Gross, F. Guldner, C. Hepp, D. Steinmetz, E. Zscherpel, S. Ghodbane, H. Sternschulte, D. Steinmüller-Nethl, Y. Liang, A. Krueger and C. Becher, *Appl. Phys. Lett.*, 2011, **98**, 243107.
9. D. Tan, S. Zhou, B. Xu, P. Chen, Y. Shimotsuma, K. Miura and J. Qiu, *Carbon*, 2013, **62**, 374.
10. S. L. Hu, J. Sun, X. W. Du, F. Tian and L. Jiang, *Diamond Relat. Mater.*, 2008, **17**, 142.
11. J. Sun, S. L. Hu, X. W. Du, Y. W. Lei and L. Jiang, *Appl. Phys. Lett.*, 2006, **89**, 183115.
12. L.-W. Yin, M.-S. Li, J.-J. Cui, Y.-S. Song, F.-Z. Li and Z.-Y. Hao, *Appl. Phys. A*, 2001, **73**, 653.
13. C.-L. Cheng, C.-F. Chen, W.-C. Shaio, D.-S. Tsai and K.-H. Chen, *Diamond Relat. Mater.*, 2005, **14**, 1455.
14. Yu. V. Butenko, V. L. Kuznetsov, E. A. Paukshtis, A. I. Stadnichenko, I. N. Mazov, S. I. Moseenkov, A. I. Boronin and S. V. Kosheev, *Fullerenes, Nanotubes, Carbon Nanostruct.*, 2006, **14**, 557.
15. A. V. Tyurina, I. A. Apolonskaya, I. I. Kulakova, P. G. Kopylova and A. N. Obraztsov, *Poverkhnost*, 2010, 106.
16. V. Pichot, M. Comet, E. Fousson, C. Baras, A. Senger, F. Le Normand and D Spitzer, *Diamond Relat. Mater.*, 2008, **17**, 13.
17. D. Mitev, R. Dimitrova, M. Spassova, Ch. Minchev and S. Stavrev, *Diamond Relat. Mater.*, 2007, **16**, 776.
18. S. Osswald, G. Yushin, V. Mochalin, S. O. Kucheyev and Y Gogotsi, *J. Am. Chem. Soc.*, 2006, **128**, 11635.
19. A. Gaisinskaya, R. Akhvlediani and A. Hoffman, *Diamond Relat. Mater.*, 2010, **19**, 1183.
20. E. Pehrsson, T. W. Mercer and J. A. Chaney, *Surface Sci.*, 2002, **497**, 13.
21. B. A. Fox, in *Thin Film Technology Handbook*, A. Elshabini-Riad and F. D. Barlow, III (eds), McGraw Hill Professional, New York 1997, and references cited therein.
22. M. I. Landstrass and K. V. Ravi, *Appl. Phys. Lett.*, 1989, **55**, 975.
23. M. I. Landstrass and K. V. Ravi, *Appl. Phys. Lett.*, 1989, **55**, 1391.

24. E. Neu, F. Guldner, C. Arend, Y. Liang, S. Ghodbane, H. Sternschulte, D. Steinmüller-Nethl, A. Krueger and C. Becher, *J. Appl. Phys.*, 2013, **113**, 203507.
25. T. Jiang and K. Xu, *Carbon*, 1995, **33**, 1663.
26. N. A. Skorik, A. L. Krivozubov, A. P. Karzhenevskii and B. V. Spitsyn, *Protection Metals Phys. Chem. Surf.*, 2011, **47**, 54.
27. Q. Zou, M. Z. Wang and Y. G. Li, *J. Exp. Nanosci.*, 2010, **5**, 319.
28. M. Comet, V. Pichot, B. Siegert, F. Britz and D. Spitzer, *J. Nanosci. Nanotechnol.*, 2010, **10**, 4286.
29. V. Mochalin, S. Osswald and Y. Gogotsi, *Chem. Mater.*, 2009, **21**, 273.
30. J.-S. Tu, E. Perevedentseva, P.-H. Chung and C.-L. Cheng, *J. Chem. Phys.*, 2006, **125**, 174713.
31. T. Enoki, K. Takai, V. Osipov, M. Baidakova and A. Vul, *Chem. Asian J.*, 2009, **4**, 796.
32. T. Petit, J.-C. Arnault, H. A. Girard, M. Sennour and P. Bergonzo, *Phys. Rev. B*, 2011, **84**, 233407.
33. I. D. Morrison, S. Ross, *Colloidal Dispersions*, Wiley Interscience, New York, 2002, p. 383.
34. R. A. Sperling and W. J. Parak, *Phil. Trans. Royal Soc., A*, 2010, **368**, 1333.
35. T. Tadros, in *Electrical Phenomena at Interfaces and Biointerfaces*, ed. H. Ohshima, John Wiley & Sons, Hoboken, New Jersey, 2012, p. 153.
36. I. D. Morrison, S. Ross, *Colloidal Dispersions*, Wiley Interscience, New York, 2002, p. 316.
37. T. Petit, J.-C. Arnault, H. A. Girard, M. Sennour, R.-Y. Kang, C.-L. Cheng and P. Bergonzo, *Nanoscale.*, 2012, **4**, 6792.
38. A. Krüger, M. Ozawa, F. Kataoka, T. Fujino, Y. Suzuki, A. E. Aleksenskii, A Ya. Vul' and E. Osawa, *Carbon*, 2005, **43**, 1722.
39. B. V. Spitsyn, M. N. Gradoboev, T. B. Galushko, T. A. Karpukhina, N. V. Serebryakova, I. I. Kulakova and N. N. Melnik, *NATO Science Series, II: Mathematics, Physics and Chemistry*, 2005, **192**, 241.
40. M. V. Avdeev, N. N. Rozhkova, V. L. Aksenov, V. M. Garamus, R. Willumeit and E. Osawa, *J. Phys. Chem. C*, 2009, **113**, 9473.
41. A. M. Panich and A. E. Aleksenskii, *Diamond Relat. Mater.*, 2012, **27–28**, 45.
42. L.-Y. Chang, E. Osawa and A. S. Barnard, *Nanoscale.*, 2011, **3**, 958.
43. M. Ozawa, M. Inakuma, M. Takahashi, F. Kataoka, A. Krueger and E. Osawa, *Adv. Mater.*, 2007, **19**, 1201.
44. Y. Liang, M. Ozawa and A. Krueger, *ACS Nano.*, 2009, **3**, 2288.
45. A. Pentecost, S. Gour, V. Mochalin, I. Knoke and Y. Gogotsi, *ACS Appl. Mater. Interfaces*, 2010, **2**, 3289.
46. C.-C. Li and C.-L. Huang, *Colloids Surfaces A*, 2010, **53**, 52.
47. Y. L. Hsin, H.-Y. Chu, Y.-R. Jeng, Y.-H. Huang, M. H. Wang and C. K. Chang, *J. Mater. Chem.*, 2011, **21**, 13213.
48. K. Y. Niu, H,-M. Zheng, Z.-Q. Li, J. Yang, J. Sun and X.-W, Du, *Angew. Chem., Int. Ed.*, 2011, **50**, 4099.

49. *Römpp Lexikon Chemie*, ed. J. Falbe, M. Regitz, 10th edn, Thieme, Stuttgart, 1997.
50. I. Barin, O. Knacke, *Thermodynamic Properties of Inorganic Substances*, Springer, Berlin, 1973.
51. R. S. Young, H. R. Simpson and D. A. Benfield, *Anal. Chim. Acta.*, 1952, **6**, 510.
52. J. Shirafuji and T. Sugino, *Diamond Relat. Mater.*, 1996, **5**, 706.
53. H. Schwertfeger, A. A. Fokin and P. R. Schreiner, *Angew. Chem., Int. Ed.*, 2008, **47**, 1022.
54. T. Rander, M. Staiger, R. Richter, T. Zimmermann, L. Landt, D. Wolter, J. E. Dahl, R. M. K. Carlson, B. A. Tkachenko, N. A. Fokina, P. R. Schreiner, T. Möller and C. Bostedt, *J. Chem. Phys.*, 2013, **138**, 024310.
55. M. B. Smith, *March's Advanced Organic Chemistry: Reactions, Mechanisms, and Structure*, 7th edn, John Wiley & Sons, New York, 2013, p. 883.
56. W. Howard, D. Huang, J. Yuan, M. Frenklach, K. E. Spear, R. Koba and A. W. Phelps, *J. Appl. Phys.*, 1990, **68**, 1247.
57. T. Jiang, K. Xu and S. Ji, *J. Chem. Soc., Faraday Trans.*, 1996, **92**, 3401.
58. B. V. Spitsyn, S. A. Denisov, N. A. Skorik, A. G. Chopurova, S. A. Parkaeva, L. D. Belyakova and O. G. Larionov, *Diamond Relat. Mater.*, 2010, **19**, 123.
59. S. Ida, T. Tsubota, O. Hirabayashi, M. Nagata, Y. Matsumoto and A. Fujishima, *Diamond Relat. Mater.*, 2003, **12**, 601.
60. O. A. Williams, private communication, Diamond Conference 2012, Granada.
61. O. A. Williams, J. Hees, C. Dieker, W. Jäger, L. Kirste and C. E. Nebel, *ACS Nano*, 2010, **4**, 4824.
62. I. I. Obraztsova and N. K. Eremenko, *Russ. J. Appl. Chem.*, 2008, **81**, 603.
63. B. A. Fox, in *Thin Film Technology Handbook*, A. Elshabini-Riad, F. D: Barlow, III (eds), McGraw Hill Professional, New York, 1997 and references cited therein.
64. S. Ghodbane, T. Haensel, Y. Coffinier, S. Szunerits, D. Steinmuller-Nethl, R. Boukherroub, S. I.-U. Ahmed and J. A. Schaefer, *Langmuir*, 2010, **26**, 18798.
65. W. Kulisch, C. Popov, D. Gilliland, G. Ceccone, F. Rossi and J. P. Reithmaier, *Surf. Interface Anal.*, 2010, **42**, 1152.
66. W. S. Yeap, S. Chen and K. P. Loh, *Langmuir*, 2009, **25**, 185.
67. M. Yeganeh, P. R. Coxon, A. C. Brieva, V. R. Dhanak, L. Šiller and Yu. V. Butenko, *Phys. Rev. B*, 2007, **75**, 155404.
68. H. A. Girard, T. Petit, S. Perruchas, T. Gacoin, C. Gesset, J. C. Arnault and P. Bergonzo, *Phys. Chem. Chem. Phys.*, 2011, **13**, 11517.
69. H. A. Girard, J. C. Arnault, S. Perruchas, S. Saada, T. Gacoin, J.-P. Boilot and P. Bergonzo, *Diamond Relat. Mater.*, 2010, **19**, 1117.
70. J.-C. Arnault, T. Petit, H. Girard, A. Chavanne, C. Gesset, M. Sennour and M. Chaigneau, *Phys. Chem. Chem. Phys.*, 2011, **13**, 11481.
71. Y. L. Zhong and K. P. Loh, *Chem. Asian J.*, 2010, **5**, 1532.

72. S. Ida, T. Tsubota, M. Nagata, Y. Matsumoto, M. Uehara and J. Hojo, *Hyomen Gijutsu*, 2003, **54**, 764.
73. W.-W. Zheng, Y.-H. Hsieh, Y.-C. Chiu, S.-J. Cai, C.-L. Cheng and C. Chen, *J. Mater. Chem.*, 2009, **19**, 8432.
74. S. Ciftan Hens, G. Cunningham, T. Tyler, S. Moseenkov, V. Kuznetsov and O. Shenderova, *Diamond Relat. Mater.*, 2008, **17**, 1858.
75. A. Krüger, Y. Liang, G. Jarre and J. Stegk, *J. Mater. Chem.*, 2006, **16**, 2322.
76. A. Krueger, M. Ozawa, G. Jarre, Y. Liang, J. Stegk and L. Lu, *Phys. Stat. Sol. A*, 2007, **204**, 2881.
77. R. Martín, M. Álvaro, J. R. Herance and H. García, *ACS Nano*, 2010, **4**, 65.
78. R. Martín, P. C. Heydorn, M. Alvaro and H. García, *Chem. Mater.*, 2009, **21**, 4505.
79. S. Turner, O. I. Lebedev, O. Shenderova, I. I. Vlasov, J. Verbeeck and G. Van Tendeloo, *Adv. Funct. Mater.*, 2009, **19**, 2116.
80. Yu. V. Butenko, V. L. Kuznetsov, E. A. Paukshtis, A. I. Stadnichenko, I. N. Mazov, S. I. Moseenkov, A. I. Boronin and S. V. Kosheev, *Fullerenes, Nanotubes, Carbon Nanostruct.*, 2006, **14**, 557.
81. A. V. Tyurina, I. A. Apolonskaya, I. I. Kulakova, P. G. Kopylova and A. N. Obraztsov, *Poverkhnost*, 2010, 106.
82. V. Pichot, M. Comet, E. Fousson, C. Baras, A. Senger, F. Le Normand and D Spitzer, *Diamond Relat. Mater.*, 2008, **17**, 13.
83. D. Mitev, R. Dimitrova, M. Spassova, Ch. Minchev and S. Stavrev, *Diamond Relat. Mater.*, 2007, **16**, 776.
84. S. Osswald, G. Yushin, V. Mochalin, S. O. Kucheyev and Y. Gogotsi, *J. Am. Chem. Soc.*, 2006, **128**, 11635.
85. H. Xie, W. Yu and Y. Li, *J. Phys. D*, 2006, **42**, 095413.
86. L. Rondin, G. Dantelle, A. Slablab, F. Grosshans, F. Treussart, P. Bergonzo, S. Perruchas, T. Gacoin, M. Chaigneau, H.-C. Chang, V. Jacques and J.-F. Roch, *Phys. Rev. B*, 2010, **82**, 115449.
87. J. Cheng, J. He, C. Li and Y. Yang, *Chem. Mater.*, 2008, **20**, 4224.
88. V. I. Anikeev and V. I. Zaikovskii, *Russ. J. Appl. Chem.*, 2010, **83**, 1202.
89. A. Kruger, E. Osawa and F. Kataoka, *Jpn. Patent App.*, 2002, **2002**, 205973.
90. B. R. Smith, D. Gruber and T. Plakhotnik, *Diamond Relat. Mater.*, 2010, **19**, 314.
91. S. K. Gordeyev and S. B. Korchagina, *J. Superhard Mater.*, 2004, **26**, 32.
92. T. Gaebel, C. Bradac, J. Chen, P. Hemmer and J. Rabeau, *Diamond Relat. Mater.*, 2012, **21**, 28.
93. S. K. Gordeev and S. B. Korchagina, *J.Superhard Mater.*, 2007, **29**, 124.
94. S. Osswald, M. Havel, V. Mochalin, G. Yushin and Y. Gogotsi, *Diamond Relat. Mater.*, 2008, **17**, 1122.
95. O. Shenderova, A. Koscheev, N. Zaripov, I. Petrov, Y. Skryabin, P. Detkov, S. Turner and G. Van Tendeloo, *J. Phys. Chem. C*, 2011, **115**, 9827.

96. G. Cunningham, A. M. Panich, A. I. Shames, I. Petrov and O. Shenderova, *Diamond Relat. Mater.*, 2008, **17**, 650.
97. S. L. Goertzen, K. D. Thériault, A. M. Oickle, A. C. Tarasuk and H. A. Andreas, *Carbon*, 2010, **48**, 1252.
98. Y. M. Wang, K. W. Wong, S. T. Lee, M. Nishitani-Gamo, I. Sakaguchi, K. P. Loh and T. Ando, *Phys. Rev. B*, 1999, **59**, 10347.
99. M. P. Schwartz, D. E. Barlow, J. N. Russell, Jr., J. E. Butler, M. P. D'Evelyn and R. J. Hamers, *J. Am. Chem. Soc.*, 2005, **127**, 8348.
100. J.-Y. Raty and G. Galli, *J. Electroanal. Chem.*, 2005, **584**, 9.
101. S. Okada, *Chem. Phys. Lett.*, 2009, **483**, 128.
102. V. L. Kuznetsov and Yu. V. Butenko, *Ultrananocrystalline Diamond: Synthesis, Properties and Applications, NATO Science Series*, 2005, **192**, 199.
103. V. L. Kuznetsov, A. L. Chuvilin, Yu. V. Butenko, I. Yu. Mal'kov and V. M. Titov, *Chem. Phys. Lett.*, 1994, **222**, 343.
104. O. O. Mykhaylyk, Yu. M. Solonin, D. N. Batchelder and R. Brydson, *J. Appl. Phys.*, 2005, **97**, 074302.
105. L. C. Qin and S. Iijima, *Chem. Phys. Lett.*, 1996, **262**, 252.
106. V. V. Roddatis, V. L. Kuznetsov, Yu. V. Butenko, D. S. Su and R. Schloegl, *Phys. Chem. Chem. Phys.*, 2002, **4**, 1964.
107. J. Hiraki, H. Mori, E. Taguchi, H. Yasuda, H. Kinoshita and N. Ohmae, *Appl. Phys. Lett.*, 2005, **86**, 223101.
108. I. P. Chang, K. C. Hwang, J.-A. A. Ho, C.-C. Lin, R. J.-R. Hwu and J.-C. Horng, *Langmuir*, 2010, **26**, 3685.
109. Y. Liang, T. Meinhardt, G. Jarre, M. Ozawa, P. Vrdoljak, A. Schöll, F. Reinert and A. Krueger, *J. Colloid Interface Sci.*, 2011, **354**, 23.
110. S. Zeppilli, J. C. Arnault, C. Gesset, P. Bergonzo and R. Polini, *Diamond Relat. Mater.*, 2010, **19**, 846.
111. X. W. Fang, J. D. Mao, E. M. Levin and K. Schmidt-Rohr, *J. Am. Chem. Soc.*, 2009, **131**, 1426.
112. A. M. Panich, A. I. Shames, H.-M. Vieth, E. Osawa, M. Takahashi and A. Ya. Vul, *Eur. Phys. J. B*, 2006, **2**, 397.
113. M. Dubois, K. Guerin, E. Petit, N. Batisse, A. Hamwi, N. Komatsu, J. Giraudet, P. Pirotte and F. Masin, *J. Phys. Chem. C*, 2009, **113**, 10371.
114. A. S. Chiganov, *Phys. Solid State*, 2004, **46**, 620.
115. H. Koch, W. Kulisch, C. Popov, R Merz, B. Merz and J. P. Reithmaier, *Diamond Relat. Mater.*, 2011, **20**, 254.
116. G.-J. Zhang, K.-S. Chong, Y. Nakamura, T. Ueno, T. Funatsu, I. Ohdomari and H. Kawarada, *Langmuir*, 2006, **22**, 3728.
117. K.-I. Sotowa, T. Amamoto, A. Sobana, K. Kusakabe and T. Imato, *Diamond Relat. Mater*, 2004, **13**, 145.
118. S. Schietinger, M. Barth, T. Aichele and O. Benson, *Nano Lett.*, 2009, **9**, 1694.
119. T. Nakamura, T. Ohana, Y. Hagiwara and T. Tsubota, *Phys. Chem. Chem. Phys.*, 2009, **11**, 730.
120. Y. Liu, Z. Gu, J. L. Margrave and V. N. Khabashesku, *Chem. Mater.*, 2004, **16**, 3924.

121. H. Huang, Y. H. Wang, J. B. Zang and L. Y. Bian, *Appl. Surf. Sci.*, 2012, **258**, 4079.
122. M. A. Ray, T. Tyler, B. Hook, A. Martin, G. Cunningham, O. Shenderova, J. L. Davidson, M. Howell, W. P. Kang and G. McGuire, *Diamond Relat. Mater.*, 2007, **16**, 2087.
123. B. V. Spitsyn, S. A. Denisov, N. A. Skorik, A. G. Chopurova, S. A. Parkaeva, L. D. Belyakova and O. G. Larionov, *Diamond Relat. Mater*, 2010, **19**, 123.
124. W. H. Saunders, Jr., *J. Org. Chem.*, 1999, **64**, 861.
125. J. S. Hovis, S. K. Coulter, R. J. Hamers, M. P. D'Evelyn, J. N. Russell, Jr. and J. E. Butler, *J. Am. Chem. Soc.*, 2000, **122**, 732.
126. K.-K. Liu, W.-W. Zheng, C.-C. Wang, Y.-C. Chiu, C.-L. Cheng, Y.-S. Lo, C. Chen and J.-I. Chao, *Nanotechnol.*, 2010, **21**, 315106.
127. A. Krueger and T. Boedeker, *Diamond Relat. Mater.*, 2008, **17**, 1367.
128. Q. Zhang, V. N. Mochalin, I. Neitzel, I. Y. Knoke, J. Han, C: A. Klug, J. G. Zhou, P. I. Lelkes and Y. Gogotsi, *Biomater.*, 2011, **32**, 87.
129. V. N. Mochalin and Y. Gogotsi, *J. Am. Chem. Soc.*, 2009, **131**, 4594.
130. W. Yang, O. Auciello, J. E. Butler, W. Cai, J. A. Carlisle, J. E. Gerbi, D. M. Gruen, T. Knickerbocker, T. L. Lasseter, J. N. Russell, L. M. Smith and R. J. Hamers, *Nature Mater.*, 2002, **1**, 253.
131. T. Nakamura, T. Ohana, Y. Hagiwara and T. Tsubota, *Appl. Surf. Sci.*, 2010, **257**, 1368.
132. T. Tsubota, S. Ida, O. Hirabayashi, S. Nagaoka, M. Nagata and Y. Matsumoto, *Phys. Chem. Chem. Phys.*, 2002, **4**, 3881.
133. T. Takimoto, T. Chano, S. Shimizu, H. Okabe, M. Ito, M. Masaaki, T. Kimura, T. Inubushi and N. Komatsu, *Chem. Mater.*, 2010, **22**, 3462.
134. T. Nakamura, M. Ishihara, T. Ohana and Y. Koga, *Chem. Commun.*, 2003, 900.
135. J. Lokajova, J. Havlik, I. Rehor, J. Slegerova, M. Ledvina, F. Treussart, A Wen, S. Shukla, N. Steinmetz, P. Cigler, *Abstracts of Papers, 243 rd ACS National Meeting*, San Diego, United States, 2012.
136. A. Krueger, J. Stegk, Y. Liang, L. Lu and G. Jarre, *Langmuir*, 2008, **24**, 4200.
137. J. Tisler, R. Reuter, A. Laemmle, F. Jelezko, G. Balasubramanian, P. R. Hemmer, F. Reinhard and J. Wrachtrup, *ACS Nano*, 2011, **5**, 7893.
138. E. J. Hwang, S. K. Lee, M. G. Jeong, Y. B. Lee and D. S. Lim, *J. Nanosci. Nanotechnol.*, 2012, **12**, 5875.
139. W. S. Yeap, Y. Y. Tan and K. P. Loh, *Anal. Chem.*, 2008, **80**, 4659.
140. A. P. Wight and M. E. Davis, *Chem. Rev.*, 2002, **102**, 3589.
141. N. Goyal, *Synlett*, 2010, 335.
142. C.-L. Park, A. Y. Jee, M. Lee and S.-g. Lee, *Chem. Comm.*, 2009, 5576.
143. A. Barras, S. Szunerits, L. Marcon, N. Monfilliette-Dupont and R. Boukherroub, *Langmuir*, 2010, **26**, 13168.
144. A. Radzicka and R. Wolfenden, *J. Am. Chem. Soc.*, 1996, **118**, 6105.
145. L. Zhao, T. Takimoto, M. Ito, N. Kitagawa, T. Kimura and N. Komatsu, *Angew. Chem.*, 2011, **123**, 1424.

146. J. L. Bahr, J. Yang, D. V. Kosynkin, M. J. Bronikowski, R. E. Smalley and J. M. Tour, *J. Am. Chem. Soc.*, 2001, **123**, 6536.
147. J.-Y. Raty, G. Galli, C. Bostedt, T. W. van Buuren and L. J. Terminello, *Phys. Rev. Lett.*, 2003, **90**, 037401.
148. G. Jarre, Y. Liang, P. Betz, D. Lang and A. Krueger, *Chem. Comm.*, 2011, **47**, 544.
149. D. Tasis, N. Tagmatarchis, A. Bianco and M. Prato, *Chem Rev.*, 2006, **106**, 1105.
150. D. Lang and A. Krueger, *Diamond Relat. Mater.*, 2011, **20**, 101.
151. C. Bingel, *Chem. Ber.*, 1993, **126**, 1957.
152. C. Bingel and H. Schiffer, *Liebigs Ann.*, 1995, 1551.
153. X. Camps and A. Hirsch, *J. Chem. Soc., Perkin Trans.*, 1997, **1**, 1595.
154. P. Betz and A. Krueger, *Chem. Phys. Chem.*, 2012, **13**, 2578.
155. E. K. Chow, X.-Q. Zhang, M. Chen, R. Lam, E. Robinson, H. Huang, D. Schaffer, E. Osawa, A. Goga and D. Ho, *Sci. Transl. Med.*, 2011, **3**, 73ra21.
156. N. M. Gibson, T. J. M. Luo, D. W. Brenner and O. Shenderova, *Biointerphases*, 2011, **6**, 210.
157. R. A. Shimkunas, E. Robinson, R. Lam, S. Lu, X. Xu, X.-Q. Zhang, H. Huang, E. Osawa and D. Ho, *Biomater.*, 2009, **30**, 5720.
158. V. N. Mochalin, I. Neitzel, B. J. M. Etzold, A. Peterson, G. Palmese and Y. Gogotsi, *ACS Nano*, 2011, **5**, 7494.
159. L.-C. L. Huang and H.-C. Chang, *Langmuir*, 2004, **20**, 5879.
160. H. C. Kolb, M. G. Finn and K. B. Sharpless, *Angew. Chem. Int. Ed.*, 2001, **40**, 2004.
161. J. E. Moses and A. D. Moorhouse, *Chem. Soc. Rev.*, 2007, **36**, 1249.
162. H. Li, F. Cheng, A. M. Duft and A. Adronov, *J. Am. Chem. Soc.*, 2005, **127**, 14518.
163. C. Schieber, A. Bestetti, J. P. Lim, A. D. Ryan, T.-L. Nguyen, R. Eldridge, A. R. White, P. A. Gleeson, P. S. Donnelly, S. J. Williams and P. Mulvaney, *Angew. Chem., Int. Ed.*, 2012, **51**, 10523.
164. T. Meinhardt, D. Lang, H. Dill and A. Krueger, *Adv. Funct. Mater.*, 2011, **21**, 494.
165. A. Barras, F. A. Martin, O. Bande, J.-S. Baumann, J.-M. Ghigo, R. Boukherroub, C. Beloin, A. Siriwardena and S. Szunerits, *Nanoscale*, 2013, **5**, 2307.
166. G. Dördelmann, T. Meinhardt, T. Sowik, A. Krueger and U. Schatzschneider, *Chem. Commun.*, 2012, 11528.
167. M. Hartmann, P. Betz, Y. Sun, S. N. Gorb, T. K. Lindhorst and A. Krueger, *Chem. Eur. J*, 2012, **18**, 6485.
168. K. Goldberg, A. Krueger, T. Meinhardt, W. Kroutil, B. Mautner and A. Liese, *Tetrahedron Asymm.*, 2008, 1171.
169. J. K. Lee, M. W. Anderson, F. A. Gray, P. John and J.-Y. Lee, *Diamond Relat. Mater.*, 2005, **14**, 675.

CHAPTER 4

Nanodiamond Purification

SEBASTIAN OSSWALD

Department of Physics, Naval Postgraduate School, 833 Dyer Road, Monterey, CA 93943, USA
Email: sosswald@nps.edu

4.1 Purification of Nanodiamond Powders

4.1.1 Introduction

Ultra-dispersed diamond (UDD), commonly referred to as detonation nanodiamond (DND), or simply, nanodiamond (ND), has stirred rapidly growing interest in the materials community.[1] Owing to its relatively simple detonation synthesis, which utilizes expired explosives as energy and carbon source for diamond growth, ND has become one of only a few nanomaterials produced on an industrial scale.[2,3]

The potential of ND for widespread application is rooted in its unique structure, combining a chemically active surface with an inert diamond core that exhibits the favorable properties of macroscopic diamonds, including high thermal conductivity, electrical resistivity, Young's modulus, an extreme hardness, and biocompatibility. The as-produced powders contain 4–5 nm-sized diamond crystals, covered by layers of graphitic and amorphous carbon. Like other carbon nanomaterials, ND crystals exhibit the tendency to agglomerate, forming larger aggregates up to several hundred nanometers in size.[2,3]

ND already serves as an additive for composite materials, cooling fluids,[4] lubricants,[5] and electroplating baths,[6] but numerous applications, including optical coatings, catalyst support, and drug delivery, remain largely under-explored. One of the primary factors prohibiting a direct use of the as-produced powders is the fact that the diamond-bearing soot collected after the detonation contains large amounts of non-amorphous and graphitic carbon, as well as metals, metal oxides, and other impurities coming from the detonation chamber or the explosives in use. Due to the high content of non-diamond species, additional purification steps are required before the ND powders can be utilized. In fact, it is this complex composition that renders purification the most complicated and expensive stage of ND production, accounting for almost half of the material cost.[7] Currently, there exists no single purification treatment that allows for a simultaneous removal of all impurities present in ND powders, and methods in use typically consist of two or more steps. While the process of purification often simply refers to the removal of non-diamond carbon and metal catalyst, it can also be used to adjust the average crystal size in ND powders[8] or to reduce agglomeration.[9]

This chapter provides an overview of the chemical methods currently employed to purify ND powders. Section 4.1 reviews the composition of as-produced ND powders and the contained impurities, summarizes the most common characterization methods used to assess ND purity, and discusses the oxidation behavior of these powders. Existing purification techniques and the corresponding treatment conditions are then reviewed in detail in section 4.2, followed by a brief discussion of the effects of the purification on ND characteristics and properties in section 4.3. The chapter closes with a brief summary and some remarks on future challenges and opportunities in ND purification.

4.1.2 Nanodiamond Structure and Composition

A major drawback of bulk synthesis methods, although critical for widespread application, is the reduced control over structural uniformity and the composition of the produced nanomaterial. This is particularly true in the case of ND, where the detonation process yields a powder that contains up to 70–80 wt% of non-diamond species.[2,3] The detonation soot is primarily composed of amorphous and graphitic carbon, but also contains significant amounts of non-carbon impurities, such as metals and metal oxides.[6] This makes purification a crucial step in ND production. The purity levels (sp^3 content) of commercially available ND powders range from less than 30 wt%, as in the case of the unpurified detonations soot, to more than 90 wt% after multi-stage purification.[10] Figure 4.1 shows the transmission electron microscopy (TEM) image of an as-produced, unpurifed ND powder. TEM analysis allows researchers to directly observe the structures and compounds contained in the sample at the nanoscale. The benefits and limitations of this, and other characterization techniques, is discussed in more detail in section 4.1.3.

The unpurified powders contain a variety of carbon nanostructures of different shapes and sizes alongside the nanodiamond crystals, including

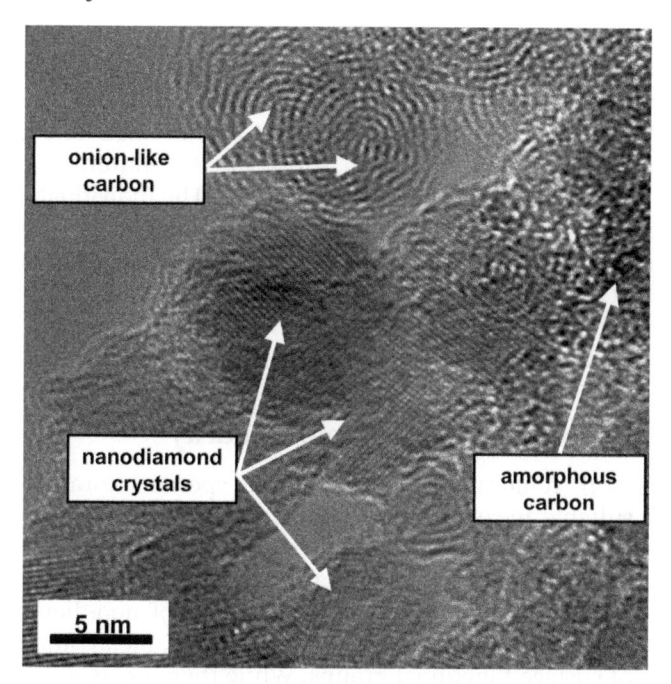

Figure 4.1 Transmission electron microscopy (TEM) image of an unpurified ND sample. The as-produced powders contain large amounts of non-diamond carbon alongside the diamond nanocrystals, including carbon onions and amorphous carbon.

amorphous carbon, carbon onions, and graphite nanocrystals and nanoribbons. Non-carbonaceous impurities include trace amounts of metals, such as Fe, Al, Ca, Mg, Zn, and Cu, and non-metal species, particularly H, N, O, and S, which typically exist in the form of surface functional groups. The exact composition of commercially available powders and their purity grades vary substantially and are strongly dependent on the chemical history of the sample (*e.g.*, synthesis and purification conditions). Low purity powders are dark in color (black or dark grey) and are characterized by high contents of amorphous and graphitic carbon. In addition, they typically contain notable amounts of metal impurities. In contrast, purified ND powders appear lighter in color (grey or light grey), exhibiting higher diamond contents and reduced amounts of metal. Unfortunately, purified ND powders are often subject to increased levels of non-metal impurities, resulting from the chemical treatments used during purification. The methodology and effectiveness of the various purification techniques is discussed in section 4.2.

The average crystal size in ND powders is around 4–5 nm, but the size of the individual diamond crystals can range from 3 to 30 nm or more. A single ND crystal consists of a diamond core that is partially or fully covered by amorphous and/or graphitic carbon, depending on the purity of the sample. Even the purest diamond samples exhibit a certain sp^2 content, resulting from the partial reconstruction of the diamond surface. This effect is negligible for macroscopic diamonds, but becomes significant in the case of ND due to the exceptionally high surface-to-volume atom ratio of nanomaterials. The majority of the ND crystals exhibit high crystallinity, but lattice defects, including vacancies, substitutional impurities, twins, dislocations, and stacking faults, are also observed. The individual ND crystals, also referred to as primary particles, tend to agglomerate and form larger aggregates, known as secondary particles, that range from hundreds to thousands of nanometers in size. Both the primary and secondary particle size is important and must be taken into consideration when selecting ND powders for the various applications.

4.1.3 Assessment of Nanodiamond Purity

The successful removal of non-diamond species from ND powders requires a suitable means for a qualitative and quantitative evaluation of ND purity. Existing characterization methods can be categorized with respect to the species one aims to analyze, focusing on: 1) structural characterization of the diamond phase, 2) evaluation of non-diamond carbon species, 3) determination of type and content of metal impurities, and 4) analysis of surface functional groups. While there exists a wide variety of adequate characterization methods for the above analyte categories, the techniques most commonly employed are electron microcopy, vibrational spectroscopy, thermal analysis, and absorption spectroscopy.

Electron microscopy, particularly scanning electron microscopy (SEM) and transmission electron microscopy (TEM), is an almost indispensable technique for nanoscale characterization as it allows for a direct observation of the ND crystals and the impurities. Two major drawbacks, however, particularly in the case of TEM, are the limited statistical reliability, due to the small probing volumes, and the relatively high cost and complex sample preparation.

Vibrational spectroscopy techniques, such as Raman and Fourier transform infrared (FTIR) spectroscopy, are particularly powerful tools for the characterization of carbon nanomaterials. While being fast, inexpensive, and relatively easy to use, these methods provide a wealth of information on sample composition, ND structure, and surface chemistry. Both techniques are highly complementary. For example, Raman spectroscopy can be used to distinguish between diamond and non-diamond carbon species and, to some extent, to quantify the sp^3 content of ND powders; whereas FTIR spectroscopy probes the ND surface chemistry, determining the nature and quantity of existing functional groups. Unfortunately, both methods share the same disadvantage as they are unable to directly identify metal impurities.

Another frequently used method for purity evaluation in ND powders is thermogravimetric analysis (TGA). This technique measures the weight loss of the sample during oxidation (burn-off), providing details on the relative amounts of the carbon species contained within the powder and their oxidation behavior (*e.g.*, oxidation rate, activation energies). While a quantitative assessment of the coexisting carbon species in ND powders is difficult and requires a more thorough analysis, TGA can readily be used to estimate the total amount of metal impurities by evaluating the ash content after burn-off.

Other characterization techniques used to evaluate ND composition and purity include X-ray photoelectron spectroscopy (XPS), energy-dispersive X-ray spectroscopy (EDS), X-ray absorption near edge structure (XANES), X-ray diffraction (XRD), and ultraviolet–visible–near infrared (UV-VIS-NIR) absorption spectroscopy. The capabilities, advantages and disadvantages of the most common techniques are summarized in Table 4.1.

Several of these characterization techniques have been used to optimize the purification process of ND powders. While, in most cases, ND samples were analyzed before and after oxidation, some characterization methods, particularly Raman spectroscopy, can also be used *in situ*, allowing for deeper insights into the structural and compositional changes that

Table 4.1 Comparison of different material characterization techniques commonly used to evaluate the purity of ND powders.

Technique	Analyte	Advantages	Limitations
Electron Microscopy	ND crystals, carbon impurities, metal impurities	direct visualization of nanostructures	low statistical reliability, costly, requires special sample preparation
Raman/FTIR Spectroscopy	ND crystals, carbon impurities, surface chemistry	extensive information on carbon species and functional groups	cannot detect metal impurities
Thermogravimetric Analysis (TGA)	carbon impurities (q), metal impurities (q)	measures metal content, provides information on oxidation behavior of carbon species	sample is destroyed during measurement
X-ray Diffraction (XRD)	ND crystals, carbon impurities, metal impurities	measures average ND crystal size	compositional information becomes inaccurate for trace quantities or small crystal sizes
X-ray Photoelectron Spectroscopy (XPS)	surface chemistry (q)	accurate assessment of functional groups	no quantitative data on diamond and metal content
Energy-Dispersive X-ray Spectroscopy (EDS)	metal impurities (q)	trace analysis of metal impurities	difficult to measure carbon content

q – provides quantitative information.

occur in the samples during purification.[10] Although generally considered non-destructive, the majority of the techniques listed in Table 4.1 irradiate the ND powders using electron beams, lasers, or other intense light sources, which increases the energy state of the probed material. This can lead to instantaneous or progressing changes in the sample, which likely interfere with the oxidation-related changes to be studied. Thermal analysis methods are very sensitive to heating rates and mass transport, and are strongly affected by experimental parameters, such as the sample size, the size of the heating chamber, and the gas flow used during analysis, to mention a few. Therefore, great care must be taken when selecting the measurement conditions, as the analysis of nanostructures often requires a high level of expertise with regards to the analysis method. Even if the higher energy state of the sample during characterization is insufficient to modify the ND samples under standard conditions, it may affect the sample at elevated temperatures, for example, during *in situ* studies.

4.1.4 Thermal Stability and Oxidation Behavior of Nanodiamond

The oxidation behavior of ND and, thus, the required purification conditions, strongly depend on the composition of the powders. Figure 4.2a depicts the weight loss curves of three commercially available ND powders obtained during TGA analysis performed in air. The corresponding weight loss rate, dw/dT, is shown in Figure 4.2b. While all three samples are considered ND powders, their oxidation behavior is noticeably different. The burn-off process can be divided into three temperature ranges of interest.[10–12] At temperatures below ~ 350 °C, referred to as range I, oxidation of carbon does not occur as the thermal energy is insufficient to

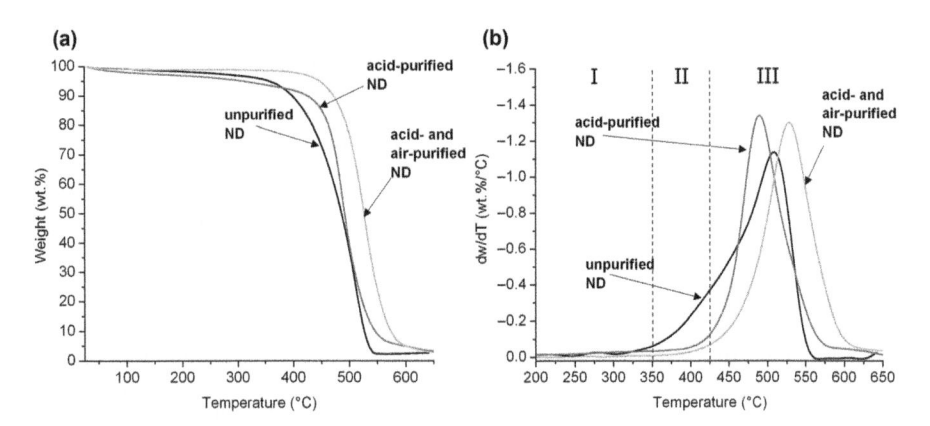

Figure 4.2 Weight loss curves (a) and corresponding weight loss rate (b) of the unpurified (detonation soot), acid-purified, and acid- and air-purified ND powders. The highest oxidation rate of each sample is determined by the maximum in dw/dT.

break the carbon–carbon bonds of amorphous, graphitic, and diamond-like carbon. In the intermediate temperature range between 350 and 425 °C (range II) only amorphous and disordered sp^2 carbon species are oxidized. Finally, when reaching temperatures above 425 °C (range III), both non-diamond carbon and ND crystals are oxidized simultaneously. While range I is similar across studies reported in literature, the transition temperature separating range II and range III varies widely.[10–12] As will be discussed in greater detail in section 4.2.3, the oxidation behavior of ND strongly depends on the composition, particularly the presence of metal impurities, which in turn is dictated by the synthesis and purification conditions of the powders.

In this study oxidation of the as-produced, unpurified ND starts at ~350 °C. The highest weight loss occurs at ~510 °C, as indicated by a maximum in the weight loss rate (Figure 4.2b). In contrast, the oxidation of purified ND, which exhibits a remaining sp^2 content of approximately ~30%, does not start below 400 °C. However, the maximum weight loss takes place around 490 °C and thus at lower temperatures, as compared to the unpurified detonation soot. This discrepancy adequately demonstrates the effect of the ND composition on the oxidation behavior. The unpurified sample consists primarily of amorphous and graphitic carbon (>70 wt%). Fullerene-like shells encapsulate the ND crystals, protecting them against any interactions with the environment. Although amorphous and disordered carbon start to oxidize around 350 °C, the graphitic shells surrounding the ND exhibit a higher resistance towards oxidation and prevent diamond oxidation at these conditions, effectively shifting the maximum weight loss rate to higher temperatures. Metal impurities are enclosed in a similar way, inhibiting catalytic reactions at lower temperatures. The catalytic activity of metal impurities in ND powders is discussed in more detail in section 4.2.3. Due to the larger structural variety of non-diamond carbon species, the as-produced powder is subject to a broader distribution of oxidation temperatures, as compared to purified ND. The burn-off occurs in two distinct temperature ranges, attributed to the removal of amorphous carbon between 350 and 420 °C (range II) and the simultaneous oxidation of graphitic carbon and diamond at temperatures above ~420 °C (range III). In the case of high-purity ND powders (acid- and air-purified), both the onset of the oxidation (~420 °C), as well as the maximum weight loss (~530 °C), occur at higher temperatures. The oxidation is dominated by the diamond phase since the sp^2 content has been reduced to less than 7 wt% during purification. However, the presence of different amounts of metal catalyst and variations in ND surface chemistry further complicate the oxidation kinetics.

The measured weight loss data can also be used to determine important kinetic parameters, such as activation energy (E_A) and frequency factor (A) by employing, for example, the Arrhenius equation or the Achar–Brindley–Sharp–Wendeworth; method.[13,14] The activation energy can be understood as a measure of the energy that is required for the oxygen–carbon reaction to occur. The frequency factor reflects the probability of a species to become activated. Therefore, it is also a measure of the number of species that possess the activation energy. Temperature range II is dominated by the oxidation of amorphous and graphitic sp^2 carbon, which account for roughly 70% of the total sample weight. The corresponding activation energy and frequency factor are 90 kJ mol^{-1} and 1.9×10^4 min^{-1}, respectively. In range III, E$_A$ and A increase to about 190 kJ mol^{-1} and 2.2×10^{11} min^{-1} due to oxidation of the remaining diamond phase, which exhibits higher resistance towards oxidation, as expected. The acid-purified ND powder exhibits slightly higher activation energy ($\sim$225.0 kJ mol^{-1}) in range III, likely due to the lower Fe content and, thus, reduced catalytic activity. The corresponding frequency factor was measured as 5.6×10^{13} min^{-1}. While the encapsulated Fe particles in the detonation soot are catalytically inactive in the low-temperature range (range II), the surrounding graphitic shells are removed at higher temperatures, allowing the Fe to catalyze the oxidation reaction. Another reason for differences in the oxidation behavior, particularly when comparing acid-treated and acid- and air-purified ND powders, are changes in the surface chemistry resulting from the purification treatment. As discussed in more detail in section 4.3, treatments in oxidizing agents, such HNO$_3$ or H$_2$SO$_4$, increase the concentration of oxygen-containing functional groups on the surface of the diamond crystals, which in turn lowers the activation energy.

A study by Chiganov *et al.* on oxidation of ND powders reported activation energies of 160 kJ mol^{-1} and 180 kJ mol^{-1} for the graphitic (range II) and the diamond phase (range III), respectively.[15] Pichot and co-workers also investigated the oxidation behavior of ND powders and reported values of 142 ± 5 kJ mol^{-1} and 189 ± 26 kJ mol^{-1} for range II and III, respectively.[16] The variations in the activation energies of non-diamond carbons in these studies result from differences in the structure and composition of the sp^2 phase. The unpurified ND analyzed in this work is the as-produced detonation soot, whereas the ND powders investigated by Pichot *et al.* and Chiganov *et al.* were pre-purified. More importantly, none of the above studies reported a complete removal of metal catalysts, making a direct comparison of the various studies difficult. The above results demonstrate that the reported activation energies (140–225 kJ mol^{-1}) are distorted by various factors and may not accurately represent the oxidation behavior of diamond phase. They are, however, a characteristic of the respective ND powders. The development of efficient purification procedures, therefore, requires a comprehensive understanding of the oxidation behavior of ND powders and the impurities contained within.

4.2 Chemical Purification of Nanodiamond

The unpurified ND collected after the detonation contains large amounts of amorphous carbon and a variety of graphitic nanostructures, co-produced during the synthesis. Powders also comprise traces of metals, metal oxides, and other foreign elements, which result either directly from the precursor formulation or constitute contaminants from the detonation chamber and other sources. An industrial-scale production and widespread use of ND, therefore, relies on the availability of simple and efficient routes to selectively remove non-diamond carbon and other impurities from the as-produced material. Unfortunately, at this point in time, there exists no single purification step that removes all carbon and non-carbon impurities from ND powders. Instead, the purification typically consists of a multi-step process involving different treatments and chemicals.

The maturity of the purification techniques currently in use can be divided into liquid-phase (wet chemistry) and gas-phase (dry chemistry) processes, both of which utilize the differences in reactivity of diamond and non-diamond species towards oxidation. The required treatment times, temperatures, and concentrations, however, depend on the reactant in use and the desired purity levels (Table 4.2).

Liquid-phase purification utilizes treatments in oxidizing acids and/or bases. In this case, reaction rates, and thus the selectivity of the process, are controlled by adjusting the concentration of the reactants. Non-diamond carbon and metal impurities can be removed either simultaneously using a suitable reactant mixture, or consecutively by employing a multi-step process. Unfortunately, liquid-phase oxidation techniques are subject to several drawbacks. Acids and bases must be handled and stored using corrosion-resistant equipment as they are environmentally harmful. This also leads to costly waste disposal and stricter operation regulations. Oxidation with liquid reactants also introduces new impurities, such as

Table 4.2 Advantages and limitations of existing liquid-phase and gas-phase purification methods used to remove non-diamond species from ND powders.

Method	Reactant	Advantages	Limitations
Liquid-Phase Purification	HNO_3, H_2SO_4, H_3PO_4, HCl, H_2O_2, KOH, NaOH, $KMnO$, NH_4OH	homogeneous process, low process temperature, high selectivity with respect to sp^2 and sp^3 carbon	need for corrosion-resistant equipment, environmentally harmful (requires extensive washing and costly waste disposal), limited process control can lead to high sample loss or reduced selectivity
Gas-Phase Purification	air, O_2, O_3, CO_2, H_2O vapor	high process control using *in situ* characterization, fast and inexpensive, easy to scale up no filtration/separation required	require high temperatures, not able to remove of metal impurities, may require additional supplements (*e.g.* catalyst, oxidation inhibitors)

nitrogen-, sulfur-, chlorine-, and chromium-containing compounds, which, depending on the application, need to be removed in additional purification steps.

Alternatively, the use of gas-phase reactants, such as air, O_2, O_3, or CO_2, for ND purification has also been studied. While these treatments have successfully been employed to remove non-diamond carbon from ND powders, the loss of ND can be substantial without sufficient process control. Unlike oxidation in HCl or HNO_3, most gaseous oxidizers do not allow for a removal of metal and metal oxide impurities. This section provides an overview of the wet-chemistry and dry-chemistry approaches currently in use, and gives insight into both the mechanisms of the purification process and the required conditions.

4.2.1 Liquid-phase Purification

The removal of amorphous and graphitic carbon species and most metal impurities has been achieved using a multi-step process, commonly involving HCl treatments for the dissolution of metals, and exposure to acidic and/or basic reactants to oxidize non-diamond carbons. Frequently employed purification agents include hydrochloric acid (HCl);[17] nitric acid (HNO_3),[18,19] sulfuric acid (H_2SO_4), perchloric acid ($HClO_4$), and mixtures thereof, with HNO_3 being the most common oxidizer in use today. Processes involving hydrogen peroxide (H_2O_2)[20] in combination with H_2SO_4 (piranha water)[21] or HNO_3, aqueous and acidic solutions of sodium perchlorate ($NaClO_4$), mixtures of potassium hydroxide (KOH) and potassium nitride (KNO_3), sodium peroxide (Na_2O_2), and carbon tetrachloride (CCl_4) have also been utilized to selectively remove amorphous and graphitic carbon from ND powders.[2,3] Addition of chromic anhydride (CrO_3) or potassium dichromate ($K_2Cr_2O_7$) to the above reactant mixtures was found to further enhance the oxidation power of these liquids.[22,23]

Purification of ND in an aqueous solution of HNO_3 for 20–30 min at elevated temperatures (230–240 °C) and pressures (6–10 MPa) was reported to remove up to 99.5% of non-diamond species.[19] HNO_3 treatments under milder conditions were found to be less suitable for ND purification as both the thermal energy and the solubility of nitrogen oxides are too low to effectively break the C=C bonds. For oxidation rates to become significant, temperatures must reach at least 160 °C. Under these conditions, HNO_3 decomposes, forming free radicals that initiate the oxidation reaction. As the free radicals are consumed during the oxidation, the concentration of HNO_3 decreases over time, which in turn reduces the oxidation rate. Therefore, in order to maintain sufficiently high oxidation rates over the duration of the process, initial acid concentrations should be at least 50–60%. The high pressures are required to suppress gas evolution during the oxidation, ensuring a true liquid-phase process, ($\sim$7–8 MPa) and to further increase the oxidation rate.[22]

The mechanism of HNO₃ purification has been described in detail by Sushchev et al.[19] The reaction process contains the following steps: (1) solvation of ND particles by the H₂O/HNO₃ mixture, which causes swelling and slow homogenization of the mixture; (2) diffusion of solvent (H₂O) and oxidizer (HNO₃) into ND agglomerates and removal of adsorbed heteroatoms from the ND surface; (3) thermal decomposition of nitric acid and formation of free radicals capable of oxidizing carbon fragments; (4) etching of carbon starting at defective sites and interparticle bonds (total surface area available for oxidation increases and becomes saturated with reaction products); (5) etch removal of the loose surface, gasification, and removal of surface oxidation products in the liquid phase volume; and (6) formation of surface oxygen-containing functional groups and surface reconstruction that offsets the excessive free energy of the ND particles. Agitation of the ND–reactant mixture during purification improves the mass transport, leading to more homogenous treatments and shorter purification times.

Although the HNO₃ purification under high pressure and high temperature conditions has successfully been used to remove non-diamond carbon and other impurities from detonation ND and is currently the most economical liquid-phase process, it suffers from several drawbacks and safety concerns. First, the use of aggressive oxidizers, such as acids and bases, necessitates corrosion-resistance equipment able to withstand the high pressures and temperatures. In many cases the reactants are environmentally harmful, requiring special handling and storage, and lead to byproducts that are subject to costly waste disposal. The purification of 1 kg of as-produced ND powder requires a staggering 35 L of concentrated acid.[24] Finally, oxidation with acids, bases, salts, and oxides introduces additional impurities. ND powders were found to exhibit increased levels of nitrogen-, sulfur-, chlorine-, and/or chromium-containing compounds after purification, which alter physical and chemical properties and, therefore, must be removed in separate purification steps.

4.2.2 Gas-phase Purification

In order to overcome the challenges associated with liquid-phase purification and to gain more process control, researchers have explored a variety of gas-phase methods, including ozone[9] and air oxidation,[23,25–27] catalyst assisted oxidation,[23,28] and oxidation using boric anhydride as an inhibitor of diamond oxidation.[15]

Gubarevich et al. successfully removed non-diamond carbon species from the detonation soot by bubbling air through an aqueous suspension of ND containing a metal catalyst.[29] The necessity for an additional catalyst is the primary drawback of this method, as catalysts are expensive and further contaminate the sample. Pavlov et al. have proposed a process utilizing an ozone–air mixture at temperatures between 150 and 400 °C for ND purification. A slight modification of this technique was employed by Petrov and co-workers, who used lower process temperatures (150–200 °C) and longer treatment times.[7] While these methods require ozone, a toxic and aggressive substance that requires special handling, they yield purified ND with a unique surface chemistry along with other benefits, such as reduced agglomeration.[9]

The treatments discussed above were all shown to substantially reduce the amount of non-diamond carbon from as-produced powders, but similar to liquid-phase methods, they require the use of either toxic and aggressive substances, or supplementary catalysts, which result in additional contamination or extensive sample loss.

Alternatively, although first thought of as being unfeasible,[15] it was recently demonstrated that air purification of ND—a simple, inexpensive, and environmentally friendly process—is a suitable alternative to ozone treatments.[10–12,23,27,30,31] Using *in situ* Raman spectroscopy to optimize the oxidation process, Osswald *et al.* selectively removed graphitic carbons from the detonation soot, eliminating the need for additional contaminants. In their study the optimal temperature range for ND purification was identified as 400–430 °C, producing purity levels as high as 96 wt% without notable losses in the diamond phase.[10] While several studies yielded similar results,[11] some authors reported higher temperatures for the selective removal of non-diamond carbon. Tyurnina *et al.* recommended oxidation temperatures as high as 550 °C.[12] In contrast, Cataldo *et al.* found that ND powders were stable towards oxidation below 450 °C, but burn rapidly above this temperature.[32] These inconsistencies are a direct consequence of the differences in the structure and composition of commercially available ND powders.[12]

Oxidation in air has also successfully been used to control the average crystal size in ND powders.[8] This approach utilizes the differences in the oxidation rates of small and large ND crystals. During air oxidation, smaller NDs (<10 nm) are subject to higher oxidation rates as compared to larger crystals, thus shifting the average ND crystal size towards higher values.[8,33] Air oxidation, therefore, provides an effective route to narrow the crystal size distribution in detonation ND by selectively removing smaller ND crystals.

4.2.3 Effect of Metal Impurities on Oxidation Behavior

While purification aims to remove non-diamond species from ND powders, the oxidation process itself is sometimes affected by the impurities. Several of the trace metals found in ND powders, particularly Fe, are known to catalyze the carbon–oxygen reaction. Metal catalysts lower the activation energy for carbon oxidation by providing alternative reaction pathways,[34] enabling reactions that would be thermodynamically impossible without its presence, or accelerate the oxidation by increasing the reaction rates. Since metal catalysts participate in multiple catalytic cycles and are not consumed by the reaction, even small amounts alter the oxidation behavior of ND powders substantially.

Nanodiamond Purification

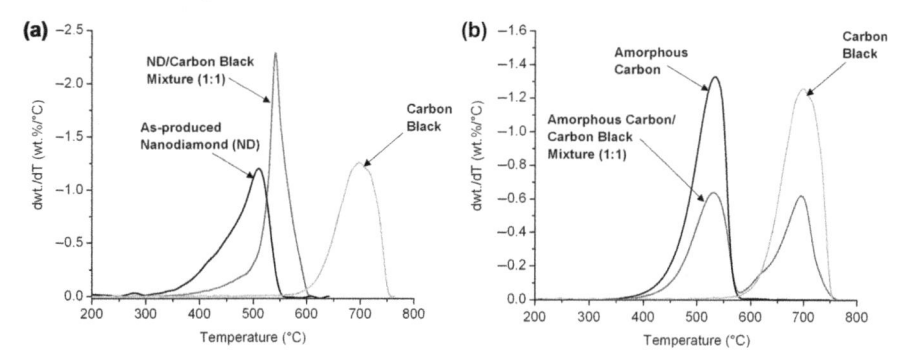

Figure 4.3 Thermogravimetric analysis (weight loss rate) of metal-free carbon black with and without the addition (mixture 1:1) of as-produced ND (a) and amorphous carbon (b), demonstrating the catalytic activity of Fe impurities in ND powders. The weight loss rates of the pristine powders are shown for comparison.

Figure 4.3 demonstrates the catalytic effect of metal impurities in ND powders during TGA analysis performed in air. An unpurified ND sample with an Fe content of approximately 0.65 wt% was mixed with carbon black, a graphitic carbon material that is essentially metal free, at a mass ratio of 1:1. Therefore, the Fe content of the mixture was only 0.325 wt%. While the pristine carbon black does not oxidize below ∼600 °C, the addition of ND results in a complete sample burn-off before even reaching this temperature (Figure 4.3a). In order to rule out other possible contributions, such as internal self-heating, resulting from the exothermic oxidation of carbon, the ND powder was replaced by a metal-free amorphous carbon (Figure 4.3b). The oxidation of the amorphous carbon/carbon black mixture occurs in two distinct temperature regions that correspond well with the oxidation range of the individual carbon powders. The recorded thermograms thus show that the reduction in oxidation temperature in Figure 4.3a was indeed caused by a catalytic process rather than internal heating effects. This relatively simple experiment illustrates that, in the case of ND powder, metal impurities are catalytically active and strongly affect the oxidation of the contained carbon species with gaseous reactants, such as O_2.

Several of the metals found in ND powders exhibit high catalytic activity. For example, lead (Pb), copper (Cu), silver (Ag), and iron (Fe) were reported to lower the oxidation temperature of a graphite powder from 740 to 382, 570, 585, and 593 °C, respectively. In all of these cases, the metal concentration did not exceed 0.2 wt%.[35] It should be mentioned that the metals do not have to exist in their elemental form to be catalytically active. Salts and oxides of alkali and alkaline earth metals were found to also catalyze the carbon–oxygen reaction.[36] It is, therefore, important to consider the nature of the existing impurities when selecting the purification conditions for

ND powders, particularly in the case of gas-phase reactions, which are sensitive to the presence of catalytic metal impurities.

4.3 Properties of Purified Nanodiamond

4.3.1 Structure, Composition, and Surface Chemistry

The purification treatments discussed in section 4.3 significantly alter the composition of ND powders. Figure 4.4a and 4.4b show the TEM images of an acid-purified and an acid- and air-purified ND powder, respectively. The acid treatment reduces the content of non-diamond carbon to less than 20 wt%, removing the majority of the amorphous and graphitic carbon impurities. The remaining sp^2 content is associated with partially reconstructed ND surfaces exhibiting mixtures of sp^1-, sp^2-, and sp^3-type bonding, and some remaining graphitic species that resisted the oxidation treatment.

By employing additional purification steps, the sp^2 content can be reduced further to less than 5 wt%. A combination of acid and air oxidation was reported to yield clean crystal surfaces (Figure 4.4b) and powders virtually free of non-diamond carbon impurities.[10]

Figure 4.5a shows the UV Raman spectra of these powders in comparison to an unpurified ND sample. UV lasers are generally preferred over excitation in the visible spectral range, as they provide a stronger diamond Raman signal and cause less fluorescence.[37,38] The Raman spectrum of the unpurified ND powder is dominated by the Raman features of amorphous and graphitic carbon, namely the disorder-induced D band (1400 cm^{-1}) and the G band (1590 cm^{-1}), which result from the in-plane vibrations of the carbon–carbon bonds.[10] Due to the low diamond content, as-produced powders exhibit only a weak or no diamond Raman signal. After purification, the intensity of the diamond peak (1327 cm^{-1}) increases, whereas the

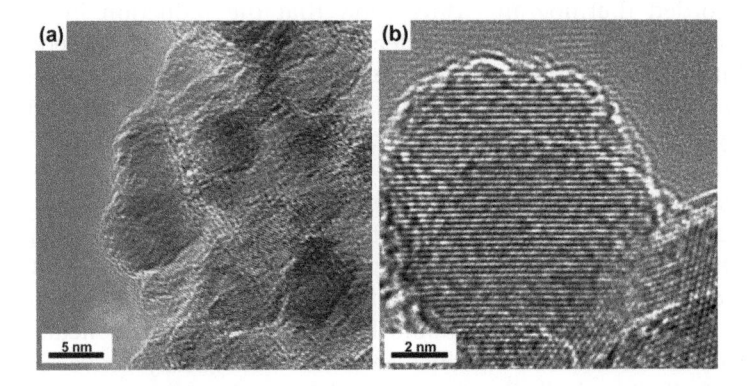

Figure 4.4 Transmission electron microscopy (TEM) image of a commercially available ND powder after acid purification (a), and acid and air purification (b).

Nanodiamond Purification

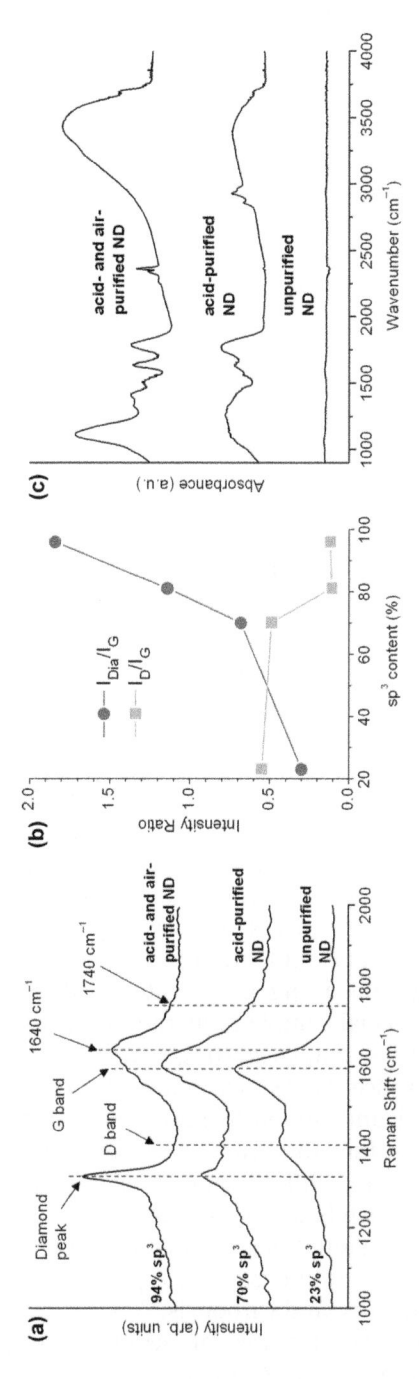

Figure 4.5 UV Raman and FTIR spectra of unpurified and purified ND powders. Raman spectra (a) and corresponding peak intensity ratios (b) allow for an evaluation of ND purity, while FTIR (c) provides valuable information on surface chemistry. Raman spectra were recorded using 325 nm laser excitation.

D band weakens. The higher the sp³ content, the lower the D band and the higher the diamond Raman intensity.[10] The spectral changes observed between 1500 and 1800 cm^{-1} are more complex.[39] In the case of unpurified ND powder with high sp² contents, this feature can be labeled as the G band. However, the contribution of graphitic carbon to the Raman spectra of well-purified ND powders (sp³ >80%) is rather small. For well-purified samples (*e.g.*, after intensive acid and/or air purification), the broad asymmetric Raman feature is composed of at least three separate peaks assigned to sp² carbon ($\sim$1590 cm^{-1}), O–H ($\sim$1640 cm^{-1}), and C=O ($\sim$1740 cm^{-1}) species.[39] The relative intensities of the 1640 (O–H) and 1740 cm^{-1} (C=O) peaks strongly depend on the surface chemistry of the ND crystals and may vary for ND powders purified by different oxidation techniques (*e.g.*, acid treatment *vs.* air oxidation).

Similar to carbon nanotubes, where the Raman intensity ratio between D and G band (I_D/I_G) is used to measure purity and structural ordering in the sample,[40–42] the diamond-to-G band intensity ratio (I_{Dia}/I_G) may be used to evaluate the purity of ND powders.[43] However, great care must be taken when using this approach as the Raman spectrum of ND is more complex and depends strongly on the composition of the sample. In general, the I_{Dia}/I_G intensity ratio increases with increasing sp³ content, as shown in Figure 4.5b. The observed increase in the slope for samples containing more than 70% sp³ carbon likely results from the removal of non-diamond species shielding the diamond core. A similar effect was observed during the oxidation of carbon nanotube samples, in which amorphous carbon on the outer walls of the nanotubes weakened their Raman signal. In contrast, the I_D/I_G intensity ratio decreases with increasing sp³ content. Although both Raman features are ascribed to sp² carbon, the D band intensity decreases at a larger rate as it is proportional to the number of hexagonal carbon rings, while the G band intensity reflects the number of sp² pairs.[44]

While Raman spectroscopy is exceptionally powerful in differentiating between various carbon species, FTIR spectroscopy is better suited for the analysis of functional groups and adsorbed molecules at ND surface. Unpurified ND is strongly absorbing, as indicated by the dark black color of the powder. Even with special sample preparation, measurements typically reveal no detectable FTIR signal, suggesting that the number of functional groups on the ND surface is relatively low. In contrast, purification treatments such as acid and air purification, yield ND powders rich in surface functional groups, including carboxylic acids, anhydrides, lactones, carbonyls, aldehydes, esters, and phenols. The amount of the various functionalities strongly depends on the purification treatment and the prevailing reaction conditions. Liquid-phase oxidation using nitric acid typically yields more carboxylic acids, while gas-phase purification results in more anhydrides, lactones, phenols, and carbonyls.[45]

The most prominent FTIR features of acid-purified powders can be assigned to O–H vibrations (3280–3675 cm^{-1} stretch and 1630–1660 cm^{-1} bend), C–H (2853–2962 cm^{-1}), C=O and C=C in aromatic rings

(1550–1760 cm^{-1}), and C–O in various chemical surroundings (950–1300 cm^{-1}), originating from –COOH, –OH –CH$_2$–, and –CH$_3$ groups of chemically bonded or adsorbed surface species.[46–48] Air oxidation largely removes –CH$_2$– and –CH$_3$ groups, but increases the amount of C–O and –OH groups. The frequency of C=O vibrations is upshifted by 20–40 cm^{-1}, indicating a conversion of aldehydes, ketones, and ester groups into carboxylic acids, anhydrides, or cyclic ketones. A recent study by Schmidlin et al. quantified the surface species of purified ND powders using the Boehm titration method, reporting COOH concentrations up to 0.81 sites per nm^2. Based on these results, the authors also suggested that the functional groups are preferentially located at the edges of ND crystals, occupying all the available sites.[49] Purification using ozone or ozone-enriched air was found to yield a similar, but more acidic, oxygen-rich surface chemistry, as compared to air oxidation.[7]

Therefore, upon removal of amorphous and graphitic carbon during purification, the surface of the ND crystals becomes accessible to the chemical reactants in use, and is immediately saturated with oxygen or oxygen-containing functional groups, which are often the basis for further functionalization.[50]

Oxidation also improves the dispersion of ND in aqueous suspensions. The oxygen-containing functional groups lead to a more negative zeta potential, which prevents the ND crystals from agglomerating. In particular, purification in ozone-enriched air has been reported to yield exceptionally low zeta potentials ($-$35 mV) over a wide pH range (3–11).[7] Stable ND suspensions can also be achieved by highly positive zeta potentials. Williams et al. modified the ND surface using a hydrogenation process, which led to a stable zeta potential of >40 mV in the pH range 3–7.[51] Zeta potentials lower than $-$30 mV or higher than 30 mV are typically required for stable colloidal suspensions in polar media.

4.3.2 Optical and Electrical Properties

The purification of ND powder also affects physical properties, such as optical absorption and electrical conductivity. The insets in Figure 4.6a show the photographic images of as-produced, acid-purified, and acid- and air-purified ND powders. The color of the powders is related to the content of amorphous and/or graphitic carbon. The unpurified detonation soot is velvet black, while the purified ND powders appear lighter in color, ranging from grey after acid purification, to light grey when employing a combination of acid and air purification. In some cases, powders exhibit faint shades of red or brown due to the presence of iron oxide.

The color of the powders is determined by their absorption characteristics. Figure 4.6a shows the UV–VIS–NIR absorption spectra of the powders, recorded using an Ulbricht sphere. The black unpurified ND powder strongly absorbs light over the entire wavelength range studied (200–2000 nm) due to the high sp^2-carbon content. Acid-purified ND exhibits a significant decrease

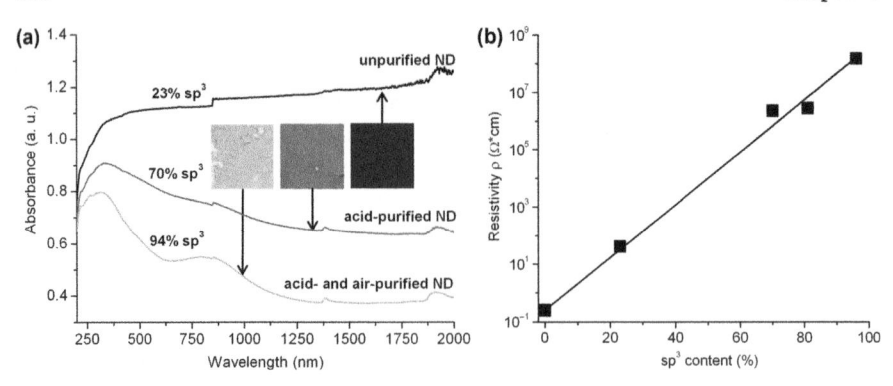

Figure 4.6 UV–VIS–NIR absorption spectra (a) and electrical resistivity (b) of ND powders with different sp³ contents. The insets show the photographic images of the respective powders.

in absorbance, especially in the VIS–NIR spectral range (600–2000 nm), which is reduced further after an additional air-purification step. Air-oxidized ND shows a 50% reduction in absorbance as compared to samples purified solely by acid treatments. The acid-purified powder revealed broad absorption peaks around 200–500 nm and 700–1000 nm. The feature between 200 and 500 nm can be ascribed to sp²-bonded carbon[52,53] and possible contributions from functional groups, which are known to strongly absorb in the lower wavelength range.[54] The absorption feature in the higher wavelength range may be assigned to contributions from lattice defects and elemental impurities (*e.g.*, nitrogen, sulfur) that absorb light in the visible spectral range.[55] While a detailed description of the absorption features of ND powders containing sp¹, sp² and sp³ carbon species, a variety of surface functional groups, and catalyst impurities is complex and outside the scope of this chapter, the results demonstrate that the purification of ND reduces absorbance by lowering the sp² content, leading to a higher transparency in the VIS–NIR range. Even small amounts of sp² carbon in acid-purified powders significantly alter the optical properties, revealing the importance of air oxidation for ND purification. The optical transparency of ND is critical for many applications, including scratch-resistant optics, windows, and displays. The absorption properties of oxidized ND may also be of great benefit for sunscreen formulations, as it is highly transparent in the visible spectral range and, at the same time, exhibits remarkable UV shielding.

The removal of non-diamond carbon impurities also has significant impact on the electrical properties of ND powders, which is of importance for applications where ND is used as an additive or reinforcement agent. Figure 4.6b shows changes in the resistivity of the ND powders with increasing sp³ content. Unpurified ND is a good electrical conductor because of the high content of graphitic carbon (~23% sp³). The resistivity of ND powder can be even further reduced by vacuum annealing (graphitization) at temperatures above 1500 °C, which leads to a full conversion of the diamond

(0% sp^3).[56] Acid-purified ND powders, with diamond contents ranging from 70 to 80%, are insulators, with electrical resistivities several orders of magnitude higher than the as-produced detonation soot. Further decreases in the sp^2 content upon air oxidation lead to resistivities as high as $>10^8$ Ω cm ($>94\%$ sp^3).

4.4 Summary and Outlook

Detonation nanodiamond, or simply nanodiamond (ND), is one of only a few nanomaterials that can be produced on an industrial scale. However, the high content of non-diamond species in the detonation soot (*e.g.*, amorphous and graphitic carbon, metal, and metal oxides) inhibits the direct use of the as-produced material in most applications. As a consequence, purification has become one of the most crucial steps in ND production, accounting for almost half of the current material cost. Existing purification treatments include liquid-phase and gas-phase methods, each of which has its unique strengths and weaknesses. The majority of these methods utilize differences in the reactivity of diamond and non-diamond species towards oxidation in order to selectively remove non-diamond carbon from the ND powders. Unfortunately, at this point in time, there exists no single purification step able to simultaneously remove all impurities from ND powders. Methods in use typically employ a series of different process steps involving a variety of chemicals. The development of effective purification treatments requires suitable characterization methods, such as TEM and Raman spectroscopy, which are able to assess and quantify the purity of ND powders. The selection of proper purification conditions also requires detailed knowledge of the powder's composition, particularly the content of metal impurities, which are known to catalyze carbon oxidation. Data reported in the literature suggest that a combination of acid and air purification currently yields the highest purity levels with respect to the remaining content of sp^2 carbon and metal impurities. The obtained powders are characterized by a light-grey color, comprising clean crystal surfaces and sp^3-carbon contents exceeding 95%.

Although much progress has been made in developing efficient and economical purification methods for ND during recent years, several challenges remain on the path toward widespread utilization of high-purity ND. The efficiencies of current synthesis processes leave room for improvement, but an evaluation of the detonation process is outside the scope of this chapter. We will, therefore, focus our concluding remarks on the purification treatments and purity assessment.

4.4.1 Purity Assessment

The evaluation of ND purity is a critical part of the purification process and relies heavily on the availability of suitable analysis methods. Although there exists a wide variety of characterization methods that have successfully been used to assess the composition and purity of ND powders, a standardized purity evaluation protocol has yet to be established. This causes difficulties when comparing the effectiveness of the various purification methods, as characterization techniques differ with respective to their selectivity and sensitivity towards the different impurity species. Currently, there exists no single characterization method that provides a quantitative assessment of all carbonaceous and non-carbonaceous impurities. Instead, a combination of different techniques is commonly employed to fully characterize ND powders. Even a slight modification in measurement conditions of a particular technique may yield different results. In order to overcome these challenges a standardized purity assessment protocol for ND powders should be developed, containing a series of selected characterization techniques and predetermined measurement conditions. The test standard must be able to quantify existing impurities in their elemental form, as well as within related compounds. It must also provide a reliable evaluation of the diamond phase and should be able to distinguish between the various non-diamond carbon species. Finally, the test protocol must be able to identify and quantify the various surface functional groups. While these are rather demanding requirements, the development of application-driven test protocols can simplify the test criteria. For example, some applications may not necessitate a complete removal of all metal impurities, while others may tolerate a certain amount of residual sp^2 species.

4.4.2 Purification Treatment

Although a wide variety of purification treatments have been shown to reduce the amount of non-diamond carbon and metal impurities, none of the existing methods has been found to effectively remove all impurities from ND powders. In fact, most reactants are selective towards a particular impurity category (*e.g.*, metals, carbon, *etc.*), and often introduce additional contaminants, for example, sulphur and nitrogen-containing compounds. Given the complexity of both the ND composition and the chemical reactions during purification, the establishment of an application-focused purification process appears to be the economically most feasible approach. However, considering the increasing number of ND manufacturers, standard purification procedures should be implemented to minimize sample-to-sample variations in ND powder composition and facilitate widespread use.

At this point in time, a standardization of both purity assessment and purification treatment seems challenging. In particular, the purification process and the required conditions are strongly dependent on the composition of the ND powder, which varies from manufacturer to manufacturer. Therefore, in the near term, the primary focus should be on the development of a purity evaluation standard. The availability of a reliable purity evaluation procedure will also allow for a direct comparison of existing purification treatments, identifying possible candidates for future treatment standardization.

Finally, as the number of ND applications grows, control over ND crystal size and surface state will become more crucial. While it has already been shown than oxidation can be used to modify the average size of ND crystals, further research is needed to fully utilize oxidation treatments for size control. In addition, surface control and adjustment of the sp^2/sp^3 carbon ratio of the individual ND crystals will play a key role in future applications, particularly in optics, electronics, and biomedicine.

References

1. V. N. Mochalin, O. Shenderova, D. Ho and Y. Gogotsi, *Nature Nanotechnology*, 2012, **7**, 11.
2. V. Y. Dolmatov, *Ultradisperse Diamonds of Detonation Synthesis: Production, Properties and Applications*, State Polytechnical University, St. Petersburg, 2003.
3. D. M. Gruen, O. A. Shenderova, A. Y. Vul, in NATO Science series. *Series II: Mathematics, Physics and Chemistry*, vol. 192, Springer, Dordrecht, Berlin, Heidelberg, New York, 2005, p. 401.
4. J. L. Davidson, D. T. Bradshaw, Vanderbilt University, USA, *US Pat.*, US6858157B2, 2005, p. 18.
5. V. E. Red'kin, *Chemistry and Technology of Fuels and Oils*, 2004, **40**, 164.
6. V. Y. Dolmatov, *Russian Chemical Reviews*, 2001, **70**, 607.
7. I. Petrov, O. Shenderova, V. Grishko, V. Grichko, T. Tyler, G. Cunningham and G. McGuire, *Diamond and Related Materials*, 2007, **16**, 2098.
8. S. Osswald, M. Havel, V. Mochalin, G. Yushin and Y. Gogotsi, *Diamond and Related Materials*, 2008, **17**, 1122.
9. O. Shenderova, A. Koscheev, N. Zaripov, I. Petrov, Y. Skryabin, P. Detkov, S. Turner and G. Van Tendeloo, *J.Phys. Chem. C*, 2011, **115**, 9827.
10. S. Osswald, G. Yushin, V. Mochalin, S. Kucheyev and Y. Gogotsi, *J. Am. Chem. Soc.*, 2006, **128**, 11635.
11. II Kulakova, *Physics of the Solid State*, 2004, **46**, 636.
12. A. V. Tyurnina, I. A. Apolonskaya, Kulakova, II, P. G. Kopylov and A. N. Obraztsov, *Journal of Surface Investigation: X-Ray Synchrotron and Neutron Techniques*, 2010, **4**, 458.
13. B. N. Achar, G. W. Brindley, J. H. Sharp, in *International Clay Conference*, ed. L. Heller, Israel Universities Press, Jerusalem, vol. 1, 1966, p. 67.
14. J. H. Sharp and S. A. Wentworth, *Anal. Chem.*, 1969, **41**, 2060.
15. A. S. Chiganov, *Physics of the Solid State*, 2004, **46**, 595.
16. V. Pichot, M. Comet, E. Fousson, C. Baras, A. Senger, F. Le Normand and D. Spitzer, *Diamond and Related Materials*, 2008, **17**, 13.
17. Z. Spitalsky, C. Aggelopoulos, G. Tsoukleri, C. Tsakiroglou, J. Parthenios, S. Georga, C. Krontiras, D. Tasis, K. Papagelis and C. Galiotis, *Materials Science and Engineering B-Advanced Functional Solid-State Materials*, 2009, **165**, 135.

18. G. Post, V. Y. Dolmatov, V. A. Marchukov, V. G. Sushchev, M. V. Veretennikova and A. E. Sal'ko, *Russian Journal of Applied Chemistry*, 2002, **75**, 755.
19. V. G. Sushchev, V. Y. Dolmatov, V. A. Marchukov and M. V. Veretennikova, *Journal of Superhard Materials*, 2008, **30**, 297.
20. T. M. Gubarevich, V. F. Pyaterikov, I. S. Larionova, V. Y. Dolmatov, R. R. Sataev, A. V. Tyshetskaya and L. I. Poleva, *Journal of Applied Chemistry of the USSR*, 1992, **65**, 2075.
21. L. Rondin, G. Dantelle, A. Slablab, F. Grosshans, F. Treussart, P. Bergonzo, S. Perruchas, T. Gacoin, M. Chaigneau, H. C. Chang, V. Jacques and J. F. Roch, *Phys. Rev. B*, 2010, **82**.
22. O. A. Shenderova, D. M. Gruen, *Ultrananocrystalline Diamond: Synthesis, Properties, and Applications*, William Andrew, 2006.
23. D. Mitev, R. Dimitrova, M. Spassova, C. Minchev and S. Stavrev, *Diamond and Related Materials*, 2007, **16**, 776.
24. J. R. Maze, P. L. Stanwix, J. S. Hodges, S. Hong, J. M. Taylor, P. Cappellaro, L. Jiang, M. V. G. Dutt, E. Togan, A. S. Zibrov, A. Yacoby, R. L. Walsworth and M. D. Lukin, *Nature*, 2008, **455**, 644.
25. E. V. Pavlov and Y. A. Skryabin, Method for Removal of Impurity of Non-Diamond Carbon and Device for Its Realization, *Rus. Pat.*, RU2019502, 1994.
26. S. Osswald, G. Yushin, V. Mochalin, S. O. Kucheyev and Y. Gogotsi, *J. Am. Chem. Soc.*, 2006, **128**, 11635.
27. X. Y. Xu, Z. M. Yu, Y. W. Zhu and B. C. Wang, *J. Solid State Chem.*, 2005, **178**, 688.
28. T. M. Gubarevich, R. R. Sataev, V. Y. Dolmatov, in *5th All-Union Meeting on Detonation*, vol. 1, Krasnoyarsk USSR, 1991, p. 135.
29. T. M. Gubarevich, R. R. Sataev, V. Y. Dolmatov, in *Proceedings of the 5th All-Union Meeting on Detonation*, vol. 1, Krasnoyarsk USSR, 5–15 August 1991, p. 135.
30. A. S. Chiganov, *Physics of the Solid State*, 2004, **46**, 620.
31. A. S. Chiganov, G. A. Chiganova, Y. V. Tushko and A. M. Staver, Method for Cleaning Detonation Diamonds, *Rus. Pat.*, RU2004491, 1993.
32. F. Cataldo and A. P. Koscheev, *Fullerenes Nanotubes and Carbon Nanostructures*, 2003, **11**, 201.
33. S. Gordeev and S. Korchagina, *Journal of Superhard Materials*, 2007, **29**, 124.
34. M. A. Vannice, *Kinetics of catalytic reactions* Springer, New York, 2005.
35. D. W. McKee, *Carbon*, 1970, **8**, 623.
36. Y. Kobayashi and M. Sano, *ChemCatChem*, 2010, **2**, 397.
37. G. N. Yushin, S. Osswald, V. I. Padalko, G. P. Bogatyreva and Y. Gogotsi, *Diamond and Related Materials*, 2005, **14**, 1721.
38. O. O. Mykhaylyk, Y. M. Solonin, D. N. Batchelder and R. Brydson, *J. Applied Phys.*, 2005, 97.
39. V. Mochalin, S. Osswald and Y. Gogotsi, *Chem. Mater.*, 2009, **21**, 273.

40. S. Osswald, M. Havel and Y. Gogotsi, *J. Raman Spectro.*, 2007, **38**, 728.
41. S. Osswald, E. Flahaut, H. Ye and Y. Gogotsi, *Chem. Phys. Lett.*, 2005, **402**, 422.
42. S. Osswald, E. Flahaut and Y. Gogotsi, *Chem. Mater.*, 2006, **18**, 1525.
43. S. Osswald, Y. Gogotsi, in *Raman Spectroscopy for Nanomaterials Characterization*, ed. C. S. R. Kumar, Springer Berlin Heidelberg, 2012, p. 291.
44. A. C. Ferrari and J. Robertson, *Philosophical Transactions of the Royal Society of London Series A*, 2004, **362**, 2477.
45. J. L. Figueiredo, M. F. R. Pereira, M. M. A. Freitas and J. J. M. Orfao, *Carbon*, 1999, **37**, 1379.
46. S. F. Ji, T. L. Jiang, K. Xu and S. B. Li, *Applied Surface Science*, 1998, **133**, 231.
47. T. Jiang and K. Xu, *Carbon*, 1995, **33**, 1663.
48. V. L. Kuznetsov, M. N. Aleksandrov, I. V. Zagoruiko, A. L. Chuvilin, E. M. Moroz, V. N. Kolomichuk, V. A. Likholobov, P. M. Brylyakov and G. V. Sakovitch, *Carbon*, 1991, **29**, 665.
49. L. Schmidlin, V. Pichot, M. Comet, S. Josset, P. Rabu and D. Spitzer, *Diamond and Related Materials*, 2012, **22**, 113.
50. A. Krueger and D. Lang, *Advanced Functional Materials*, 2012, **22**, 890.
51. O. A. Williams, J. Hees, C. Dieker, W. Jager, L. Kirste and C. E. Nebel, *ACS Nano*, 2010, **4**, 4824.
52. A. E. Aleksenskii, V. Y. Osipov, A. Y. Vul', B. Y. Ber, A. B. Smirnov, V. G. Melekhin, G. J. Adriaenssens and K. Iakoubovskii, *Physics of the Solid State*, 2001, **43**, 145.
53. O. Shenderova, V. Grichko, S. Hens and J. Walch, *Diamond and Related Materials*, 2007, **16**, 2003.
54. D. R. Lide, *CRC Handbook of Chemistry and Physics: a Ready-reference Book of Chemical and Physical Data*, CRC, Taylor & Francis, 2006.
55. A. M. Zaitsev, *Optical Properties of Diamond*, Springer, Heidelberg, 2001.
56. V. L. Kuznetsov, A. L. Chuvilin, Y. V. Butenko, I. Y. Mal'kov and V. M. Titov, *Chem. Phys. Lett.*, 1994, **222**, 343.

CHAPTER 5

Pure Nanodiamonds Produced by Laser-assisted Technique

BORIS ZOUSMAN AND OLGA LEVINSON*

Ray Techniques Ltd, The Hebrew University of Jerusalem, P.O.B. 39162, 91391, Israel
*Email: olga.levinson@nanodiamomd.co.il

5.1 Introduction, Nanodiamond Market, Problems and Prospects

Nanodiamond powder belongs to the carbon nanomaterial family, which, along with graphene, fullerene and nanotubes, has attracted great interest in recent years due to its unique physical and chemical properties.[1–3] Initially, nanodiamond applications were rooted in the defense industry but have now reached into a variety of fields, such as fine polishing, lubricating, coatings and polymers. Currently, nanodiamonds have entered biomedicine, thermal management in electronics, photovoltaics and energy storage applications.

Nanodiamond powder is composed of diamond nanoparticles with average size of 4–5 nm, usually collected in aggregates of a few hundred nanometers and even microns. Each primary particle consists of a nanocrystal of tetrahedral bonded carbon atoms collected in a three-dimensional cubic lattice, which determines the unique properties of diamond, and an onion-like carbon shell with a chemically active "coat" of functional groups on the surface.[1] This coat enables interaction with various molecules, which transfers to them the unique properties of diamond and makes possible the creation of novel, unique and economically viable materials and objects with desired properties.

Nanodiamond powder is one of a few nanomaterials being produced at a commercial scale. Traditionally, it is synthesized by detonation of solid explosives in metal chambers, then isolated from the obtained blend and purified in boiling acid.[1–4] Having been discovered in 1963 in the Soviet Union, nanodiamonds have been studied in depth and a lot of applications exploiting the unique nanodiamond features have been developed in the fields of fine polishing, lubrication, coatings and polymers. The potential market of nanodiamond powder is huge[4] and is estimated at 300 billion dollars annually. However, wide use of nanodiamonds is currently impeded by some problems, including:

- Low quality consistency of nanodiamonds produced by various manufacturers.
- Absence of standards for the regulation of nanodiamond quality.
- Problem of nanodiamond aggregation for most advanced applications and the absence of industrial technologies for nanodiamond dispersion within various media.
- Lack of fundamental understanding of the mechanisms influencing the structure of nanodiamonds (crystalline shape and size, thickness of sp^2 cover and type and quantity of functional groups on the surface) and the effect on the nanodiamond performance (mechanical, thermal, electrical, magnetic and optical properties of nanodiamond composites and objects).
- Insufficient quantity of already designed final nanodiamond-based products.

At the same time, recent achievements in the development of advanced nanodiamond applications[5,6] in biomedicine, photovoltaics, energy and optics present new demands on the quality of diamond nanopowder, such as the purity and homogeneity of the primary particle dimensions, as well as the surface chemistry. Due to differences in synthesis conditions, inside the detonation charge volume detonation nanodiamonds (DND) usually contain primary particles of different sizes, as well as various ratios of functional groups and sp^2 shells on the surface, which define the wide range of particle behavior in diverse reactions. An additional problem is metal impurities from the detonator and chamber materials. Insufficient levels of purity and homogeneity in DND limit the applicability and efficiency of this unique material in many important fields, such as heat conductive insulating compounds and fine polishing materials for electronics, nuclear fuel, electrodes for efficient energy storage, cold emitters for displays, photovoltaic elements, terahertz radiation sensors, agents for high performance liquid chromatography, diagnostic kits and drugs. Moreover, detonation technology is dangerous, polluting and requires additional expenses for safety and security. Apart from the non-controllable character of the detonation process, insufficient quality, safety and regulatory issues, the cost of DND is relatively expensive for some applications.

Special techniques for DND unification and fractionalization should be developed, as well as special standards ensuring ND quality. These activities may subsequently raise the price of already expensive DND. Therefore, novel technologies for nanodiamond fabrication, enabling control of the synthetic process, to increase the quality of the product and to reduce its cost are currently of particular importance.

5.2 Synthesis Nanodiamonds by Laser Ablation in Liquid

Laser ablation has been studied from the creation of ruby lasers in the sixties of the last century and is currently widely used for analytical spectroscopy,[6] surface treatment, coatings and the synthesis of nanomaterials.[7] The synthesis of nanoparticles by laser ablation in liquid (PLAL), named also liquid-phase pulse laser ablation (LP-PLA) and the pulsed-laser-induced liquid–solid interfacial reaction (PLIIR), was first reported in 1987 when a metastable form of iron oxide was obtained by high-power pulsed-laser radiation of liquid–solid interfaces.[8] Since then, interest in this method of synthesis has been steadily increasing.[9,10] The synthesis of nanoscale particles of silver,[11] gold,[12] cubic-boron nitride,[13] titanium dioxide,[14] cobalt oxide,[15] cubic-carbon nitride[16] and other nanostructures[8] has been reported in various publications. Special interest regarding this method for nanoparticle fabrication was ignited by the possibility to control the parameters of the obtained nanostructures, such as their size, morphology and shape, by adjusting the parameters of laser radiation, the liquid and the target. This control is highly important for most nanomaterial applications, such as nano-biosurgery and light emission.

In this method a pulsed laser beam of high density is focused onto the surface of a solid target, which is placed in a liquid. The interaction of the laser beam and the target surface can result in vaporization of the target in the form of an ablation plume. Atoms of the target and liquid interact under high pressure and high temperature conditions, allowing the formation of nanoparticles dispersed in suspension.

The PLAL technique for producing nanodiamonds has been studied by several academic groups.[15–18] Theoretical and experimental justifications for the growth of diamond nanocrystals *via* PLAL have been performed in order to predict the size of primary particles and the yield of the process.[7,9,10,26,27] It was shown that the growth of diamond crystals is followed by the formation of single nanocrystals and twin structures, including a single twin and triple twin, as well as four- and five-fold twin structures.[25,26] Nanodiamond particles of various sizes were obtained using laser Nd-YAG radiation with different parameters and in diverse liquids (Table 5.1). However, to the best of our knowledge, attempts to synthesize nanodiamonds by this method proved to be economically impractical.

Table 5.1 Nanodiamond synthesis by PLAL reported earlier.

#	Wavelength, nm	Repetition frequency, Hz	Power density, W cm^{-2}	Pulse width	Graphite target	Liquid	ND size, nm	Author, year	Ref.
1	532	5	10^{11}	10 nc	Poly-crystalline	Water	300–400	G. W. Yang, 1998, 2000	19, 20
2	532	5	10^{11}	10 nc	Poly-crystalline	Acetone	30–40	G. W. Yang, 2002	21
3	532	10	66 J cm^{-2} Spot: 0.5 mm 6.6×10^9	10 nc	Graphite disk	Water	?	P. W. May, 2004	22
4	532	10	66 J cm^{-2} Spot: 0.5 mm 6.6×10^9	10 nc	Graphite disk	Cyclo-hexane (in Ar)	?	P. W. May, 2004	22
5	532	?	28 J cm^{-2} Spot: 0.5 mm	15 nc	Graphite	Water	~100	P. W. May, 2006	15
6	1064	20	4×10^6 W cm^{-2}	1.2 ms	Graphite particles <2.0 μm	Water	3–6	Xi-Wen Du, 2006	23
7	355	10	40–1000 J cm^{-2} Spot: 0.1–0.5 mm	5 nc	Pyrolytic graphite	Water?	5–15	Olivier Guillois, 2008	18
8	1064	20	4×10^6 W cm^{-2}	0.4 ms	Graphite particles <2.0 μm	Water	2–7	P. Bai, 2010	25
9	1064	20	4×10^6 W cm^{-2}	1.2 ms	Graphite particles <2.0 μm	Water	2–13	P. Bai, 2010	25

5.3 Light Hydro-dynamic Pulse for Nanodiamond Fabrication

In this work we describe a novel approach to produce nanodiamonds by a controlled synthesis. The proposed method is a form of PLAL; however, at the same time, two main innovations change the physics of the process, which allow us to describe this as a new method of synthesis, named light hydro-dynamic pulse (LHDP). The new approach is based on the treatment of a specially prepared multi-component solid target, containing a carbon non-diamond source, by a radiation beam focused in a transparent liquid at some predetermined distance from the target surface. In this case the formation of diamond nanocrystals is not the result of the plasma treatment of graphite as happens in reported experiments with PLAL, but as a consequence of the impact of an acoustic shock wave created by plasma on the surface of the composite material containing carbon black and hydrocarbons.

The treatment of specially prepared multi-component targets, rather than graphite, as was done earlier, by acoustic shock waves and not by plasma has led to a considerable increase in productivity, which offers the prospect of industrial implementation of the laser nanodiamond synthesis.

5.3.1 Technological Process

Presently, nanodiamonds are produced by LHDP synthesis under laboratory conditions. The process consists of the following operations:

1. Formation of a special target from pure carbon soot and a hydrocarbon binder.
2. Laser treatment of targets in liquid accompanied by ND synthesis and the production of a carbon blend that is free of metals.
3. Removal of hydrocarbons; isolation of synthesized nanodiamonds from the blend.
4. Washing and drying.

The targets are prepared with commercially available soot and wax in a 50/50 weight content by heating and mixing. Various additives (stearic acid, fullerenes, urea, naphthalene, *etc.*) can be used to increase the productivity or provide specific surface chemistry. Then, the mixture is poured into a glass bowl and, after drying, the bowl with the prepared target is filled with glycerin or water.

The treatment of the targets is conducted by a series of laser pulses of specific parameters in liquid media at room temperature and under normal pressure. We use a YAG solid-state laser with a wavelength of 1064 nm and laser pulse intensity of 10^{10} W cm^{-2}. The target is fixed on the table, which moves automatically so that the laser spot scans in the fluid at a distance of 3 mm above the target surface.

After removal of the hydrocarbons with an organic solvent, the synthesized NDs are isolated from the non-diamond carbon particles by flotation. Then, the NDs are rinsed with deionized water and dried.

5.3.2 Physical Mechanism

The specially prepared target is placed in a liquid media and a laser beam of specific parameters is focused at the some predetermined distance from the target surface. The laser beam leads to the emergence of acoustic shock waves of high power that impact on the target surface and provide specific conditions (*e.g.*, temperature, pressure, *etc.*) that are sufficient for the formation of the diamond cubic crystal structure of carbon. This happens when the specific ratio of the liquid refractive index and the light flow intensity provides a self-focusing effect and a laser beam energy of high density is absorbed at the focus. This energy concentrated at the focus area causes a rapid jump of the temperature and liquid evaporation, which results in a shock wave of high power (light hydraulic effect) and sound excitation.[30] This is the acoustic shock wave in front of which carbon atoms of the target are collected in a cubic diamond structure, forming nanocrystals. This happens due to the thermo-mechanical instability characterized by the highly inhomogeneous space distribution of both the pressure (P) and temperature (T) of the target mixture. As the duration of the laser pulses is rather short, the time derivatives of both the pressure (dP/dt) and the temperature (dT/dt) are extremely high and both P and T rise dramatically in certain micro-regions, reaching values in which the thermodynamically stable form of carbon is a diamond.

Contrary to the existing technology for nanodiamond fabrication by detonation, LHDP is environmentally friendly and not dangerous. No explosives are needed and the crystalline sizes can be controlled. Depending on the parameters of laser radiation, nanodiamonds with an average size of 4–5 nm size or 250–300 nm can be obtained by this method. The crystal sizes have been confirmed by transmission electron microscopy (Figure 5.7).

5.4 Characterization of Nanodiamonds Obtained by Laser Synthesis

5.4.1 X-Ray Diffraction (XRD) Analysis

The crystalline structure of laser nanodiamonds (LND) was determined by X-ray diffractometry (D8 Advance, Bruker AXS) in Cu Kα radiation (Figure 5.1). The XRD pattern obtained for the samples indicates that the powder consists of only a crystalline phase, *i.e.*, the diamond crystal lattice (identified with JCPDS card 6–675).

A lattice constant of 0.35687 Å was calculated with a Rietveld refinement using TOPAS software, which also indicated a cubic diamond structure.

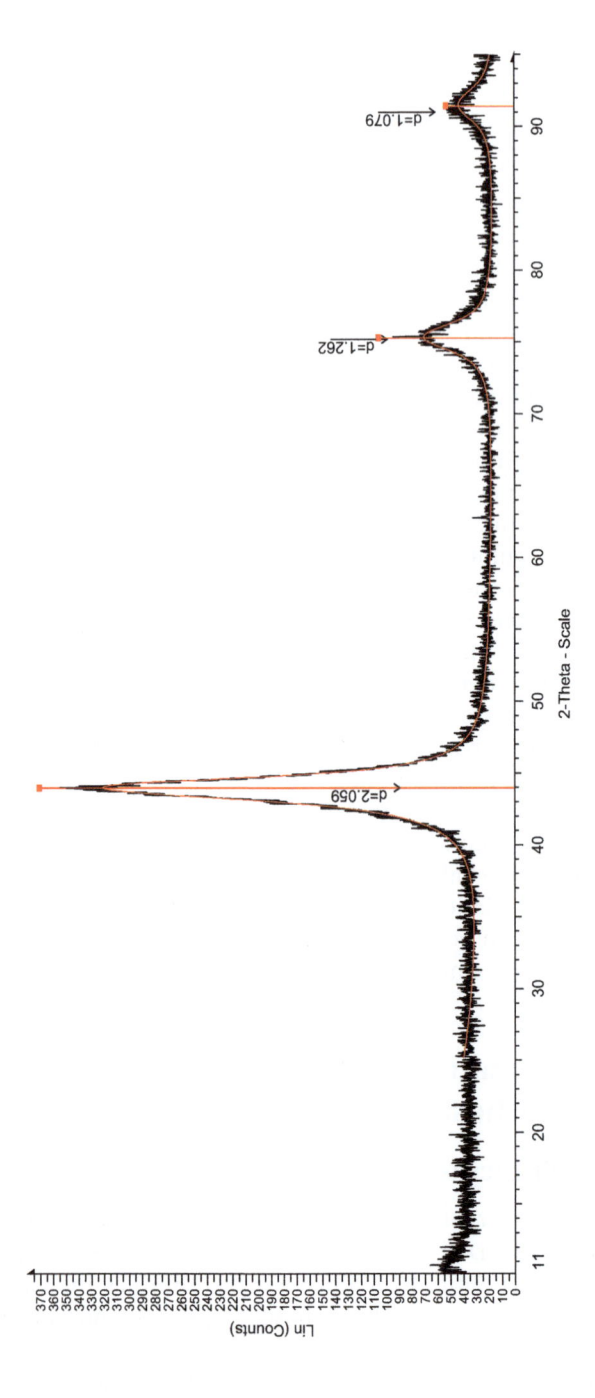

Figure 5.1 XRD analysis.

Pure Nanodiamonds Produced by Laser-assisted Technique

The absence of non-diamond peaks and troughs is evidence of the high purity and fine crystallinity of the LND powder. The average diamond crystallite size was evaluated from the full width half maxima (FWHM) of the peaks, which appeared on the diffraction pattern at angles of 44.01 and 75.34° (2Theta scale) and grain size values of 4.3 and 3.7 nm, respectively, were calculated.

5.4.2 Scanning Electron Microscopy

Scanning electron microscopy (SEM) was performed using a high-resolution microscope (SEM Serion) with energy dispersive X-ray spectroscopy (EDS) and Magellan™ 400L (Figure 5.2). The Serion image (Figure 5.2a) shows that the powder consists of agglomerates with a size of more 1 μm. Microanalysis EDS (Figure 5.2b) indicated the high purity of the local area and did not identify any other elements, except for carbon, oxygen and nitrogen. To see the samples in high resolution, dry nanodiamond powder was placed on a special grid (Figure 5.2c) and observed with the Magellan. The three-dimensional images show aggregates with a size of 25–400 nm (Figure 5.2d).

Figure 5.2 Scanning electron microscopy.

The aggregates consist of nanoparticles with an average size of around 4–5 nm. They are usually collected in small, highly porous aggregates, which in turn form big agglomerates, as shown in the Serion image.

5.4.3 Transmission Electron Microscopy

The dry nanodiamond powder was investigated by high resolution transmission electron microscopy (HRTEM; Tecnai F20 G2). The obtained TEM images (Figure 5.3 and Figure 5.4) are similar to DND.[28] They show that the material consists of porous agglomerates in which primary particles with a nanoscale size from 2–3 to 15–17 nm have a clearly observable crystalline lattice and contact with one another through some kind of disordered material, with no evidence of periodicity. Crystalline regions are noted to sometimes consist of twinned structures with a size of 10–20 nm. The size distribution of the crystallites with visible lattice fringes was determined by TEM point counting. The maximum of the distribution was at around 4–5 nm, which coincides with the results obtained from the XRD data.

Electron energy-loss spectroscopy (EELS) was used to determine the bonding of LND. It was found (Figure 5.5) that the EELS spectrum looks very similar to the results of the DND analysis: a pre-peak in the region of 285–290 eV, a core-loss peak at 300 eV, which characterizes ND at the core-loss range, a main peak in the region of 300 eV and "sole" and twin peaks in the regions of 315 and 335 eV, respectively. The small peak in the region of 410 eV[1] shows some slight presence of nitrogen vacancy (NV) defects in the nanodiamond grains.[31] The small dimensions of the pre-peak at 285–290 eV are indicative of the very low content of non-diamond structures on the surface compared with DND.[1]

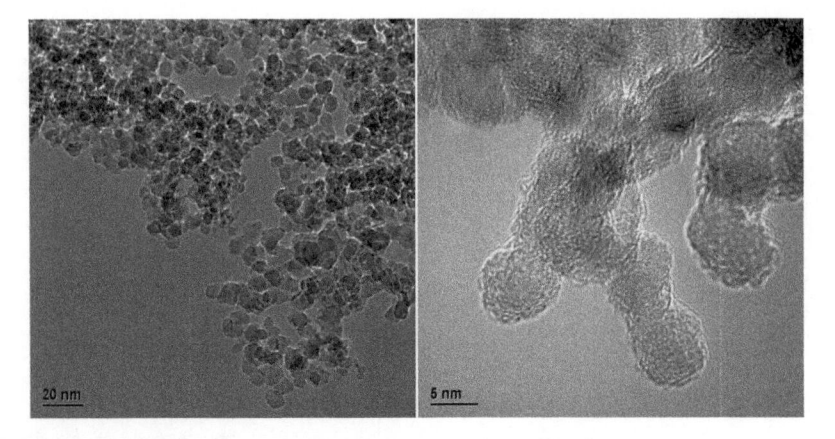

Figure 5.3 TEM images with 20 nm and 5 nm scale bars.

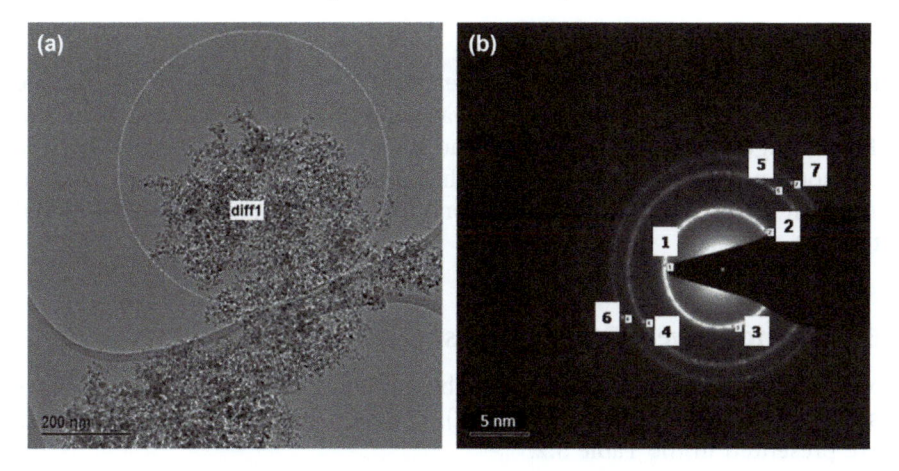

Pure Nanodiamonds Produced by Laser-assisted Technique 121

(c)	Measured spacing, nm	Miller index
1	0.2026	111
2	0.2101	111
3	0.2108	111
4	0.1261	220
5	0.1274	220
6	0.1072	311
7	0.1089	311

Figure 5.4 Selected area electron diffraction (SAED): a) selected area, b) electron diffraction pattern, c) measured interplanar spacings.

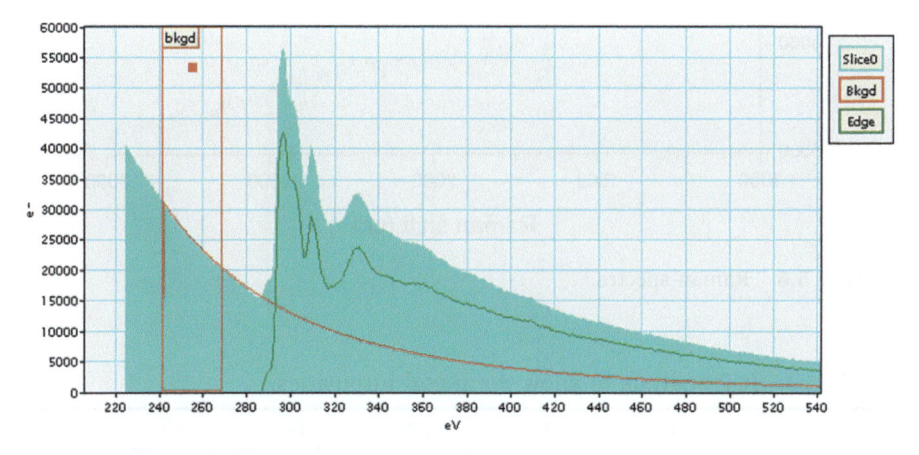

Figure 5.5 EELS.

5.4.4 Raman Spectroscopy

Raman spectroscopy analysis was performed using a Raman microscope (WITec GmbH) with an excitation wavelength of 488 nm and a power of 1 mW (Figure 5.6). In contrast to the Raman shift of purified DND,[5] the diamond peak at 1323 cm^{-1} is much higher than that of the sp^2 peak at 1590 cm^{-1}. This indicates a higher content of the diamond phase compared to DND.

5.4.5 Inductively Coupled Plasma Mass Spectrometry

Metal impurities were determined by inductively coupled plasma mass spectrometry (ICP-MS; Agilent 7500cx). The averages of three measurements are presented in the Table 5.2.

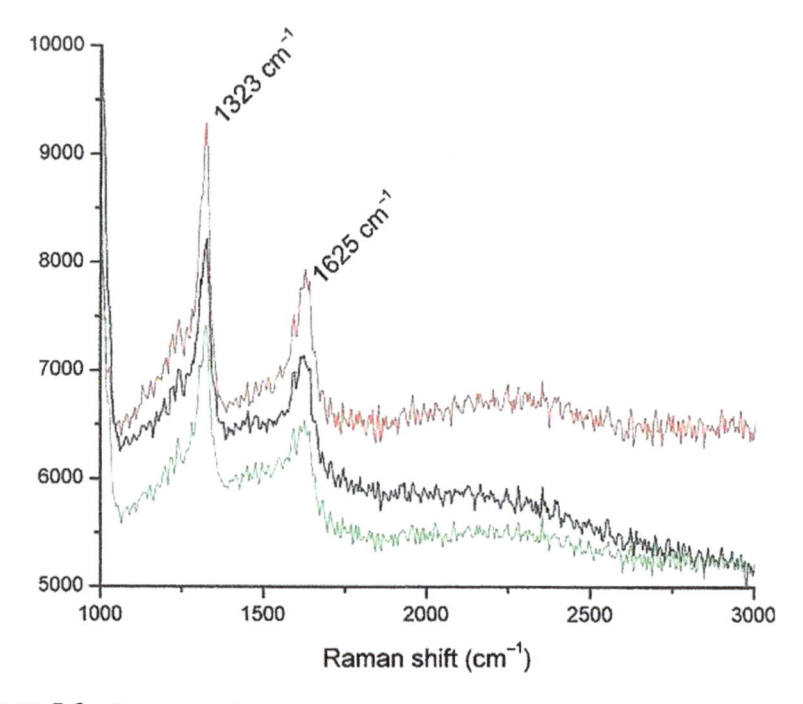

Figure 5.6 Raman spectra.

Table 5.2 ICP analysis to show the metal impurities.

Element	Zn	Ni	Mn	Fe	Cr	Al
Blank, ppm	<1	<2	<1	<1	<2	<5
LND, ppm	44	20	35	45	114	19

5.5 Controlled Nanodiamond Synthesis

The LHDP process of LND synthesis can be controlled by varying at least one of the following parameters:

- Content of the specially prepared carbon source target.
- Optical and mechanical characteristics of the liquid media.
- Width and/or shape of the laser pulse.
- Energy flux.
- The distance between the focusing plane and the surface of the solid carbon-source target.

Changing these parameters affects the size of the obtained diamond nanocrystals, the size of the shell and the character of functional groups on the crystal surface. These features define the interactions that nanodiamonds have with each other (aggregation) and with other particles, as well as their performance in applications.

As has been reported in previous investigations using PLAL,[24] the pulse width affects the size of the synthesized nanodiamonds. LNDs of different sizes (Figure 5.7) were obtained by LHDP when the target, liquid, focus location and light were kept constant. Just enhancing the pulse width resulted in a significant increase in the size of the nanodiamonds and in their magnetic resonance (Figure 5.8), which was studied earlier.[28,29] These special LNDs were analyzed by electron paramagnetic resonance (EPR) spectroscopy and compared with DND. The EPR spectra are shown in Figure 5.8. LNDs demonstrate enhanced paramagnetism compared with DND. The special LND spectrum, in contrast to normal LND and DND, shows a hyperfine pattern, usually attributed to NV paramagnetic centers.[29,30]

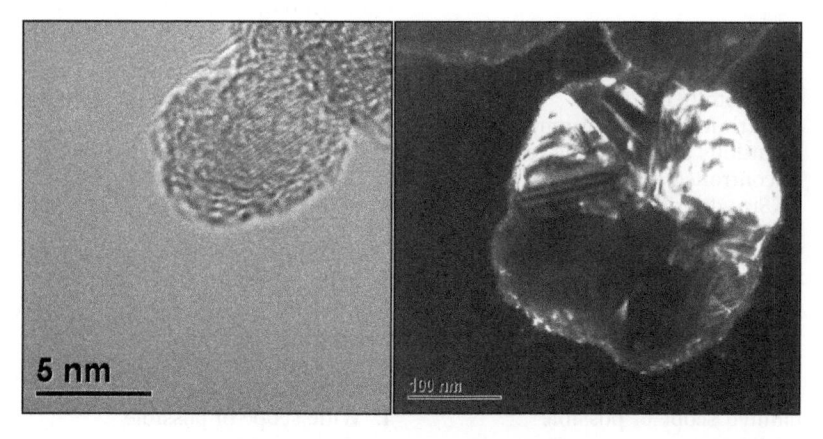

Figure 5.7 HRTEM images of LND on the left and LND obtained under an enhanced pulse width on the right.

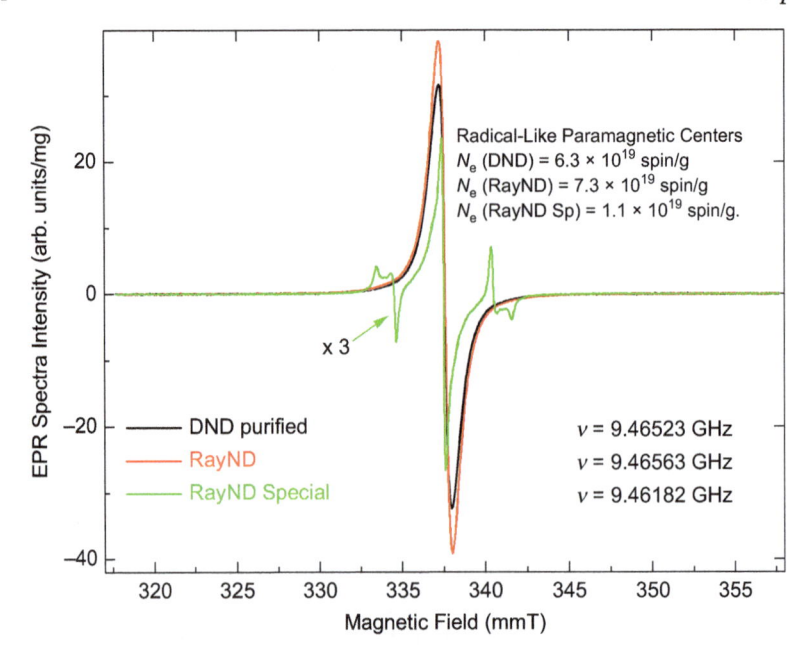

Figure 5.8 EPR spectra of LND (RayND) and DND samples.

5.6 LHDP vs. Existing Technology for Detonation Nanodiamond Synthesis

Table 5.3 summarizes the main advantages of LHPS over the detonation synthesis approach.

Table 5.3 Comparison of detonation synthesis and light hydrodynamic pulse synthesis.

Detonation Synthesis: mass production	*Light Hydrodynamic Pulse Synthesis: laboratory fabrication*
1. Raw material: explosives (trinitrotoluene, TNT; cyclotrimethylenetrinitramine, RDX); hazardous and polluting technology	1. Raw material: carbon soot and hydrocarbon binder; environmentally friendly technology

2. Uncontrolled synthesis
   a) Size distribution non-constant
   b) High variability of surface chemistry
	2. Controlled synthesis
	   a) Size distribution can be controlled (Figure 5.7)
	   b) Controlled surface chemistry

3. Insufficient (non-constant) quality:
   a) Incombustible residue: 0.2–1.4 wt.%
   b) Metal impurities
   c) Low homogeneity
   d) Presence of polycrystals
	3. High quality:
	   a) Incombustible residue: <0.01 wt.%
	   b) Near metal free
	   c) High homogeneity
	   d) Polycrystal free

4. Limited scope of possible applications	4. Wide scope of possible applications
5. Difficulties for implementation	5. Ease for implementation
6. Problematic to reduce cost; additional security expenses	6. Highly competitive when produced on an industrial scale

5.7 Conclusion and Prospects in the Development of the Nanodiamond Industry

A novel laser synthesis technology for nanodiamond powder fabrication has been developed. The technology is environmentally friendly and non-hazardous, enables control over the diamond nanocrystal dimensions and defects, and allows nanodiamonds of high purity to be obtained, which are metal- and graphite-free with high homogeneity.

The possibility to control the characteristics of nanodiamond are very important for most advanced applications, such as:

- Optics: better surface control important for IR and visible spectra.
- Thermal management in heat capacitors and heat conductors: LND is more pure and has a narrower size distribution and, as a result, holds more promise for high thermo-conductivity.
- Biomedical and biochemical fields: better promise due to the purity and enhanced size control.
- Mechanical (production of polymers, lubricants, polish components and additives for coatings): LND better due to the higher homogeneity and dispersion.

After transition to a mass production, the cost of LND powder is expected to drop considerably. The low price, together with the possibility of controlling the dimensions and surface chemistry, will ensure the rapid development of advanced ND applications, such as drug and gene delivery agents, biosensors and diagnostic kits, optical filters and photovoltaic elements, thermal interface materials and heat sinks for electronics, field emission displays, and quantum computers.

Acknowledgments

We thank to Dr Elena Perevedentseva for the analysis of LND samples and who kindly provided the Raman spectroscopy results. We also thank to Dr Alexander Shames for the comparative EPR analysis of LND and DND samples.

References

1. O. Shenderova and D. Gruen, *Ultrananocrystalline Diamond: Synthesis, Properties and Applications*, William Andrew, 2006.
2. A. M. Schrand, S. A. Ciftan Hens and O. A. Shenderova, *Critical Rev. Solid State Mater. Sci.*, 2009, **34**, 18.
3. M. Baidakova and A. Vul, New prospects and frontiers of nanodiamond clusters, *J. Phys. D: Appl. Phys.*, 2007, **40**, 6300–6311.
4. V. V. Danilenko, Detonation nanodiamonds: problems and prospects, *Superhard Materials*, 2010, **N5**.

5. V. N. Mochalin, O. Shenderova, D. Ho and Y. Gogotsi, The properties and applications of nanodiamonds, *Nat Nanotechnol.*, 2012, 7(1), 11–23.
6. B. C. Windom and D. W. Hahn, Laser ablation—laser induced breakdown spectroscopy (LA-LIBS): A means for overcoming matrix effects leading to improved analyte response, *J. Anal. At. Spectrom*, 2009, **24**, 1665–1675.
7. G. W. Yang, Laser ablation in liquids: Applications in the synthesis of nanocrystals, *Prog. Mater. Sci.*, 2007, **52**(5), 648–698.
8. P. P. Patil, D. M. Phase, S. A. Kulkarni, S. V. Ghaisas, S. K. Kulkarni, S. M. Kanetkar, S. B. Ogale and V. G. Bhide, Pulsed-laser-induced reactive quenching at liquid-solid interface: aqueous oxidation of iron, *Phys. Rev. Lett.*, 1987, **58**, 238.
9. H. Zeng, X.-W. Du, S. C. Singh, S. A. Kulinich, S. Yang, J. He and W. Cai, Nanomaterials via Laser Ablation/Irradiation in Liquid: A Review, *Adv. Funct. Mater.*, 2012, **22**, 1333–1353.
10. G. Yang (ed.), *Laser Ablation in Liquids, Principles and Applications in the Preparation of Nanomaterials*, Pan Stanford Publishing, Singapore, 2012.
11. A. Pyatenko, K. Shimokawa, M. Yamaguchi, O. Nishimura and M. Suzuki, Synthesis of silver nanoparticles by laser ablation in pure water, *Appl. Phys. A.*, 2004, **79**, 803.
12. T. E. Itina, On Nanoparticle Formation by Laser Ablation in Liquids, *J. Phys. Chem. C*, 2011, **115**, 5044.
13. J. B. Wang, G. W. Yang, C. Y. Zhang, X. L. Zhong and Z. H. A. Ren, Symmetric organization of self-assembled carbon nitride, *Chem. Phys. Lett.*, 2003, **367**, 10.
14. C. H. Liang, Y. Shimizu, T. Sasaki and N. Koshizaki, Preparation of ultrafine TiO2 nanocrystals via pulsed-laser ablation of titanium metal in surfactant solution, *Appl. Phys. A*, 2005, **80**, 819.
15. L. Yang, P. W. May, L. Yin, J. A. Smith and K. N. Rosser, Growth of diamond nanocrystals by pulsed laser ablation of graphite in liquid, *Diamond & Related Materials*, 2007, **16**, 725–729.
16. J. Sun, S.-L. Hu, X.-W. Du, Y.-W. Lei and L. Jiang, Ultrafine diamond synthesized by long-pulse-width laser, *Appl. Phys. Lett.*, 2006, **89**, 183115.
17. S. Hu, J. Sun, X. Du, F. Tian and Lei, The formation of multiply twinning structure and photoluminescence of well-dispersed nanodiamonds produced by pulsed-laser irradiation, *Diamond & Related Materials*, 2008, **17**, 142.
18. D. Amans, A. Chenus, G. Ledoux, C. Dujardin, C. Reynaud, O. Sublemontier, K. Masenelli-Varlot and O. Guillois, Nanodiamond synthesis by pulsed laser ablation in liquids, *Diamond & Related Materials*, 2009, **18**, 177–180.
19. G.-W. Yangyz, J.-B. Wangy and Q.-X. Liuy, Preparation of nano-crystalline diamonds using pulsed laser induced reactive quenching, *J. Phys.: Condens. Matter.*, 1998, **10**, 7923–7927.

20. G. W. Yang and J. B. Wang, Pulsed-laser-induced transformation path of graphite to diamond via an intermediate rhombohedral graphite, *Appl. Phys. A*, 2001, **72**, 475.
21. J. B. Wang, C. Y. Zhang, X. L. Zhong and G. W. Yang, Cubic and hexagonal structures of diamond nanocrystals formed upon pulsed laser induced liquid–solid interfacial reaction, *Chemical Physics Letters*, 2002, **361**, 86.
22. S. R. J. Pearce, S. J. Henley, F. Claeyssens, P. W. May, K. R. Hallam, J. A. Smith and K. N. Rosser, Production of nanocrystalline diamond by laser ablation at the solid liquid interface, *Diamond and Related Materials*, 2004, **13**, 661.
23. J. Sun, S.-L. Hu, X.-W. Du, Y.-W. Lei and L. Jiang, Ultrafine diamond synthesized by long-pulse-width laser, *Appl. Phys. Lett.*, 2006, **87**, 183115.
24. P. Bai, S. Hua, T. Zhang, J. Sun and S. Cao, Effect of laser pulse parameters on the size and fluorescence of nanodiamonds formed upon pulsed-laser irradiation, *Materials Research Bulletin*, 2010, **45**, 826.
25. C. X. Wang, Y. H. Yang and G. W. Yang, Thermodynamical predictions of nanodiamonds synthesized by pulsed-laser ablation in liquid, *J. Appl. Phys.*, 97, 2005, **6**, 066104.
26. C. X. Wang, P. Liu, H. Cui and G. W. Yang, Nucleation and growth kinetics of nanocrystals formed upon pulsed-laser ablation in liquid, *Appl. Phys. Lett.*, 2005, **87**, 201913.
27. Q. Zou, Y. G. Li, L. H. Zou and M. Z. Wang, Characterization of structures and surface states of the nanodiamond synthesized by detonation, *Materials Characterization*, 2009, **60**(11), 1257–1262.
28. L. B. Casabianca, A. I. Shames, Alexander M. Panich, O. Shenderova and L. Frydman, Factors affecting DNP NMR in Polycrystalline Diamond Samples, *J. Phys. Chem. C*, 2011, **115**(39), 19041–19048.
29. A. Panich, A. Shames, B. Zousman and O. Levinson, Magnetic resonance study of nanodiamonds prepared by laser-assisted technique, *Diamond and Related Materials*, 2012, **23**, 150–153.
30. "Science Discovery" under number No. 65 in the name of; G. A. Askar'yan, A. M. Prokhorov, G. F. Chanturiya and G. P. Shipulo, The effects of a laser beam in a liquid, *Sov. Phys. JETP*, 1963, **17**(6), 1463–1465.
31. I. I. Vlasov, O. Shenderova, S. Turner, O. I. Lebedev, A. A. Basov, I. Sildos, M. Rähn, A. A. Shiryaev and G. Van Tendeloo, *Small*, 2010, **6**(5), 687.

CHAPTER 6
Electrochemistry of Nanodiamond Particles

KATHERINE B. HOLT

Department of Chemistry, University College London, 20, Gordon Street, London, WC1H 0AJ, UK
Email: k.b.holt@ucl.ac.uk

6.1 Introduction

This chapter is concerned with the electrochemical (redox) activity of undoped diamond in its nanoparticle form; however, undoped diamond is perhaps one of the best known examples of an insulating material, with a band gap of *ca.* 5.5 eV. In principle one should expect no electrochemical activity from such a material but, as explained below, the small size of the particles and the resulting high surface-atom-to-bulk-atom ratio allows us to probe surface electronic states using standard electrochemical techniques. In the case of 5 nm detonation nanodiamond the surface properties begin to dominate over those of the bulk and the rich surface chemistry of the nanodiamond particles reveals a complex redox response.

This chapter will start with an overview of diamond in electrochemistry followed by a brief description of the chemistry of the nanodiamond surface as it is presently understood. This will be followed by a summary of investigations illustrating how the surface chemistry of nanodiamond gives rise to redox chemistry that can be probed by immobilising the diamond nanoparticles onto electrodes or treating them with solution redox species. Finally, a comprehensive survey of the literature on nanodiamond in electrochemistry is given, including applications in fuel cell catalysis, supercapacitors, sensors, biosensors and photochemistry.

6.2 Diamond in Electrochemistry

Although undoped diamond is an insulator, in its boron-doped form diamond makes an excellent electrode material. Grown as a polycrystalline film using chemical vapour deposition, the diamond can be substitutionally doped with boron to a sufficient level to introduce broad impurity bands into the band gap, allowing for metallic conductivity. Boron-doped diamond (BDD) electrodes are now commercially available and have found use in a range of electroanalytical and electrolysis applications.[1]

Undoped diamond films grown by chemical vapour deposition also demonstrate interesting electrochemical properties. Hydrogen-terminated diamond exhibits p-type surface conductivity due to charge transfer doping of valance band electrons into the lowest unoccupied electronic levels of a solution redox species.[2] Undoped nanocrystalline diamond films also exhibit conductivity and have been used as electrodes.[3] Conduction has been shown to be n-type and is believed to occur through overlapping band gap impurity states of π and π^* character.[4]

6.3 Structure and Surface Chemistry of Nanodiamond

The surface chemistry of nanodiamond has been investigated using infrared (IR) spectroscopy,[5] Raman spectroscopy,[6] X-ray photoelectron spectroscopy (XPS),[7] nuclear magnetic resonance (NMR)[8] and zeta-potential measurements[9] among other techniques. The commercial 5 nm powders are relatively high purity as-received (*i.e.*, free of non-diamond graphitic content) and can be further purified by heating in air or exposure to acid solutions. The core of each nanoparticle is crystalline diamond, with trace nitrogen incorporation from the precursor material. The outer several atomic layers are more complex in structure, with the carbon found in both sp^3 (saturated) and sp^2 (unsaturated) bonding environments. Raman spectroscopy suggests that extended graphitic carbon is not present in this material after purification and that sp^2 carbon is instead present in isolated defect sites.[6] Some models of the nanodiamond structure have suggested an outer fullerenic shell (so-called 'bucky-diamond'), but in most nanodiamond materials this structure is not observed unless it is annealed to high temperatures in a vacuum.

Most importantly for redox or catalytic activity, the outer surface of each nanodiamond particle is highly oxidised due to the purification treatment. An array of oxygen-containing functional groups is present, as shown by the complex IR absorption spectrum in Figure 6.1.

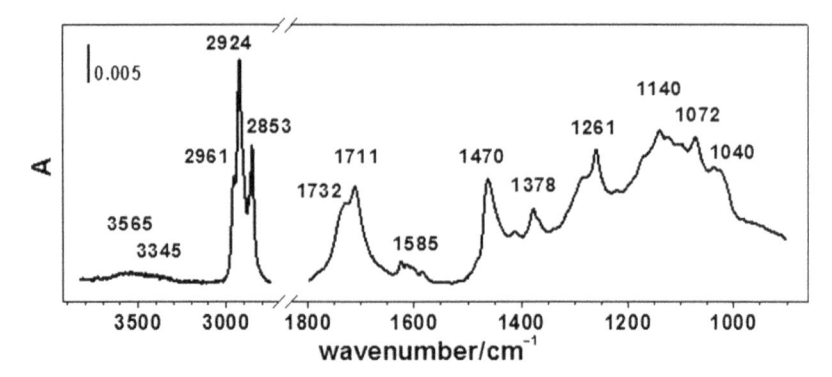

Figure 6.1 IR absorption spectrum of dry, untreated 5 nm nanodiamond powder drop-coated from ethanol onto a diamond ATR prism (256 scans, 4 cm^{-1} resolution, 2700–1800 cm^{-1} prism cut-out region omitted). Reproduced with permission from *Chem. Commun.*, 2011, **47**, 12140.

Oxygen is present as carbonyl (C=O) and carboxyl (COOH) at *ca.* 1680–1770 cm^{-1}, representing the most oxidised carbon moieties, as well as alcohol (C–OH) at 3340 cm^{-1} and ether (C–O–C) at 1500–900 cm^{-1}. Carbon–hydrogen (C–H) bonding is noted at 2970–2850 cm^{-1}, but interestingly little evidence for the graphitic (C=C) functionality is observed in the 3100–3000 cm^{-1} region. The range of carbon–oxygen bond types is confirmed by XPS and negative zeta-potential measurements indicate that such functionalities are maintained in solution and can readily undergo deprotonation.

The unsaturated nature of the bonding on the nanodiamond surface gives rise to opportunities for redox chemistry at relatively mild potentials (*i.e.*, at energies within the diamond band gap). Moreover, delocalisation and electronic communication between neighbouring moieties may allow electron transfer to take place over a broad range of potentials.

6.4 Electrochemical Behaviour of Diamond Nanoparticles

6.4.1 Electrode Preparation Methods for the Study of Nanodiamond Particles

Early studies of nanodiamond as electrodes were carried out using pressed compacts sintered at high temperature and pressure to form disks.[10] The electrochemical response of the compacts was obtained in aqueous solutions containing a KCl background electrolyte and different redox couples. However, the high resistance of the undoped nanodiamond leads to an unsatisfactory, sloped electrochemical response. Electrodes have also been prepared by mixing nanodiamond powder with Nafion polymer and applying the paste to electrode surfaces.[11] Composite electrodes have been formed with other non-diamond carbons to enhance the otherwise low conductivity of nanodiamond electrodes.[12] Stable homogeneous nanodiamond particle films have been deposited on conducting p-type Si surfaces using electrophoresis of nanodiamond suspended in isopropanol and 240–300 V applied potential.[13] Another approach is to use cavity electrodes filled with powder to investigate the response of nanodiamond. These may be constructed by etching away platinum from the tip of a platinum-in-glass microelectrode using *aqua regia* and then filling the cavity with nanodiamond powder.[14] Alternatively, a simple plastic micropipette can be used, where the tip is filled with the powder and a Pt wire introduced to provide an electrical contact.[15] The simplest method to explore the electrochemistry of nanodiamond is to form a suspension in ethanol or water and to drop-coat onto an electrode surface. When the solvent evaporates, a thin adherent film of nanodiamond powder is left on the electrode and is stable to repeated electrochemical investigations in solution.[16] The nanodiamond adheres to the electrode surface by electrostatic interaction with the electrode surface, but coverage is difficult to control and not always uniform and is almost certainly in the form of agglomerates rather than discrete monolayers of 5 nm particles. Layer-by-layer deposition of larger 500 nm high temperature high pressure (HTHP) diamonds onto electrodes to form two-dimensional arrays has been demonstrated by modifying electrodes with a thin layer of positively charged polyelectrolyte with which the negatively charged diamond particles interact through electrostatic attractions.[17] There are advantages and disadvantages to all of these preparation methods and the technique adopted depends on the ultimate application of the nanodiamond electrode, as discussed further below.

6.4.2 Electrochemical Response of Electrode-immobilised Nanodiamond Particles

6.4.2.1 Influence of surface modification on electrochemical response of nanodiamond

Early reports of the electrochemical activity of nanodiamond particles were followed with more in-depth studies to determine the extent and origins of the electrochemical response. Electrodes were prepared by binding the nanodiamond powder with mineral oil and immobilising a thin layer onto a glassy carbon electrode.[18] Commercial detonation nanodiamond samples were used either untreated (as-received), oxidised by heating to 420 °C in air, or hydrogenated by heating in a H_2/N_2 gas flow at 800 °C. None of the powders exhibited significant conductivity in their dry form. The untreated, oxidised and hydrogenated powders were characterised using transmission electron microscopy (TEM), Raman spectroscopy, IR spectroscopy and XPS before electrochemical investigations. Raman spectroscopy indicated small concentrations of sp^2-bonded carbon incorporated in the particles, presumably in the surface layers. However the characteristic D and G bands associated with extended graphitic bonding were not observed, allowing the sp² bands at 1620 cm⁻¹ to be assigned to localised sp² bonding sites on the nanodiamond surface. IR analysis indicated that the surfaces of the untreated and oxidised samples were both highly oxidised with a range of C–O and C=O moieties present. In contrast, the hydrogenated sample contained –OH bonding on the surface but there was no evidence of higher oxidation state C=O species. This was confirmed by XPS, which also confirmed there was no measurable metal (*e.g.*, iron) contaminant on the nanodiamond surfaces.

The oxidation response of the three nanodiamond samples is shown in Figure 6.2. Differential pulse voltammetry was carried out and the electrode potential swept from −0.6 V to 1.2 V *vs.* Ag/AgCl while the current at the nanodiamond-coated electrode was recorded. Similar responses were obtained for the untreated and oxidised powders, with broad peaks noted at −0.4 V and −0.15 V due to the oxidation of the powders. For the untreated nanodiamond, a further oxidation peak is observed at 0.65 V that is not present for the oxidised sample. This peak is absent on subsequent scans, indicating that it corresponds to some moiety that is oxidised irreversibly on the first scan and not regenerated. The hydrogenated sample shows very different voltammetry to the other samples, with no clear oxidation peaks observed below 0.95 V.

These results suggest that the electrochemical response of nanodiamond is dependent on the surface chemistry of the particles, with the oxidised particles (including the untreated particles, which are also oxidised) showing

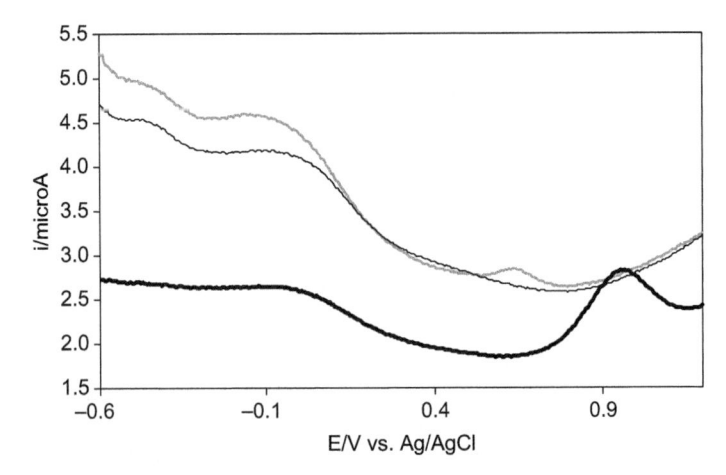

Figure 6.2 Differential pulse voltammetry of nanodiamond powders bound in mineral oil and immobilised on a glassy carbon electrode. The electrode potential was swept from −0.6 V to 1.2 V, step potential 0.005 V, modulation amplitude 0.025 V: thick black line = hydrogenated nanodiamond; thin black line = oxidised nanodiamond; thick grey line = untreated nanodiamond.
Reproduced with permission from *Phys. Chem. Chem. Phys.*, 2008, **10**, 303.

Electrochemistry of Nanodiamond Particles 133

increased redox activity in comparison to the hydrogenated powder. This indicates strongly that the origin of the redox activity is the surface of the particle with no contribution from the bulk, as would be anticipated given that the bulk is insulating, crystalline and undoped diamond. This is an interesting example of how surface properties of a material begin to dominate those of the bulk as nanometer dimensions are reached. It illustrates that properties of a bulk material cannot necessarily be extrapolated to nanoparticle dimensions and that care must be exercised in making such assumptions.

Further detailed studies were carried out on the electrochemical response of surface-oxidised nanodiamond using BDD electrodes modified with a thin layer of nanodiamond by drop-coating from an ethanol suspension. Again the electrode-immobilised nanodiamond was probed using differential pulse voltammetry and the stability of the nanodiamond electrochemical response to repeated cycling was probed.[16] A typical response of a nanodiamond layer in pH 4 buffer solution is shown in Figure 6.3. The red trace indicates the response of a fresh layer of nanodiamond and the black trace represents the response of the same nanodiamond layer on subsequent scans.

The voltammetry reveals a complex response as the nanodiamond can undergo both oxidation and reduction, although there are more oxidation peaks than reduction peaks. Some of the redox processes are reversible, as indicated by an asterisk (*) on the figure, but other processes are irreversible.

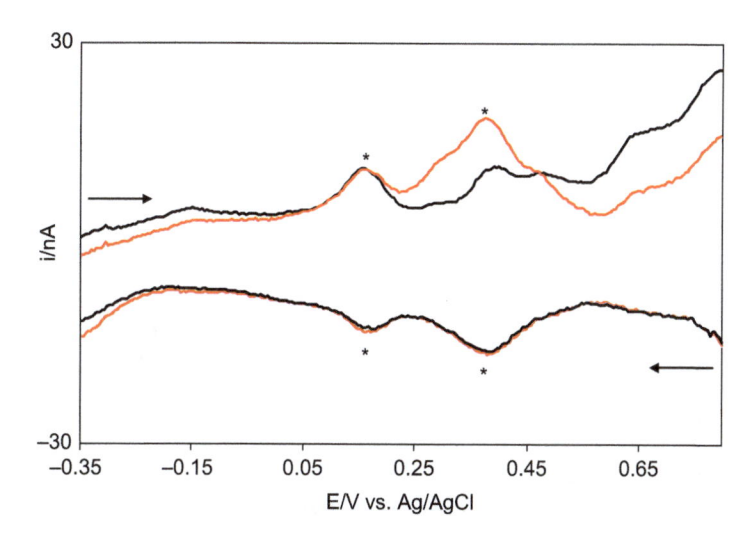

Figure 6.3 Differential pulse voltammetry response of immobilised nanodiamond on a boron-doped diamond electrode in 0.2 M pH 4 phosphate buffer solution. Red lines indicate the first scan and the black line shows the second scan. Asterisk denotes the position of the reversible redox couple. Reproduced with permission from *Phys. Chem. Chem. Phys.*, 2010, **12**, 2048.

For example, oxidation peaks are noted above 0.5 V that have no corresponding reductions peaks, indicating that the oxidation processes in the potential range are not reversible. Redox processes take place at discrete and defined potentials as indicated by the well-resolved and peak-like responses. However the peaks are not evenly distributed and some of the peaks overlap. Additionally, the relative heights of the peaks vary. The oxidation response changes from the first scan to the second; although the same peaks seem to be present, they vary in height. However, the reduction response is very stable to repeated cycling and is identical on the first and subsequent scans.

6.4.2.2 Influence of Solution pH on the Electrochemical Response of Nanodiamond

The effect of solution pH was investigated by carrying out differential pulse voltammetry on nanodiamond drop-coated onto a BDD electrode, as shown in Figure 6.4.[19] It is immediately obvious that the response changes dramatically with solution proton concentration. At pH 4 and 5 the response is very well defined and complex over the potential range. Clear peaks are noted that are overlapping yet well resolved. Peaks between −0.1 and 0.5 V become much smaller in magnitude and far less resolved as the solution pH increases. Oxidation of the nanodiamond above 0.5 V begins to dominate the response at pH 7 and above. The reversibility of the oxidation processes

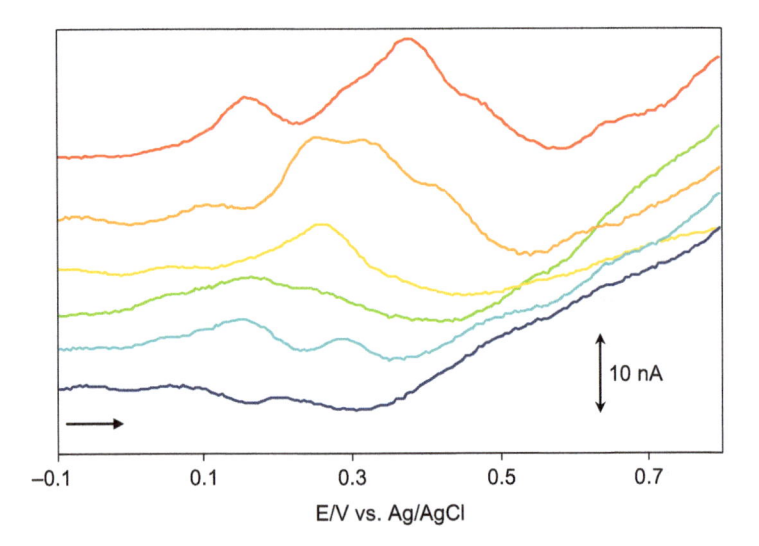

Figure 6.4 Differential pulse voltammograms of electrode-immobilised nanodiamond in different pH solutions. Second scans in each case. Red = pH 4; orange = pH 5; yellow = pH 6; green = pH 7; light blue = pH 8 and dark blue = pH 9.
Reproduced with permission from *Phys. Chem. Chem. Phys.*, 2010, **12**, 2048.

and stability to repeated cycling were investigated in the different solutions, where it was found that some processes were reversible and stable over many cycles. Some peaks showed a shift in position of close to −60 mV per increase in pH unit, which is indicative of electron transfer coupled to proton transfer. This is more clearly illustrated in Figure 6.7, discussed further below.

6.5 Electronic Structure of the Nanodiamond Surface

As the nanodiamond particles used in these studies are undoped, any redox activity must be attributed to the surface of the particles as confirmed by the pH dependence of the electrochemical response. As described in section 6.3, the surface of the nanodiamond is very complex in terms of the range of chemical functionalities present. The bonding is highly unsaturated with sp^2 bonding evident in both C=C and C=O motifs. At a molecular level it is easy to predict redox processes that could take place under relatively mild oxidative or reductive conditions, for example, the oxidation of alcohol (C–OH) groups to ketones (C=O) or carboxylic acids (COOH). These transformations have been probed using IR spectroscopy and are described in some detail in section 6.7.[5a] An alternate and equivalent model is to consider how the rich surface chemistry contributes to significant heterogeneity in the electronic character of the nanodiamond surface. The cartoon depicted in Figure 6.5

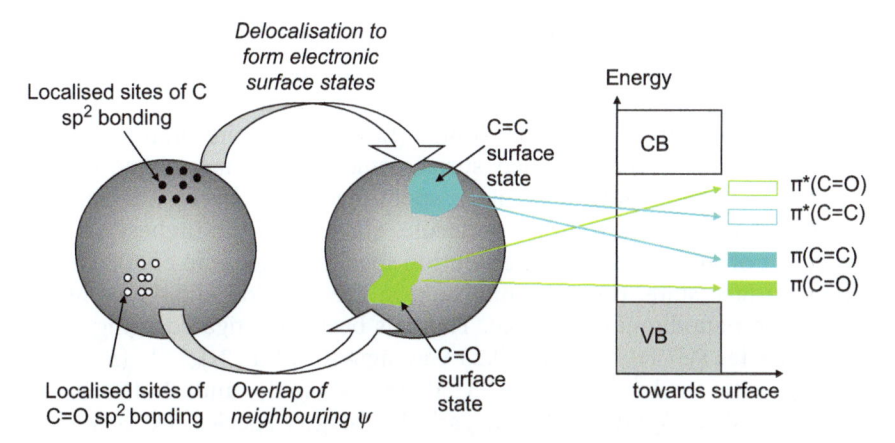

Figure 6.5 Cartoon showing the formation of surface states on the surface of a nanodiamond particle. Localised sites of specific chemical identity (sp^2 carbon and C=O shown as examples) from regions of differing electrochemical potential due to the overlap of neighbouring wavefunctions and electron delocalisation to form surface states. These surface states have energies distributed in the diamond band gap (impurity bands) and can be occupied or empty depending on their bonding or antibonding character.
Reproduced with permission from *Phys. Chem. Chem. Phys.*, 2010, **12**, 2048.

illustrates how localised sites of unsaturated bonding may contribute to regions on the surface that have different electronic character to bulk diamond.

Overlap of neighbouring wavefunctions for the different surface functional groups will give rise to surface states that can be occupied by electrons or not depending on their bonding or antibonding character. These states are likely to have energies within the band gap of bulk diamond, as shown in Figure 6.5, and those that are occupied by electrons can be oxidised (donate electrons to an electrode at the corresponding potential) and those that are empty can be reduced (accept electrons from an electrode). States formed from different types of neighbouring atoms with different bonding environments will have different energies, resulting in discrete electronic states within the band gap. The capacity of these states (*i.e.*, how many electrons they are able to provide or accept) will depend on the number of constituent atomic orbitals that contribute to the state. The simplest experimental methods to detect such states are photoluminescence measurements, where, in the gas phase, transitions attributed to surface band gap states have been reported.[20] However, as yet, no solution-phase experiments have been carried out in redox environments to probe this model.

6.6 Interaction of Nanodiamond Particles with Solution Redox Species

Assuming that the redox activity of nanodiamond can be attributed to surface band gap electronic states, the particles should also be able to transfer electrons to solution redox species, as well as to an underlying electrode. If the lowest unoccupied molecular orbital (LUMO) of a solution redox species is lower in energy than a filled surface state of the nanodiamond, then electron transfer should take place spontaneously between the nanodiamond surface and the solution molecule. Likewise, if the highest occupied molecular orbital (HOMO) of the solution species is higher in energy than an empty surface state, then the molecule in the solution should be able to reduce the nanodiamond surface. This has been investigated using the solution species $Fe(CN)_6^{3-/4-}$ ($E = 0.23$ V *vs.* Ag/AgCl), $Ru(CN)_6^{3-/4-}$ ($E = 0.75$ V *vs.* Ag/AgCl), $IrCl_6^{2-/3-}$ ($E = 0.75$ V *vs.* Ag/AgCl) and $Ru(NH_3)_6^{3+/2+}$ ($E = -0.15$ V *vs.* Ag/AgCl). The negatively charged species are very useful for these investigations as there are no complicating adsorption effects as they are not electrostatically attracted to the negatively charged surface of the nanodiamond.

6.6.1 Interaction with $Fe(CN)_6^{4-/3-}$

Figure 6.6a shows cyclic voltammograms for the reversible reduction of $Fe(CN)_6^{3-}$ at a clean BDD electrode and at the same electrode modified with a thin layer of oxidised nanodiamond.

Electrochemistry of Nanodiamond Particles

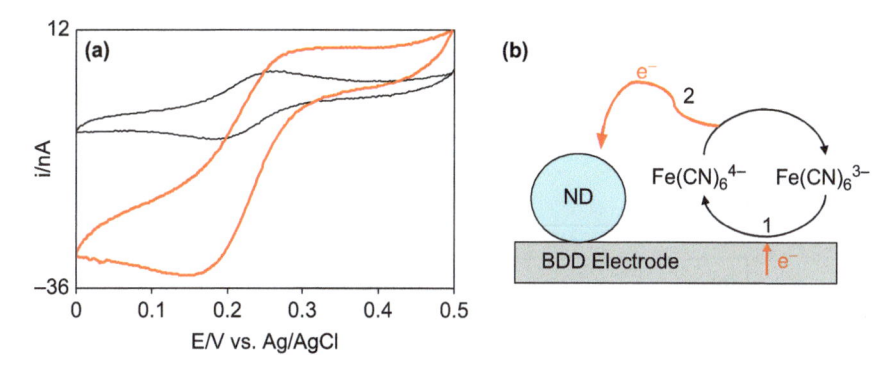

Figure 6.6 a) Cyclic voltammograms of $Fe(CN)_6^{3-}$ reduction at clean boron-doped diamond electrode (black) and at nanodiamond-modified boron-doped diamond electrode (red). 1 µM $Fe(CN)_6^{3-}$ in 0.2 M pH 7 phosphate buffer solution, 10 mV s^{-1}. b) Schematic to show the posited mechanism of reduction current enhancement at the nanodiamond-modified electrode. Red arrows indicate direction of electron transfer. Labels 1 and 2 are explained in the text.
Adapted from *J. Am. Chem. Soc.*, 2009, **113**, 11272 with permission. Copyright 2009 American Chemical Society.

At a clean BDD electrode, the peaks for the reduction and oxidation of the redox couple can be seen, where both forward and reverse peaks are equal in magnitude, indicating a reversible electron transfer process. When nanodiamond is present, the magnitude of the reduction peak is considerably enhanced on the first scan but, importantly, the oxidation currents are not greatly enhanced. Figure 6.6b shows a schematic for the suggested mechanism for current enhancement during the reduction process. It is assumed that the dominant electron transfer process is the reduction of $Fe(CN)_6^{3-}$ at the underlying electrode (process 1). The product of the reduction is $Fe(CN)_6^{4-}$ and this is formed at the electrode interface. If the nanodiamond has unfilled surface states of a suitable energy then electron transfer can then take place between the $Fe(CN)_6^{4-}$ species and the nanodiamond; the nanodiamond will be reduced (accept electrons) and the $Fe(CN)_6^{4-}$ will become oxidised back to the $Fe(CN)_6^{3-}$ species. This is process 2 in Figure 6.6b. The result of this step is that $Fe(CN)_6^{3-}$ is regenerated at the electrode surface and can once again be reduced; this leads to enhanced (catalytic) reduction currents as a positive feedback loop is established.

This mechanism suggests that nanodiamond can spontaneously oxidise $Fe(CN)_6^{4-}$ under these conditions. To investigate this process further, the voltammetric response of nanodiamond to $Fe(CN)_6^{4-}$ in a range of different pH solution was investigated. The responses are shown in Figure 6.7 along with the electrochemical response of the nanodiamond in the absence of the redox molecules in the same pH solutions.

In Figure 6.7a–e the black cyclic voltammogram responses in each case show the reversible oxidation of the $Fe(CN)_6^{4-}$ at a clean BDD electrode and

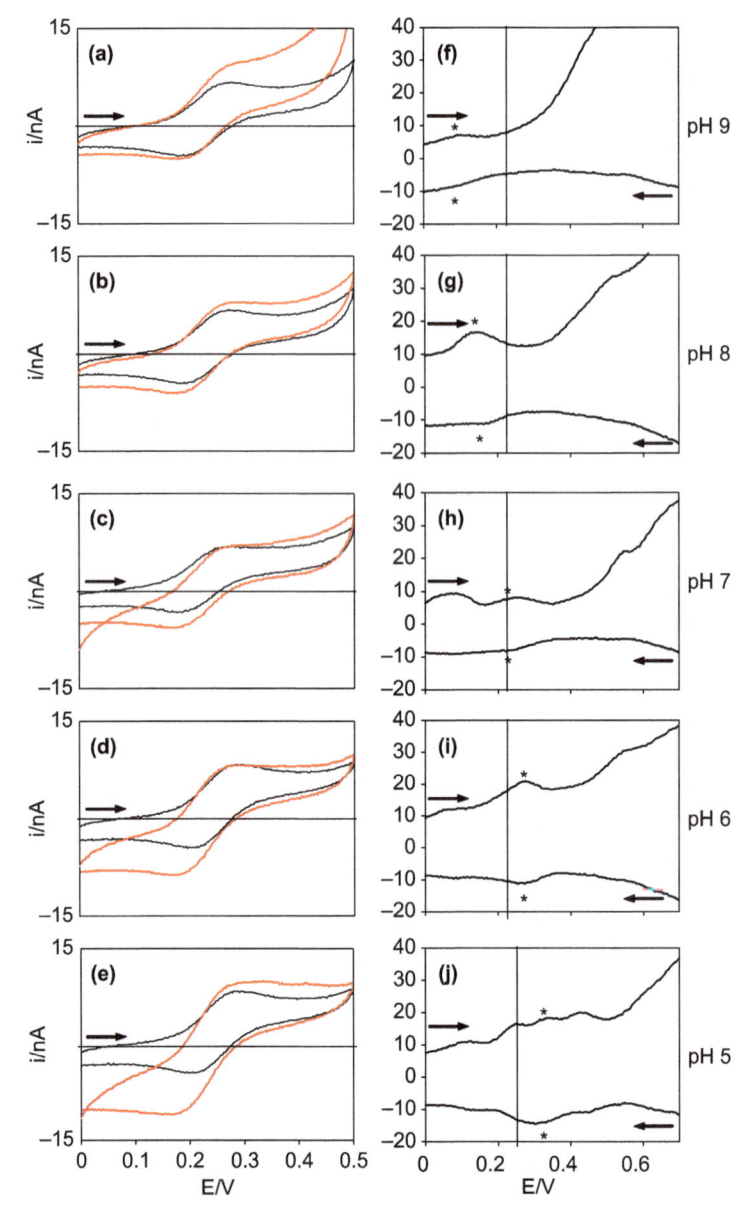

Figure 6.7 (a)–(e): Cyclic voltammograms of 1 µM Fe(CN)$_6^{4-}$ in 0.2 M phosphate buffer solution at 10 mV s^{-1} at a clean boron-doped diamond electrode (black) and an electrode modified with nanodiamond (red). (a) pH 9; (b) pH 8; (c) pH 7; (d) pH 6; (e) pH 5. (f)–(j): Differential pulse voltammograms in 0.2 M phosphate buffer solution (modulation potential = 45 mV; modulation time = 0.05 s; step potential = 5 mV; interval time = 0.5 s; scan rate = 10 mV s^{-1}) of nanodiamond immobilised onto a boron-doped diamond electrode (background response of boron-doped diamond electrode has been subtracted). (f) pH 9; (g) pH 8; (h) pH 7; (i) pH 6; (j) pH 5. Grey vertical line indicates E^0 for the Fe(CN)$_6^{4-/3-}$ couple. Asterisk labels explained in the text. Reproduced from *J. Am. Chem. Soc.*, 2009, **113**, 11272 with permission. Copyright 2009 American Chemical Society.

they are independent of solution pH. In contrast, the red responses are those of the nanodiamond-modified electrode and these are clearly highly pH dependent. At pH 9, oxidation currents are greatly enhanced in comparison to reduction currents. In contrast, at pH 5 and 6 only the reduction currents exhibit significant enhancement. Importantly, at pH < 7 reduction currents flow as soon as a potential is applied at 0 V. This indicates that there is a species in solution that is being reduced at this potential; however, only $Fe(CN)_6^{4-}$ should be present and this can only be oxidised. The observed reduction currents can only be explained if somehow $Fe(CN)_6^{3-}$ is being spontaneously generated in solution before the potential is applied. This could occur if $Fe(CN)_6^{4-}$ is oxidised by the nanodiamond surface by a spontaneous electron transfer as described in step 2 of Figure 6.6b and discussed above.

It seems that the spontaneous oxidation of $Fe(CN)_6^{4-}$ by the nanodiamond surface only takes place below pH 7 and this can be understood by comparing the electrochemical response of the nanodiamond itself at the different pH values, shown in Figure 6.7f–j. In all of the figures a reversible redox process is observed, indicated by an asterisk (*). This couple moves approximately 60 mV towards more negative potentials for each increase in pH unit, indicating an electron transfer coupled to a proton transfer. At pH 5, 6, and 7 this couple lies positive of the potential at which $Fe(CN)_6^{4-}$ can be oxidised (indicated by the grey line at 0.23 V on the figure). As electron transfer must always take place from a low potential state to a high potential state, electron transfer from the $Fe(CN)_6^{4-}$ to the nanodiamond state marked by * will be spontaneous at pH < 7. Above this pH the surface state has moved to more negative potentials than that of the $Fe(CN)_6^{4-/3-}$ couple, so electron transfer will no longer be allowed. These observations provide evidence that electron transfer can take place between the nanodiamond surface and solution redox couples, but only if the surface state/solution molecular orbital energies are of the suitable relative energies.

Scanning electrochemical microscopy (SECM) was used to further investigate the direction of electron transfer at the nanodiamond surface for the $Fe(CN)_6^{4-/3-}$ redox couple.[21] In an experiment where $Fe(CN)_6^{3-}$ was generated at the microelectrode tip, very little current enhancement (positive feedback) was recorded at the tip as it approached the nanodiamond layer. This indicates that the nanodiamond surface does not readily transform $Fe(CN)_6^{3-}$ to $Fe(CN)_6^{4-}$. In contrast, when $Fe(CN)_6^{4-}$ was generated at the tip, a greater increase in tip current was noted as it was moved towards the surface; this confirmed that $Fe(CN)_6^{4-}$ was spontaneously converted to $Fe(CN)_6^{3-}$ at the powder surface, consistent with the observations from voltammetry studies.

The response to the $Fe(CN)_6^{4-/3-}$ of nanodiamond annealed in a vacuum and in air at 1100 °C has been studied using a cavity electrode.[22] It was found that the electrochemical response was highly dependent on the presence of oxygen-containing functional groups. When these groups were removed from the surface by heating above 850 °C the electrochemical response was diminished and electron transfer kinetics became slower. When the powder was annealed in air to restore the oxygen groups, the electrochemical response was once again enhanced. These observations confirm that the electrochemical response of the nanodiamond can be attributed to surface oxygen-containing groups that can undergo pH-dependent oxidation and reduction.

6.6.2 Interaction of Nanodiamond Particles with Other Solution Redox Species

Similar studies have been carried out with redox couples of different potential to the $Fe(CN)_6^{4-/3-}$ couple.[50,16,19] As expected, the redox couples interact with different electronic states of the nanodiamond depending on their potential relative to the energies of the diamond surface states. The availability of surface states at a particular pH will dictate whether oxidation or reduction currents of a redox species are enhanced. If no surface states are present at that potential then no current enhancement is seen and currents are, in fact, suppressed as the nanodiamond acts only as an inert blocking layer on the electrode.

An enhancement of oxidation currents in the presence of nanodiamond is observed for the $IrCl_6^{3-/2-}$ and $Ru(CN)_6^{4-/3-}$ couples, which both have a potential at *ca.* 0.75 V *vs.* Ag/AgCl. The electrochemical response of the nanodiamond itself (Figure 6.4) shows that it undergoes oxidation at this potential at pH 5–9. Thus, as shown in Figure 6.8, oxidation currents for the $IrCl_6^{3-/2-}$ couple show enhancement attributed to the illustrated feedback mechanism over the full pH range.

Studies using different surface terminations of nanodiamond were also carried out using the $Ru(NH_3)_6^{3+/2+}$ redox couple.[21] Interpretation of the

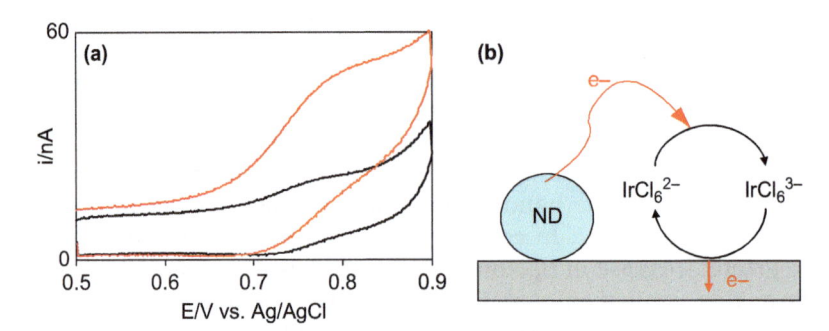

Figure 6.8 a) Cyclic voltammogram of 1 μM $IrCl_6^{3-}$ oxidation at a clean boron-doped diamond (black) and a nanodiamond-modified electrode (red) at 10 mV s^{-1}, 0.2 M pH 7 phosphate buffer solution. b) Postulated mechanism for oxidation current enhancement. Red arrows indicate direction of electron transfer.
Adapted from *J. Am. Chem. Soc.*, 2009, **113**, 11272 with permission. Copyright 2009 American Chemical Society.

current enhancements using this redox species are less straightforward as it is positively charged and so can be electrostatically adsorbed on to the surface of the nanodiamond by interaction with negatively charged oxygen functionalities. Interestingly, for the as-received and oxidised powders, significant current enhancements were observed at the potentials for the redox species and also in the potential region corresponding to oxygen reduction. An enhancement of oxygen reduction current was not noted for the nanodiamond-modified electrode in the absence of the $Ru(NH_3)_6^{3+/2+}$, which suggests the nanodiamond is somehow mediating electron transfer between the $Ru(NH_3)_6^{3+/2+}$ species and the dissolved oxygen. The more oxidised powder exhibited higher current enhancements than the as-received powder and the hydrogenated powder caused very little enhancement in the currents for both the $Ru(NH_3)_6^{3+/2+}$ species and oxygen reduction. These results confirm that oxygen functionalities are necessary for the redox activity of this material.

6.7 *In Situ* Spectroscopy Studies of the Nanodiamond Surface in Redox Solutions

To investigate the processes taking place on the nanodiamond surface on exposure to a redox solution containing $IrCl_6^{2-/3-}$, an *in situ* infrared spectroscopy experiment was designed.[5a] A layer of the nanodiamond powder was drop-coated onto an attenuated total reflectance (ATR) prism. ATR IR spectroscopy directs an IR beam through a prism, at which point it undergoes total internal reflection. Some of the IR beam penetrates the region just above the prism as an evanescent wave and so can probe the IR vibrational spectrum of material deposited on the prism. The large surface area of the nanodiamond allows high resolution spectra of the nanodiamond surface to be obtained. The nanodiamond layer was equilibrated in electrolyte solution (0.1 M KCl) and then the redox species $IrCl_6^{2-}$ was introduced. Changes in the spectrum of the nanodiamond were recorded over time after addition of the redox molecule to allow temporal analysis of changes to surface functional group chemistry.

IR difference spectra are shown in Figure 6.9a; peaks showing a decrease in absorbance are loss peaks indicating surface functionalities that are lost during the redox treatment, and peaks showing an increase in absorbance are gain peaks indicating an increase in the presence of surface groups absorbing at that energy during the redox treatment. The loss peak at 1072 cm^{-1} is attributed to surface alcohol groups (C–OH) and the concomitant gain at 1665 cm^{-1} is an unsaturated ketone moiety (C=C–C=O), for example, from a quinone group. A cartoon illustrating the observed surface transformations is shown in Figure 6.9b. This study demonstrates that changes in the solution redox environment are able to induce transformations in surface functional groups and proves that the nanodiamond surface is redox active under relatively mild redox conditions. It also helps us

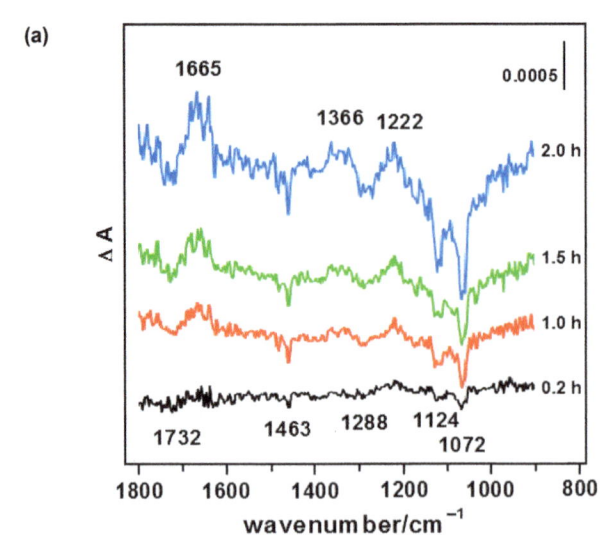

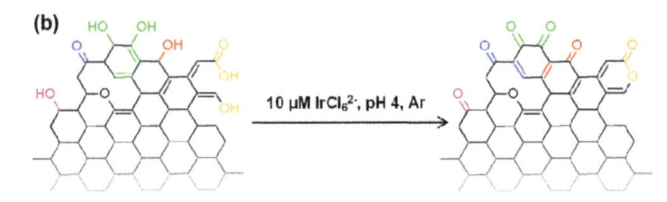

Figure 6.9 a) IR difference absorption spectra of nanodiamond film exposed to 10 µM IrCl$_6^{3-}$ vs. film equilibrated in 0.1 M NaCl solution. From 64 scans, resolution 4 cm^{-1}. Spectra offset on absorbance scale for clarity. b) Scheme of nanodiamond surface functional groups and their transformation upon contact with an aqueous IrCl$_6^{2-}$ solution in 0.1 M NaCl. Reproduced with permission from *Chem. Commun.*, 2011, **47**, 12140.

to attribute the observed heterogeneous electronic nature of the diamond surface to specific chemistries and enables us to design surfaces with more or less redox character.

6.8 Electrochemical Impedance Spectroscopy Studies of Nanodiamond Particles

Electrochemical impedance spectroscopy provides information on the resistive and capacitive elements of the nanodiamond powders. Using a cavity powder electrode, the obtained Nyquist plot for as-received nanodiamond in 0.1 M KCl solution gave a resistance of 15 kΩ for the powder, with significant porosity.[14] In the presence of a redox species the same value for resistance was obtained and was potential independent. The charge transfer resistance (*i.e.*, the resistance associated with electron transfer across the nanodiamond—solution interface) was too large to be accurately measured.

A more systematic impedance spectroscopy study was carried out using a specially designed pressure cell to compare dry as-received and hydrogenated (hydrogen microwave plasma, 800 °C, 10 min) nanodiamond powders.[23] As-received powder exhibited a near perfect semi-circular Cole–Cole plot and could be simulated to a simple RC parallel circuit. Resistance was determined at 5×10^9 Ω. This emphasises the non-conducting nature of the as-received nanodiamond powder. After hydrogen treatment the resistance fell several orders of magnitude to 5×10^5 Ω and also showed more porosity. This porosity and the decreased resistance were attributed to adsorbed surface water with other adsorbates as the effect lessened once the sample was heated to 100 °C. The increased p-type conductivity of diamond films is also noted after hydrogenation[2] and also observed for 100 nm diamond particles (see section 6.10.3).[17]

6.9 Applications of Nanodiamond in Electrochemical Technologies

6.9.1 Nanodiamond and Electrochemical Sensors

The motivation for incorporating nanodiamond into electrochemical sensors is to take advantage of the catalytic properties of the nanodiamond surface towards particular reactions; for example, nanodiamond powder electrodes have been shown to be catalytic towards nitrite oxidation.[15] In most cases the reasons for enhanced activity of nanodiamond towards different analytes has not been further explored and explained, but the reason is likely to be the presence of specific surface functional groups, as described earlier in the chapter, that interact with the analytes and catalyse their oxidation or reduction.

A mixture of nanographite and nanodiamond in a chitosan matrix was used to detect the immunosuppressant drug azothioprine in pharmaceutical samples and blood serum.[24] The nanodiamond-modified electrode showed a much improved overpotential for the reduction of the drug molecule, as well as enhanced electron transfer kinetics—all hallmarks of a catalytic electrochemical response. The sensor also showed long-term stability and reproducibility. A sensor prepared by drop-coating the nanodiamond/graphite aqueous suspension onto a glassy carbon electrode likewise showed an excellent electrocatalytic response to epinephrine and uric acid, allowing their simultaneous detection without interference from ascorbic acid present in the sample.[25] The same authors also showed this electrode could be used for the voltammetric determination of tryptophan and 5-hydroxytryptophan.[26]

6.9.2 Nanodiamond in Biosensors

Nanodiamond has been incorporated into biosensors both as a benign, high-surface-area enzyme support and as an active electron transfer mediator. An example of the former application is its use in an alcohol biosensor based on alcohol dehydrogenase adsorbed to the surface of oxidised diamond nanoparticles.[27] The nanodiamond provides a biocompatible surface on which the enzyme can adsorb without denaturing and, therefore, maintains a high activity within the device. The enzyme converts alcohols to aldehydes with the concomitant reduction of NAD^+ to NADH. The generated NADH is then detected at an electrode and the nanodiamond does not appear to play a role in the electrochemical response. In contrast, a glucose biosensor formed from a gold electrode–polyelectrolyte–nanodiamond–glucose; oxidase composite showed much enhanced current responses when nanodiamond was incorporated.[28] This device relies on the detection of the oxygen reduction current. It was found that pre-treating the electrode at a potential of +0.7 V *vs.* SCE resulted in enhanced oxygen reduction peaks and so greater device sensitivity. It became clear the anodically-treated ND was mediating and enhancing the oxygen reduction reaction, presumably through interaction of the oxygen molecule with specific surface functional groups generated during the oxidative pre-treatment. IR spectroscopy did not reveal any obvious changes in the nanodiamond surface chemistry so the exact mechanism responsible for the current enhancement remains unknown.

In another study, nanodiamond was found to act as a direct mediator of electron transfer between a redox protein and an electrode by the construction of a chitosan–nanodiamond–haemoglobin assembly.[29] It was shown by UV–visible spectroscopy that haemoglobin does not denature on the nanodiamond surface. It was speculated that the protein interacts with the nanodiamond surface through both hydrophobic and hydrophilic interactions. In the presence of nanodiamond, reversible and stable voltammetry of the haemoglobin redox centre was observed with evidence of fast electron transfer kinetics. This very interesting result illustrates that nanodiamond is able to mediate and promote electron transfer between an electrode surface and an adsorbed protein molecule. Finally, a biosensor for nitrite detection was constructed from electropolymerised polyaniline grafted onto $-NH_2$-modifed nanodiamond.[30] Gold nanoparticles were electrodeposited onto this composite and cytochrome c was then immobilised to the array. Direct electrochemistry of cytochrome c was detected and the assembly was found to be a sensitive and stable biosensor for nitrite ions.

6.9.3 Nanodiamond as Fuel Cell Catalyst Electrode Supports

One of the most widely reported electrochemical applications of nanodiamond is as a support for fuel catalysts, such as Pt or PtRu, particularly for methanol oxidation. The motivation behind the use of nanodiamond is that it offers greater stability to oxidation potentials than the sp^2 carbon supports used presently. Platinum nanoparticles have been deposited on nanodiamond using electrodeposition from H_2PtCl_6[31,32] and by the chemical reduction of H_2PtCl_6 using sodium borohydride.[33] XPS studies have suggested that the Pt precursor molecules interact with the –OH and –CH$_{2/3}$ groups of the nanodiamond surface prior to reduction.[32] Microwave-assisted polyol synthesis has been used to deposit Pt and PtRu nanoparticles on nanodiamond[34] and core–shell 'bucky-diamond'[35] for methanol oxidation catalysis. This process involves the microwave heating of metal precursor with the nanodiamond in ethylene glycol, which acts simultaneously as a solvent and reducing agent. For all of the preparations used, the metal nanoparticles were found to be adherent and catalytic for methanol oxidation with high activity and stability. Using 'bucky diamond', which is nanodiamond with a graphitic outer shell (formed by annealing in a vacuum), gave still greater activity due to the higher conductivity from the sp^2 content. As well as greater stability inferred from the use of nanodiamond, the increased surface area available for deposition of Pt or PtRu catalyst was also cited as an advantage of using nanodiamond in this application.

6.9.4 Nanodiamond as a Supercapacitor Electrode Material

The electrochemical performance of nanodiamonds in electrical double layer capacitors (supercapacitors) has been compared to the performance of carbon onions, carbon black and multiwalled nanotubes using galvanostatic cycling measurements, electrochemical impedance spectroscopy and cyclic voltammetry in acetonitrile and water.[36,37] The main focus of this study was to understand the influence of pore size on capacitive charge/discharge events as nanodiamond and carbon onions provide materials with controlled pore sizes between the particles. In comparison to carbon onions, nanodiamond showed smaller capacitance due to a smaller specific surface area and higher resistance due to the decreased sp^2 content. In an attempt to increase capacitance of these materials, their surfaces were modified with a layer of phosphomolybdate, which adds a pseudocapacitance (Faradaic) response.[38] However, this was found to decrease the capacitance of the ND soot as it covered the surface functional groups that give rise to its redox chemistry.

Ungraphitised nanodiamond does not perform particularly well as a supercapacitor material as it is too resistive. However, it has been incorporated into polyaniline to produce electrodes with high capacitance and pseudocapacitance. For example, polyaniline has been electropolymerised from aniline onto the surface of nanodiamond in a cavity electrode.[39] Nanodiamond–polyaniline composites were also synthesised by chemical oxidation.[40] Addition of 3–28% by weight nanodiamond resulted in electrodes with excellent charge–discharge characteristics that were stable to cycling over 10 000 cycles. This is in contrast to polyaniline without the nanodiamond, which entirely degraded during this time. The nanodiamond composites also showed more stable responses than those made with other carbon additives, such as onion-like carbon. The nanodiamond was believed to act as a toughening agent by minimising the volume changes in polyaniline during cycling that lead to degradation. The root of this stability is traced to strong interactions between the aniline and surface groups on the nanodiamond—in particular, hydrogen bonding between the aniline amino groups and the nanodiamond carboxyl groups.

6.9.5 Nanodiamond in Electrode Coatings and Composites

Following from the investigations of nanodiamond–polyaniline composites for supercapacitor applications, the same materials have been investigated as corrosion inhibition films on electrodes.[41] Cyclic voltammograms of these films showed redox peaks for both polyaniline and nanodiamond surface groups and the nanodiamond appeared to have an influence on the proportion of emeralidine to emeralidine salt in the polyaniline. Electrochemical polymerisation of polyaniline was also investigated as a catalytic electrode coating for the I^-/I_2 redox couple with possible future applications in dye-sensitised solar cells.[42] In this study the nanodiamond was found not to simply blend with the polymer but to provide a surface for controlled adsorption of the monomer and subsequent nucleation and growth. Nanodiamond has also been incorporated into gold electrode coatings during the electrodeposition process, where it was found to have an influence on grain size and hardness of the coating.[43]

6.9.6 Nanodiamond in Photocatalysis

Related to electrochemical applications, nanodiamond has also been shown to have interesting photo-induced redox chemistry. As received, oxygenated and hydrogenated nanodiamond samples in deionised water were irradiated with 532 nm laser pulses, resulting in the evolution of hydrogen.[44] This was shown to be a multiphoton process, where valance electrons excited into the diamond conduction band were collected at hydrogen-terminated surface sites, acting as electron reservoirs. Proton reduction at these sites results in hydrogen evolution and IR spectroscopy revealed a loss of nanodiamond surface C=O groups and a concomitant gain of C–H bonds. The nanodiamond was also used as a photocatalyst to reduce graphene oxide. Nanodiamond has also been incorporated into a donor–acceptor composite as the electron donor for photo-induced electron transfer and photocurrent generation.[45] Hydrogen bonding interactions were exploited to form supramolecular porphyrin–nanodiamond assemblies, the photodynamics of which were investigated using spectroscopy and photoelectrochemistry. The composite showed an efficient photo-induced electron transfer process in the visible light region due to efficient electron transfer between the porphyrin and the nanodiamond.

6.10 Electrochemistry of Other Diamond Nanoparticles

6.10.1 Boron-doped Nanodiamond

A disadvantage of using nanodiamond in electrochemical applications is its low conductivity as it is usually undoped. Attempts have been made to dope the particles with boron by solid-state diffusion methods.[46] Boron powder was mixed in a 2 : 1 ratio with nanodiamond powder and compressed to make a pellet. The pellet was then heated at 1000 °C for 1 day under a nitrogen atmosphere. The pellet was then refluxed in nitric acid to remove unreacted boron as boric acid. This treatment appears to change the outer morphology of the nanodiamond, as seen by high resolution TEM. XPS of the boron region showed a B^0 peak at 190 eV, which the authors attributed to substitutionally doped boron within the diamond lattice. An additional peak was noted at 187 eV, attributed to a different boron species. The degree of penetration of B into the particles is unclear, as are the actually sites occupied by the boron atoms. These treated particles show lower capacitance and higher conductivity than undoped nanodiamond and a featureless voltammetric response over −0.3 to 1.8 V *vs.* SCE.

6.10.2 Electrochemistry with 100 nm Diamond Particles

Many of the above applications, particularly in fuel cell catalyst supports, have also explored the use of larger diamond particles of 50–100 nm, produced through mechanical crushing.[47,48] As the electrochemical activity of detonation nanodiamond is attributed to surface groups, one would expect that large diamond particles should show decreased electrochemical activity due to the smaller surface-area-to-bulk ratio and this, in general, is observed.

6.10.3 Electrochemical Studies with HTHP Type 1b Diamond

An interesting series of experiments has been carried out on HTHP type 1b diamonds, which were immobilised onto an indium tin oxide (ITO) electrode using electrostatically driven layer-by-layer assembly.[17] The redox response of as-received diamond particles revealed a reversible redox couple around 0.1 V, attributed to a surface species. As this response was similar to those noted on glassy carbon it was attributed to quinone-type species and was calculated to correspond to only 0.1% coverage of the surface. In further studies the electrochemical responses of oxygenated and hydrogenated diamond particles were investigated using the same layer-by-layer deposition technique to construct the two-dimensional assembly of diamonds.[49] Interestingly, despite the different surface terminations for the treated diamonds, they all still exhibited a negative zeta-potential. The oxidised diamond showed a featureless voltammetric response but the hydrogenated diamond showed current increases above 0.2 V that seemed a consequence of reversible capacitive charging in the region 0.2 to 0.8 V, above which an irreversible oxidation of the surface took place. The region of reversible capacitive charging appeared stable over repeated cycling and was posited to be due to hole accumulation at the hydrogenated diamond surface, *i.e.*, a p-type surface-doping process like that noted at *H*-terminated chemical vapour deposition (CVD) diamond films. The same insulating diamond particles were used as supports for Pd nanoparticles for the oxidation of CO and methanol.[50] The particles were modified by mixing a Pd precursor salt with sodium borohydride in the presence of the diamond particles. The modified particles were then mixed with Nafion and drop-coated onto a glassy carbon electrode. The fact that these insulating particles could support electrocatalysis was attributed to partial hydrogenation of the diamond during reduction of the metal precursor, inducing p-type conductivity of the diamond surface.

6.11 Conclusions and Outlook

Although one may not expect much exciting redox behaviour from an undoped, insulating material, detonation nanodiamond and diamond nanoparticles formed through other synthetic methods show interesting electrochemical activity. The two important mechanisms for electron transfer seem to be through surface functional group chemistry, giving rise to band gap electronic states in the case of oxygenated diamond and, for hydrogenated diamond, a surface hole-accumulation mechanism, inducing p-type conductivity. Fundamental studies of these behaviours are still at an early stage and further work is required using coupled electrochemical and spectroscopic techniques to elucidate the underlying mechanisms. It can be seen that nanodiamond is posited in a range of electrochemical applications. If nanodiamond is to achieve a lasting role in any of these, rather than merely remain a passing novelty, a significant optimisation of surface chemistry is required along with deeper understanding of its reactivity. This is an exciting interdisciplinary research field, encompassing chemistry, physics, engineering, materials processing and even biology. Researchers in all disciplines need to collaborate to ensure the potential of diamond nanoparticles in electrochemical applications is fulfilled.

References

1. See for example: R. G. Compton, J. S. Foord and F. Marken, *Electroanalysis*, 2003, **15**, 1349.
2. (a) D. Shin, H. Watanabe and C. E. Nebel, *J. Am. Chem. Soc.*, 2005, **127**, 11236; (b) V. Chakrapani, J. C. Angus, A. B. Anderson, S. D. Wolter, B. R. Stoner and G. U. Sumanasekera, *Science*, 2007, **318**, 1424.
3. (a) L. C. Hian, K. J. Grehan, R. G. Compton, J. S. Foord and F. Marken, *Diam. Relat. Mater.*, 2003, **12**, 590; (b) L. C. Hian, K. J. Grehan, R. G. Compton, J. S. Foord and F. Marken, *J. Electrochem. Soc.*, 2003, **150**, E59.
4. O. A. Williams, *Semicond. Sci. Technol.*, 2006, **21**, R49.
5. (a) J. Scholz, A. J. McQuillan and K. B. Holt, *Chem. Commun.*, 2011, **47**, 12140; (b) J.-S. Tu, E. Perevedentseva, P.-H. Chung and C.-L. Cheng, *J. Chem. Phys.*, 2006, **125**, 174713.

6. (a) V. Mochalin, S. Osswald and Y. Gogotsi, *Chem. Mater.*, 2009, **21**, 273; (b) S. Prawer, K. W. Nugent, D. N. Jamieson, J. O. Owra, L. A. Bursill and J. L. Peng, *Chem. Phys. Lett.*, 2000, **332**, 93.
7. (a) A. Dementjev, K. Maslakov, I. Kulakova, V. Korolkov and V. Dolmatov, *Diam. Relat. Mater.*, 2006, **15**, 1813; (b) O. Shenderova, A. M. Panich, S. Moseenkov, S. C. Hens, V. Kuznetsov and H.-M. Vieth, *J. Phys. Chem. C.*, 2001, **115**, 19005.
8. (a) A. M. Panich, H.-M. Vieth and O. Shenderova, *Fullerenes, Nanotubes and Carbon Nanostructures*, 2012, **20**, 579; (b) X. W. Fang, J. D. Mao, E. M. Levin and K. Schmidt-Rohr, *J. Am. Chem. Soc.*, 2009, **131**, 1426.
9. M. Ozawa, M. Inaguma, M. Takahashi, F. Kataoka, A. Kruger and E. Osawa, *Adv. Mater.*, 2007, **19**, 1201.
10. I. A. Novoselova, E. N. Fedoryshena, E. V. Panov, A. A. Bochechka and L. A. Romanko, *Phys. Solid State*, 2004, **46**, 748.
11. G. P. Bogatyreva, M. A. Marinich, E. V. Ishchenko, V. L. Gvyazdovskaya, G. A. Bazalii and N. A. Oleinik, *Phys. Solid State*, 2004, **46**, 738.
12. T. L. Kulova, Yu. E. Evstefeeva, Yu. V. Pleskov, A. M. Skundin, V. G. Ral'chenko, S. B. Korchagina and S. K. Gordeev, *Phys. Solid State*, 2004, **46**, 726.
13. L. La-Torre-Riveros, D. A. Tryk and C. R. Cabrera, *Rev. Adv. Mater. Sci.*, 2005, **10**, 256.
14. J. B. Zang, Y. H. Wang, S. Z. Zhao, L. Y. Bian and J. Lu, *Diam. Relat. Mater.*, 2007, **16**, 16.
15. L. H. Chen, J. B. Zang, Y. H. Wang and L. Y. Bian, *Electrochim. Acta*, 2008, **53**, 3442.
16. K. B. Holt, *Phys. Chem. Chem. Phys.*, 2010, **12**, 2048.
17. W. Hongthani and D. J. Fermin, *Diam. Relat. Mater.*, 2010, **19**, 680.
18. K. B. Holt, C. Ziegler, D. J. Caruana, J. B. Zang, E. J. Millan-Barrios, J. P. Hu and J. S. Foord, *Phys. Chem. Chem. Phys.*, 2008, **10**, 303.
19. K. B. Holt, D. J. Caruana and E. J. Millan-Barrios, *J. Am. Chem. Soc.*, 2009, **131**, 11272.
20. Y. D. Glinka, K.-W. Lin, H.-C. Chang and S. H. Lin, *J. Phys. Chem. B.*, 1999, **103**, 4251.
21. K. B. Holt, C. Ziegler, J. B. Zang, J. P. Hu and J. S. Foord, *J. Phys. Chem. C*, 2009, **113**, 2761.
22. J. B. Zang, Y. H. Wang, L. Y. Bian, J. H. Zhang, F. W. Meng, Y. L. Zhao, S. B. Ren and X. H. Qu, *Electrochim. Acta*, 2012, **72**, 68.
23. S. Shi, J. Li, V. Kundrat, A. M. Abbot and H. Ye., *Diam. Relat. Mater.*, 2012, **24**, 49.
24. S. Shahrokhian and M. Ghalkhani, *Electrochim. Acta*, 2010, **55**, 3621.
25. S. Shahrokhian and M. Khafaji, *Electrochim. Acta*, 2010, **55**, 9090.
26. S. Shahrokhian and M. Bayat, *Microchim. Acta*, 2011, **174**, 361.
27. E. Nicolau, J. Mendez, J. J. Fonseca, K. Griebenow and C. R. Cabrera, *Bioelectrochemistry*, 2012, **85**, 1.
28. W. Zhao, J.-J. Xu, Q.-Q. Qiu and H.-Y. Chen, *Biosens. Bioelectron.*, 2006, **22**, 649.

29. J.-T. Zhu, C.-G. Shi, J.-J. Xu and H.-Y. Chen, *Bioelectrochemistry*, 2007, **71**, 243.
30. A. I. Gopalan, K.-P. Lee and S. Komathi, *Biosens. Bioelectron.*, 2010, **26**, 1638.
31. L. Y. Bian, Y. H. Wang, J. B. Zang, J. K. Hu and H. Huang, *J. Electroanal. Chem.*, 2010, **644**, 85.
32. L. La-Torre-Riveros, K. Soto, M. A. Scibioh and C. R. Cabrera, *J. Electrochem. Soc.*, 2010, **157**, B831.
33. L. La-Torre-Riveros, R. Guzman-Blas, A. E. Mendez-Torres, M. Prelas, D. A. Tryk and C. R. Cabrera, *ACS Appl. Mater. Interfaces*, 2012, **4**, 1134.
34. R. Lu, J. B. Zang, Y. H. Wang and Y. L. Zhao, *Electrochim. Acta*, 2012, **60**, 329.
35. J. B. Zang, Y. H. Wang, L. Y. Bian, J. H. Zhang, F. W. Meng, Y. L. Zhao, X. H. Qu and S. B. Ren, *Int. J. Hydrogen Energy*, 2012, **37**, 6349.
36. C. Portet, G. Yushin and Y. Gogotsi, *Carbon*, 2007, **45**, 2511.
37. C. Portet, J. Chmiola, Y. Gogotsi, S. Park and K. Lian, *Electrochim. Acta*, 2008, **53**, 7675.
38. S. Park, K. Lian and Y. Gogotsi, *J. Electrochem. Soc.*, 2009, **156**, A921.
39. J. B. Zang, Y. H. Wang, X. Y. Zhao, G. X. Xin, S. P. Sun, X. H. Qu and S. B. Ren, *Int. J. Electrochem. Sci.*, 2012, **7**, 1677.
40. I. Kovalenko, D. G. Bucknall and G. Yushin, *Adv. Funct. Mater.*, 2010, **20**, 3979.
41. H. Gomez, M. K. Ram, F. Alvi, E. Stefanakos and A. Kumar, *J. Phys. Chem. C*, 2010, **114**, 18797.
42. E. Tamburri, S. Orlanducci, V. Guglielmotti, G. Reina, M. Rossi and M. L. Terranova, *Polymer*, 2011, **52**, 5001.
43. F. Wunsche, A. Bund and W. Plieth, *J. Solid State Electrochem.*, 2004, **8**, 209.
44. D. M. Jang, Y. Myung, H. S. Im, Y. S. Seo, Y. J. Cho, C. W. Lee, J. Park, A.-Y. Jee and M. Lee, *Chem. Commun.*, 2012, **48**, 696.
45. M. Ohtani, P. V. Kamat and S. Fukuzumi, *J. Mater. Chem.*, 2010, **20**, 582.
46. L. Cunci and C. R. Cabrera, *Electrochem. Solid State Lett.*, 2011, **14**, K17.
47. L. Y. Bian, Y. H. Wang, J. B. Zang, F. W. Meng and Y. L. Zhao, *Int J. Hydrogen Energy*, 2012, **37**, 1220.
48. J. B. Zang, Y. H. Wang, L. Y. Bian, J. H. Zhang, F. W. Meng, Y. L. Zhao, R. Lu, X. H. Qu and S. B. Ren, *Carbon*, 2012, **50**, 3032.
49. W. Hongthani, N. A. Fox and D. J. Fermin, *Langmuir*, 2011, **27**, 5112.
50. A. Moore, V. Celorrio, M. Montes de Oca, D. Plana, W. Hongthani, M. J. Lazaro and D. J. Fermin, *Chem Commun.*, 2011, **47**, 7656.

CHAPTER 7

Nanodiamonds for Drug Delivery and Diagnostics

HAN MAN,[a] JOSHUA SASINE,[b] EDWARD K. CHOW*[c,d] AND DEAN HO*[d]

[a] Department of Mechanical Engineering, Northwestern University, Evanston, IL 60208, USA; [b] Department of Medicine, David Geffen School of Medicine, Los Angeles, CA 90095, USA; [c] Cancer Science Institute and Department of Pharmacology, National University of Singapore, Singapore; [d] Divisions of Oral Biology and Medicine, Advanced Prosthodontics, and Center for Oral, Head and Neck Cancer Research, The Jane and Jerry Weintraub Center for Reconstructive Biotechnology, UCLA School of Dentistry, Jonsson Comprehensive Cancer Center, California NanoSystems Institute, University of California, Los Angeles, CA 90095, USA
*Email: dean.ho@ucla.edu; csikce@nus.edu.sg

7.1 Introduction

The major diseases and physiological disorders of our generation include cancer, heart disease, regenerative medicine, and several other challenges. Their successful diagnosis and treatment require the development of novel technologies with improved diagnostic sensitivity and specificity, as well as therapeutic delivery with enhanced efficacy and safety. Several classes of nanomaterials have been previously explored. These have included metallic (*e.g.*, gold, TiO_2, silver, *etc.*), polymer, lipid, and several others. Emerging as a promising class of therapeutic and imaging agents are nanocarbons. In addition to the widely studied nanotubes and graphene, nanodiamonds

(NDs) combine several important and, in some cases, uniquely beneficial properties for translationally relevant drug release and imaging.[1–10] In addition to NDs, ultra/nanocrystalline diamond has also gained important traction for biomedical applications particularly in the areas of implant coating and diagnostic interfaces. NDs are promising due to the range of critically important properties they possess. This chapter will focus primarily on the application of detonation NDs and high pressure, high temperature (HPHT) diamond to examine several innovative applications in the therapeutics, imaging, and diagnostics arenas.

7.2 Nanodiamond Systemic Drug Delivery

One of the most promising applications of NDs is their use as drug delivery agents. The initial demonstration of ND-mediated chemotherapeutic delivery[1] pertained to using doxorubicin (Dox) as the model therapeutic. Additional studies utilized NDs to deliver water insoluble therapeutics, nucleic acids (*e.g.*, DNA plasmids, siRNA), and proteins (*e.g.*, therapeutic antibodies, such as anti-TGF-beta, insulin). Requirements for clinically relevant enhancements to drug delivery include scalable processing parameters for the material to rapidly yield uniform particles that are lower in cost relative to the current classes of nanomaterials. In addition, the versatility of being able to carry multiple classes of compounds individually or in a combinatorial fashion is important given the expanding work exploring the role of multi-drug delivery towards improving efficacy and safety in the treatment of cancer, or for other applications, such as guiding stem cell differentiation.

7.2.1 Nanodiamond–Doxorubicin Complexes

Following the initial demonstration of ND–Dox (NDX) synthesis, the ND–drug complex was harnessed for pre-clinical validation in multiple drug-resistant tumor models. Drug resistance is a major cause of failure in metastatic cancer therapy. Under these conditions, proteins with endogenous roles, such as enabling toxin removal (*e.g.*, through the liver), that are also present within tumors also result in chemotherapeutic efflux following drug administration. This can result in little to no efficacy and, depending on the drug that is administered, significant levels of toxicity. An LT2-M drug resistant liver tumor model and a 4T1 drug-resistant breast cancer model have previously been utilized to investigate NDX as a possible therapeutic.[1] In both models NDX administration resulted in clearly improved intratumoral retention compared to unmodified Dox administration. Furthermore, NDX administration resulted in a ten-fold increase in the circulatory half-life of Dox compared to unmodified Dox administration. One of the most important findings from this work was the fact that there was *virtually no myelosuppression* following NDX administration, while unmodified Dox administration caused significant myelosuppression. Myelosuppression is the major dose-limiting side effect of chemotherapy that can cause superinfections, and often necessitates the stoppage of treatment. *This is a critically important finding because this demonstrates that premature or early Dox release does not appear to occur with the NDX complex.* This further highlights that the reversible nature of ND–Dox binding simultaneously enables potent binding during circulation *in vivo*, which resulted in reduced toxicity compared to the Dox clinical standard, while subsequent release resulted in major improvements in drug efficacy. More specifically, in the 4T1 model, unmodified Dox administration resulted in virtually no treatment efficacy, a similar outcome observed with phosphate buffered saline (PBS). However, following NDX administration, tumor sizes were markedly reduced. One of the most important findings from this study was the effect of doubling the Dox dosage and the delivery of this dosage *via* NDX. The doubled Dox dosage, when administered alone, was lethal, resulting in markedly accelerated animal mortality (Figure 7.1). However, when this lethal dose was delivered in the form of NDX, animal survival was significantly improved, and the efficacy of treatment was the most efficient observed compared to all study conditions (Figure 7.1). This demonstrated that NDX could potentially be both a safe and highly efficacious form of Dox that may be translationally relevant and is a candidate for further development.

7.2.2 Targeted Nanodiamond Drug Delivery

In addition to passively delivered NDX, recent studies have also demonstrated that multimodal NDs carrying the epidermal growth factor receptor (EGFR) antibody, paclitaxel, and fluorescein were capable of integrated targeting, imaging, and therapy.[2] These complexes were used to address a triple negative breast cancer line. Triple negative breast cancer is characterized as having among the highest rates of recurrence, as well as metastasis. Therefore, a significant effort has been placed on the development of new therapeutics to address this challenging disorder. The ND surfaces were modified with fluorescein-labeled oligonucleotides and a hetero-bifunctional cross-linker sulfosuccinimidyl 6-(3-[2-pyridyldithio]propionamido)hexanoate (sulfo-LC-SPDP). This multifunctional ND complex enabled simultaneous targeting of the epidermal growth factor receptor (EGFR) over-expressing MDA-MB-231 cell lines *in vitro*, imaging of the ND complexes using the fluorescein-functionalized oligonucleotides, and therapeutic intervention using covalently conjugated paclitaxel. MCF-7 was utilized as the control cell line. Confocal microscopy clearly indicated increased ND uptake in the MDA-MB-231 cell line compared to MCF-7. In addition, epidermal growth factor (EGF) saturation studies using gradually increasing concentrations indicated gradually decreasing targeting efficacy, demonstrating that the increased ND localization was indeed due to the EGFR antibody. IC_{50} studies demonstrated that, while there was a subtle improvement in MCF-7 cell death following targeted ND administration, which was likely due to the delivery of a larger dose of paclitaxel from the nanoparticle surface, the paclitaxel efficacy was enhanced

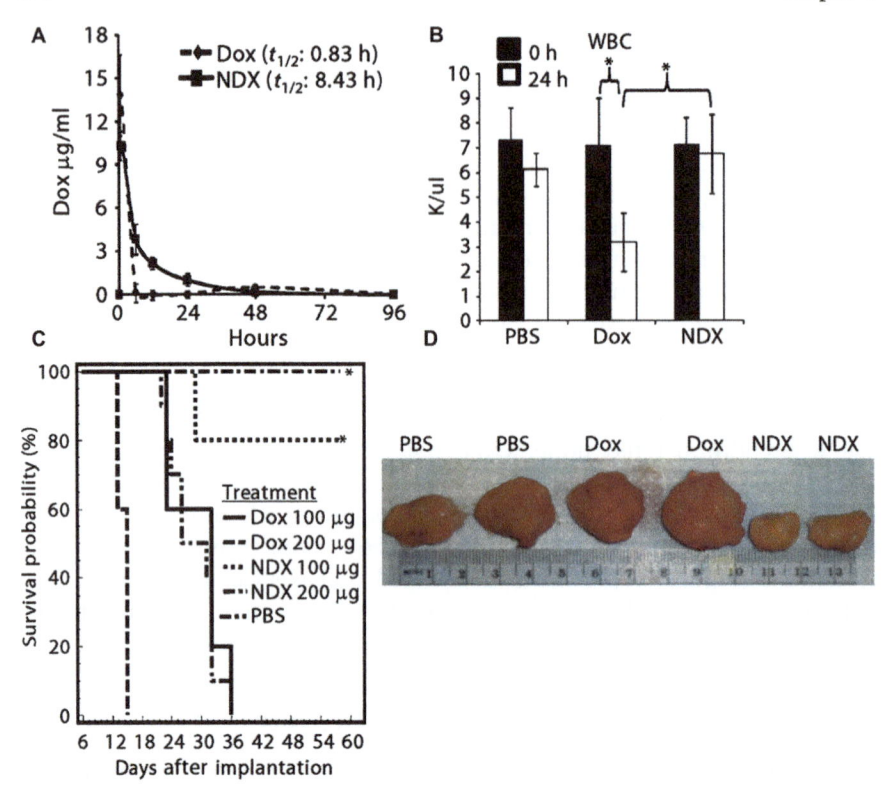

Figure 7.1 (A) A ten-fold increase in circulation half-life was observed with NDX administration compared to unmodified drug administration. (B) Compared to Dox administration, which resulted in myelosuppresion, NDX administration resulted in no apparent myelosuppression. (C) NDX administration markedly improved drug tolerance and animal survival, and also resulted in marked improvements in impaired tumor growth. (D) Tumor images demonstrate clear improvements in efficacy *via* NDX administration.
Reprinted with permission from Science Translational Medicine/ American Association for the Advancement of Science.[1]

nearly two-fold using the targeted ND complexes in the MDA-MD-231 cells. This study demonstrated the immense potential that multimodal NDs may have towards the therapy of hard-to-treat cancers.

7.3 Nanodiamond-based Implantable Devices

In addition to systemic ND administration, the localized implantation of ND–drug complexes or ND-embedded devices may impact several facets of biology and medicine. Localized drug delivery has the potential to significantly decrease drug toxicity and increase efficacy in a specific area of treatment.[3,5,11–14] In addition, the versatile ND surface chemistry and mechanical properties may be applicable towards tissue engineering. Therefore, the continued development of ND-based scaffolds and implantable devices may impact new areas of medical treatment.

7.3.1 Nanodiamond–Chemotherapeutic Tablets

Polyethylene glycol diacrylate (PEGDA) hydrogels embedded with ND conjugates (ND–PEGDA) were compared with PEGDA hydrogels containing unmodified drug molecules. For the release of the chemotherapeutic doxorubicin (Dox), hydrogels containing Dox, which had been previously loaded onto ND vehicles, were able to demonstrate a robust drug sequestering capacity compared with hydrogels containing unmodified Dox. ND-embedded hydrogels were capable of displaying slow and controlled administration of therapeutic compounds while mitigating the harmful effects of burst release.[5] Two chemotherapeutics, daunorubicin (DNR) and temozolomide (TMZ), were simultaneously embedded into multi-reservoir patches. After incubation, these patches were able to elute both drugs sequentially (Figure 7.2). Comparing the release of unmodified and ND-conjugated therapeutics, NDs were able to facilitate an improvement over standard diffusion-based release profiles. Following its release, the therapeutics were incubated with cells *in vitro* to confirm the retention of its therapeutic efficacy. The ability to customize both the structure and content of PEGDA hydrogels, as well as its scalable synthesis, make this material an effective multi-therapeutic complement to the drug delivery abilities of NDs. A hybrid material that combines the advantages of both PEGDA and ND components has the potential for application in a wide range of therapeutic challenges.

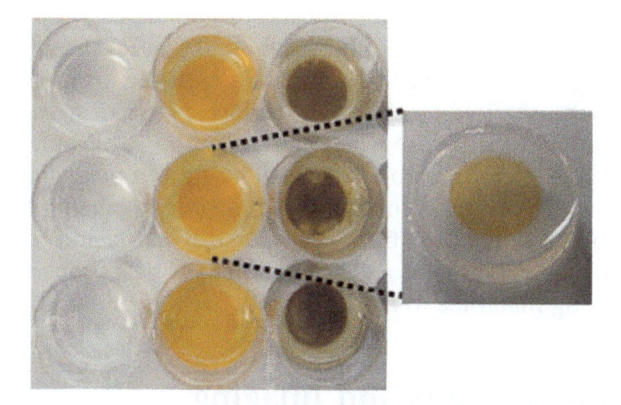

Figure 7.2 An image of a multi-layered ND–PEGDA tablet is shown. This platform is capable of eluting multiple therapeutic compounds in a sequential manner and may serve as a combinatorial drug delivery implant. Reprinted with permission from Wiley-VCH.[5]

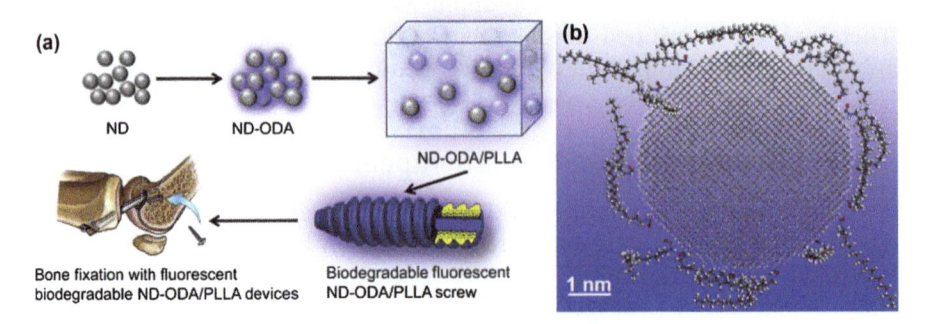

Figure 7.3 Nanodiamond–PLLA hybrids are being explored as potential tissue engineering platforms.
Reprinted with permission from Elsevier.[14]

7.3.2 Nanodiamond–PLLA Complexes for Tissue Engineering

NDs integrated with poly(L-lactic acid) (PLLA), a biodegradable polymer, have demonstrated potential in biomaterial applications.[14] Because of their positive mechanical properties, complex surface chemistry, and high biocompatibility, octadecylamine-functionalized NDs (ND-ODA) are promising candidates for use in bone scaffold materials. The hybrid ND–polymer material exhibited an increased Young's modulus and hardness, achieving similar properties to human cortical bone. It is the addition of NDs uniformly dispersed within the biodegradable polymer that enhances the mechanical properties. In fact, a 10% by weight supplement of ND-ODA was able to achieve a two-fold increase in Young's modulus and an eight-fold increase in hardness (Figure 7.3).

ND-ODA constructs provide additional advantages for use as a biomaterial and the scaffolds containing NDs were assessed for their biocompatibility by incubation with murine osteoblast cells (7F2). They were found to demonstrate biocompatibility for up to a week. In addition, the NDs displayed bright fluorescence, which could be used to monitor the integration of bone scaffolding. The ND surface properties also allow facile functionalization, in this case with octadecylamine. This surface treatment allowed the NDs to form stable suspensions in non-polar liquids, such as chloroform, which is a good solvent of PLLA. Functionalizing NDs with ODA made possible an enhanced affinity between the matrix and filler. ND-ODA/PLLA composites also possess a combination of critical properties that make them promising materials to aid in bone tissue regeneration.

7.4 Nanodiamonds and Imaging

In the area of imaging, NDs have demonstrated significant promise both as cellular imaging and potential clinically-relevant imaging modalities.[15–21] HPHT NDs possess the advantage of being resistant to photobleaching, while still remaining biocompatible. This has enabled their widespread use in cellular uptake assays, and they have recently been explored in *in vivo* studies as well. In addition to HPHT NDs, detonation NDs have also been explored as imaging agents. Examples include conjugating octadecylamine (ODA) to the ND surface, which resulted in a blue fluorescent ND. In addition, as ND development continues towards a translational roadmap, ND–gadolinium(III) (ND–Gd(III)) complexes have been synthesized for magnetic resonance imaging (MRI) applications.

7.4.1 Fluorescent Nanodiamonds for Bio-imaging Applications

Fluorescent nanodiamonds (FNDs) have received significant attention in recent years for their potential in bio-imaging applications.[15–17,19–21] This is related to the fact that their negatively charged nitrogen–vacancy centers allow them to serve as fluorophores. Mohan and colleagues were one of the first groups to demonstrate the imaging potential for FNDs *in vivo*.[20,21] Utilizing the model organism *Caenorhabditis elegans* (*C. elegans*), a colloidal FND solution was used to image the digestion patterns of FNDs in *C. elegans*. Bare FNDs were taken up by *C. elegans* and subsequently passed through the digestive tract for excretion (Figure 7.4). Interestingly, FNDs functionalized with either dextran or bovine serum albumin (BSA) were absorbed into intestinal cells, suggesting that future FND bioconjugates can be used for drug delivery or imaging applications through oral delivery. In addition to the uptake of FNDs by feeding, Mohan and colleagues used microinjection to analyze the potential of FNDs in the imaging of specific organs. Injection of FNDs into the distal gonads of gravid hermaphrodites allowed the researchers to image the reproductive organs, as well as their offspring at various stages of embryonic development. While this work demonstrates the potential for FNDs in bio-imaging applications, translation of FNDs into the clinic also requires that they are safe for use. In order to address this, Mohan and colleagues also analyzed the *C. elegans* response to FNDs with respect to the stress response, life span, and other markers of toxicity. In all tests, FNDs appeared to be non-toxic with no apparent effects on the feeding response, stress response, life span, and other factors. These experiments demonstrate

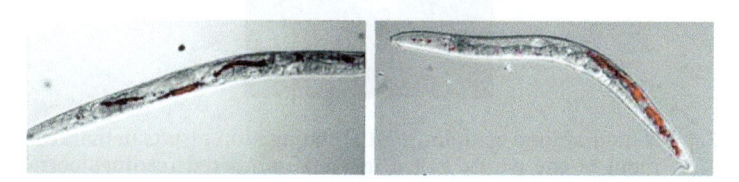

Figure 7.4 Fluorescent nanodiamond (FND) uptake within the *C. elegans* model. The FNDs resulted in no observable toxicity and were cleared through the digestive system.
Reprinted with permission from the American Chemical Society.[20]

that, in addition to being capable of imaging *C. elegans*, FNDs were non-toxic and thus have potential to be evaluated in larger organisms. Overall, this work was important in demonstrating *in vivo* both the bio-imaging capabilities of FNDs as well as the biocompatibility of these fluorescent particles.

7.4.2 Nanodiamonds for Magnetic Resonance Imaging

A spectrum of imaging approaches for disease diagnosis and monitoring are being developed amidst a need for enhanced imaging efficacy, specificity, and safety. Among the spectrum of approaches being developed, nanomaterial loading with gadolinium (Gd) for magnetic resonance imaging (MRI) applications has been widely explored because of the high loading capacity of the nanomaterials, which can improve relaxivity (contrast efficiency). However, an increased loading can also increase toxicity. Recently, a novel ND–Gd complex has been developed that has resulted in a twelve-fold per-Gd relaxivity increase compared to clinical agents and other nanoparticle Gd(III) contrast agents (Figure 7.5).[18] Instead of relying on excess Gd loading on the ND surface, the ND itself utilizes its uniquely faceted surface to mediate potent water binding, thereby enhancing Gd(III) performance. The major advantage of this powerful combination is the possibility of reducing patient Gd dosages by at least one order of magnitude, while still obtaining vital diagnostic and monitoring information.

7.5 Nanodiamond Safety and Biocompatibility Studies

A major precursor to the translation of NDs into the clinic is confirmation of their safety and biocompatibility. Several studies, using both *in vitro* and

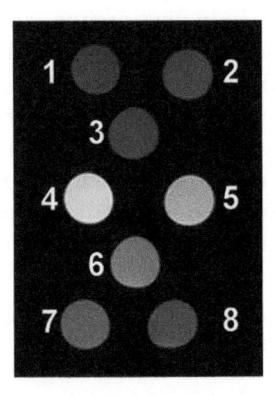

Figure 7.5 Nanodiamond–gadolinium(III) conjugation results in marked enhancement of the per-Gd relaxivity. The numbered readings correspond to varying Gd concentrations and other conditions. The conditions were as follows: 1, water; 2, 1 m/mL^{-1} undecorated ND; 3, undecorated ND + coupling reagents; 4, 48 µM Gd(III); 5, 38 µM Gd(III); 6, 22 µM Gd(III); 7, 10 µM Gd(III); 8, 5 µM Gd(III).
Reprinted with permission from the American Chemical Society.[18]

in vivo models, have sought to comprehensively provide this information. These studies have ranged from conventional (3-[4,5-dimethylthiazol-2-yl]-2,5 diphenyl tetrazolium bromide) MTT assays to more in-depth polymerase chain reaction (PCR) studies. More recent studies have compared ND safety with other carbon nanomaterials, while ND safety *in vivo* has also been evaluated.[1,20,22–24] These studies have revealed that there is no apparent toxicity following ND administration. While continued studies are needed, these findings serve as a promising foundation for the continued development of NDs for medical applications.

7.5.1 Pre-clinical Evaluation of Nanodiamond Safety

Conventional cell viability assays (*e.g.*, MTT), as well as quantitative real time polymerase chain reaction (qRTPCR) studies, have been previously conducted to evaluate preliminary ND safety. Recent studies in mice have evaluated localized ND organ safety and systemic toxicity.[1] For example, systemic administration of NDs was revealed to have virtually no negative impact on serum alanine transaminase (ALT) secretion (a measurement of liver toxicity). In addition, systemic interleukin-6 levels were virtually unchanged following ND administration. Hematoxylin and eosin staining also demonstrated that there were no changes to organ histology. As NDs have also been shown to improve cytotoxic drug tolerance, their continued development may yield a promising clinical platform for improved and safe drug release.

7.5.2 *In Vitro* Validation of Nanodiamond Safety

Some of the most widely used carbon-based nanomaterials (CNMs), including carbon nanotubes (CNTs), nanodiamond (ND), and grapheme oxide (GO), have been extensively studied for their physicochemical and biological properties.[24] Determining the intrinsic biochemical toxicity of CNMs is essential prior to their therapeutic use. Preliminary data have suggested that numerous factors can influence toxicity, such as metal contaminants, surface properties, aggregation state, and the cell type under investigation. In a study by Zhang *et al.* the cell uptake ratios of three types of CNMs (CNTs, GO, ND) were quantitatively determined using a radiolabeling technique. All of the CNMs were readily internalized by HeLa cells through nonspecific cellular uptake and the uptake ratios showed significant differences in the following order: ND > CNTs > GO. Afterwards, the overall induced cytotoxicity was evaluated by a series of assays, including an MTT assay, which depends on the mitochondrial enzyme reduction of a tetrazolium dye to detect and determine cell viability, and the levels of malondialdehyde, superoxide dismutase, lactate dehydrogenase, and reactive oxygen species (ROS). They all exhibited dose- and time-dependent cytotoxicity toward HeLa cells. However, the cytotoxicity of CNMs was not associated with their cell uptake ratios. NDs showed relatively low cytotoxicity compared with CNTs and GO, and, importantly, NDs exhibited the highest cell uptake ratio along with the low cytotoxicity. Furthermore, compared with the control, there was a 2.5-fold ROS increase induced by CNTs in HeLa cells, and a 3.6-fold increase for GO at the same concentration. Surprisingly, when cells were incubated with different concentrations of ND, the generation of ROS was found to be inversely proportional to the concentration of ND. This may be due to the ability of NDs to scavenge ROS (unless a large amount of ND taken up by cells could interfere with the detection of ROS, although this would be unexpected). Ultimately, for biomedical applications, the high uptake and low toxicity of NDs suggests that they may be the most promising candidate in this field.

7.6 Nanodiamonds and Diagnostics

Nanodiamonds are more commonly utilized as cellular/animal imaging agents, as well as drug delivery platforms. However, recently, NDs have also been explored as potential diagnostic platforms for bacterial detection. Due to their versatile surface chemistry, NDs can be modified to bind specific bacteria, which mediate changes in particle aggregation, producing an immediate readout for positive detection. This promising approach may serve as a gateway to the use of NDs as diagnostic reagents for rapid bacterial detection.[25]

7.6.1 Novel Approaches for Nanodiamond-based Detection

NDs have also been incorporated into a sandwich assay sensing system for the detection of pathogenic bacteria.[25] NDs were covalently attached to carbohydrates, which were able to facilitate the precipitation of type 1 fimbriated uropathogenic *Escherichia coli* bacteria. The ability to detect and remove dangerous bacteria, such as *E. coli*, from aqueous environments is a very important technique with regards to ensuring safety (Figure 7.6).

The readily modified surface of ND allows the covalent grafting of glycosides following a Diels–Alder reaction. The glycosylated NDs mimic the surface of host cells that bacteria typically attach to. Following the binding reaction, the ND–glycoside conjugate (glycol–ND) formed stable dispersions in media environments and the final carbohydrate loading on the nanoparticle surface reached 0.1 mmol g^{-1}. A new sandwich assay, utilizing dual layers of bacterial strains, was used to test the specificity of the glycol–ND interactions. Ultimately, glycol–NDs were able to form mechanically stable agglutinates through interactions with specific target strains of bacteria. This result confirmed the successful detection of appropriate bacterial strains. A further benefit was the ability to remove bacteria using scalable, low-cost filtration methods. After interaction with the modified NDs, the captured bacteria agglomerates were trapped by filtration through 10 µm-pore filters. By adding glycol–ND concentrations of 0.08–0.8 mg mg^{-1}, high bacterial removal rates of between 93–97% were achieved. Remarkably, the glycol–NDs were also recovered by the addition of an appropriate carbohydrate to

Nanodiamonds for Drug Delivery and Diagnostics

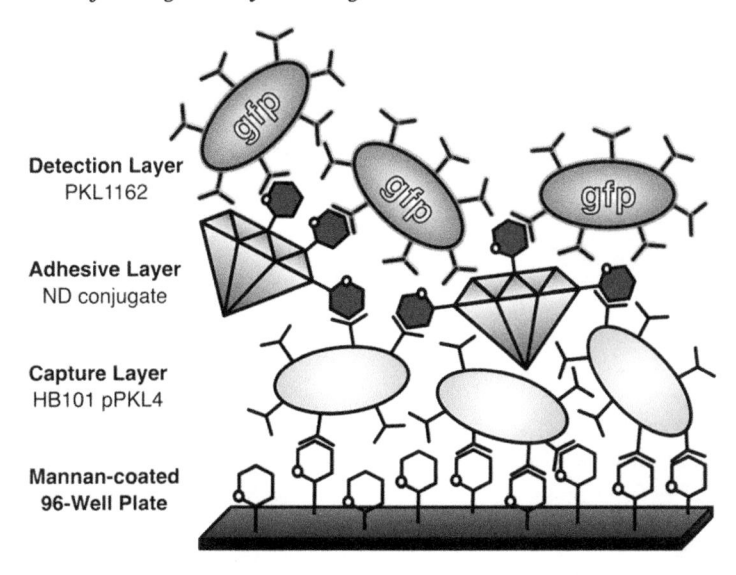

Figure 7.6 Multifunctional nanodiamond particles targeted towards a specific strain of bacteria were synthesized to serve as diagnostic agents. The sandwich assay resulted in a highly specific detection modality that can be expanded towards additional types of pathogens.
Reprinted with permission from Wiley-VCH.[25]

reversibly remove the bacteria from the ND surface. This ND-based detection technique has wide-ranging applications since it can be tuned to capture other types of bacteria by varying the carbohydrate attachment.

7.7 Modeling of Nanodiamond Vehicles for Optimized Applications

Several promising experimental studies have been conducted that demonstrate the promise of NDs as drug delivery and imaging agents. To further the development and optimization of ND platforms, modeling and simulation techniques have been employed to better understand the surface composition of NDs and how they interact with therapeutic payloads. These studies are vital as NDs are continually developed for pre-clinical and clinical applications.[26-32]

7.7.1 Nanodiamond-based Gene Delivery

NDs have been shown to effectively facilitate the delivery of a wide range of therapeutic modalities. Not only have NDs been able to augment the efficacy of chemotherapeutics, but they have also been engineered as a vehicle for gene delivery in the treatment of cancer. NDs functionalized with polyethylenimine (PEI) have demonstrated *in vitro* transfection of both plasmid DNA (pDNA) and small interfering RNA (siRNA) (Figure 7.7). The

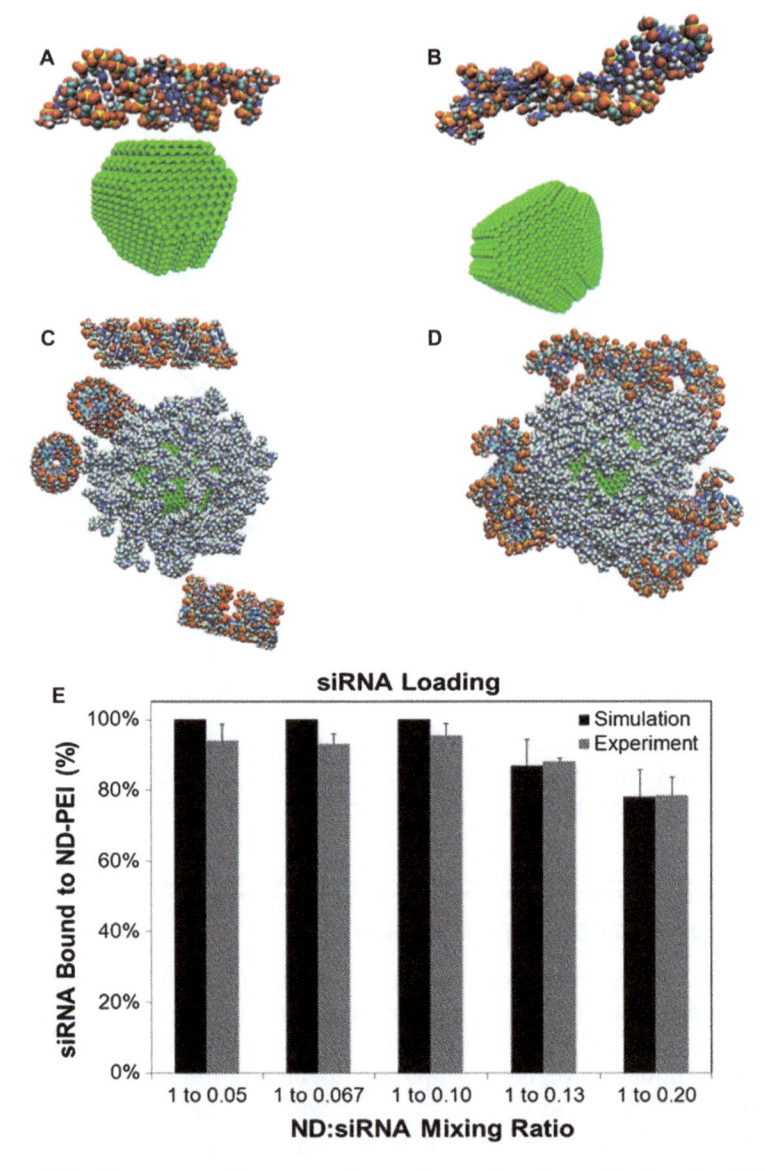

Figure 7.7 The sequential process of nanodiamond–polymer binding in relation to functionalization with siRNA. Complementary experimental validation with simulation/computation can mediate the optimization of gene delivery. A–D demonstrate gradual interaction of polyethylenimine (PEI) and siRNA coordination around the nanodiamond surface. E gives confirmation of material safety.
Reprinted with permission from the American Chemical Society.[32]

combination of high efficiency and low cytotoxicity makes ND-based gene delivery systems attractive. However, *in vivo* gene delivery is currently confronted with many challenges; for example, siRNA is easily degraded in serum, and some carriers, such as polymer-based transfection agents, exhibit a tradeoff between transfection efficiency and toxicity.

Recently, the use of multiscale modeling to simulate ND interactions has provided a framework to improve the design of ND–PEI platforms, with the capability to be extended to other ND vehicles.[32] A comprehensive model of a ND was constructed from first principles to accurately account for the ND surface properties. Quantum-scale simulations were performed to optimize the shape of the structure and the state of the surface graphitization. Titration experiments provided a comprehensive view of the ionization of surface functional groups by determining the effective pKa of an ND solution. Using these inputs, a molecular dynamics (MD) model of NDs was constructed, and its interaction with PEI and siRNA was tested. Various binding ratios of ND and PEI and ND–PEI and siRNA were tested and compared against experimental results. The simulations were able to predict the saturation limits and loading trends of the siRNA sequence for the c-Myc gene, a commonly overexpressed gene in cancer. The comparisons between the modeling results and experimental results demonstrated remarkable similarity, validating the use of simulations as a tool to inform ND–platform design. Utilizing this type of comprehensive model, one of the most accurate simulations of NDs to date, has opened the door for improved efficiency in the understanding and development of novel ND conjugates.

7.8 Concluding Remarks

Nanodiamonds are versatile platforms that can be applied towards drug delivery, imaging, and diagnostics.[33–77] Their scalable processing parameters, rich surface chemistry, biocompatibility, ability to deliver nearly any type of therapeutic, and unique surface facet architectures have resulted in compelling improvements to both the efficacy and safety of a spectrum of applications. By combining experimental and theoretical/computational validation studies to better understand the composition of the ND surface, further optimization of applications, such as chemotherapeutic loading and elution, and gadolinium-based magnetic resonance imaging, may be possible. These findings may further reduce the dosages of drugs or imaging compounds needed for marked improvements in cancer therapy or high-resolution imaging. Furthermore, continued safety/biocompatibility studies will further forge a developmental roadmap for the continued translation of NDs towards a clinical development pathway. It is envisioned that the fruition of ND agents will result in substantial improvements towards the management of human health in the coming generations.

Acknowledgements

D.H. gratefully acknowledges support from the National Science Foundation CAREER Award (CMMI-0846323), Center for Scalable and Integrated Nano-Manufacturing (DMI-0327077), CMMI-0856492, DMR-1105060, V Foundation for Cancer Research Scholars Award, Wallace H. Coulter Foundation Translational Research Award, Society for Laboratory Automation and Screening (SLAS) Endowed Fellowship, Beckman Coulter, National Cancer Institute grant U54CA151880 (The content is solely the responsibility of the authors and does not necessarily represent the official views of the National Cancer Institute or the National Institutes of Health), and European Commission funding program FP7-KBBE-2009-3. E.C. gratefully acknowledges support from the NUS CSI RCE Main Grant. H.B.M. gratefully acknowledges support from the Northwestern University Mechanical Engineering Department for the Walter P. Murphy fellowship, terminal year Cabell fellowship, and Predictive Science and Engineering Design (PSED) fellowship.

References

1. E. K. Chow, X.-Q. Zhang, M. Chen, R. Lam, E. Robinson, H. Huang, *et al.*, Nanodiamond therapeutic delivery agents mediate enhanced chemoresistant tumor treatment, *Sci. Transl. Med.*, 2011, **3**, 73ra21.
2. X-Q. Zhang, R. Lam, X. Xu, E. K. Chow, H.-J. Kim and D. Ho, Multimodal nanodiamond drug delivery carriers for selective targeting, imaging, and enhanced chemotherapeutic efficacy, *Advanced Materials*, 2011, **23**, 4770–4775.
3. R. Lam, M. Chen, E. Pierstorff, H. Huang, E. Osawa and D. Ho, Nanodiamond-embedded microfilm devices for localized chemotherapeutic elution, *ACS Nano*, 2008, **2**, 2095–2102.
4. A. Alhaddad, M.-P. Adam, J. Botsoa, G. Dantelle, S. Perruchas, T. Gacoin, *et al.*, Nanodiamond as a vector for sirna delivery to Ewing sarcoma cells, *Small*, 2011, **7**, 3087–3095.
5. H. B. Man, R. Lam, M. Chen, E. Osawa and D. Ho, Nanodiamond-therapeutic complexes embedded within poly(ethylene glycol) diacrylate hydrogels mediating sequential drug elution, *Physica Status Solidi (A)*, 2012, **209**, 1811–1818.
6. M. Chen, E. D. Pierstorff, R. Lam, S.-Y. Li, H. Huang, E. Osawa, *et al.*, Nanodiamond-mediated delivery of water-insoluble therapeutics, *ACS Nano*, 2009, **3**, 2016–2022.
7. D. Ho, Beyond the sparkle: The impact of nanodiamonds as biolabeling and therapeutic agents, *ACS Nano*, 2009, **3**, 3825–3829.
8. H. Huang, E. Pierstorff, E. Osawa and D. Ho, Active nanodiamond hydrogels for chemotherapeutic delivery, *Nano Letters,*, 2007, **7**, 3305–3314.
9. H. Huang, E. Pierstorff, E. Osawa and D. Ho, Protein-mediated assembly of nanodiamond hydrogels into a biocompatible and biofunctional multilayer nanofilm, *ACS Nano*, 2008, **2**, 203–212.
10. N. Gibson, T.-J. Luo, O. Shenderova, A. Koscheev and D. Brenner, Electrostatically mediated adsorption by nanodiamond and nanocarbon particles, *J. Nanoparticle Research*, 2012, **14**, 1–12.

11. M. Chen, X.-Q. Zhang, H. B. Man, R. Lam, E. K. Chow and D. Ho, Nanodiamond vectors functionalized with polyethylenimine for sirna delivery, *J. Phys. Chem. Lett.*, 2010, **1**, 3167–3171.
12. L. Lai and A. S. Barnard, Interparticle interactions and self-assembly of functionalized nanodiamonds, *J. Phys. Chem. Lett.*, 2012, **3**, 896–901.
13. V. N. Mochalin, I. Neitzel, B. J. M. Etzold, A. Peterson, G. Palmese and Y. Gogotsi, Covalent incorporation of aminated nanodiamond into an epoxy polymer network, *ACS Nano*, 2011, **5**, 7494–7502.
14. Q. Zhang, V. N. Mochalin, I. Neitzel, I. Y. Knoke, J. Han, C. A. Klug, et al., Fluorescent plla-nanodiamond composites for bone tissue engineering, *Biomaterials*, 2011, **32**, 87–94.
15. Y. R. Chang, H. Y. Lee, K. Chen, C. C. Chang, D. S. Tsai, C. C. Fu, et al., Mass production and dynamic imaging of fluorescent nanodiamonds, *Nature Nanotechnology*, 2008, **3**, 284–288.
16. V. N. Mochalin and Y. Gogotsi, Wet chemistry route to hydrophobic blue fluorescent nanodiamond, *J. Am. Chem. Soc.*, 2009, **131**, 4594.
17. Y.-R. Chang, H.-Y. Lee, K. Chen, C.-C. Chang, D.-S. Tsai, C.-C. Fu, et al., Mass production and dynamic imaging of fluorescent nanodiamonds, *Nature Nanotechnology*, 2008, **3**, 284–288.
18. L. M. Manus, D. J. Mastarone, E. A. Waters, X.-Q. Zhang, E. A. Schultz-Sikma, K. W. MacRenaris, et al., Gd(iii)-nanodiamond conjugates for MRI contrast enhancement, *Nano Letters*, 2009, **10**, 484–489.
19. O. Faklaris, V. Joshi, T. Irinopoulou, P. Tauc, M. Sennour, H. Girard, et al., Photoluminescent diamond nanoparticles for cell labeling: Study of the uptake mechanism in mammalian cells, *ACS Nano*, 2009, **3**, 3955–3962.
20. N. Mohan, C.-S. Chen, H.-H. Hsieh, Y.-C. Wu and H.-C. Chang, In vivo imaging and toxicity assessments of fluorescent nanodiamonds in caenorhabditis elegans, *Nano Letters*, 2010, **10**, 3692–3699.
21. N. Mohan, Y.-K. Tzeng, L. Yang, Y.-Y. Chen, Y. Y. Hui, C.-Y. Fang, et al., Sub-20-nm fluorescent nanodiamonds as photostable biolabels and fluorescence resonance energy transfer donors, *Advanced Materials*, 2010, **22**, 843.
22. A. M. Schrand, H. J. Huang, C. Carlson, J. J. Schlager, E. Osawa, S. M. Hussain, et al., Are diamond nanoparticles cytotoxic? *J. Phys. Chem. B*, 2007, **111**, 2–7.
23. Y. Yuan, Y. Chen, J.-H. Lui, H. Wang and Y. Liu, Biodistribution and fate of nanodiamonds in vivo, *Diamond and Related Materials*, 2009, **18**, 95–100.
24. X. Zhang, W. Hu, J. Li, L. Tao and Y. Wei, A comparative study of cellular uptake and cytotoxicity of multi-walled carbon nanotubes, graphene oxide, and nanodiamond, *Toxicology Research*, 2012, **1**, 62–68.
25. M. Hartmann, P. Betz, Y. Sun, S. N. Gorb, T. K. Lindhorst and A. Krueger, Saccharide-modified nanodiamond conjugates for the efficient detection and removal of pathogenic bacteria, *Chemistry – A European Journal*, 2012, **18**, 6485–6492.

26. A. S. Barnard and M. Sternberg, Substitutional nitrogen in nanodiamond and bucky-diamond particles, *J. Phys. Chem. B*, 2005, **109**, 17107–17112.
27. A. S. Barnard and M. Sternberg, Crystallinity and surface electrostatics of diamond nanocrystals, *J. Mater. Chem.*, 2007, **17**, 4811–4819.
28. L. Lai and A. S. Barnard, Modeling the thermostability of surface functionalisation by oxygen, hydroxyl, and water on nanodiamonds, *Nanoscale*, 2011, **3**, 2566–2575.
29. A. S. Barnard, Self-assembly in nanodiamond agglutinates, *J. Mater. Chem.*, 2008, **18**, 4038–4041.
30. E. Ōsawa, D. Ho, H. Huang, M. V. Korobov and N. N. Rozhkova, Consequences of strong and diverse electrostatic potential fields on the surface of detonation nanodiamond particles, *Diamond and Related Materials*, 2009, **18**, 904–909.
31. A. Adnan, R. Lam, H. Chen, J. Lee, D. J. Schaffer, A. S. Barnard, *et al.*, Atomistic simulation and measurement of ph dependent cancer therapeutic interactions with nanodiamond carrier, *Molecular Pharmaceutics*, 2010, **8**, 368–374.
32. H. Kim, H. B. Man, B. Saha, A. M. Kopacz, O.-S. Lee, G. C. Schatz, *et al.*, Multiscale simulation as a framework for the enhanced design of nanodiamond-polyethylenimine-based gene delivery, *J. Phys. Chem. Lett.*, 2012, **3**, 3791–3797.
33. K. D. Behler, A. Stravato, V. Mochalin, G. Korneva, G. Yushin and Y. Gogotsi, Nanodiamond-polymer composite fibers and coatings, *ACS Nano*, 2009, **3**, 363–369.
34. G. Dantelle, A. Slablab, L. Rondin, F. Lainé, F. Carrel, P. Bergonzo, *et al.*, Efficient production of nv colour centres in nanodiamonds using high-energy electron irradiation, *J. Luminescence*, 2010, **130**, 1655–1658.
35. C. Bradac, T. Gaebel, N. Naidoo, M. J. Sellars, J. Twamley, L. J. Brown, *et al.*, Observation and control of blinking nitrogen-vacancy centres in discrete nanodiamonds, *Nature Nano.*, 2010, **5**, 345–349.
36. X. Q. Zhang, M. Chen, R. Lam, X. Y. Xu, E. Osawa and D. Ho, Polymer-functionalized nanodiamond platforms as vehicles for gene delivery, *ACS Nano,*, 2009, **3**, 2609–2616.
37. J.-P. Boudou, P. A. Curmi, F. Jelezko, J. Wrachtrup, P. Aubert, M. Sennour, *et al.*, High yield fabrication of fluorescent nanodiamonds, *Nanotechnology*, 2009, **20**, 235602.
38. V. Petráková, A. Taylor, I. Kratochvílová, F. Fendrych, J. Vacík, J. Kučka, *et al.*, Luminescence of nanodiamond driven by atomic functionalization: Towards novel detection principles, *Adv. Func. Mater.*, 2012, **22**, 812–819.
39. V. Bondar, I. Pozdnyakova and A. Puzyr', Applications of nanodiamonds for separation and purification of proteins, *Phys. Solid State*, 2004, **46**, 758–760.
40. N. Gibson, O. Shenderova, T. J. M. Luo, S. Moseenkov, V. Bondar, A. Puzyr, *et al.*, Colloidal stability of modified nanodiamond particles, *Diamond and Related Materials,*, 2009, **18**, 620–626.

41. S. C. Hens, G. Cunningham, T. Tyler, S. Moseenkov, V. Kuznetsov and O. Shenderova, Nanodiamond bioconjugate probes and their collection by electrophoresis, *Diamond and Related Materials*, 2008, **17**, 1858–1866.
42. C. C. Fu, H. Y. Lee, K. Chen, T. S. Lim, H. Y. Wu, P. K. Lin, *et al.*, Characterization and application of single fluorescent nanodiamonds as cellular biomarkers, *Proc. Natl Acad. Sci. U S A*, 2007, **104**, 727–732.
43. B. Guan, F. Zou and J. Zhi, Nanodiamond as the ph-responsive vehicle for an anticancer drug, *Small*, 2010, **6**, 1514–1519.
44. P. Zhang, J. Yang, W. Li, W. Wang, C. Liu, M. Griffith, *et al.*, Cationic polymer brush grafted-nanodiamond via atom transfer radical polymerization for enhanced gene delivery and bioimaging, *J. Mater. Chem.*, 2011, **21**, 7755–7764.
45. A. M. Kopacz, N. A. Patankar and WK. Liu, The immersed molecular finite element method, *Comp. Method, Appl. Mech. Eng.*, 2012, **233–236**, 28–39.
46. I. Kovalenko, D. G. Bucknall and G. Yushin, Detonation nanodiamond and onion-like-carbon-embedded polyaniline for supercapacitors, *Adv. Func. Mater.*, 2010, **20**, 3979–3986.
47. I. Kratochvilova, A. Kovalenko, F. Fendrych, V. Petrakova, S. Zalis and M. Nesladek, Tuning of nanodiamond particles' optical properties by structural defects and surface modifications: Dft modelling, *J. Mater. Chem.*, 2011, **21**, 18248–18255.
48. A. Krueger, New carbon materials: Biological applications of functionalized nanodiamond materials, *Chemistry – An European Journal*, 2008, **14**, 1382–1390.
49. R. A. Shimkunas, E. Robinson, R. Lam, S. Lu, X. Xu, X.-Q. Zhang, *et al.*, Nanodiamond–insulin complexes as pH-dependent protein delivery vehicles, *Biomaterials*, 2009, **30**, 5720–5728.
50. A. M. Schrand, S. A. C. Hens and O. A. Shenderova, Nanodiamond particles: Properties and perspectives for bioapplications, *Critical Rev. Solid State Mater. Sci.*, 2009, **34**, 18–74.
51. A. Kruger, Y. J. Liang, G. Jarre and J. Stegk, Surface functionalisation of detonation diamond suitable for biological applications, *J. Mater. Chem.*, 2006, **16**, 2322–2328.
52. Y. Zhu, J. Li, W. Li, Y. Zhang, X. Yang, N. Chen, *et al.*, The biocompatibility of nanodiamonds and their application in drug delivery systems, *Theranostics*, 2012, **2**, 302–312.
53. Y. Liang, M. Ozawa and A. Krueger, A general procedure to functionalize agglomerating nanoparticles demonstrated on nanodiamond, *ACS Nano*, 2009, **3**, 2288–2296.
54. X. Y. Xu, Z. M. Yu, Y. W. Zhu and B. C. Wang, Influence of surface modification adopting thermal treatments on dispersion of detonation nanodiamond, *J. Solid State Chem.*, 2005, **178**, 688–693.
55. K. K. Liu, C. L. Cheng, C. C. Chang and J. I. Chao, Biocompatible and detectable carboxylated nanodiamond on human cell, *Nanotechnology*, 2007, **18**, 10.

56. U. Maitra, K. E. Prasad, U. Ramamurty and C. N. R. Rao, Mechanical properties of nanodiamond-reinforced polymer-matrix composites, *Solid State Communications,*, 2009, **149**, 1693–1697.
57. H. B. Man and D. Ho, Diamond as a nanomedical agent for versatile applications in drug delivery, imaging, and sensing, *Physica Status Solidi (A)*, 2012, **209**, 1609–1618.
58. O. A. Williams, M. Nesladek, M. Daenen, S. Michaelson, A. Hoffman, E. Osawa, *et al.*, Growth, electronic properties and applications of nanodiamond, *Diamond and Related Materials*, 2008, **17**, 1080–1088.
59. S.-J. Yu, M.-W. Kang, H.-C. Chang, K.-M. Chen and Y.-C. Yu, Bright fluorescent nanodiamonds: No photobleaching and low cytotoxicity, *J. Am. Chem. Soc.*, 2005, **127**, 17604–17605.
60. H. J. Huang, L. M. Dai, D. H. Wang, L. S. Tan and E. Osawa, Large-scale self-assembly of dispersed nanodiamonds, *J. Mater. Chem*, 2008, **18**, 1347–1352.
61. V. N. Mochalin, O. Shenderova, D. Ho and Y. Gogotsi, The properties and applications of nanodiamonds, *Nature Nano.*, 2012, **7**, 11–23.
62. M. Ozawa, M. Inaguma, M. Takahashi, F. Kataoka, A. Krüger and E. Ōsawa, Preparation and behavior of brownish, clear nanodiamond colloids, *Adv. Mater.*, 2007, **19**, 1201–1206.
63. L. Schmidlin, V. Pichot, M. Comet, S. Josset, P. Rabu and D. Spitzer, Identification, quantification and modification of detonation nanodiamond functional groups, *Diamond and Related Materials*, 2012, **22**, 113–117.
64. A. Krueger, J. Stegk, Y. Liang, L. Lu and G. Jarre, Biotinylated nanodiamond: Simple and efficient functionalization of detonation diamond, *Langmuir*, 2008, **24**, 4200–4204.
65. O. Shenderova, C. Jones, V. Borjanovic, S. Hens, G. Cunningham, S. Moseenkov, *et al.*, Detonation nanodiamond and onion-like carbon: Applications in composites, *Physica Status Solidi (A)*, 2008, **205**, 2245–2251.
66. O. A. Shenderova, V. V. Zhirnov and D. W. Brenner, Carbon nanostructures, *Critical Rev. Solid State Mater. Sci.*, 2002, **27**, 227–356.
67. A. Krueger, Beyond the shine: Recent progress in applications of nanodiamond, *J. Mater. Chem.*, 2011, **21**, 12571–12578.
68. A. H. Smith, E. M. Robinson, X.-Q. Zhang, E. K. Chow, Y. Lin, E. Osawa, *et al.*, Triggered release of therapeutic antibodies from nanodiamond complexes, *Nanoscale*, 2011, **3**, 2844–8.
69. J. Tisler, G. Balasubramanian, B. Naydenov, R. Kolesov, B. Grotz, R. Reuter, *et al.*, Fluorescence and spin properties of defects in single digit nanodiamonds, *ACS Nano,*, 2009, **3**, 1959–1965.
70. V. Vaijayanthimala, Y.-K. Tzeng, H.-C. Chang and C.-L. Li, The biocompatibility of fluorescent nanodiamonds and their mechanism of cellular uptake, *Nanotechnology*, 2009, **20**, 425103.
71. Y. Xing and L. Dai, Nanodiamonds for nanomedicine, *Nanomedicine*, 2009, **4**, 207–218.
72. K.-K. Liu, W.-W. Zheng, C.-C. Wang, Y.-C. Chiu, C.-L. Cheng, Y.-S. Lo, *et al.*, Covalent linkage of nanodiamond-paclitaxel for drug delivery and cancer therapy, *Nanotechnology*, 2010, **21**, 315106.

73. X. Y. Xu, Y. W. Zhu, B. C. Wang, Z. M. Yu and S. Z. Xie, Mechanochemical dispersion of nanodiamond aggregates in aqueous media, *J. Mater. Sci. Technol.*, 2005, **21**, 109–112.
74. J. Yan, Y. Guo, A. Altawashi, B. Moosa, S. Lecommandoux and NM. Khashab, Experimental and theoretical evaluation of nanodiamonds as ph triggered drug carriers, *New J. Chem.*, 2012, **36**, 1479–1484.
75. W. S. Yeap, Y. Y. Tan and K. P. Loh, Using detonation nanodiamond for the specific capture of glycoproteins, *Analytical Chem.*, 2008, **80**, 4659–4665.
76. L. C. L. Huang and H.-C. Chang, Adsorption and immobilization of cytochrome c on nanodiamonds, *Langmuir*, 2004, **20**, 5879–5884.
77. L.-Y. Chang, E. Osawa and A. S. Barnard, Confirmation of the electrostatic self-assembly of nanodiamonds, *Nanoscale*, 2011, **3**, 958–962.

CHAPTER 8

Biophysical Interaction of Nanodiamond with Biological Entities In Vivo

J. MONA,[a] E. PEREVEDENTSEVA[a,b] AND C.-L. CHENG*[a]

[a] Department of Physics, National Dong Hwa University, No.1 Sec. 2 Da-Hsueh Rd., Shoufeng, Hualien, 97401, Taiwan;
[b] P. N. Lebedev Physics Institute, Rus. Acad. Sci., Moscow, Russia
*Email: clcheng@mail.ndhu.edu.tw

8.1 Introduction

Day-by-day improvements in medical science depend on the discovery of new methodologies, which essentially require new materials for diagnosis through to treatment. Amongst the various nanostructures identified, nanometer-sized diamond particles (nanodiamonds, NDs) stand ahead due to their special structure and surface, high chemical stability, biocompatibility and widespread physical–chemical properties.[1–6] The unique surface of NDs, with large amounts of surface atoms, makes them amenable to functionalization/conjugation with various biologically and medically relevant molecules/drugs through physical adsorption and chemical linking.[7,8] NDs do not induce significant cytotoxicity in a variety of animal and human cells, such as lung,[9,10] neuronal,[11] renal,[5,12] cervical,[13] immune (macrophage),[14] stem,[10,15] blood[16] cells and whole blood.[17] Functionalizing the ND surface with cellular targeting elements (*e.g.*, antibodies, aptamers, *etc.*) adds another layer of advantages to this drug carrier technology, which helps to drive targeted and slow drug release, imaging and biosensing. NDs are promising for diverse biological applications, such as in bioprobes, photoluminescent and Raman biomarkers, or biolabels, biosensors, bio-chips, and for drug delivery (*i.e.*, cancer therapeutics); they have also been explored for use in analytical diagnostics and biomolecule target capturing, *etc.*[18–26] Uptake of NDs by living cells found has facilitated their use as drug carriers, for bio-imaging and as delivery vehicles.[27] Overall, the collective capabilities of NDs have significantly contributed to the rapid progress of research in an ever increasing number of real applications in different developed systems. Nevertheless, due to obvious reasons, consideration of the risks involved in research concerning health is necessary. At present, some of the major concerns are that NDs interact with various biological entities and their cytotoxicity and long-term effects on cells/animals have not been fully elucidated for clinical use. Furthermore, nanoparticle uptake by cells and the mechanisms involved in their cellular uptake and subsequent pathways are not fully understood. In this chapter recent research related to the biophysical interaction of various carbon-based nanoparticles, including NDs, with various cells, tissues and unicellular organisms are highlighted and illustrated.

For the study of any type of biophysical interaction concerning NDs with biological tissues, a first consideration is the surface functionalizations/conjugations of the NDs. So, we start with a discussion of the surface functionalizations/conjuagations of NDs and how this relates to their interaction with biomolecules.

8.2 Nanodiamond Surface Functionalization/Conjugation with Biomolecules

Complex arrays of functional groups are present on the ND surface due to the extreme environment during the production process. These surface functional groups can be modified using chemical or physical approaches, which further provide a unique platform for connecting biological molecules of interest for specific or non-specific interactions with a target.[6,7,19,26,28–37] Physical adsorption has been achieved on NDs of different sizes using proteins, DNA, polypeptides, *etc.* through electrostatic interactions or ionic bonds.[7] For example, adsorption of lysozyme and alpha-bungarotoxin proteins on the surface of carboxylated NDs (cNDs) was achieved *via* non-covalent chemical bonding[38] and the adsorption of apoobelin and luciferase on the surface of NDs was used for the efficient separation and purification of proteins.[39] In addition, hydrophilic ND powders coated with poly-L-lysine for the non-covalent surface amination of proteins and fluorescent labeling has been demonstrated.[7] Covalent grafting has also been demonstrated, for example, with biotin (a biologically active moiety), which was attached to the surface of NDs with a high surface loading. In this case an increased stability of the resulting conjugates was also achieved.[8] Covalent (chemical) conjugation was also attained when the NDs were functionalized with immunoglobulin, bovine serum albumin, rabbit anti-mouse antibody, etc.[40]

New approaches for ND functionalization/conjugation include the "click" chemistry approach, which has shown scope for attaching organic moieties on ND surfaces, as well as the grafting of larger proteins, DNA strands, glycomoieties, antibodies, etc.[41–43] For a non-hindered drug release, surface functionalization has been achieved by coupling azomethine ylides using the Prato reaction.[44] For the stable grafting of organic moieties onto the surface of NDs, different types of o-quinodimethanes have been used and the possibility of subsequent modifications with reactions on the aromatic rings formed during grafting has been shown.[45] Functionalization for the purpose of the selective capture of a certain kind of organic molecule with aminophenylboronic acid to selectively capture glycoproteins from unfractionated protein mixtures has been demonstrated.[46] A hydrogel of NDs with chemotherapeutic drugs, such as doxorubicin (DOX), was developed by adsorption and introduced into a murine liver tumor, as well as mammary carcinoma models, facilitating enhanced chemotherapeutic efficacy through improved drug delivery.[47,48] The effect of pH on the delivery of DOX from fluorescein-labeled NDs (Fc-NDs) has been investigated using experimental and theoretical methods. Cell viability results have shown that Fc-NDs are safe to use as drug carriers in both neutral and acidic conditions and Fc-NDs have a promising dual potential of imaging and controlled drug release at the same time.[49] DOX has also been coated on fluorescent NDs (FNDs) by physical adsorption to study drug delivery and as a fluorescent probe to investigate the interactions and pathways with cells.[50] Chen et al. developed a novel route using water-insoluble compounds in treatment-relevant scenarios using several therapeutics, such as purvalanol A, 4-hydroxytamoxifen and dexamethasone, to enhance the dispersive properties of NDs in water, thereby preserving their functionality.[51] Functional covalent conjugation of ND–paclitaxel, which could be delivered into lung carcinoma cells, was found to preserve the anticancer activity of paclitaxel (mitotic blockage, apoptosis and anti-tumor genesis) by Liu et al.[52] Recently, heterofunctional NDs were developed by attaching fluorescently labeled paclitaxel–DNA conjugates and anti-human epidermal growth factor monoclonal antibodies onto NDs for multimodal imaging and therapeutically relevant applications by Zhang et al.[53] The enhanced therapeutic efficacy and specific internalization within breast cancer cell lines was observed.

The anticancer drug cis-dichlorodiamminplatinum (II) (CDDP) was conjugated to NDs by adsorption and complexation. Results demonstrated that the released CDDP retained its anticancer activity, as observed from the inhibition of proliferation in human cervical cancer (HeLa) cells.[54] NDs have been covalently functionalized with amine groups using (3-aminopropyl)-trimethoxysilane or the surface immobilization of 800 Da polyethyleneimine to introduce NDs as delivery vehicles for gene therapy.[55] Furthermore, detonation NDs have been coated with DNA, a synthetic ethylene precursor or an ethylene antagonist, or were labeled with fluorescent dyes for their use as "nanobullets" for ballistic delivery in the work by Grichko et al.[56] Controlled delivery of functionalized NDs into cells has also been demonstrated by Loh et al.[57] They used a broadly applicable nanofountain probe, which is a tool for direct-write nanopatterning, with sub-100 nm resolution and also investigated its use in direct in vitro single-cell injections. Attempts using diamond nanofilms as biosensors and bio-arrays have been reported by Huang et al.[30] Immobilization of antibodies and bacterial binding on NDs for biosensor applications has been discussed by Smirnov et al.[58] They fabricated aligned diamond nanowires for these investigations. Zhang et al. reported the role of ND surface functionalization to achieve uniform dispersion and high affinity between the components of the composite, leading to improved mechanical properties, which allowed NDs to be used for bone tissue engineering.[59] Methods have been developed to produce NDs embedded within polyethylene glycol diacrylate hydrogels for localized drug release.[60] Continued research into the development of new methods and strategies to functionalize/conjugate NDs will provide prospects to attach an increasing range of therapeutic compounds.

8.3 Essential Characterizations for the Biophysical Interactions of Nanodiamond

During detonation or shock wave synthesis, NDs are found to agglomerate and surface reconstruction takes place giving rise to sp^2 hybridization and graphitization. Some agglomeration can be overcome by ultrasonic treatments, milling methods, oxidation of detonated ND in air, etc.[61,62] Additionally, chemical functional groups, such as carbonyl groups, ethers, hydroxyl structures, carboxylic acids, ketones, lactones, simple carbonyls and, in some cases, amides and esters, are present at the ND surface.[29,63-67] These functional groups can be controlled by oxidation/carboxylation (with a mixture of concentrated H_2SO_4 and HNO_3), reduction (with $LiAlH_4$ in a solution of $BH_3 \cdot THF$) or hydrogenation of the ND surface with hydrogen-plasma, followed by oxidation, photochemical functionalization or a combination of these approaches. The surface groups can be analyzed using IR spectroscopy.[19,29,31,32,63,68,69] The IR spectra of carboxylated/oxidized NDs measured under ambient conditions and in a low vacuum (10^{-4} torr) are presented in Figure 8.1.[19] Spectra are plotted using the absorbance signal and can be directly compared with standard molecular vibrations.

Molecular groups existing on the ND surface allow biomolecules to be conjugated. The ND spectroscopic signals (Raman and fluorescence) are used for detection and allow these interactions with the target to be mapped.[70-78] The ND lattice structure provides a unique Raman signal (~ 1332 cm^{-1}) for the phonon mode of sp^3-bonded carbon, which is sharp and isolated, and can be excited with a wide range of wavelengths. NDs can interact with cells and the distribution of NDs in the cell can be observed via Raman mapping. Figure 8.2 shows examples of the typical Raman spectra

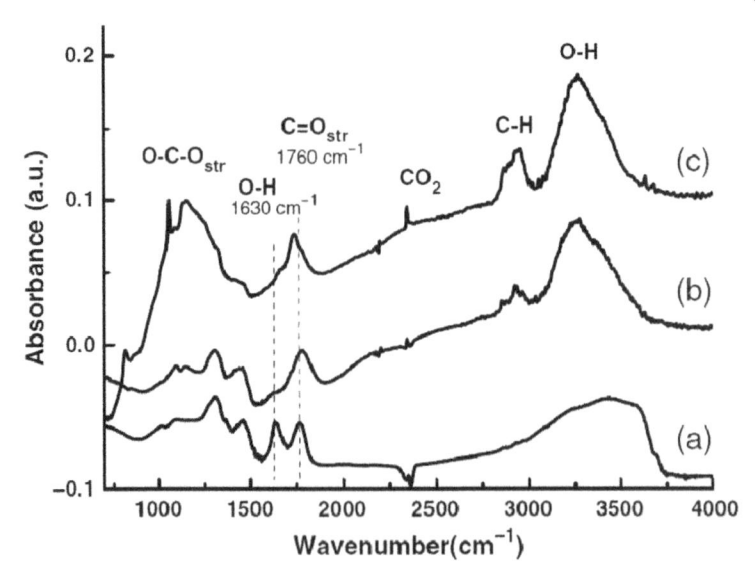

Figure 8.1 IR spectra of carboxylated/oxidized NDs: (a) 100 nm; initial concentration in suspension = 20 mg/60 μL; spectrum taken in ambient air; (b) concentration = 1.4 mg/200 μL; spectrum taken in a vacuum; (c) 5 nm; concentration = 1 mg/200 μL, spectrum taken in a vacuum.[19]

measured in every pixel of a mapped area of a sample and the spatial distribution of the intensity of some characteristic peaks. This mapping of the Raman signal was in the range of 1300–1350 cm^{-1}.

The diamond peak at 1332 cm^{-1} is characteristic of the NDs distribution in the cell, while other wavenumber ranges are characteristic of the cell's molecular structures. The ease with which ND penetration inside the cells can be determined and the intensity of the spectroscopic signal makes NDs promising for bio-labeling and tracers of drug delivery. Attempts to use NDs as a Raman label were initially successfully demonstrated by Cheng et al.[34] The interaction between the surface growth hormone receptor of A549 human lung epithelial cells and growth hormone chemically attached to 100 nm NDs was observed *via* confocal Raman mapping. The Raman spectroscopic signal of the NDs provided direct observation of the growth hormone receptor localization under physiological conditions on the single-cell level.

The photoluminescence (PL) of NDs is also a subject of extensive study. It is determined by a variety of luminescent centers associated with the crystal lattice defect admixtures and the nanosize effect.[75–78] Methods to increase the luminescence of NDs using high energy beam treatment followed by high temperature annealing is under development,[79–83] and various methods to produce FNDs have been reported.[78,82] So far, the possibility of using NDs as fluorescent cellular labels through the intrinsic fluorescence of pristine NDs with enhanced fluorescence has been demonstrated using confocal scanning microscopy. Chao et al. demonstrated that

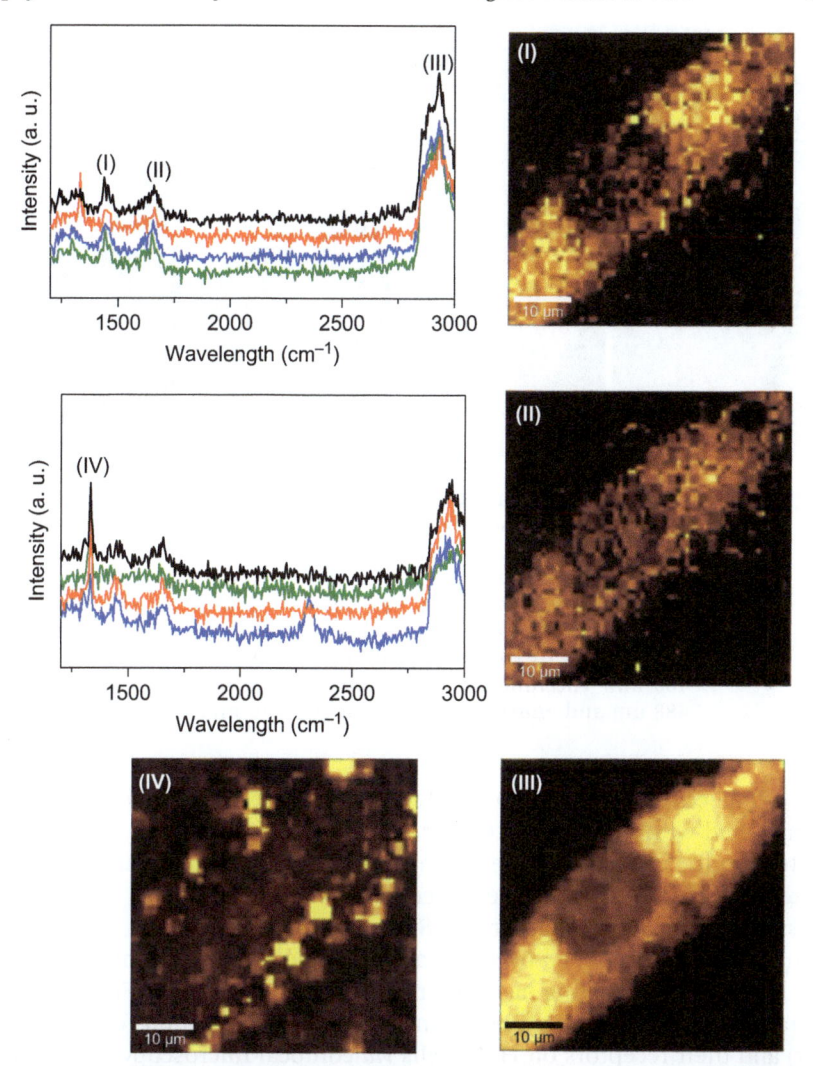

Biophysical Interaction of Nanodiamonds with Biological Entities In Vivo 175

Figure 8.2 Raman mapping images with distribution of the signal in different characteristic spectral ranges and the corresponding spectra of human lung fibroblast (HFL-1) cells with NDs (concentration = 10 µg mL^{-1}; incubation = 4 h; Witec α-SNOM (Germany); excitation = 488 nm).

nanometer-sized diamonds could be used as probes for detection of the interactions of nanoparticles with bio-objects, such as cells and bacteria.[84] Figure 8.3 shows the fluorescence images of localized 100 nm NDs in human foetal lung fibroblast (HFL1) cells.

It can be seen that 100 nm cNDs penetrate the cell membrane and are located in the cytoplasm of HFL1 cells, which proves that the inherent fluorescence of cNDs can be used as a probe for detection. Considering the fluorescence properties of NDs, many reports have showed and discussed

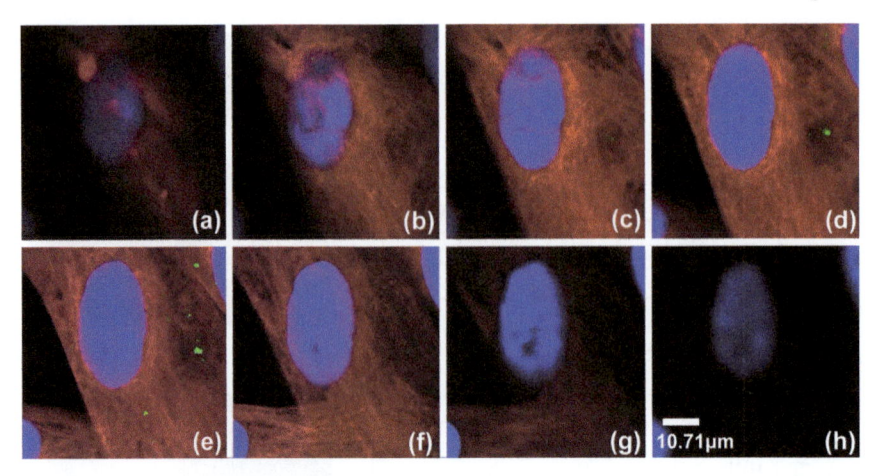

Figure 8.3 Cross-sectional scan of a single HFL1 cell. The figures represent a series of confocal fluorescence images at changing position in the z direction from the top (a) to the bottom (h). Confocal fluorescence images of HFL1 cells after incubation with 100 nm cNDs (1 μg mL^{-1} in DMEM) for 48 h. The cytoplasm has been died with anti-β-tubulin (Cy3). Excitation = 543 nm; emission = 550–615 nm. The nuclei have been died with Hoechst 33258. Excitation = 351 nm; emission = 355–460 nm. The inherent fluorescence of 100 nm NDs was excited with 488 nm and emission collected at 500–530 nm.

the perspectives of using ND and FND for biolabeling. NDs have been bioconjugated with target intracellular structures, such as actin filaments and mitochondria, and transfected into HeLa cells. This conjugation has been detected using the ND fluorescence, showing that they can be utilized as possible cellular biomarkers.[85] Super-resolution imaging of albumin-conjugated FNDs in cells by stimulated emission depletion microscopy and for homogenous labeling has been reported by Tzeng et al.[86] FNDs was used as targeted fluorescence probes to investigate the interactions of transferrins (Tf) and their receptors on HeLa cells *via* confocal microscopy.[87] Optiz et al. reported the receptor-mediated uptake of FND-Tf bioconjugates into HeLa cells, which was confirmed by confocal microscopy, where the NDs were conjugated with antibody for membrane binding.[88] Long-term *in vivo* imaging of FNDs in wild-type *Caenorhabditis elegans* was achieved by Mohan et al.[89] For mammalians, ND entrapment in the liver was observed following intravenous injection in mice and was tracked using spectroscopic methods by Yuan et al.[90] and Zhang et al.[91]

Additionally, new methods to detect NDs in cells have also been developed. Smith et al. used NDs as scattering optical labels in a biological environment.[92] NDs were efficiently transfected into cells using cationic liposomes and imaged using differential interference and Hoffman modulation "space" contrast microscopy techniques. An innovative approach in the field of bio-imaging using NDs was presented by

Balasubramanian et al.[93] They showed that magneto-optical spin detection, with nanometer resolution under ambient conditions, can be used to determine the location of a spin associated with a single nitrogen–vacancy (NV) center in diamond. This shows that a single NV center in a ND crystal, which is highly sensitive to magnetic fields, has great potential for use as a magnetometer for nanobiosensing. The use of NDs as single-atom quantum probes for nanoscale processes, utilizing the quantum behavior of the NV centers in ND, has been discussed theoretically in detail.[94] The quantum dynamics of a NV probe have been explored when in proximity to ion channels and the lipid bilayer, as well as its surrounding aqueous environment. The sensitivity of the NV centers' decoherence to various magnetic field sources has been estimated and indicated the ability to detect single ion-channel switch-on/off events. Decoherence refers to the loss of quantum coherence between magnetic sublevels of the NV atomic system due to interactions with an environment. Theoretical results indicated that the real-time detection of an ion-channel in operation at millisecond resolution is possible by directly monitoring the quantum decoherence of the NV probe with nanometer spatial resolution. The ND surface and adsorption properties have allowed these nanostructures to effectively capture proteins for mass spectrometry analysis. Matrix-assisted laser desorption/ionization (MALDI) time of flight (TOF) mass spectrometry has been used for analysis. Due to the high affinity of NDs towards proteins, their solutions can be analyzed with very low protein concentrations and the adsorbed protein molecules do not need to be separated from the NDs. Yeap et al. analyzed the attachment of glycoproteins with NDs using the MALDI-TOF-MS method[46] and, to achieve this, NDs were chemically functionalized to specifically bind glycoproteins for separation in a protein matrix. Hens et al. demonstrated the electrophoretic collection of ND–protein complexes for analytical purposes.[95] It has been reported that the high electrophoretic mobility and controlled surface charges of the NDs can make them available for electrophoretic applications and allows them to be used as target-specific capture probes to bind and collect their target using electrophoresis at low electric field strengths.

8.4 Interactions of Nano-carbon with Unicellular Microorganisms

The investigations associated with the effect of NDs on various kinds of unicellular organisms are important to understand the interactions of NDs with living organisms *in vivo*, along with the problems related to bio-applications or nano-biosafety. In the following sections recent investigations that have considered carbon-based nanomaterials, including ND interactions with bacteria and the very first investigation concerning the interaction of ND with ciliated protozoa microorganisms, are discussed.

8.4.1 Interactions with Bacteria

Nanomaterials present themselves as promising agents to counteract the emergence of antibiotic resistant bacteria by preventing and treating infectious diseases. Bacteria, especially those found in diverse environments, provide a particularly valuable model for exploring how single-celled organisms respond to environmental stressors. In order to design effective strategies against bacteria, antibacterial properties of many nanomaterials have been explored. Such nanomaterials include metal- and carbon-based nanomaterials, silver, titanium oxide and zinc oxide nanoparticles, as well as nano-carbons, like carbon nanotubes (CNTs), fullerene derivatives, graphene, *etc.*[96–98] In addition, NDs were also recently recognized as suitable antibacterial agents. These antibacterial agents are important as they have been found vital for a wide range of applications, such as disinfection/microbial control of medical devices, home appliances and also to remove contaminants from water or air, *etc.*[96,99,100] Conversely, ND particles have been explored as antibacterial agents due to their biocompatibility with animal cells and some other microorganisms. The ND biocompatibility differs for various biological constructs and it has also been demonstrated that the cytotoxic mechanisms involved in human/animal cell and microorganism models may be different.

The cytotoxicity of different nanoparticles, in particular, iron oxide nanoparticles, silver nanoparticles, TiO_2, cerium oxide (CeO_2), magnesium oxide (MgO) and zinc oxide (ZnO), *etc.*, has been investigated on gram-negative bacteria, *e.g.*, *Escherichia coli* (*E. coli*) and different gram-positive bacteria.[101–115] Metal and metal–polymer composites as antibacterial agents have also been discussed.[116] Few nanomaterials have been found to be appropriate as antimicrobial agents and these have been used for water disinfection and microbial control and found to be promising. Their potential applications and implications have also been discussed.[96] The antibacterial activity of two water-dispersible graphene derivatives (graphene oxide and reduced graphene oxide nanosheets) was reported to effectively inhibit the growth of *E. coli* bacteria with mild cytotoxicity.[97] The mass production of graphene derivatives and their use for the production of freestanding and flexible paper have previously been discussed. Kang *et al.* reported that the direct contact of *E. coli* K12 bacteria with carbon nanotubes (CNTs) seriously impacted the cellular membrane integrity and metabolic activity of the bacteria.[117] Single-walled nanotubes (SWNTs) were also investigated and exhibited much pronounced antibacterial activity compared to multi-walled CNTs (MWCNTs). Interestingly, the size of the nanotube played an important role in the inactivation of the bacteria. Mechanical damage of the gram-negative bacteria *Acidothiobacillus ferrooxidans* by electroporation of the bacterial cell membrane using highly ordered CNTs has been demonstrated.[118] Although, for many nanoparticles, the antibacterial activity and the multiple factors associated with specific antibacterial activity has been investigated, the underlying mechanisms still remain unclear. Some of the mechanisms used to explain the antibacterial activity of the nanoparticles are described below.

Photoactivated antimicrobial agents generally function by generating lethal reactive oxygen species (ROS) when exposed to light. ROS can be generated in the form of superoxide anions or hydroxyl radicals (type I) or singlet oxygen (type II).[99] ROS acts on multiple targets within microbes, thereby reducing the probability of the emergence of resistance against ROS. There are numerous photoactivate antimicrobial agents, which include titania and its alloys, and titanium oxide, as well as a few other materials and their nanostructures. Metal nanoparticles can directly damage bacterial cell walls and this has been explained using three hypothetical mechanisms: *via* formation of ROS on the surface of particles, by release of metal ions followed by increased membrane permeability and dysfunction, and cytolitic damage.[116] Furthermore, this damage results in morphological changes[119] and changes in the hardness or elasticity can alter cell division and cell motility. Sorption of the nanoparticles on the cell itself, as well as surface charge variations, may alter the adhesion characteristics, *e.g.*, disruption of bacterial colonization.

The mechanisms of the antibacterial action of nanoparticles without formation of ROS has been discussed by Cui *et al.*[120] Figure 8.4 shows a schematic diagram of the mechanism of action of bactericidal gold nanoparticles on *E. coli*.[120] It was reported that the gold nanoparticles induce the down-regulation of the oxidative phosphorylation pathway (F-type ATP synthase and ATP level) and ribosome pathways, and the transient upregulation of chemotaxis.

Figure 8.4 shows that the gold NPs do not induce a change in ROS-related processes. It has been shown that antibacterial agents can target the energy metabolism and transcription of bacteria without triggering the ROS reaction, which may be, at the same time, harmful for the host when killing bacteria.

Compared to various nanoparticles, the mechanisms used to explain the antibacterial activity of carbon-based materials and composite materials,

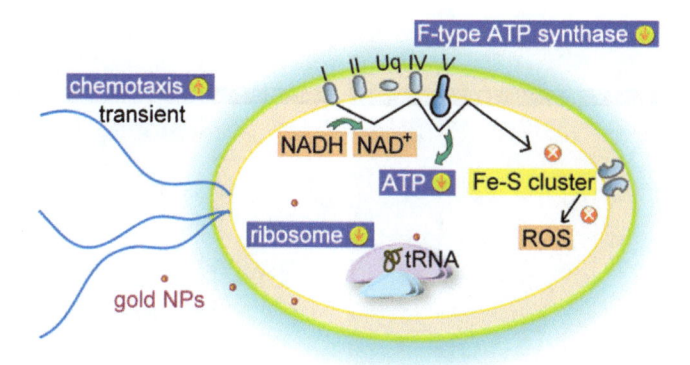

Figure 8.4 Schematic diagram of the mechanism of action of bactericidal gold nanoparticles on *E. coli*. Gold NPs induce the down-regulation of the oxidative phosphorylation pathway (F-type ATP synthase and ATP level) and ribosome pathways, and the transient upregulation of chemotaxis. Gold NPs do not induce a change in ROS-related processes.[120]

such as Ag-CNT, can be interesting as they exert both oxidative and mechanical damage on bacteria.[121,122] Bacterial cell damage and the alteration of gene expression have been observed and compared for single-walled and multi-walled CNTs.[117] Significantly pronounced results were achieved with SWNTs than with MWNTs. The enhanced bacterial toxicity of SWNTs was attributed to (1) a smaller nanotube diameter that facilitates the partitioning and partial penetration of nanotubes through the cell wall, (2) a larger surface area for contact and interaction with the cell surface, and (3) unique chemical and electronic properties conveying greater chemical reactivity. Although some researches suppose that the observed antibacterial effect is because of ROS formation. The antibacterial activity of fullerene water suspensions (nC_{60}) has been reported to not be due to ROS-mediated damage or ROS production.[123] Instead, an alternative hypothesis has been reported, whereby nC_{60} behaves as an oxidant and exerts ROS-independent oxidative stress.

Several reports have shown the antibacterial activity of carbon-based materials, such as CNTs, fullerenes, *etc.*; however, work regarding ND particles/films are rare and to study NDs as antibacterial agents may be promising. This inspired us to study the antibacterial effect of detonated NDs on *E. coli*. Figure 8.5 shows the SEM images demonstrating the interaction of 5 nm NDs with the *E. coli*.

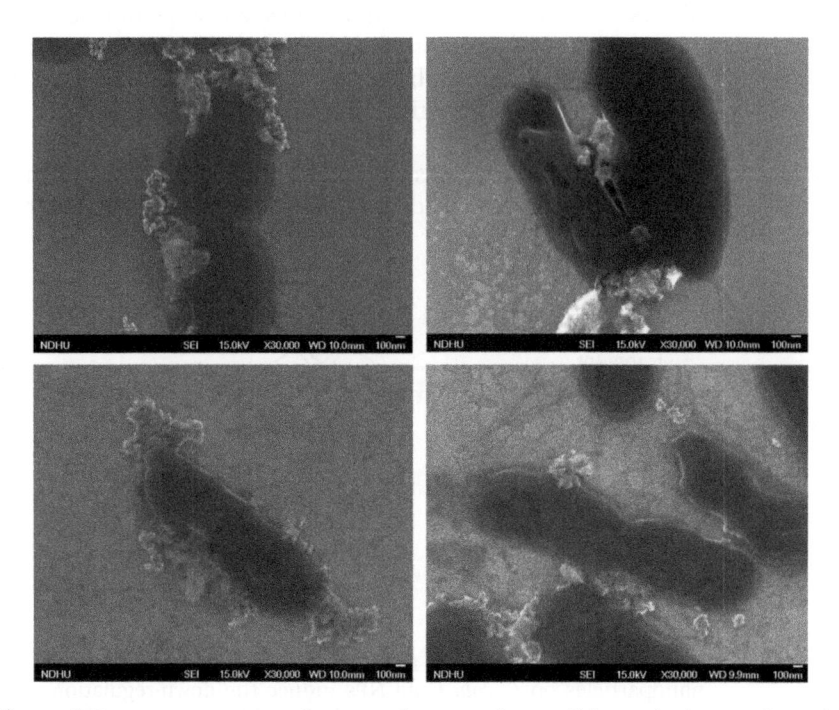

Figure 8.5 Four representative SEM images of *E. coli* bacteria interacting with 5 nm detonation NDs.

The images reveal that the NDs are adsorbed on the bacterial cell wall and substantial changes in the bacterial morphology have occurred. According to our preliminary results, the small sized NDs affect bacteria division and colony formation.

In contrast to what is observed with detonation ND and bacteria, it has been shown that diamond-like carbon coatings (and films) result in low bacterial adhesion and resistance to bacteria colonization compared to medical steel.[124] To further decrease the bacterial adhesion, Si was doped in the carbon coatings.[125] Significant improvement in the antibacterial properties was achieved by improving the diamond structure of the coatings. Diamond surfaces with a nanocrystalline structure have been found to exhibit the highest resistance to bacterial colonization.[126] Medical stainless steel coated with nanocrystalline diamond, titanium and plain medical steel were analyzed for their use in surgery and implant technology thanks to the biocompatibility of NDs with human cells and the antibacterial properties. Overall, the bactericidal acitivity of the diamond coating/film, with a crystalline size of about 20 nm, was sufficient and better than medical steel, Ti and Ag, but not as good as Cu (which, in any case, is less biocompatible than ND).[127] However, microcrystalline diamond with a grain size of more than 500 nm did not show significant antibacterial activity.

In addition to the antibacterial activity of carbon-based nanoparticles, research into other applications, such as the detection and sorption of bacteria, have been demonstrated. CNTs have a high affinity for bacterial cells, which makes them a key candidate for bacteria adsorption. Free-standing monolithic uniform macroscopic hollow cylinders, with radially aligned carbon nanotube walls with diameters and lengths up to several centimeters, have been fabricated by Srivastava *et al.*[100] These cylindrical membranes were used as filters, demonstrating their utility in the filtration of bacterial contaminants, such as *E. coli*, or the nanometer-sized polio-virus (~25 nm) from water. These macro-filters were reportedly cleaned repeatedly after each filtration process *via* ultrasonication and autoclaving to regain their full filtering efficiency. It was mentioned that these prepared carbon nanotube filters have exceptional thermal and mechanical stability, high surface area, easy and cost-effective fabrication, which may allow them to compete with commercially used ceramic- and polymer-based separation membranes. Furthermore, owing to the high binding affinity of CNTs to bacteria and their absorbance in the visible to near-infrared (NIR) region (most bio-tissues are relatively transparent to NIR radiation), CNTs have been developed as photo-thermal laser-activated antibacterial agents.[128] Another example is the potent photo-induced antimicrobial effect of porphyrin-conjugated MWNTs.[98] The light-activated antimicrobial activity of porphyrins against some pathogens and a facile way to incorporate porphyrins into coatings may lead to more effective use. The conjugates were reported to effectively deactivate *Staphylococcus aureus* bacteria in solution upon irradiation with visible light. Other examples include biosensors using SWCNTs for electrochemical detection of bacteria.[129] Taking advantage of the electrical properties of metallic CNTs, these constructs have been developed for the localized electroporation of bacteria through the cell membrane.[118]

Recently, Hartmann et al. reported the synthesis of ND covalently modified with specific carbohydrates (glycol-ND or saccharide-modified ND) for capture, detection and agglutination of pathogenic bacteria.[130] The glycosylated ND interacted with bacterial cells through specific protein–carbohydrate interactions. Depending on the respective surface termination, the ND conjugates can either interact specifically or nonspecifically with the surface of different pathogenic bacterial strains. The agglutination experiments with E. coli allowed corroboration that ND particles are a useful means to flocculate bacteria and remove them by filtration. It was reported to be a low-cost and high-throughput filter material with a large pore size. Moreover, via decoration with specific carbohydrates, it was possible to use NDs for the detection and removal of bacteria in a contaminated sample by the formation of mechanically stable agglutinates. Highly specific anti-salmonella and anti-S. aureus antibodies were immobilized on hydrophobic and hydrophilic NDs and carbon-nanotube-coated silicon substrates.[30] Bacteria binding was observed and analyzed for biosensor applications and express-diagnostics. The labeling and treatment of E. coli using lysozyme-conjugated NDs was demonstrated.[21] cND particles interacted with lysozyme via an electrostatic attraction between the surface-terminated anionic groups (-COO-) and the positively charged amino groups (-NH^{+3}) in the protein to form stable lysozyme–carboxylated ND complexes (lysozyme–cND). The surface bonded protein did not change its conformation upon adsorption to a substrate and no surface-induced changes of the protein function were observed. Figure 8.6 displays the SEM images of the lysozyme–cND complex with bacteria.[21] From the images, it can be seen that E. coli are destroyed by the lysozyme–cND complexes.

The ND Raman signal has been used to label the interaction of protein lysozyme with E. coli via Raman mapping for further confirmation.[21] Figure 8.7(a) shows the optical images for the cND and lysozyme–cND complex interacting with E. coli. Figure 8.7(b) shows the three-dimensional Raman mapping images; the bright spots correspond to the ND signal. The signal from lysozyme is too weak to be detected, thus it is convenient to use the Raman mapping of the ND signal to locate the positions of the lysozyme–cND complexes. Figure 8.7(c) shows the merged images from (a) and (b) and clearly shows that, in contrast to lysosyme–cND, cNDs do not interact with E. coli. The results confirm that NDs can be used efficiently as nano-biolabels (or nano-bioprobes).

A number of research studies have demonstrated the interaction of bacteria and nanoparticles for the development of micro/nano biorobotic systems,[131] which are unicellular organisms integrated with engineered micro- or nano-structures focused on controlled behavior and tracking.[132] In these studies fluorescent or bioluminescent genes are loaded onto the polymer nanoparticles, which adhere to the bacteria surface, allowing the

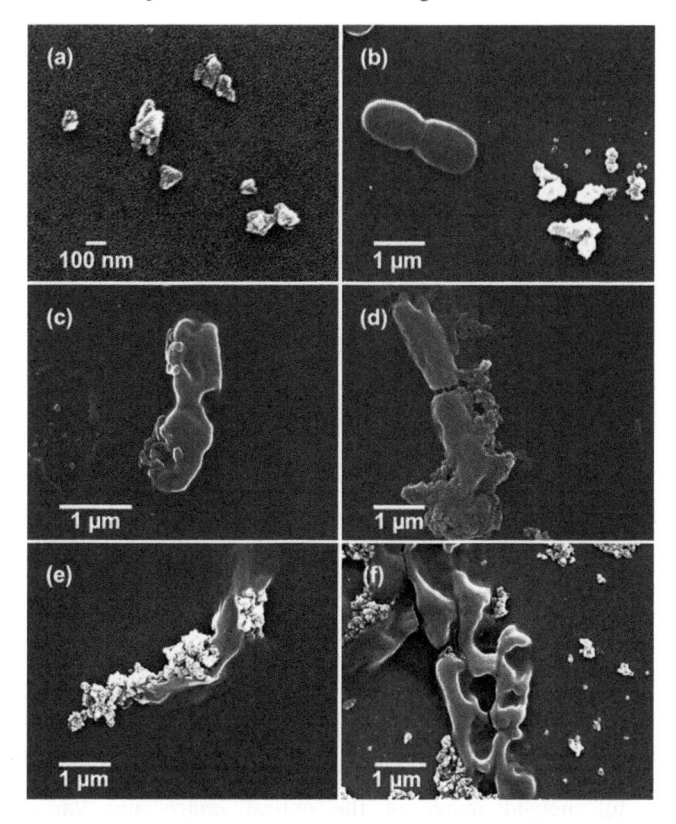

Figure 8.6 SEM images of (a) 100 nm cND, (b) *E. coli* mixed with ND, (c, d) *E. coli* treated with lysozyme–cND (interaction for 30–90 s), (e, f) *E. coli* treated with lysozyme–cND (interaction for 150–200 s). The ND concentration was 25 μg mL^{-1}.[21]

delivery of DNA-based model drug molecules.[131] In the same way, more recently Lin *et al.* have demonstrated for the first time the interaction of NDs with protists ciliated protozoa microorganisms, *P. caudatum* and *T. thermophila* (*in vivo*), for bio-imaging and cargo delivery.[133] Detonation NDs with a crystallite size of 5 nm were adsorbed on the surface of *E. coli* that was further used to feed the *T. thermophila*, resulting in delivery into the food vacuole. The ND luminescence was used to observe the food vacuoles. Figure 8.8 shows a 5 nm ND localized in a food vacuoles of *T. thermophila*, which correspondingly shows the position of *E. coli*.[133] Even though the fluorescence intensity of the ND is low, the number of NDs delivered by *E. coli* was high enough for fluorescence detection. ND localization corresponds to the process of *E. coli* digested by the *T. thermophila*. Since NDs attach on the carrier (*i.e.*, the bacterial cell wall), allowing entry into the target system (*i.e.*, the microorganism) through feeding, these processes could be engineered for future development of a micro/nano biorobotic systems.

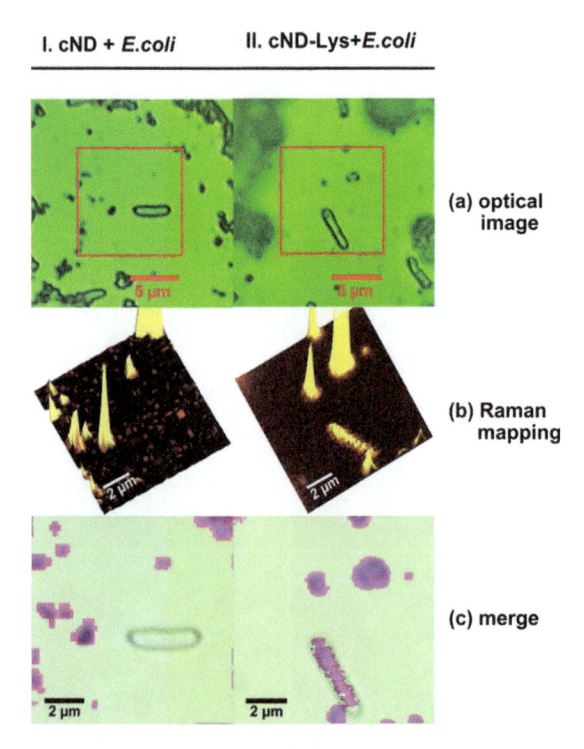

Figure 8.7 The interaction of lysozyme with *E. coli*. (a) Optical images, (b) Raman mapping in three dimensions of the square box area in (a), and (c) merged image of the optical image and Raman mapping. (I) *E. coli* with cND, (II) *E. coli* with lysozyme–cND complex. The cND and lysozyme–cND concentrations were 10 μg mL^{-1}.[21]

8.4.2 Interactions with Protozoa

The protists unicellular ciliated protozoa are key members of many ecosystems. These microorganisms are widely used as model systems for studying the toxicity of environmental agents. In fact ciliated protozoa are very well studied and extensively used as a model organism in cell biology (particularly for phagocytosis and exocytosis studies), genetics studies, *etc*. Protozoa can feed on suspended particles in the size range of nanometers to microns and, as a consequence, they are excellent model organisms for studying, *in situ*, the bio-distribution of internalized nanoparticles, their toxic effects and can help develop methods for *in vivo* imaging.

The *in situ* observation of unicellular microorganisms is a step forward that may drive research from the cellular level to studies related to organisms. This is more evident when the microorganism motility and functional polarity are considered, *i.e.*, it has a complex structure with the cell mouth at the anterior and the anus or cytoproct at the posterior. On the other hand, the ND compatibility with unicellular ciliate protozoa is interesting and allows exploration of the mechanisms by which NDs exert their effect on the

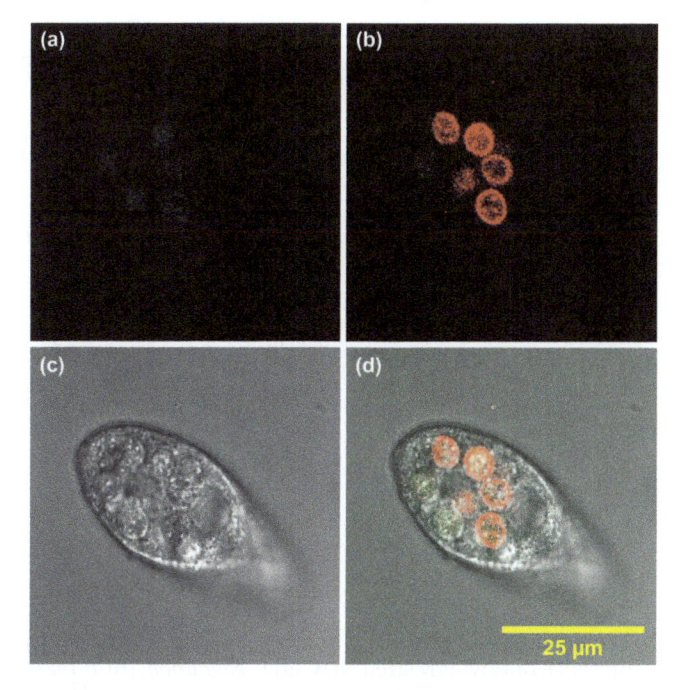

Biophysical Interaction of Nanodiamonds with Biological Entities In Vivo   185

Figure 8.8 T. thermophila fed with E. coli with adsorbed 5 nm NDs. (a) ND fluorescence (excitation = 488 nm; emission = 500–545 nm), (b) fluorescence of LysoTracker™, which was used to stain the food vacuoles in T. thermophila (excitation = 488 nm; emission = 585–595 nm), (c) T. thermophila optical image, (d) merged image of (a), (b) and (c).[133]

cell/organism, as well as the ND bio-distribution, nanoparticle transfer by a moving carrier and the problems of eco- and nano-safety, *etc.*

Presently, only limited studies exists where protozoa microorganisms have been used for the evaluation of biological effects using engineered carbon nanoparticles. For example, the toxicity of CNTs in *P. caudatum* and their influence on proliferation and interaction in culture medium has been determined.[134,135] The influence of SWNTs on the ability of *T. thermophila* to ingest and digest bacteria has been discussed.[136] It was observed that SWNTs can be internalized (eaten) by protozoa and affect the protozoa ability for bacteria killing. In contrast to SWNTs, the protozoans do not appear to be particularly sensitive to C_{60} fullerene[137] and a negligibly small effect on the proliferation was observed using fullerenol $C_{60}(OH)_x$.[138] In fact, investigations using ND are still lacking and have been carried out just recently.

The interaction of ciliated eukaryotic unicellular organisms, such as *P. caudatum* and *T. thermophila*, with commercially available ND and cND with an average diameter of 5 nm and 100 nm have been studied.[133] Carboxylation of the NDs was performed to remove the metallic impurities, non-diamond carbon and to create carboxyl groups on the surface using standard chemical methods. *P. caudatum* and *T. thermophila* were cultivated

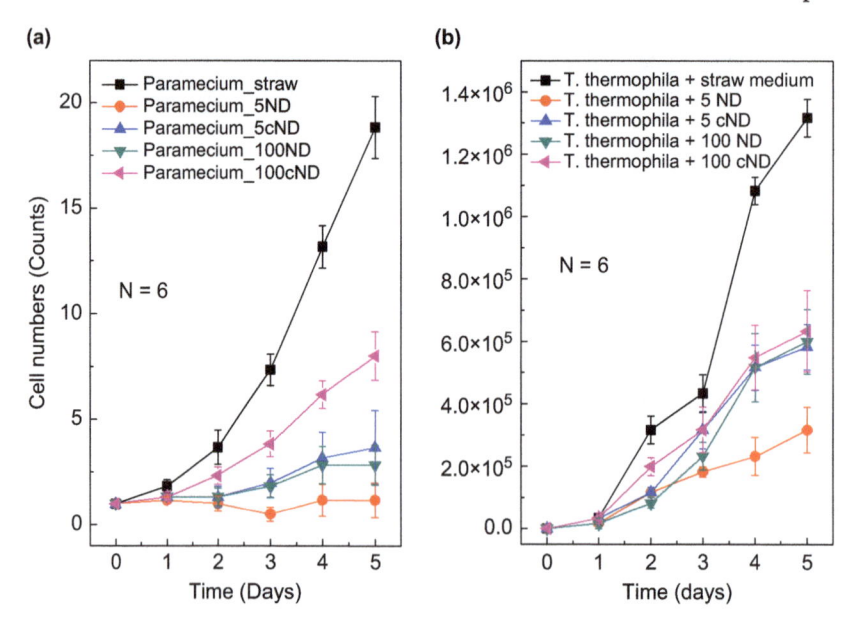

Figure 8.9 The dependence of the cell number against time of cultivation in the presence of varyingly sized NDs. (a) *P. caudatum*; (b) *T. thermophila*. The presented results are the average of six experiments.[133]

in straw medium. The ND influence on protozoal growth, microorganisms feeding with NDs and excretion, protozoa division and their interaction with *E. coli* in the presence of NDs was observed and analyzed. The effect of NDs on the growth of *P. caudatum* and *T. thermophila* can be seen in Figure 8.9.[133] The plot represents the number of cells cultivated in straw medium in the presence of NDs with time. It can be seen that 5 nm NDs terminate the growth of *P. caudatum* and decrease the growth rate of *T. thermophila* after 2 days. This indicates that the effect of the NDs is more pronounced in *P. caudatum* than in *T. thermophila*. Both NDs and cNDs decrease the viability of the protozoa. Furthermore, small-sized NDs (5 nm) are more toxic than large-sized NDs (100 nm) and NDs (5 and 100 nm) are more toxic than cNDs (5 and 100 nm). It was assumed that disorder and loosely bonded carbons on the surface of the NDs, especially 5 nm NDs, could be the source of toxicity.

At the same time, microorganisms that are able to survive in the presence of NDs are still able to function and the microorganisms are seen to "eat" the NDs during the *in vivo* observations. In this regard, we have observed that NDs enter and localize in the food vacuoles of living microorganisms. Figure 8.10 shows *P. caudatum* and *T. thermophila* with 100 nm cND in the food vacuoles.[133] *T. thermophila* can also excrete NDs and divide in the presence of NDs.

The combined research that has been carried out to date using NDs and the other already proven concepts show that NDs provide a suitable platform for drug delivery too.

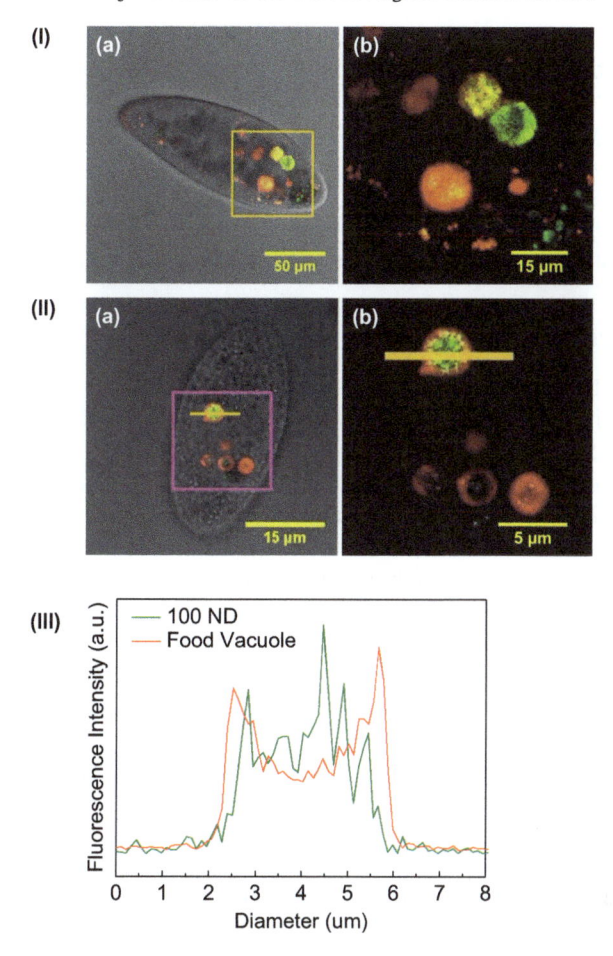

Biophysical Interaction of Nanodiamonds with Biological Entities In Vivo 187

Figure 8.10 Observation of ND internalization into microorganisms. (I) *P. caudatum* and (II) *T. thermophila* with 100 nm NDs in food vacuoles. (a) Merged optical image of the microorganism with the fluorescence image of 100 nm NDs (shown in green; excited at 488 nm; collected in the 510–540 nm range) and the fluorescence image of LysoTracker™-stained food vacuoles (shown in red; excited at 543 nm; collected in the 580–600 nm range). (b) Fluorescence images of the marked area; the color of the vacuoles depends on the microscope focus. (III) Distribution of intensity of fluorescence along the yellow line in the vacuole shown in II(b). Distribution of fluorescence intensity of 100 nm ND is shown by the green line and the food vacuole is shown by the red line.[133]

8.5 Summary

In summary we have exemplified research works related to the functionalizations/conjugations of NDs and the characterizations tools used to examine them. We have discussed nano-carbon (including ND particle) interactions with bacteria and the very first investigations concerning the interaction of NDs with ciliated protozoa microorganisms. These investigations allow us to further develop new methods for *in vivo* imaging and to study the various mechanisms of nanostructure interactions with different bio-systems. We believe that this type of study is necessary and important as it is an intermediate step that interlinks cells and organisms for the development of effective applications and helps in understanding more deeply the interaction of NDs with living organisms *in vivo*.

References

1. V. Yu. Dolmatov, *Russ. Chem. Rev.*, 2001, **70**, 607.
2. O. A. Shenderova, V. V. Zirnov and D. W. Brenner, *Crit. Rev. Solid State Mater. Sci.*, 2002, **27**, 227.
3. K. Iakoubovskii, M. V. Baidakova, B. H. Wouters, A. Stesmans, G. J. Adriaenssens, A. Y. Vul' and P. J. Grobet, *Dia. & Rel. Mater.*, 2000, **9**, 861.
4. S. V. Kidalov, F. M. Shakhov and A. Ya. Vul', *Dia. & Rel. Mater.*, 2008, **17**, 844.
5. S.-J. Yu, M.-W. Kang, H.-C. Chang, K.-M. Chen and Y.-C. Yu, *J. Am. Chem. Soc.*, 2005, **127**, 17604.
6. E. Perevedentseva, P.-J. Cai, Y.-C. Chiu and C.-L. Cheng, *Langmuir*, 2011, **27**, 1085.
7. L.-C. L. Huang and H.-C. Chang, *Langmuir*, 2004, **20**, 5879.
8. A. Krueger, *Chem. Eur. J.*, 2008, **14**, 1382.
9. K. K. Liu, C.-L. Cheng, C. C. Chang and J. I. Chao, *Nanotechnology*, 2007, **18**, 325102.
10. K. K. Liu, C. C. Wang, C.-L. Cheng and J. I. Chao, *Biomaterials*, 2009, **30**, 4249.
11. A. M. Schrand, H. Huang, C. Carlson, J. J. Schlager, S. E. Omacr, S. M. Hussain and L. Dai, *J. Phys. Chem. B*, 2007, **111**, 2.
12. T. Lechleitner, F. Klauser, T. Seppi, J. Lechner, P. Jennings, P. Perco, B. Mayer, D. Steinmüller-Nethl, J. Preiner, P. Hinterdorfer, M. Hermann, E. Bertel, K. Pfaller and W. Pfaller, *Biomaterials*, 2008, **29**, 4275.
13. I. P. Chang, K. C. Hwang and C. S. Chiang, *J. Am. Chem. Soc.*, 2008, **130**, 15476.
14. V. Thomas, B. A. Halloran, N. Ambalavanan, S. A. Catledge and Y. K. Vohra, *Acta Biomaterialia*, 2012, **8**, 1939.
15. C.-Y. Fang, V. Vaijayanthimala, C.-A. Cheng, S.-H. Yeh, C.-F. Chang, C.-L. Li and H.-C. Chang, *Small*, 2011, 7, 3363.
16. Y.-C. Lin, L.-W. Tsai, E. Perevedentseva, H.-H. Chang, C.-H. Lin, D.-S. Sun, A. E. Lugovtsov, A. Priezzhev, J. Mona and C.-L. Cheng, *J. Biomed. Optics*, 2012, **17**, 101512.
17. A. P. Puzyra, A. V. Baron, K. V. Purtov, E. V. Bortnikov, N. N. Skobelev, O. A. Mogilnaya and V. S. Bondar, *Dia. & Rel. Mater.*, 2007, **16**, 2124.

18. J. C. Sung and J. Lin, *Diamond Nanotechnology: Syntheses and Applications*, Pan Stanford Publishing Ltd, Singapore, 2009.
19. P.-H. Chung, E. Perevedentseva, J.-S. Tu, C. C. Chang and C.-L. Cheng, *Dia. & Rel. Mat.*, 2006, **15**, 622.
20. C. Bradac, T. Gaebel, N. Naidoo, M. J. Sellars, J. Twamley, L. J. Brown, A. S. Barnard, T. Plakhotnik, A. V. Zvyagin and J. R. Rabeau, *Nat. Nanotech.*, 2010, **5**, 345.
21. E. Perevedentseva, C.-Y. Cheng, P.-H. Chung, J.-S. Tu, Y.-H. Hsieh and C.-L. Cheng, *Nanotech.*, 2007, **18**, 315102.
22. R. Lam and D. Ho, *Expert opinion on Drug Delivery*, 2009, **6**, 883.
23. A. Qureshi, Y. Gurbuz, M. Howell, W. P. Kang and J. L. Davidson, *Dia. and Rel. Mater.*, 2010, **19**, 457.
24. R. S. Schoenfeld and W. Harneit, *Phys. Rev. Lett.*, 2011, **106**, 030802.
25. Y. L. Liu and K. W. Sun, *Nanoscale Research Lett.*, 2010, **5**, 1045.
26. A. P. Puzyr, I. O. Pozdnyakova and V. S. Bondar, *Phys. Solid State*, 2004, **46**, 76.
27. O. Faklaris, V. Joshi, T. Irinopoulou, P. Tauc, M. Sennour, H. Girard, C. Gesset, J.-C. Arnault, A. Thorel, J.-P. Boudou, P. A. Curmi and F. Treussar, *ACS Nano*, 2009, **3**, 3955.
28. E. Katz and I. Willner, *Angew. Chem. Int. Ed.*, 2004, **43**, 6042.
29. K. Ushizawa, Y. Sato, T. Mitsumori, T. Machinami, T. Ueda and T. Ando, *Chem. Phys. Lett.*, 2002, **351**, 105.
30. T. S. Huang, Y. Tzeng, Y. K. Liu, Y. C. Chen, K. R. Walker, R. Guntupalli and C. Liu, *Dia. & Rel. Mater.*, 2004, **13**, 1098.
31. W. Yang, O. Auciello, J. E. Butler, W. Cai, J. A. Carlisle, J. E. Gerbi, D. M. Gruen, T. Knickerbocker, T. L. Lasseter, J. N. Russell, Jr., L. M. Smith and R. J. Hamers, *Nat. Mater.*, 2002, **1**, 253.
32. T. Knickerbocker, T. Strother, M. P. Schwartz, J. N. Russell, Jr., J. Butler, L. M. Smith and R. J. Hamers, *Langmuir*, 2003, **19**, 1938.
33. X. L. Kong, L. C. L. Huang, C.-M. Hsu, W.-H. Chen, C.-C. Han and H.-C. Chang, *Anal. Chem.*, 2005, **77**, 259.
34. C. Y. Cheng, E. Perevedentseva, J. S. Tu, P. H. Chung, C. L. Cheng, K. K. Liu, J. I. Chao, P. H. Chen and C. C. Chang, *Appl. Phys. Lett.*, 2007, **90**, 163903.
35. T. T. B. Nguyen, H.-C. Chang and V. W.-K. Wu, *Dia. & Rel. Mater.*, 2007, **16**, 872.
36. N. Kossovsky, A. Gelman, H. J. Hnatyszyn, S. Rajguru, R. L. Garrell, S. Torbati, S. S. F. Freitas and G.-M. Chow, *Bioconjugate Chem.*, 1995, **6**, 507.
37. P. John, N. Polwart, C. E. Troupe and J. I. Wilson, *J. Am. Chem. Soc.*, 2003, **125**, 6600.
38. K.-K. Liu, M.-F. Chen, P.-Y. Chen, T. J. F. Lee, C.-L. Cheng, C.-C. Chang, Y.-P. Ho and J.-I. Chao, *Nanotech.*, 2008, **19**, 205102.
39. V. S. Bondar, I. O. Pozdnyakova and A. P. Puzyr, *Phys. Solid State*, 2004, **46**, 758.

40. K. V. Purtov, A. I. Petunin, A. E. Burov, A. P. Puzyr and V. S. Bondar, *Nanoscale Res. Lett.*, 2010, **5**, 631.
41. A. Krueger, Y. J. Liang, G. Jarre and J. Stegk, *J. Mater. Chem.*, 2006, **16**, 2322.
42. A. Krueger, J. Stegk, Y. Liang, L. Lu and G. Jarre, *Langmuir*, 2008, **24**, 4200.
43. T. Meinhardt, D. l. Lang, H. Dill and A. Krueger, *Adv. Funct. Mater.*, 2011, **21**, 494.
44. D. Lang and A. Krüger, *Dia. & Rel. Mater.*, 2010, **20**, 101.
45. G. Jarre, Y. Liang, P. Betz, D. Lang and A. Krueger, *Chem. Commun.*, 2011, **47**, 544.
46. W. S. Yeap, Y. Y. Tan and K. P. Loh, *Anal. Chem.*, 2008, **80**, 4659.
47. H. Huang, E. Pierstorff, E. Osawa and D. Ho, *Nano Lett.*, 2007, 7, 3305.
48. E. K. Chow, X.-Q. Zhang, M. Chen, R. Lam, E. Robinson, H. Huang, D. Schaffer, E. Osawa, A. Goga and D. Ho, *Cancer*, 2011, **3**, 73ra21.
49. J. Yan, Y. Guo, A. Altawashi, B. Moosa, S. Lecommandouxc and N. M. Khashab, *New J. Chem.*, 2012, **36**, 1479.
50. Y. Li, X. Zhou, D. Wang, B. Yang and P. Yang, *J. Mater. Chem.*, 2011, **21**, 16406.
51. M. Chen, E. D. Pierstorff, R. Lam, S. Y. Li, H. Huang, E. Osawa and D. Ho, *ACS Nano*, 2009, **3**, 2016.
52. K. K. Liu, W. W. Zheng, C. C. Wang, Y. C. Chiu, C. L. Cheng, Y. S. Lo, C. Chen and J. I Chao, *Nanotech.*, 2010, **21**, 315106.
53. X.-Q. Zhang, R. Lam, X. Xu, E. K. Chow, H.-J. Kim and D. Ho, *Adv. Mater.*, 2011, **23**, 4770.
54. B. Guan, F. Zou and J. F. Zhi, *Small*, 2010, **6**, 1514.
55. X. Q. Zhang, M. Chen, R. Lam, X. Xu, E. Osawa and D. Ho, *ACS Nano*, 2009, **3**, 2609.
56. V. Grichko, V. Grishko and O. Shenderova, *Nanobiotech.*, 2007, **2**, 37.
57. O. Loh, R. Lam, M. Chen, N. Moldovan, H. Huang, D. Ho and H. D. Espinosa, *Small*, 2009, **5**, 1667.
58. W. Smirnov, A. Kriele, N. Yang and C. E. Nebel, *Dia. & Rel. Mater.*, 2010, **19**, 186.
59. Q. Zhang, V. N. Mochalin, I. Neitzel, I. Y. Knoke, J. Han, C. A. Klug, J. G. Zhou, P. I. Lelkes and Y. Gogotsi, *Biomaterials*, 2011, **32**, 87.
60. H. B. Man, R. Lam, M. Chen, E. Osawa and D. Ho, *Phys. Status Solidi A*, 2012, **209**, 1811.
61. A. Krüger, *J. Mater. Chem.*, 2008, **18**, 1485.
62. S. Osswald, G. Yushin, V. Mochalin, S. O. Kucheyev and Y. Gogotsi, *J. Am. Chem. Soc.*, 2006, **128**, 11635.
63. A. Krüger, M. Ozawa, F. Kataoka, T. Fujino, Y. Suzuki, A. E. Aleksenskii, A. Ya. Vul' and E. Osawa, *Carbon*, 2005, **43**, 1722.
64. A. E. Aleksenskii, M. V. Baidakova, A. Ya. Vul' and V. I. Siklitskii, *Phys. Solid State*, 1999, **41**, 668.

65. V. L. Kuznetsov, M. N. Aleksandrov, I. V. Zagoruiko, A. L. Chuvilin, E. M. Moroz, V. N. Kolomiichuk, V. A. Likholobov, P. M. Brylyakov and G. V. Sakovich, *Carbon*, 1991, **29**, 665.
66. T. Jiang and K. Xu, *Carbon*, 1995, **33**, 1663.
67. Yu. V. Butenko, V. L. Kuznetsov, E. A. Paukshtis, A. I. Stadnichenko, I. N. Mazov, S. I. Moseenkov, A. I. Boronin and S. V. Kosheev, *Fullerenes, Nanotubes, Carbon Nanostruct.*, 2006, **14**, 557.
68. T. Strother, T. Knickerbocker, J. N. Russell, Jr., J. E. Butler, L. M. Smith and R. J. Hamers, *Langmuir*, 2002, **18**, 968.
69. J. Mona, J.-S. Tu, T.-Y. Kang, E. C.-Y. Tsai, E. Perevedentseva and C.-L. Cheng, *Dia. & Rel. Mater.*, 2012, **24**, 134.
70. A. Aleksenskiy, M. Baidakova, V. Osipov, A. Vul', The fundamental properties and characteristics of Nanodiamonds, in *Nanodiamonds: Applications in Biology and Nanoscale Medicine*, ed. D. Ho, Springer, New York Dordrecht Heidelberg London, 2010, p. 55.
71. A. M. Panich, *Dia. & Rel. Mater*, 2007, **16**, 2044.
72. D. S. Knight and W. B. White, *J. Mater. Res.*, 1989, **4**, 385.
73. A. C. Ferrari and J. Robertson, *Phys. Rev. B*, 2011, **60**, 121405.
74. P.-H. Chung, E. Perevedentseva and C.-L. Cheng, *Surface Science*, 2007, **601**, 3866.
75. F. Jelezko, C. Tietz, A. Gruber, I. Popa, A. Nizovtsev, S. Kilin and J. Wrachtrup, *Single Molecules*, 2001, **2**, 255.
76. K. Iakoubovskii, A. Stesmans, B. Nouwen and G. J. Adriaenssens, *Phys. Rev. B*, 2000, **62**, 16587.
77. A. E. Aleksenskii, Yu. V. Osipov, A. Ya. Vul, B. Ya. Ber, A. B. Smirnov, V. G. Melekhin, G. J. Adriaenssens and K. Iakoubovskii, *Phys. Solid State*, 2001, **43**, 145.
78. M. E. Kompan, E. I. Terukov, S. K. Gordeev, S. G. Zhukov and Yu. A. Nikolaev, *Phys. Solid State*, 1997, **39**, 1928.
79. C.-C. Fu, H.-Y. Lee, K. Chen, T.-S. Lim, H.-Y. Wu, P.-K. Lin, P.-K. Wei, P.-H. Tsao, H.-C. Chang and W. Fann, *Proc. Natl. Acad. Sci. USA*, 2007, **104**, 727.
80. T.-L. Wee, Y.-W. Mau, C.-Y. Fang, H.-L. Hsu, C.-C. Han and H.-C. Chang, *Dia. & Rel. Mater.*, 2009, **18**, 567.
81. Y. R. Chang, H. Y. Lee, K. Chen, C. C. Chang, D. S. g Tsai, C. C. Fu, T. S. Lim, Y. K. Tzeng, C. Y. Fang, C. C. Han, H. C. Chang and W. Fann, *Nat. Nanotech.*, 2008, **3**, 284.
82. Y.-P. Sun, X. Wang, F. Lu, L. Cao, M. J. Meziani, P. G. Luo, L. Gu and L. M. Veca, *J. Phys. Chem. C*, 2008, **112**, 47.
83. A. B. Bourlinos, A. Stassinopoulos, D. Anglos, R. Zboril, V. Georgakilas and E. P. Giannelis, *Chem. Mater.*, 2008, **20**, 4539.
84. J. I. Chao, E. Perevedentseva, P. H. Chung, K. K. Liu, C. Y. Cheng, C. C. Chang and C. L. Cheng, *J. Biophys.*, 2007, **93**, 2199.
85. M. Mkandawire, A. Pohl, T. Gubarevich, V. Lapina, D. Appelhans, G. Rodel, W. Pompe, J. Schreiber and J. Opitz, *J. Biophoton.*, 2009, **2**, 596.

86. Y.-K. Tzeng, O. Faklaris, B.-M. Chang, Y. Kuo, J.-H. Hsu and H.-C. Chang, *Angew. Chem. Int. Ed.*, 2011, **50**, 2262.
87. M. F. Weng, S. Y. Chiang, N. S. Wang and H. Niu, *Dia. & Rel. Mater.*, 2009, **18**, 587.
88. J. Opitz, M. Mkandawire, M. Sorge, N. Rose, M. Rudolph, P. Krüger, I. Hannstein, V. A. Lapina, D. Appelhans, W. Pompe, J. Schreiber and G. Roede, *Proc. of SPIE*, 2010, **7759**, 775914.
89. N. Mohan, C. S. Chen, H. H. Hsieh, Y. C. Wu and H. C. Chang, *Nano Lett.*, 2010, **10**, 3692.
90. Y. Yuan, Y. Chen, J.-H. Liu, H. Wang and Y. Liu, *Dia. & Rel. Mater.*, 2009, **18**, 95.
91. X. Zhang, J. Yin, C. Kang, J. Li, Y. Zhu, W. Li, Q. Huang and Z. Zhu, *Toxicology Lett.*, 2010, **198**, 237.
92. B. R. Smith, M. Niebert, T. Plakhotnik and A. V. Zvyagin, *J. Lumines.*, 2007, **127**, 260.
93. G. Balasubramanian, I. Y. Chan, R. Kolesov, M. Al-Hmoud, J. Tisler, C. Shin, C. Kim, A. Wojcik, P. R. Hemmer, A. Krueger, T. Hanke, A. Leitenstorfer, R. Bratschitsch, F. Jelezko and J. Wrachtrup, *Nature*, 2008, **455**, 648.
94. L. T. Halla, C. D. Hill, J. H. Cole, B. Städler, F. Caruso, P. Mulvaney, J. Wrachtrup and L. C. L. Hollenberg, *Proc. Natl. Acad. Sci. USA*, 2010, **107**, 18777.
95. S. C. Hens, G. Cunningham, T. Tyler, S. Moseenkov, V. Kuznetsov and O. Shenderova, *Dia. & Rel. Mater.*, 2008, **17**, 1858.
96. Q. Li, S. Mahendra, D. Y. Lyon, L. Brunet, M. V. Liga, D. Li and P. J. J. Alvarez, *Water Res.*, 2008, **42**, 4591.
97. W. Hu, C. Peng, W. Luo, M. Lv, X. Li, D. Li, Q. Huang and C. Fan, *ACS Nano*, 2010, **4**, 4317.
98. I. Banerjee, D. Mondal, J. Martin and R. S. Kane, *Langmuir*, 2010, **26**, 17369.
99. K. Page, M. Wilson and I. P. Parkin, *J. Mater. Chem.*, 2009, **19**, 3819.
100. A. Srivastava, O. N. Srivastava, S. Talapatra, R. Vajtai and P. M. Ajayan, *Nature Mater.*, 2004, **3**, 610.
101. H. Schwegmann, A. J. Feitz and F. H. Frimmel, *J. Coll. & Interface Science*, 2010, **347**, 43.
102. J. R. Morones, J. L. Elechiguerra, A. Camacho, K. Holt, J. B. Kouri, J. T. Ramirez and M. J. Yacaman, *Nanotechnology*, 2005, **16**, 2346.
103. S. Pal, Y. K. Tak and J. M. Song, *Appl. Environ. Microbiol.*, 2006, **73**, 1712.
104. I. Sondi and B. Salopek-Sondi, *J. Colloid Interface Sci.*, 2004, **275**, 177.
105. C. Wei, W. Y. Lin, Z. Zainal, N. E. Williams, K. Zhu, A. P. Kruzic, R. L. Smith and K. Rajeshwar, *Environ. Sci. Technol.*, 2004, **28**, 934.
106. P. K. Stoimenov, R. L. Klinger, G. L. Marchin and K. J. Klabunde, *Langmuir*, 2002, **18**, 6679.
107. L. Huang, D.-Q. Li, Y. J. Lin, M. Wei, D. G. Evans and X. Duan, *J. Inorg. Biochem.*, 2005, **99**, 986.

108. O. B. Koper, J. S. Klabunde, G. L. Marchin, K. J. Klabunde, P. Stoimenov and L. Bohra, *Curr. Micobiol.*, 2002, **44**, 49.
109. S. Makhluf, R. Dror, Y. Nitzan, Y. Abramovich, R. Jelinek and A. Gedanken, *Adv. Funct. Mater.*, 2005, **15**, 1708.
110. L. Zhang, Y. Jiang, Y. Ding, M. Povey and David York, *J. Nanopart. Res.*, 2007, **9**, 479.
111. L. K. Adams, D. Y. Lyon and P. J. J. Alvarez, *Water Res.*, 2006, **40**, 3527.
112. V. Shah, S. Shah, H. Shah, F. J. Rispoli, K. T. McDonnell, S. Workeneh, A. Karakoti, A. Kumar and S. Seal, *Plos ONE*, 2012, 7, e47827.
113. A. Thill, O. Zeyons, O. Spalla, F. Chauvat, J. Rose, M. Auffan and A. M. Flank, *Environ. Sci. Technol.*, 2006, **40**, 6151.
114. Z. Huang, X. Zheng, D. Yan, G. Yin, X. Liao, Y. Kang, Y. Yao, D. Huang and B. Hao, *Langmuir.*, 2008, **24**, 4140.
115. M. Li, L. Zhu and D. Lin, *Environ. Sci. Technol.*, 2011, **45**, 1977.
116. J. Díaz-Visurraga, C. Gutiérrez, C. von Plessing and A. García, Metal nanostructures as antibacterial agents, in *Science Against Microbial Pathogens: Communicating Current Research and Technological Advances*, ed. A. Méndez-Vilas, Formatex, 2011, pp. 201–218.
117. S. Kang, M. Herzberg, D. F. Rodrigues and M. Elimelech, *Langmuir.*, 2008, **24**, 6409.
118. J. A. Rojas-Chapana, M. A. Correa-Duarte, Z. Ren, K. Kempa and M. Giersig, *Nano Lett.*, 2004, **4**, 985.
119. A. E. Nel, L. Mädler, D. Velegol, T. Xia, E. M. V. Hoek, S. Somasundaran, F. Klaessig, V. Castranova and M. Thompson, *Nature Mater.*, 2009, **8**, 543.
120. Y. Cui, Y. Zhao, Y. Tian, W. Zhang, X. Lü and X. Jiang, *Biomaterials*, 2012, **33**, 2327.
121. V. K. Rangari, G. M. Mohammad, S. Jeelani, A. Hundley, K. Vig, S. R. Singh and S. Pilla, *Nanotechnology*, 2010, **21**, 095102.
122. J. H. Jung, G. B. Hwang, J. E. Lee and G. N. Bae, *Langmuir*, 2011, **27**, 10256.
123. D. Y. Lyon, L. Brunet, G. W. Hinkal, M. R. Wiesner and P. J. J. Alvarez, *Nano Lett.*, 2008, **8**, 1539.
124. A. Soininen, V.-M. Tiainen, Y. T. Konttinen, H. C. van der Mei, H. J. Busscher and P. K. Sharma, *J Biomed. Mater. Res. Part B: Appl. Biomater.*, 2009, **90B**, 882.
125. W. Shao, Q. Zhao, E. W. Abel and A. Bendavid, *J. Biomed. Mater. Res. Part A*, 2009, **93A**, 133.
126. W. Jakubowski, G. Bartosz, P. Niedzielski, W. Szymanski and B. Walkowiak, *Dia. & Rel. Mater.*, 2004, **13**, 1761.
127. O. Medina, J. Nocua, F. Mendoza, R. Gómez-Moreno, J. Ávalos, C. Rodríguez and G. Morell, *Dia. & Rel. Mater*, 2012, **22**, 77.
128. J.-W. Kim, E. V. Shashkov, E. I. Galanzha, N. Kotagiri and V. P. Zharov, *Lasers Surg. Med.*, 2007, **39**, 622.
129. G. A. Zelada-Guill, J. Riu, A. Düzgün and F. X. Rius, *Angew. Chem. Int. Ed.*, 2009, **48**, 7334.

130. M. Hartmann, P. Betz, Y. Sun, S. N. Gorb, T. K. Lindhorst and A. Krüger, *Chem. Eur. J.*, 2012, **18**, 6485.
131. D. Akin, J. Sturgis, K. Rageb, D. Sherman, K. Burkholder, J. P. Robinson, A. K. Bhunia, S. Mohhamed and R. Bashir, *Nature Nanotech*, 2007, **2**, 441.
132. M. S. Sakar, E. B. Steager, D. H. Kim, A. A. Julius, M. Kim, V. Kumar and G. J. Pappas, *Int. J. Robot. Res.*, 2011, **30**, 647.
133. Yu.-C. Lin, E. Perevedentseva, L.-W. Tsai, K.-T. Wu and C.-L. Cheng, *J. Biophotonics*, 2012, **5**, 838.
134. N. Haga and K. Haneda, *Jpn. J. Protozool.*, 2007, **40**, 139.
135. Y. Zhu, T. Ran, Y. Li, J. Guo and W. Li, *Nanotechnology*, 2006, **17**, 4668.
136. P. Ghafari, C. H. St-Denis, M. E. Power, X. Jin, V. Tsou, H. S. Mandal, N. C. Bols and X. Tang, *Nat. Nanotech.*, 2008, **3**, 347.
137. A. Johansen, A. Pedersen, U. Karlson, B. M. Hansen, J. Scott-Fordsmand and A. Winding, *Environ. Toxicol. Chem.*, 2008, **27**, 1895.
138. Q.-F. Zhao, Y. Zhu, T.-C. Ran, J.-G. Li, Q.-N. Li and W.-X. Li, *Nuclear Sci. Tech.*, 2006, **17**, 280.

CHAPTER 9

Neuron Growth on Nanodiamond

ROBERT EDGINGTON AND RICHARD B. JACKMAN*

London Centre for Nanotechnology, 17–19 Gordon Street, UCL, London, WC1H 0AH, UK
*Email: r.jackman@ucl.ac.uk

9.1 Introduction: Neuronal Biomaterials

Neurons are one of the most environmentally demanding cell types. Due to their remarkable ability to generate action potentials and electrically communicate with their neighbours, their extreme differentiated phenotype leaves them with little resilience to physiological stress. For this reason, neurons are protected behind the blood–brain barrier and have their own hypersensitive immune system that is tended by glial cells. Unlike the sophisticated targeted immune system of the rest of our bodies, the glial immune system aggressively attacks any foreign body it encounters,[1–4] and implanted materials and electrodes often incite such an immune response and result in a process called gliosis.[5,6] Gliosis, or glial scarring, is where insulating glial tissue surrounds and progressively electrically isolates implanted devices from neurons. This increase in impedance renders the implant incapacitated and gliosis accounts for a significant contribution to most neural prosthesis failures.[1–4] Therefore, neuronal biomaterials must meet very high specifications if they are to achieve neuronal biocompatibility, and only select materials can promote the formation of functional neuronal networks on their surfaces (e.g., flat or unmodified glass, polystyrene, sapphire, iridium oxide, silicon, metal oxides, polymers,[6,5]

RSC Nanoscience & Nanotechnology No. 31
Nanodiamond
Edited by Oliver Williams
© The Royal Society of Chemistry 2014
Published by the Royal Society of Chemistry, www.rsc.org

diamond[7] and platinum, as reviewed herein).[8] For those that can promote neuronal networks, many of the surface properties of these biomaterials are paramount to achieving biocompatibility.

First and foremost, neuronal biomaterials must have an inert chemistry in physiological media in order to not be toxic to cells. Beyond this, specific surface functionalities have been found to promote cellular attachment on materials.[9] For example, Patel et al. showed that silane moieties have an adverse effect on neuronal attachment.[10] In the case of neurons growing on diamond, typical hydrogen- or oxygen-based functionalities have been found to be of little difference to cell proliferation,[7,11,12] except for flat homo-epitaxial diamond (as opposed to nanocrystalline diamond, NCD), where hydrophilic oxygenated diamond has been found to be more conducive to neuronal attachment.[11] Chen et al.[13] found that neural stem cells grown on hydrogen-terminated ultra-nanocrystalline diamond (UNCD) spontaneously differentiated into neurons, while the cells grown on oxygen-terminated UNCD films exhibited a preference towards oligodendrocyte (insulating glia cells) differentiation. Although surface chemistry affects cellular attachment—and positively charged surfaces can cause initial (but short lived[14]) cell adhesion *via* electrostatic bonding with negatively charged cell membranes[15]—a direct cell–biomaterial interaction has not been observed, and supposed surface chemical dependencies are often dependent on a mediatory protein adsorption on a biomaterial that brings about cellular attachment. In this context, in order to perform neuronal cell cultures *in vitro*, a coating of extracellular matrix (ECM) proteins is routinely used for cellular attachment.[8,16,17] ECM proteins, such as laminin and fibronectin, form the molecular scaffolding of cells and contain specific peptide sequences (*e.g.*, RGD) from which cells obtain biochemical cues to initiate various developmental processes *via* integrin binding.[18–20] Such sequences can be artificially synthesised and immobilised onto biomaterial surfaces in order to promote neuronal adhesion.[21,22]

With respect to the common surface terminations (H/O) and ECM protein adsorption on diamond, Rezek et al.[23] have shown the orientation of serum proteins to be dependent on surface termination, which could explain the slight chemical selectivity of neurons on diamond, but could also explain the more apparent selectivity of other cells, such as osteoblasts on hydrophilic oxygenated diamond surfaces.[24] Curtis and Wilkinson[25] state that when both chemical patterns and topographic ones are offered to cells, topography tends to have a greater effect than chemical patterns, and the nano- and micro-structure of surfaces has been established as a decisive factor affecting neuron morphology and adhesion. Fan et al. found that surface roughness of 20–50 nm promoted neuronal adhesion (central neural cells, substantia nigra) on silicon substrates, and surfaces of below 10 nm and above 70 nm roughness adversely affected adhesion.[26,27] Similarly, Khan et al.[28] found rat primary cortical neurons adhere well onto surfaces of 10–100 nm roughness. Whilst nano-roughness is known to affect neuronal adhesion, micro-roughness determines neuron morphology. Clark et al.

showed chick cerebral neurons to align to 2 μm groves,[29] and various other studies on neuronal guidance using micro-scale topography have been carried out showing neurons to be guided by micro-scale ridges,[30] grids[31] and grooves as extensively reviewed herein.[8,32] Aside from the absolute roughness or relatively large micro-scale structures of biomaterials, the shape and curvature of surfaces can strongly affect neuronal attachment. This is most likely due to the functional conformation and orientation of proteins adsorbed onto nano-textured surfaces being affected by the curvature of the surface[9,33–35] *via* the balance of electrostatic and hydrophobic adsorption forces, which in turn affect the degree of cell attachment on biomaterial surfaces. In fact, out of the surface metrics of diamond as a biomaterial, Rezek *et al.*[23] suggest wetting angle and hydrophobicity to be the most important metric for cell growth on diamond, which incorporates fundamental properties, such as surface chemistry, surface roughness and porosity. Other properties, such as high porosity[36] and high elasticity,[37] have been shown to enhance neuronal proliferation on biomaterials, both of which mimic the properties of neuronal tissue.

Considering the metrics of modern biomaterials, diamond is a promising candidate for chronic neuronal biomaterials. Its extreme chemical inertness yet ability to be stably functionalised with a variety of molecular and macromolecular functionalities (H/O, DNA, proteins, *etc.*)[38–41] allow the tailoring of its biocompatibility towards cells. Moreover, compared to single- and poly-crystalline diamond coatings, nanodiamond coatings are a promising biomaterial due to their additional properties of increased curvature and nano-roughness, ease of deposition, porosity and a high affinity for protein adsorption,[42,43] wherein adsorbed proteins remain in their functional state.[44,45]

9.2 The Use of Nanodiamond Monolayer Coatings to Promote the Formation of Functional Neuronal Networks

9.2.1 Introduction

It is becoming increasingly clear that the interaction of nano-sized particles with living cells transgresses the framework laid down by previously known macroscopic interactions.[46] The excellent biocompatibility of diamond in different biological environments[13,40,47] and its excellent electrical properties[48] make this material a prime candidate for chronic electrical interfaces with neurons. Whilst nanodiamonds (NDs) created by a detonation process have been available for some years,[49] the primary particles, which have a size typically within the range 5–10 nm, aggregate during the formation process into particles with sizes on the micron scale. These particles have excellent electrical properties[50] but are not readily attached to substrate materials without complex chemical procedures. In contrast, recently developed techniques for the dispersion of the primary particles in solution[51,52] yield individual detonation ND (DND) particles and can be used to coat a given substrate by attachment through Van der Waals, electrostatic and polar forces without the need for hazardous chemical species or aggressive protocols.

Primary cultures of rodent neurons require a specifically treated surface for initial attachment and survival, and this is routinely provided by the deposition of extracellular matrix (ECM) proteins. Patterned deposition of ECM proteins has been shown to direct ordered outgrowth of neurons on single-crystal diamonds,[7] exemplifying the necessity of protein-coating for neuronal growth. Whilst protein coating is sufficient for *in vitro* experiments, *in vivo* implants, such as in neural prosthetics, will require biocompatible coatings that preferably will not introduce foreign proteins into the body.

9.2.2 Neuronal Cell Attachment on ND Layers

Newly emerging biomaterials aim to mimic the physicochemical properties of ECM proteins and surrounding tissue, and the ND properties compare favourably to the metrics of modern biomaterials.[9] Their tissue equivalence (carbon-based), non-cytotoxicity,[53] versatile organic surface functionality,[54] high affinity for protein adsorption[54] and their particle curvature and size[9,46] all present themselves as possible contributing agents to providing a nurturing environment for neuronal networks. Given the outstanding properties of NDs for biological application, DND monolayers have been tested by us[55] for their ability to support functional neuronal network formation even in the absence of otherwise routine extracellular matrix (ECM) protein coatings.

Thalhammer *et al.*[55] showed ND layering to be sufficient for promoting neuronal attachment and growth on four different substrate materials. Neurons grown on such ND-coated materials display intrinsic electrical excitability, synaptic transmission and network function comparable to those grown under standard ECM protein coating conditions.

Primary neuronal cultures were seeded onto glass, mechanically polished polycrystalline diamond (PCD), nanocrystalline diamond (NCD) and silicon (Si). All four substrates had been treated with monodispersed NDs in an ultrasonication bath to form a monolayer of NDs on the substrate surfaces (0.05 g L^{-1} DND for 10 min from New Metals & Chemicals, Japan). The resulting surface coatings promoted the attachment of primary murine neurons, which were absent on untreated substrate surfaces. Confocal microscopy of immunostained neuronal cultures showed comparable attachment on all ND-coated substrate materials (Figure 9.1, column 2), confirming that it is the NDs themselves that are responsible for the successful attachment and growth of the neurons. None of the surfaces led to cell attachment and growth in the absence of the NDs (Figure 9.1, column 1).

Neuron Growth on Nanodiamond 199

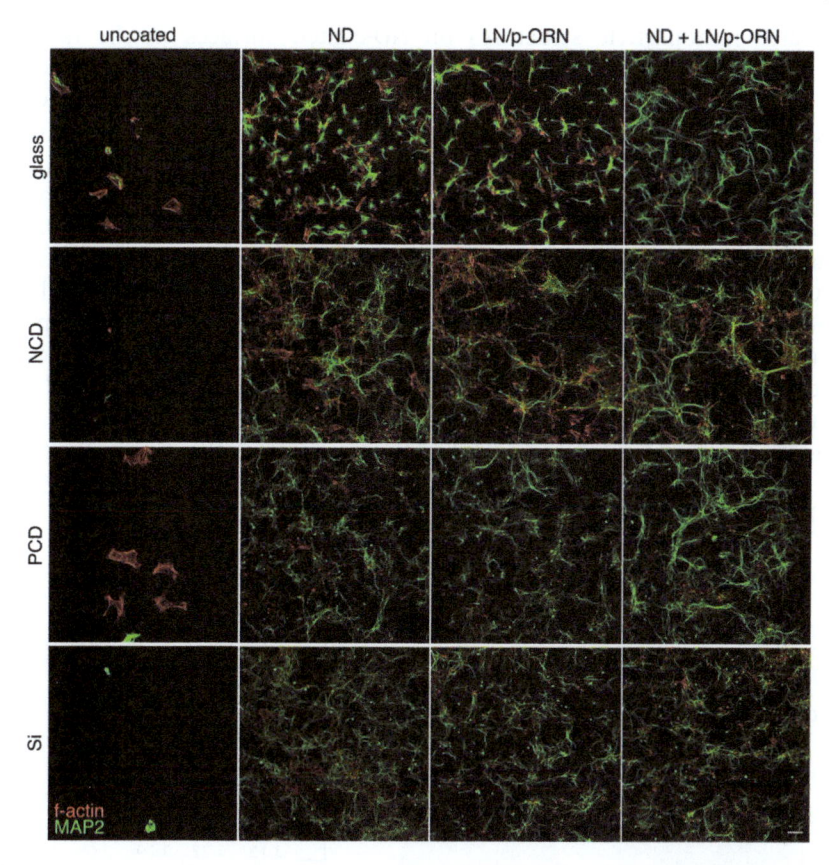

Figure 9.1 Neuronal growth on ND-coated substrates. Confocal analysis of growth of hippocampal neurons after 2 days in culture (DIV2). Substrates of glass, nanocrystalline diamond (NCD), polycrystalline diamond (PCD), and silicon (Si) were coated with nanodiamonds (ND) and/or laminin (LN)/poly-DL-ornithine (p-ORN); surfaces of uncoated substrates are oxygen terminated. Immunostaining revealeds neurons via the dendrite-specific marker MAP2 (shown in green), while the cytoskeletal filaments of actin (f-actin) were stained for with rhodamine-phalloidin (red), highlighting structures rich in f-actin, such as growth cones at tips of neurites. ND-coating (2nd column) promoted neuronal attachment and outgrowth similar to conventional protein coating (3rd column), whereas the uncoated substrates displayed no significant attachment and growth of neurons, where only few cells of apparent non-neuronal origin can be detected (1st column). Scale bar: 50 μm. 3 independent batches of cultures and coatings were tested.
Reproduced with permission from Elsevier copyright 2010, Ref. 55.

9.2.3 Substrate Dependence of Surface Roughness Following ND Layering

Atomic force microscopy (AFM) was used to investigate the physical nature of NDs on the various substrate materials. AFM scans of the sample surfaces revealed nanoparticle structures on ND-coated surfaces in the range of 10–20 nm, which was comparable for all underlying substrates tested and did not vary substantially between different batches of ND solution used for coating (Figure 9.2a). The uncoated glass, PCD and Si surfaces were very flat with mean roughness of less than 2 nm. NCD was rougher than the other three materials with a mean roughness of approximately 8 nm. Coating the materials with NDs resulted in an increase in roughness on the glass, PCD and Si by 1.5–2 nm, whilst the NCD layer experienced a reduction in mean roughness by approximately 2 nm following ND coating. The NDs filled in and smoothed the roughness of the NCD substrate that had topographical features in excess of the ND particle size and increased the roughness of the flatter surfaces. Interestingly, all uncoated substrates showed only slight differences in mean roughness to their ND-coated equivalents (Figure 9.2c) and no clear correlation to neuronal attachment could be

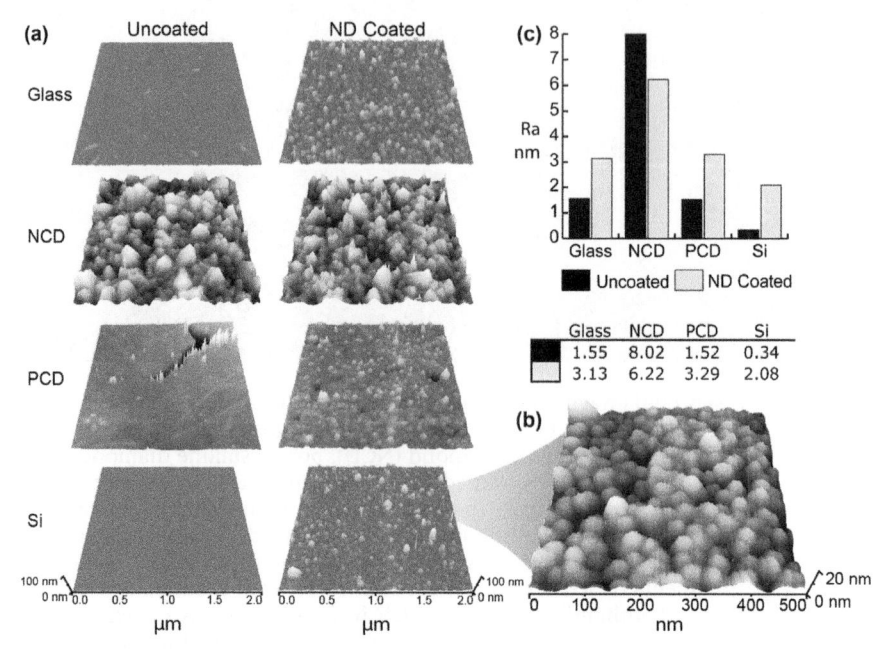

Figure 9.2 Surface roughness of ND-coated substrates that promote neuronal growth. a) AFM images of glass, NCD, PCD and Si surfaces with and without the ND coating. Scans are 2 μm×2 μm sample areas and are presented in 25% perspective. The scans were representative of the whole substrate areas and were selected at random over the entire substrate surface. Layer coverings of ND particles can be seen on all substrates. b) A random 500 nm×500 nm area from the Si ND sample showed the ND layer to have particles ranging in size from 10 to 20 nm in diameter. Particle size was determined from the z-axis measurement in order to avoid AFM tip convolution effects. c) Mean roughness (R_a) data of the 500 nm square sample areas of all substrates for uncoated and ND coated samples.
Reproduced with permission from Elsevier copyright 2010, Ref. 55.

observed (see Figure 9.1 and Figure 9.5, first two columns). This suggests that the observed neuron adhesion and growth could likely be ascribed to other properties linked specifically to the NDs themselves. The surface functionalisation of the DNDs was predominantly oxygen-terminated with partial hydrogen functionalities as determined by Fourier transform infrared (FTIR) spectroscopy.

9.2.4 Neuronal Attachment and Outgrowth on Different Materials

Our observation of ND-promoted neuronal attachment prompted an investigation of these neuronal cultures with respect to neurite outgrowth, direct interaction with the substrate, long-term survival and functional properties. On ND-coated substrates, hippocampal neurons displayed branched neurite outgrowth comparable to those on conventional protein-coated materials—a mixture of ECM proteins of laminin (LN) and poly-DL-ornithine (p-ORN), (Figure 9.1, columns 2 and 3). Sequential coating with NDs and protein did not result in any further improvement of cell attachment (Figure 9.1, right column) or negative interference. The ND layer supported growth of both neurons and glia cells (Figure 9.3). This raised the possibility that neurons might not be growing directly on the ND monolayer but on glia cells attached to the ND coating. To distinguish between these possibilities we stained the cultures with neuron- and glia-specific markers. Importantly, neurons were found to be able to grow in direct contact with the ND-coated surfaces and without interpositioned glia cells (Figure 9.3, arrow heads).

9.2.5 Formation of Neuronal Networks

The capability of ND coatings to support the establishment of electrically active networks in more mature cultures was next investigated. The

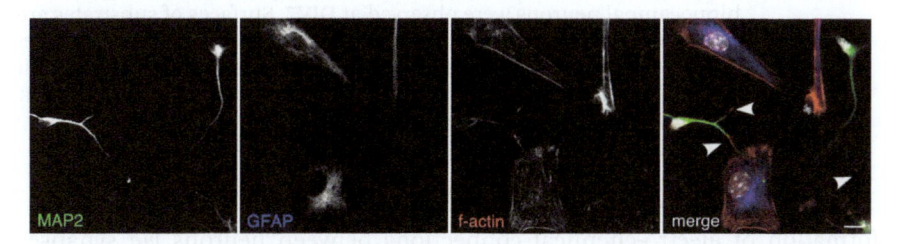

Figure 9.3 Neuronal outgrowth on ND-layered NCD. Confocal imaging of cell types and cellular compartments revealed that neuronal outgrowth could take place in direct contact with ND-layer. Dendrites of neurons (MAP-2, green in false colour merge), glia cells (GFAP, blue), f-actin (rhodamine-phalloidin, red), cell nuclei (Hoechst33258, grey values in merge). Arrowheads indicate neuronal growth cones. Scale bar: 10 µm.
Reproduced with permission from Elsevier copyright 2010, Ref. 55.

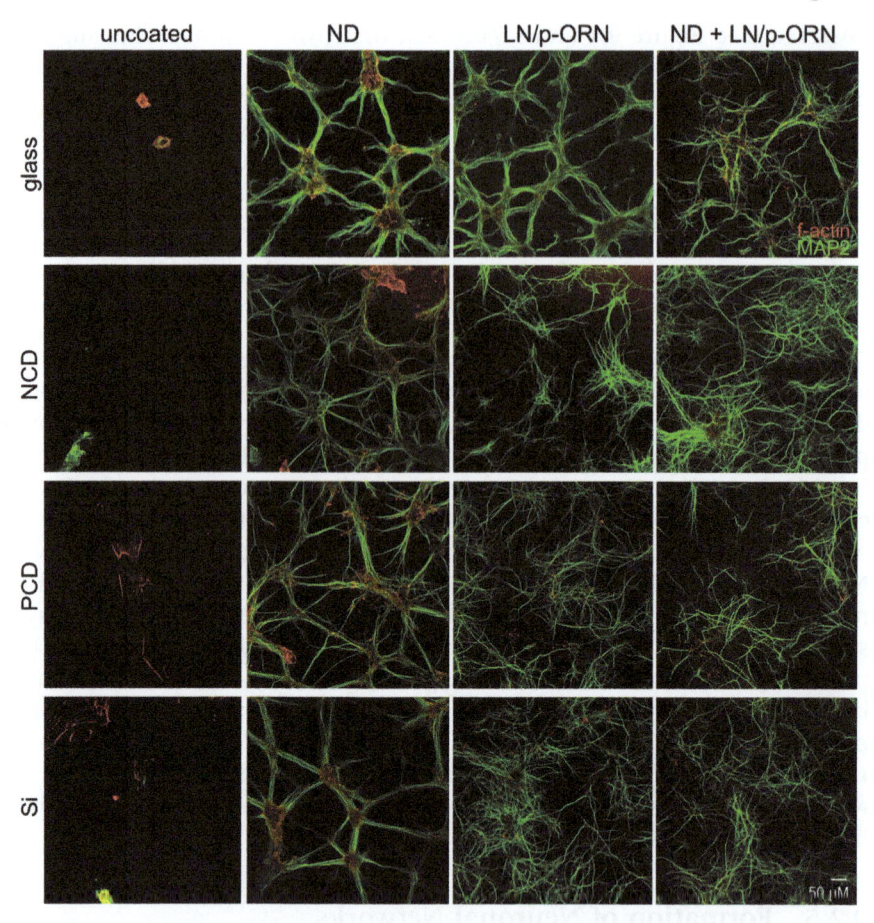

Figure 9.4 Neuronal growth on differently coated substrates after 7 days *in vitro* (DIV7). Substrates of glass, nanocrystalline diamond (NCD), polycrystalline diamond (PCD), and silicone (Si) were coated with nanodiamonds (ND) and/or laminin (LN)/poly-DL-ornithine (p-ORN) and the growth of hippocampal neurons were observed at DIV7. Surfaces of substrates are oxygen-terminated. Immunostaining reveals neurons via the dendrite-specific marker MAP2 (shown in green), while f-actin is stained for with rhodamine-phalloidin. Scale bar: 50 μm. The neurons have made clear physical connections between neighbouring cells.
Reproduced with permission from Elsevier copyright 2010, Ref. 55.

formation of electro-chemical connections between neurons *via* synapses takes place from about five days in culture onwards.[56] Confocal microscopy of immunostained neuronal cultures on all four substrates showed robust neuronal arborisation at seven days *in vitro* (DIV7) (Figure 9.4). Scanning electron microscopy (SEM) confirmed formation of well-developed neuronal networks by day seven and, in addition, revealed little or no degradation of the ND coating (Figure 9.5).

Neuron Growth on Nanodiamond 203

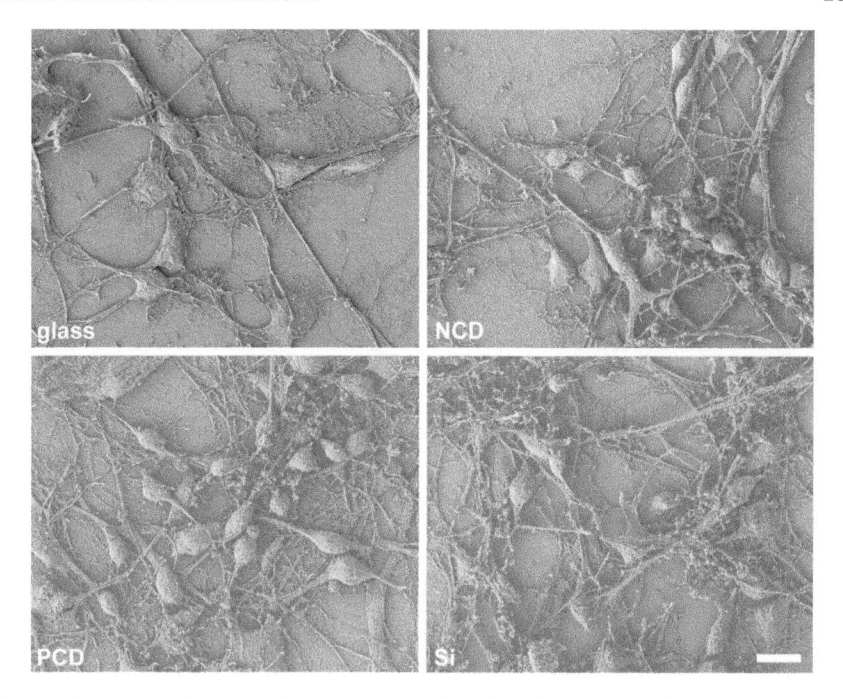

Figure 9.5 SEM images of neurons growing directly on ND-only coated material surfaces. Neurons appear highly branched and networked at DIV7. The ND layer appears intact on all supporting materials. Scale bar 10 μm. Reproduced with permission from Elsevier copyright 2010, Ref. 55.

9.2.6 Intrinsic Electric Excitability of ND-grown Neurons

To investigate whether neurons grown on ND layers displayed normal cell-autonomous electrical excitability, whole-cell patch-clamp recordings were performed on individual neurons in the current-clamp configuration. Graded somatic current injections successfully elicited trains of action potentials (APs) in neurons grown on ND-coated glass (Figure 9.6, right panel). This denotes that ND coating supports the development of electrically functional neurons. A comparison with recordings performed in parallel on sister cultures grown on conventional laminin and poly-DL-ornithine-coated glass (LN/p-ORN) indicates that neurons grown on the two substrates behave similarly in terms of AP frequency for injected currents up to 150 pA (Figure 9.6). Average AP frequency at 150 pA injected currents was 28.00 ± 10.04 Hz for neurons on LN/p-ORN coatings and 33.75 ± 6.44 Hz for the ND coating, $p = 0.65$, unpaired two-tailed Student's t-test. In order to probe to what extent the two conditions were comparable, increasingly stronger and saturating currents were injected. Only for the very high current injections, especially at the end of the pulses, did neurons grown on ND layers display a reduced capacity to fire APs (compare lower panels of Figure 9.6). As there were no significant differences in input resistance (R_{in}) and

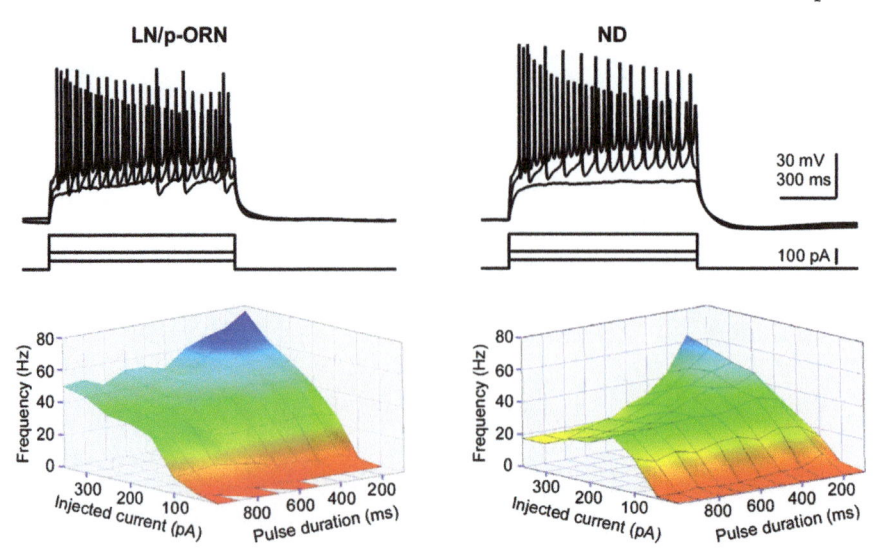

Figure 9.6 Intrinsic excitability of neurons on LN- vs. ND-coated glass. Neurons from sister cultures grown on laminin ('control', left) or ND-only coated glass (right) were recorded in current-clamp mode after 7 DIV. 1 s-long rectangular pulses of current were injected via the patch pipette at the soma, with increments of 50 pA from 0 to 350 pA. Example traces recorded at 50, 100 and 200 pA are shown in top panels; middle panels depict the corresponding stimulus protocols. Summary data (n = 3 cells for control and 4 for ND) are shown in bottom panels, where action potential (AP) frequencies are plotted against pulse duration (binned at 100 ms) and intensity. Note the similarity in AP frequencies between control and ND-grown neurons for current injections up to 150 pA. Reproduced with permission from Elsevier copyright 2010, Ref. 55.

membrane capacitance (C_m) between the two groups ($R_{in} = 778 \pm 246$ MΩ and $C_m = 172 \pm 15$ pF for LN/p-ORN coating vs. $R_{in} = 735 \pm 118$ MΩ and $C_m = 156 \pm 31$ pF for ND coating, $p = 0.88$ and 0.67, respectively, unpaired two-tailed Student's t-test), the reduced number of APs in ND-grown neurons for the highest current injections (250–350 pA) is likely due to ion channel differences that make it more difficult for neurons grown on NDs to cope with strong depolarising inputs.

9.2.7 Synaptic Connectivity

Having established the individual neuron's ability to generate APs, the ND coating's ability to support the formation of functional synapses was investigated. To this end, whole-cell patch-clamp measurements in the voltage-clamp configuration were performed to record spontaneous miniature excitatory postsynaptic currents (mEPSCs) from individual neurons. mEPSCs are thought to represent the activation of postsynaptic glutamate receptors in response to the spontaneous release of a single synaptic vesicle of

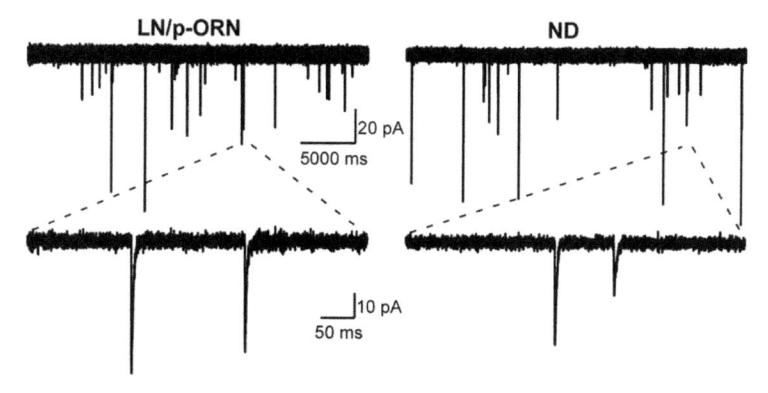

Figure 9.7 ND-coating supports formation of functional synaptic connections. Spontaneous synaptic transmission of neurons on LN- vs. ND-coated glass. Spontaneous mEPSCs were recorded from neurons of sister cultures grown on laminin ('control', left) or ND-only coated glass (right) for 7 days. Note the similar size and frequency of mEPSCs in the two conditions.
Reproduced with permission from Elsevier copyright 2010, Ref. 55.

neurotransmitter and hence provide invaluable information on the mechanism of synaptic transmission.[57] As shown in the right panel of Figure 9.7, neurons grown on ND layers displayed large and frequent mEPSCs only after seven days. A comparison with mEPSCs recorded in parallel from sister cultures grown on LN/p-ORN (Figure 9.7, left panel) revealed no differences in terms of mEPSC size and frequency between the two groups of cells (32.8 ± 9.8 pA and 1.18 ± 0.48 Hz for ND coating vs. 22.6 ± 4.0 pA and 1.73 ± 0.95 Hz for LN/p-ORN coating, $n=3$ cells per condition, $p=0.41$ and 0.64, respectively, unpaired two-tailed Student's t-test), suggesting that ND coating supports normal synaptogenesis and synapse maturation, and possibly the formation of functional neuronal networks.

9.2.8 Electrochemical Network: Calcium Oscillations

The electrophysiological experiments discussed above addressed the cell-autonomous electrical properties of neurons on the ND-coated surfaces, as well as the synaptic input an examined neuron received from contacting cells. In addition, the formation of the full neuronal network formed in the presence of ND coating *via* calcium imaging with the calcium dye fura-2 was examined. Following the pharmacological blockade of intrinsic inhibitory inputs, cultures grown on protein-coated glass showed the expected slow oscillatory calcium waves that are associated with the epileptic-like state induced by this treatment. Similar oscillatory activity was observed in cultures grown on the ND coating, which confirmed the formation of interconnected, electrically active networks of neurons on substrates coated with NDs (Figure 9.8).

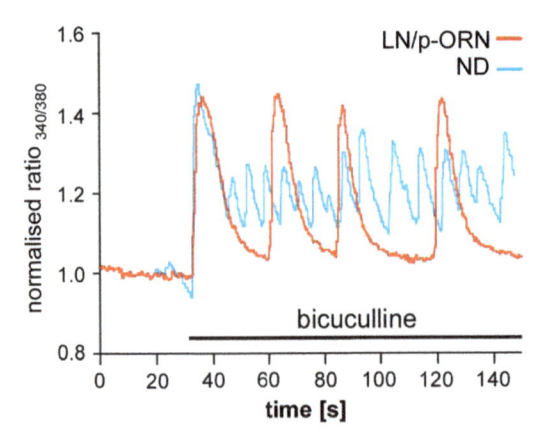

Figure 9.8 ND-coating supports formation of functional neuronal networks. Calcium measurements of neurons grown on protein or ND-only coated glass samples demonstrated network activity in primary hippocampal cultures at DIV12. Global calcium transients were elicited by bath application of 50 μM bicuculline. Fura-2 ratio$_{340/380}$ values are shown normalized to baseline.
Reproduced with permission from Elsevier copyright 2010, Ref. 55.

9.2.9 Discussion

9.2.9.1 Simplicity and Universality of ND Coating

Previous studies have revealed that nanodiamonds are, in general, much more biocompatible than other carbon-based nanostructured materials, such as carbon nanotubes (CNTs) or carbon blacks.[58] Nanodiamonds are chemically inert, mechanically robust[59] and non-cytotoxic[60] in contrast to CNTs, which have been shown to induce apoptosis and, therefore, appear less suitable for use *in vivo*.[61]

In this study, coating all the four different substrates tested (glass, NCD, PCD and Si) with ND layers was found to convert them from unfavourable surfaces for cell attachment, growth and maintenance into most suitable platforms that compared well to the established culture conditions. This suggests that ND coating may be a widely applicable surface treatment to confer cell-interacting properties to various materials. Such properties are desirable for certain medical applications, such as chronic implants. This is particularly interesting since ND coating is an amazingly simple process that can coat most materials in three dimensions (3D) and is only limited by the size of the ultrasonic bath being used, although the long-term stability (*i.e.*, >2 weeks) of the ND coating by this process has not yet been tested. Importantly, the ultrasound-mediated layering does not rely on specialised equipment or chemically–thermally aggressive protocols that are commonly used with the deposition and post-growth cleaning of other carbon materials, such as CNTs.[62]

9.2.9.2 Comparison with Other Growth Platforms

Whilst other studies have shown increased cell proliferation on oxygen- and hydrogen-terminated surfaces of NCD in the absence of further coating,[13,63-65] significant cell attachment of essentially post-mitotic neurons on the oxygen-terminated NCD substrates was not detected on the NCD used in this study. This suggests that cultures of primary murine hippocampal neurons are more demanding with respect to a suitable growth surface than cultures of renal epithelial and bone marrow cell cultures, or mesenchymal and neural stem cells. Alternatively, subtle differences in the surface properties of the uncoated NCD substrates from the studies compared to those used here, such as roughness or curvature of the surface topography, or propensity to protein adsorption may account for the differences in cell attachment.

Neurons on ND layers were remarkably similar to those grown on traditional ECM-protein-coated samples. Electrophysiological recordings suggest that there might be only minor differences in neuronal excitability. Equally, the differences in calcium wave oscillation frequency of induced epileptic activity could have been caused by these small differences in intrinsic excitability of neurons, but might reflect, in addition, differences in network connectivity. Such variations are in line with the known variability of primary neuronal cultures maintained under slightly variant conditions.

In contrast to electrical shortcuts observed in neurons grown on CNTs,[66] the electrical excitability, and hence the ability to generate APs, seems not to be influenced by NDs, which remain electrically inert. This would allow the usage of patterned NDs on conductive biomaterials for generation of devices with spatially defined electrical properties for biological and medical application.

9.2.9.3 Direct Attachment of Neurons to ND Layers

In the case of cell cultures containing different cell types, it is of interest to know whether the various cell types attach and grow directly on the ND layer or whether the more demanding cell types grow only on top of the less demanding ones. In the case of neuronal cultures, glia cells are known to adhere to many surfaces and can serve as an efficient substrate layer for neurons. We observed direct attachment and growth of neurons on the ND layer without interpositioned glia cells. This suggests that low-capacitance and sensitive ion-sensing electrical coupling between neurons and substrate could be achieved on ND-coated materials, which is desirable for electrical interfacing between an implant and neuronal tissue.

9.3 The Effect of Varying Nanodiamond Properties on Neuronal Adhesion and Outgrowth

9.3.1 Introduction

DND monolayers show a remarkable ability to support neuronal networks in stark contrast to nanocrystalline diamond (NCD) thin films, despite the seemingly similar bulk and surface properties of these materials. DND and NCD coatings differ mostly in their morphological properties; however, the observed adhesion of neurons on DND *vs.* NCD is not sufficiently described by the established surface metric of average roughness. The average roughness of both DND and NCD lie below the reported window of neuronal adhesion on silicon[26–28] (10–70 nm average roughness). The protein-sized dimensions, chemical inertness and organic surface functionalisations of monodispersed nanodiamonds and their coatings suggest DNDs have an influence on neuronal adhesion on a nanoscale subcellular level. In this context the mechanisms responsible for cell adhesion on DND surfaces have yet to be identified. In order to investigate the effect of varying nanodiamond coating properties on neuronal adhesion, we[67] varied the size, surface chemistry, source and deposition method of nanodiamond coatings and infer possible mechanisms of neuronal adhesion on ND-coated substrates.

9.3.2 Coating of Different Nanodiamond Coatings and Characterisation

To investigate the dependence of ND coating type on neuronal adhesion, ten different types of coatings have been prepared on glass slides using different ND sources and deposition methods with or without the addition of a chemically modifying plasma-enhanced chemical vapour deposition (PECVD) treatment. Firstly, monodispersed DNDs were deposited on glass cover slips using ultrasonic seeding (US), as performed in section 9.2.2 (Figure 9.9a, column 1). The same DNDs were next deposited by spin coating (SC), resulting in more aggregated, sparse DND coatings (Figure 9.9a, column 2). In order to investigate the effect of ND particle size on neuronal adhesion, as well as differing morphology and surface functionalisation, *ca.* 20 nm high pressure high temperature (HPHT) ND hydrocolloids were then prepared and deposited *via* the methods of spin coating (Figure 9.9a, column 3), electrostatic binding using ND/polyvinyl alcohol (PVA) hydrocolloids on poly(diallydimethylammonium) chloride (PDDAC)-coated glass substrates, as developed by Girard *et al.*[68] (Figure 9.9a, column 4), and finally dip coating (Figure 9.9a, column 5). PDDAC was used to deposit both HPHT NDs and electrostatically bound aggregated NDs. PDDAC was removed by a short (20–40 s) hydrogen microwave plasma.

For each coating type described above, additional duplicate samples were then prepared, which, following deposition, were subjected to a short microwave plasma chemical vapour deposition (MWPECVD) treatment in order to result in better-adhered ND coatings and also hydrogenated ND surfaces, as opposed to untreated, predominantly oxygenated ND surface functionalisation (see below). Following MWPECVD treatment, the increased adhesion of ND coatings was observed *via* the improved resistance of the coatings to mechanical abrasion. MWPECVD processing is commonly

Neuron Growth on Nanodiamond 209

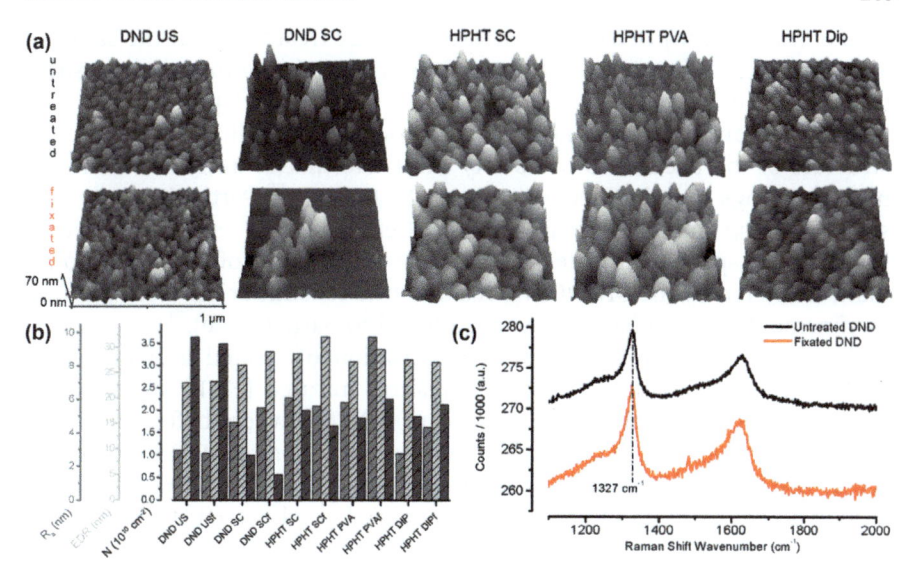

Figure 9.9 a) 1 µm square AFM images of ND-coated substrates coated using a variety of ND types and deposition methods. Row 1 show untreated coatings and Row 2 show MWPECVD fixated resulting in a hydrogenated surface. No differences in coating were apparent before and after fixing. Scans were representative of the whole sample surface and were selected at random over the entire substrate surface. b) Corresponding average roughness, R_a; equivalent disc radius, EDR; and seeding density, N of AFM images. EDR measurements were subject to AFM tip convolution contributions resulting in an overestimate of particle size by ca. 10 nm. c) Raman spectra of untreated and fixated DNDs show no significant differences in spectra. Pertinently the 1327 cm^{-1} peak was not shifted, indicating no change in core particulate size from the fixing treatment.

used to grow nanocrystalline diamond (NCD) films[69] from ND seeded coatings; therefore, in order to not result in an accrued mass of diamond on the NDs or significantly alter their morphology, the fixing step was kept short (11 min) and at relatively low power in comparison to typical diamond growth parameters. Atomic force microscopy (AFM) image pairs of most coatings (Figure 9.9a and 9.9b) revealed no discernible differences in the morphology of the coatings before and after fixation outside of the heterogeneity expected from the employed deposition methods, except for perhaps the sample 'fixed HPHT PVA', which had a skewed average roughness (R_a) measurement that could be ascribed to an atypical coating heterogeneity for the fixed sample area measured. The average roughness (R_a) and ND particle statistics of equivalent disc radius (EDR) and seeding density (N) were measured using image processing and grain analysis of the AFM images, and parameterised ND coating properties are displayed in Figure 9.9b.

In order to confirm the lack of increase in ND particle size after MWPECVD fixation treatment, Raman spectroscopy was performed on thicker (*ca.* 100 nm) DND layers before and after treatment. All the samples were not tested due to the smallest diamond seeds being the most sensitive to changes in Raman spectra with respect to increased particle size (see below). Raman spectra in Figure 9.9c show the crystalline carbonaceous structure of the fixed DND to be very similar to that of untreated nanodiamond coatings and distinct from grown NCD films.[70] Pertinently, the redshift of the sp^3 diamond peak from 1333 cm^{-1} to 1327 cm^{-1} observed in both Raman spectra indicates the primary particle size of the DNDs to remain unaltered by the MWPECVD fixation (the redshift arises from phonon confinement in nanodiamond and is related to particle size).[71]

To characterise the surface functionalisation of treated and untreated DND and HPHT coatings, Fourier transform infrared (FTIR) spectroscopy measurements were taken from DND and HPHT powders before and after MWPECVD treatment. Figure 9.10 shows that, prior to MWPECVD treatment, DNDs are predominantly hydroxyl (broad band *ca.* 3400 cm^{-1}) and carboxylic acid (*ca.* 1750 cm^{-1}) functionalised,[72] whilst HPHT NDs show increased anhydride functionalities[73] (*ca.* 1800 cm^{-1}). MWPECVD treatment resulted in increased hydrogenation (C–H, *ca.* 3000-2900 cm^{-1}) of both DNDs and HPHT NDs, as well as reduced coverage of the aforementioned oxygen moieties.

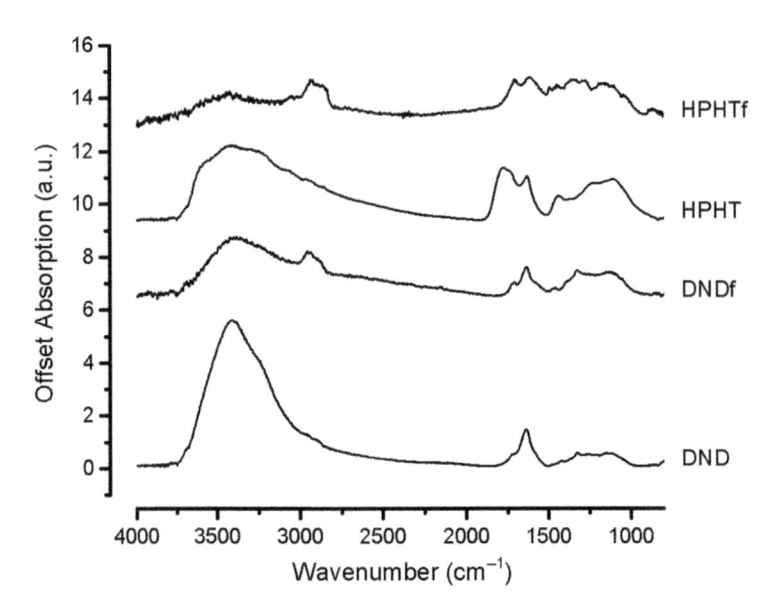

Figure 9.10 FTIR scans of DND and HPHT samples and their corresponding fixated sample spectra, DNDf and HPHTf. DND and HPHT show predominantly oxygen moiety functionalisation prior to MWPECVD fixation, after which increased hydrogenation is measured for both ND types. Intensity magnitude is arbitrary between samples.

Neuron Growth on Nanodiamond

9.3.3 Neuronal Cell Attachment on ND Coatings

Primary neuronal cultures were seeded onto the ten different ND-coated samples and a control sample comprised of a glass cover slip coated with LN/p-ORN. Confocal microscopy images of immunostained neuronal cultures after two days *in vitro* (DIV2) revealed attachment and outgrowth of neurons on all samples tested (Figure 9.11). In order to quantify neuronal outgrowth on the different ND coatings the total neurite length (TNL) was measured using NeuriteTracer. Average TNL per quadrant for all samples is shown in the bottom left hand panel of Figure 9.11. For ND-coated substrates, neurite extension was similar or in excess to the control LN/p-ORN-coated substrate.

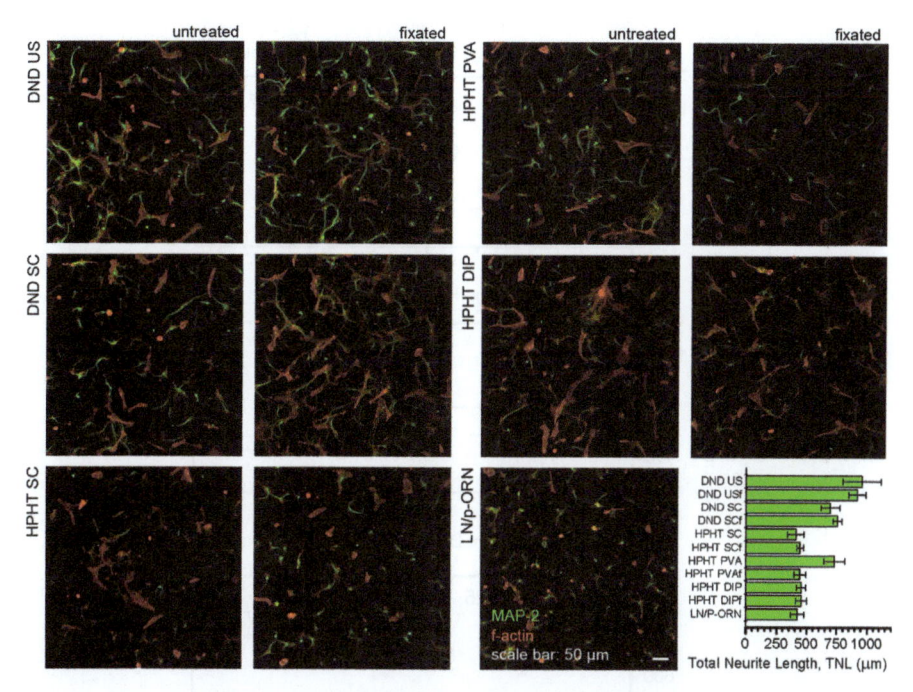

Figure 9.11 Confocal microscopy images of immunostained neuronal cell cultures after DIV2 on ND-coated and laminin (LN)/poly-DL-ornithine (p-ORN) coated glass substrates. Immunostaining reveals neurons via the dendrite-specific marker MAP2 (shown in green) and cytoskeletal filaments of actin (f-actin) are stained for using rhodamine-phalloidin (red), which highlights structures rich in f-actin, such as growth cones at the tips of neurites. All ND-coated samples promoted neuronal adhesion and neurite outgrowth in a similar or favourable comparison to protein coated substrates. Notably, hydrogenated and untreated samples showed very similar levels of attachment and outgrowth. Scale bar 50 μm. The quadrant-averaged Total Neurite Length (TNL) of each sample (measured using the NeuriteTracer ImageJ plugin) is shown in the bottom left panel, with error bars corresponding to the standard error of the mean (s.e.m.).

Notably, no discernable differences between untreated and hydrogenated, fixed ND coatings were observed, except for untreated and fixed HPHT PVA samples that showed a reduced neurite extension post-fixation (Figure 9.11, forth row).

Having parameterised the degree of neurite extension on the different ND-coated samples, the Pearson product-moment correlation coefficients, r, and their significance, p, were calculated in order to test for any significant linear correlation between TNL and the associated ND coating parameters of R_a, EDR and N. Scatter plots of R_a, EDR and N vs. TNL are shown in Figure 9.12, accompanied by their associated r and p values inset in each panel. The largest and only statistically significant ($r = -0.77$; $p = 0.01$) correlation observed was EDR vs. TNL, suggesting that the size of particles has a

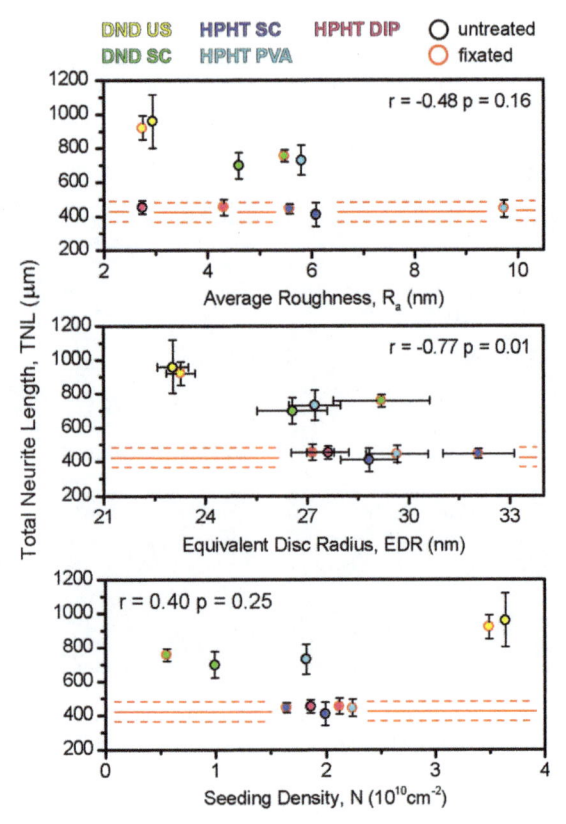

Figure 9.12 Scatter plots showing the relationships between the AFM-determined morphological properties (Figure 9.9b) of the ND-coated samples against the total neurite extension measured from confocal microscopy images shown in Figure 9.11. Data point fill colour: ND coating type. Line border colour: untreated/fixated. TNL and EDR error bars are both s.e.m.. The solid horizontal line across each panel indicates the LN/p-ORN TNL with the flanking dashed lines indicating ± s.e.m.. Inset r and p-values report the Pearson correlation coefficient and 2-tailed significance for each data set.

determining factor on neuronal adhesion and development in the nanoscale range tested. R_a also showed a negative but insignificant correlation coefficient over the R_a range tested. No significant correlation to TNL is observed for the N range tested; however, the correlation coefficient for N vs. TNL was positive ($r = 0.4$; $p = 0.25$), as expected (as N tends to zero so should TNL). Cultures were maintained up to at least DIV7, forming apparent neuronal networks on all substrates tested.

9.3.4 Mechanism of Neuronal Adhesion with Respect to ND Surface Properties

Whilst cellular adhesion is regularly achieved on artificial substrates *via* biomolecule and peptide lamination, some artificial substrates have been shown to be able to promote cellular adhesion. The most primitive mechanism known to promote cellular adhesion on artificial substrates is the electrostatic binding of negatively charged cell membranes to positively charged surfaces of materials such as polylysine.[74] NDs could also have a similar effect on neuronal adhesion, especially considering their polar surfaces; however, a common fate for cells adhering in such a manner is apoptosis, which is caused by the cells' subsequent inability to secrete ECM proteins.[14]

Considering the remarkable ability of NDs to support and direct neuronal adhesion for extended periods of time, it is tempting to entertain the possibility that NDs are mimicking the functional properties of ECM proteins and are directly interfacing with cells and initiating cellular attachment and development. After all, the size, charge, shape and the organic functionality of ND particles that could mimic the attachment peptide motifs of ECM proteins (*e.g.*, RGD motif)[16] is similar to that of globular proteins, and other nanoparticles have already been shown to mimic other biological processes.[75] In the particular case of NDs being anchors for the development of focal adhesions with neurons, NDs would have to mimic the function of adhesion molecules (AMs) of the ECM by mimicking AM peptide motifs, such as (for laminin) RGD, YIGSR, IKVAV, *etc.*,[16] which in turn interface with cells *via* membrane proteins called integrins.[19] In this context similarities can be drawn between ND surface functionalities and the functionalities of the R groups of amino acids of AM motifs, such as the amine groups on arginine (R) and carboxylic groups on aspartic acid (D) of the common RGD motif. Whilst it is not inconceivable that functionalities on NDs could be serendipitously arranged as to occasionally mimic the motifs of AMs (especially considering organic RGD mimetics exist[76]), a direct ND–cell interaction is still highly unlikely. Without integrin signalling, cells die in a process called anoikis, and even slight changes in integrins or AM motif sequences cause anoikis or severe loss of function in cells.[77,78] Therefore, it would be surprising that NDs—crudely forged in detonation chambers—could mimic the highly specific, evolved interactions of AMs with integrins. In this context a direct cell–ND interaction is refuted by the observation of neuronal adhesion on both untreated and hydrogenated NDs, the latter of which could not have such AM motif mimetic surface chemistry.

Instead, it is much more likely that NDs provide a substrate for the optimal adsorption of AMs (*e.g.*, vitronectin and fibronectin from serum proteins and subsequent ECM proteins produced by cells themselves)[19] as a result of the NDs' high affinity for protein adsorption[43] and, moreover, their small radii of curvature that is reported[35] to promote the functional adsorption of (in particular) fragile globular proteins, such as vitronectin and fibronectin. Indeed, in the absence of serum, neurons do not attach to ND-coated substrates (*note*: serum also contains other growth factors essential to cell attachment that are not AMs) and the slight fasciculation of neurons observed in Figure 9.4, column 2 and Figure 9.13c (v) (after DIV7) can be ascribed to neurons attaching onto their own matrix of endogenous AMs and not the underlying surface. One should note, however, that such observations can be ascribed to varying culture conditions and is evident on standard protein coated substrates, such as in Figure 9.4, column 3, row 1.

In this context the correlation coefficient analysis in Figure 9.12 between the neurite extension and parameterised morphological variables of the ND coatings shows a significant correlation between the radii of curvature (proportional to EDR) of the ND particles and the total neurite length measured ($r = -0.77$; $p = 0.01$), whereas the more commonly studied biomaterial metric of average roughness is only weakly correlated to neurite extension ($r = -0.48$; $p = 0.16$). The negative correlation measured agrees with the above hypothesis that as the ND particles tend towards the size of proteins and increase in curvature, the surface becomes more biocompatible and more neurite extension is observed. Furthermore, it can explain the increased adhesion of neurons on ND-coatings and reduced adhesion on facetted, low curvature NCD.

9.4 Patterned Neuronal Networks Using Nanodiamonds

In view of the promising nature of ND monolayers for *in vitro* and *in vivo* brain-machine interface applications, due to their ease of deposition in 3D on most materials,[79] their porous nature and electrode functionality,[80] the usage of micro-patterning for the directed growth of hippocampal neurons has been demonstrated by us[67] using patterned DND coatings of varying line widths from 5 to 70 µm.

Patterned ND coatings were deposited as described by Bergonzo *et al.*[81] using the photolithography mask pictured in Figure 9.13a. The ND type and deposition method of untreated High-Pressure, High Temperature (HPHT) grown single crystal (SC) diamond was used due to its spin coating process being readily incorporated into photolithography processing and it also having a high seeding density for ease of feature identification using AFM. ND patterns were etched with negligible residual ND particulates and

Neuron Growth on Nanodiamond 215

Figure 9.13b shows an AFM image of a crossroad junction of 15 μm thick ND tracks. Neuronal cultures were carried out on top of patterned substrates. Confocal microscopy images after DIV2 revealed NDs are able to directly pattern neuronal adhesion down to line widths of at least 10 μm with high contrast (Figure 9.13c). Inspecting Figure 9.13c (i) one can see the majority of neuron nuclei sat over the junction of the ND tracks for 10 μm line widths, whereas for larger widths (Figure 9.13c (ii), (iii)) nuclei were dispersed seemingly irrespective of the patterned features. For the unpatterned area of Figure 9.13c (iv), less neurite branching was observed than on the patterned areas, with the neurons taking on a more circular morphology, as also

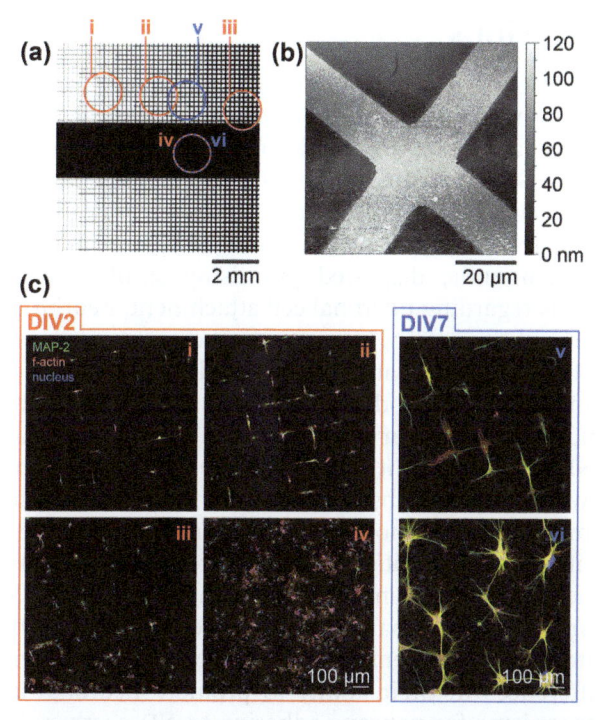

Figure 9.13 a) Diagram of photolithography mask used for ND patterning (9 mm square). Black areas were ND filled. Grid pitch was 200 μm and incremental gradated line widths of 5, 10, 20, 25, 30, 35, 40, 50, 60 and 70 μm increased from left to right every 4th to 5th grid unit. The central unpatterned region was used as an ND control surface for each sample. The indicated circular regions correspond to the confocal images presented in panel (c). b) 70 μm square AFM image of patterned HPHT SC tracks on glass substrate. Width: 15 μm. Height: ca. 35 nm. c) Confocal microscopy images of immunostained neuronal cultures on untreated HPHT SC patterned glass substrates. Cells have MAP2 and f-actin staining as in Figure X.10 and an additional nuclei stain using Hoechst 33258. After DIV2 neurons could be seen to be following ND-tracks down to at least 10 μm (c.i) and for larger line widths as well (c.ii, c.iii). After DIV7, neurons formed interconnected networks as directed by the ND patterning (c.v). Scale bar 100 μm.

observed in HPHT SC in Figure 9.11. After DIV7 (Figure 9.13c (v)), neurons were seen to have formed physically connected, ordered networks on top of the ND tracks and, in comparison to the unpatterned mask area (Figure 9.13c (vi)), the neurons were not fasciculated. The ability to directly pattern neuronal adhesion using artificial nanodiamond coatings presents itself as a valuable tool for incorporating neuronal patterning into *in vitro* neuronal devices (*e.g.*, multi-electrode recording arrays, MEAs) and *in vivo* neural prosthetics (*e.g.*, artificial retina) at the micro-processing stage, without the need for post-process biomolecular printing.

9.5 Conclusions

Nanodiamond monolayers provide an exceptional substrate for supporting functional neuronal networks with electrophysiological characteristics in remarkable comparison to traditional protein-coated substrates, and even *in vivo* neurons. Section 9.2 describes the first demonstration of neuronal growth on nanodiamond, where the surfaces of four different materials (glass, PCD, NCD and Si), coated with monolayers of monodispersed detonation nanodiamonds, displayed promising similarity to the protein-coated materials regarding neuronal cell attachment, neurite outgrowth and functional network formation. Importantly, the neurons were able to grow in direct contact with the ND-coated material and could be easily maintained in culture for an extended period equal to those on protein-coated substrates. Given the biocompatibility of NDs and their potential for surface functionalisation, ND layering might prove a valuable material technique for implants on a wide range of substrates. Section 9.3 describes further research by us, wherein various nanodiamond coatings of different origin, size, functionalisation and deposition method were investigated for their ability to support neuronal adhesion. All ND coatings were shown to universally promote murine hippocampal neuronal adhesion. Predominantly oxygenated (untreated) and hydrogenated, fixed nanodiamond coatings showed no apparent differences in their ability to promote neuronal adhesion. The possible mechanisms for neurons adhering to ND coatings have been discussed and, *via* parametric correlation coefficient analysis of the morphological properties of ND coatings and their corresponding neuronal cultures, a significant correlation between neurite extension and nanodiamond size and curvature has been found, wherein smaller sized nanodiamonds promote greater neurite extension. This effect is most probably ascribed to functional protein adsorption onto ND layers. Finally, Section 9.4 describes the use of nanodiamond patterns to pattern neuronal adhesion with high contrast, which demonstrates ND patterning to be a promising technique for the *in vitro* study of ordered neuronal networks without using additional biomolecules to promote adhesion. Patterned ND coatings show great potential for directing neurons towards electrodes of neuroprosthetics, such as artificial retina microelectrode arrays, cortical or cochlear implants.

References

1. M. P. Ward, P. Rajdev, C. Ellison and P. P. Irazoqui, *Brain Research*, 2009, **1282**, 183–200.
2. W. J. Streit, *Toxicologic Pathology*, 2000, **28**, 28–30.
3. K. Cheung, *Biomedical Microdevices*, 2007, **9**, 923–938.
4. W. J. Streit and C. A. Kincaid-Colton, *Sci. Am.*, 1995, **273**(54–5), 58–61.
5. C. Hassler, T. Boretius and T. Stieglitz, *J. Polym. Sci. Pol. Phys.*, 2011, **49**, 18–33.
6. V. Polikov, P. Tresco and W. Reichert, *J. Neurosci. Methods*, 2005, **148**, 1–18.
7. C. Specht, O. Williams, R. B. Jackman and R. Schoepfer, *Biomaterials*, 2004, **25**, 4073–4078.
8. S. Khan and G. Newaz, *J. Biomed. Mater. Res. A*, 2010, **93A**, 1209–1224.
9. P. Roach, D. Eglin, K. Rohde and C. C. Perry, *J. Mater. Sci. Mater. Med.*, 2007, **18**, 1263–1277.
10. K. Patel, H. Tang, W. Grever, K. Ng, J. Xiang, R. Keep, T. Cao and J. McAllister, *Biomaterials*, 2006, **27**, 1519–1526.
11. P. Ariano, O. Budnyk, S. Dalmazzo, D. Lovisolo, C. Manfredotti, P. Rivolo and E. Vittone, *European Phys. J. E*, 2009, **28**, 1–8.
12. P. Ariano, P. Baldelli, E. Carbone, A. Gilardino, A. Lo Giudice, D. Lovisolo, C. Manfredotti, M. Novara, H. Sternschulte and E. Vittone, *Diamond and Related Materials*, 2005, **14**, 669–674.
13. Y.-C. Chen, D.-C. Lee, C.-Y. Hsiao, Y.-F. Chung, H.-C. Chen, J. P. Thomas, W.-F. Pong, N.-H. Tai, I.-N. Lin and I.-M. Chiu, *Biomaterials*, 2009, **30**, 3428–3435.
14. L. Bacakova, E. Filova, F. Rypacek, V. Svorcik and V. Stary, *Physiol Res*, 2004, **53**, S35–S45.
15. E. Yavin and Z. Yavin, *J. Cell. Biol.*, 1974, **62**, 540–546.
16. S. S. Rao and J. O. Winter, *Frontiers in Neuroengineering*, 2009, **2**, 1–14.
17. R. O. Hynes, *Science*, 2009, **326**, 1216–1219.
18. A. J. Garcia, *Adv. Polym. Sci.*, 2006, **203**, 171–190.
19. A. García, *Biomaterials*, 2005, **26**, 7525–7529.
20. C. Bokel and N. Brown, *Dev Cell*, 2002, **3**, 311–321.
21. N. H. Romano, D. Sengupta, C. Chung and S. C. Heilshorn, *Bba-Gen Subjects*, 2011, **1810**, 339–349.
22. K. S. Straley and S. C. Heilshorn, *Frontiers in Neuroengineering*, 2009, **2**, 9.
23. B. Rezek, E. Ukraintsev, L. Michalíková, A. Kromka, J. Zemek and M. Kalbacova, *Diamond and Related Materials*, 2009, **18**, 918–922.
24. B. Rezek, E. Ukraintsev, A. Kromka, M. Ledinsky, A. Broz, L. Noskova, H. Hartmannova and M. Kalbacova, *Diamond and Related Materials*, 2010, **19**, 153–157.
25. A. Curtis and C. Wilkinson, *Biochem Soc Symp*, 1999, **65**, 15–26.
26. Y. Fan, F. Cui, L. Chen, Y. Zhai, Q. Xu and I. Lee, *Appl Surf Sci*, 2002, **187**, 313–318.
27. Y. Fan, F. Cui, S. Hou, Q. Xu, L. Chen and I. Lee, *J. Neurosci. Methods*, 2002, **120**, 17–23.

28. S. P. Khan, G. G. Auner and G. M. Newaz, *Nanomedicine-UK*, 2005, **1**, 125–129.
29. P. Clark, P. Connolly, A. S. Curtis, J. A. Dow and C. D. Wilkinson, *Development*, 1990, **108**, 635–644.
30. D. Y. Fozdar, J. Y. Lee, C. E. Schmidt and S. Chen, *Biofabrication*, 2010, **2**, 035005.
31. N. M. Dowell-Mesfin, M.-A. Abdul-Karim, A. M. P. Turner, S. Schanz, H. G. Craighead, B. Roysam, J. N. Turner and W. Shain, *J. Neural Eng.*, 2004, **1**, 78–90.
32. D. Hoffman-Kim, J. A. Mitchel and R. V. Bellamkonda, *Annu. Rev. Biomed. Eng.*, 2010, **12**, 203–231.
33. P. Roach, D. Farrar and C. C. Perry, *J. Am. Chem. Soc.*, 2005, **127**, 8168–8173.
34. P. Asuri, S. S. Bale, S. S. Karajanagi and R. S. Kane, *Curr. Opin. Biotech.*, 2006, **17**, 562–568.
35. P. Roach, D. Farrar and C. C. Perry, *J Am Chem Soc*, 2006, **128**, 3939–3945.
36. K. A. Moxon and N. M. Kalkhoran, *Proceedings of the 25th Annual International Conference of the IEEE*, 2003, vol. 4, pp. 3698–3699, DOI:10.1109/IEMBS.2003.1280961.
37. P. C. Georges, *J. Appl. Physiol.*, 2005, **98**, 1547–1553.
38. R. J. Hamers, J. E. Butler, T. L. Lasseter, B. Nichols, J. N. Russell Jr, K. Tse and W. Yang, *Diamond and Related Materials*, 2005, **14**, 661–668.
39. A. Härtl, E. Schmich, J. A. Garrido, J. Hernando, S. C. R. Catharino, S. Walter, P. Feulner, A. Kromka, D. Steinmueller-Nethl and M. Stutzmann, *Nat Mater*, 2004, **3**, 736–742.
40. W. Yang, O. Auciello, J. Butler, W. Cai, J. Carlisle, J. Gerbi, D. Gruen, T. Knickerbocker, T. Lasseter, J. Russell, L. Smith and R. Hamers, *Nat Mater*, 2002, **1**, 253–257.
41. C. Stavis, T. L. Clare, J. E. Butler, A. D. Radadia, R. Carr, H. Zeng, W. P. King, J. A. Carlisle, A. Aksimentiev, R. Bashir and R. J. Hamers, *Proc. Natl Acad. Sci*, 2011, **108**, 983–988.
42. X. Kong, L. Huang, C. Hsu, W. Chen, C. Han and H.-C. Chang, *Anal Chem*, 2005, 77, 259–265.
43. S. C. Wasdo, D. S. Barber, N. D. Denslow, K. W. Powers, M. Palazuelos, S. M. J. Stevens, B. M. Moudgil and S. M. Roberts, *Int J Nanotechnol*, 2008, **5**, 92–115.
44. T. Nguyen, H.-C. Chang and V. Wu, *Diamond and Related Materials*, 2007, **16**, 872–876.
45. H.-D. Wang, C. H. Niu, Q. Yang and I. Badea, *Nanotechnology*, 2011, **22**, 145703.
46. A. E. Nel, L. Maedler, D. Velegol, T. Xia, E. M. V. Hoek, P. Somasundaran, F. Klaessig, V. Castranova and M. Thompson, *Nature*, 2009, **8**, 543–557.
47. K.-K. Liu, C.-L. Cheng, C.-C. Chang and J.-I. Chao, *Nanotechnology*, 2007, **18**, 325102.
48. *Diamond; Electronic Properties and Applications*, ed. L. S. Pan and D. R. Kania, Springer, USA, 1995, pp. 241–284.

49. O. Shenderova, V. Zhirnov and D. Brenner, *Critical Reviews in Solid State and Materials Sciences*, 2002, **27**, 227–356.
50. M. Bevilacqua, S. Patel, A. Chaudhary, H. Ye and R. B. Jackman, *Appl. Phys. Lett.*, 2008, **93**, 132115-1-3.
51. A. Krüger, F. Kataoka, M. Ozawa, T. Fujino, Y. Suzuki, A. E. Aleksenskii, A. Y. Vul and E. Osawa, *Carbon*, 2005, **43**, 1722–1730.
52. O. A. Williams, O. Douhéret, M. Daenen, K. Haenen, E. Osawa and M. Takahashi, *Chem. Phys. Lett.*, 2007, **445**, 255–258.
53. A. M. Schrand, H. Huang, C. Carlson, J. J. Schlager, E. Osawa, S. M. Hussain and L. Dai, *J. Phys. Chem. B*, 2007, **111**, 2–7.
54. A. Krueger, *Chemistry – A European Journal*, 2008, **14**, 1382–1390.
55. A. Thalhammer, R. J. Edgington, L. A. Cingolani, R. Schoepfer and R. B. Jackman, *Biomaterials*, 2010, **31**, 2097–2104.
56. W. M. Cowan, T. C. Südhof, and C. F. Stevens, *Synapses*, Johns Hopkins University Press, Baltimore, USA, 2003.
57. J. E. Lisman, S. Raghavachari and R. W. Tsien, *Nat Rev Neurosci*, 2007, **8**, 597–609.
58. Y. Xing and L. Dai, *Nanomedicine-UK*, 2009, **4**, 207–218.
59. O. A. Shenderova and D. W. Brenner, *Solid State Phenomena*, 2002, **87**, 205–214.
60. A. M. Schrand, L. Dai, J. J. Schlager, S. M. Hussain and E. Osawa, *Diamond and Related Materials*, 2007, **16**, 2118–2123.
61. C. A. Poland, R. Duffin, I. Kinloch, A. Maynard, W. A. H. Wallace, A. Seaton, V. Stone, S. Brown, W. Macnee and K. Donaldson, *Nature Nanotech*, 2008, **3**, 423–428.
62. Y. Abdi, S. Mohajerzadeh, J. Koohshorkhi, M. D. Robertson and C. M. Andrei, *Carbon*, 2008, **46**, 1611–1614.
63. T. Lechleitner, F. Klauser, T. Seppi, J. Lechner, P. Jennings, P. Perco, B. Mayer, D. Steinmueller-Nethl, J. Preiner, P. Hinterdorfer, M. Hermann, E. Bertel, K. Pfaller and W. Pfaller, *Biomaterials*, 2008, **29**, 4275–4284.
64. M. Amaral, A. G. Dias, P. S. Gomes, M. A. Lopes, R. F. Silva, J. D. Santos and M. H. Fernandes, *J. Biomed. Mater. Res. A*, 2008, **87A**, 91–99.
65. W. C. Clem, S. Chowdhury, S. A. Catledge, J. J. Weimer, F. M. Shaikh, K. M. Hennessy, V. V. Konovalov, M. R. Hill, A. Waterfeld, S. L. Bellis and Y. K. Vohra, *Biomaterials*, 2008, **29**, 3461–3468.
66. G. Cellot, E. Cilia, S. Cipollone, V. Rancic, A. Sucapane, S. Giordani, L. Gambazzi, H. Markram, M. Grandolfo and D. Scaini, *Nature Nanotech*, 2008, **4**, 126–133.
67. R. Edgington, A. Thalhammer, J. O. Welch, A. Bongrain, P. Bergonzo, E. Scorsone, R. B. Jackman, and R. Schoepfer, *J. Neural Eng.*, 2013, **10**, 056022.
68. H. A. Girard, S. Perruchas, C. Gesset, M. Chaigneau, L. Vieille, J.-C. Arnault, P. Bergonzo, J.-P. Boilot and T. Gacoin, *Acs Appl Mater Inter*, 2009, **1**, 2738–2746.
69. O. A. Williams, *Diamond and Related Materials*, 2011, **20**, 621–640.

70. O. A. Williams, A. Kriele, J. Hees, M. Wolfer, W. Müller-Sebert and C. E. Nebel, *Chemical Physics Letters*, 2010, **495**, 84–89.
71. K. W. Sun, J. Y. Wang and T. Y. Ko, *Applied Physics Letters*, 2008, **92**, 153115.
72. T. Jiang and K. Xu, *Carbon*, 1995, **33**, 1663–1671.
73. O. Shenderova, A. Koscheev, N. Zaripov, I. Petrov, Y. Skryabin, P. Detkov, S. Turner and G. Van Tendeloo, *J. Phys. Chem. C*, 2011, **115**, 9827–9837.
74. G. Rainaldi, A. Calcabrini and M. Santini, *J Mater Sci-Mater M*, 1998, **9**, 755–760.
75. N. A. Kotov, *Science*, 2010, **330**, 188–189.
76. M. Shimaoka and T. Springer, *Nat Rev Drug Discov*, 2003, **2**, 703–716.
77. M. D. Mager, V. LaPointe and M. M. Stevens, *Nat Chem*, 2011, **3**, 582–589.
78. R. O. Hynes, *Cell*, 2002, **110**, 673–687.
79. J. Hees, A. Kriele and O. A. Williams, *Chem. Phys. Lett.*, 2011, **509**, 12–15.
80. J. Zang, Y. Wang, S. Zhao, L. Bian and J. Lu, *Diamond and Related Materials*, 2007, **16**, 16–20.
81. P. Bergonzo, A. Bongrain, E. Scorsone, A. Bendali, L. Rousseau, G. Lissorgues, P. Mailley, Y. Li, T. Kauffmann, F. Goy, B. Yvert, J. A. Sahel and S. Picaud, *Irbm*, 2011, **32**, 91-94.

CHAPTER 10

Diamond Nucleation and Seeding Techniques: Two Complementary Strategies for the Growth of Ultra-thin Diamond Films

J. C. ARNAULT* AND H. A. GIRARD

CEA, LIST, Diamond Sensors Laboratory, F-91191 Gif-sur-Yvette, France
*Email: Jean-Charles.ARNAULT@cea.fr

10.1 Introduction

Continuous diamond films with a thickness lower than 150 nm grown on heterosubstrates are required for current technological applications, such as heat dissipation[1,2] (silicon on diamond technology) or nano/micro-electromechanical systems (NEMS/MEMS).[3] The main limitation to overcome remains the low diamond nucleation densities obtained on untreated substrates, typically between 10^3–10^5 cm^{-2} for silicon.[4,5] Growth of ultrathin coalesced diamond layers requires a very high density of nucleation sites (typically >10^{11} cm^{-2}). Depending on the expected microstructure of diamond (polycrystalline, nanocrystalline or highly oriented), different experimental approaches have been developed over the last decades.

Two main pathways have been explored:

(i) **A nucleation pathway**: *ex situ* or *in situ* substrate pre-treatments, which induce the formation of diamond *nuclei*, *i.e.*, the smallest thermodynamically stable diamond crystal on the substrate during the early stages of CVD.
(ii) **A seeding pathway**: different surface treatments, which provide stable diamond seeds at the substrate surface that directly grow under CVD conditions.

In the present chapter we will focus first on diamond nucleation (section 10.2), especially on the widely used bias enhanced nucleation (BEN) process (section 10.2.4). The second part will detail seeding techniques: surface abrasion using diamond particles (section 10.3.1) and how to control the homogeneous deposition of diamond nanoparticles on substrates (section 10.3.2). To conclude, we will draw a comparison between BEN and seeding approaches to underline their complementarities in terms of substrates (properties and size), diamond film adhesion and microstructure of the diamond films grown.

10.2 Diamond Nucleation

10.2.1 What is Nucleation?

Nucleation is the formation of the smallest thermodynamically stable aggregate, which is called a *nucleus*. Indeed, below this critical size, aggregates have significant probability to disappear, while bigger aggregates can only grow.

For ***homogeneous nucleation***, *i.e.*, direct condensation, no external system, such as a surface, is involved. For diamond, homogeneous nucleation could occur directly within the gas phase during chemical vapor deposition (CVD). The critical size could be obtained considering the capillary theory of Volmer and Weber.[6] The supersaturation rate is defined as the ratio of the real vapor pressure to the equilibrium vapor pressure. For an aggregate containing *n* atoms and a positive supersaturation rate, the free enthalpy ΔG_n could be expressed as the sum of the bulk and surface terms:

$$\Delta G_n = n\Delta\mu_c + \sum_{j=1}^{k} \sigma_j \cdot S_j, \qquad (10.1)$$

with $\Delta\mu_c$ the difference of atomic free enthalpy and σ_j the surface tension of the j crystal facet of surface S_j.

The critical size n^* corresponds to the maximum value of ΔG_n. Addition of atoms to the stable aggregate leads to a decrease of the free enthalpy, *i.e.*, to thermodynamically stable aggregates.

In most cases diamond deposition is taking place on a surface. It corresponds to ***heterogeneous nucleation***. Here, the interface AB between an aggregate (A) on a surface (B) should be taken into account *via* its interfacial energy γ_{AB} *via* the Dupray relation.[7] This implies two additional terms for the free enthalpy ΔG_n[8]:

$$\Delta G_n = n\Delta\mu_c + \sum_{j=1}^{k-1} \sigma_j \cdot S_j + \gamma_{AB}S_{AB} - \sigma_B S_{AB}, \qquad (10.2)$$

with $\Delta\mu_c$ the difference of atomic free enthalpy, σ_j the surface tension of the crystallographic j facet of surface S_j, γ_{AB} the interfacial energy, S_{AB} the interface surface and σ_B the surface tension of the substrate B.

10.2.2 Diamond Homogeneous Nucleation

Homogenous nucleation, *i.e.*, the formation of diamond *nuclei* in a gas phase, was predicted by classical nucleation theory[9] and diamondoids from the adamantane family[10] were first proposed as precursors. However, they were finally found to be highly unstable under CVD plasma conditions[11] and graphite precursors, like polycyclic aromatics, may constitute possible precursors due to their higher stability.[12] In particular cases, a few experimental studies reported diamond formation in the gas phase[13–15] and some hexagonal diamond was even collected.[15] More recently, homogeneous diamond nucleation was reported at low pressure (<10^2 Pa compared to 10^3–10^4 Pa for conventional CVD) and a low electron temperature (2.6 eV instead of 5 eV) using surface microwave plasma (MPCVD) for diamond growth on plastic substrates at a temperature lower than 100 °C.[16] Transmission electron microscopy (TEM) investigations have demonstrated the presence of nanometric diamond crystals, probably formed in the gas phase (Figure 10.1). These findings are supported by calculations, which predicted suitable plasma conditions.[17,18]

A recent article has review the state-of-art of diamond homogenous nucleation models from bulk liquid. These calculations may explain why the homogeneous diamond nucleation is likely in carbon-rich stars and planets, such as Uranus and Neptune.[19] Indeed, the presence of 5 nm diamond particles in meteorites was first reported by Lewis *et al.*[20] The Raman signature of nanodiamonds extracted from the meteorite Orgueil (averaged size of 2–3 nm) was then compared to synthetic 5 nm diamonds.[21] Finally, nanometric diamond particles (5–10 nm) were also formed by detonation synthesis under non-equilibrium thermodynamics conditions.[22]

10.2.3 Diamond Heterogeneous Nucleation

10.2.3.1 Models and Theory

A review of models was reported in 1995. The reaction involved in the formation of a critical diamond *nucleus* was analyzed in detail.

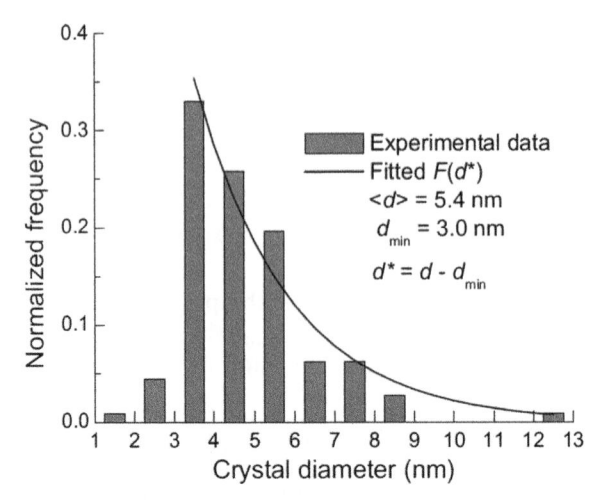

Figure 10.1 Size distribution of diamond crystals from TEM observations. The minimum size is 3 nm.[16]

A computational method to obtain the size distribution function was checked with both experimental distribution functions and nucleation densities extracted from scanning electron microscopy (SEM).[23] Avrami formalism was developed to model diamond nucleation and growth.[24] The ultimate aim was to describe, by means of rate equations, the kinetics of film formation over the whole range of coverage. Kinetic quantities were calculated such as i) the film perimeter ii) the evolution of the fractional surface covered by the film (through the KJMA† model[25–28]) and iii) the behavior of the island density during the growth up to the closure of the film (when direct impingement of carbon species from the gas phase to the surface cannot be neglected). More recently, the particle size distribution function was analyzed in terms of particle volume rather than, as usually done, the diameters of particles.[29,30] The model was then employed to describe experimental data.[31]

10.2.3.2 Experimental Observations of Diamond Nuclei

Theoretically, diamonds of 3–5 nm in size are more stable than graphite.[32,33] As previously shown, the critical size is substrate dependent (via the interfacial energy γ_{AB} and the surface tension σ_B, eqn 10.2). In the literature experimental observations of very small diamond *nuclei* have been reported. For example, diamond nanocrystals (2–6 nm) were observed by high resolution transmission electron microscopy (HRTEM) grown directly on Si surfaces[34,35] (Figure 10.2) or on a silicon carbide interlayer,[36] sometimes even embedded in an amorphous carbon matrix.[37] For diamond nucleation

---

†This model corresponds to the one initially developed by Kolmogorov and, independently, two years later, by Johnson and Mehl, and Avrami. It is called the KJMA model.

Diamond Nucleation and Seeding Techniques 225

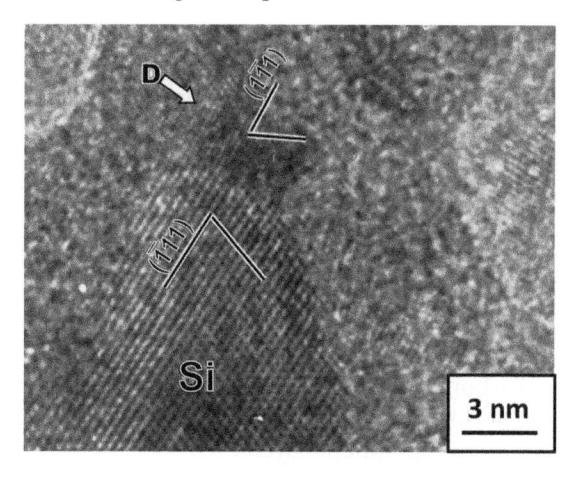

Figure 10.2 HRTEM picture of a 3 nm diamond island grown epitaxially on silicon.[34]

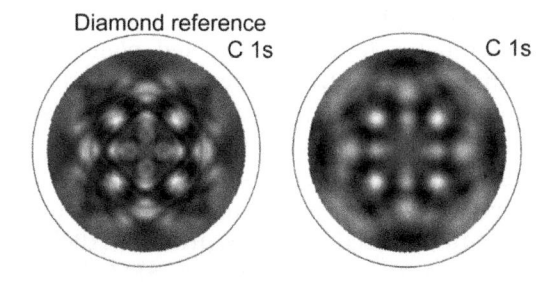

Figure 10.3 XPD patterns of a diamond reference and after nucleation on iridium.[40]

on iridium, it was impossible to observe the smallest diamond *nuclei* using HRTEM before short CVD growth.[38] Nevertheless, the diamond signature was extracted from X-ray photoelectron diffraction (XPD)[39,40] (Figure 10.3) and high resolution electron energy loss spectroscopy (HREELS).[41] Indeed, these two techniques are particularly sensitive to the order at short distances.

10.2.3.3 Surface Mechanisms Competing with Heterogeneous Diamond Nucleation

Since 1980, CVD diamond has been grown on numerous heterosubstrates.[4,42] In some studies heteroepitaxial growth was achieved.[43] This research field will be further discussed (section 10.2.4). The intrinsic reactivity of each heterosubstrate, with energetic species coming from the CVD plasma, is an essential parameter to consider. Indeed, interactions with the reactive species may induce surface or subsurface mechanisms, which strongly influence the nucleation pathways. As a consequence, the nature of the diamond/substrate interface could strongly differ. Let us illustrate this with heterosubstrates exhibiting very different reactivities toward CVD plasma.

A previous review has focused on the mechanisms taking place on silicon,[44] like surface etching by atomic hydrogen and silicon carbide formation, and their consequences on diamond nucleation. A sequential XPS analysis performed in a UHV set-up connected to a MPCVD reactor has shown the removal of the native silicon oxide after H_2/CH_4 CVD exposure of 5 min while a silicon carbide 2.5 nm thick overlayer is stabilized.[45] With nickel or platinum substrates, the high carbon solubility[46] leads to carbon dissolution into the substrate under CVD plasma and induces the formation of a graphitic overlayer before diamond nucleation can occur.[47–51] Under CVD conditions, carbon graphitic nanostructures, like carbon onions, were formed on copper surfaces.[52] More recently, this specific behavior was used to grow high quality CVD graphene layers.[53]

On the other hand, cubic silicon carbide (100) oriented surfaces were found to be very inert towards CVD exposure. Their surface stoichiometry within the first atomic planes was weakly modified. A silicon rich surface (three reconstructed planes) is slightly etched under H_2/CH_4 plasma and becomes silicon-terminated (one reconstructed plane), as shown by low energy electron diffraction (LEED) (Figure 10.4) and XPS investigations.[54] These modifications of the extreme surface were, however, sufficient to significantly influence diamond nucleation.[55] Lastly, on iridium, a weak amount of carbon may be dissolved into the substrate during plasma exposure,[56] which could segregate at the surface after cooling. The formation of furrows along <110> directions at the iridium (100) surface has been observed.[57,58] According to high resolution SEM investigations, these surface structures are closely related to the formation of oriented diamond nucleation[59] (see section 10.2.4).

10.2.3.4 Role of Structural Defects on Heterogeneous Nucleation

A few papers have discussed the role of extended defects on diamond nucleation. Nevertheless, it is a well-established fact that surface defects play a fundamental role on heterogeneous nucleation processes.[60,61] Defects are needed to induce the nucleation process by lowering the enthalpy of *nuclei* formation.[62–64] This effect was particularly underlined on silicon at the

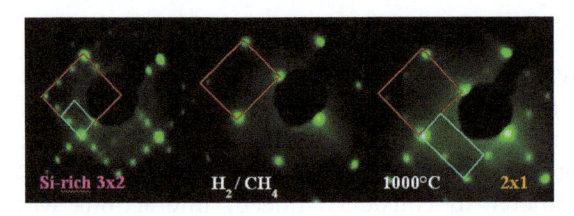

Figure 10.4 LEED patterns of (3×2) Si rich 3C-SiC; after H_2/CH_4 plasma; and after annealing: (2×1) Si-terminated.[54]

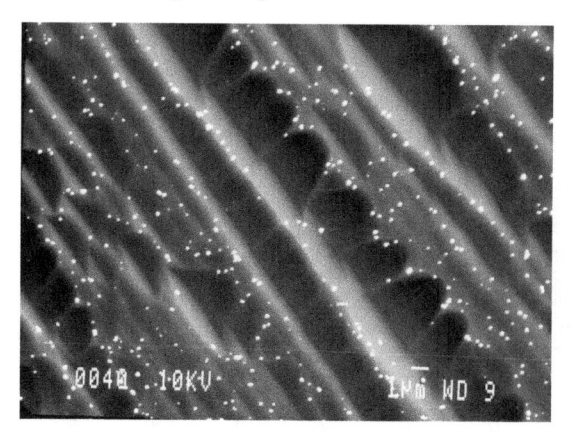

Figure 10.5 SEM picture of diamond crystals along steps in a silicon dimple.[71]

apexes of pyramids,[65] sharp edges,[66] whiskers[67] and the edges of terraces on cleaved silicon.[68] Several explanations have been attempted to justify the preferential nucleation on these sites, such as the minimization of the interfacial energy,[66] the presence of more dangling bonds enhancing the chemisorption energy and the faster carbon saturation on these sites.[69] In the presence of line defects on silicon (*i.e.*, dislocations) the nucleation density was enhanced.[70] The nucleation reached completion after a relatively short period of time and the nucleation rate could be modeled according to a Dirac delta function. Indeed, nucleation sites were available at the Si surface at the beginning of CVD.

Another approach consisted of transforming reconstructed vicinal surfaces into a silicon dimple by ultra-high vacuum (UHV) annealing. SEM investigations (Figure 10.5) revealed a preferential location of diamond *nuclei* in the vicinity of step edges.[71] The link between structural defects and diamond nucleation was also established on a tungsten carbide (WC) substrate.[72] This study showed a correlation between broadening of the substrate X-ray diffraction (XRD) peaks and the nucleation density N_d, *i.e.*, the number of diamond crystals per cm^2 (Figure 10.6). Nucleation after etching of the WC-strained material leads to lower N_d values (by two orders of magnitude). The higher the full-width at half-maximum (FWHM) of the XRD peaks, the larger the nucleation density.

10.2.4 Different Approaches for Enhanced Diamond Nucleation

Several reports have reviewed in detail the different *ex situ* techniques developed to promote diamond nucleation.[4,5] Two main objectives are important: (i) first, the enhancement of the nucleation density N_d (up to 10^{11} diamond crystals per cm^2) on substrates as large as possible; (ii) second, heteroepitaxial diamond nucleation. Let us briefly summarize the methods

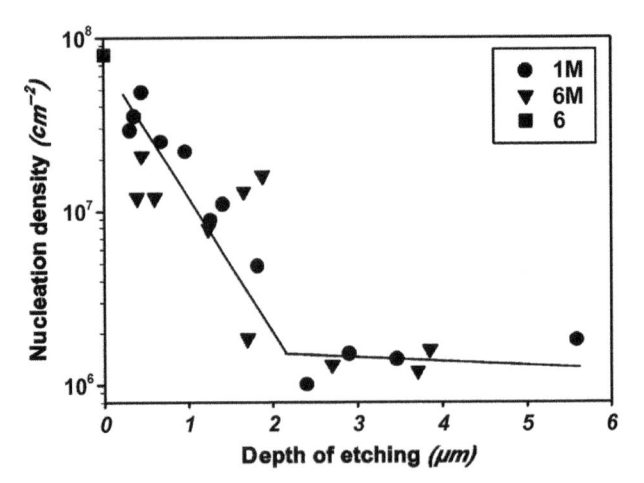

Figure 10.6 Evolution of the nucleation density *versus* etched depth of WC substrate.[72]

used to obtain very high N_d (section 10.2.4.1). Then, the bias enhanced nucleation process (BEN), which permits us to reach both objectives, will be discussed (section 10.2.4.2). As an alternative to nucleation methods, seeding techniques, which directly provide diamond seeds at the surface, will be further described (section 10.3).

10.2.4.1 *Techniques for Diamond Nucleation Enhancement*

Very low nucleation densities ($N_d = 10^3$–10^5 cm^{-2}) are obtained on non-treated substrates due to the high surface energy of diamond relative to heterosubstrates, the low sticking coefficient of gaseous precursors and the competition of non-diamond phases.[4,73]

The first widely used method to enhance diamond nucleation was mechanical scratching of the substrates with abrasives (BN, SiC, Al$_2$O$_3$, *etc.*).[74,75] A better efficiency, with N_d values up to 10^{10} cm^{-2}, was observed using diamond powder for scratching. An alternative method used an ultrasonic treatment with abrasive particles suspended in methanol or acetone.[76,77] Three different mechanisms for nucleation enhancement were proposed:[78] (i) the mechanical treatment generated **surface defects** (edges, steps, dislocations) with lowered interfacial energy. It created available dangling bonds and a faster carbon saturation could happen at these defects; this effect was previously discussed (section 10.2.3). (ii) A **seeding** effect was also considered to occur, especially using diamond, c-BN or SiC powders. (iii) Finally, the production of **non-volatile graphitic species** was suggested, which could act as precursors for nucleation.[42] A surface study on scratched Si (100) by X-ray photoelectron spectroscopy (XPS), Auger electron spectroscopy (AES) and SEM well demonstrated the coexistence of two pathways: a seeding pathway occurring by direct growth from part of the diamond seeds left by the mechanical pretreatment, and a nucleation pathway taking place through a stepwise process, including the formation of nucleation sites suitable for diamond *nuclei*.[79] These aspects will be further described in the section on seeding techniques (section 10.3). Chemical etching or passivation surface treatments were also reported to create etch pits or to remove the surface oxide on silicon.[80,81] Moderate N_d enhancements were obtained using surface coatings[82] or ion implantation (Si^+, C^+).[4,83]

10.2.4.2 Bias Enhanced Nucleation

10.2.4.2.1 Principle. Compared to previous surface treatments, bias enhanced nucleation (BEN) is an *in situ* pretreatment, which consists of applying a DC bias voltage to the substrate with respect to the grounded reactor walls or to a second internal electrode. This technique was first used for diamond nucleation by Yugo *et al.*[80,84] in an MPCVD reactor. A significant enhancement (N_d values up to 10^{10} per cm^2) was reported. Then, BEN was developed in different reactor geometries for microwave[85,86] or hot filament[87–89] CVD techniques, applying DC and AC voltages,[90] or pulse bias.[91] Negative and positive bias voltages led to nucleation enhancement.[92,93] Nevertheless, negative biasing is currently the most used technique. Under particular BEN conditions, heteroepitaxial nucleation was achieved on Si,[94] 3C-SiC[95] and Ir surfaces.[96]

10.2.4.2.2 Relevant Parameters. The CH_4/H_2 MPCVD plasma is basically composed of C_xH_y species in their neutral, radical and ionic forms. The CH_4/H_2 plasma chemistry was recently investigated in detail.[97] Under BEN, a negative polarization is applied in most cases, leading to the formation of a cathodic sheath above the substrate.[98] This induces two main phenomena: electron emission from the surface to the plasma, and surface bombardment by H_y^+ and $C_xH_y^+$ ions. At first glance, the applied bias voltage and the duration seem to be pertinent parameters for BEN. Frequently used bias voltages are included between ¬100 V and ¬300 V. However, due to the diverse reactor geometries, the bias voltage value by itself is quite meaningless. Indeed, the energy provided by the bias voltage is not totally transferred to the positively charged species as reported in the literature.[99,100] On the other hand, the **kinetic energy** of accelerated H_y^+ and $C_xH_y^+$ ions is a critical parameter. *In situ*, real-time, mass-selective energy analysis of the incoming ions has been carried out during BEN by Kátai *et al.*[99] For example, the ion energy distributions were measured for a bias voltage of ¬200 V and a methane content of 4%. All the involved species (H_y^+, $C_xH_y^+$) have kinetic energies lower than 100 eV.[100] The second pertinent parameter to consider is the **ion flux**, which bombards the surface. According to Kátai *et al.*[99] it is closely related to the bias voltage and the methane content (Figure 10.7).

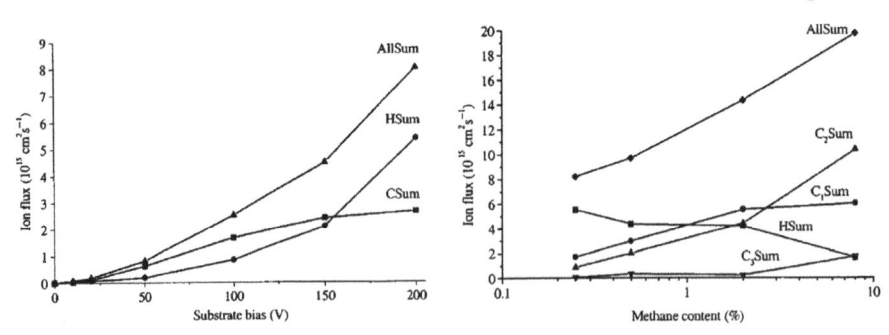

Figure 10.7 Left: Evolution of ion fluxes *versus* bias voltage; right: evolution of ion flux *versus* methane content.[99]

Moreover, the abundance of $C_xH_y^+$ ions is strongly dependent on the methane content. $C_2H_y^+$ ions become the major ion species for methane contents higher than 3% (Figure 10.7). This explains why higher methane contents (2–10%) are commonly used for BEN compared to CVD growth. The importance of **ion flux** explains the very different BEN durations used in the literature. On silicon, Barrat *et al.* used ultra-short BEN steps of 10–90 s,[101–103] while BEN procedures up to 15 min have been reported.[104] On iridium, longer BEN steps from 60 min up to 180 min[57,105,106] were applied using MPCVD plasma, whereas durations lower than 60 s were used with DC plasma.[107] In this latter case, stronger ion bombardment is expected. In conclusion, concentrations, kinetic energies and the flux of ionic species impinging the substrate during BEN are the most relevant parameters. Unfortunately, very little experimental data is available in the literature.

10.2.4.2.3 BEN-induced Mechanisms.
Surface modifications induced by MPCVD plasma on different substrates (Si, 3C-SiC, Ir, *etc.*) were previously discussed (section 10.2.3). We underlined that the substrate's intrinsic reactivity plays an important role in diamond nucleation pathways. During BEN, the induced mechanisms are also substrate dependent.[108] Different competing phenomena have been observed (Figure 10.8).

The ion bombardment induced under BEN conditions led to **subplantation** of carbon species. This mechanism, first proposed by Gerber *et al.*,[109] was investigated in detail on silicon by Lifshitz *et al.*[110] According to calculations,[111] a minimum kinetic energy of 30 eV is required for carbon ions to lead to subplantation into the silicon substrate. Experimental kinetic energy distributions[99] confirmed that a significant part of $C_xH_y^+$ ions exhibit higher kinetic energy.

Surface bombardment can also lead to an **enhanced diffusion** of carbon species at the surface. Indeed, impinging ions transfer their kinetic energy to surface atoms. This mechanism was well identified on silicon by Jiang *et al.*,[112] who demonstrated that the spatial distribution of diamond islands generated by BEN was not random. A depletion zone was evidenced around existing diamond crystals, which is bias-voltage dependent. The two

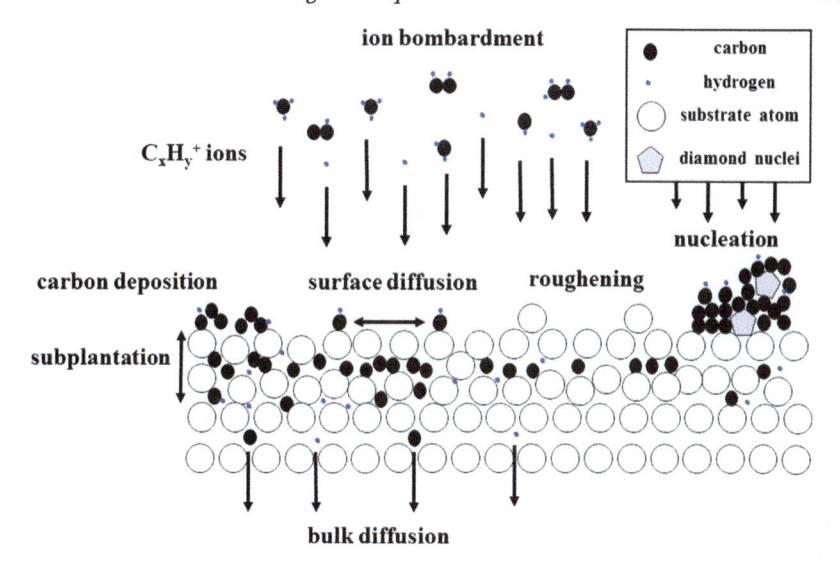

Diamond Nucleation and Seeding Techniques 231

Figure 10.8 Different mechanisms reported or expected under bias enhanced nucleation.

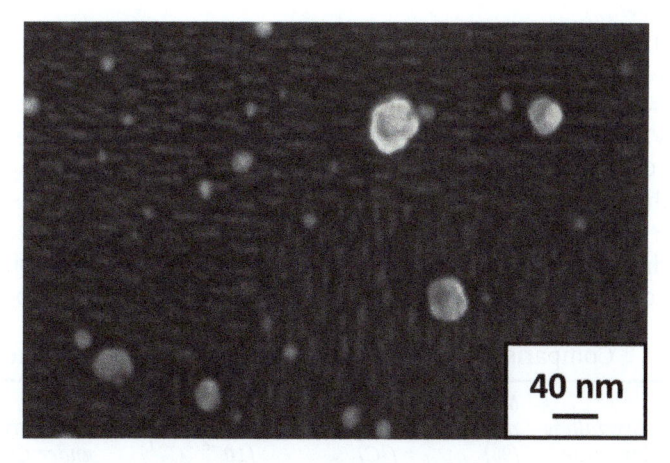

Figure 10.9 Surface roughening observed after BEN on 3C-SiC (001).[115]

previous mechanisms yield a **higher carbon supersaturation** at the surface, which is a required condition for diamond nucleation. The formation of diamond crystals under BEN then increases the electronic emission towards the plasma leading to a stronger sheath.[113]

Another consequence of ion bombardment is the **surface roughening**. Under proper BEN conditions, organized nanostructures could be formed at the surface. Perpendicular stripes corresponding to different crystallographic domains were observed after BEN on 3C-SiC surfaces[114,115]

(Figure 10.9). A surface roughening was also reported on iridium surfaces.[57,116] It leads to the formation of furrows along ⟨110⟩ directions, which may have an important role in the unique nucleation pathway on iridium.[57,58] These surface nanostructures generated during BEN may significantly influence diamond nucleation, modifying the surface diffusion of species.

10.2.4.2.4 Heteroepitaxy. Bias enhanced nucleation treatments have revealed a powerful process to obtain diamond heteroepitaxial nucleation. In this case diamond crystals showed a crystallographic relationship with the substrate lattice. This property was achieved using a narrow window of BEN parameters. Among the wide variety of heterosubstrates, several are of interest for the growth of highly oriented diamond films (Table 10.1). The ideal heterosubstrate should have the lowest misfit parameter, a crystalline structure and a thermal expansion coefficient close to those of diamond, a high melting temperature and a weak reactivity under CVD. Among suitable substrates, c-BN exhibits the lowest misfit parameter and a weak difference of thermal expansion coefficient with diamond. Some studies reported epitaxial diamond crystals on (111) boron-terminated faces and (001) faces.[117] Nevertheless, the availability of c-BN substrates remains low.

On (001) silicon, polar misorientations‡ close to 5° were obtained for diamond films of several tens of microns thick.[42] The diamond crystalline quality is strongly limited by the formation of a silicon carbide interlayer under CVD (section 10.2.3). A significant improvement was observed using directly 3C-SiC (001) layers epitaxied on silicon substrates, yielding diamond films with a polar misorientation of 0.6°.[95] This improved result is related to the weak reactivity of 3C-SiC under CVD plasma (section 10.2.3). At the present time, the best (001) diamond films have been grown on iridium with polar and azimuthal misorientations of 0.3° and 0.2°, respectively.[118]

Table 10.1 Comparison of substrate properties for diamond heteroepitaxy.

	Crystalline structure	Misfit parameter (%)	Melting temperature (°C)	Therm. exp. coefficient (10^{-6} °C^{-1})	Reactivity under CVD
Diamond	Fd3m	–	3550	1.5	Weak
Silicon	Fd3m	52.0	1410	2.6	Carbide formation etching
3C-SiC	F43m	22.0	2830	4.6	Surface roughening
Platinum	Fm3m	10.0	1770	8.8	Carbon dissolution
c-BN	F43m	1.5	2727	0.6	Weak
Iridium	Fm3m	7.6	2410	6.4	Surface roughening

‡For (001) heteroepitaxy, the polar misorientation (or tilt) corresponds to the angular difference between the [001] direction of diamond lattice and the normal to the (001) oriented substrate. The azimuthal misorientation (or twist) is defined as the angular difference between ⟨100⟩ directions of diamond lattice *versus* the ⟨100⟩ directions of the substrate surface plane.

For iridium, a unique nucleation pathway is occurring. While on other heterosubstrates, diamond crystals are formed *via* a 3D Volmer–Weber growth pathway, a *pseudo*-2D growth takes place on iridium in specific areas called domains.[57,119,120] Within domains, a very high ratio of oriented diamond crystals (>90%) is observed. A recent SEM/atomic force microscopy (AFM) combined study suggested that furrows created by iridium roughening could be involved in domain formation.[58] For (111) heteroepitaxy, Tachibana *et al.* achieved the growth of diamond films on platinum with a polar misorientation close to 1° using either BEN or a seeding/annealing treatment.[50]

10.3 Nanodiamond Seeding

Among the methods developed over the last decades to promote the growth of ultrathin diamond layers, a so-called "seeding" technique appeared as the more versatile approach to enable diamond coating on heterosubstrates. It relies on the deposition of diamond nanoparticles (nanodiamonds, NDs) on a substrate, which act as seeds for diamond growth. Unlike the BEN technique previously described (section 10.2.4), seeding is not truly a nucleation technique as growth starts from already deposited nanometric diamond phases, even if the confusion is often encountered in the literature. However, this technique enables diamond growth on a large variety of substrates as no electrical conductivity is required. Moreover, it is theoretically not limited by the shape or the dimension of the substrate. Indeed, in the case of diamond growth initiated by nanodiamond seeds, the main limitations come from the CVD process, hardly achievable on substrate larger than six inches.

10.3.1 Abrasion Techniques Involving Particles

Development in the 90s of nanodiamond seeding toward CVD growth partly resulted from the evolution of abrasion methods, originally developed in 80s to enhance the nucleation density on a substrate.[121–124] These abrasion techniques rely on the formation of defects on the substrate by the mean of diamond or non-diamond particles applied on the substrate by mechanical strength (polishing or scratching) or by ultrasonic agitation.[124] Typically, a nucleation density N_d of 10^7–10^8 cm^{-2} is routinely obtained on a scratched silicon substrate, which compares to 10^4 cm^{-2} on non-scratched substrates.[125] In this approach the enhancement of N_d is expected to be linked to the minimization of interfacial energy on sharp convex surfaces, the breaking of a number of surface bonds and the rapid carbon saturation (fast carbide formation) at sharp edges.[124] Different abrasive materials have been studied and ranked, such as oxides, silicides, carbides and nitrides. For instance, aluminum oxides or cubic boron nitrides were shown to be much more efficient at reaching a high nucleation density than silicon or zinc oxides.[41] However, if diamond particles are used, an even higher nucleation density is generally measured as the embedding of small residual diamonds in the surface contributes to the nucleation enhancement through the growth occurring on these diamond seeds.[76,126,127] In ultrasonic treatments with diamond particles the predominant nucleation mechanism has even been shown to be mainly related to seeding by "diamond dust" on the substrate surface.[75] Consequently, in order to maximize this effect, the use of diamond particle slurries became rapidly predominant instead of non-diamond particles.[128,129] In parallel, the particle diameter used in the diamond slurries reduced,[76] but significant differences between both abrasion methods (mechanical polishing and ultrasonic treatment) appeared. For mechanical abrasion, the use of smaller diamond particles rapidly appeared as more efficient in enhancing the nucleation density by the combined effect of finer defects and more embedded seeds.[123,129,130] In contrast, the use of smaller particles, with nanometric diameters and dispersed in slurries, for ultrasonic agitation, did not directly enhance the nucleation density on the substrate. Several authors emphasize that the smallest particles have insufficient momentum to induce damage on the substrate or embedding.[76,131,132] To increase this momentum, the use of larger particles in the ND slurries has been shown to be efficient, a so-called "hammering" effect. Note that these larger particles can be made of diamond or non-diamond materials, as reported by Akhvlediani et al.[131] with the use of micrometric diamond, alumina and titanium particles. If these abrasion/seeding mixed methods are still in use to produce thick diamond layers, they are limited by the damage induced on the substrate and the high probability of leaving large particles on the substrate. These drawbacks drastically impact the diamond quality of the grown layer and are incompatible with the quality standards required for MEMS processing, and optical or tribological devices.

10.3.2 Controlled Deposition of Nanodiamonds on Substrates

The progressive availability on the market of much smaller diamond particles enabled the development of methods only devoted to the deposition of diamond crystals on a substrate without the formation of defects. The main advantage concerns the possibility to not damage the substrate, an essential requirement in some applications (notably, optics and sensing), and the possibility to deposit a homogenous size distribution of diamond seeds on the substrate. The quality of the resulting diamond film is thus directly dependent on the quality of the seeding. Three major parameters drive the characteristics of the grown layer: (i) the density of seeds N_s deposited on the substrate; (ii) the spatial homogeneity of the deposit; and (iii) the size distribution of the particles deposited on the substrate. The former will mainly impact the thickness of the resulting coalesced layer, while the two others factors drive the homogeneity and the roughness of the grown layer. Therefore, an efficient seeding process will rely on the quality of the

Diamond Nucleation and Seeding Techniques 235

deposition technique, but also on the quality of the nanoparticles suspension. Prior to the description of the major seeding methods reported so far in the literature, a short introduction on NDs suspensions will be given below.

10.3.2.1 Nanodiamond Suspensions

Nanodiamond Synthesis. The formation of clusters and the superficial chemistry of NDs are closely related and mainly depend on their origin. Among the different synthesis methods of NDs, detonation[133] and milling processes[134] are maybe the most widely used. In the detonation method, particles are synthesized from carbon provided by the explosive itself. During a detonation and under a lack of oxygen, the conditions of temperature and pressure required for the formation of a diamond phase can be met for a short time, leading to the synthesis of nanodiamonds.[135] The diameter is thus determined by the duration of the detonation wave, which gives rise to uniform primary particles generally around 5 nm. However, non-diamond phases continue to be synthesized after the pressure drops, which generally cover the particles with sp^2 or amorphous carbon species, which surround small clusters of a few nanodiamonds (Figure 10.10). Thorough purification steps are thus required to remove all these non-diamond phases and dedicated techniques to deagglomerate the clusters have been developed, such as bead milling[136,137] or air oxidation.[138] However, access to primary particles is still a challenge and detonation nanodiamonds usually carry on their surfaces a small amount of non-diamond phases.

Another approach to synthesize diamond nanoparticles consists in the milling of micrometric crystallites (natural or synthetic).[134] The resulting nanodiamonds are less contaminated with non-diamond phases on their surface and are less sensitive to aggregation. Nevertheless, they exhibit a

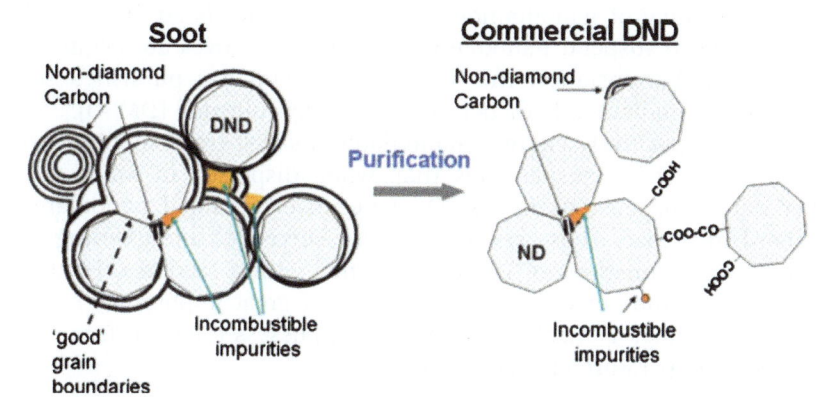

Figure 10.10 Major structural components of detonation soot (left) and commercial NDs (right).[139]

broader size distribution than detonation particles and diameters below 10 nm are not yet commercially available. Finally, as a major difference between both production routes, if detonation nanodiamonds are roughly spherical, nanodiamonds synthesized by a milling process exhibit the typical sharp edges of a diamond crystallite, even at the nanometer scale.

Surface Charge. A key parameter for the colloidal stability of nanodiamonds is their surface potential E_{ZETA} and the resulting electrostatic forces. If this surface charge is high enough in absolute value ($E_{ZETA} > |30|$ mV), repulsive forces between the particles are able to ensure good colloidal stability. The diamond core is surrounded by a versatile carbon layer on which different chemistries can be found. The use of oxidative treatments during the purification steps gives rise to an oxidized surface, mainly composed of carboxylic, alcoholic or etheric groups.[140] Polar groups ensure a hydrophilic surface, but also may confer a charge to the particle, which will depend on the pH. For instance, carboxylic groups can be found on nanodiamonds exposed to a strong oxidative treatment (air annealing or an acid treatment). Above pH 4, the carboxylate form prevails, leading to a negative zeta potential, generally below − 40 mV if well dispersed. This method is commonly used to prepare stable colloidal suspension of nanodiamonds. Conferring a positive charge to nanodiamonds is more tedious. Surface functionalization can be performed to cover the particle with cationic groups (*e.g.*, ammonium groups). However, it has been recently reported in the literature that "unfunctionalized" particles (*e.g.*, without noticeable acid/base groups) can exhibit such positive zeta potentials, especially after a reduction treatment under hydrogen or a vacuum annealing.[141,142] A recent hypothesis on the origin of this positive charge involves adsorption of oxygen on the surface of the particles, allowing hole doping at the surface of the nanodiamonds.[141]

Solvent. The choice of solvent is also a major factor when assessing the colloidal stability of nanodiamonds. Ultrapure water is so far the wildest solvent used to suspend nanodiamonds, but alternatives are beginning to be reported. For instance, the Shenderova group has published several studies on nanodiamonds dispersed in dimethylsulfoxide (DMSO).[143] They showed that positively charged nanodiamonds dispersed in DMSO often lead to more stable suspensions than when dispersed in water. On the other hand, suspensions in DMSO of negatively charged nanodiamonds are hard to realize. These researchers also succeeded in achieving stable suspensions in mixtures of DMSO combined with a second solvent, such as alcohol, acetone or water. This aspect is of great importance for nanodiamond seeding, as alcohols, for instance, possess a low surface tension and a low temperature of vaporization.

To summarize, nanodiamonds can differ in many aspects, such as: (i) spherical or faceted shape; (ii) surrounded or not with non-diamond carbon species; (iii) scattered surface chemistry; (iv) surface charge; (v) sizes from a few nanometers to tens of nanometers; and (vi) polydispersed (or not) distributions. This diversity will directly impact the characteristics of the colloidal suspension of nanodiamonds and, consequently, the seeding method. As will be described, the dispersion method is mainly chosen according to these characteristics: the sign of the zeta potential (which depends on the surface chemistry and the solvent), the tendency of the suspension to agglomerate (depends on detonation or milling synthesis) or the polydispersity of the suspension.

10.3.2.2 Stability of the Nanodiamonds Under CVD Plasma

The aim of nanodiamond seeding is the growth promotion of CVD diamond layers. However, the chemical or thermal stability of 5 nm diamond cores under CVD plasma conditions is not obvious, as well as the interaction between the seeds and the substrate under harsh conditions. Arnault *et al.* reported on a detailed XPS analysis of detonation particles seeded on a silicon substrate under pure hydrogen MPCVD plasma.[144] They revealed that sp^2 species are entirely etched by atomic hydrogen, without altering the sp^3 seeds even after a high temperature treatment ($T = 1213$ K). At the same time, the formation of silicon carbide at the interface between the seed and the silicon substrate has been identified, more predominant at 1213 K than at 993 K (Figure 10.11). This interlayer has been predicted by the authors to significantly enhance the adhesion of nanodiamonds on the substrate and consequently the adhesion of the grown diamond layer. Indeed, beyond the roughness of the substrate known to enhance the adhesion of the diamond layer, specific studies also reported the impact of such interlayers on the adhesion[145,146] or, conversely, on the possibility to realize spontaneously detaching self-standing diamond films.[147] The stability of the nanodiamonds under CVD plasma was also confirmed by Sumant *et al.*,[148] who studied the underside of a diamond nanocrystalline film grown from detonation nanodiamonds. Surface analysis showed that 98% of the underside was composed of diamond, the rest being mainly composed of sp^2 carbon attributed to the detonation nanodiamonds used for the seeding.

10.3.2.3 Methods for Seeding

10.3.2.3.1 Evaporation of Nanodiamond Suspensions. The easiest way to transfer and deposit nanodiamonds from a colloid onto a surface obviously consist in the deposition of a drop of nanodiamond suspension, the solvent being evaporated naturally. Unfortunately, such a deposit generally leads to a poor spatial homogeneity due to convection forces during the drying process, combined with a progressive aggregation of the nanodiamonds during the concentration increase. Shenderova *et al.*[143] illustrated this last point by allowing nanodiamond suspensions to dry and be reconstituted in different solvent. They revealed that a dried aqueous suspension that was reconstituted in water promoted a strong

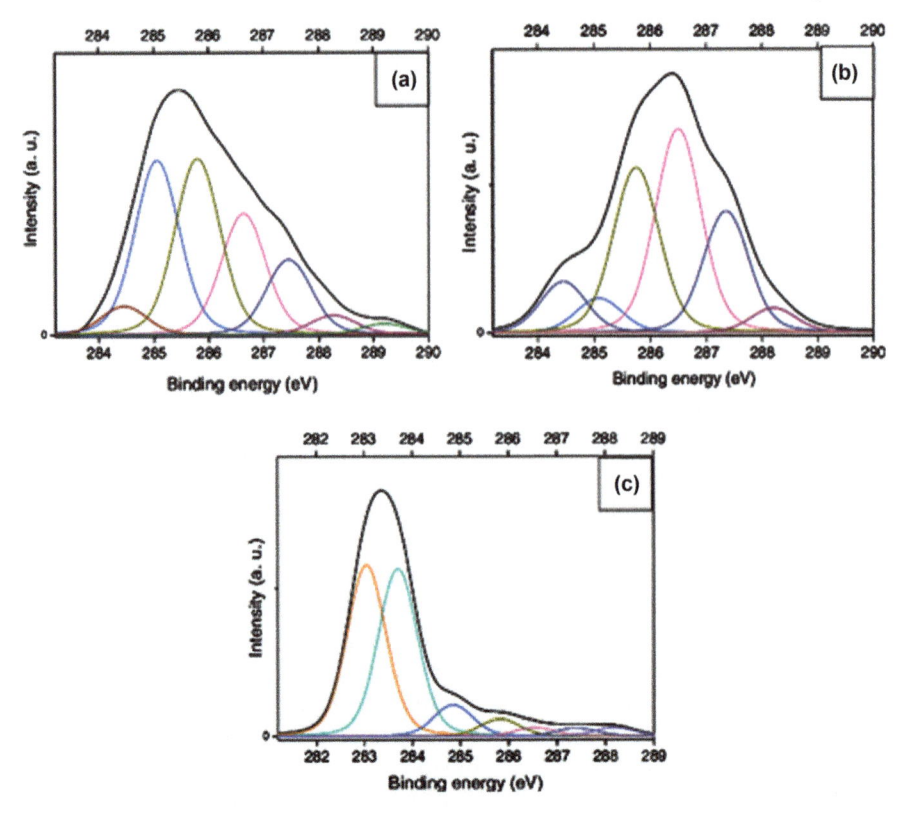

Figure 10.11 XPS C1s core level spectra; a) initial state, after a UHV treatment; b) at 1073 K; c) at 1173 K. The detailed assignation of components is discussed in ref. 144.

agglomeration of the NDs (Figure 10.12). To underline the effect of the solvent, a similar experiment, performed from a DMSO suspension, which was dried and the resulting particles reconstituted in water, maintains the initial average diameters of the nanodiamonds.

To overcome these issues, different strategies to spread and dry nanodiamond suspensions homogeneously on a substrate have been explored in the literature. One of the most efficient is the use of a spin coater to ensure a good dispersion of NDs on a flat substrate and enhance the drying step. The technique consists of the deposition of a volume of ND suspension on the substrate, which is then spin-coated at a defined rotation speed. This basic approach can be used either with nanodiamonds suspended in water or alcohol,[149,150] or with composite suspensions, in which organics can be added to modulate some parameters, such as the viscosity. For instance, the addition of polyvinyl alcohol (PVA), a water-soluble polymer, in a water-based suspension of nanodiamonds was shown to be highly efficient in realizing dense seeding. PVA is expected to help the formation of a homogenous polymeric layer on the substrate as it exerts a stabilizing effect on the nanodiamonds.[151,152] Scorsone et al. reported the use of such composite

Sizes of ND in initial suspensions as well as after drying and re-dispersion.

Initial solvent	ND size in initial solvent, nm	ND size after drying and resuspension, nm	Solvent for resuspension
DMSO	35	35	DMSO
DMSO	35	37	H_2O
DMSO	35	50	DMSO:methanol[a] (1:9)
H_2O	30	98	H_2O
DMSO/H_2O[b] (1:9)	40	55 (res.)[c]	H_2O
H_2O + DMSO[d] (1:1)	30	48 (res.)[c]	H_2O

[a] ND powder was first dispersed in DMSO, then methanol added.
[b] ND was fractionated in a mixture DMSO/H_2O.
[c] res. – Small residue presents.
[d] ND was fractionated in water, then DMSO added.

Figure 10.12 Sizes of NDs in their initial suspensions, as well as after drying and re-dispersion in different solvents.[143]

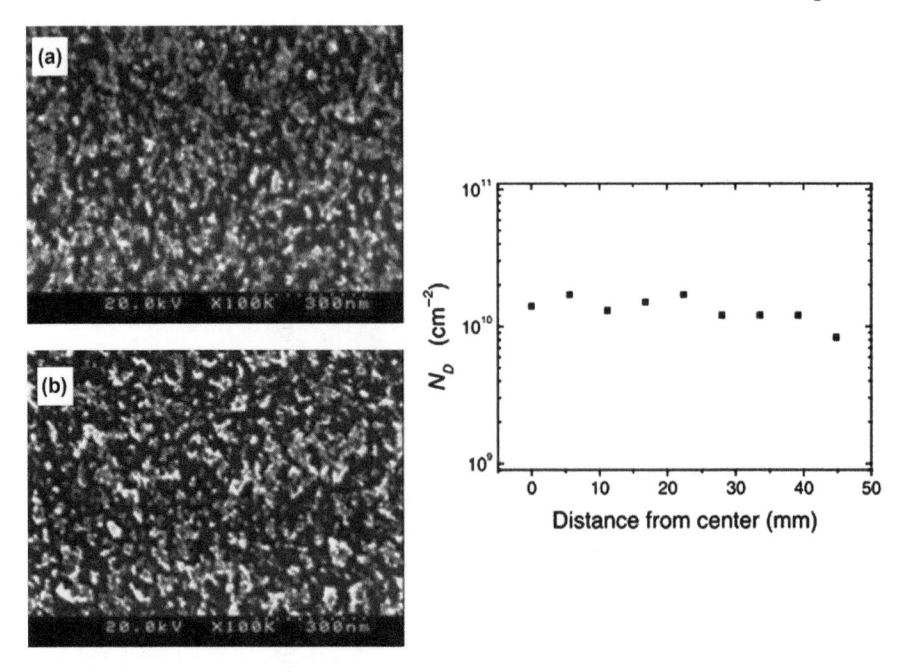

Figure 10.13 (Left) SEM images of seeded silicon substrate prepared from a 0.1% wt/wt nanoparticle with 1.5% wt/wt PVA aqueous solution after (a) pyrolysis under hydrogen plasma and (b) pyrolysis in air at 350 °C. (Right) Density of seeds across the radius of a 4 inch silicon wafer.[32]

mixtures (PVA/nanodiamonds) to finely control the density of seeds deposited on a substrate by manipulating the nanodiamond concentration.[151] High density seed layers ($N_0 > 10^{10}$ seeds cm^{-2}) were deposited over 4-inch substrates (Figure 10.13), matching the required characteristics to grow ultrathin diamond layers. In this study a thorough XPS study was also conducted to exclude any effect of the polymer on the diamond growth, showing that the polymer matrix in which the nanodiamonds are embedded instantly burns away under the plasma conditions at the beginning of the growth. Such methods involving spin-coating of nanodiamond suspensions give rise to excellent results on flat surfaces, but may encounter limitations on 3D shapes due to screening effects.

10.3.2.3.2 Seeding by Adhesion Transfer. An alternative to solvent evaporation to spread nanodiamonds on a substrate consists of the exploitation of specific interactions between both surfaces (substrate and nanodiamonds). As detailed above, nanodiamonds carry on their surface a complex chemistry, which yield adequate surface properties for adhesion. Among them, electrostatic forces can be used to promote the self-adhesion of nanoparticles onto an oppositely charged substrate. For instance, silicon, which is the most common substrate on which polycrystalline diamond is grown, exhibits a negative surface charge due to the native oxide layer (thus also including silica-based substrates, like quartz, glass, annealed silicon, *etc.*). In this case electrostatic adhesion will require positively charged nanodiamonds, which can be found either commercially or prepared by different methods (see section 10.3.2.1). Note that such an approach based on electrostatic interactions requires careful handling of the nanodiamond suspension, especially regarding the pH and the solvent chosen.

To achieve electrostatic seeding, the easiest method consists of dipping the silicon substrate in a positively charged nanodiamond slurry under ultrasonic agitation.[142,148,153–156] Ultrasonic agitation is used here to maintain an optimized dispersion of the nanodiamonds, especially when detonation nanoparticles are used. Agitation also enhances the diffusion of the particles over the substrate to promote their self-adhesion and the seeding homogeneity. In this case the adhesion of the nanodiamonds is expected to be due to electrostatic forces and not to mechanical embedding of the seeds in the substrate, for which larger particles in the slurry are required to confer enough momentum to the nanodiamonds ("hammering" effect see section 10.3.1). This method, so-called "ultrasonic seeding", is not limited to pure aqueous suspensions of nanodiamonds. DMSO-based suspensions were also reported in the literature to work efficiently with this approach[143] (Figure 10.14). However, a small fraction of methanol had to be added to the DMSO slurries because of its high surface tension, high boiling point and slow evaporation. With this *ultrasonic seeding* approach based on electrostatic adhesion, highly dense seeding is achieved with N_s densities regularly reported above 10^{11} seeds cm^{-2}.[153,155] Furthermore, as the adhesion of the first layer of nanodiamonds will largely screen the substrate charge, the formation of several layers of nanodiamonds is limited. Using highly dispersed detonation nanodiamonds (typically with diameters below 10 nm), Williams *et al.* reported the growth of an ultrathin diamond coating

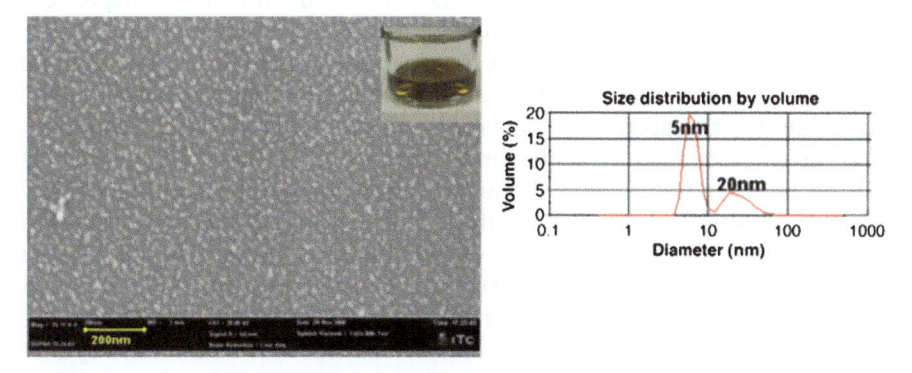

Figure 10.14 (Left) SEM image of NDs seeded on silicon using a mixture of 0.5% NDs in DMSO (1 part) and methanol (3 parts), and treated in an ultrasonic bath. The inset illustrates the ND/DMSO/MeOH seeding slurry. (Right) Particle size distribution.[143]

with a thickness below 70 nm.[153] Furthermore, this technique also allows the seeding of complex surfaces with 3D shapes. For instance, Si tips, carbon nanotubes[157] and polymeric stripes[158] have been covered with nanodiamonds by ultrasonic seeding.

Despite the efficiency of this "ultrasonic seeding", which is driven by electrostatic forces, a limitation can come from the undesired charge of the substrate or of the nanodiamonds. For instance, carboxylated nanodiamonds exhibiting a negative zeta potential at neutral pH will not stick spontaneously on a silicon substrate, even with a highly dispersed suspension. A method to control and adapt the surface charge of the substrate is thus required. Polarization of the substrate to initiate an electrophoretic deposition is possible and has been explored by different authors.[157,159,160] For instance, silicon field emission tips for cold cathode fabrication have been seeded by such an approach with nanodiamonds of different charge by adapting the applied voltage. Selective seeding on polarized gold strips was also demonstrated by Hens et al.[157] using positively charged nanodiamonds functionalized with amino groups.

However, electrophoretic deposition implies a conductive substrate. To overcome this limitation, an efficient alternative resides in the use of polyelectrolytes to coat the substrate and confer to them the required charge. Polycationic polymers, such as poly(diallyldimethylammonium) chloride (PDDAC), will give a positive charge to the surface, while polyanionic polymers, such as poly(styrenesulfonate) (PSS), will confer a negative charge. These polymers, which are soluble in water, can be deposited as a very thin layer of a few nanometers on almost all materials, even with complex 3D features. By this way, insulating materials or patterned substrates exhibiting different charges can be coated and have a homogenous charge on their entire surface. The seeding will thus be achieved *via* a layer-by-layer approach,[161,162] with the first layer of polyelectrolyte deposited on the substrate by dipping the substrate in a polyelectrolyte solution. After thorough rinsing, the second layer, made of nanodiamonds, is deposited, also by dipping the coated substrate in a nanodiamond suspension of opposite charge.[163–167] A first monolayer of nanodiamonds is thus achieved, and multilayers of nanodiamonds can also be deposited on the substrate by repeating this sequence. Note that the use of this very thin polymer coating has no consequence on the diamond growth as all the polymer is immediately burned away by the plasma. The density reached by this technique is equivalent to those achieved by ultrasonic seeding (10^{11} seeds cm^{-2}). As for the other techniques based on the electrostatic adhesion of seeds, 3D complex structures can also be seeded through this approach. Girard et al.[164] reported the fabrication of all-diamond nanostructures grown in nanoscaled silicon molds seeded by this method (Figure10.15).

10.3.2.3.3 Patterned Seeding. 3D micro- or nano-structures are often required in the design of diamond-based microelectromechanical systems

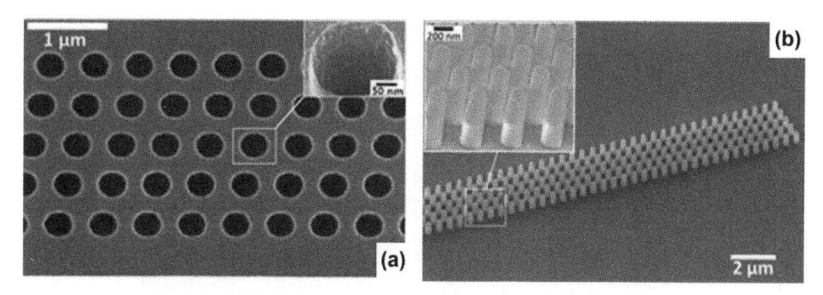

Figure 10.15 SEM pictures of (a) a mold made of silicon structured with wells (200 × 800 nm²) after seeding; (b) diamond nanostructures grown in the molds after silicon dissolution.[164]

(MEMS) or nanoelectromechanical systems (NEMS), optical systems or biomedical devices. Depending on the targeted properties, all-diamond structures may have to be realized as a simple coating of nanocrystalline diamond (as described above) may not be sufficient. Top-down approaches can be considered, mostly based on the etching of a thick diamond layer through a mask.[168,169] Almost perfect structures can be achieved with this approach for photonic applications, either in monocrystalline[170] or polycrystalline diamond.[171] However, etching of diamond is time consuming and post-processing can be cumbersome. An alternative bottom-up approach is also possible, based on the patterning of the seeds, to obtain a selective area deposition (SAD) of the diamond layer. However, a reliable selective growth will be achieved only if several orders of magnitude exist between voluntary seeded areas and the residual density of nucleation of the substrate. Here, two strategies have been reported in the literature. The first relies on the seeding of a substrate on which the selective removal of the nanodiamond has been performed.[172] For instance, the patterning of the nanodiamonds can be achieved by a plasma etching of unwanted nanoparticles through the protection of an aluminum hard mask deposited by lithography.[173] Babchenko *et al.* described a technique in which patterning of the seeds is obtained through the wet chemical removal of nanodiamonds with a SiO_2 layer on pre-defined areas by a buffer oxide etchant.[174,175] Selective growth on patterned seeding can also be achieved by a selective deposition of the seeds. Fox *et al.* developed a technique based on water ink containing nanodiamonds.[176] By standard ink jet technology, local seeding has been realized by Fox *et al.* on glass, silicon, copper and fused quartz (Figure 10.16) with micrometric patterns. An alternative at the nanometric scale has been recently proposed by Zhuang *et al.*, who used microcontact printing with a nanodiamond-based ink.[177] Using the electrostatic seeding method, a pre-patterning of the polyelectrolyte layer can also be used to drive the deposition of the nanodiamonds only on the regions of opposite charge.[163,165]

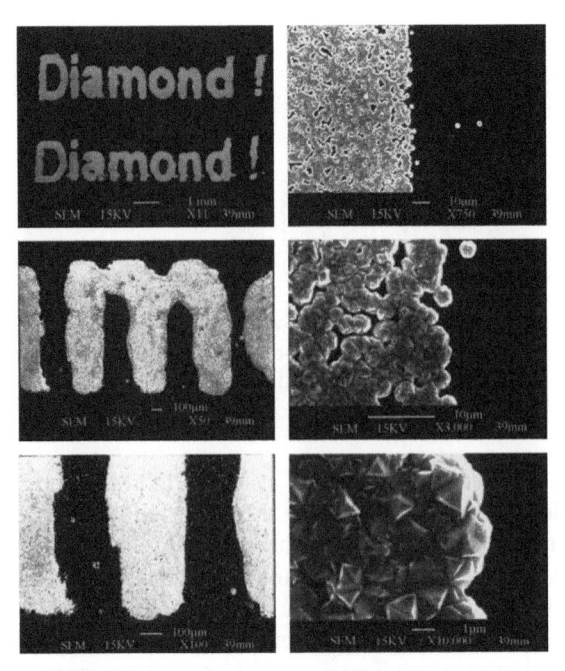

Figure 10.16 SEM pictures, at various magnifications, of ink-jet seeded CVD-grown diamond on a silicon substrate.[176]

10.4 Conclusion

The control of ultrathin diamond films (<150 nm) is a major challenge for applications (heat dissipation, NEMS and MEMS). The present chapter focuses on the two experimental approaches investigated in the literature: the nucleation and seeding pathways.

Firstly, we have shown that, for diamond, both homogeneous (*i.e.*, nucleation in a gas phase) and heterogeneous (*i.e.*, nucleation onto a surface) nucleation has been experimentally observed. Focusing on heterogeneous nucleation, we underlined that the intrinsic reactivity of a given heterosubstrate under CVD conditions is an essential parameter. It leads to different mechanisms competing with diamond nucleation and participates in the interface formation. We have also mentioned the role of structural defects on heterogeneous diamond nucleation. Then, we focused on the BEN process, which remains the most controllable *in situ* nucleation technique. The most relevant parameters (concentration, kinetic energy and the fluxes of ionic species) were detailed. A brief state of the art was given for heteroepitaxy comparing the promising substrates.

In the second part of this chapter we reviewed the seeding techniques, focusing on the controllable deposition of nanodiamonds on surfaces (evaporation, adhesion transfer, patterning). Describing the main characteristics of a nanodiamond suspension (particle shape, surface chemistry

Table 10.2 Comparison of BEN and seeding.

Bias enhanced nucleation	Seeding
Advantages	
– In situ technique	– Large substrates (up to six inches)
– Diamond heteroepitaxy	– 3D substrates
– Good film adhesion	– Variety of substrates
Drawbacks	
– Non-conductive substrates	– Film adhesion to be improved
– 3D substrates[179]	– Not usable for heteroepitaxy
– Substrate size limited by CVD reactor geometry	

and charge, size distribution, solvent nature), we underlined the relevance of the colloidal chemistry on the seeding quality. The stability of the diamond nanoparticles under plasma conditions was also mentioned.

Both pathways (nucleation and seeding) were successfully applied to obtain diamond layers with thicknesses lower than 150 nm. Nevertheless, each pathway possesses its own advantages and drawbacks (Table 10.2). Seeding is powerful for diamond deposition on large or non-conductive substrates, while BEN remains the best approach for heteroepitaxy and well-adherent films.[178]

References

1. J. P. Mazellier, O. Faynot, S. Cristoloveanu, S. Deleonibus and P. Bergonzo, *Diam. Relat. Mater.*, 2008, **17**, 1248.
2. M. Lions, S. Saada, B. Bazin, M.-A. Pinault, F. Jomard, F. Andrieu, O. Faynot and P. Bergonzo, *Diam. Relat. Mater.*, 2010, **19**, 413.
3. W. Zhu, G. P. Kochanski and S. Jin, *Science*, 1998, **282**, 1471.
4. H. Liu, D. S. Dandy, *Diamond Chemical Vapor Deposition: Nucleation and Early Growth Stages*, 1995, Noyes Publications, New Jersey.
5. O. A. Williams, *Diam. Relat. Mater.*, 2011, **20**, 621.
6. M. Volmer and A. Weber, *Z. Physik Chem.*, 1926, **199**, 277.
7. R. Kern, G. de Lay, J. J. Métois in *Current Topics in Materials Science*, ed. E. Kaldis, North-Holland Publ. Co, 1979, vol. 3, pp. 135–419.
8. S. Barrat, *Vide: Science, technique et applications*, 2001, **2/4**, 320.
9. B. V. Derjaguin and D. V. Fedoseev, *Scientific American*, 1975, **233**, 102.
10. H. Fujimoto, Y. Kitagawa, H. Hao and K. Fukui, *Bulletin of the Chemical Society of Japan*, 1970, **43**, 52.
11. S. A. Godleski, P. V. Schleyer, E. Osawa and W. T. Wipke, *Prog. Phys. Org. Chem.*, 1981, **13**, 63.
12. M. Frenklach and K. Spear, *J. Mater. Res.*, 1988, **3**, 133; G. A. Raiche and J. B. Jeffries, *Carbon*, 1990, **28**, 796.
13. S. Mitura, *J. Cryst. Growth*, 1987, **80**, 417.

14. W. Howard, D. Huang, J. Yuan, M. Frenklach, K. E. Spear, R. Koba and A. W. Phelps, *J. Appl. Phys.*, 1990, **68**, 1247.
15. M. Frenklach, R. Kematick, D. Huang, W. Howard, K. E. Spear, A. W. Phelps and R. Koba, *J. Appl. Phys.*, 1989, **66**, 395.
16. K. Tsugawa, M. Ishihara, J. Kim, Y. Koga and M. Hasegawa, *Phys. Rev. B*, 2010, **82**, 125460.
17. H. M. Jang and N. M. Hwang, *J. Mater. Res.*, 1998, **13**, 3527.
18. N. M. Hwang, *J. Cryst. Growth*, 1999, **198–199**, 945.
19. L. M. Ghiringhelli, C. Valeriani, J. H. Los, E. J. Meijer, A. Fasolino and D. Frenkel, *Molecular Physics*, 2008, **106**, 2011.
20. R. S. Lewis, T. Ming, J. F. Wacker, E. Anders and E. Steel, *Nature*, 1987, **326**, 160.
21. A. P. Jones, L. B. d'Hendecourt, S.-Y. Sheu, H.-C. Chang, C.-L. Cheng and H. G. M. Hill, *Astronomy & Astrophys*, 2004, **416**, 235.
22. V. N. Mochalin, O. Shenderova, D. Ho and Y. Gogotsi, *Nature Nanotech*, 2012, 7, 11.
23. M. Tomellini and B. Bunsenges, *Phys. Chem.*, 1995, **99**, 838.
24. M. Fanfoni and M. Tomellini, *J. Phys.: Condens. Matter*, 2005, **17**, R571.
25. A. N. Kolmogorov, *Bull. Acad. Sci. URSS (cl. Sci. Math. Nat.)*, 1937, **3**, 355.
26. W. A. Johnson and R. F. Mehl, *Trans. Am. Inst. Min., Metall. Pet. Eng.*, 1939, **135**, 416.
27. M. J. Avrami, *Chem. Phys.*, 1939, **7**, 1103.
28. M. J. Avrami, *Chem. Phys.*, 1940, **8**, 212.
29. M. Tomellini and M. Fanfoni, *Int. J. Nano Science*, 2010, **9**, 1.
30. M. Tomellini and M. Fanfoni, *Diam. Relat. Mater.*, 2010, **19**, 1135.
31. R. Polini, M. Tomellini, M. Fanfoni and F. Le Normand, *Surf. Science*, 1997, **373**, 230.
32. P. Badziag, W. S. Verwoerd, W. P. Ellis and N. R. Greiner, *Nature*, 1990, **343**, 244.
33. J. Y. Raty and G. Galli, *Nature Mater.*, 2003, **2**, 792.
34. S. T. Lee, H. Y. Peng, X. T. Zhou, N. Wang, C. S. Lee, I. Bello and Y. Lifshitz, *Science*, 2000, **287**, 104.
35. S. Pecoraro, J. C. Arnault and J. Werckmann, *Diam. Relat. Mater.*, 2004, **13**, 342.
36. D. Wittorf, W. Jager, C. Dieker, A. Floter and H. Guttler, *Diam. Relat. Mater.*, 2000, **9**, 1696.
37. Y. Lifshitz, X. M. Meng, S. T. Lee, R. Akhveldiany and A. Hoffman, *Phys. Rev. Lett.*, 2004, **93**, 056101.
38. R. Brescia, M. Schreck, S. Gsell, M. Fischer and B. Stritzker, *Diam. Relat. Mater.*, 2008, **17**, 1047.
39. S. Kono, M. Shiraishi, N. I. Plusnin, T. Goto, Y. Ikehima, T. Abukawa, M. Schimomura, Z. Dai, C. Bernarski-Meinke and B. Golding, *New Diamond and Frontier Carbon Technology*, 2005, **15**, 363.
40. S. Gsell, S. Berner, T. Brugger, M. Schreck, R. Brescia, M. Fischer, T. Greber, J. Osterwalder and B. Stritzker, *Diam. Relat. Mater.*, 2008, **17**, 1029.

41. A. Hoffman, Sh. Michaelson, R. Akhvlediani, N. K. Hangaly, S. Gsell, R. Brescia, M. Schreck, B. Stritzker, J. C. Arnault and S. Saada, *Phys. Stat. Solidi*, 2009, **206**, 1972.
42. D. Das and R. N. Singh, *Inter. Mater. Reviews*, 2007, **52**, 29.
43. M. Schreck in *Heteroepitaxial Growth in CVD Diamond for Electronic Devices and Sensors*, ed. Ricardo S. Sussmann, John Wiley & Sons, Chichester, UK, 2009, p. 125.
44. J. C. Arnault, *Surf. Rev. Lett.*, 2003, **10**, 127.
45. S. Saada, J. C. Arnault, N. Tranchant, M. Bonnauron and P. Bergonzo, *Phys. Stat. Sol.*, 2007, **204**, 2854.
46. T. B. Massalski, H. Okamoto, P. R. Subramanian, L. Kacprzak (eds), *Binary Alloy Phase Diagramms*, vol. 1, ASM International, Materials Park, OH, 1990, p. 871.
47. D. N. Belton and S. J. Schmieg, *Thin Solid Films*, 1992, **212**, 68.
48. A. Lindlbauer, R. Haubner and B. Lux, *Diam. Films Technol.*, 1992, **2**, 81.
49. P. C. Yang, W. Zhu and J. T. Glass, *J. Mater. Res.*, 1993, **8**, 1773.
50. T. Tachibana, Y. Yokota, K. Kobashi and Y. Shintani, *J. Appl. Phys.*, 1997, **82**, 4327.
51. T. Tachibana, Y. Yokota, K. Hayashi and K. Kobashi, *Diam. Relat. Mater.*, 2001, **10**, 1633.
52. L. Constant, C. Speisser and F. Le Normand, *Surf. Science*, 1997, **387**, 28.
53. X. Li, W. Cai, J. An, S. Kim, J. Nah, D. Yang, R. Piner, A. Velamakanni, I. Jung, E. Tutuc, S. K. Banerjee, L. Colombo and R. S. Ruoff, *Science*, 2009, **324**, 1312.
54. J. C. Arnault, S. Delclos, S. Saada, N. Tranchant and P. Bergonzo, *J. Appl. Phys.*, 2007, **101**, 014904.
55. J. C. Arnault, L. Intiso, S. Saada, S. Delclos, P. Bergonzo and R. Polini, *Appl. Phys. Lett.*, 2007, **90**, 044101.
56. A. Chavanne, J. C. Arnault, J. Barjon and J. Arabski, *Surf. Science*, 2011, **605**, 564.
57. Th. Bauer, S. Gsell, F. Hormann, M. Schreck and B. Stritzker, *Diam. Relat. Mater.*, 2004, **13**, 335.
58. A. Chavanne, J. Barjon, B. Vilquin, J. Arabski and J. C. Arnault, *Diam. Relat. Mater.*, 2012, **22**, 52.
59. N. Vaissière, S. Saada, M. Bouttemy, A. Etcheberry, P. Bergonzo and J. C. Arnault, *Diam. Relat. Mater.*, 2013, **36**, 16.
60. J. C. Angus, A. Argoitia, R. Gat, Z. Li, M. Sunkara, L. Wang and Y. Wang, *Trans. Phil. Mag., A*, 1993, **342**, 195.
61. W. A. Yarbrough, *J. Am. Ceram. Soc.*, 1992, **75**, 3179.
62. B. Lewis and J. C. Anderson, *Nucleation and Growth of Thin Films*, Academic Press, New York, 1978.
63. J. A. Venables, G. D. T. Spiller and M. Hanbücken, *Rep. Progr. Phys.*, 1984, **47**, 399.
64. Y. Bar Yam and T. D. Moustakas, *Nature*, 1989, **342**, 786.
65. R. Ramesham and C. Ellis, *J. Mat. Res.*, 1992, **7**, 1189.
66. P. A. Dennig and D. A. Stevenson, *Appl. Phys. Lett.*, 1991, **59**, 1562.

67. E. I. Givargizov, V. V. Zhirnov, A. V. Kuznetsov and P. S. Plekhanov, *Mater. Letters*, 1993, **18**, 61.
68. R. Polini, *J. Appl. Phys.*, 1992, **72**, 2517.
69. K. Kobayashi, N. Mutsumura and Y. Machi, *Mater. Manufacturing Processes*, 1992, 7, 395.
70. R. Polini, M. Tomellini, M. Fanfoni and F. Le Normand, *Surf. Science*, 1997, **373**, 230.
71. J. C. Arnault, S. Pecoraro, J. Werckmann and F. Le Normand, *Diam. Relat. Mater.*, 2001, **10**, 1612.
72. R. Polini, P. D'Antonio and E. Traversa, *Diam. Relat. Mater.*, 2003, **12**, 340.
73. W. S. Yang and J. H. Je, *J. Mater. Research*, 1996, **1**, 1787.
74. K. Mitsuda, Y. Kojima, T. Yoshida and K. Akashi, *J. Mater. Sci.*, 1987, **22**, 1557.
75. P. E. Pehrsson, F. G. Celii, J. E. Butler, in *Diamond Films and Coatings*, ed. R. F. Davis, Noyes Publications, Park Ridge, 1993, p. 68.
76. S. Iijima, Y. Aikawa and K. Baba, *Appl. Phys. Lett.*, 1990, **57**, 2646.
77. P. Ascarelli and S. Fontana, *Appl. Surf. Science*, 1993, **64**, 307.
78. B. Lux and R. Haubner, in *Diamond and Diamond-like Films and Coatings*, R. E. Clausing *et al.* (eds), 1991, New York, Plenum Press, pp. 579–609.
79. J. C. Arnault, L. Demuynck, C. Speisser and F. Le Normand, *Europ. Phys. Journal B*, 1999, **11**, 327.
80. S. Yugo, A. Izumi, T. Kanai, T. Muto and T. Kimura, *Proc. 2nd Int. Conf. New Diamond Science and Technology*, 1991, Pittsburg, PA, MRS, pp. 385–389.
81. C. H. Lee, Z. D. Lin, N. G. Shang, L. S. Liao, I. Bello, N. Wang and S. T. Lee, *Phys. Rev. B*, 2000, **62**, 17134.
82. A. A. Morrish and P. E. Pehesson, *Appl. Phys. Lett.*, 1991, **59**, 417.
83. J. Yang, X. W. Su, Q. J. Chen and Z. D. Lin, *Appl. Phys. Lett.*, 1995, **66**, 3284.
84. S. Yugo, T. Kanai, T. Kimura and T. Muto, *Appl. Phys. Lett.*, 1991, **58**, 1036.
85. B. R. Stoner, G. H. M. Ma, S. D. Wolter and J. T. Glass, *Phys. Rev. B*, 1992, **45**, 11067.
86. X. Jiang and C. P. Klages, *Diam. Relat. Mater.*, 1993, **2**, 1112.
87. F. Stubhan, M. Ferguson, H.-J. Füsser and R. J. Behm, *Appl. Phys. Lett.*, 1995, **66**, 1900.
88. Q. Chen, J. Yang and Z. Lin, *Appl Phys. Lett.*, 1995, **67**, 1853.
89. X. T. Zhou, H. L. Lai, H. Y. Peng, C. Sun, W. J. Zhang, N. Wang, I. Bello, C. S. Lee and S. T. Lee, *Diam. Relat. Mater.*, 2000, **9**, 134.
90. S. D. Wolter, T. H. Borst, A. Vescan and E. Kohn, *Appl. Phys. Lett.*, 1996, **68**, 3558.
91. A. Flöter, H. Güttler, G. Schulz, D. Steinbach, C. Lutz-Elsner, R. Zachai, A. Bergmaier and G. Dollinger, *Diam. Relat. Mater.*, 1998, 7, 283.
92. M. Katoh, M. Aoki and H. Kawarada, *Jpn. J. Appl. Phys.*, 1994, **33**, L194.

93. S.-T. Lee, Z. D. Lin and X. Jiang, *Mater. Sci. Eng.*, 1999, **25**, 123.
94. X. Jiang, K. Schiffmann, C. P. Klages, D. Wittorf, C. L. Jia, K. Urban and W. Jaeger, *J. Appl. Phys.*, 1998, **83**, 2511.
95. H. Kawarada, C. Wild, N. Herres, R. Locher and P. Koidl, *J. Appl. Phys.*, 1997, **81**, 3490.
96. M. Schreck, F. Hörmann, H. Roll, T. Bauer and B. Stritzker, *New Diamond and Frontier Carbon Technology*, 2001, **11**, 189.
97. K Hassouni, F Silva and A Gicquel, *J. Phys. D: Appl. Phys.*, 2010, **43**, 153001.
98. M. Schreck, T. Bauer and B. Stritzker, *Diam. Relat. Mater.*, 1995, **4**, 553.
99. S. Kátai, Z. Tass, G. Hars and P. Deak, *J. Appl. Phys.*, 1999, **86**, 5549.
100. S. Kátai, A. Kováts, I. Maros and P. Deák, *Diam. Relat. Mater.*, 2000, **9**, 317.
101. S. Barrat, S. Saada, J. M. Thiebaut and E. Bauer-Grosse, *Diam. Relat. Mater.*, 2001, **10**, 1637.
102. A. Guise, S. Barrat and E. Bauer-Grosse, *Diam. Relat. Mater.*, 2007, **16**, 695.
103. C. Sarrieu, N. Barth, A. Guise, J. C. Arnault, S. Saada, S. Barrat and E. Bauer-Grosse, *Phys. Stat. Sol. (A)*, 2009, **206**, 1967.
104. X. Jiang, K. Schiffmann and C. P. Klages, *Phys. Rev. B*, 1994, **50**, 8402.
105. T. Bauer, S. Gsell, M. Schreck, J. Goldfuß, J. Lettieri, D. G. Schlom and B. Stritzker, *Diam. Relat. Mater.*, 2005, **14**, 314.
106. C. Bednarski, Z. Dai, A. P. Li and B. Golding, *Diam. Relat. Mater.*, 2003, **12**, 241.
107. T. Aoyama, N. Amano, T. Goto, T. Abukawa, S. Kono, Y. Ando and A. Sawabe, *Diam. Relat. Mater.*, 2007, **16**, 594.
108. J. C. Arnault, S. Saada, S. Delclos, S. Pecoraro, P. Bergonzo, *Mater. Res. Soc. Symp. Proc.* 2008, Warrendale, PA, 1039.
109. J. Gerber, S. Sattel, K. Jung, H. Ehrhard and J. Robertson, *Diam. Relat. Mater.*, 1995, **4**, 559.
110. Y. Lifshitz, T. Kohler, T. Frauenheim, I. Guzmann, A. Hoffman, R. Q. Zhang, X. T. Zhou and S. T. Lee, *Science*, 2002, **297**, 1531.
111. S. Sattel, J. Gerber and H. Ehrhardt, *Phys. Status Solidi A*, 1996, **154**, 141.
112. X. Jiang, K. Schiffmann and C. P. Klages, *Phys. Rev. B*, 1994, **50**, 8402.
113. K. G. Perng, K. S. Liu and I. N. Lin, *Appl. Phys. Lett.*, 2001, **79**, 3257.
114. T. Suesada, N. Nakamura, H. Nagasawa and H. Kawarada, *Jap. J. Appl. Phys.*, 1995, **34**, 4898.
115. J. C. Arnault, S. Saada, S. Delclos, L. Intiso, N. Tranchant, R. Polini and P. Bergonzo, *Diam. Relat. Mater.*, 2007, **16**, 690–694.
116. T. Bauer, S. Gsell, F. Hormann, M. Schreck and B. Stritzker, *Diam. Relat. Mater.*, 2004, **13**, 335.
117. S. Koizumi and T. Inuzuka, *Jpn J. Appl. Phys.*, 1993, **32**, 3920.
118. M. Fischer, S. Gsell, M. Schreck, R. Brescia and B. Stritzker, *Diam. Relat. Mater.*, 2008, **17**, 1035.

119. M. Schreck, Th. Bauer, S. Gsell, F. Hörmann, H. Bielefeldt and B. Stritzker, *Diam. Relat. Mater.*, 2003, **12**, 262.
120. B. Golding, C. Bednarski-Meinke and Z. Dai, *Diam. Relat. Mater.*, 2004, **13**, 545.
121. N. Jiang, B. W. Sun, Z. Zhang and Z. Lin, *J. Mater. Res.*, 1994, **9**, 2695.
122. X. Jiang, K. Schiffmann and C.-P. Klages, *Physical Review B*, 1994, **50**, 8402.
123. K. Suzuki, A. Sawabe, H. Yasuda and T. Inuzuka, *Applied Physics Letters*, 1987, **50**, 728.
124. H. Liu and D. S. Dandy, *Diamond and Related Materials*, 1995, **4**, 1173.
125. S.-T. Lee, Z. Lin and X. Jiang, *Materials Science and Engineering: R: Reports*, 1999, **25**, 123.
126. S. Iijima, Y. Aikawa and K. Baba, *J. Mater. Res.*, 1991, **6**, 1491.
127. W. S. Yang and J. H. Je, *J. Mater. Res.*, 1996, **11**, 1787.
128. C.-P. Chang, D. L. Flamm, D. E. Ibbotson and J. A. Mucha, *J. Appl. Phys.*, 1988, **63**, 1744.
129. B. V. Spitsyn, L. L. Bouilov and B. V. Derjaguin, *J.Cryst. Growth*, 1981, **52**, 219.
130. P. A. Dennig and D. A. Stevenson, *Appl. Phys. Lett.*, 1991, **59**, 1562.
131. R. Akhvlediani, I. Lior, S. Michaelson and A. Hoffman, *Diamond and Related Materials*, 2002, **11**, 545.
132. M. Varga, T. Ižák, A. Kromka, M. Veselý, K. Hruška and M. Michalka, *Central European J. Phys.*, 2011, **10**, 218.
133. O. A. Shenderova, V. V. Zhirnov and D. W. Brenner, *Critical Rev. Solid State Mater. Sci.*, 2002, **27**, 227.
134. K. Niwase, T. Tanaka, Y. Kakimoto, K. N. Ishihara and P. H. Shingu, *Mater. Trans. JIM*, 1995, **36**, 282.
135. A. Krueger, *Adv. Mater.*, 2008, **20**, 2445.
136. M. Ozawa, M. Inaguma, M. Takahashi, F. Kataoka, A. Krueger and E. Osawa, *Adv. Mater.*, 2007, **19**, 1201.
137. A. Krueger, F. Kataoka, M. Ozawa, T. Fujino, Y. Suzuki, A. E. Aleksenskii, A. Y. Vul' and E. Osawa, *Carbon*, 2005, **43**, 1722.
138. S. Osswald, G. Yushin, V. Mochalin, S. O. Kucheyev and Y. Gogotsi, *J. Am. Chem. Soc.*, 2006, **128**, 11635.
139. A. M. Schrand, S. A. C. Hens and O. A. Shenderova, *Critical Rev. Solid State Mater. Sci.*, 2009, **34**, 18.
140. A. Krueger, Y. Liang, G. Jarre and J. Stegk, *J. Mater. Chem.*, 2006, **16**, 2322.
141. T. Petit, J. C. Arnault, H. A. Girard, M. Sennour, T.-Y. Kang, C.-L. Cheng and P. Bergonzo, *Nanoscale*, 2012, **4**, 6792.
142. O. A. Williams, J. Hees, C. Dieker, W. Jäger, L. Kirste, and C. E. Nebel, *ACS Nano*, 2010, **4**, 4824.
143. O. Shenderova, S. Hens and G. McGuire, *Diamond and Related Materials*, 2010, **19**, 260.
144. J. C. Arnault, S. Saada, O. A. Williams, K. Haenen, P. Bergonzo, M. Nesladek, R. Polini and E. Osawa, *Phys. Stat. Sol. (A)*, 2008, **205**, 2108.

145. M. Amaral, F. Almeida, A. J. S. Fernandes, F. M. Costa, F. J. Oliveira and R. F. Silva, *Surface Coatings Technol.*, 2010, **204**, 3585.
146. T. Hartnett, R. Miller, D. Montanari, C. Willingham and R. Tustison, *J. Vacuum Sci. Technol. A*, 1990, **8**, 2129.
147. D. Varshney, A. Kumar, M. J.-F. Guinel, B. R. Weiner and G. Morell, *Diamond and Related Materials*, 2012, **21**, 99.
148. A. V. Sumant, P. U. P. A. Gilbert, D. S. Grierson, A. R. Konicek, M. Abrecht, J. E. Butler, T. Feygelson, S. S. Rotter and R. W. Carpick, *Diamond and Related Materials*, 2007, **16**, 718.
149. K. Hanada, K. Matsuzaki and T. Sano, *Surface Science*, 2007, **601**, 4502.
150. A. Bongrain, E. Scorsone, L. Rousseau, G. Lissorgues, C. Gesset, S. Saada and P. Bergonzo, *J. Micromech. Microeng.*, 2009, **19**, 74015.
151. E. Scorsone, S. Saada, J. C. Arnault and P. Bergonzo, *J. Appl. Phys.*, 2009, **106**, 14908.
152. A. Kromka, O. Babchenko, H. Kozak, K. Hruska, B. Rezek, M. Ledinsky, J. Potmesil, M. Michalka and M. Vanecek, *Diamond Relat. Mater.*, 2009, **18**, 734.
153. O. A. Williams, O. Douhret, M. Daenen, K. Haenen, E. Osawa and M. Takahashi, *Chem. Phys. Lett.*, 2007, **445**, 255.
154. V. V. Danilenko, *Phys. Solid State*, 2004, **46**, 595.
155. A. Kromka, B. Rezek, Z. Remes, M. Michalka, M. Ledinsky, J. Zemek, J. Potmesil and M. Vanecek, *Chem. Vap. Deposition*, 2008, **14**, 181.
156. L. Ondic, K. Dohnalova, M. Ledinsky, A. Kromka, O. Babchenko and B. Rezek, *ACS Nano*, 2011, **5**, 346.
157. S. C. Hens, G. Cunningham, T. Tyler, S. Moseenkov, V. Kuznetsov and O. Shenderova, *Diamond and Related Materials*, 2008, **17**, 1858.
158. A. Kromka, O. Babchenko, H. Kozak, K. Hruska, B. Rezek, M. Ledinsky, J. Potmesil, M. Michalka and M. Vanecek, *Diamond Relat. Mater.*, 2009, **18**, 734.
159. I. Zhitomirsky, *Materials Letters*, 1998, **37**, 72.
160. A. N. Alimova, N. N. Chubun, P. I. Belobrov, P. Y. Detkov, and V. V. Zhirnov, in *Papers from the 11th International Vacuum Microelectronics Conference*, AVS, Asheville, North Carolina (USA), 1999, vol. 17, p. 715.
161. A. Mamedov, J. Ostrander, F. Aliev and N. A. Kotov, *Langmuir*, 2000, **16**, 3941.
162. W. Xue and T. Cui, *Nanotechnology*, 2007, **18**, 145709.
163. S.-K. Lee, J.-H. Kim, M.-G. Jeong, M.-J. Song and D.-S. Lim, *Nanotechnology*, 2010, **21**, 505302.
164. H. a. Girard, E. Scorsone, S. Saada, C. Gesset, J. C. Arnault, S. Perruchas, L. Rousseau, S. David, V. Pichot, D. Spitzer and P. Bergonzo, *Diamond and Related Materials*, 2012, **23**, 83.
165. H. A. Girard, S. Perruchas, C. Gesset, M. Chaigneau, L. Vieille, J.-C. Arnault, P. Bergonzo, J.-P. Boilot and T. Gacoin, *ACS Appl. Mater. Interfaces*, 2009, **1**, 2738.
166. E. Chevallier, E. Scorsone, H. A. Girard, V. Pichot, D. Spitzer and P. Bergonzo, *Sensors Actuators B*, 2010, **151**, 191.

167. J. H. Kim, S. K. Lee, O. M. Kwon, S. I. Hong and D. S. Lim, *Diamond Relat. Mater.*, 2009, **18**, 1218.
168. H. Masuda, M. Watanabe, K. Yasui, D. Tryk, T. Rao and A. Fujishima, *Advanced Materials*, 2000, **12**, 444.
169. N. Yang, H. Uetsuka, E. Osawa and C. E. Nebel, *Nano Letters*, 2008, **8**, 3572.
170. T. M. Babinec, B. J. M. Hausmann, M. Khan, Y. Zhang, J. R. Maze, P. R. Hemmer and M. Loncar, *Nature Nanotechnology*, 2010, **5**, 195.
171. X. Checoury, D. Néel, P. Boucaud, C. Gesset, H. Girard, S. Saada and P. Bergonzo, *Applied Physics Letters*, 2012, **101**, 171115.
172. H. Liu, C. Wang, C. Gao, Y. Han, J. Luo, G. Zou and C. Wen, *J. Physics: Condensed Matter*, 2012, **14**, 10973.
173. A. Bongrain, E. Scorsone, L. Rousseau, G. Lissorgues and P. Bergonzo, *Sensors Actuators B*, 2011, **154**, 142.
174. O. Babchenko, E. Verveniotis, K. Hruska, M. Ledinsky, A. Kromka and B. Rezek, *Vacuum*, 2012, **86**, 693.
175. A. Kromka, O. Babchenko, B. Rezek, M. Ledinsky, K. Hruska, J. Potmesil and M. Vanecek, *Thin Solid Films*, 2009, **518**, 343.
176. N. A. Fox, M. J. Youh, J. W. Steeds and W. N. Wang, *J. Appl. Phys.*, 2000, **87**, 8187.
177. H. Zhuang, B. Song, T. Staedler and X. Jiang, *Langmuir*, 2011, **27**, 11981.
178. Y. C. Lee, S. J. Lin, C. Y. Lin, M. C. Yip, W. Fang and I. N. Lin, *Diam. Relat. Mater.*, 2006, **15**, 2046.
179. M. Bonnauron, S. Saada, L. Rousseau, G. Lissorgues, C. Mer and P. Bergonzo, *Diam. Relat. Mater.*, 2008, **17**, 1399.

CHAPTER 11

The Microstructures of Polycrystalline Diamond, Ballas and Nanocrystalline Diamond

ROLAND HAUBNER

Institute of Chemical Technologies and Analytics, University of Technology Vienna, Getreidemarkt 9/164-CT, A-1060 Vienna, Austria
Email: rhaubner@mail.zserv.tuwien.ac.at

11.1 Introduction

The history of chemical vapour deposition (CVD) diamond synthesis is based on the deposition of well-faceted diamond layers with high crystal quality, and on the growth of polycrystalline materials containing many defects. This happens because the deposition conditions and the amount of atomic hydrogen vary strongly.

The initially grown half-spherical and polycrystalline diamond deposits are called ballas.[1] However, because of their morphologies, the CVD ballas layers are also called cauliflower-like,[2] ball-shaped,[3] *etc.* by some authors.

Later, nanocrystalline (NCD)[4,5] and ultra-nanocrystalline (UNCD)[6,7] diamond films were described. For these types, the diamond growth must be hindered to create very small individual crystals. These coatings can be grown by adding fullerenes to the deposition gas or using Ar/CH$_4$ mixtures with small additions of H$_2$. Small crystals surrounded by grain boundaries contain amorphous carbon, making it difficult to distinguish between diamond (crystalline) and amorphous carbon.

Optically clear diamond is used in optical and heat sink applications and the ballas, NCD and UNCD coatings show good performance in wear applications.[8]

11.2 Polycrystalline Diamond

The notation "polycrystalline diamond" suggests that several diamond crystals are located side by side, forming a compact material. The questions to be answered are: "what grain sizes do the diamond crystals have in the compacts?" and "what is the nature of the grain boundaries?"

Polycrystalline diamonds are found in nature, and can be synthesised by high-pressure high-temperature (HP-HT) processes, as well as by CVD. In the polycrystalline diamond aggregates that are found in nature the individual diamond crystals are large and, at the grain boundaries, consist of almost no amorphous carbon[9] (Figure 11.1). Examples of compact polycrystalline diamonds are the cubic aggregates found in the Congo. Ballas are special forms of globular polycrystalline aggregates and do not contain any second phases. Some of them are nearly spherical and range in size up to several millimetres and more.[10–12] Carbonado is a polycrystalline aggregate of individual diamond grains whose microstructure, *i.e.*, the grain boundary structure, resembles that of a sintered diamond ceramic with varying degrees of bonding, porosity, and secondary phases.[13,14]

HP-HT processes are primarily used to produce diamond powders in various grain sizes. Diamond powders can be sintered at HP-HT conditions to "polycrystalline diamond" (PCD).[15,16] PCD can be described as "synthetic carbonado", and can contain an interpenetrating metallic phase. PCD is used in numerous applications, such as cutting tools, drills, wire dies, anvils, wear parts, *etc*. At the diamond grain boundaries, no amorphous carbon is found.

In the case of low-pressure diamond deposition, which is carried out by a CVD process, mainly polycrystalline diamond is produced.[9] From the surface morphology, various crystal facets and also unfaceted ballas diamonds can

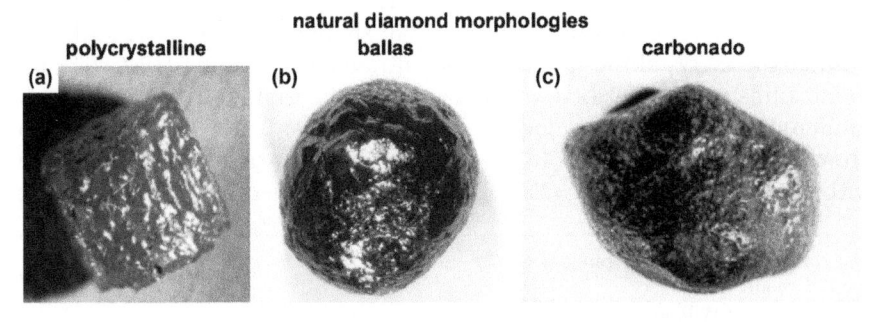

Figure 11.1 Natural polycrystalline diamond aggregates. (a) Polycrystalline cube, (b) spherical ballas diamond, (c) carbonado.

be observed. The quality of the deposited diamond is predetermined by the deposition conditions, and mainly the amount of atomic hydrogen.

For NCD and UNCD deposition, the diamond growth must be interrupted and therefore secondary nucleation occurs.[17] Because of the circumstances under which the amount of amorphous grain boundary material increases with a decreasing grain size of the diamond crystals, we should ask the question: "is it still a diamond layer?"

11.3 Low-pressure Diamond

11.3.1 Characterisation of the Diamond Deposits

The morphology of diamond in various coatings, deposited by variations in the process parameters, can easily be observed by scanning electron microscopy (SEM). However, because of the inhomogeneous distribution of the nanocrystalline regions, it is not possible to correlate the observed surface morphology with the real grain size.

Raman spectroscopy is the most frequently used technique to characterise carbon bonding in diamond films[18] to distinguish between diamond, graphite, microcrystalline diamond, and amorphous carbon.[19] The peak at 1332 cm^{-1} is characteristic of diamond and decreases because of defects and impurities in the crystals. In micro- and nano-crystalline materials, a peak at 1140 cm^{-1} is observed. With increasing amounts of micro-twinned areas, the intensities of the D-band and the 1470 cm^{-1} peak increase. The G-band also increases but, in addition, its position changes from 1539 cm^{-1} (faceted area) to 1585 cm^{-1} (ballas area).[19]

In principal, two different types of Raman spectra were observed by analysing the different types of diamond morphology:[19]

- diamond layers with amorphous carbon contents
- ballas layers with graphitic carbon contents

Figure 11.2 compares the relevant spectra and gives information about the various peaks observed.[19] Correlations between the different diamond morphologies and the Raman spectra are shown in Figure 11.3.

For Raman spectroscopy it has to be noted that investigations with different laser wavelengths result in different Raman spectra. In Figure 11.4 the Raman spectra of selected diamond coatings are shown, measured with three different lasers (472.681 nm/blue, 532.1 nm/green, 632.81 nm/red).[20]

For detailed information about the microstructure and crystallinity of diamond deposits, transmission electron microscope (TEM) investigations are necessary.[21]

Some typical pictures of defects and its distribution in the diamond are shown in Figure 11.5. Faceted CVD diamond depositions can form nearly perfect diamond crystals, or crystals containing many defects.[21] Micro-twins are mainly observed in the (111) and dislocations in the (100) growth sectors.[21]

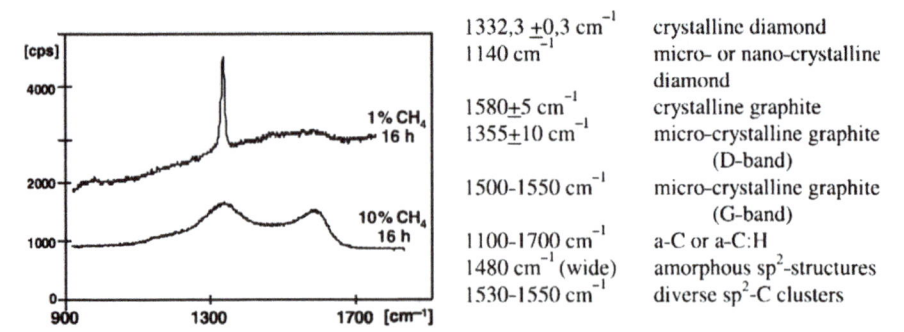

Figure 11.2 Raman spectra and assignment of the various peaks to carbon microstructures.[19]

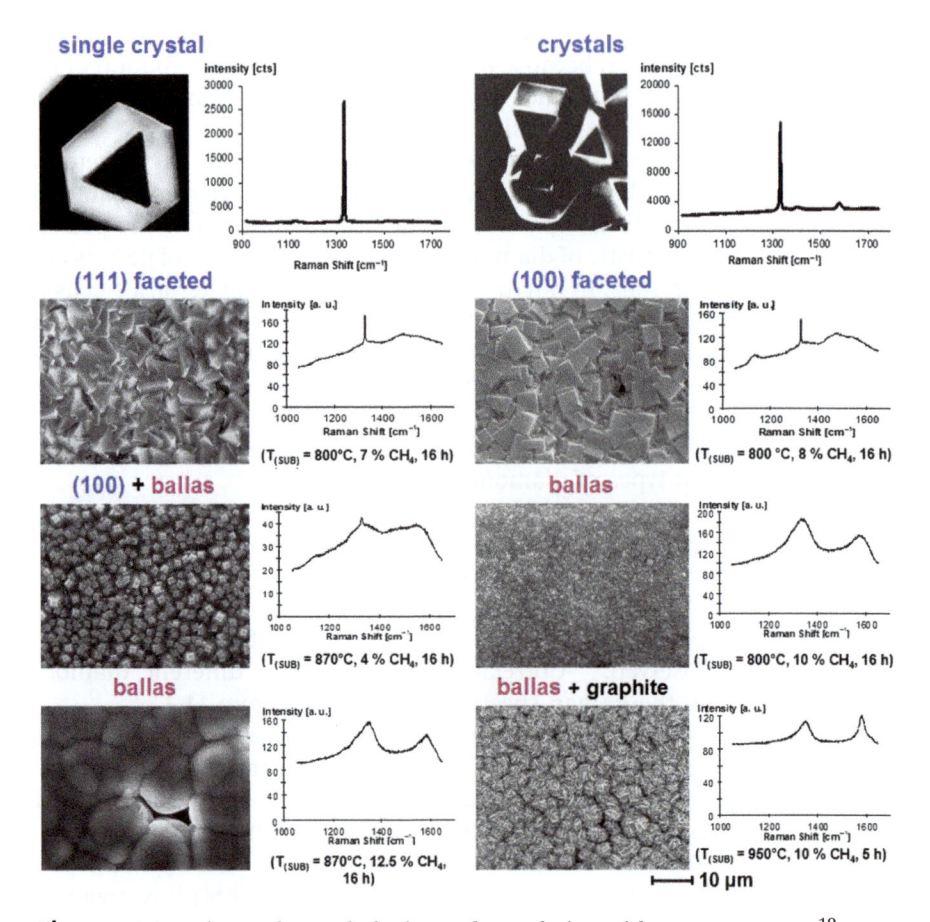

1332.3 ±0.3 cm^{-1}	crystalline diamond
1140 cm^{-1}	micro- or nano-crystalline diamond
1580±5 cm^{-1}	crystalline graphite
1355±10 cm^{-1}	micro-crystalline graphite (D-band)
1500-1550 cm^{-1}	micro-crystalline graphite (G-band)
1100-1700 cm^{-1}	a-C or a-C:H
1480 cm^{-1} (wide)	amorphous sp^2-structures
1530-1550 cm^{-1}	diverse sp^2-C clusters

Figure 11.3 Diamond morphologies and correlation with Raman spectra.[19]

As it can be seen in larger diamond grains, vacancies are formed. In the ballas layers, twin grain boundaries and normal grain boundaries are observed. Depending on the deposition conditions, the twinning can be very

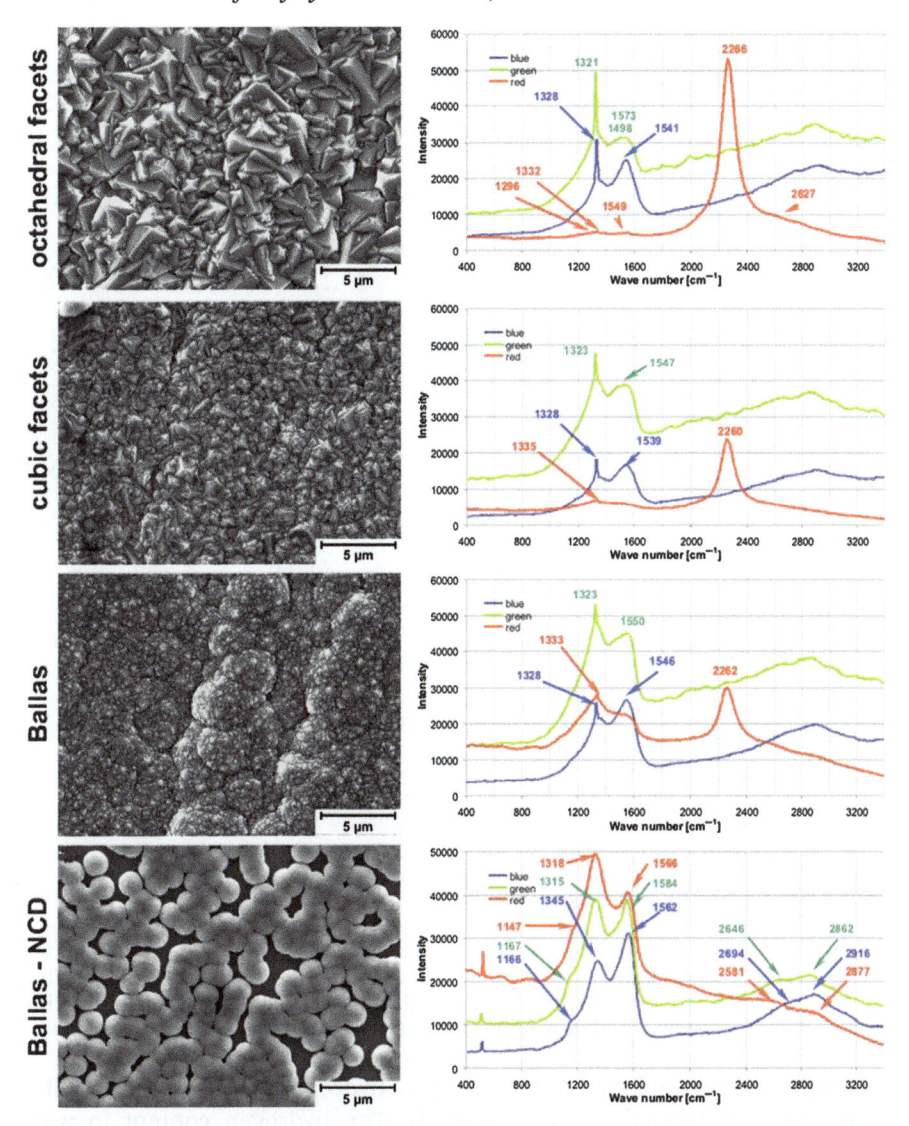

Figure 11.4 Raman spectroscopy with different laser wavelength on selected diamond coatings (472.681 nm/blue, 532.1 nm/green, 632.81 nm/red).[20]

high, resulting in a nanocrystalline material. Amorphous carbon inclusions are observed at normal grain boundaries and not at the twin grain boundaries.

In the case of NCD and UNCD only small crystalline areas, in the nm range, are observed, which are surrounded by amorphous carbon.[17,22]

To obtain further information about the impurities in the diamond, secondary ion mass spectrometry (SIMS),[23] Rutherford backscattering (RBS), and nuclear reaction analysis (NRA) measurements are useful.[24,25] An indicator of the amount of amorphous impurities in the coating is

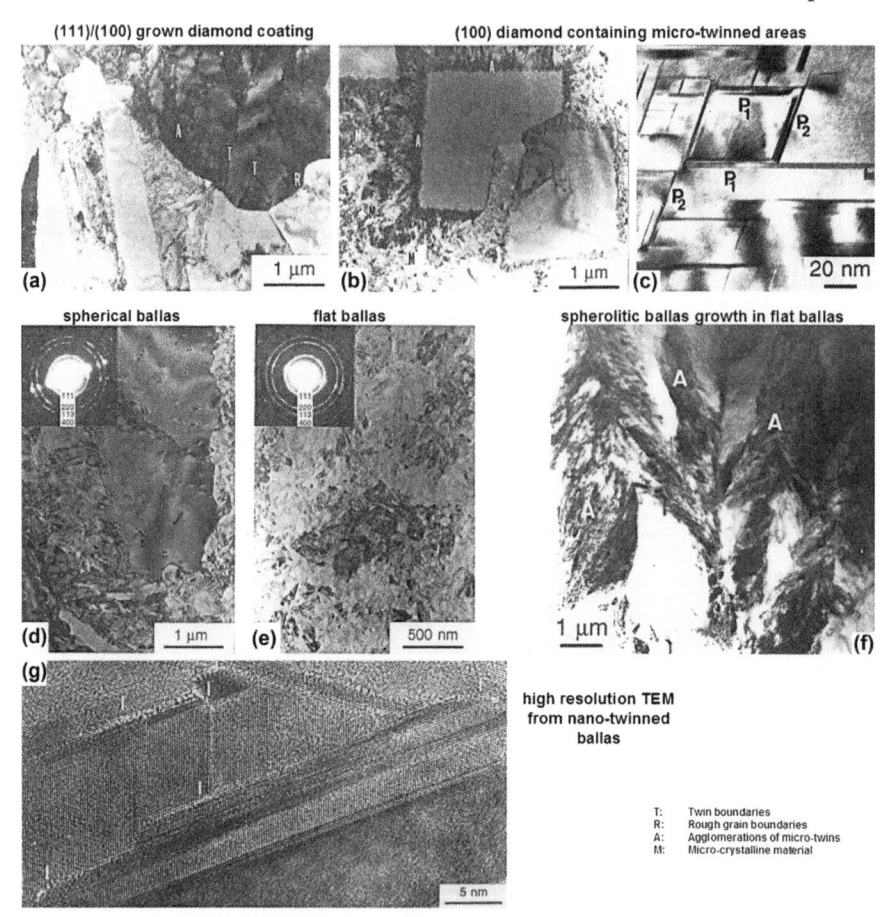

Figure 11.5 TEM investigations on various diamond coatings. (a) (111)/(100) grown diamond; (b) (100) grown diamond; (c) micro-twins region; (d) spherical ballas; (e) flat ballas; (f) flat ballas showing spherolitic growth; (g) high resolution TEM of micro-twinned area.[21]

hydrogen. All methods reveal much higher hydrogen contents in the ballas layers than in faceted diamond crystals. The hydrogen content in well-faceted layers with coarse crystals is below 0.2 at.% and about 1.2 at.% in ballas type layers.

11.3.2 Growth Mechanism for the Different Diamond Morphologies

Experimental results show an increasing growth rate with increasing hydrocarbon concentrations, while the diamond growth facets change from (111) to (100). After passing a maximum, a decreased growth rate for ballas deposition occurs (Figure 11.6). Similar trends are observed by varying other deposition parameters (*e.g.*, gas pressure and substrate temperature).

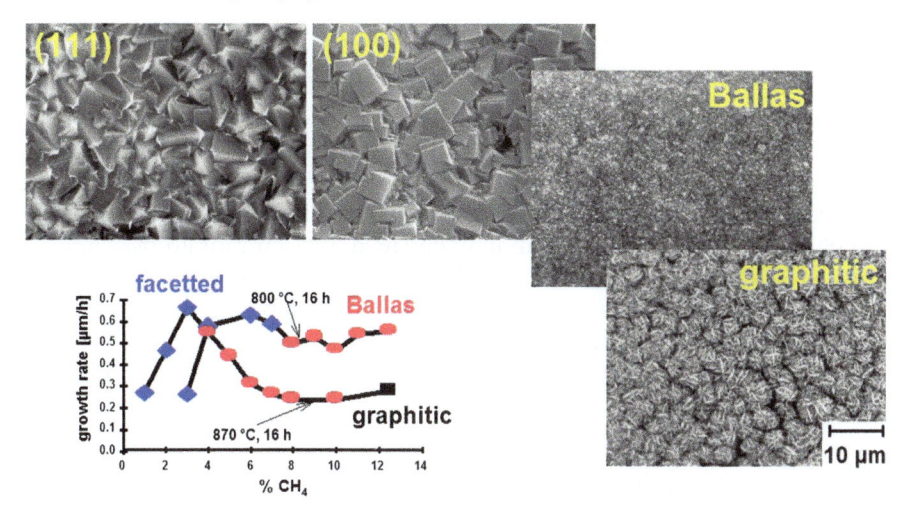

The Microstructures of Polycrystalline Diamond, Ballas and NCD 259

Figure 11.6 Correlations between diamond growth rate, methane concentration, and coating morphology.[19]

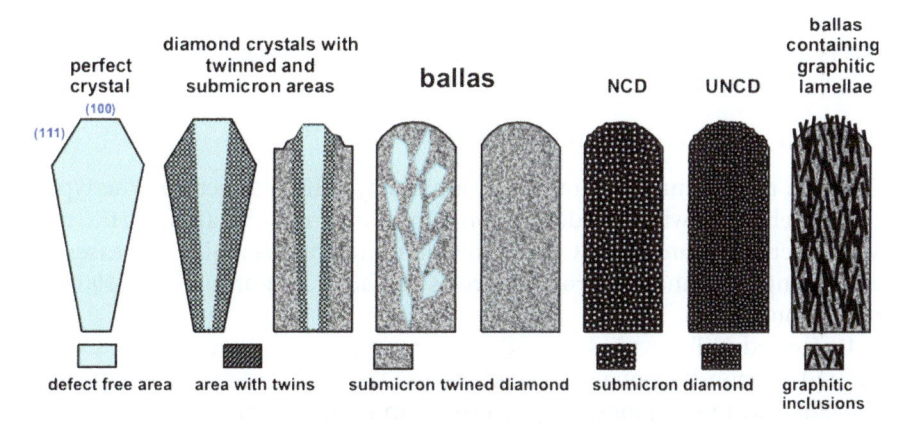

Figure 11.7 Schematic drawings of the various defect distributions observed in diamond layers with different morphologies.

The growth of single crystals seems to be disturbed by too high hydrocarbon super-saturation, too low atomic hydrogen or by a decreasing surface mobility of the growth species at low temperatures. The increased layer growth rate observed during a change in the deposition conditions towards ballas can be explained by the formation of twinned areas, which allow a quicker incorporation of the growth species on the surface. For ballas growth, but also for NCD and UNCD, the layer growth rates decrease. An alternative explanation for the decreased growth rate could be the preferred etching of amorphous carbon by atomic hydrogen. The various defect distributions observed in columnar structures of the diamond layers with different morphologies are summarised in Figure 11.7.

11.3.2.1 The (111) Crystal Facet Growth

The diamond (111) surface has the lowest surface energy and is, therefore, thermodynamically stable. For the growth of faceted crystals with a (111) morphology, re-entrant corners play a critical role by providing sites for rapid growth. Where three atoms are required to form a stable nucleus on a smooth (111) surface, only two atoms suffice to form a stable nucleus at the tip of a re-entrant corner.[26] Twin lamellae lead to re-entrant grooves and are formed by defects or inclusions on the growing (111) facet.

Well-faceted diamond crystals, with octahedral habits, grown under "optimal" deposition conditions, contain large areas, which are virtually free of defects. Mainly near the grain boundaries, the formation of defects, like micro-twins, starts.[9]

11.3.2.2 The (100) Crystal Facet Growth

During growth of (100) facets, the separation of defect-free and micro-twinned areas can be observed clearly. Defect-free columns located in the centre of a crystal (formed by the growth of a (100) facet) are surrounded by micro-twinned areas and grain boundaries. By SEM investigations, the large (100) facets and disturbed areas in between the faceted crystals are observed.[27]

11.3.2.3 Ballas Diamond Growth

Ballas is nearly pure diamond with a nanocrystalline structure. The type of defects changes with the deposition conditions and the atomic H/C ratio. With decreasing atomic H/C ratio, the diamond grain size also decreases and the strongly twinned microstructure is replaced by normal[21] or (100) Σ29 grain boundaries.[21,28]

Unfaceted ballas-type diamonds grown under conditions adjacent to the faceted diamond region were called "coarse ballas".[19,21] In the layers between the nano-twinned regions, larger (several μm in diameter) regions with defect-free diamond crystals were observed. The Raman spectra did not show the diamond 1332 cm^{-1} peak, while the D-band and the graphite peak appeared clearly. It must be noted that the Raman intensity (background) increases strongly due to the morphological change from faceted to ballas diamond.

The ballas-type diamond, grown at deposition conditions near the formation of graphite or amorphous carbon, is characterised by a homogenous micro-twinned matrix . A slight formation of graphite lamellae is possible. In this "flat ballas" the D-band is smaller and the graphite peak larger than in the "coarse ballas". By investigating a flat ballas layer by high resolution electron microscopy (HREM), multiple twinning with the grains misaligned relative to the exact twin orientation across several boundaries was observed. Some of the boundaries showed a bright contrast, indicating incoherent grain boundaries. This indicated the existence of, for example, amorphous or platelet-like intergranular phases, but neither HREM nor electron energy loss spectrometry (EELS) measurements indicated the presence of large volume fractions of amorphous or graphitic carbons.[21]

11.3.2.4 Graphitic Carbon Morphology

Transition from the ballas-like morphologies to layers containing graphitic carbon inclusions is initialised by implementation of fine lamellae. The lamellae appear to be graphitic in nature since the Raman spectra show a sharp graphite peak at about 1585 cm^{-1} when such fine lamellae are present. With an increasing CH_4 concentration, these lamellae grow until the whole microstructure becomes exclusively lamellar.

11.3.2.5 NCD and UNCD

For NCD and UNCD deposition, the crystal growth of diamond has to be hindered and secondary nucleation has to be favoured. Fullerenes can serve as a carbon source. It is also possible to work with $H_2/Ar/CH_4$ mixtures to reduce the amount of atomic hydrogen in comparison with the carbon content. By reducing the substrate temperature, the surface mobility decreases and NCD and UNCD can be deposited. Because such deposition parameters are subject to high losses compared to high-quality diamond growth, the layer growth rate is low.[29–32]

Characterisation of NCD and UNCD coatings is expensive and time-consuming because only TEM can verify the real microstructure. The TEM images show the crystalline nanodiamonds embedded in the grain boundary material.[4,6,21] On the whole, Raman spectra are used for characterisation, but it is not characteristic enough. In Raman, mainly the amorphous parts of the deposited layer are visualised but not the nanocrystalline diamond particles.

11.4 Correlation of Growth Conditions with Diamond Morphology

As shown in Figure 11.6, the diamond growth rate goes through a maximum with variations in the deposition parameters. At the conditions of maximal growth rate, the (100) facets are mainly observed. If the deposition conditions are optimised for diamond (*e.g.*, more atomic hydrogen), the quality of the deposited diamond increases. On the other hand, if many defects are formed in the diamond, the growth rate decreases. The decrease in growth rate can be explained by an increased etching of sp^2 carbons by atomic hydrogen. In other words, for the growth of high quality diamonds, high amounts of atomic hydrogen are necessary to stabilise the sp^3 carbon.[33,34]

11.4.1 Gas Activation (Atomic Hydrogen and Carbon Species)

Diamond in faceted morphologies can be deposited by various deposition techniques)[35,36] and within a wide range of parameters (*e.g.*, a temperature range between 500 and 1000 °C). The geometry of the deposition area depends on the activation method; *e.g.*, in a hot-filament apparatus[36] the optimal area is more linear under the filament and, in the case of microwave plasma,[37,38] the area is circular.

All of these methods are based on the generation of large amounts of atomic hydrogen compared with the amount of carbon applied.[37]

With increasing carbon super-saturation (*e.g.*, the amount of CH_4 in the gas phase), and decreasing atomic hydrogen content, the quality of the diamond decreases and a high amount of twins and stacking faults are introduced into the deposited diamond.[21,39] At this stage the radial growth of polycrystalline, unfaceted diamond (ballas-type morphology) occurs.[1,9] The ballas diamond itself is not a homogeneous, well-defined material because its microstructure and the diamond grain size can vary greatly. Finally, if the amount of atomic hydrogen is low, graphite lamellae are deposited instead of ballas diamond.[40]

11.4.2 Gas-phase Composition

The overall gas-phase composition determines the concentration of the growth species.[41] The deposition of diamond is possible in a wide range of different gas mixtures,[42] whereby CH_4–H_2 and CH_4–O_2–H_2 are the basic systems.[43] In general, an increasing concentration in CH_x compounds leads to ballas deposition. Large differences in the carbon content have been reported for the transition from faceted diamond crystals to ballas deposition (1.5% CH_4[44] and 11% CH_4[45]). The differences are caused by the filament temperature, the filament–substrate distance, the gas speed in the reactor, *etc.*, which can also vary widely.

C_{60} fullerene has been used as a carbon source to grow NCD.[46] Experiments were carried out in an Ar atmosphere with varying hydrogen additions (10^{-2} Torr C_{60}, 90–99% Ar, 1–10% H_2).[47,48] As, under such deposition conditions, the H/C atomic ratio is low, the pure crystalline diamond growth is strongly disabled and nanocrystalline diamond with amorphous grain boundaries is deposited. This effect is not specific for C_{60} as a carbon source because, with CH_4/H_2/Ar mixtures, similar results were observed.[49] The experiments with Ar addition showed the transition from faceted to ballas growth in gas mixtures with 10–20% H_2.

Similar to the deposition in H_2 carrier gas, a decrease in the layer growth rate was observed at high Ar contents (NCD deposition).[52] Figure 11.8 shows depositions grown at different Ar/H_2 ratios and the corresponding Raman spectra.[30]

11.4.3 Influence of Impurities

Impurities are known to hinder crystal growth and, in the low-pressure synthesis of diamond, they can also influence the diamond crystal morphology. Impurities are usually introduced by additions to the source gas or by

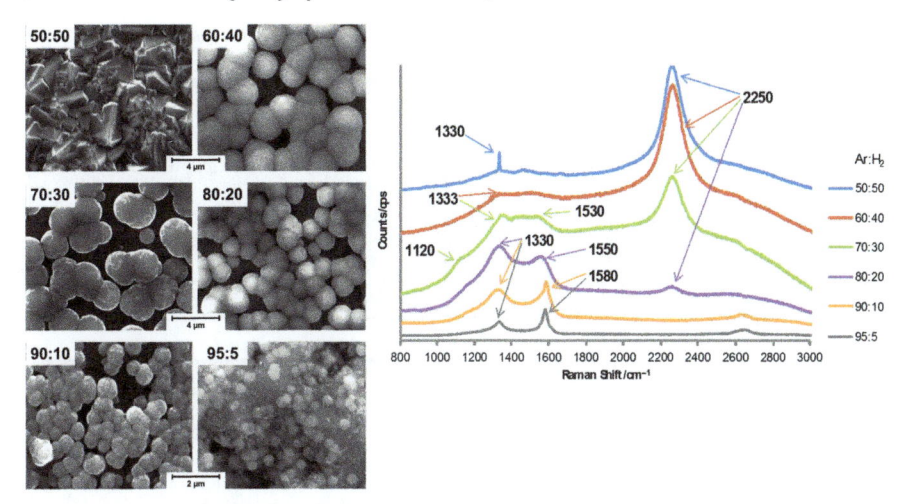

Figure 11.8 Diamond depositions grown at different Ar/H$_2$ ratios and the corresponding Raman spectra.[30]

evaporation from the substrate.[50] The addition of foreign gases, like N$_2$, PH$_3$ or boron compounds, *etc.*, can change the diamond morphology from faceted to ballas with increasing concentrations.[51–54] In all cases Raman and morphological investigations showed the formation of the ballas diamond (Figure 11.9).

The occurrence of impurities during the deposition process has been observed during diamond deposition on iron substrates. Due to the high vapour pressure of iron, ballas is formed at methane concentrations lower than 0.4%.[55] Impurities can also evaporate from the substrate when chemical reactions with atomic hydrogen occur. Such effects were observed on ceramic substrates, like h-BN or SiO$_2$.[56]

11.5 Summary and Conclusions

The microstructure of polycrystalline diamond types is given by the grain size of the individual crystals and the structure of the grain boundaries.

In faceted diamond and ballas diamond the small grain sizes are caused by micro-twinned regions. NCD and UNCD contain nanocrystalline diamond particles and the grain boundaries contain amorphous carbon. Furthermore, the microstructures shown by TEM for the nanocrystalline and ultra-nanocrystalline diamond can also be observed in ballas diamond. Because the notation of ballas diamond includes all diamond layers without facets (from the faceted region to graphitic deposits), the nanocrystalline and ultra-nanocrystalline coatings are a subgroup of ballas also.

The formation of ballas, NCD and UNCD are observed with increasing carbon super-saturation (CH$_4$ content) in the gas phase with a decreasing atomic H/C ratio. These conditions can be reached by variation of the carbon

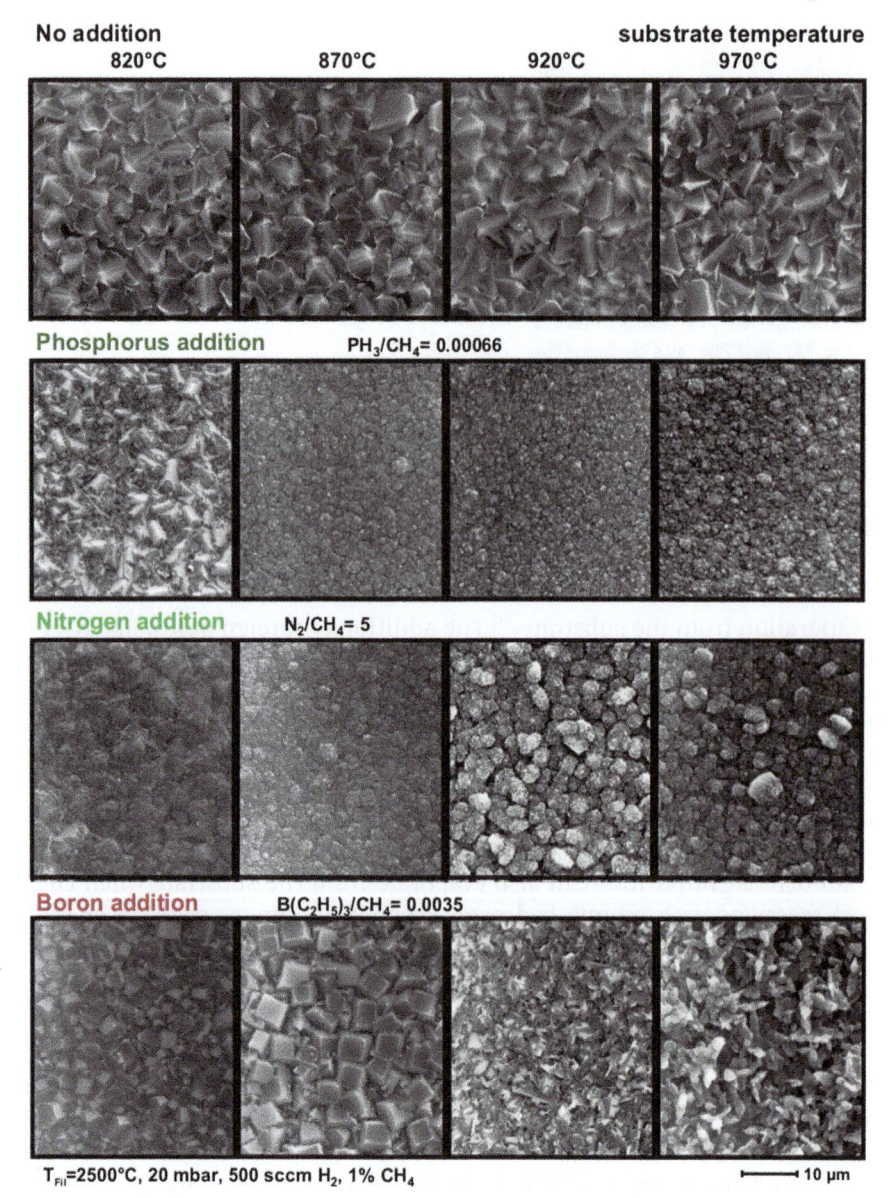

Figure 11.9 Deposition of ballas and NCD coatings by adding foreign gases, like N_2, PH_3 or boron. Influence of the surface temperature at constant concentrations of the added compounds.[51–53]

source, *e.g.*, with fullerenes or by reduction of the atomic hydrogen concentration in the case of Ar additions. By variation of the other diamond deposition parameters, ballas can be deposited also (*e.g.*, plasma intensity, surface temperature, impurities, *etc.*).

In order to use the high potential of the unique ballas structure in industrial applications, further investigations should be performed with regards to low-pressure ballas synthesis. NCD and UNCD are difficult to characterise but, nevertheless, these coatings show interesting properties for various applications.

References

1. R. Bichler, R. Haubner and B. Lux, *High Temperature-High Pressure*, 1989, **21**, 576.
2. P. N. Barnes and R. L. C. Wu, *Appl.Phys.Lett.*, 1993, **62**, 37.
3. P. K. Bachmann, D. U. Wiechert, in *Diamond and Diamond-Like Films and Coatings*, ed. R. E. Clausing, *et al.*, Plenum Press, New York, 1991, p. 677.
4. D. M. Gruen, in *Properties, Growth and Application of Diamond*, ed. M. H. Nazare and A. J. Neves, 2001, INSPEC, London, p. 299.
5. O. A. Williams and M. Nesladek, *Physica Status Solidi A*, 2006, **203**, 3375.
6. D. M. Gruen, in *Properties, Growth and Application of Diamond*, ed. M. H. Nazare and A. J. Neves, 2001, INSPEC, London, p. 303.
7. O. A. Williams, M. Daenen, J. D'Haen, K. Haenen, J. Maes, V. V. Moshchalkov, M. Nesladek and D. M. Gruen, *Diamond and Related Materials*, 2006, **15**, 654.
8. A. M. Bogus, I. Gebeshuber, A. Pauschitz, M. Roy and R. Haubner, *Diamond and Related Materials*, 2008, **17**, 1998.
9. B. Lux, R. Haubner, H. Holzer and R. C. DeVries, *Refractory Metals & Hard Materials*, 1997, **15**, 263.
10. A. F. Williams, *The Genesis of Diamond*, vol. 2, Ernest Benn Ltd, London 1932, University Microfilms, Ann Arbor, MI, 1971.
11. F. V. Kaminskii, Yu. A. Klyuev, B. I Prokopchuk, S. A Shchekam, V. I. Smirnov and I. N. Ivanovskaya, *Dokl. Akad. Nauk SSSR*, 1978, **242**, 687.
12. V. P. Martovitskii, N. A. Bul'enkov and Yu. P. Solodova, *Izv. Akad. Nauk SSSR, Ser. Geol.*, 1985, **6**, 71.
13. L. F. Trueb and E. C. deWys, *Science*, 1969, **165**, 799.
14. L. F. Trueb and E. C. deWys, *Amer. Min.*, 1971, **56**, 1252.
15. R. H. Wentorf, R. C. DeVries and F. P. Bundy, *Science*, 1980, **208**, 873.
16. J. C. Walmsley, *Mater. Sci. Engr. A*, 1988, **105/106**, 549.
17. C.-S. Wang, G.-H. Tong, H.-C. Chen, W.-C. Shih and I.-N. Lin, *Diamond and Related Materials*, 2010, **19**, 147.
18. R. J. Nemanich, L. Bergman, Y. M. LeGrice, R. E. Schroder, *Proc. 2nd Int. Conf. New Science and Technology*, Sept 23–27, 1990, Washington, DC, eds R. Messier, J. T. Glass, J. E. Butler and R. Roy, MRS, p. 741.
19. S. Bühlmann, E. Blank, R. Haubner and B. Lux, *Diamond and Related Materials*, 1999, **8**, 194.
20. M. Rudigier and R. Haubner, *Anal. Bioanal. Chem.*, 2012, **403**, 675.
21. M. Joksch, P. Wurzinger, P. Pongratz, R. Haubner and B. Lux, *Diamond and Related Materials*, 1994, **3**, 681.

22. C.-S. Chuan-Sheng, H.-C. Chen, H.-F. Cheng and I.-N. Lin, *J. Appl. Phys.*, 2009, **105**, 124311/1.
23. H. Spicka, M. Griesser, H. Hutter, M. Grasserbauer, S. Bohr, R. Haubner and B. Lux, *Diamond and Related Materials*, 1996, **5**, 383.
24. W. Wagner, F. Rauch, R. Haubner and B. Lux, *Thin Solid Films*, 1992, **207**, 257.
25. B. Lux, R. Haubner, M. Griesser and M. Grasserbauer, *Microchimica Acta*, 1997, **125**, 197.
26. J. C. Angus, M. Sunkara, S. R. Sahaida and J. T. Glass, *J. Mater. Res.*, 1992, **7**, 3001.
27. P. Wurzinger, P. Pongratz, P. Hartmann, R. Haubner and B. Lux, *Diamond and Related Materials*, 1997, **6**, 763.
28. D. M. Gruen, in *Properties, Growth and Application of Diamond*, ed. M. H. Nazare and A. J. Neves, 2001, INSPEC, London, p. 307.
29. M. Lessiak: *Diploma Thesis*, University of Technology, Vienna, 2012.
30. O. A. Williams and M. Nesládek, *Phys. Stat. Sol. (A)*, 2006, **203**, 3375.
31. J. E. Gerbia, J. Birrella, M. Sardelab and J. A. Carlislea, *Thin Solid Films*, 2005, **473**, 41.
32. T. Lin, G. Y. Yu, A. T. S. Wee and Z. X. Shen, *Appl. Phys. Lett.*, 2000, **77**, 2692.
33. R. Brunsteiner, R. Haubner and B. Lux, *Diamond and Related Materials*, 1993, **2**, 1263.
34. A. Lindlbauer, R. Haubner and B. Lux, *Refractory Metals & Hard Materials*, 1992, **11**, 247.
35. P. K. Bachmann and W. van Enckevort, *Diamond and Related Materials*, 1992, **1**, 1021.
36. K. E. Spear, *J. Am. Ceram. Soc.*, 1989, **72**, 171.
37. R. Haubner and B. Lux, *Refractory Metals & Hard Materials*, 1987, **12**, 210.
38. R. Haubner, A. Lindlbauer and B. Lux, *Refractory Metals & Hard Materials*, 1996, **14**, 119.
39. J. Michler, J. Stiegler, Y. von Kaenel, P. Moeckli, W. Dorsch, D. Stenkamp and E. Blank, *J. Crystal Growth*, 1997, **172**, 404.
40. J. Stiegler, T. Lang, M. Nygard-Ferguson, Y. von Kaenel and E. Blank, *Diamond and Related Materials*, 1996, **5**, 226.
41. R. Beckmann, B. Sobisch and W. Kulisch, *Diamond and Related Materials*, 1995, **4**, 256.
42. P. K. Bachmann, D. Leers and H. Lydtin, *Diamond and Related Materials*, 1991, **1**, 1.
43. R. Brunsteiner, R. Haubner and B. Lux, *Refractory Metals & Hard Materials*, 1996, **14**, 127.
44. K. Kobashi, K. Nishimura, Y. Kawate and T. Horiuchi, *Physical Review B*, 1988, **38**, 4067.
45. J. Brückner and T. Mäntylä, *Diamond and Related Materials*, 1993, **2**, 373.
46. D. M. Gruen, S. Liu, A. R. Krauss, J. Luo and X. Pan, *Appl. Phys. Lett.*, 1994, **64**, 1502.

47. L. C. Qin, D. Zhou, A. R. Krauss and D. M. Gruen, *Nanostruct. Mater.*, 1998, **10**, 649.
48. H. G. Busmann, U. Brauneck and H.-E. David, *Carbon*, 1998, **36**, 529.
49. D. Zhou, D. M. Gruen, L. C. Quin, T. G. McCauley and A. R. Krauss, *J. Appl. Phys. Lett.*, 1998, **84**, 1981.
50. R. Haubner and B. Lux, *Diamond Films and Technology*, 1994, **3**, 209.
51. S. Bohr, R. Haubner and B. Lux, *Diamond and Related Materials*, 1995, **4**, 133.
52. S. Bohr, R. Haubner and B. Lux, *Appl. Phys. Lett.*, 1996, **68**, 1075.
53. P. Hartmann, S. Bohr, R. Haubner, B. Lux, P. Wurzinger, M. Griesser, A. Bergmaier, G. Dollinger, H. Sternschulte and R. Sauer, *Refractory Metals & Hard Materials*, 1998, **16**, 223.
54. R. Haubner, S. Bohr and B. Lux, *Diamond and Related Materials*, 1999, **8**, 171.
55. A. Lindlbauer, R. Haubner and B. Lux, *Diamond Films and Technology*, 1992, **2**, 81.
56. A. Lindlbauer, R. Haubner and B. Lux, *Wear*, 1992, **159**, 67.

CHAPTER 12

Low-temperature Growth of Nanocrystalline Diamond Films in Surface-wave Plasma

KAZUO TSUGAWA AND MASATAKA HASEGAWA*

Nanotube Research Center, National Institute of Advanced Industrial Science and Technology (AIST), 1-1-1 Higashi, Tsukuba, Ibaraki 305-8565, Japan
*Email: hasegawa.masataka@aist.go.jp

12.1 Introduction

A nanocrystalline diamond (NCD) film is a polycrystalline diamond layer constructed from diamond crystals of a size from several nm to several tens of nm. Recently, NCDs have been generating interest in the materials field of chemical vapor deposition (CVD) diamond due to its potential industrial applications.

The major advantage of the NCD film in practical use is its small crystal size, which gives a smooth surface and transparency to the film. The surface roughness of a polycrystalline diamond film is comparable to its crystal size. The rough surface of a film with a large crystal size causes high light scattering, poor friction and inadequate wear performance. Consider the use of conventional polycrystalline diamond films comprising micrometer-sized diamond crystals for optical or tribological coatings as an example: the required transparency or friction and wear performance are compromised by the large surface roughness. Thus, mechanical or chemical polishing is required for these applications. However, this results in a significant cost due to the nature of diamond, *i.e.*, the hardness and the chemical stability. NCD films, therefore, have advantages in applications for optical and tribological coatings.

Since a number of groundbreaking works,[1–3] the plasma CVD technique has been widely used for the synthesis of diamond thin films. In particular, microwave-excited plasma is mostly used for the diamond CVD process at present. For diamond CVD growth, the following five processes go on simultaneously: 1) decomposition of the carbon source gas (generation of precursors), 2) deposition, 3) nucleation of diamond, 4) sp^3 bonding, and 5) removal of amorphous and graphite components by atomic hydrogen.

The primary requirement of plasma parameters for diamond synthesis is a low electron temperature T_e. It is considered that the most important precursor in the above process (1) is the CH_3 radical. Suppose, for instance, methane (CH_4) is used as the carbon source gas; the increasing T_e leads to the accelerated decomposition of CH_4. This means that radicals, such as CH_2 and C_2, which promote the formation of non-diamond components, increase in the plasma. In addition, a higher T_e invokes a higher energy of ion bombardment,[4] which damages the deposited diamond. Thus, a diamond film is effectively synthesized in low-electron temperature plasmas. However, the complete eradication of deposition of the non-diamond components is still difficult. Diamond deposition with selective removal of the depositing non-diamond components is therefore necessary for the formation of diamond films. Atomic hydrogen etches the sp^2-bonded non-diamond carbon, such as amorphous carbon and graphite, more rapidly than the sp^3 bonding network of carbon (diamond). In order to eliminate the non-diamond components this property of atomic hydrogen is utilized as the above process (5). Hence a massive amount of atomic hydrogen is fundamental for the quality improvement of the synthesized diamond films.

Here, the problem is that the efficient generation of atomic hydrogen needs high electron energies, *i.e.*, high T_e. That is to say, processes (1) and (5) require different plasma conditions and T_e. We have to work out the differences between these dilemmatic requirements in order to obtain high quality diamond films. Since the fundamental properties of plasma are roughly determined by the frequency of the electromagnetic wave used for its excitation, T_e cannot be controlled in such a wide range. Thus, the optimum conditions are obtained by adjusting the mixture ratio of carbon source gas and hydrogen under the given conditions. The microwave frequency of the 2.45 GHz industry–science–medical (ISM) band is frequently used for diamond plasma CVD, in which methane (CH_4) is mostly employed as the carbon source gas. In this case the pressure of the reaction gas mixture is roughly[2–5] kPa and the gas mixture ratio of methane to hydrogen is approximately 1:100. In this way diamond microwave plasma CVD (MWPCVD) needs a hydrogen-rich ambience. In addition, the substrate temperature is also an important parameter of diamond CVD. Since the etching effect of the non-diamond components by atomic hydrogen is more effective at higher temperatures, diamond CVD requires a substrate temperature from 700 to 1000 °C. This high process temperature has limited the substrate material selection for diamond deposition to materials such as Si. From this viewpoint, many efforts to lower the substrate temperature of diamond CVD have been made.

12.2 Surface-wave Plasma CVD

As a measure to overcome the problem of the high substrate temperature, we have utilized surface-wave plasmas for diamond MWPCVD. For typical MWPCVD processes of diamond, ball-shaped plasmas[5] in a bulk-heating mode[6] are used. In this case the microwaves introduced into the reaction chamber generate a ball-shaped plasma of the reaction gas mixture, where the microwave penetrates the plasma, and heats and excites its whole body, resulting in a high T_e ($\sim$10 eV) throughout the plasma.[6] In contrast to this, in a surface-heating mode including a surface-wave mode, the microwaves penetrate only a thin layer, heating and exciting just the surface of the plasma due to its high electron density near the plasma-generation region. To generate surface-wave plasmas, microwaves are introduced into the reaction chamber through a dielectric window[7,8] or a tube,[9] mostly made of quartz, and the plasma is generated on them. It is extended and maintained by a surface wave generated and propagating along with the interface between the window (tube) and the plasma as illustrated in Figure 12.1.[7]

The features of the surface-wave plasma are: 1) a high plasma density of 10^{11}–10^{12} cm^{-2} even at a low pressure of less than 100 Pa, 2) a low T_e of $\sim$1 eV (T_e decreases with increasing distance from the window and reaches $\sim$1 eV), and 3) befitting for large-area processes as the plasma is spread by the low pressure. These are preferable features for the diamond

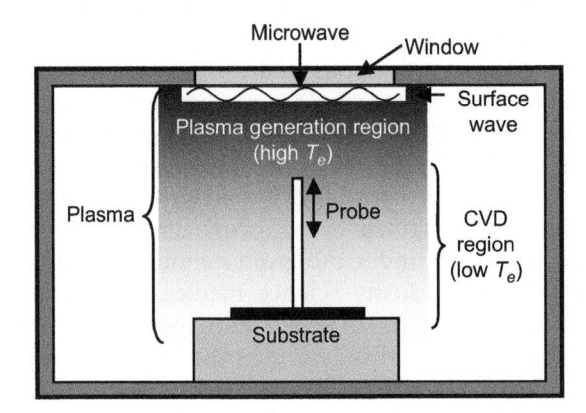

Figure 12.1 Schematic illustration of a typical surface-wave plasma in a reaction chamber with a dielectric window.
Reprinted from Tsugawa et al.[7] Copyright (2011), with permission from Elsevier.

film growth, where both high and low T_e are needed and a low-temperature process is desirable as aforementioned. Atomic hydrogen is generated in the high T_e region near the window, while the substrate is placed in the low T_e region far from the window, where CH_3 radicals are effectively created. Furthermore, substrate heating is suppressed owing to the low T_e and the low pressure. This enables the low process temperature.

In Figure 12.1, the plasma generated under the window diffuses in the chamber downward to the substrate, which is placed at some distance from the window (indicated as "CVD region" in Figures 12.1 and 12.2). Figure 12.2 shows the vertical distribution of electron densities n_e and electron temperatures T_e along with the distance from the window in one of our surface-wave plasma CVD chambers measured by a Langmuir probe. The probe measurements were performed in hydrogen plasma at pressures from 10 to 50 Pa without any carbon source gas in order to prevent carbon deposition on the probe.

In Figure 12.2(a), at any pressure, n_e near the window exceeds the cut-off density of surface waves (7.4×10^{10} cm^{-3}), indicating the generation of a surface-wave plasma under the window. For pressures over 50 Pa, probe measurements are hardly achieved due to the scattered data with large variances, especially in estimations of electron temperature T_e. This is due to a rapid decrease in the electron density n_e with distance from the window and instabilities of the plasma at the relatively high pressure for maintaining the plasma. In Figure 12.2(b), the electron temperature T_e decreases along with the distance from the window and then becomes nearly constant (2–3 eV for 10 and 20 Pa) over the CVD region.

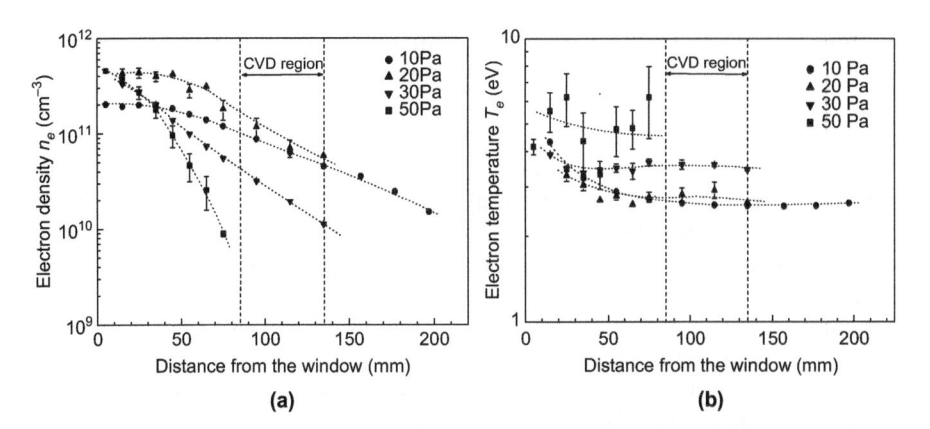

Figure 12.2 Electron density n_e and electron temperature T_e distributions of the surface-wave plasma as a function of the distance from the quartz window. The substrate is put within the region between two vertical dashed lines (CVD region). Dashed lines on the plots are guides for the eye.
Reprinted from Tsugawa et al.[7] Copyright (2011), with permission from Elsevier.

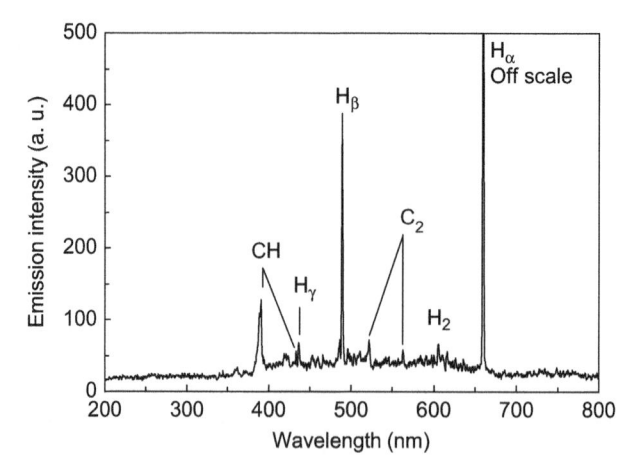

Figure 12.3 Optical emission spectrum from the surface-wave plasma of a H_2 (90%) + CH_4 (5%) gas mixture at 20 Pa. The spectrum was taken within the CVD region (around 100 mm from the quartz window). Reprinted from Tsugawa et al.[7] Copyright (2011), with permission from Elsevier.

Figure 12.3 shows the optical emission spectrum (OES) from the CVD region of the surface-wave plasma.[7] In the spectrum the Balmer series of atomic hydrogen, including H_α (656 nm), H_β (486 nm) and H_γ (434 nm) lines, is clearly observed. The intensity of H_α is approximately six times larger than H_β. Signals from CH radicals corresponding to the transitions of $B^2\Sigma^- \to X^2\Pi$ at 389 nm and $A^2\Delta \to X^2\Pi$ at 431 nm also emerge in the spectrum.

The microwave plasma of a hydrogen-rich gas mixture exhibits a continuous band in the visible and near UV regions in its OES spectrum, from 340 to 520 nm in wavelength, generated by radiative relaxations from the higher energy, bound $a^3\Sigma_g^+ 2s\sigma$ state to the lower, repulsive $b^3\Sigma_u^+ 2p\sigma$ excited state in the dissociation processes of H_2 molecules by electron collisions. This continuum is observed for microwave plasmas of $T_e > 5$ eV, while it is not present for a thermal plasma of $T_e < 1$ eV due to lack of hot electrons with sufficient energy to invoke excitation to the upper levels associated with the radiative transitions corresponding to the continuum. It is also not observed in Figure 12.3, reflecting the low T_e of the surface-wave plasma.

Atomic hydrogen, other radicals, and ions in the surface-wave plasma are generated mainly in the plasma-generation region with high T_e near the window. Under a low pressure (~10 Pa) of the reaction gas mixture, they diffuse into the CVD region, where dissociation processes occur with statistically much lower rates than the plasma-generation region. Atomic hydrogen plays significant roles in chemical reactions, as well as in the formation of sp^3 bonds. Once it is provided, various chemical reactions, and activating and cycling hydrocarbon species, are set off in the gas phase to grow diamond. The spectrum in Figure 12.3 is of emission from the CVD region with nearly homogeneous low T_e, which indicates the presence of atomic hydrogen and CH radicals near the substrate. This may permit the growth of diamond with low activation energy (0.07 eV) as discussed in section 12.4

Intense emission peaks of the Swan system are observed in OES spectra from hydrogen-poor plasmas for ultrananocrystalline diamond (UNCD) growth,[10] where C_2 is considered to be an important growth species. In contrast, in the spectrum shown in Figure 12.3, weak bands at around 519 and 561 nm are present, which we assign to $\Delta v = 0$ and $\Delta v = -1$ sequences of the C_2 Swan system ($d^3\pi_g \rightarrow a^3\pi_u$),[11] respectively. The intensity of the Swan system is weaker than those of the CH radical, indicating that C_2 is present in relatively small amounts compared to CH in surface-wave plasmas of hydrogen-rich feed gas conditions. This indicates that C_2 is not a major growth species for the NCD growth in the present work, although we cannot discuss radicals from only emission spectra. Nevertheless, continuous re-nucleation of diamond occurs in the hydrogen-rich surface-wave plasma as discussed section 12.3 and 12.4, just like UNCD growth in a hydrogen-poor plasma. This is a totally different feature of the surface-wave plasma compared to others.

12.3 Nanocrystalline Diamond Grown in Surface-wave Plasmas

Figure 12.4 shows the Raman spectrum of a NCD film deposited on a Pyrex glass substrate excited by a UV light of 244 nm wavelength.[9] The UV Raman spectrum indicates the formation of a diamond film by the sharp peak at 1333 cm^{-1} in Figure 12.4. The broader band with lower intensity at around 1590 cm^{-1} represents sp^2-bonded carbon, which is often observed together with NCD. At the nanocrystalline scale, the sp^2-bonded carbon coexists in

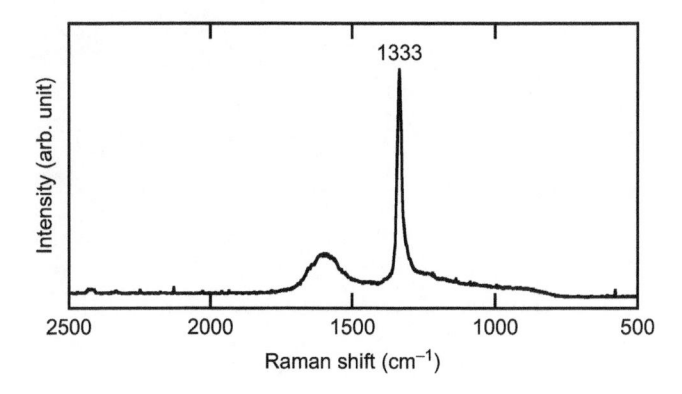

Figure 12.4 UV-excited Raman spectrum of a NCD film deposited on a Pyrex glass substrate (excitation wavelength = 244 nm).
Reprinted from Tsugawa et al.[9] Copyright (2006), with permission from MYU K.K.

the diamond film at the boundaries between the diamond grains. For this film, the relative intensity of the peak at 1333 cm^{-1} was much larger than that at 1590 cm^{-1} and the film was highly transparent because of the low sp^2 component.

Figure 12.5 exhibits a typical high-resolution cross-sectional transmission electron microscope (TEM) image of a NCD film on Pyrex glass.[9] The TEM observation reveals that the film consists of continuous NCD grains with sizes in the range from 5 to 20 nm. The diamond crystal size is almost constant over the cross section from the surface to the substrate. This indicates that continuous re-nucleation of diamond occurs during the NCD growth. The inset of Figure 12.5 shows the electron diffraction patterns of the diamond (111), (220) and (311) reflections. Their ring patterns also indicate the random orientation of the NCD grains.

The optical properties, *i.e.*, the transmittance for visible light, have been investigated for the NCD films on the Pyrex glass substrate.[9] The optical transmittance of a NCD film with a thickness of 500 nm is shown in Figure 12.6.[9] The average transmittance of the visible light (wavelength 400 to 800 nm) is 90%. This high transmittance in the visible-light region makes it possible to coat glass windows with NCD films for use in abrasive or other extreme environments.

Figure 12.7(a) shows a plastic substrate, polyphenylene sulfide (PPS), after the CVD treatment at a substrate temperature of around 80 °C[8] with a color pattern due to optical interference, indicating the formation of a transparent thin film. Figure 12.7(b) shows typical Raman-scattering spectra of films

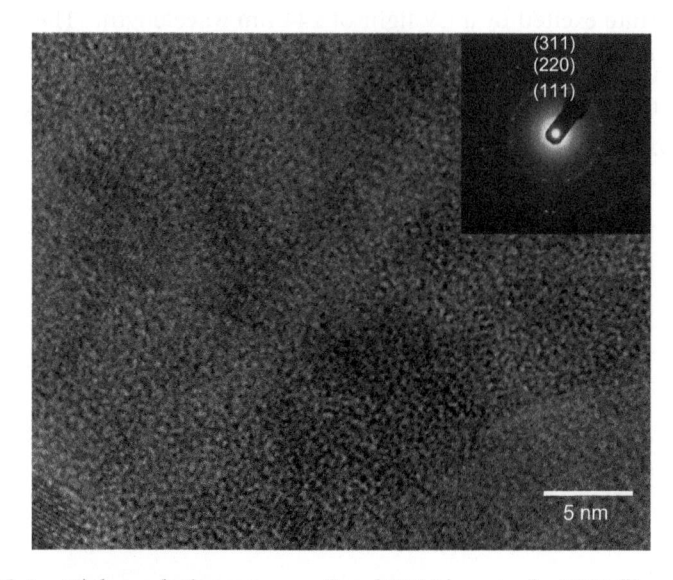

Figure 12.5 High-resolution cross-sectional TEM image of a NCD film on a Pyrex glass substrate.
Reprinted from Tsugawa *et al.*[9] Copyright (2006), with permission from MYU K.K.

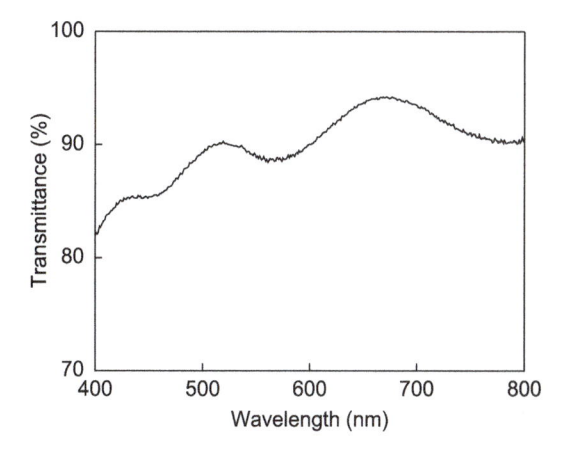

Figure 12.6 Wavelength dependence of the transmittance of a NCD film of 500 nm thickness deposited on a Pyrex glass substrate. The average transmittance for visible light ($\lambda = 400$–800 nm) is 90%.
Reprinted from Tsugawa et al.[9] Copyright (2006), with permission from MYU K.K.

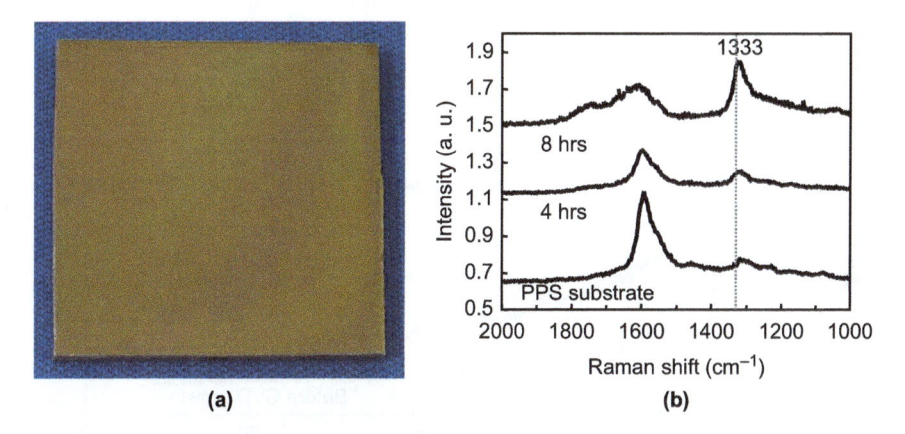

Figure 12.7 (a) PPS substrate after the CVD process at a substrate temperature around 80 °C and (b) Raman spectra of CVD-treated PPS substrates using an excitation wavelength of 244 nm. In (a), a color pattern is observed on the surface, indicating the formation of a transparent film. In (b), spectra from two different durations of the CVD growth are displayed, i.e., 4 and 8 h. A spectrum of a bare PPS substrate before the CVD treatment is also shown. The peak around 1333 cm^{-1} in Raman shift indicates the diamond signature, the intensity of which increases with increasing duration of the CVD treatment.
Reprinted from Tsugawa et al.[8] Copyright (2010), with permission from the American Physical Society.

deposited on the PPS substrates using an excitation wavelength of 244 nm.[8] Spectra from two films deposited for different durations of CVD treatment are demonstrated, i.e., 4 and 8 h. For comparison, a Raman spectrum of a

PPS substrate without the CVD treatment is also shown. The signature of diamond at 1333 cm^{-1} in Raman shift is clearly observed. The intensity of the diamond peak increases with deposition time, indicating the growth of the diamond film in the CVD process.

Figure 12.8 demonstrates a typical X-ray diffraction pattern using Cu Kα radiation of the diamond film on PPS.[8] In the pattern the peak of reflection from diamond (111) planes is observed at $2\theta = 43.9°$, while an untreated PPS substrate exhibits no diffraction peak near that position. The average crystallite size of the film is estimated from the broadening of the diffraction peak of (111) using the Scherrer equation.[12] The inset of Figure 12.8 depicts the (111) diffraction peak fitted by the Pearson VII function. We have used the full-width at half-maximum of the fitted Pearson VII function, and consequently, the average crystallite size is approximately 5.5 nm. This indicates that the film on the PPS substrate consists of NCDs.

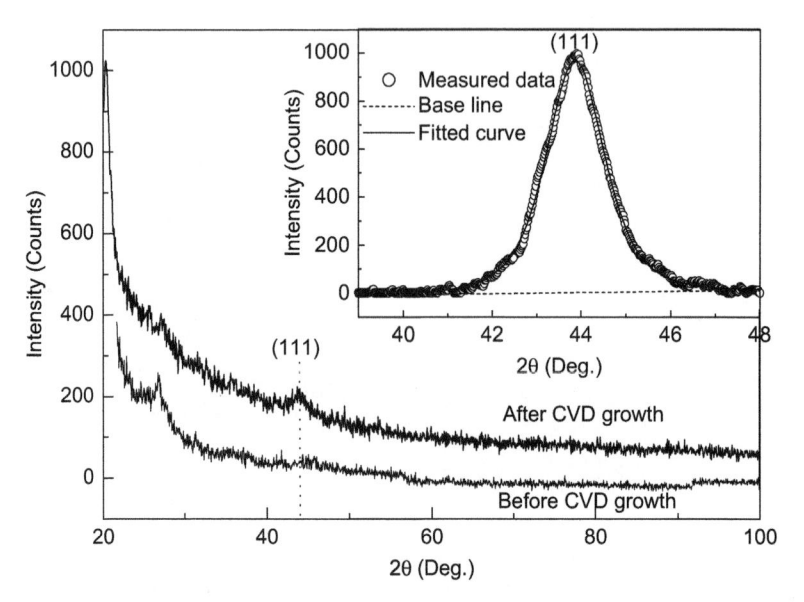

Figure 12.8 X-Ray diffraction pattern of an NCD film deposited on a PPS substrate using Cu Kα radiation. The duration time of the CVD growth is 8 h and the thickness of the NCD film is around 500 nm. A diffraction pattern from a PPS substrate without the CVD treatment is also shown for comparison. A diffraction peak at $2\theta = 43.9°$ from the reflections of diamond (111) is observed in the pattern from the CVD-treated PPS substrate (indicated by "after CVD growth"), while no peak is observed around that position in the pattern from the PPS substrate without the CVD treatment (indicated by "before CVD growth"). The inset shows the magnified (111) peak with a fitted curve with background elimination. The dashed line in the inset denotes a base line. Reprinted from Tsugawa et al.[8] Copyright (2010), with permission from the American Physical Society.

12.4 Growth Mechanism

The growth mechanism of NCD films grown in the surface-wave plasma is discussed in this section. We focus on the growth rate and the nucleation rate, as well as the crystallite size distribution in the NCD film that is the fingerprint of the growth mechanism. For example, rare nucleation leads to large columnar grains and frequent, continuous nucleation results in films of uniform small grains.

In conventional diamond CVD a hydrogen-rich plasma suppresses diamond nucleation. In this case a polycrystalline diamond film with micrometer-sized columnar grains is grown (microcrystalline diamond: MCD). Furthermore, stopping deposition by a thickness of thinner than ~100 nm gives nanometer-sized grains. This type of diamond film grown in hydrogen-rich plasma is generally referred to as nanocrystalline diamond (NCD). On the other hand, UNCD indicates a nanocrystalline diamond film of less than 10 nm in crystallite size caused by continuous nucleation in hydrogen-poor plasma.[13] In contrast to this, hydrogen-rich surface-wave plasma allows frequent, continuous nucleation of diamond, leading to the deposition of an NCD film, which is similar to a UNCD film. In fact, in this work, the average crystallite size obtained from X-ray diffraction is independent of the film thickness, *i.e.*, the average crystallite size is almost constant regardless of the film thickness. This indicates that the size distribution of diamond crystals does not change considerably from near the substrate to the surface of the film like UNCD films. This fact suggests the constant occurrence of diamond nucleation on the surface of the growing NCD film during the CVD process, even in hydrogen-rich plasma.

In order to analyze the mechanism of the NCD deposition, first, growth rates are estimated.[8] Here, we define growth rate r as film thickness grown per unit time. When a film of thickness L is deposited for a process time t, its growth rate $r = L/t$. Figure 12.9 shows the growth rates of NCD films on PPS, stainless steel, silicon (Si) and sintered tungsten carbide (WC) substrates at various substrate temperatures, deposited by the microwave surface-wave plasma CVD system using the same gas mixture and pressure (20 Pa).[8] The substrate temperatures range from below 100 °C for PPS, around 470 °C for stainless steel, 500–600 °C for the Si, to around 700 °C for WC. Typical growth rates of MCD and UNCD are also indicated in the figure.

We assume that the growth rate does not depend on substrate material so that the activation energy of the growth rate can be derived from the temperature dependence of the growth rates (Arrhenius plots). Suppose the growth rate r is related to the substrate temperature T by an Arrhenius-type form $r = \exp(-E_a/k_B T)$, where E_a is derived from the gradient of a least-square fit among the growth rates. Here k_B denotes the Boltzmann constant. In the case of Figure 12.9 the derived E_a is 0.069 ± 0.012 eV. This value is much lower than typical activation energies of UNCD (0.254 ± 0.019 eV [ref. 14]), conventional MCD (1.00 ± 0.05 eV [ref. 15]) and homoepitaxial single-crystalline diamond (0.3 ± 0.1, 0.78 ± 0.09, and 0.52 ± 0.17 eV for [100], [110],

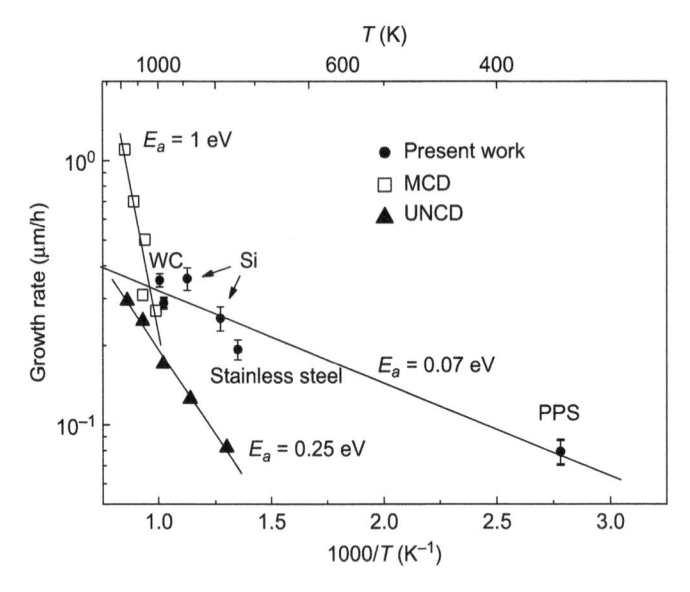

Figure 12.9 Temperature dependence of the growth rates of three types of diamond. In addition to the growth rate of the NCD in this work, typical growth rates of MCD (ref. 15) and UNCD (ref. 14) are also shown for comparison.
Reprinted from Tsugawa et al.[8] Copyright (2010), with permission from the American Physical Society.

and [111] growth, respectively [ref. 16]). In a similar style to Robertson,[17] Figure 12.9 shows a comparison of the temperature dependence of growth rates among MCD, UNCD and the NCD in the present work. The differences of the activation energies are clearly observed in the figure. This result implies that the growth mechanism of the NCD in the present work is different from that of MCD or UNCD growth.

For further analysis of the growth mechanism, the nucleation rate is estimated by using a simple growth model.[8] To simplify the estimation, we assumed a constant nucleation rate per unit area per unit time, ν, for a given substrate temperature. Furthermore, each nucleated diamond crystal on the growing diamond film surface is assumed to grow up to the mean crystal diameter $\langle d \rangle$, i.e., the diamond film is constructed from diamond crystals with a uniform size of $\langle d \rangle$. In addition, nucleation is assumed to occur randomly on the growing surface and the diamond film grows layer by layer on average.

Using this simple model, the nucleation density per unit area n is described as $n = 1/\langle d \rangle^2 = \nu t_0$. Here t_0 is the time for which a layer with a thickness of $\langle d \rangle$ grows. In addition, the growth rate of a film r can be written as $r = \langle d \rangle/t_0 \, (= L/t)$. Hence the nucleation rate ν is given by:

$$\nu = \frac{r}{\langle d \rangle^3}. \tag{12.1}$$

Here, L, t and $\langle d \rangle$ are obtained experimentally. Thus, the growth rates r and the nucleation rates v are evaluated using these experimental values. Estimated nucleation rates v versus reciprocal temperature are shown in Figure 12.10(a), along with the growth rates, which were obtained using average crystallite sizes obtained from X-ray diffraction measurements as $\langle d \rangle$.[8] Generally speaking, the mechanism of crystal growth can be divided sequentially into two stages, namely, nucleation and growth. In the present work and from Figure 12.10(a), the nucleation rate decreases with increasing substrate temperature T. This tendency is in contrast with conventional nucleation, where the nucleation rate increases along with temperature. In fact, in UNCD growth for example, the nucleation rate increases with increasing substrate temperature, as demonstrated in Figure 12.10(b).[8]

Figure 12.10(b) shows Arrhenius plots of growth rates and nucleation rates of UNCD film growth in HFCVD.[18] The plotted data were calculated from experimental data in the same manner as the present work. The nucleation rate of UNCD growth in Figure 12.10(b) increases with increasing substrate temperature with an activation energy of 0.64 eV, exhibiting a similar trend to the growth rate. This type of temperature dependence of the nucleation rate coincides with conventional nucleation. On the other hand, in the NCD growth in the present work shown in Figure 12.10(a), the nucleation rate seems not to be thermally activated. It decreases with increasing substrate temperature. In other words, NCD growth is hardly explained by a model based on conventional nucleation at the substrate surface. This suggests that nucleation occurs in the surface-wave plasma *via* a different mechanism from that of conventional nucleation.

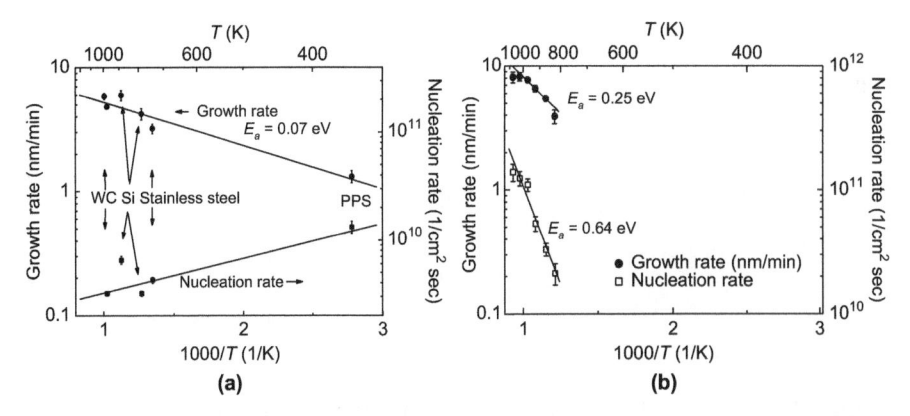

Figure 12.10 Temperature dependence of growth rates and nucleation rates of (a) NCD films in this work and of (b) an UNCD film in ref. 18, plotted on the same scales for comparison. In (a) three kinds of substrates were used for specific substrate temperature ranges, namely, PPS for below 100 °C, stainless steel for around 470 °C and silicon for around 700 °C. In (b) the plotted data are calculated using experimental data taken from ref. 18.
Reprinted from Tsugawa et al.[8] Copyright (2010), with permission from the American Physical Society.

We have investigated crystal size distribution of a NCD film synthesized on a PPS substrate at a substrate temperature of ∼86 °C in order to analyze the nucleation and growth mechanism further.[8] The size distribution has been obtained from the high-resolution TEM image shown in Figure 12.11(a).[8] The obtained size distribution is presented in Figure 12.11(b).[8] From Figure 12.11, the NCD film is found to consist of spherical diamond crystals of diameters ranging from 3 to 10 nm.

In the current investigation it has been determined experimentally that the surface roughness does not depend on the film thickness, which indicates that there is no correlation between successive nucleation events. Therefore, a Poisson process, with nucleation at a constant rate ν, can be utilized. In this case the distribution of a time interval τ between two nucleation events follows the exponential distribution:

$$f(\tau) = \nu \langle d \rangle^2 e^{-\nu \langle d \rangle^2 \tau}. \tag{12.2}$$

Here, a diamond crystal grows at a growth rate r within a time interval τ. We therefore obtain $d = r\tau$. Since r is a constant and $d\tau = dd/r$, using eqn (12.1) and (12.2), the distribution function of the crystal size $F(d)$ is given as:

$$F(d) = \frac{\nu \langle d \rangle^2}{r} e^{-\nu \langle d \rangle^2 / rd} = \frac{1}{d} e^{-d/\langle d \rangle}. \tag{12.3}$$

The distribution function $F(d)$ is normalized since $f(\tau)$ is normalized.

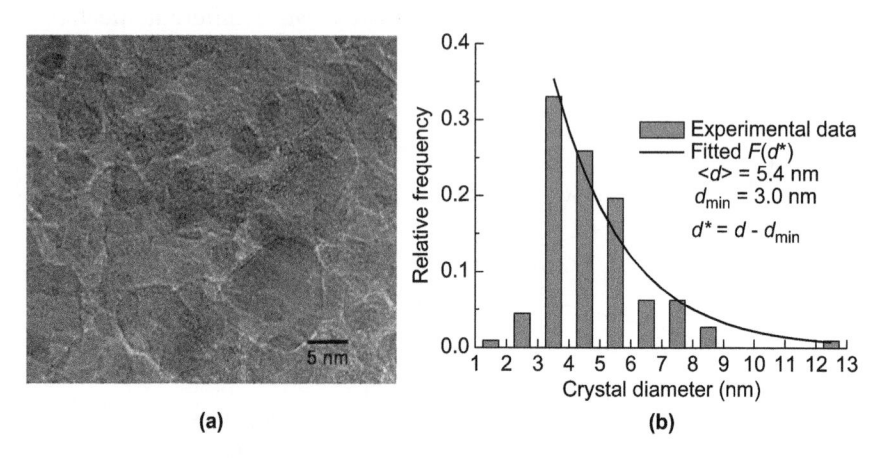

Figure 12.11 TEM characterization to obtain the size distribution of diamond crystals. (a) A high-resolution TEM image of a NCD film on PPS. Diamond (111) lattice fringes are observed in some particles. (b) The size distribution of diamond crystals in the NCD film, the data of which were taken from the TEM image (a). Each frequency for the respective crystal diameter is normalized by the total sample number of 112. The solid line indicates a theoretically derived distribution function of crystal size d, which is fitted to the experimental data. Reprinted from Tsugawa et al.[8] Copyright (2010), with permission from the American Physical Society.

As shown in Figure 12.11(b), few diamond crystals smaller than 3 nm in diameter are present. This fact suggests the existence of a minimum size of diamond crystals. Here, the minimum diameter is defined as d_{min}. Substitution of $d^* = d - d_{min}$ for d yields a distribution function $F(d^*)$ as:

$$F(d*) = F(d - d_{min}). \tag{12.4}$$

Fitting $F(d^*)$ to the experimental size distribution in Figure 12.11(b) using $\langle d \rangle$ and d_{min} as fitting parameters, $\langle d \rangle = 5.4$ nm and $d_{min} = 3.0$ nm, have resulted in the best fit. This value of $\langle d \rangle$ is quite close to the average crystallite size obtained from the X-ray diffraction (5.5 nm).

Diamond is a stable phase of carbon (more stable than graphite) at a pressure of over 2 GPa at room temperature. The pressure at which diamond is stable increases with increasing temperature. On the other hand, responding to a demonstration of the presence of small diamonds (~5 nm) in meteorites, Nuth[19] has pointed out the possibility that nanometer-sized diamond is stable in that particular size regime even at low pressures, where bulk diamond is metastable. In this size regime, thermodynamic predictions based on the bulk properties are not necessarily applicable to nanometer-sized particles. He attributed the reversal in the relative stability between diamond and graphite to the increasing effects of surface free energy on the stability with decreasing dimensions of carbon particles.

After Nuth's indication, more elaborated investigations supporting his discussion have been made to date. For example, Raty and Galli,[20] in particular, have applied the discussion to diamond CVD growth that requires the presence of hydrogen using first principles calculations. They found that the size distribution should exhibit a sharp peak at around 3 nm in the presence of hydrogen on the surfaces of diamond nanoparticles for a broad range of pressures and temperatures. They also suggested that diamonds start to grow from particles of this size to a film. In the present work few crystals less than 3 nm are found in the crystal size distribution shown in Figure 12.11(b), which indicates that the minimum size is 3 nm for our CVD process. In other words, the diamond crystals start to grow from 3 nm. This result supports these theoretical discussions.

In several experiments, nucleation of diamond in the gas phase at atmospheric or sub-atmospheric pressures has been observed. Frenklach et al.[21] have extracted diamond aggregates from deposits collected from microwave plasma without using any substrate. The aggregates were typically 50 nm in diameter. They revealed that each aggregate consisted of many smaller diamond crystallites using selected-area electron diffraction.

The above results and discussion indicate the possibility of diamond nucleation in the gas phase under certain conditions. Once the gas-phase nucleation is assumed, the extremely low-temperature growth of diamond and the temperature dependences of the growth rate and the nucleation rate in the present work become understandable. That is to say, the substrate temperature is not responsible for the diamond nucleation in the gas phase. A nucleated diamond cluster in plasma precipitates onto the substrate and grows there until another diamond cluster falls on it. The substrate temperature affects only the growth rate of the precipitated diamonds on the substrate. This assumption naturally leads to NCD film growth at substrate temperatures below 100 °C and the weak temperature dependence of the growth rate. Moreover, under gas-phase nucleation, the substrate temperature is also considered not to be responsible for the nucleation rate. However, the nucleated diamonds are etched by atomic hydrogen in plasma, the rate of which increases with increasing substrate temperature. This explains the reciprocal temperature dependence of the nucleation rate. If this is true, then nucleation pretreatments of substrates, such as mechanical abrasion, ultrasonic particle treatment, seeding with diamond nanoparticles, and so on,[22] are not necessary for diamond deposition. In fact, this has been confirmed in surface-wave plasma CVD of NCD on stainless steel without any nucleation pretreatments.[23]

In summarizing the foregoing discussion we have reached the following model for NCD growth in surface-wave plasmas. (1) Diamond crystals nucleate and grow into nuclei clusters of ~3 nm in size in the plasma. The diamond clusters precipitate on the substrate. (2) The precipitated clusters grow, or are removed by hydrogen plasma etching, which depends on the substrate temperature. The decreasing nucleation rate with increasing temperature, which is shown in Figure 12.10(a), is attributed to the increasing etching effect of the hydrogen plasma on the clusters, which increases with temperature. The nucleation in the plasma enables low-temperature deposition of NCD films on plastic substrates at below 100 °C.

Dielectric particles in plasma, such as the diamond clusters nucleated in the plasma, are bombarded by ions. The energy of the ions is proportional to the electron temperature of the plasma. The high-energy ion bombardment, caused by a high electron temperature, blasts the nucleated diamonds in the plasma, viz., diamonds hardly nucleate in a high-electron-temperature plasma, including conventional microwave plasmas for diamond growth (over 5 eV). The low-electron temperature of the surface-wave plasma enables the gas-phase nucleation with a nucleation rate around 10^{10} cm^{-2} s^{-1} [Figure 12.10(a)] even in a hydrogen-rich plasma, whereas a conventional hydrogen-rich plasma reduces diamond re-nucleation so that MCD films are grown. Besides, the low-electron temperature suppresses excessive decomposition of carbon precursors, which causes nondiamond components in the deposited film. In brief, the low-electron temperature of the plasma is essential for gas-phase nucleation and low-temperature growth of diamond, as well as for the reduction of the nondiamond-component formation. The gas-phase nucleation is indispensable for diamond deposition at such low temperatures.

12.5 Tribological Properties

We have performed friction coefficient measurements of these smooth NCD films on AISI 440C stainless steel by a reciprocating sliding test with a pin-on-plate configuration under a normal load of 1 N in air at room temperature without any lubrication. We selected an AISI 440C stainless steel ball of $\frac{3}{16}$ inches diameter as the counterpart of sliding. A relatively low sliding speed (10 mm s^{-1}, sliding amplitude: 10 mm, frequency: 0.5 Hz) was chosen in order to reflect the intrinsic friction property by avoiding external effects, such as lifting with speed.[23]

Figure 12.12 shows the dynamic friction behavior of a film with a thickness of 1.7 µm up to 3600 cycles.[23] Two results from the different parts of the film are indicated in the figure. The friction coefficients for both the measurements tend to converge at values smaller than 0.1. The average friction coefficient from further measurements is around 0.06, which is lower than that of generic diamond-like carbon (DLC).

Figure 12.13 shows a summary of the results of the friction tests of a pin-on-plate configuration using AISI 440C pin counterparts with a spherical sliding contact end of 30 mm radius with a normal load of 1 N.[23] We have also tested pristine stainless steel substrates without any coating and commercially available DLC films deposited on the same substrates for comparison. We selected two types of DLC films for the friction tests, i.e., hydrogen-free tetrahedral amorphous carbon (ta-C) and DLC with hydrogen inclusion (a-C:H). Figure 12.13(a) indicates the results of the friction coefficients of the films. The friction coefficient of the NCD film is less than 0.1 and is nearly equal to that of the ta-C film. In the friction tests the specific wear rates of the samples, excluding AISI 440C, are immeasurably small compared to their counterparts since the NCD, ta-C and a-C:H films are much harder than the AISI 440C stainless steel. Figure 12.13(b) shows the specific wear rates of the counterparts, which indicate the aggression strength of the films. In the figure, a marked feature of the NCD film is its low

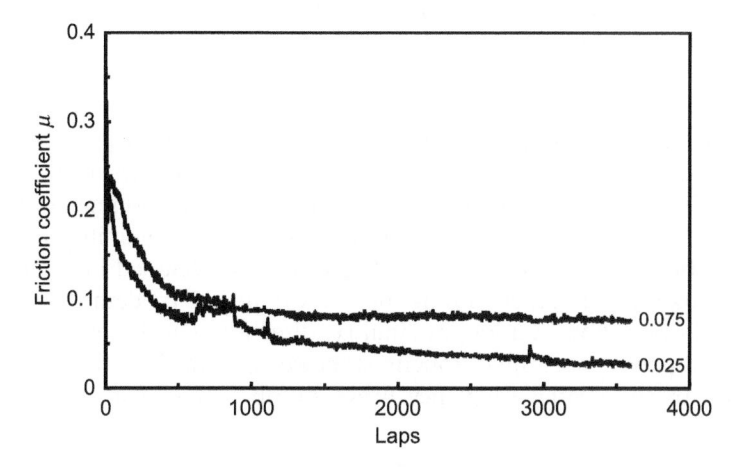

Figure 12.12 Dynamic friction behavior of an NCD film on stainless steel in a reciprocating sliding test with a pin-on-plate configuration under a normal load of 1 N in air at room temperature (vs AISI 440C stainless steel ball of $\frac{3}{16}$ inches diameter, 26–27 °C, relative humidity: 50–60%). Reprinted from Tsugawa et al.[23] Copyright (2012), with permission from the Japan Society of Applied Physics.

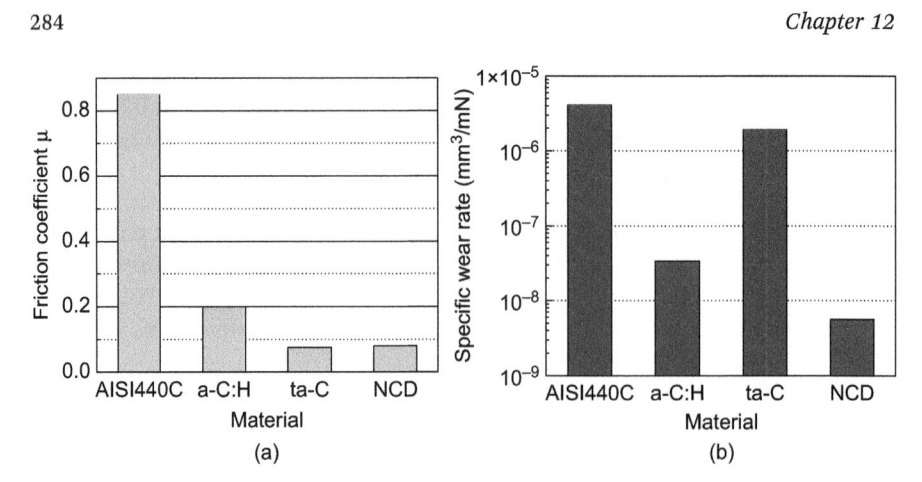

Figure 12.13 (a) Friction coefficients and (b) specific wear rates of AISI 440C stainless steel counterparts on tested materials. The measurements were performed in air at room temperature at 30% relative humidity by a pin-on-plate reciprocating sliding test.
Reprinted from Tsugawa *et al.*[23] Copyright (2012), with permission from the Japan Society of Applied Physics.

aggression strength against the stainless counterpart. The specific wear rate of the counterpart for the NCD film is one or three orders of magnitude less than those for the a-C:H and ta-C films. These tribological properties in air without lubrication of the low-temperature-grown NCD films on steel substrates, such as the low friction coefficient and the low aggression strength against counterparts of sliding, are advantageous for mechanical applications.

Surface roughness involves a variety of phenomena, including friction and wear. The most important property of rough surfaces is the surface roughness power spectrum.[24] We have found that the low aggression property of the NCD film is ascribed to the power spectrum of its surface roughness. Roughness parameters, the arithmetic average of absolute values R_a, the root mean square R_q and the maximum height R_t, are $R_a = 8.9$ nm, $R_q = 12.4$ nm and $R_t = 142$ nm, where the roughness of the substrate before the CVD growth is 3–4 nm in R_a. This small roughness of the NCD film originates from the nanosized grains on the surface, indicating a spatially-uniform constant re-nucleation of diamond during CVD growth, as discussed in section 12.4. These roughness parameters only include the height (z) data of the profile with no information in a longitudinal (x) direction. On the other hand, the roughness power spectrum or roughness power spectral density (PSD) $C(q)$ is a function of the spatial frequency (wave number) q of surface roughness, defined by:

$$C(q) = \frac{1}{2\pi} \int \langle h(x)h(0) \rangle e^{-iqx} dx. \tag{12.5}$$

Here, $z = h(x)$ is the surface profile function (the height from the average level, defined so that $\langle h \rangle = 0$) and $\langle \ldots \rangle$ stand for ensemble averaging.

We assume that the surfaces of the samples are isotropic, such that the one-dimensional analysis [using $h(x)$ instead of $h(x, y)$] is adequate. Equation (12.5) is also expressed as:

$$C(q) = \frac{2\pi}{L} \langle |h_L(q)|^2 \rangle, \quad (12.6)$$

where L is the length under study. $h_L(q)$ is the Fourier transform of $h(x)$, i.e.:

$$h_L(q) = \frac{1}{2\pi} \int_L h(x) e^{-iqx} \mathrm{d}x. \quad (12.7)$$

We used the fast Fourier transform (FFT) method to obtain $h_L(q)$. Using the discrete Fourier transform H_m by FFT eqn (12.7) can be written as:

$$h_L \approx \frac{a}{2\pi} H_m, \quad (12.8)$$

where a is the lattice constant of discretization. Using eqn (12.6)–(12.8), the discretized version of $C(q)$ is obtained.

Figure 12.14 shows the obtained power spectra from the roughness profiles of the samples, i.e., the AISI 440C substrate without coatings and the

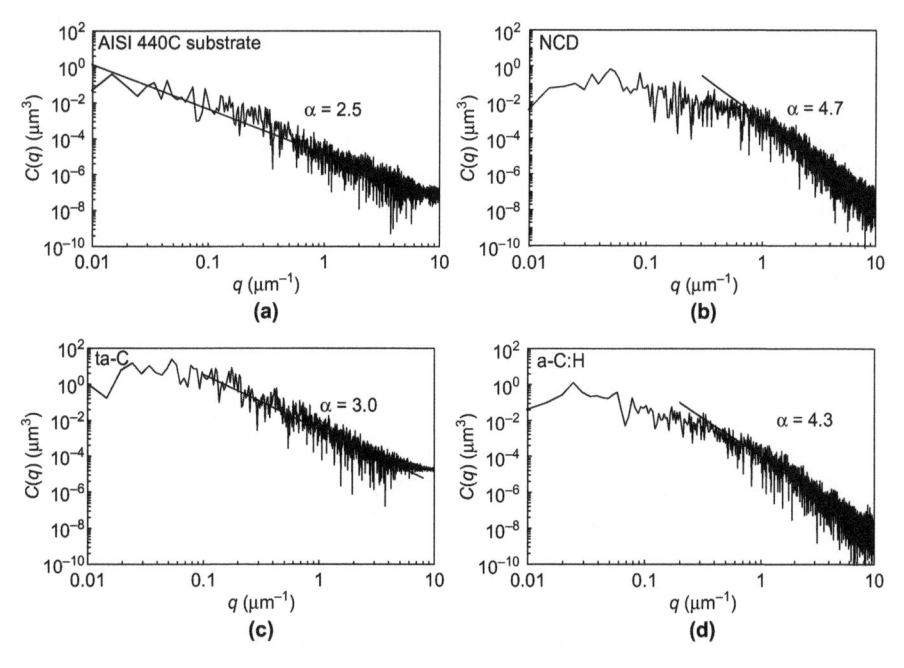

Figure 12.14 Power spectra $C(q)$ of surface roughness of samples for the tribological tests: (a) the AISI 440C substrate without coatings and (b) NCD, (c) ta-C, and (d) a-C:H films on the AISI 440C substrates. They obey the power law in a specific region of wave number q, indicated by solid lines. α denotes the power law exponent.
Reprinted from Tsugawa et al.[23] Copyright (2012), with permission from the Japan Society of Applied Physics.

NCD, ta-C and a-C:H films on the substrates.[23] The power spectrum obeys the power law in a specific region of wave number q for each sample, i.e., $C(q) = q^{-\alpha}$, as indicated by the solid lines. For the as-received AISI 440C stainless steel substrate with $R_a = 3-4$ nm in Figure 12.14 (a), $C(q)$ obeys the power law in almost the entire region under study ($0.01 < q < 10$ µm^{-1}). In this case, the least-squares fit in the entire region indicated by the solid line gives a power law exponent α of 2.5. Likewise, the power law exponents of 4.7, 3.0 and 4.3 are obtained for NCD, ta-C and a-C:H-coated surfaces, respectively. In these cases, however, the regions governed by the power law are narrower than the substrate surfaces. The power spectra $C(q)$ exhibit a shape typical to a self-affine fractal surface with long distance roll-off wave number q_0 and short distance cut-off wave number q_1. In Figure 12.14(b) and (d), q_0 is around 1 µm^{-1} and q_1 is 1 µm^{-1} or larger. In Figure 12.14(c), q_0 and q_1 are respectively 0.1–0.2 and around 3 µm^{-1}. The power law exponents have been obtained in the region of $q_0 < q < q_1$.

For a self-affine fractal surface, the roughness power law exponent α is translated into the Hurst exponent (roughness exponent) H as $\alpha = 2(H+1)$, which is related to the fractal dimension D_f as $H = 2 - D_f$. The roughness exponent H varies within the range of $0 < H < 1$ for a self-affine fractal surface. In the cases shown in Figure 12.7, the values of H are 0.25, 1.35, 0.50 and 1.15 for the substrates and the NCD, ta-C and a-C:H films, respectively. For the NCD and a-C:H films, H is larger than unity, indicating that the film surfaces cannot be described as fractal at any scale.

Figure 12.15 shows the correlation between the specific wear rate w of pins (Figure 12.13) and the power law exponent α (Figure 12.14).[23] The plots in the figure suggest a linear correlation between log w and α, which implies

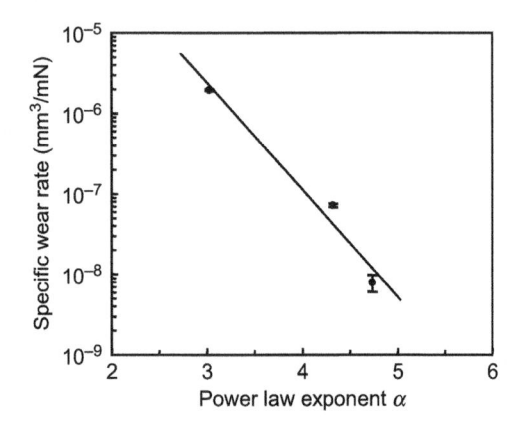

Figure 12.15 Log–linear plots of the specific wear rates of the opposing pin counterparts in the tribology tests, plotted against the power law exponents of carbon films.
Reprinted from Tsugawa et al.[23] Copyright (2012), with permission from the Japan Society of Applied Physics.

that $w = p^{-a\alpha}$, where p and α are constants. This comes from the power law in the surface roughness profile of the carbon films on the stainless steel substrates. The low specific wear rate of the sliding counterpart of the NCD film on stainless steel originates from the large power law exponent of its surface roughness profile. In other words, the NCD film exhibits a lower aggression strength against the frictional counterpart than DLC films, owing to its large power law exponent.

Figure 12.15 shows that the specific wear rates of the stainless steel pins sliding on the carbon films under study are determined by the surface roughness profiles of the harder carbon films. This is a feature of the abrasive wear between a soft surface and a hard surface. In this case the soft surface is unilaterally abraded away by the rough hard surface, where the wear rate depends on the roughness profile of the hard surface. A surface with power-law roughness implies that the amplitude of roughness protrusion decreases with increasing wave number q. The rapidity of the amplitude reduction is determined by the power law exponent α. That is, the roughness amplitude decreases with increasing α more rapidly on a smaller length scale. Figure 12.15 shows that the rapidity of the roughness reduction with length scale defines the wear rate. A low amplitude of protrusion of a wave number indicates a more obtuse apex of protrusion. This is an intuitive explanation for the relationship between the specific wear rates and the power-law exponents in Figure 12.15. However, this experimental result does not agree with the assumption that the surface of an amorphous material is smoother than that of a crystalline material. Although the reason for this is not clear, it may be due to the nanometer-sized crystals and the characteristic crystal size distribution of the NCD film. A feature of this distribution is the exponentially decreasing number with increasing size, which is related to the growth mechanism of the NCD film in surface-wave plasmas.

12.6 Summary

In the CVD growth of diamond films, microwave-excited plasmas are widely used. The properties of the deposited diamond films are essentially affected by the characteristics of the plasma used, as well as other growth parameters, including substrate temperature and reaction gas mixture. The revealed plasma chemistry and physics are connected to the mechanism of the diamond CVD growth. In conventional diamond CVD processes substrate temperatures are typically 700–1000 °C. These high substrate temperatures limit substrate selection to materials such as silicon and limit the application fields of diamond.

We have developed MWPCVD techniques using surface-wave plasmas for NCD film deposition and it has realized low-temperature growth of diamond. In the surface-wave plasma, NCD films are grown at substrate temperatures less than 500 °C and they still grow below 100 °C. This technology has enabled a further wide-ranging selection of substrate materials, including silicon, steel, glass and even plastics.

Surface-wave plasmas provoke a low electron temperature of 2–3 eV in the diamond CVD region, while it exceeds 5 eV in a conventional diamond MWPCVD process using the ball-shaped plasma. Diamond films deposited in the surface-wave plasma are NCD films consisting of nanometer-sized (<100 nm) diamond crystals, which are formed by continuous re-nucleation of diamond.

We have investigated the temperature dependence of the growth rate and the nucleation rate of NCD films in the surface-wave plasma. They exhibit different trends from other diamond CVD growth processes, *i.e.*, the temperature-insensitive growth rate and the nominally negative activation energy of the nucleation rate. This fact, and the analysis of the size distribution of diamond crystals in the NCD film, suggests diamond nucleation in a stable phase in the plasma, which is invoked by the low-electron temperature of the surface-wave plasma. The gas-phase nucleation enables an explanation of the mechanism of the low-temperature growth of NCD films.

The tribological properties of NCD films directly deposited on stainless steel substrates using low-temperature growth in surface-wave plasma have been investigated. The nanodiamond films on stainless steel exhibit good tribological properties in air at room temperature without any lubrication, particularly a low friction coefficient less than 0.1 and a low aggression strength against the counterpart. Surface roughness analysis of the NCD film reveals that the low aggression strength originates from the large power-law exponent of the power spectral density of the surface-roughness profile.

References

1. B. V. Spitsyn, L. L. Bouilov and B. V. Derjaguin, *J. Cryst. Growth*, 1981, **52**, 219.
2. S. Matsumoto, Y. Sato, M. Kamo and N. Setaka, *Jpn. J. Appl. Phys.*, 1982, **21**, L183.
3. M. Kamo, Y. Sato, S. Matsumoto and N. Setaka, *J. Cryst. Growth*, 1983, **62**, 642.
4. J. Kim, K. Tsugawa, M. Ishihara, Y. Koga and M. Hasegawa, *Plasma Sources Sci. Technol.*, 2010, **19**, 015003.
5. J. E. Butler, Y. A. Mankelevich, A. Cheesman, Jie Ma and M. N. R. Ashfold, *J. Phys.: Condens. Mater.*, 2009, **21**, 364201.
6. H. Sugai, I. Ghanashev and K. Mizuno, *Appl. Phys. Lett.*, 2000, **77**, 3523.
7. K. Tsugawa, S. Kawaki, M. Ishihara, J. Kim, Y. Koga, H. Sakakita, H. Koguchi and M. Hasegawa, *Diamond Relat. Mater.*, 2011, **20**, 833.
8. K. Tsugawa, M. Ishihara, J. Kim, Y. Koga and M. Hasegawa, *Phys. Rev. B*, 2010, **82**, 125460.
9. K. Tsugawa, M. Ishihara, J. Kim, M. Hasegawa and Y. Koga, *New Diamond Frontier Carbon Technol.*, 2006, **16**, 337.
10. D. Zhou, D. M. Gruen, L. C. Qin, T. G. McCauley and A. R. Krauss, *J. Appl. Phys.*, 1998, **84**, 1981.

11. A. Tanabashi, T. Hirao, T. Amano and P. F. Bernath, *Astrophys. J. Suppl. Ser.*, 2007, **169**, 472.
12. H. P. Krug and A. L. Elbert, *X-ray Diffraction Procedures for Polycrystalline and Amorphous Materials*, 2nd edn., Wiley, New York, 1974, Chap. 9.
13. O. A. Williams, M. Nesladek, M. Daenen, S. Michaelson, A. Hoffman, E. Osawa, K. Haenen and R. B. Jackman, *Diamond Relat. Mater.*, 2008, **17**, 1080.
14. T. G. McCauley, D. M. Gruen and A. R. Krauss, *Appl. Phys. Lett.*, 1998, **73**, 1646.
15. E. Kondoh, T. Ohta, T. Mitomo and K. Ohtsuka, *Appl. Phys. Lett.*, 1991, **59**, 488; *J. Appl. Phys.*, 1993, **73**, 3041.
16. C. J. Chu, R. H. Hauge, J. L. Margrave and M. P. D'Evelyn, *Appl. Phys. Lett.*, 1992, **61**, 1393.
17. J. Robertson, *Phys. Status Solidi A*, 2008, **205**, 2233.
18. D. C. Barbosa, F. A. Almeida, R. F. Silva, N. G. Ferreira, V. J. Trava-Airoldi and E. J. Corat, *Diamond Relat. Mater.*, 2009, **18**, 1283.
19. J. A. Nuth III, *Nature*, 1987, **329**, 589.
20. J. Y. Raty and G. Galli, *Nature Mater.*, 2003, **2**, 792.
21. M. Frenklach, R. Kematick, D. Huang, W. Howard, K. E. Spear, A. W. Phelps and R. Koba, *J. Appl. Phys.*, 1989, **66**, 395.
22. O. A. Williams, *Diamond Relat. Mater.*, 2011, **20**, 621.
23. K. Tsugawa, S. Kawaki, M. Ishihara and M. Hasegawa, *Jpn. J. Appl. Phys.*, 2012, **51**, 090122.
24. B. N. J. Persson, O. Albohr, U. Tartaglino, A. I. Volokitin and E. Tosatti, *J. Phys.: Condens. Matter*, 2005, **17**, R1.

CHAPTER 13

Low Temperature Diamond Growth

TIBOR IZAK,* OLEG BABCHENKO, STEPAN POTOCKY, ZDENEK REMES, HALYNA KOZAK, ELISSEOS VERVENIOTIS, BOHUSLAV REZEK AND ALEXANDER KROMKA

Institute of Physics, Academy of Sciences of the Czech Republic, v.v.i., Cukrovarnická 10, Praha 6, 162 00, Czech Republic
*Email: izak@fzu.cz

13.1 Introduction

Diamond used to be a material used merely in polishing pastes and in cutting tools. Due to tremendous progress in the science and technology of diamond, diamond thin films represent today a promising material for many advanced applications in electronics, biology and medicine. Monocrystalline diamond is employed more often than nanocrystalline diamond (NCD) in fundamental studies because monocrystalline diamond exhibits higher thermal conductivity in comparison to nanocrystalline or polycrystalline diamond. On the other hand, NCD films are more likely to be widely applicable as they are inexpensive in comparison to monocrystalline diamond and they can be easily fabricated from methane using microwave plasma on arbitrary substrates (silicon, glass, metals, plastic)[1] and on large areas.[2,3] However, from the electronic point of view, NCD is a complicated system due to the presence of sp^2 phases and grain boundaries, and the role of these features still needs to be fully elucidated.[4,5]

The first application of diamond in the electronic industry was motivated by its outstanding thermal properties.[6] Diamond has become a suitable material for thermal management applications in high power and high frequency devices due to its high thermal conductivity compared to other applicable materials (two to four fold greater than SiC or Cu and ten fold greater than Si or thick AlN). Devices producing high heat flux are, for example, lasers, high brightness LEDs, RF power transistors and power switching devices. It has been shown that the use of even a few micrometers of a diamond film enhances the sustainable power density of electronic devices by silicon on diamond (SOD) technology.[7] The combination of diamond with GaN leads to even greater possible power density increases. All-diamond power devices are being intensively researched, in particular in Japan, and will likely soon become reality.[8,9]

Diamond is recognized also as a promising active artificial substrate for biosensors and biointerfaces in the life sciences and medicine.[10] This is because, unlike other semiconductors, diamond exhibits a unique combination of favorable electronic, optical and mechanical properties with chemical and biocompatible properties. The diamond surface can be relatively easily, yet reliably, functionalized with organic molecules, DNA or enzymes[11–13] and it is suitable for the attachment of cells, such as osteoblasts, fibroblasts, cervical carcinoma cells (HeLaG)[14,15] or cardiomyocyte cells.[16]

The biological, as well as electronic properties, of intrinsic diamond can be significantly altered by hydrogen and oxygen atomic surface termination, which results in different properties, such as electrical conductivity, electron affinity and surface wettability. Oxygen-terminated surfaces are hydrophilic and highly resistive, while hydrogen-terminated surfaces are hydrophobic and induce p-type surface conductivity even on undoped diamond.[17–19] These properties can be used to create microscopic human cell arrays[15] on hydrophilic patterns or to fabricate solution-gated field effect transistors (SGFET) based on nanocrystalline diamond films on glass[5] employing the surface conductivity of *H*-terminated diamond.[20] It is notable that diamond SGFET can operate without a gate oxide layer because the gate is insulated by hydrogen atoms, hence it allows a direct contact between biomolecules and the surface of the FET channel. These transistors enable miniaturization and the direct transduction of signals in biosensing. This is indispensable for high density, high sensitivity and high-speed sensors.[21]

Both the intrinsic NCD films and the boron-doped NCD films accelerate the proliferation of human cells in culture on these films in the early exponential phase of growth, increase their numbers and enhance their differentiation, as manifested by the increased concentration of osteocalcin and collagen I per mg of cellular protein in comparison with cells grown in standard polystyrene dishes.[10] At the same time, the NCD films do not evoke the increased immune activation of cells. These beneficial effects can be explained by the generally good biocompatibility of diamond and the nano-roughness of NCD films,[22] which can be further enhanced by boron doping.

Thus, diamond thin films can be understood as a class of materials whose morphological, chemical and electronic properties can be tailored on demand for specific applications. Nevertheless, this inherently requires a high degree of control and understanding of diamond growth technology, especially at low temperatures because of the substrates used in bioapplications.

13.2 Synthesis of Diamond Thin Films

Presently, diamond growth is routinely performed all over the world. For successful growth, two crucial steps must be performed: i) proper pretreatment of a non-diamond substrate (nucleation or seeding) and ii) diamond chemical vapor deposition. The nucleation/seeding process, often called the pre-growth treatment, is required due to the high surface energy of diamond. In other words, initial diamond clusters do not want to be formed on foreign substrates. The nucleation/seeding process influences (controls) the final character of the diamond film, especially its crystallographic character (size of crystals and their preferential orientation), the chemical purity (ratio of diamond phases to non-diamond carbons—amorphous and graphitic carbon), the film roughness, the adhesion of the diamond layer to the substrate, *etc.* Detailed reviews on the nucleation/seeding techniques can be found in refs 23–26.

The second step—diamond chemical vapor deposition (CVD) from the gas phase—is a complex process, which has to i) stabilize the formed diamond nuclei/seeds, ii) enlarge their size into well faceted crystals up to the formation of a continuous layer and iii) keep a thermodynamic equilibrium for the development of the diamond phase over non-diamond carbon phases. The CVD process can be divided into five stages: 1) introduction of an appropriate gas mixture (often 1% methane diluted in hydrogen), 2) activation of gases by an external source of energy, 3) production of various gas species through chemical reactions, 4) transport of gas species to the substrate, and 5) their interaction with/on the substrate surface to form the diamond layer.[24,27] The diamond CVD growth can be further classified by external activation sources (plasma, hot wire, irradiation of substrate) through the use of gas mixtures, by the deposition area or by the total process pressure.

13.2.1 CVD Diamond Process Temperature: Limiting Factor for Substrates

The most common feature of all CVD techniques is an exposure of the substrate material to relatively high substrate temperatures of 700–950 °C. The high temperature process limits the variety of substrate materials over which CVD diamond films can be grown. This limitation can be specified by substrates divided into three groups: a) low temperature melting/softening substrates, b) substrates with high coefficients of thermal expansion and c) substrates with high diffusion coefficients/or high chemical reactivity.

13.2.1.1 Low Temperature Melting/Softening Substrates

For example, heat spreaders have to be deposited on complementary metal–oxide–semiconductor (CMOS) devices and electrical contacts, which are often stable up to a maximum temperature of 400 °C.[28,29] In addition, some semiconductor devices implement robust solders, which can degrade at high temperatures. In biological applications (like tissue engineering) cheap glass substrates with adequate optical transparency are commonly used. In this case microscope glass slides are often made of soda lime glass or borosilicate glass, which becomes soft at high temperatures (600–800 °C). From this point of view, fused quartz silica glass is more compatible with the diamond technology due to its high melting/softening temperature. Moreover, this glass is also suitable for fluorescence microscopy that works in the ultraviolet spectral regions. Coating of glass substrates with diamond brings an added value. In this case, the diamond film is used either as a passive functional material for immobilization of various biomolecules[11] or as an electrically active part of the biosensing element.[13] In addition, implementing diamond-coated metal mirrors and silicon prisms for grazing angle reflection[30] and attenuated total reflectance FTIR spectroscopy is an attractive solution for the identification of chemical groups on the nano-sized diamond crystals (more detailed data can be found in section 13.8).[31]

Plastics represent an additional material group that has attracted the attention of diamond technology. Plastics are widely used as cheap, transparent, lightweight and robust substitutes for glass in various applications, including the automotive and soft drinks industries, as well as for organic light emitting diode (OLED) display applications. However, the use of polymers is limited because of their low hardness, poor scratch resistance and high susceptibility to gases and chemicals. Alternatives have been developed by modifying the polymer surfaces to improve their properties. A challenging technological issue is the diamond CVD growth on such substrates with extremely low melting points (<200 °C). The low temperature deposition of diamond makes possible the use of diamond films instead of other carbon forms (diamond-like carbon, amorphous carbon, *etc.*). Recently, several attempts at diamond deposition on polymer substrates have been made by Piazza *et al.*,[32] Tsugawa *et al.*[3] and Kromka *et al.*[33] However, even today, this field is still emerging in laboratories rather than being practical for industrial uses.

13.2.1.2 Substrates with High Coefficients of Thermal Expansion

Diamond is known as a material with a very low coefficient of thermal expansion (CTE) of $\sim 0.8 \times 10^{-6}$ K^{-1}. The large difference in the CTEs of diamond and substrate materials induces high tensile or compressive stresses in the layers/substrate. Such a situation occurs when the standard high temperature (700–950 °C) process is applied to substrates with a

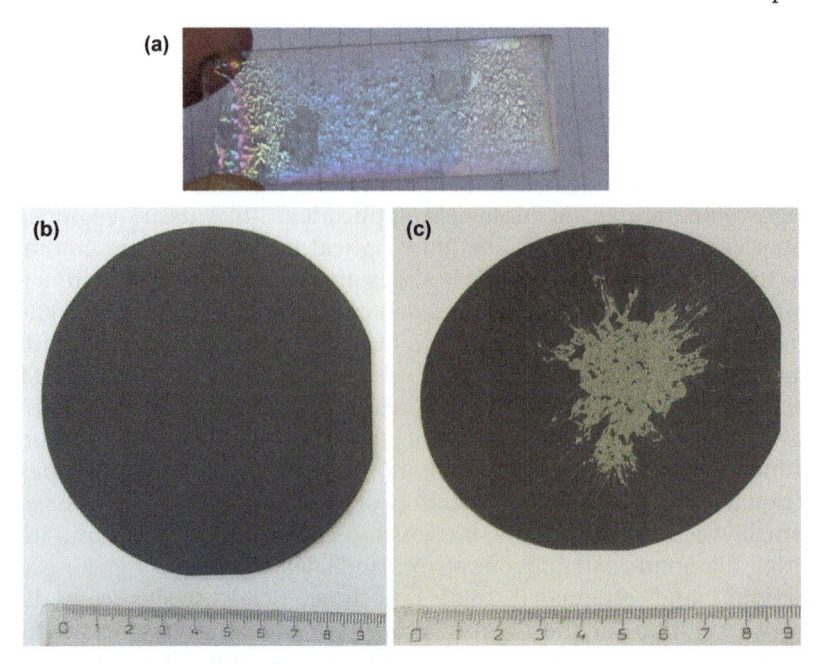

Figure 13.1 Delamination of a diamond layer deposited at ∼700 °C by PLAMWP system[2] a) on an alkali-free glass substrate, b) and c) on a four inch Si wafer due to internal stress. Note: the diamond film on the Si wafer shown in b) is a few hours after deposition and in c) is a few days after deposition.

high CTE mismatch with diamond (*e.g.* alkali-free glass with a CTE of 4.5×10^{-6} K^{-1}). The high CTE mismatch results in development of increased stress during cooling of the sample to room temperature (Figure 13.1). As a consequence, bending of the substrate material, cracking of the deposited diamond film and its delamination from the substrate can occur. One solutions to this is low temperature diamond growth, which reduces the large thermal stresses in the films[34] and allows diamond deposition also on temperature-sensitive substrates. Figure 13.2 compares the thermal conductivity and thermal expansion coefficient for diamond with a variety of substrate materials.

13.2.1.3 Substrates with High Diffusion Coefficients/High Chemical Reactivity

Silicon still represents the most cost effective solution for the semiconductor industry. Electronics based on silicon are often noted to be functional up to temperatures of 125 °C. This temperature is defined as the upper limit for the proper functionality of the pn junctions.[35] Improvement of the thermal management has motivated researchers to combine silicon with other

Low Temperature Diamond Growth

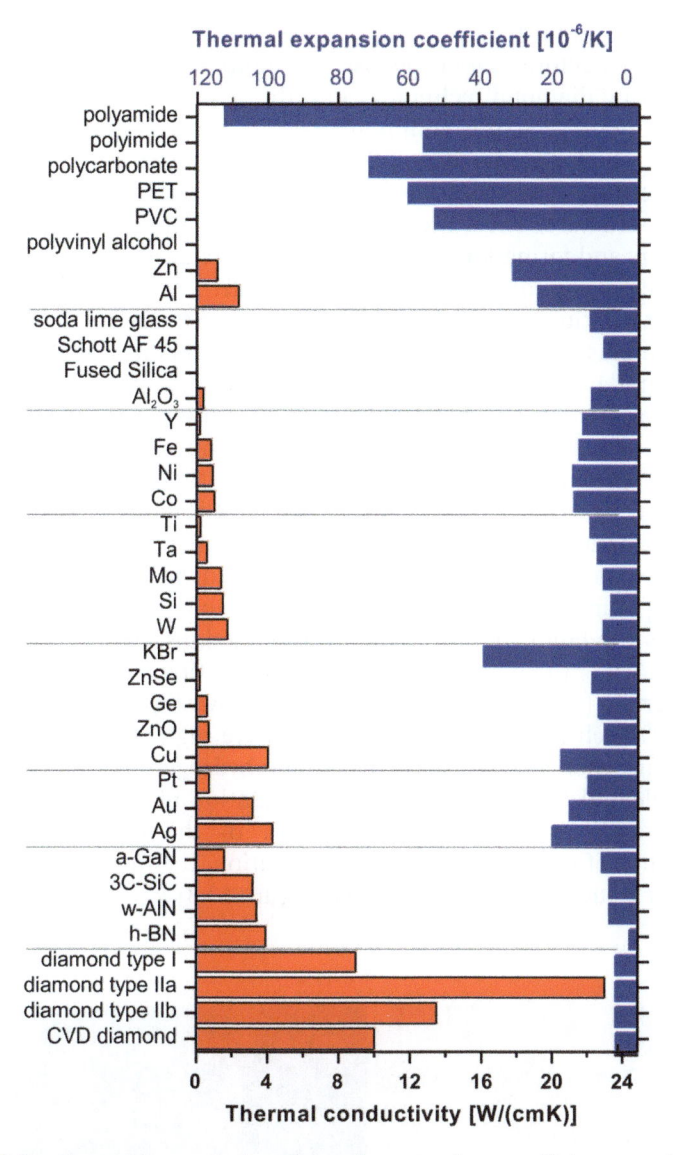

Figure 13.2 Comparison of the thermal expansion coefficients and thermal conductivities of various materials.

materials, like diamond. Diamond is an extremely good thermal conductor (*i.e.*, a property that is welcomed for heat spreading) and exhibits high resistivity (*i.e.*, intrinsic diamond). By adding properties such as chemical and mechanical resistivity, the diamond thin film becomes a very promising candidate for encapsulating standard silicon chips for their use in harsh environments (higher temperatures, acids, *etc.*). However, the deposition of diamond at standard temperatures for several hours can drastically change the electronic performance of the final device due to out-diffusion of the pn junctions. Therefore, lower deposition temperatures are preferable for implementing diamond technology into the semiconductor industry.

CVD diamond growth is commonly performed in the presence of atomic hydrogen, which can detrimentally react with most metals. The situation becomes critical once multi-layered ultrathin metals are combined in a sandwich structure. Then, the effects of delamination, cracking of the thin metal layers and formation of metal nano-sized clusters, or even full damage and etching of the metal layers, can play a role.

Another limiting factor of the high temperature process in combination with harsh plasma conditions can be the higher chemical reactivity and the faster degradation of substrates. For example, commonly used materials suitable for IR optics are ZnSe, KBr or germanium. These materials are known as soft and are unstable in harsh chemical environments and/or aggressive plasma conditions. We observed that, even though a low temperature process was used, out-diffusion of Ge atoms occurred, resulting in void formation in the Ge substrate (see section 13.8.3).

Finally, diamond is recognized as an excellent material for high power/high temperature electronics. Unfortunately, n-type doping of diamond still limits the fabrication of diamond pn-junctions. Fabrication of heterostructures using the p-type diamond in combination with n-type wide band gap materials (GaN, AlN or ZnO) has been investigated as an alternative solution.[36] Such wide band gap materials have high melting temperatures, but they are chemically unstable in atomic hydrogen. For example, oxygen will be released from ZnO following degradation of the n-ZnO/p-diamond interface and, resulting in deterioration of the electronic or optical performance of the heterostructure. In addition, the mismatch in the CTEs is another issue that has to be solved (Figure 13.3).

Figure 13.3 Optical (left) and SEM (right) image of diamond film grown on a GaN substrate. Darker regions in the SEM image represent areas of diamond delamination from the GaN substrate.

13.3 Low Temperature Diamond Growth

As mentioned above, low temperature diamond CVD growth (LTDG) is a crucial objective for the practical use of diamond thin films as semiconducting or optically emitting materials, a passivation coating of optically transparent materials or semiconductors, bio-encapsulating layers for biochemistry and the life science, *etc.*

Standard CVD diamond growth is realized in hydrogen-rich gas mixtures containing 1–2% methane, which represents the source of carbon and hydrogen that contributes to the stabilization of sp^3 dangling bonds (characteristic for diamond) on the diamond surface plane. Without this stabilizing effect, these bonds would not be maintained and the diamond plane would collapse to the graphite structure.[37] The presence of atomic hydrogen is very important; it reacts with hydrocarbon gases and creates active radicals containing carbon atoms, *e.g.*, $CH_4 + H = CH_3 + H_2$. These radicals diffuse onto the substrate surface and form C–C bonds necessary for diamond lattice formation. Moreover, in contrast to molecular hydrogen, atomic hydrogen is more chemically reactive and selectively etches graphite (non-diamond carbon). Therefore, atomic hydrogen prevents the formation of sp^2 graphitic bonds. The etching rate of graphite is at least twenty times faster than that of diamond. However, the CVD growth from hydrogen-rich gas mixtures is effective at relatively high substrate temperatures (>600 °C). Any lowering of the substrate temperature needs a proper process modification.

LTDG is technologically not a trivial or simple task. The first diamond growth at room temperature was demonstrated by graphite sputtering using either an ion beam of Ar or hydrogen in 1985/1988; however, it was achieved with very poor film quality.[38] Similarly, attempts at decreasing the substrate temperature whilst keeping all other deposition parameters constant failed due to: i) a drastic decrease of the growth rate to a few nanometers per hour and ii) incorporation of non-diamond carbon phases as unwanted impurities over the film.[39] Both these limitations make diamond thin film technology uneconomical and unviable for industrial uses.

A systematic effort toward LTDG can be dated back to the early 1990s. Several alternative processes have been investigated and experimentally tested. These processes can be classified into two strategies. The first main strategy is related to the modification of CVD deposition systems and processes and the second strategy is based on a change of the gas chemistry (Figure 13.4).

13.3.1 Strategy 1: Modification of CVD Deposition Systems

Polycrystalline diamond films are commonly deposited by chemical vapor deposition techniques from hydrocarbon gas sources. The two most used CVD methods are hot filament (HFCVD) and microwave plasma CVD (MWCVD).

Synthesis of Diamond Thin Films

A) before-growth treatment
- nucleation from carbon species
- seeding with (nano) particles
- mechanical scratching
- thin film overcoating
- others

B) chemical vapor deposition
- process chamber (geometry, heat & gas flow, ...)
- energetic activation of gas species (heat, plasma, light, irradiation)
- gas reactions chemistry (gas mixture, catalytic effects)
- reaction at substrate surface (migration, atoms attachment, ...)
- others

Low Temperature Diamond Growth (LTDG)

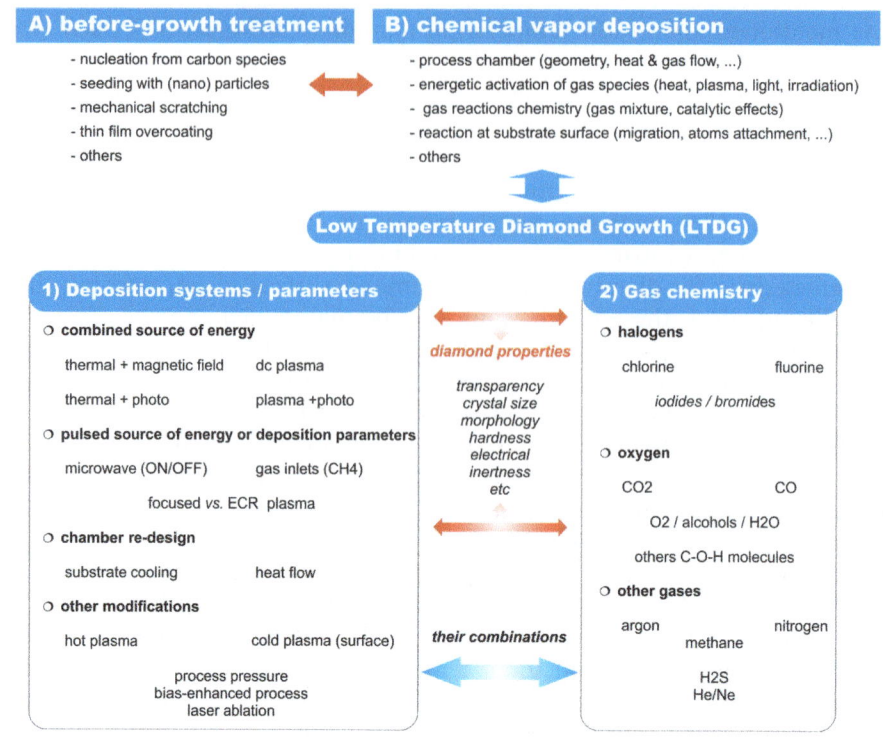

1) Deposition systems / parameters
- ○ combined source of energy
  - thermal + magnetic field   dc plasma
  - thermal + photo   plasma +photo
- ○ pulsed source of energy or deposition parameters
  - microwave (ON/OFF)   gas inlets (CH4)
  - focused vs. ECR plasma
- ○ chamber re-design
  - substrate cooling   heat flow
- ○ other modifications
  - hot plasma   cold plasma (surface)
  - process pressure
  - bias-enhanced process
  - laser ablation

diamond properties

transparency
crystal size
morphology
hardness
electrical
inertness
etc

their combinations

2) Gas chemistry
- ○ halogens
  - chlorine   fluorine
  - iodides / bromides
- ○ oxygen
  - CO2   CO
  - O2 / alcohols / H2O
  - others C-O-H molecules
- ○ other gases
  - argon   nitrogen
  - methane
  - H2S
  - He/Ne

Figure 13.4 Low temperature diamond growth processes.

In the HFCVD method the radicals are thermally activated by hot filaments (*i.e.*, W, Ta) *via* heating up to 2200–2700 °C. The filaments in the HFCVD system are very close to the substrate (typically ~1–3 cm) and, hence, strongly influence the substrate temperature T_S.

In the microwave CVD method the gas growth species are activated by plasma. In the standard MWCVD reactor (*e.g.* Figure 13.5, left), ball-shaped plasma with a temperature of 3000–5000 K is localized very close to the substrate (1–2 mm). Thus, the substrate surface is strongly heated by flow to the substrate from plasma (*i.e.*, electromagnetic radiation, charged or neutral species with kinetic energy, and the release of chemical or electronic excitation).

The primary aim in LTDG is minimizing the (over-) heating of the substrate even in the HFCVD or in MWCVD method. This could be achieved in several ways; for example, by the application of a more effective cooling system for the substrate, by moving up the filaments far from the substrate or by decreasing the filament temperatures in HFCVD. Alternatively, in MWCVD, by minimizing the power of the microwaves.[40] Ihara *et al.*[41] successfully grew diamond crystals on Si wafers at 135 °C using the HFCVD system equipped with a very effective cooling of the substrate holder

Low Temperature Diamond Growth

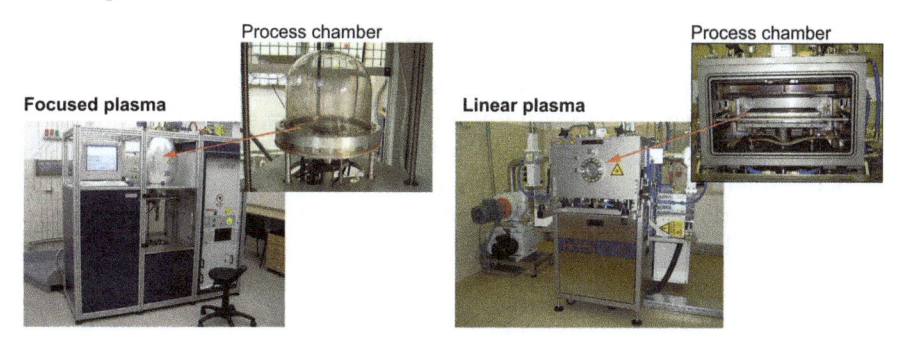

Figure 13.5 The microwave plasma systems used at the Institute of Physics.[2] Focused plasma (left) and linear antenna plasma (right).

stage. The substrate temperature was evaluated by measuring the holder temperature with a thermocouple inserted into a hole in the substrate holder and recalculated in terms of the thermal gradient of the Si substrate. However, a simple decrease in the substrate temperature generally reduced the growth rate to low values (<20 nm h^{-1}) and negatively influenced the quality of the grown diamond films. Therefore, several technological attempts have been investigated and some of them have proved their suitability for low temperature diamond growth. Related technological and process modifications are discussed further below.

13.3.1.1 Pulsed MW Plasma

In order to minimize the heating effect of the substrate from focused plasma, Ong *et al.*[42] replaced a continuous MW power with switchable MW power (*i.e.*, on/off cycles). This technology is often called *pulsed MWCVD*, which is widely used nowadays. The time duration of each cycle varies from tens of seconds up to ten minutes and has a strong influence on the diamond grain size and film continuity (*i.e.*, closed layers or diamond clusters). The lowest temperature of continuous diamond film deposited by a $CH_4/H_2/O_2$ gas mixture in pulsed mode was 400 °C. It should be noted that this was an average temperature calculated from the temperature behavior during switching cycles.

13.3.1.2 Decreasing of process pressure

In conventional MW plasma the dominant thermal flux that heats the substrate is generated *via* molecules or collisions within the high-temperature gas. Therefore, one of the natural ways to minimize the substrate heating from plasma is to lower the process pressure. However, reducing the pressure causes difficulty with the plasma stability due to the loss of charged particles, especially the recombination of electrons at the chamber walls.[43] Thus, prolonging the trajectory of charged particles in the process chamber should help in the stabilization of plasma.

13.3.1.3 Magnetic Field-assisted MW Plasma

High density plasma at low pressures can be formed by application of a magnetic field, which causes the trapping of electrons along the magnetic lines of force, keeping electrons away from the vacuum chamber walls and hence reducing recombination and increasing the path length. In 1992 Yuasa *et al.* reported large area diamond growth under low pressures of ~10 Pa and low temperatures of 500 °C using such magneto-active plasma CVD.[44]

13.3.1.4 ECR Plasma

Another method, which operates at much lower pressures and relies on magnetic confinement, is electron cyclotron resonance CVD (ECR-CVD), *i.e.*, magneto-microwave plasma. The ECR-CVD technique runs at very low pressures (3 Pa or less), adequately long electron mean free paths (>5.8 mm), which allow much more effective absorption of the microwave energy through cyclotron resonance. In fact, at 1.5 Pa, the electrons can absorb energy over many orbits, resulting in a larger fraction of very high energy (10–20 eV) electrons. These high energy electrons promote dissociation and ionization of neutral gas atoms and molecules at very efficient rates. Consequently, in the early 1990s it was confirmed that ECR-CVD is able to produce diamond films even at temperatures as low as 300 °C.[45]

13.3.1.5 Magnetic Field-assisted HF CVD

In the last two decades magnetic fields (static or periodic) have been used in combination with other techniques to obtain high quality diamond films at lower temperatures. In comparison to the conventional HF CVD method, applying a periodic magnetic field in the HFCVD reactor increases the diamond growth rate up to seven times for substrate temperatures of 520 °C.[46] Furthermore, the chemical film purity improves as confirmed by Raman measurements.

13.3.1.6 DC Plasma-assisted CVD

The change of the growth activation principle has been presented by Nakao *et al.*,[47] who used DC plasma in which the excitation of reactive species can be accomplished by electrons impinging on the reactive species at the substrate surface. Their study confirmed that the formation of diamond films is possible even at 400 °C when the discharge current is sufficiently high.

13.3.1.7 Surface Wave MW Plasma

A breakthrough in LTDG over large areas can be dated back to 2006, when surface wave plasma MWCVD systems with linear antennas were introduced[3] and later modified to become multi-slot antennas.[48] These systems are characterized by cold plasma due to the lower pressures (500 Pa) and larger distance of the hot plasma regions from the substrate surfaces (typically larger than 5 cm). A successful diamond growth was presented at extremely low temperatures (<100 °C) over areas as large as 30×30 cm^2. Recently, large area diamond films have become available using $CO_2 + H_2 + CH_4$ in a linear antenna microwave plasma CVD system. An additional advantage is its simple scalability, *i.e.*, the antenna length can be lengthened up to 1 m and the number of antennas can be multiplied.[3,48] Additional data on the LTDG by this technique are summarized in section 13.3.2 (*Strategy 2: Influence of Gas Chemistry*) and the key differences between focused (ball-shaped MW) and surface wave plasma are described in more detail in sections 13.7 and 13.8.

As demonstrated in several selected key studies (section 13.4–13.8), the growth of diamond thin films by pulsed linear antenna microwave plasma CVD processes opens new fields in which diamond can be used as a multi-functional material. The representative pulsed linear antenna microwave plasma (PLAMWP) CVD system used in our group is shown in Figure 13.5 (right). The PLAMWP reactor is based on the commercially available AK 400 platform, which is used for solar cell technology. The system was developed in a close cooperation with Roth and Rau, AG (Germany).

The system employs two microwave generators (2.45 GHz, MX4000D, Muegge) working at frequencies up to 500 Hz and a maximum power up to 4.4 kW in a pulse regime at each side of the linear conductor located in the quartz tubes, which are situated above the substrate (Figure 13.5). The distance between the quartz tubes and the substrate can be varied from 4 to 12 cm by the up/down positioning of the substrate holder. Moreover, the substrate holder is resistively heated, which allows plasma-independent control of the substrate temperature from 250 °C up to 700 °C. The deposition area is up to 20×30 cm^2 and the total gas pressure can be varied from 6 Pa to 200 Pa for various gas mixtures. A large distance between plasma and the substrate holder means that substrate overheating from plasma is minimized. This feature allows low temperature deposition on various temperature-sensitive substrates (*e.g.*, plastics, glasses, *etc.*).

13.3.1.8 Other Techniques

Laser-driven reactions in CO/H$_2$ mixtures,[49] self-bias enhanced growth (SBEG)[50] or laser ablation[51] have been also used to grow diamond thin films at lower substrate temperatures. Either due to the limited deposition area or plasma stability, they have been more generally employed for fundamental research rather than for industrial uses.

The milestones in the modified processes of LTDG are summarized in Figure 13.6.

Milestones in modified processes of LTDG

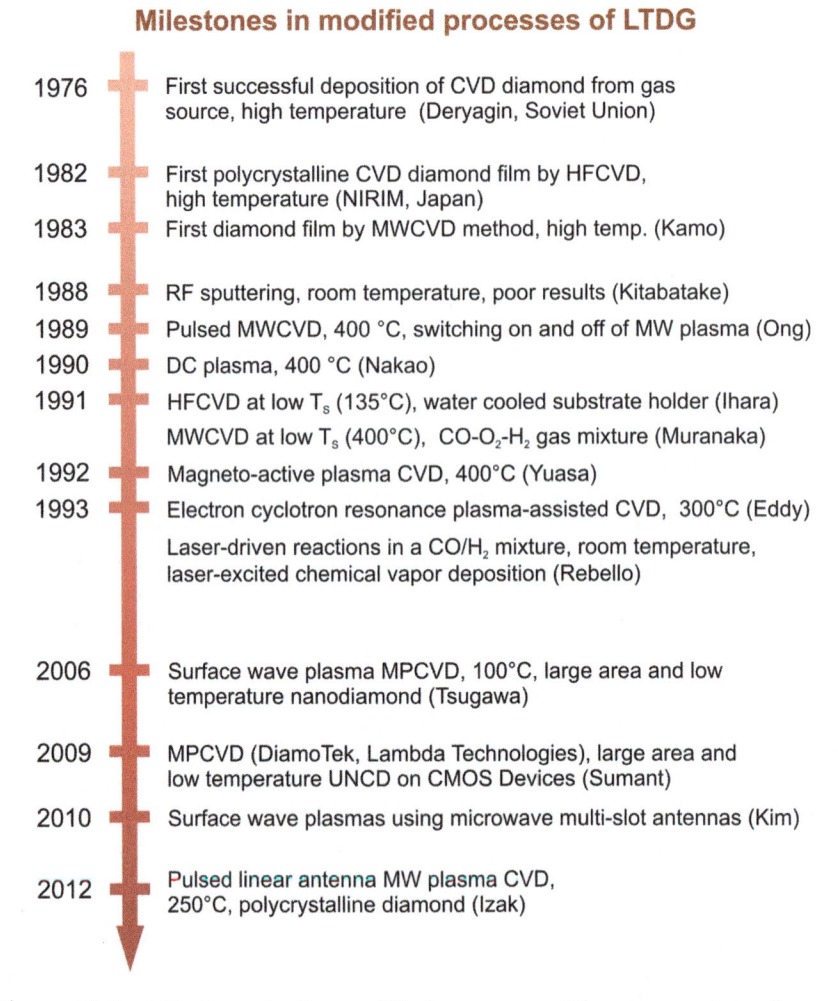

- **1976** — First successful deposition of CVD diamond from gas source, high temperature (Deryagin, Soviet Union)
- **1982** — First polycrystalline CVD diamond film by HFCVD, high temperature (NIRIM, Japan)
- **1983** — First diamond film by MWCVD method, high temp. (Kamo)
- **1988** — RF sputtering, room temperature, poor results (Kitabatake)
- **1989** — Pulsed MWCVD, 400 °C, switching on and off of MW plasma (Ong)
- **1990** — DC plasma, 400 °C (Nakao)
- **1991** — HFCVD at low T_s (135°C), water cooled substrate holder (Ihara); MWCVD at low T_s (400°C), CO-O_2-H_2 gas mixture (Muranaka)
- **1992** — Magneto-active plasma CVD, 400°C (Yuasa)
- **1993** — Electron cyclotron resonance plasma-assisted CVD, 300°C (Eddy); Laser-driven reactions in a CO/H_2 mixture, room temperature, laser-excited chemical vapor deposition (Rebello)
- **2006** — Surface wave plasma MPCVD, 100°C, large area and low temperature nanodiamond (Tsugawa)
- **2009** — MPCVD (DiamoTek, Lambda Technologies), large area and low temperature UNCD on CMOS Devices (Sumant)
- **2010** — Surface wave plasmas using microwave multi-slot antennas (Kim)
- **2012** — Pulsed linear antenna MW plasma CVD, 250°C, polycrystalline diamond (Izak)

Figure 13.6 Milestones in the modified processes of low temperature diamond growth.

13.3.2 Strategy 2: Influence of Gas Chemistry

13.3.2.1 Adding Halogens

The first LTDG attempts were based on simple requirements: i) termination of the diamond surface with atoms that have lower bonding energy with carbon than hydrogen with carbon and ii) enhancement of the chemical reactions that take place in comparison to the standard gas mixture, $CH_4 + H_2$. Halogen precursors were assumed to be one of the best candidates. Halogenated hydrocarbons were proposed to be more effective growth precursors as they form more radicals at lower temperatures (including the formation of halogen-methyl radicals), undergo more effective gas–surface reactions and by the abstraction of surface terminating groups (hence increasing the number of active surface sites). Among all halogens, chlorinated and fluorinated systems have been intensively investigated and compared.

The influence of $CHCl_3$ and CH_2Cl_2 as diamond growth precursors on the diamond growth was investigated by Nagano and Shibata.[52] They concluded that the chlorine precursor promoted the growth rate and participated in the etching of the diamond surface. Later, Rego et al. studied several halogen-based gas mixtures, including CH_2Cl_2 in H_2, CH_3Cl in H_2 and Cl_2/CH_4 in H_2, and compared them with standard $CH_4 + H_2$ gas mixtures.[53] It was found that too high Cl contents decreased the diamond quality at elevated temperatures. For lower temperatures, the growth was enhanced, which was attributed to the greater efficiency of surface hydrogen abstraction by chloride ions. They proposed that the gas-phase chemistry of chlorohydrocarbon radicals does not play a key role in the enhanced growth rates at lower T_S. The growth kinetics of LTDG was studied for CCl_4/H_2 mixtures in the temperature range of 280 to 820 °C.[54] The threshold temperature for crystalline diamond was 420 °C. Concurrent with the well-known role of atomic hydrogen, chlorine participates in hydrogen abstraction from the growing diamond surface. It was proposed that abstraction of surface H atoms by chlorine is 60 times faster than by atomic H at 670 °C. They concluded that the concentration of atomic hydrogen is a key factor in LTDG.[54] Taking into account that the C–H bond energy in methane (CH_4) is 439 kJ mol^{-1}, chlorine should play a positive role in LTDG as the C–Cl bond energy in $CHCl_3$ is much smaller (349 kJ mol^{-1}).[55] Evidence that chlorine is a key factor for LTDG was also shown in practice,[56] where diamond growth was shown to nearly stop for temperatures lower than 370 °C when a methane/hydrogen gas mixture was used. Using C_2H_5Cl as a precursor diluted in H_2 resulted in diamond growth at temperatures below 370 °C.

Fluorine-based precursors have also been intensively studied. Fluorine-based precursors, such as CH_3F and CF_4, have been added to hydrogen, and diamond films grown by a microwave plasma CVD process.[57] The calculated activation energies were found to be 12–13 kcal mol^{-1} for a temperature range of 600–900 °C. Thus, there was no evident influence of fluorine precursors on the growth kinetics. Maeda et al. presented similar findings in a comparative study of CF_4 and CHF_3 precursors as alternative growth precursors to CH_4.[58] The activation energy for a temperature range of 640–850 °C increased with increasing fluorine concentration. The highest activation energy was found for CHF_3 (28 kcal mol^{-1}); it was lower for CF_4 (21 kcal mol^{-1}) and the lowest was for CH_4 (10 kcal mol^{-1}). The origin of this observation was assigned to fluorine termination of the active sites on diamond. For temperatures higher than 850 °C, fluorine-based diamond growth exhibited a near constant growth rate. However, contrary results were observed for CH_4 gas mixtures, where the growth rate decreased when the temperature was increased to 950 °C. These differences may originate from different pretreatment techniques (nucleation or seeding) and slight differences in the deposition systems (gas injection, control of substrate temperature by microwave power, varied distance between filament and substrate, *etc.*).

There are complex discussions regarding the role of fluorine in diamond growth. Although, the above studies can be classified as high temperature, it is interesting to note that fluorine effectively etches non-diamond carbon phases at high temperatures. Discussions regarding this complexity have considered that the C–F bond is stronger (446 kcal mol^{-1} in CHF$_3$) than the H–H bond (436 kcal mol^{-1}).[55] LTDG was achieved at 380 °C for a gas mixture of CH$_4$/CF$_4$.[56] It was proposed that fluorine atoms captured by atomic hydrogen can form very stable HF molecules and the system should be free of fluorine.[54] Remaining CF$_x$ radicals can take part in surface hydrogen abstraction.

Other methyl-halogen precursors, such as CH$_3$I and CH$_3$Br, were tested with regards to diamond thin films, but they have not been adopted for diamond technology.[59]

13.3.2.2 Adding Oxygen

In oxygen-containing gas chemistry it is supposed that oxygen partially replaces the role of hydrogen in etching of the amorphous components, the inhibition of carbon π bonding at the surface and the suppression of impurity incorporation. At lower temperatures, atomic oxygen exhibits stronger preferential etching behavior than atomic hydrogen and more effective removal of hydrogen surface atoms.[60] Such behavior suggests that oxygen-containing gases are promising candidates for the synthesis of high quality diamond films at low substrate temperatures.

In 1991 Bachmann *et al.*[61] summarized the results of diamond deposition experiments with oxygen containing gases and constructed a so-called atomic C–H–O phase diagram (Figure 13.7). The low pressure synthesis of high quality diamond was possible within only a well-defined area close to the H–CO "tie line". It was supposed that the Bachmann ternary phase diagram was most likely only appropriate for high energy density systems that are controlled by the concentrations of the active species, determined by equilibrium considerations, where the border can be shifted by the temperature, pressure and gas composition.[62] Experimental and theoretical studies have confirmed such a variable diamond deposition window.[63–65]

Several research groups have confirmed the enhanced growth rates and improved diamond film quality (resulting in lower non-diamond sp^2-bonded carbon content) by the addition of oxygen-containing gases into the CH$_4$/H$_2$ gas mixture.[2,3] However, diamond growth is possible even without supplying additional hydrogen with gas mixtures of CH$_4$/CO$_2$[39,40,66] or C$_2$H$_2$/O$_2$.[67] In CH$_4$/CO$_2$ plasma, the gas phase concentration of CO is ∼100 times higher that of atomic H and adsorbed CO molecules have also been measured by X-ray photoelectron spectroscopy on the surface of deposits. According to this result, CO was considered to be the growth species.[68] However, another study concluded that, although CO species are dominant in the gas-phase,

Low Temperature Diamond Growth 305

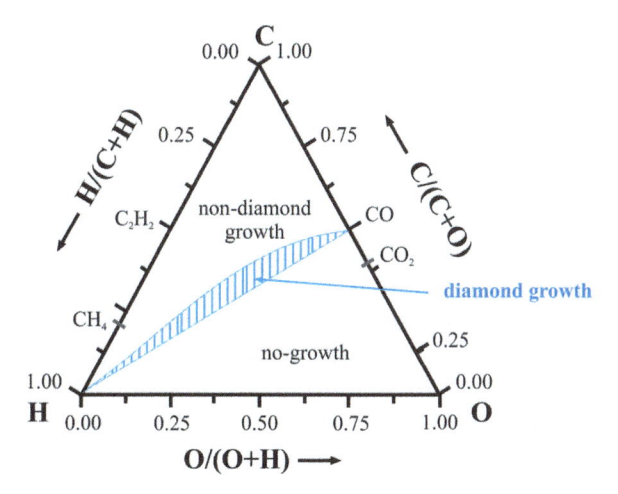

Figure 13.7 Simplified model of the Bachmann triangle with highlighted diamond growth region.[6]

they do not participate in gas-surface chemistry and CH_3 species are responsible for diamond growth rather than CO or C_2H_2.[66]

Interesting results were also achieved for diamond growth using CO/H_2[69] or CH_3CHO/H_2[70] gas mixtures. Acetaldehyde has a weak C–H bond and reacts with hydrogen atoms faster than methane and, therefore, it was considered that it might provide an initially "cleaner" source of methyl radicals involving the following reaction:

$$H + CH_3CHO \rightarrow CH_3 + H_2 + CO \quad (13.1)$$

The rate limiting step of this conversion is:

$$H + CH_3CHO \rightarrow CH_3CO + H_2, \quad (13.2)$$

which is about 30 times faster than the $H + CH_4$ reaction at 650 °C and about 75 times faster at 500 °C.[70] That means acetaldehyde could also be a very promising carbon source for low temperature diamond growth.

LTDG was achieved at temperatures as low as 200 °C by a magneto-active microwave plasma CVD process using a mixture of CH_3OH (15%) diluted in hydrogen. As the process temperature decreased from 600 °C to 300 °C, the growth rate steeply decreased from 120 nm h^{-1} to approximately 20 nm h^{-1} and it was constant at this value when the temperature was further decreased to 200 °C.[71] The diamond character of the film was confirmed by Raman measurements. Another LTDG was achieved at the National Institute of Advanced Industrial Science and Technology (AIST) using surface microwave plasma also, often labeled as cold plasma (*i.e.*, electron temperature of 1–2 eV at the substrate surface).[72] In this case, CO_2 was added to the $CH_4 + H_2$ gas mixture and the substrate temperature was as low as 100 °C. Formed diamond films consisted of diamond nanocrystals with an average size of 5 nm. The origin for the growth of nanocrystals was assigned to gas-phase nucleation.

13.3.2.3 Other Gas Mixtures

According to the abovementioned theories of diamond growth, the rate-limiting step in the low temperature range is the hydrogen abstraction from the growing diamond surface. Except for the addition of halogens or oxygen-containing gases, abstraction of hydrogen at low temperatures can be improved also by the addition of other gases, such as argon, nitrogen or hydrogen sulfide. For example, carbon dimers (C_2) were proposed as the dominant growth species in CH_4/Ar plasma.[73] It was considered that argon helps not only the excitation of plasma at low pressures, but it also controls the discharge state, which can increase the density and activity of reaction radicals and hence improve the quality of diamond films. That means argon can cool the plasma and maintain the low temperature of the substrate due to its large ionization range and high collision probability with gas molecules.[74] However, in most cases, using argon in the gas mixtures resulted in the growth of nano- or even ultra-nanocrystalline (UNCD) diamond films. Such films are usable only for specific applications, often tribologically or mechanically oriented uses (*i.e.*, cutting tools, drills, passivation of seals, microelectromechanical system (MEMS) parts, *etc.*).[75]

Similarly, nitrogen has been added to CH_4/H_2 gas mixtures.[39] First, nitrogen was found as a possible way to control the relative growth velocities of {100} and {111} facets, but at high concentrations it deteriorates the film morphology by enhancing defect formation. UNCD-like films have been grown from 17% CH_4 in N_2 gas within the temperature range of 520–770 °C.[76] It was shown that the size of the formed diamond crystals is not affected by the substrate temperature, which was assigned to the high rate of secondary nucleation.

LTDG through the addition of 500 ppm H_2S to a 0.3% CH_4 in H_2 gas mixture was reported by Piazza *et al.*[77] The diamond film was successfully formed on glass at a temperature of 440 °C by a hot filament CVD technique. The proposed growth mechanism involved several steps: i) CS is produced at the hot filaments ($\sim$2500 °C) from H_2S and CH_4, ii) CS diffuses to the substrate (the colder part), iii) H_2S and C atoms are produced after a set of chemical reactions, and iv) H_2S is stable due to the low temperature region, while C is a further source of carbon at the growing film surface.[77,78] Later, Piazza *et al.* reported the successful growth of diamond on a polyimide film at 360 °C[79] and on thick Kapton® VN foil at temperatures as low as 250 °C.[32]

The influence of adding other inert gases (He, Ne and Kr) to the $CH_4 + H_2$ gas mixture was investigated also. Adding Kr to the gas mixture did not result in the growth of UNCD films. However, UNCD films were grown from He- or Ne-containing gas mixtures.[80] In contrast to the well-established UNCD growth model based on C_2 dimers in Ar-rich plasma, no C_2 dimers were detected in He or Ne plasmas. This observation indicates that different growth pathways can dominate during the UNCD growth.

The well-known change of diamond morphology from a poly- to nanocrystalline character for higher methane concentrations should be noted in nearly any gas mixture due to the increasing renucleation rate, which can be helpful for maintaining the growth rate at low substrate temperatures. Unfortunately, films produced at high CH_4 concentrations are of substantially lower optical and electronic quality and such materials are applicable only for limited studies and/or industrial uses.

13.4 Determination of Substrate Temperature and Calculation of Activation Energy

13.4.1 Substrate Temperature

In the chemical vapor deposition of diamond films, the temperature is the most difficult parameter to determine and control because of the extreme environment (high concentration of atomic hydrogen) and high-power flux to the substrate surface due to plasma. Heating of the substrate by plasma strongly influences the lowest temperature limit. There are several kinds of energy flow onto the substrate from plasma, *i.e.*, infrared, visible or ultraviolet radiation, electromagnetic waves, charged or neutral species with kinetic energy, or the release of chemical or electronic excitation. These energies will heat up the substrate and contribute to the surface reactions, which are also exothermic themselves.[43]

A reliable measurement of the substrate temperature (T_s) is a basic requirement for LTDG. There are several methods that may be used to determinate the T_s, such as a thermocouple, pyrometry, interferometry[81,82] or spectroscopic ellipsometry[69,83] (see Figure 13.8).

In general, measurement methods can be separated into two groups: (i) contact and (ii) non-contact methods. Contact methods use thermocouples for the determination of the temperature. This method consists of two conductors composed of different materials and its principle is based on the thermoelectric (or so-called Seebeck) effect when the temperature gradient between the two metals leads to the generation of a voltage drop. It can be used with a wide range of temperatures; however, it does not show the real substrate temperature (T_S) because, in most cases, it is attached to the substrate holder from the bottom side (see Figure 13.8a) and commonly measures the temperature of the substrate holder, which is obviously lower than the substrate temperature due to the thermal gradient. However, thermal losses in the substrate can be calculated by eqn 13.3:

$$\Delta T = \frac{j}{(\lambda/l)}, \qquad (13.3)$$

where λ is the thermal conductivity of the substrate (in the case of silicon it is 150 J m^{-2} s K), l is the thickness of the substrate and j is the heat flux. The

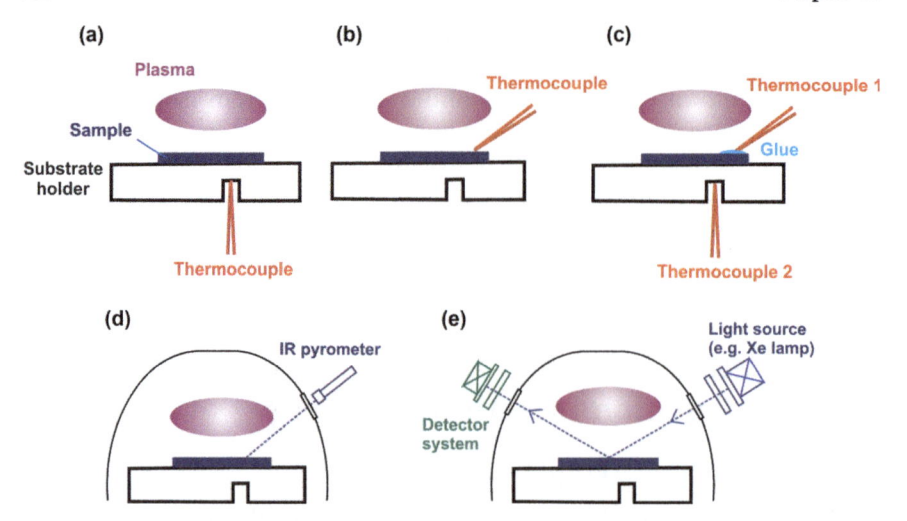

Figure 13.8 Schematic illustrations of the various methods that can be used to measure the substrate temperature *i.e.* with thermocouple attached a) to the bottom of the substrate holder or b) to the top of the sample; c) with two thermocouples. Figure d) shows a representative measurement setup of applying pyrometry or interferometry for temperature evaluation and e) shows the measurement setup for ellipsometry.

heat flux in the case of HFCVD can be calculated from the equation described in ref. 84. Of course, there is always a questionable thermal contact between the substrate and the substrate holder. To eliminate the problem of the thermal gradient, the thermocouple can be attached to the topside of the substrate (Figure 13.8b).[85] In such a case the measured value can be strongly influenced by plasma radiation. One way to solve this is the protection of the thermocouple tip from plasma by an insulator (Figure 13.8c).[43] However, this concept can be time consuming and sensitive to manipulation with the sample(s). In some cases two or more thermocouples are recommended.

Optical methods, such as pyrometry or interferometry, are non-contact methods and thus do not need thermal contact with the substrates and measurement tip (Figure 13.8d and e). They can be performed in emission, transmission or reflection modes in the near-infrared to visible range. The proper identification of emission constants from the substrate and/or from the grown diamond films, and background subtraction from plasma radiation are the most limiting factors for a proper temperature determination.

Commonly, infrared (IR) pyrometers are used because they exclude any interference from plasma emission. IR pyrometer records the thermal radiation from the substrate at IR wavelengths (depending on its type) through the plasma. However, most pyrometers determine the temperature from an absolute irradiance measurement and, therefore, require frequent calibration. This calibration should involve the difference in emissivity of diamond and the Si substrate (or other materials used as the substrate), then optical losses that may occur upon transmission through the reactor window or due to errors in the assumed emissivity of the surface (*e.g.*, from changes in emissivity that may occur during film growth).[83] During the growth, the emitted radiation intensity of the diamond film is modulated in time by interference phenomena, causing a periodic variation of the pyrometers apparent temperature reading. The growth rate of the diamond films can be determined from the period of these oscillations *via* eqn 13.4:

$$dG/dt = (\lambda_1 + \lambda_2)\cos\theta/(4n\tau), \qquad (13.4)$$

where λ_1 and λ_2 are the measuring wavelengths of the pyrometer, n is the refractive index of the film, τ is the period of the apparent temperature oscillations and θ is the viewing angle of the pyrometer, as measured from the sample normal (it is assumed that $n(\lambda_1) = n(\lambda_2)$).[67]

A similar method, interferometric thermometry, determines the temperature from the thermal expansion and refractive index changes of a transparent substrate of known thickness, whose front and back faces are polished and approximately parallel. Temperature changes are measured by counting the oscillations (fringes) in the reflectance signal. It is a suitable technique for measurements both near and above room temperature and, in contrast to pyrometric techniques, for cryogenic measurements as well.[81]

Ellipsometry employs the temperature dependence of the optical properties of the substrate to determine the temperature. In general it measures a change in the polarization as light reflects or transmits from a material structure. The measured response depends on the optical properties and thickness of the individual materials. Thus, ellipsometry is primarily used to determine film thickness and optical constants and, together with pyrometry and interferometry, provides *in situ* (real time) monitoring of growth.[86,87] It is also a well-used method for the calibration of thermocouples during the growth process.[69] Based on rapid-scanning spectroscopic ellipsometry, Wakagi *et al.* developed and utilized methods for precise ± 5 °C Si substrate temperature calibration.[83] Moreover, ellipsometry has an advantage over Raman spectroscopy for characterizing and optimizing NCD film quality owing to its ability to provide diamond, sp^2 C and void volume fractions with roughly equal sensitivity in very thin films (<100 nm).[69]

13.4.2 Activation Energy

A characteristic parameter, which can provide significant insight into the chemical kinetics on the surface growth, is the activation energy E_a. The activation energy represents the amount of energy needed to start the reaction, *i.e.*, it is the minimum energy needed to form an activated complex during a collision between reactants (Figure 13.9). Once the activated complex is formed, the final products are formed from it. The path from the reactants, through the activated complex and to the final products is known as the reaction path. Activation energies are usually determined experimentally by measuring the growth (or reaction) rate k at different

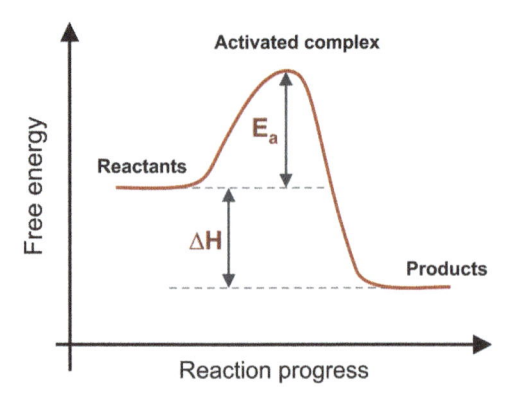

Figure 13.9 An energy profile for an exothermic reaction (E_a is the activation energy and ΔH is the enthalpy).

temperatures T, plotting the logarithm of k against $1/T$ and determining the slope of the straight line.

Physical and chemical reactions during diamond growth are temperature dependent and the activation energy of the process can be determined from the Arrhenius equation (eqn 13.5).[73]

$$k = Ae^{-E_a/RT_s}, \qquad (13.5)$$

where k is the rate constant, A is a temperature-independent constant called the Arrhenius (or frequency) factor, E_a is the activation energy in J mol^{-1}, R is the universal gas constant (8.314 J K^{-1} mol^{-1}) and T_s is the substrate temperature in K.

However, the calculated activation energy is often questionable due to difficulties in the precise measurement of the substrate temperature and in the evaluation of growth rates—it is well known that the growth rate of the diamond deposition is different to the nucleation (*i.e.*, the incubation period) and growth regimes and, in most cases, the incubation period is not taken into account when the growth rates are evaluated.[88] During our studies we observed that the nucleation induction period varies with the substrate preparation and temperature, with a period of approximately 15 s observed at 1000 °C and 30 min at 500 °C. Corrections for this induction period are necessary if accurate activation energies are to be determined from *ex situ* measurements.

A precise real-time measurement method of the growth rates seems to be real time *in situ* spectral ellipsometry, applied during the deposition process.[69] Activation energies have also been calculated from the mass growth rate, as shown by Barbosa *et al.*[89]

The standard CVD diamond process (1–3% CH$_4$/H$_2$) features a high activation energy of ~ 28 kcal mol^{-1}. That means the growth rate strongly depends on the temperature and indicates a significant reduction in the growth rate as the substrate temperature is lowered.

Small differences can be found between the activation energies of CH_4/H_2 systems when using different deposition techniques; e.g., the E_a is ~22–24 kcal mol^{-1} in the standard HF CVD systems, 28 kcal mol^{-1} in the microwave plasma jet and 20–30 kcal mol^{-1} in MW plasma CVD systems (Table 13.1). However, when activation energies are compared, it is necessary to take into account the quality of the final diamond film, i.e., if it is microcrystalline (MCD), nanocrystalline (NCD), ultrananocrystalline

Table 13.1 Activation energies of diamond growth by different techniques in various gas mixtures.

	Gas system	Activation energy (kcal mol^{-1})	Temperature range (°C)	Reactor type	Reference
Standard CH_4/H_2 system	CH_4/H_2	22–24	740–930	HFCVD	Corat (1997)
		20–30	850–1000	MPCVD	Zhu (1989)
		28	700–1180	MW plasma jet	Mitsuda (1989)
		22	420–820	HFCVD	Corat (1997)
		12 (MCD) 9 (NCD)	370–1100	Focused MWCVD	Potocky (2006)
		1–5	210–700	HFCVD	Yamaguchi (1994)
Halogen addition	1% CCl_4 in H_2	11	420–820	HFCVD	Corat (1997)
	$Cl_2/CH_4/H_2$	3.6^a	600–750	Hot-tube system	Wu (1997)
		7.9^a	400–600		
	CF_4/H_2	11	420–820	HFCVD	Corat (1997)
	CF_4/H_2	28	640–950	MWCVD	Maeda (1994)
	CHF_3/H_2	20	640–950	MWCVD	Maeda (1994)
	CHF_3/H_2	13.7	600–900	MWCVD	Fox (1995)
	CF_4/H_2	12	600–900	MWCVD	Fox (1995)
Oxygen addition	C_2H_2/O_2	16 and 23^a	444–1200	Oxygen-acetylene flame	Snail (1992)
	CO/H_2	8	400–800	PECVD	Lee (1996)
	CH_4/CO_2	6.7	435–865	MWCVD	Petherbridge (2001)
	CH_4/CO_2	5^b	345–560	MWCVD	Stiegler (1998)
	$CH_4/H_2/CO_2$	1.7 and 7.8^a	250–680	PLAMWP	Izak (2012)
	$CH_4/H_2/CO_2$	1.6 (NCD)	100–700	SWP	Tsugawa (2010, PR B)
Argon addition	$CH_4/H_2/Ar$ (1/9/90%)	5.7 (UNCD)	550–850	HFCVD	Barbosa (2009)
	1% CH_4 in Ar	5.85 (UNCD)	500–90	MWCVD	McCauley (1998)
		2–3 (UNCD)	400–800	MPCVD	Xiao (2004)
Nitrogen addition	CH_4/N_2	8.7 (NCD)	520–770	MWCVD	Kulisch (2006)
	$CH_4/CO_2/N_2$	4.7–10	430–600	MWCVD	Stiegler (1999)

^{a}Two distinct diamond growth regimes.
^{b}Under 430 °C two-dimension (2D) nucleation.

(UNCD), as well as its chemical purity (such as the sp³ to sp² ratio, *etc.*), because the growth models differ strongly and, hence, the activation energies are different. For example, in our earlier studies we showed that the activation energy for a standard CH_4/H_2 gas system in a focused MWCVD system is 12 kcal mol^{-1} for microcrystalline diamond and 9 kcal mol^{-1} for nanocrystalline diamond.[90] Low activation energies (4-5 kcal mol^{-1}) are characteristic for argon–methane gas mixtures; however, those films have UNCD character.

Low activation energies were also calculated for C–O–H gas mixtures together with increased growth rates due to the altered plasma chemistry and surface reactions. In these experiments the final diamond film morphology had, in most cases, microcrystalline character. The calculated activation energies for different deposition techniques, gas mixtures and final diamond film quality are briefly summarized in Table 13.1.

13.5 LTDG in Pulsed Linear Antenna Microwave Plasma

In the previous section we introduced that the pulsed linear antenna microwave plasma (PLAMWP) CVD system also allows low temperature deposition of diamond, even at 250 °C, over large areas and with high diamond film homogeneity.[91] Several experiments have been carried out with different microwave powers, gas mixtures and pressures to study the growth and diamond film quality at low temperatures. It was found that the addition of oxygen-containing gases (in our case, CO_2) is essential to achieve high diamond film quality at low T_S. The optimal gas mixture ratio seems to be 2.5% CH_4 and 10% CO_2 to H_2. Some interesting diamond growth features as a function of the process conditions are discussed below.

Figure 13.10 shows the SEM images of diamond thin films deposited on Si substrates at different microwave powers of 2500 W (a) and 1200 W (b and c). Samples a) and b) were deposited at a high substrate temperature (600 °C), while sample c) was deposited at a low temperature of 250 °C.[91] In this study the substrate temperature (T_S) was controlled by a resistively heated table

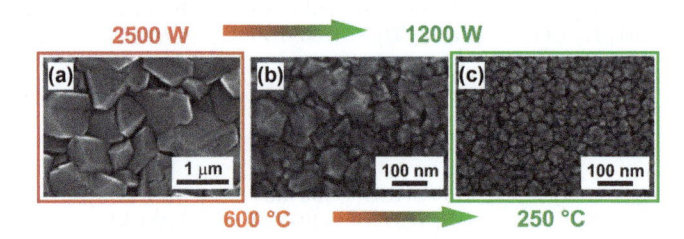

Figure 13.10 Morphology of diamond films (by SEM) deposited at different microwave powers and substrate temperatures: a) diamond film deposited at 2500 W and 600 °C, b) 1200 W and 600 °C, and c) 1200 W and 250 °C. Note: sample a) has a different magnification to b) and c).

holder. In the standard setup the lowest T_S (250 °C) was achieved by lowering the microwave power to 1200 W. The substrate temperature was measured by IR pyrometry. Diamond thin films deposited at 250 °C and 2500 W (sample a) exhibit large diamond crystals with sizes from 600 nm up to 1 μm. Keeping the substrate temperature constant and decreasing the MW power from 2500 to 1200 W leads to a decrease in the crystals size to 50–100 nm (sample b). This change in film morphology is assigned to the growth rate (Figure 13.12) caused by reduced plasma density. The grown diamond film has a thickness of 45 nm and crystals with diameters of 50–70 nm. The growth rate at 250 °C was found to be as low as 3 nm h^{-1}.[91]

The quality of the diamond film, *i.e.*, the ratio of sp^3 to sp^2 carbon bonds, was evaluated by Raman spectroscopy with an excitation wavelength of 325 nm. The Raman spectra of the grown samples are dominated by a sharp diamond peak at a frequency of 1332 cm^{-1} (sp^3 carbon bonds) and a broad band centered at a frequency of ~1580 cm^{-1} (G-band, graphitic sp^2 carbon phases) (Figure 13.11). Surprisingly, nearly no sp^2 amorphous carbon bands were detected for the diamond film deposited at 250 °C (sample c). This indicates that such an ultra-thin diamond layer is of good quality.

Activation energies (E_a), as calculated from an Arrhenius plot, reveal two different growth regimes: at lower T_S the activation energy is 1.6 to 2 kcal mol^{-1} and at higher T_S it is ~7.8 kcal mol^{-1} (Figure 13.12). The threshold temperature between these two regimes is ~550 °C. Two different growth regimes were also observed by Snail *et al.* in the oxygen–acetylene flame system[67] and Yamaguchi *et al.* in the HFCVD system.[92] These results suggest

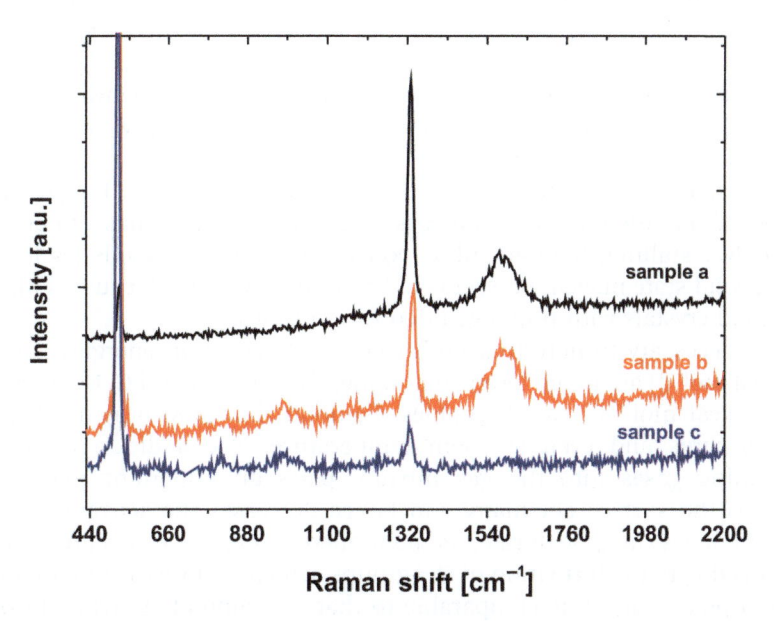

Figure 13.11 Raman spectra of diamond films deposited at different microwave powers and substrate temperatures.

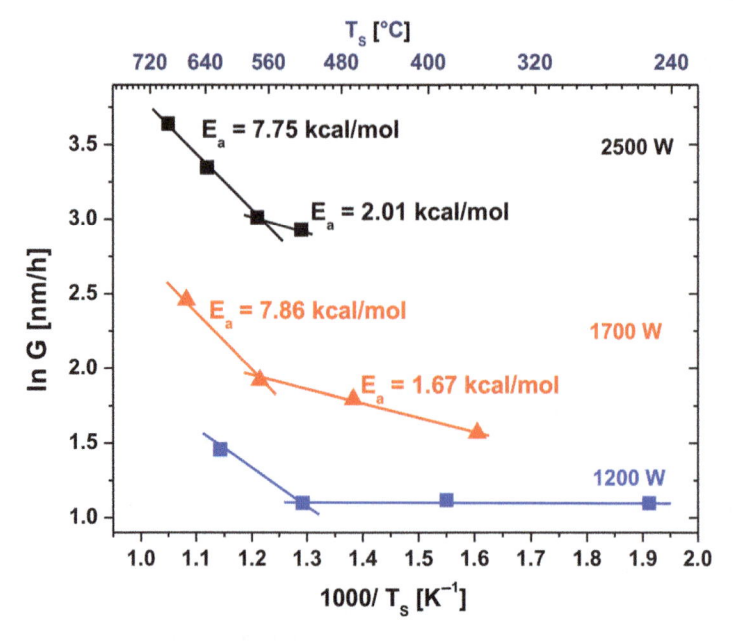

Figure 13.12 Arrhenius plots and calculated activation energies of diamond films deposited at different MW powers and substrate temperatures (with permission of *Physica Status Solidi B*).[90]

that diamond growth is not a process characterized by a single rate-determining step. We propose that two different growth mechanisms are present: active growth species and the dominance of specific crystal face development, or a combination of both these factors.[67] The activation energy at low T_S (1.6 kcal mol^{-1}) has a good agreement with Tsugawa et al.,[72] who grew NCD films at 100 °C. However, in their case, only small diamond nanocrystals were grown with an average diameter of 5.5 nm. Tsugawa et al. supposed that the nucleation process occurs from the gas phase. In our case the polycrystalline diamond films consisted of larger crystals (~50 nm). Thus, solid state nucleation seems to be dominating, which causes enlarged diamond crystals with prolonged deposition times.

It is necessary to note that both calculated activation energies (1.6 and 7.8 kcal mol^{-1}) are much lower than values for the standard C–H gas system (22–30 kcal mol^{-1}). Low E_a enables diamond film growth at much lower temperatures and even at room temperature. The addition of oxygen-containing gases into the gas mixture plays an important role, which changes the surface chemistry and hence decreases the activation energy and enhances the growth rate. As mentioned above, one explanation of this effect is due to the formation of OH groups, which remove the non-diamond carbon phases at a rate comparable to that of diamond growth.[93] In other words, oxygen at lower temperatures has a stronger preferential etching behavior than hydrogen. Another role of atomic oxygen is the effective

Low Temperature Diamond Growth 315

abstraction of surface hydrogen atoms.[41] In our case the low E_a is related to the presence of oxygen-containing gases (CO_2) and the implementation of a specific linear antenna microwave plasma system.

The present results confirm that the lowest temperature that can be achieved in our PLAMWP system (by addition of CO_2) is about 250 °C. This temperature is limited by ineffective cooling of the substrate holder. A further decrease of the microwave power is not possible due to a) instability of plasma for MW powers <1200 W and b) extremely low plasma densities, which make diamond growth nearly impossible. The dependence of the substrate temperature on the MW power and heater temperature (which is controlled by external ohmic heating) is shown in Figure 13.13. The grey area represents a situation when the external ohmic heating of the substrate holder is turned off. It should be noted that increasing the MW power shifts the lower limit of the temperature when the substrate heating is switched off to higher temperature values, starting from 108 °C (at 1200 W) up to 182 °C (for 2500 W MW power). We suppose that even lower deposition temperatures, such as 250 °C, can be achieved by re-designing the water cooling mechanism of the substrate holder stage.

At certain distances from the dielectric material (often a quartz tube), the surface wave plasma is much colder as has been shown by Tsugawa *et al.*[94] In their study the electron density n_e and electron temperature T_e as a function of the distance from the antenna and process pressure were measured by a Langmuir probe. Their findings indicated that the surface wave plasma

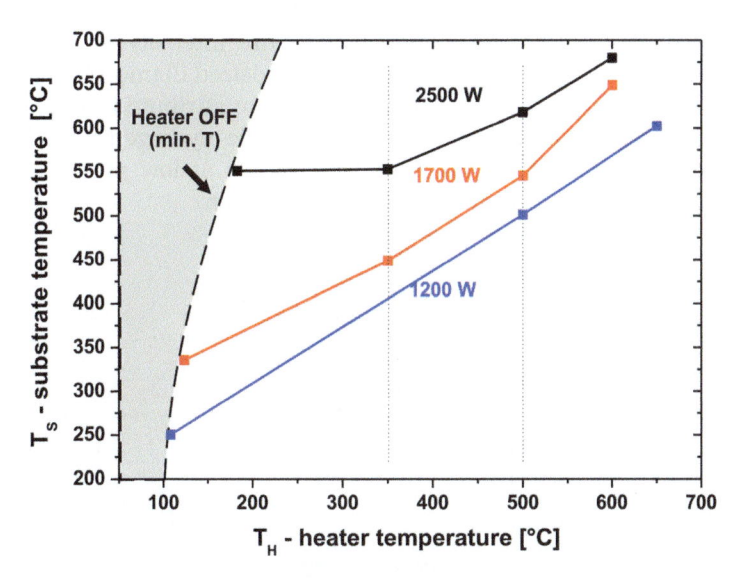

Figure 13.13 Effect of process conditions (microwave power and substrate heater ON/OFF) on the substrate temperature. Note: the heater temperature was measured by a thermocouple and the substrate temperature by IR pyrometry.

system does not penetrate far from the antenna and the microwave radiation heats and excites just the *top surface* of the plasma due to its high electron density near the plasma-generation region. The bulk of the plasma is not heated and excited by microwaves. In other words, this plasma region is *cold* and has a *low electron temperature* (<3 eV).

It should be noted that plasma is ignited at low pressure (~10 Pa) for surface wave plasma systems. Thus, the LTDG process should be reliable because, at lower pressures, the ion (T_i) and neutral gas (T_g) temperatures are approximately room temperature, $T_i \approx T_g \approx 300$ K,[94] while $T_e \approx 10^4$ K. This is due to the small kinetic energy transfer in elastic collisions between electrons and heavy particles.

For comparison, focused MW plasma is much hotter. In conventional MW plasma CVD processes, microwaves penetrate the plasma bulk, and heat and excite its whole body, which results in a high electron temperature (~5–10 eV).[95] In a ball-shaped microwave plasma the temperature of the gas can be ~3000 K within the plasma ball, and in the HFCVD process it can be ~2700 K near the filament.[94]

Our recent study showed that pressure plays a crucial role in diamond deposition.[2] Lowering of the process pressure in the PLAMWP system from 200 Pa to 6 Pa changed the growth kinetics and improved the diamond film morphology from a nano- to poly-crystalline character. The observed dependence was explained by a change of the plasma chemistry near the substrate surface. At low pressures (<10 Pa), the plasma volume expands towards the substrate, and at high pressures (≥100 Pa) it is localized near the quartz tubes (Figure 13.14). It was also found that the concentration of atomic hydrogen, indirectly deduced from the intensity of the H_α line, significantly depends on the gas pressure. Nanosized diamond crystals were formed at high pressures (≥100 Pa), which were attributed to the high re-nucleation rate at low concentrations of atomic hydrogen. On the other hand, polycrystalline diamond films were grown at low pressures due to adequately high concentrations of atomic hydrogen.

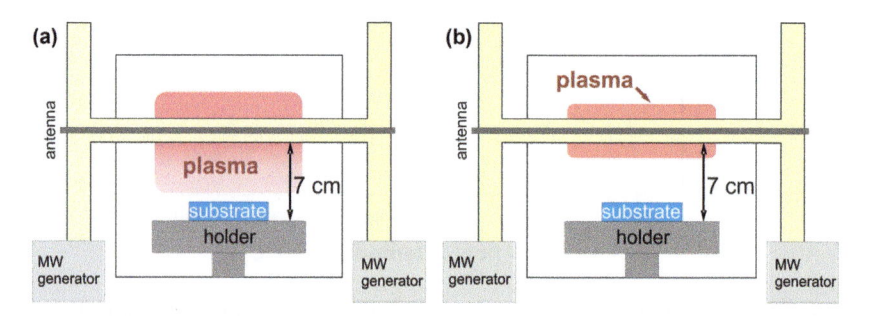

Figure 13.14 Schematic view of the plasma volume in a pulsed linear antenna MP CVD system at a) low (<10 Pa) and b) high (>100 Pa) process pressures.

13.6 Diamond Growth on Amorphous Silicon

The next potentially interesting application of LTDG is a combination of diamond nanocrystals with amorphous or crystalline silicon. Differences between low and high temperature diamond growth by linear antenna microwave plasma CVD processes can be compared (*e.g.*, the recrystallization of amorphous silicon at high temperatures).

So far, research in this field has been focused on boron-doped monocrystalline (BDD) or polycrystalline diamond (PCD) for heterojunction fabrication with a-Si:H[96–98] and SOD technology for heat dissipation.[99] Substituting BDD or PCD for intrinsic nanocrystalline diamond could generally reduce fabrication costs and enable large-area processing. In addition, implementation of nanocrystals may open novel prospects for sensing, biological or optoelectronic applications, where isolated crystals play a fundamental role.[100,101]

13.6.1 Nucleation and Growth of Diamond on a-Si

Untreated (*i.e.*, not seeded or nucleated) a-Si:H samples have been used as substrates for diamond nanocrystal growth.[102] In our study they were preferred because we focused on spontaneous nucleation (*i.e.*, the formation of individual crystals) rather than the growth of continuous diamond films.

Experiments were performed in a linear antenna plasma (PLAMWP) system at different substrate temperatures (250–750 °C) for 24 h. The pressure was 10 and 100 Pa and the $CH_4/H_2/CO_2$ flow rates were 5/200/20 sccm, respectively. The nucleation density, as well as the nanocrystal size and geometry, were evaluated by scanning electron microscope (SEM) measurements. Micro-Raman spectroscopy was used to detect the diamond and determine the phase (crystalline or amorphous) of the underlying silicon layer after diamond deposition. Atomic force microscope (AFM)/Kelvin probe force microscope (KFM) measurements were implemented for further microscopic evaluation of the nanocrystals that were undetectable by micro-Raman.

Figure 13.15a shows the SEM micrograph of individual diamond crystals deposited at 250 °C, 10 Pa. This low growth temperature was used to avoid

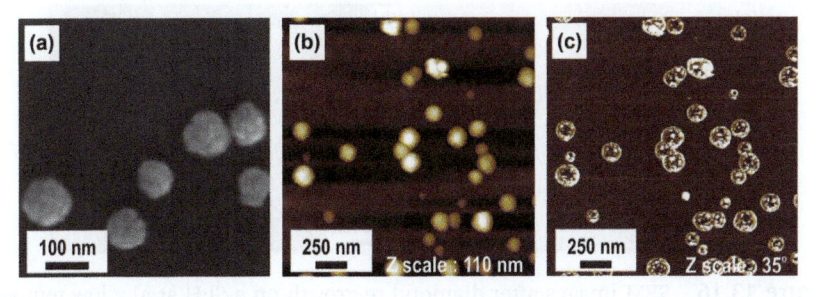

Figure 13.15 Diamonds grown on a-Si:H at 250 °C, 10 Pa: a) SEM micrograph, b) AFM topography and c) phase for the same sample.

metal-induced solid-phase crystallization. The crystals are, essentially, 100 nm (or smaller) clusters composed of several smaller nanocrystallites. Micro-Raman was unable to detect the diamond or graphitic phase of such samples. However, the simultaneous AFM topography/phase measurements (shown in Figure 13.15b and c) indicate that the deposited nanocrystals exhibit phase contrasts different from the substrate. This means that the substrate and the nanocrystals are essentially different materials.

Micro-Raman was unable to provide direct evidence of the diamond phase most likely due to resolution limitations: an isolated nanocrystal is four times (or more) smaller than the available micro-Raman focus (wavelength 442 nm, spot size ∼400 nm), meaning that it occupies, at best, only 6% of the probed spot size. In addition, the nanocrystals are on a silicon substrate 200–400 nm thick, while their height is around 100 nm (from AFM topography). Given that the Raman signal is obtained by the full sample thickness (theoretical characterization limit is ∼1 μm) and assuming that the signal is collected from a cylinder with a radius of the Raman laser and height of the film thickness, a diamond nanocrystal comprises, at best, only 1.5–3% of the total volume.

Keeping in mind the abovementioned suggestion, the samples with sub-100 nm nanocrystals were used as substrates for (re-)deposition of diamond in order to enlarge the crystals up to sizes detectable by Raman measurements. Thus, if these non-identified nano-sized features were diamond, they would already act as nucleation sites. The deposition parameters were 350 °C at 10 Pa and 750 °C at 100 Pa. The results of these secondary diamond depositions are shown in Figure 13.16.

In both cases the nanocrystal enlargement of the initial deposited features was successful. The crystals deposited with a pressure of 10 Pa appear irregular with fine, "cauliflower-like" structures. Their 100 Pa counterparts exhibit more homogeneous, spherical structures. Note that the geometric differences are mostly attributed to the process pressure[91] and not the temperature.[2] The size of each crystal (or cluster) is around 400 nm. Despite their larger size, these crystals are also comprised of small nanocrystallites.

Figure 13.17 shows micro-Raman measurements conducted on the samples illustrated in Figure 13.16. Both samples exhibit peaks, indicating

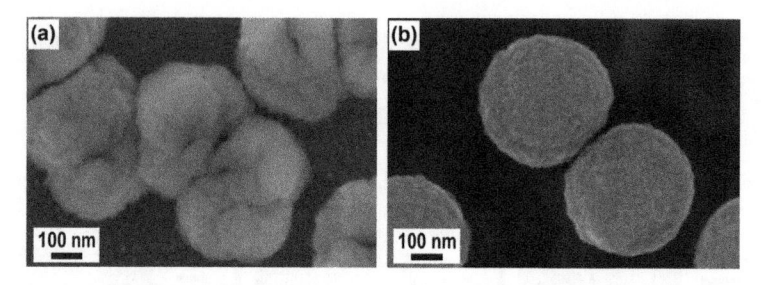

Figure 13.16 SEM images after diamond re-growth on a-Si:H at a) a low temperature (350 °C) and pressure 10 Pa, and b) a high temperature (750 °C) and pressure 100 Pa.

Low Temperature Diamond Growth 319

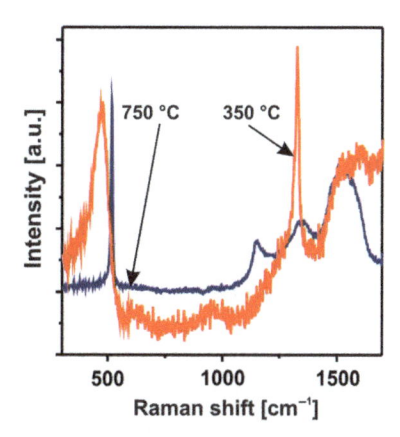

Figure 13.17 Raman spectra measured on the samples shown in Figure 13.16.

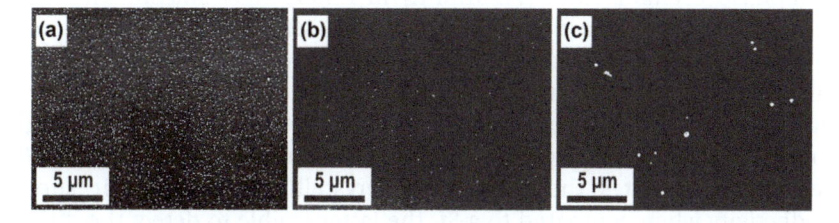

Figure 13.18 SEM images of diamonds grown on untreated a-Si after the CVD process at a) 250 °C, b) 350 °C and c) 750 °C.

the presence of diamond and non-diamond carbon phases. Nevertheless, the sample deposited at a lower temperature exhibits a sharper sp^3 peak in agreement with previous results.[103] It is noteworthy that, at the high temperature of 750 °C, the amorphous silicon substrate crystallized fully. On the contrary, deposition at a low temperature of 350 °C preserved the amorphous nature of the film, which indicates that the initial deposition at 250 °C also left the substrate amorphous.

The influence of the temperature on the spontaneous nucleation and growth of diamond on pristine amorphous silicon substrates was also studied. For this, we used untreated a-Si samples (*i.e.*, without the prior deposition at 250 °C as before) as substrates. The deposition temperature was 250 °C, 350 °C and 750 °C. The pressure was kept constant at 10 Pa for all three individual depositions.

SEM micrographs of the grown diamond crystals are shown in Figure 13.18. We can see that the density of the diamond crystals declines as the temperature rises. This is explained by kinetics at the early stage of diamond growth. Since the growth here is merely spontaneous (clean substrates before CVD), there are no nucleation sites for the carbon atoms to bond to. Elevated temperatures cause the etching agents in the reaction (atomic H and also O from CO_2) to be more active. This has a pronounced impact in the very first nucleation steps as the newly deposited carbon species are more likely to be etched away before additional carbon atoms have the chance to bond with them. This is clearly evidenced in Figure 13.18, where the highest density of surface features is observed at the lowest substrate temperature of 250 °C.

The findings in Figure 13.18 also show that the dense packing of grains observed in Figure 13.16 originate from the previous deposition at 250 °C and are not due to the adjusted deposition conditions. It is obvious that carbon atoms are more strongly bound to each other than to the substrate during CVD (Volmer–Weber island growth).[104]

13.6.2 Selective Growth of Diamond Nanocrystals

With all the above in mind, we used amorphous silicon samples for selective diamond deposition in templates created by field-enhanced, metal-induced solid-phase crystallization (FE-MISPC) in amorphous silicon. The experimental setup can be seen in Figure 13.19. FE-MISPC leads to patterning of a-Si:H. These patterns can be either crystalline (due to substrate phase transitions) or exhibit more deformations than the amorphous form. Evaluation of the process outcome is typically performed by conductive atomic force microscopy (C-AFM) and micro-Raman measurements. While the former indicates the presence of crystallites due to their superior conductivity when compared to a-Si, the latter is able to detect the different Si phases (given that the size of the features is within the detection range). For more detailed information about the process, see refs 105–107.

We have already demonstrated the selective growth of silicon nanocrystals in such patterns.[108] Here, we performed the same procedure for diamond growth. After processing of the amorphous silicon substrates, low temperature diamond CVD growth was performed at 250 °C, 10 Pa (the gas concentration was the same as in section 13.6.1) as this resulted in a better crystal size (~100 nm) and density of diamond crystals (Figure 13.18).

Figure 13.20a shows the topography of the FE-MISPC-patterned amorphous silicon sample after LTDG. Since it appeard similar to the corresponding topography before the growth, one would assume that no

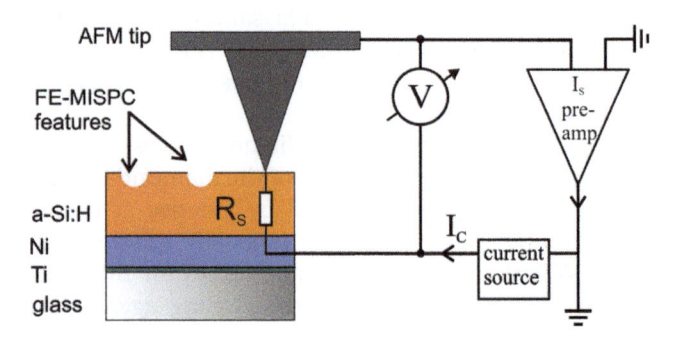

Figure 13.19 Experimental setup for FE-MISPC.

Low Temperature Diamond Growth 321

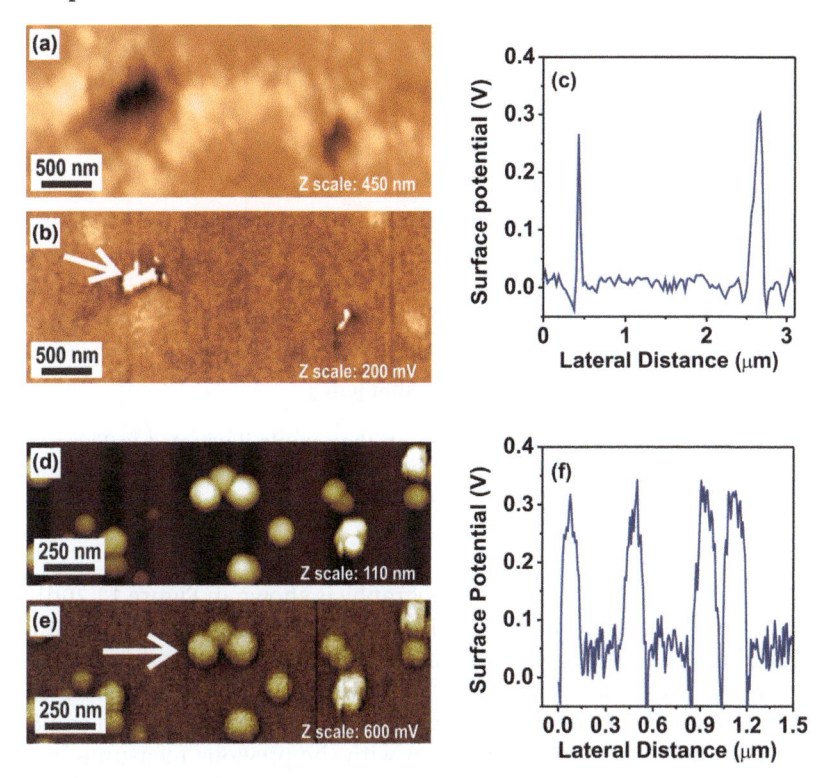

Figure 13.20 Diamond CVD on the FE-MISPC-templated a-Si:H sample: a) AFM topography, b) KFM and c) spatial profile measured on the KFM map, as indicated by the arrow. Results of diamond CVD on pristine a-Si:H: d) AFM topography, e) KFM and f) spatial profile measured on the KFM map as indicated by the arrow.

additional species were deposited. Nevertheless, AFM measurements far away from the FE-MISPC-induced features (*i.e.*, several tens of μm away) showed the deposition of diamond with a structure similar to what can be seen in Figure 13.15. In addition, the pronounced signal in the KFM image shown in Figure 13.20b indicates that there is some material accumulated inside the pits, which is different to the a-/μc-Si surroundings. The surface potential difference of that material *vs.* the surroundings is around 300 mV, as seen in the spatial profile indicated by the arrow in the KFM image, plotted in Figure 13.20c. Note that C-AFM measurements on such samples do not provide evidence for diamond growth as the differences in conductivity can occur for various reasons, such as the presence of graphite, diamond crystals or the local phase transitions of silicon due to the elevated deposition temperature. For this reason, the KFM evaluation after the diamond deposition is preferential.

A surface potential map of the diamond grains deposited at 250 °C on a clean a-Si sample (without FE-MISPC processing) can be seen in

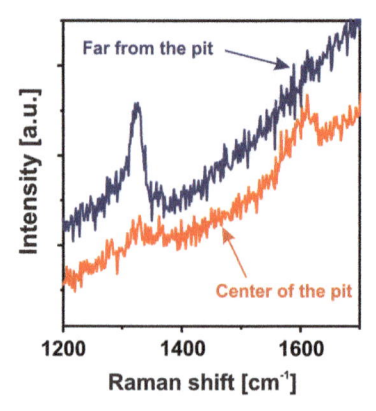

Figure 13.21 Raman spectra measured on the sample illustrated in Figure 13.20 a) and b). Note: spectra were measured after diamond re-growth to collect more resolvable signals.

Figure 13.20e (the corresponding topography is shown in Figure 13.20d). The corresponding spatial profile from this KFM image, shown in Figure 13.20f, indicates that the surface potential difference of the diamond grains *vs.* the substrate is also around 300 mV. This provides indirect evidence of deposited carbon species inside the pits. Nevertheless, micro-Raman could not detect a diamond phase, in agreement with the previous measurements on samples deposited at 250 °C. With that in mind, we performed an additional CVD at 250 °C in order to increase the diamond crystals to sizes detectable by Raman.

The micro-Raman spectra in Figure 13.21 were measured after the second diamond deposition. The red spectrum indicates the presence of diamond nanocrystals with pronounced graphitic content inside the larger pit. Measurements performed far from the FE-MISPC-induced features show a strong sp^3 bonded carbon signal, which corresponds to diamond (blue spectrum). This means that the mechanism of CVD deposition in FE-MISPC-induced features is different for silicon and diamond.

13.7 Diamond Growth on Glass

Diamond films on glass substrates have many useful applications. However, it is necessary to mention that various glasses exist that differ with regards to their bulk properties.[109] In the case of diamond deposition only some of these properties are important, particularly the thermal expansion coefficient, the softening point, the annealing point and the strain point (see Table 13.2 for selected glass substrates).

As we can notice, in most cases, the process temperature for diamond deposition has to be less than 600 °C if glass substrates are to be used. However, lowering the temperature results in a decreased growth rate.

Low Temperature Diamond Growth 323

Table 13.2 Material properties of selected glasses.

Material property	Soda lime	Schott AF 45	Corning EAGLE2000	UV-quality fused silica
CTE[a]	8.6×10^{-6} K^{-1}	4.5×10^{-6} K^{-1}	3.61×10^{-6} K^{-1}	0.57×10^{-6} K^{-1}
Strain point	514 °C	627 °C	666 °C	893 °C
Annealing point	546 °C	663 °C	722 °C	1042 °C
Softening point	726 °C	883 °C	985 °C	1585 °C

[a]Coefficient of thermal expansion.

Therefore, to achieve deposition of a fully closed layer, it is necessary to accordingly prolong the deposition time or increase the nucleation density. Moreover, optimization of the deposition process (for example the gas composition) is beneficial.[93,110] In the following, two deposition processes for diamond growth on glass substrates are compared, namely, focused and linear antenna microwave plasma CVD.

13.7.1 Diamond Growth Using Focused MW Plasma

We initially performed experiments with a Schott AF45 substrate glass slides (1 in × 3 in) and temperatures >600 °C to 370 °C. The growth was carried out in a focused microwave plasma CVD reactor (Aixtron P6).[111] The nucleation procedure of the glass substrates was realized ultrasonically using ultradispersed detonation diamond (UDD) powder solution. The standard diamond growth step was provided at a constant methane concentration (1% CH_4 in H_2). High temperatures (above 600 °C) were measured by a two-color pyrometer working at wavelengths of 1.35 and 1.55 µm (CHINO type) and low temperatures (below 600 °C) were measured by a two-color pyrometer working at wavelengths of 2.13 and 2.35 µm (Williamson type). Both pyrometers were found to be insensitive to the quartz bell jar.

We found that film adhesion was limited to a certain film thickness (400–600 nm) for our experimental setup. Thick NCD films (>600 nm) have tendency to delaminate spontaneously from the glass substrates due to the different thermal expansion coefficients of diamond and the glass substrate (see Figure 13.1).

Figure 13.22 shows a typical transmittance/reflectance spectrum of the NCD layer. The deposited NCD film is optically transparent in the broad spectral range from infrared to ultraviolet light (transmittance below 0.5 eV and above 3.5 eV is suppressed due to the glass substrate). The calculated refractive index of the NCD film is 2.34. It is interesting to note that the refractive index is not influenced by the substrate temperature. The surface scattering from the NCD layers increased with an increasing film thickness. An optical roughness (σ) of ~20–30 nm was recorded for 500 nm thick NCD films grown at high temperatures (>600 °C). This value decreased below 10 nm for ultra-thin NCD films typically grown at low substrate temperatures (<400 °C). The calculated values of the optical surface roughness were confirmed by AFM measurements.

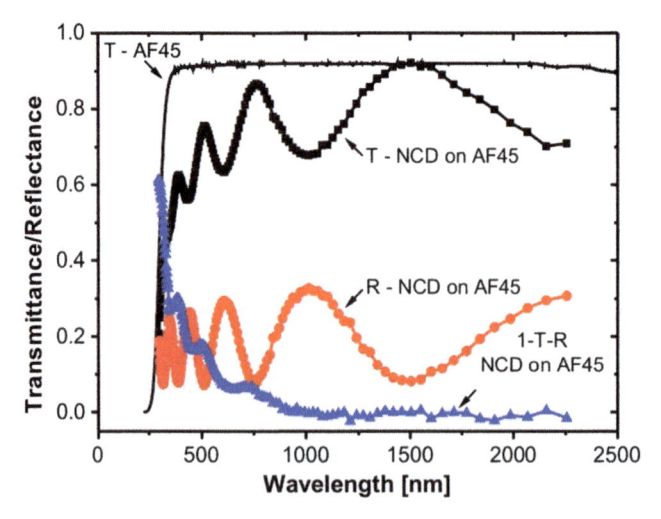

Figure 13.22 Transmittance and reflectance spectra of a 320 nm NCD film deposited on low alkaline borosilicate glass (Schott AF 45) as the substrate.

The calculated mass density was $\rho = 3.45$ g cm^{-3} (for $n = 2.34$) using the Clausius–Mossotti equation.[112] The slightly lower mass density of NCD compared to single crystal diamond ($\rho = 3.52$ g cm^{-3}) is probably due to a non-diamond content or voids between crystallites. Using FTIR reflectance–absorbance measurements, we estimated the amount of hydrogen bonded to sp^3 carbon. Generally, no measurable absorption was observed in the sp^3 CH stretching region.[113] The absorption (A) related to hydrogen bound to carbon was estimated to be less than 1% for 1 μm thick NCD films. Thus, the related absorption coefficient (α) is lower than $A/d < 100$ cm^{-1}. According to Jacob and Unger,[114] the total bound hydrogen content in the diamond layer can be roughly estimated as $N_H \approx \alpha \times 10^{19}$ cm^{-3}. Based on these facts, we concluded that the total content of bound hydrogen in our films is below 10^{21} cm^{-3} (*i.e.*, less than 1 at.%). Similar results were observed over the whole temperature range.

Based on these results and process optimizations, we were able to use soda lime glass substrates for the deposition of thin diamond films. The substrate temperature was maintained at ≤400 °C. Figure 13.23 shows the Raman spectra of the NCD films deposited at low temperatures on glass and, for comparison, also on Si substrates. We noticed that the Raman spectra of samples deposited at the high temperatures (>500 °C) exhibited peaks centered at 1332.5 cm^{-1}, which was identified as the characteristic line of diamond.[115] At low temperatures, we observed that the variation and intensity of non-diamond phases increased with decreasing substrate temperature. The quantitative percentage of the sp^3 hybridized carbon atoms in the analyzed volume of the NCD films were evaluated from the high resolution C 1s XPS spectra recorded at two emission angles, *i.e.*, 0° and 60° with respect to the surface normal.[116] The resulting percentage of

Low Temperature Diamond Growth 325

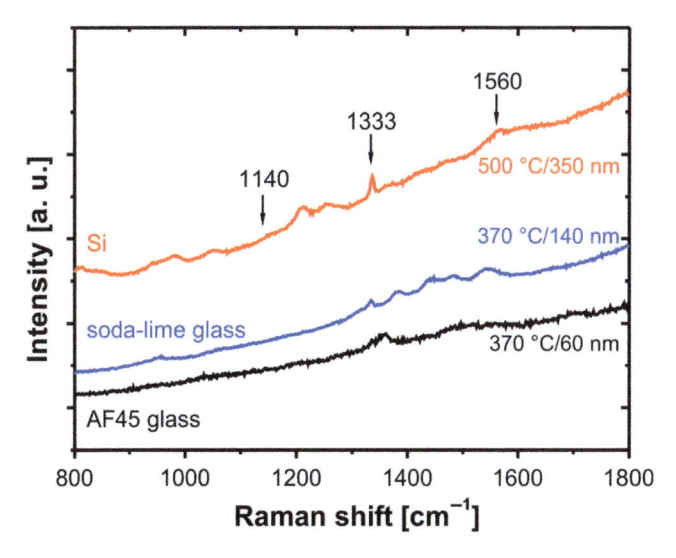

Figure 13.23 Raman spectra of NCD films deposited on low alkaline borosilicate glass (Schott AF 45), soda-lime glass and Si substrates. The corresponding deposition temperatures and final thicknesses of diamond films are also shown.

sp^3 hybridized carbon atoms was above 90%. Generally, this percentage was larger under the surface region (from 95.3% to 98.5%) than at the surface region (from 92.7% to 95.2%). No remarkable differences between the NCD films prepared in the whole temperature range were observed by XPS measurements.

Figure 13.24 shows the surface morphology of the NCD films. NCD deposited on glass exhibited a homogenous surface morphology. The deposition resulted in the growth of randomly oriented nanocrystals with sizes up to approximately 60 nm. The surface root-mean-square (RMS) roughness was less than 10 nm (measured by AFM). On the other hand, the NCD film deposited on silicon exhibited a relatively rough surface with crystal sizes from tens to hundreds nanometers and visible crystal faceting. The surface of this film exhibited some dark regions, which are most probably related to non-diamond carbon phases and is in good agreement with the Raman measurements. The observed difference in the surface morphology of the NCD films deposited on the glass and silicon substrates seems to originate from a different (BEN used for Si substrates) nucleation procedure.

The adhesion of the NCD films to the substrates was also compared. As expected, for Si substrates, very good adhesion was observed where no dependence on the growth temperature or the film thickness was registered. This was not the case for deposition on glass substrates. Due to the very different CTEs of the glass substrates (see Table 13.1) compared to diamond, the internal stress increases with the thickness of the NCD films. The 600 nm NCD film spontaneously delaminated from the glass substrate (Figure 13.1). The solution to this problem is to control the

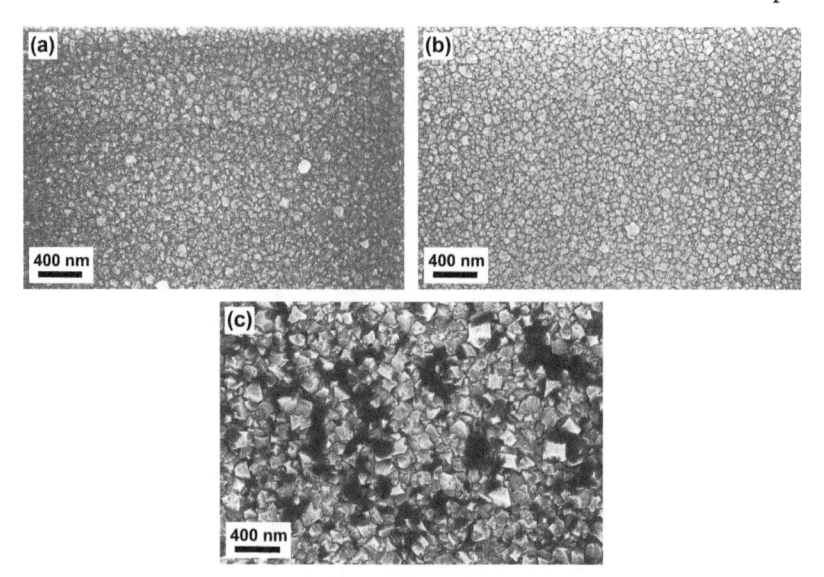

Figure 13.24 Surface morphology of NCD films grown at low substrate temperatures (370 °C) on a) low alkaline borosilicate glass (Schott AF 45), b) soda lime glass and c) a Si substrate.

temperature decrease to the strain-point temperature. Then, the temperature can be safely dropped to room temperature at the end of the growth process. This is usually very difficult in focused plasma systems, where heating of the substrate is mostly generated by plasma and only small variations in the temperature are available through variation of the process parameters. In the case of linear MW plasma the influence of plasma on the substrate temperature can be more easily suppressed.

13.7.2 Diamond Growth Using PLAMWP

Here, we present the successful growth of diamond on Schott AF 45 glass (1 in × 3 in) using pulsed linear antenna microwave plasma (PLAMWP). The deposition parameters were as follows: pressure 10 Pa, MW power 2 × 2.5 kW, substrate temperature 700 °C, deposition time 15 h, 2.5% CH_4 in H_2 gas and a varied concentration of CO_2 (0 to 80%).

Figure 13.25 shows optical images of all these samples, where samples (a) to (d) exhibit interference fringes corresponding to a certain thickness of the deposited diamond layer. Samples (e) and (f) do not reveal any interference fringes due to the higher etching rate of diamond than the deposition rate, mostly caused by the addition of oxygen-containing gases. In these cases no diamond growth was confirmed by SEM and Raman spectroscopy. On other hand, the deposited diamond layers (samples a–d) featured good adhesion without delamination on the glass substrates. We have to note another advantage of PLAMWP, which is its ability to deposit films over large areas

Low Temperature Diamond Growth 327

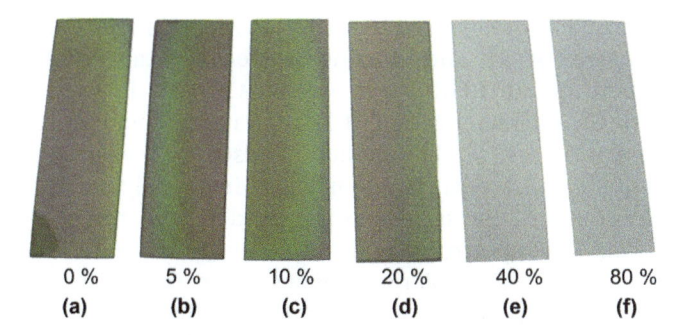

Figure 13.25 Optical images of diamond layers deposited on AF 45 glass substrates (1 in × 3 in) prepared with different CO_2 concentrations in CH_4/H_2 gas.

(in our case, 20 × 30 cm^2) with very good homogeneity. This is very important, for example, in life sciences, where a requirement for a large set of experimental substrates, preferably with a standard size of 1 in × 3 in, and high reproducibility is ordinary. Moreover, the possibility to covalently graft organic molecules to the diamond surfaces with excellent chemical stability of diamond has significantly widened its field of applications.

13.8 Diamond Growth on Optical Elements for IR Spectroscopy

Another challenge in LTDG is diamond deposition on optical elements or other substrates that are optically transparent to infrared (IR) light, which may have promising applications in IR spectroscopy. Such substrates used in IR spectroscopy are, for example, Ge, ZnSe, KBr, BaF_2 and ZnS. Most of them are thermally sensitive and/or reactive materials (in terms of diamond deposition conditions), therefore, they cannot be exposed to high temperature or aggressive plasma. Moreover, due to their optical applications, there is a greater need for the sufficient quality of the diamond layer (*i.e.*, high quality optical properties, smoothness, *etc.*), including suppression of any substrate degradation during the diamond growth process.

Traditionally, IR optical absorption spectra were measured in the transmission mode using optically transparent windows or cuvettes. When the optical spectra of *thin films* are measured, films need to be deposited on optically transparent substrates, *e.g.*, polished and slightly wedged intrinsic crystalline silicon wafers. On the other hand, the optical spectra of *powder sample* can be measured after grinding to a fine powder, which is then dispersed and pressed in a KBr pellet. However, this method suffers from inevitable reproducibility issues given the complexity of the sample preparation complicated by difficulties in obtaining sample that is homogenously dispersed throughout the KBr pellet. In addition, KBr is highly hydrophilic and drying of the KBr pellet *in situ* during spectrometry is required in order to avoid water-related bands in the final IR spectrum.[117]

The two most commonly used IR spectroscopy techniques are attenuated total reflectance (ATR) and grazing angle reflectance (GAR) Fourier transmittance infrared (FTIR) spectroscopy. *ATR-FTIR* is a well-established technique for the characterization of liquids, non-abrasive pastes, gels and deformable thin layers.[118] This method is based on measuring the optical absorption of the evanescent wave that occurs in a totally internally reflected infrared beam when the sample is attached to the prism. The second technique, *GAR-FTIR* spectroscopy, is based on measuring the reflectance–absorbance spectra of p-polarized light under the angle of incidence nearly parallel to a surface. It requires highly reflective mirror-like substrates, preferably with an aluminum or gold layer. We found that coating mirror-like substrates with a thin diamond layer not only protects the metallic surface, but also enhances the hydrophilicity of the substrate surface with other additional advantages.[30] Such advantages for both ATR and GAR methods using diamond-coated optical elements (Dc-O elements) offer a faster sample preparation than the transmission method and improve sample-to-sample reproducibility. Moreover, they offer other very important advantages, such as mechanical robustness, surface wettability that can be tuned for the study of proteins and other biomolecules[15] and inertness of the material to heat or gas treatments,[119] as well as the biocompatibility of diamond.[120] Diamond coating may also provide new insights with regards to diamond chemistry due to the possibility of *in situ* monitoring of any kind of reaction performed on surface-adsorbed nanoparticles or directly on diamond thin films. For instance, starting from hydrogenated layers, one may be able to monitor the grafting of alkenes under UV or diazonium chemistry,[12,121] as well as the intermediate species, which are usually hardly accessible with common techniques. Next, the diamond film promises to be an exceptional biointerface, hence Dc-O elements in combination with ATR can also be a valuable tool to validate and study the biochemical grafting, or even interactions, with biological species.

As mentioned in the previous section, the growth of diamond films at high temperatures (>600 °C) is an undesired technological step in the fabrication of Dc-O elements due to partial or complete damage of the metal coating. Decreasing this temperature in a standard microwave system is not a simple task as the main process principle is based on igniting a plasma ball close to the substrate (*i.e.*, approximately 1–2 mm). This arrangement results in a high thermal load on the substrate surface; therefore, low temperature deposition is needed.

Here, we report on the deposition of nanocrystalline diamond thin films on commercially available Al and Au optical mirrors, Si prisms and Ge substrates using two fundamentally different deposition techniques (focused and linear microwave plasma CVD). We show that, with an appropriate method and deposition parameters, one can achieve diamond films with high quality in terms of their optical properties, homogeneity and enhanced optical performance of the Dc-O elements.

Low Temperature Diamond Growth 329

13.8.1 Diamond Growth on Metallic Optical Mirrors for GAR-FTIR

As GAR-FTIR substrates, we used 22×22 mm^2 large and 5 mm thick $\lambda/4$ surface-protected Al mirrors on fine annealed pyrex glass (NT33-510, Edmund Optics Inc). The control sample was a 0.5 mm thick optically polished Si wafer. The substrates were first ultrasonically seeded with diamond nanoparticles,[122] then deposited with diamond using the focused (FMWP)[111,122] or linear antenna (PLAMWP) plasma CVD systems described above.[2,123] Figure 13.26 shows schematic drawings of both deposition systems and the optical images

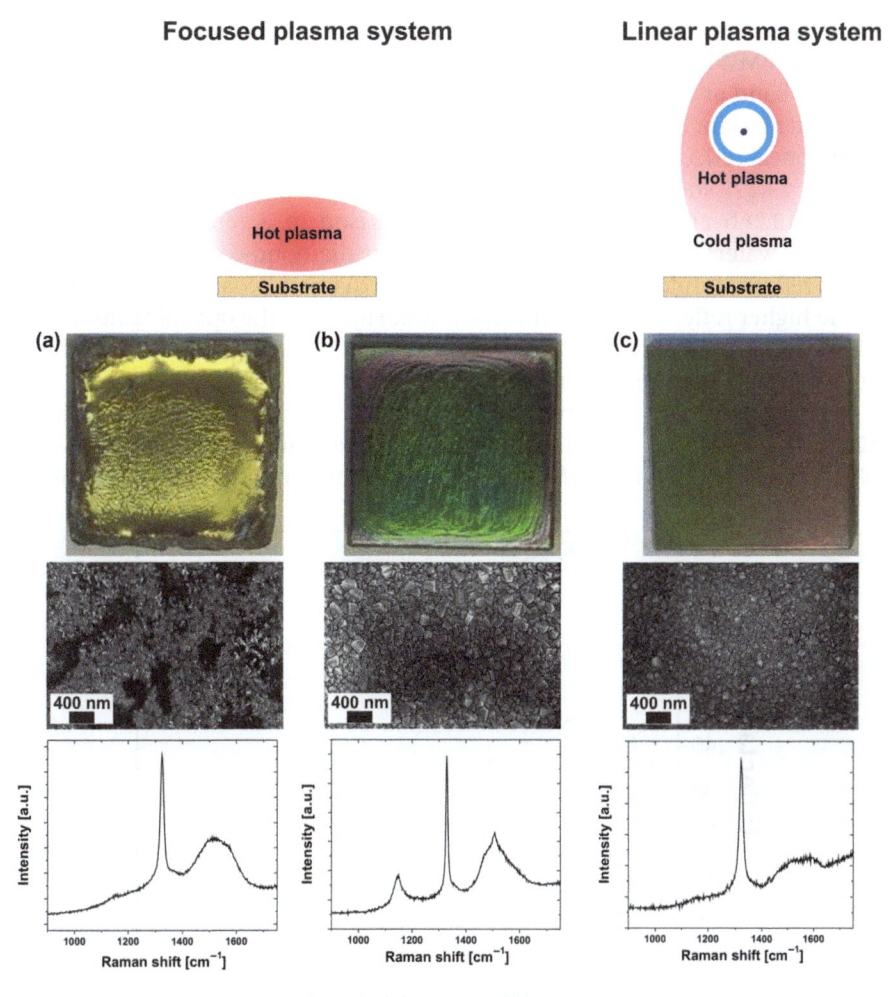

Figure 13.26 Comparison of Al mirror samples coated with diamond films using focused plasma at a) standard and b) optimized parameters, and c) using linear antenna plasma. Schematic drawings of the processes (top), optical images (second from top), SEM images (second from bottom) and Raman spectra (bottom) are given.

of the final samples, including the corresponding SEM images and Raman spectra. It is evident that applying the focused plasma system under standard deposition conditions (>800 °C) results in detrimental damage to Al mirror surface (Figure 13.26a). Lowering the deposition temperature (<550 °C) allows diamond film growth to a thickness of ~400 nm; however, noticeable changes (cracks) in the Al mirror are still optically observed (Figure 13.26b). In contrast, use of PLAMWP results in the growth of an optically uniform diamond layer (approx. 300 nm in thickness) without any damage of the mirror surface (Figure 13.26c).

The Raman spectra of the samples confirmed the high quality of the diamond films, especially that grown by PLAMWP (*i.e.*, sharp peak at 1332 cm^{-1}, which is attributed to the sp^3 diamond bonds). In the case of samples grown in FMWP, the broad graphite band (1500 cm^{-1}) had a higher intensity and the band at 1150 cm^{-1} is attributed to *trans*-polyacetylene-like structures. It is essential for the successful use of Dc-O elements in IR spectroscopy that high optical reflectance and low optical scattering in the IR region are present. Figure 13.27 shows the reflectance spectra of the samples shown in Figure 13.26b and Figure 13.26c in comparison to a reference (diamond-coated Si wafer). It is clearly confirmed that the diamond-coated Si wafer has a lower reflection (plot a) in Figure 13.27) than diamond-coated Al mirrors due to the higher reflection of the metal surface. However, the optical scattering of the sample grown in FMWP (plot b) in Figure 13.27) deteriorates the reflectance because of partially damaged substrates.

Beside Al mirrors, diamond growth on Au mirror substrates (3.2 mm thick, 12.7 mm in diameter, Thorlabs) was also studied. Similar to previous, it was observed that applying FMWP even at low temperatures resulted in damage

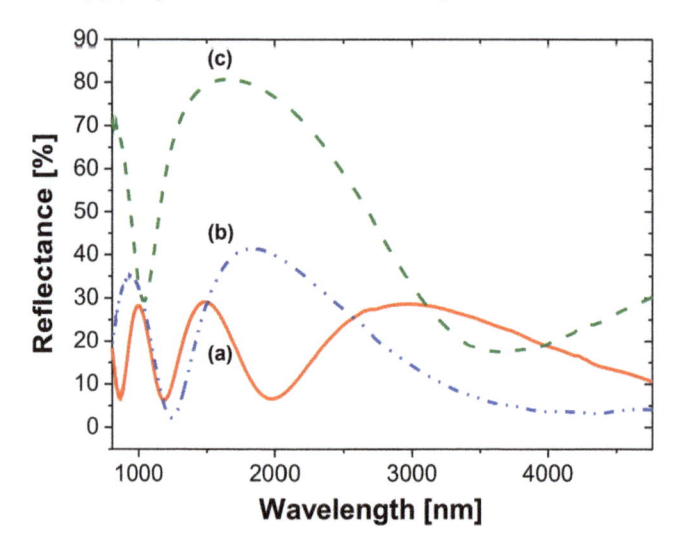

Figure 13.27 Infrared reflectance spectra of NCD films grown on a) polished Si substrate and b) Al mirror using focused plasma system, and c) on Al mirror using linear antenna system.

Low Temperature Diamond Growth 331

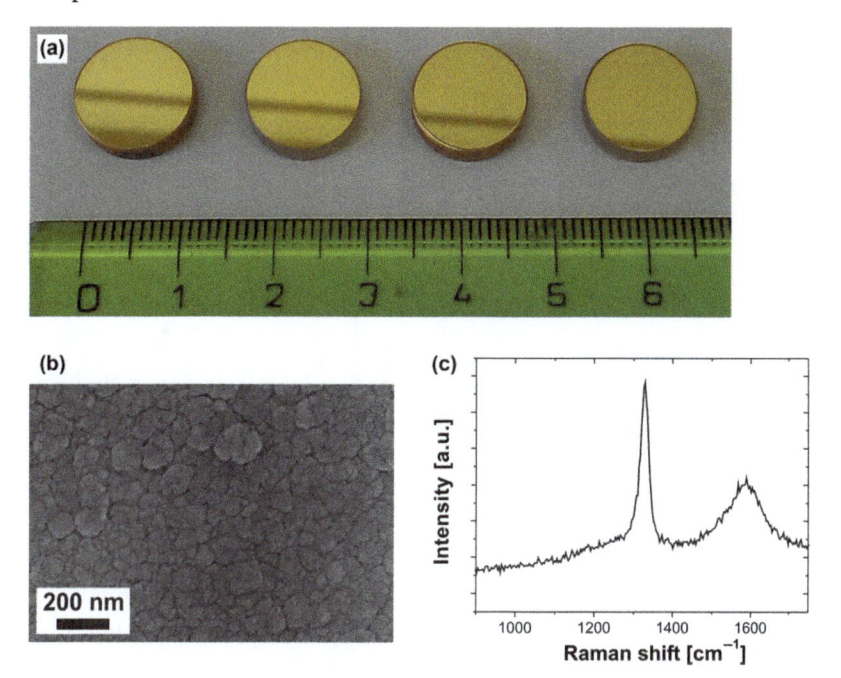

Figure 13.28 Au mirror substrates covered with a diamond film: a) optical image, b) corresponding SEM image, and c) Raman spectrum.

to the Au mirror surface. On the other hand, deposition with PLAMWP resulted in a high quality diamond film (thickness of 80 nm), as shown in Figure 13.28. Optical images revealed a homogeneous and optically transparent diamond layer with high reflectivity from the Au layer (Figure 13.28a).

13.8.2 Diamond Growth on Si Prism for ATR-FTIR Applications

As mentioned above, coating of Ge or Si ATR prisms with a diamond layer in combination with ATR-FTIR spectroscopy is a promising tool for biointerface studies. However, Si or Ge prisms are more complex 3D objects than the standardly used thin flat substrates, *e.g.*, the silicon 6-reflection ATR prism has a length 72 mm, width 10 mm and height 6 mm (Figure 13.29a). Therefore, FMWP is not a usable deposition technique and only the PLAMWP system can be applied. We successfully deposited homogeneous diamond films on such Si prisms using 0.5% CH_4 in H_2 gas using a PLAMWP system (Figure 13.29a). Before optical measurements were performed, it was oxidized in RF plasma to achieve a hydrophilic surface.

In our preliminary study diamond-coated ATR prisms were used as an active tool to validate and investigate the grafting of fetal bovine serum (FBS) to the NCD surface. In this study the diamond layer acts as a functional material, *i.e.*, a material with defined wetting properties due to oxygen termination and as an exceptional biointerface. The measured IR absorbance spectrum of FBS

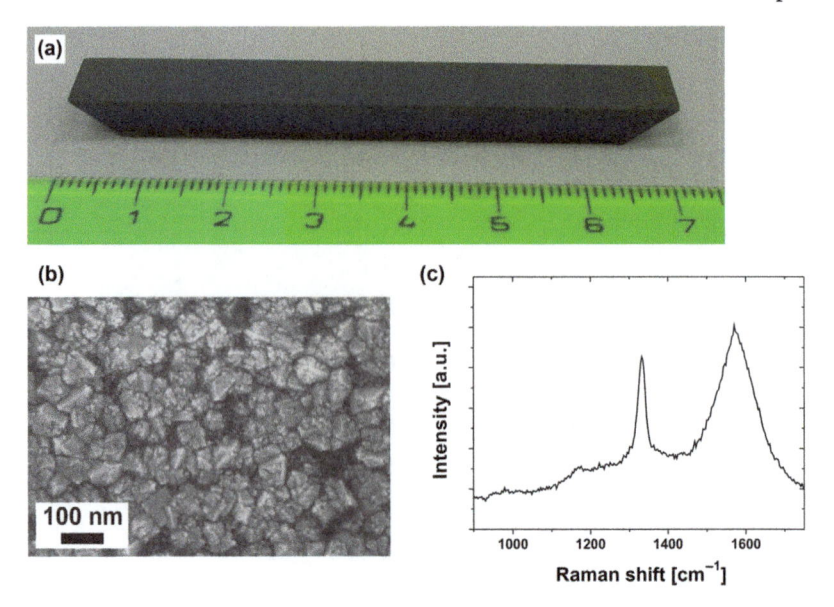

Figure 13.29 NCD-coated Si ATR prism: a) optical image, b) SEM image, and c) Raman spectrum.

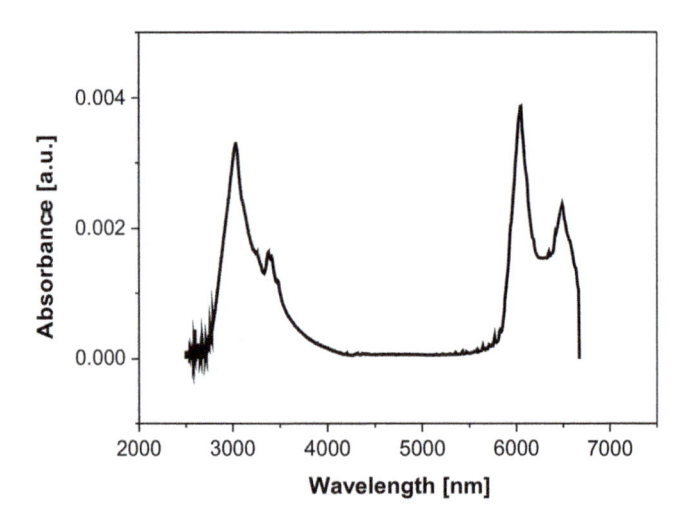

Figure 13.30 IR absorbance spectrum of FBS after 20 min adsorption on an oxidized diamond-coated Si prism.

adsorbed on an oxygen-terminated NCD surface is presented in Figure 13.30. The bands at 3030 nm and 3268 nm correspond to the N–H stretching of the proteins. The absorbance bands at 3378, 3409 and 3480 nm are related to the antisymmetric and symmetric stretching vibrations of CH_2 and CH_3 groups. The strong bands at 6053 and 6489 nm correspond to C=O stretching and N–H bending and the C–N stretching of proteins.

In addition, the functionality and suitability of the diamond-coated Si prism was also studied during characterization of the diamond nanoparticles with different surface modifications. As-received, air-annealed, plasma-oxidized and plasma-hydrogenated nanoparticles were distributed on the diamond-coated Si prism by drop-casting methanol dispersions. ATR-FTIR spectra of the diamond nanoparticles were measured in the 4000–1500 cm^{-1} spectral range.[124] The limiting point of this technique was the lack of information in the region below 1500 cm^{-1} because of absorption in the Si prism. To avoid this problem, we decided to use a diamond-coated Ge prism.

13.8.3 Diamond Growth on Ge Substrates

Germanium prisms allow IR spectroscopy measurements below 1500 cm^{-1}; however, the main disadvantages of applying unprotected Ge substrates are the low mechanical resistance, brittleness and difficulties with its chemical/mechanical cleaning for recycled measurements. As preliminary attempts, diamond-like carbon layers were applied as mechanically protective coatings. However, they reduced the transmittance properties of the germanium IR windows as the thickness required for effective protection was quite large.[125] On the other hand, diamond has almost no absorption in the IR region and possesses even better mechanical properties, *i.e.*, abrasion protection. Thus, coating diamond films onto the Ge substrates is advantageous. However, there are several limitations that prohibit the trivial deposition of diamond onto Ge, such as: (i) low melting point of Ge, which is close to the standard diamond deposition temperature, (ii) non-carbide forming material, (iii) the high thermal expansion mismatch between Ge and diamond, and (iv) the high out-diffusion constants of Ge at higher temperatures.

Under standard diamond growth conditions, these limitations become detrimental and finally result in (i) thermal damage of the Ge substrate, (ii) spontaneous delamination of the diamond layer due to low adhesion and high CTE mismatch or (iii) formation of voids due to out-diffusion of Ge atoms. To overcome these limitations, protective coatings (amorphous carbon, SiO_2 and Si_3N_4) were used to prevent Ge overheating and to improve adhesion.[126] However, cracking and spontaneous delamination of diamond limited use of diamond coatings.

Here, we present progress in LTDG on polished Ge substrates (10 × 10 mm^2 large, 0.6 mm thick) by employing a PLAMWP deposition system. SEM images of the Ge substrates before and after ultrasonic seeding are shown in Figure 13.31. In comparison to other pretreatment methods, ultrasonic seeding is gentle and does not damage the Ge substrate.

First, diamond films on Ge substrates were deposited using both PLAMWP and FMWP systems for comparison. As clearly observed from the SEM images in Figure 13.32a, focused MW plasma resulted in the formation of voids, even at low substrate temperatures of 400 °C. Similarly, just lowering the deposition temperature to 250 °C in the linear antenna MW plasma (Figure 13.32b) resulted in damage of the Ge substrate as well. However, in

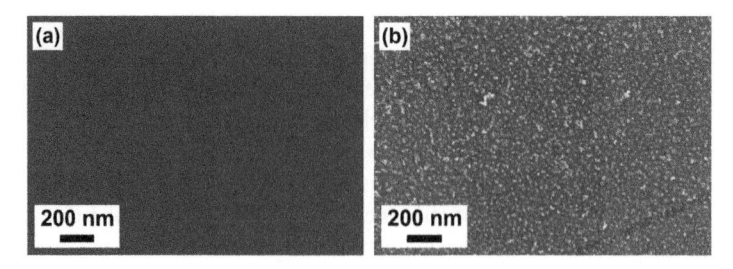

Figure 13.31 SEM images of a) a clean Ge surface and b) the GE surface nucleated by diamond powders.

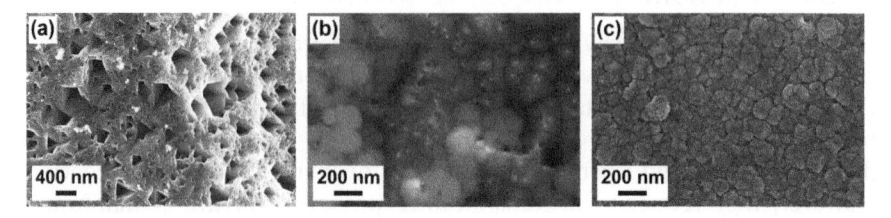

Figure 13.32 SEM images of the Ge surface after deposition using a) focused plasma and linear antenna plasma under b) standard and c) optimized process conditions.

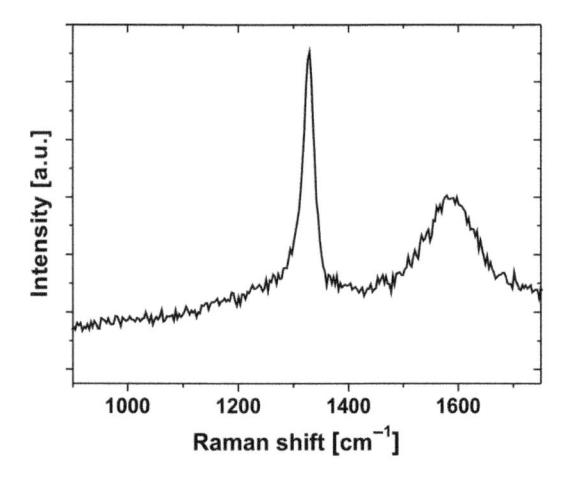

Figure 13.33 Characteristic Raman spectrum of an NCD film on a Ge substrate grown using optimized process conditions in the linear plasma system (excitation wavelength 325 nm).

this case, voids were not observed. Further optimizing the deposition process, mainly the gas mixture,[91] resulted in the successful deposition of diamond film (thickness of 100 nm) on the Ge substrates by PLAMWP (Figure 13.32c). The Raman spectrum confirms the diamond character of the film (Figure 13.33). These results indicate that coating of Ge ATR prisms with high quality diamond films is realistic and promising for perspective applications. This also shows the advantages of the linear antenna plasma system compared to the focused plasma system.

13.9 Conclusions

Diamond technology has reached a stage where optically and electronically high quality films can be grown at relatively low temperatures (<400 °C) and large areas (up to a square meter). Accurate measurements of the substrate temperature and evaluation of growth rates are crucial for fundamental understanding of the growth kinetics and chemistries involved. In harsh and aggressive diamond growth conditions, non-contact optical techniques are commonly used for determination of the substrate temperature. Nevertheless, one has to take into account specific substrate emissivity, plasma shadowing or interference with the substrate emission, as well as low emission from substrate near room temperature, *etc.*

In particular, during early research, most of the LTDG experiments were devoted to testing new gas mixtures, mainly containing halogens and oxygen gas species. Gas-phase and surface reactions involved in diamond growth from chlorine and fluorine precursors differ. These chemistries and growth models still remain partially open and specific details are under discussion. Nevertheless, the higher production of atomic hydrogen and more effective etching of non-diamond phases seem to be common features to Cl and F gas species. Both these features promote diamond growth at low temperatures. It is also assumed that both these halogens not only promote growth but also enhance the nucleation density in the early stages of growth. However, these halogens can cause significant corrosive damage to the deposition system (etching of O-rings, damaging vacuum pumps, *etc.*). Therefore, they are less favorable for industrial LTDG.

Using oxygen as the gas precursor, either as pure O_2 or in the form of oxides or alcohols (CO_2, CO, CH_3OH, CH_2COOH, *etc.*), for LTDG seems to be a more promising solution. Adding oxygen-containing molecules to the $CH_4:H_2$ gas mixture results in an increased growth rate and improved diamond film quality. Oxygen provides several key functions in LTDG: i) it increases the concentration of atomic hydrogen, ii) it leads to faster etching of non-diamond phases, iii) it produces highly reactive -OH radicals (*i.e.*, an additional effective etchant for non-diamond phases), and iv) it influences the electron temperature/density in the plasma. Still, there are some drawbacks with regards to oxygen-containing gas mixtures: i) oxygen in combination with hydrogen is explosive, ii) in hot wire CVD systems it leads to wire burning, iii) it can introduce defects into diamond and iv) hydrogen-induced surface conductivity of such diamond grains is an open issue. Nevertheless, just recently we showed that even such films are fully functional for surface-conductive field-effect transistors.

Modification of deposition systems has been shown as another route for LTDG. Recently introduced microwave surface wave plasma reactors, either with linear antennas or slots, bring diamond technology closer to industrial applications due to their scalability to square meters. In such deposition systems plasma in the sample vicinity is cold due to the larger distance (6–10 cm) of the substrate surface from the hot plasma region (*i.e.*, the region at the quartz tube). A broad range of diamond deposition parameters, for example, the process pressure, microwave pulses, gas composition and the distance of the substrate from the linear antenna(s), opens new fields for fundamental studies of diamond nucleation and growth. Phenomena, such as spontaneous nucleation, the controlled nano- or poly-crystalline character of grains or even the bottom-up growth of porous diamond layers, can be achieved.

As practical examples, we showed that the pulsed linear antenna microwave plasma system can control spontaneous nucleation of diamond nanocrystals on a-Si:H at temperatures as low as 250 °C. The employed LTDG preserved the amorphous silicon substrate and resulted in the formation of sub-100 nm grains. Even targeted localized growth of diamond nanocrystals inside the nanoscale pits in the a-Si:H matrix was achieved under appropriate conditions. We assume that diamond nanocrystals prefer to grow in these pits as energetically favorable sites. Further work is still needed to elucidate the mechanism of diamond growth in such features, for instance, by further experimentation with the diamond deposition parameters (CH_4 dilution, pressure, temperature).

We also showed that LTDG is favorable for growing diamond thin films on materials used as optical elements. The diamond films with high uniformity were grown on glass, Al and Au mirrors, Si ATR prisms and Ge substrates. These diamond-coated optical elements were successfully used in GAR and ATR FTIR spectroscopies, where they provided advantages in terms of their sensitivity, repeatability, interface properties, *etc.* Advantages of the linear antenna plasma system compared to the more common focused plasma system were clearly demonstrated.

For broad industrial uses, the generally low deposition rate (below 50 nm h^{-1}) of LTDG in the linear antenna plasma system may be a limiting factor. Simulation of chemical reactions for various gas mixtures, re-design of the process chamber, coupling of microwave radiation and increasing the microwave power densities can lead to further improvement of the growth rate and, thus, make this technology more attractive. Nevertheless, we have demonstrated that, already now, the low temperature growth of high quality diamond films or diamond nanocrystals is not only achievable but also practically applicable. Further progress in LTDG will enable broader expansion of diamond films in new fields, such as the semiconductor industry, optics, spinotronics, life sciences, and others.

Acknowledgements

We would like to gratefully thank Karel Hruska and Jitka Libertinova for SEM, Milan Vanecek and Vit Jirasek for valuable discussions, Jiri Potmesil and Ondrej Rezek for technical assistance and Zdenka Polackova for chemical treatments. There were also small contributions from undergraduate and PhD. students in our group. This work was supported by the Academy of Sciences of the Czech Republic by J. E. Purkyne Fellowship, and by the P108/12/G108 grant (GACR Excellence Centre). This work was carried out in the frame of the LNSM infrastructure.

References

1. A. Kromka, B. Rezek, Z. Remes, M. Michalka, M. Ledinsky, J. Zemek, J. Potmesil and M. Vanecek, *Chemical Vapor Deposition*, 2008, **14**, 181–186.
2. A. Kromka, O. Babchenko, T. Izak, K. Hruska and B. Rezek, *Vacuum*, 2012, **86**, 776–779.
3. K. Tsugawa, M. Ishihara, J. Kim, M. Hasegawa and Y. Koga, *New Diamond & Frontier Carbon Technology*, 2006, **16**, 337–346.
4. P. Hubík, J. J. Mareš, H. Kozak, A. Kromka, B. Rezek, J. Krištofik and D. Kindl, *Diamond and Related Materials*, 2012, **24**, 63–68.
5. B. Rezek, M. Krátká, A. Kromka and M. Kalbacova, *Biosensors and Bioelectronics*, 2010, **26**, 1307–1312.
6. P. W. May, *Philosophical Transactions of the Royal Society A: Mathematical, Physical and Engineering Sciences*, 2000, **358**, 473–495.
7. G. Chandler and J. Zimmer, *IEEE*, 2008, 101–102.
8. H. Kato, K. Oyama, T. Makino, M. Ogura, D. Takeuchi and S. Yamasaki, *Diamond and Related Materials*, 2012, **27–28**, 19–22.
9. D. Takeuchi, T. Makino, H. Kato, M. Ogura, H. Okushi, H. Ohashi and S. Yamasaki, *Japanese Journal of Applied Physics*, 2012, **51**, 090113.
10. L. Grausova, A. Kromka, Z. Burdikova, A. Eckhardt, B. Rezek, J. Vacik, K. Haenen, V. Lisa and L. Bacakova, *PLoS ONE*, 2011, **6**, e20943.
11. W. Yang, O. Auciello, J. E. Butler, W. Cai, J. A. Carlisle, J. E. Gerbi, D. M. Gruen, T. Knickerbocker, T. L. Lasseter, J. N. Russell, L. M. Smith and R. J. Hamers, *Nat Mater*, 2002, **1**, 253–257.
12. B. Rezek, D. Shin, H. Uetsuka and C. E. Nebel, *Physica Status Solidi (a)*, 2007, **204**, 2888–2897.
13. A. Hartl, E. Schmich, J. A. Garrido, J. Hernando, S. C. R. Catharino, S. Walter, P. Feulner, A. Kromka, D. Steinmuller and M. Stutzmann, *Nat Mater*, 2004, **3**, 736–742.
14. M. Kalbacova, L. Michalikova, V. Baresova, A. Kromka, B. Rezek and S. Kmoch, *Physica Status Solidi (b)*, 2008, **245**, 2124–2127.
15. B. Rezek, L. Michalíková, E. Ukraintsev, A. Kromka and M. Kalbacova, *Sensors*, 2009, **9**, 3549–3562.
16. M. Dankerl, S. Eick, B. Hofmann, M. Hauf, S. Ingebrandt, A. Offenhäusser, M. Stutzmann and J. A. Garrido, *Advanced Functional Materials*, 2009, **19**, 2915–2923.
17. H. Kawarada, *Surface Science Reports*, 1996, **26**, 205–206.
18. F. Maier, M. Riedel, B. Mantel, J. Ristein and L. Ley, *Phys. Rev. Lett.*, 2000, **85**, 3472.

19. V. Chakrapani, J. C. Angus, A. B. Anderson, S. D. Wolter, B. R. Stoner and G. U. Sumanasekera, *Science*, 2007, **318**, 1424–1430.
20. B. Rezek, D. Shin, H. Watanabe and C. E. Nebel, *Sensors and Actuators B: Chemical*, 2007, **122**, 596–599.
21. K.-S. Song, T. Hiraki, H. Umezawa and H. Kawarada, *Applied Physics Letters*, 2007, **90**, 063901.
22. M. Kalbacova, B. Rezek, V. Baresova, C. Wolf-Brandstetter and A. Kromka, *Acta Biomaterialia*, 2009, **5**, 3076–3085.
23. B. V. Spitsyn, L. L. Bouilov and B. V. Derjaguin, *Journal of Crystal Growth*, 1981, **52**, 219–226.
24. D. Das and R. N. Singh, *International Materials Reviews*, 2007, **52**, 29–64.
25. H. Liu and D. S. Dandy, *Diamond Chemical Vapor Deposition Nucleation and Early Growth Stages*, Noyes Publications, Park Ridge, N.J., 1995.
26. A. Kromka, O. Babchenko, T. Izak, S. Potocky, M. Varga, B. Rezek, A. Sveshnikov and P. Demo, Diamond nucleation and seeding techniques for tissue regeneration, in *Diamond-based Materials for Biomedical Applications*, ed. R. Narayan, Woodhead Publishing Ltd, 2009.
27. International Conference of New Diamond Science and Technology, M. Yoshikawa, O. Fukunaga, S. Saito, and Japan New Diamond Forum (JNDF), *Science and technology of new diamond: proceedings of the 1st international conference of new diamond science and technology*, Tokyo, Japan, October 24–26, 1988, KTK Scientific Publishers: Terra Scientific Pub. Co., Tokyo, 1990.
28. A. V. Sumant, O. Auciello, H.-C. Yuan, Z. Ma, R. W. Carpick, and D. C. Mancini, in *Proceedings of SPIE - The International Society for Optical Engineering*, 2009, vol. 7318, 731817.
29. M. Dipalo, Z. Gao, J. Scharpf, C. Pietzka, M. Alomari, F. Medjdoub, J.-F. Carlin, N. Grandjean, S. Delage and E. Kohn, *Diamond and Related Materials*, 2009, **18**, 884–889.
30. Z. Remes, H. Kozak, O. Babchenko, S. Potocky, E. Ukraintsev, B. Rezek and A. Kromka, *Diamond and Related Materials*, 2011, **20**, 882–885.
31. A. Kromka, O. Babchenko, T. Izak, S. Potocky, M. Davydova, N. Neykova, H. Kozak, Z. Remes, K. Hruska, and B. Rezek, *Pulsed Linear Antenna Microwave Plasma – a Step Ahead in Large Area Material Depositions and Surface Functionalization*, Tanger Ltd, Slezska, 2011.
32. F. Piazza, F. Solá, O. Resto, L. F. Fonseca and G. Morell, *Diamond and Related Materials*, 2009, **18**, 113–116.
33. A. Kromka, S. Potocky, B. Rezek, O. Babchenko, H. Kozak, M. Vanecek and M. Michalka, *Phys. Status Solidi b*, 2009, **246**, 2654–2657.
34. H. Windischmann, G. F. Epps, Y. Cong and R. W. Collins, *Journal of Applied Physics*, 1991, **69**, 2231.
35. P. L. Dreike, D. M. Fleetwood, D. B. King, D. C. Sprauer and T. E. Zipperian, *IEEE Transactions on Components, Packaging, and Manufacturing Technology: Part A*, 1994, **17**, 594–609.
36. J. Zhao, D. Wu and J. Zhi, *Bioelectrochemistry*, 2009, **75**, 44–49.

37. H. O. Pierson, *Handbook of Carbon, Graphite, Diamond, and Fullerenes: Properties, Processing, and Applications*, Noyes Publications, Park Ridge, N.J., U.S.A, 1993.
38. M. Kitabatake, *Journal of Vacuum Science & Technology A: Vacuum, Surfaces, and Films*, 1988, **6**, 1793.
39. J. Stiegler, J. Michler and E. Blank, *Diamond and Related Materials*, 1999, **8**, 651–656.
40. C.-F. Chen, S.-H. Chen, H.-W. Ko and S. E. Hsu, *Diamond and Related Materials*, 1994, **3**, 443–447.
41. M. Ihara, H. Maeno, K. Miyamoto and H. Komiyama, *Diamond and Related Materials*, 1992, **1**, 187–190.
42. T. P. Ong and R. P. H. Chang, *Applied Physics Letters*, 1989, **55**, 2063.
43. A. Hiraki, *Materials Chemistry and Physics*, 2001, **72**, 196–200.
44. M. Yuasa, O. Arakaki, J. S. Ma, A. Hiraki and H. Kawarada, *Diamond and Related Materials*, 1992, **1**, 168–174.
45. C. R. Eddy, D. L. Youchison and B. D. Sartwell, *Diamond and Related Materials*, 1994, **3**, 105–111.
46. X. You, Z. Yu, L. Shi and L. Wang, *Journal of Crystal Growth*, 2009, **311**, 4675–4678.
47. S. Nakao, S. Maruno, M. Noda, H. Kusakabe and H. Shimizu, *Journal of Crystal Growth*, 1990, **99**, 1215–1219.
48. J. Kim, K. Tsugawa, M. Ishihara, Y. Koga and M. Hasegawa, *Plasma Sources Science and Technology*, 2010, **19**, 015003.
49. J. H. D. Rebello, V. V. Subramaniam and T. S. Sudarshan, *Applied Physics Letters*, 1993, **62**, 899.
50. J. Jiang and Y. Tzeng, *AIP Advances*, 2011, **1**, 042117.
51. T. Yoshitake, T. Hara, T. Fukugawa, L. yun Zhu, M. Itakura, N. Kuwano, Y. Tomokiyo and K. Nagayama, *Japanese Journal of Applied Physics*, 2004, **43**, L240–L242.
52. T. Nagano and N. Shibata, *Japanese Journal of Applied Physics*, 1993, **32**, 5067–5071.
53. C. A. Rego, R. S. Tsang, P. W. May, M. N. R. Ashfold and K. N. Rosser, *Journal of Applied Physics*, 1996, **79**, 7264.
54. E. J. Corat, R. C. Mendes de Barros, V. J. Trava-Airoldi, N. G. Ferreira, N. F. Leite and K. Iha, *Diamond and Related Materials*, 1997, **6**, 1172–1181.
55. M. Asmann, J. Heberlein and E. Pfender, *Diamond and Related Materials*, 1999, **8**, 1–16.
56. I. Schmidt and C. Benndorf, *Diamond and Related Materials*, 1998, **7**, 266–271.
57. C. A. Fox, M. C. McMaster, W. L. Hsu, M. A. Kelly and S. B. Hagstrom, *Applied Physics Letters*, 1995, **67**, 2379.
58. H. Maeda, M. Irie, T. Hino, K. Kusakabe and S. Morooka, *Diamond and Related Materials*, 1994, **3**, 1072–1078.
59. B. J. Bai, J. C. Chu, D. E. Patterson, R. H. Hauge and J. L. Margrave, *Journal of Materials Research*, 2011, **8**, 233–236.

60. B. Sun, X. Zhang, Q. Zhang and Z. Lin, *Journal of Applied Physics*, 1993, **73**, 4614.
61. P. K. Bachmann, D. Leers and H. Lydtin, *Diam. Relat. Mater.*, 1991, **1**, 1–12.
62. J.-T. Wang, Y.-Z. Wan, D. W. Zhang, Z.-J. Liu and Z.-Q. Huang, *Journal of Materials Research*, 2011, **12**, 3250–3253.
63. M. Marinelli, E. Milani, M. Montuori, A. Paoletti, A. Tebano, G. Balestrino and P. Paroli, *Journal of Applied Physics*, 1994, **76**, 5702.
64. S. C. Eaton and M. K. Sunkara, *Diam. Relat. Mater.*, 2000, **9**, 1320–1326.
65. Š. Potocký, O. Babchenko, K. Hruška and A. Kromka, *Physica Status Solidi (b)*, 2012, **249**, 2612–2615.
66. J. R. Petherbridge, P. W. May, S. R. J. Pearce, K. N. Rosser and M. N. R. Ashfold, *Journal of Applied Physics*, 2001, **89**, 1484.
67. K. A. Snail and C. M. Marks, *Applied Physics Letters*, 1992, **60**, 3135.
68. K. Itoh and O. Matsumoto, *Thin Solid Films*, 1998, **316**, 18–23.
69. J. Lee, B. Hong, R. Messier and R. W. Collins, *Journal of Applied Physics*, 1996, **80**, 6489.
70. J. Tarr and M. Kaufman, *Diamond and Related Materials*, 1998, **7**, 1328–1332.
71. A. Hiraki, *Applied Surface Science*, 2000, **162–163**, 326–331.
72. K. Tsugawa, M. Ishihara, J. Kim, Y. Koga and M. Hasegawa, *Physical Review B*, 2010, **82**, 125460.
73. T. G. McCauley, D. M. Gruen and A. R. Krauss, *Applied Physics Letters*, 1998, **73**, 1646.
74. L. Wang, Y. Wang, J. Zhou and S. Ouyang, *Materials Science and Engineering: A*, 2008, **475**, 17–19.
75. X. Xiao, J. Birrell, J. E. Gerbi, O. Auciello and J. A. Carlisle, *Journal of Applied Physics*, 2004, **96**, 2232.
76. W. Kulisch, C. Popov, S. Boycheva, M. Jelinek, P. N. Gibson and V. Vorlicek, *Surface and Coatings Technology*, 2006, **200**, 4731–4736.
77. F. Piazza, J. A. González, R. Velázquez, J. De Jesús, S. A. Rosario and G. Morell, *Diamond and Related Materials*, 2006, **15**, 109–116.
78. R. Haubner and D. Sommer, *Diamond and Related Materials*, 2003, **12**, 298–305.
79. F. Piazza and G. Morell, *Diamond and Related Materials*, 2007, **16**, 1950–1957.
80. J. Griffin and P. C. Ray, *Nanotechnology*, 2006, **17**, 1225–1229.
81. K. L. Saenger, F. Tong, J. S. Logan and W. M. Holber, *Review of Scientific Instruments*, 1992, **63**, 3862.
82. V. M. Donnelly, *Journal of Vacuum Science & Technology A: Vacuum, Surfaces, and Films*, 1990, **8**, 84.
83. M. Wakagi, *Journal of Vacuum Science & Technology A: Vacuum, Surfaces, and Films*, 1995, **13**, 1917.
84. M. Ihara, H. Maeno, K. Miyamoto and H. Komiyama, *Applied Physics Letters*, 1991, **59**, 1473.
85. I. Schmidt, *Solid State Ionics*, 1997, **101–103**, 97–101.

86. S. Kulesza, *Surface and Coatings Technology*, 2012, **206**, 3554–3558.
87. A. J. SpringThorpe, *Journal of Vacuum Science & Technology B: Microelectronics and Nanometer Structures*, 1990, **8**, 266.
88. O. A. Williams, *Diamond and Related Materials*, 2011, **20**, 621–640.
89. D. C. Barbosa, F. A. Almeida, R. F. Silva, N. G. Ferreira, V. J. Trava-Airoldi and E. J. Corat, *Diamond and Related Materials*, 2009, **18**, 1283–1288.
90. S. Potocky, A. Kromka, J. Potmesil, Z. Remes, Z. Polackova and M. Vanecek, *Physica Status Solidi (a)*, 2006, **203**, 3011–3015.
91. T. Izak, O. Babchenko, M. Varga, S. Potocky and A. Kromka, *Physica Status Solidi (b)*, 2012, **249**, 2600–2603.
92. A. Yamaguchi, M. Ihara and H. Komiyama, *Applied Physics Letters*, 1994, **64**, 1306.
93. J. Stiegler, T. Lang, M. Nygård-Ferguson, Y. von Kaenel and E. Blank, *Diamond and Related Materials*, 1996, **5**, 226–230.
94. K. Tsugawa, S. Kawaki, M. Ishihara, J. Kim, Y. Koga, H. Sakakita, H. Koguchi and M. Hasegawa, *Diamond and Related Materials*, 2011, **20**, 833–838.
95. T. Sharda, D. Misra, D. Avasthi and G. Mehta, *Solid State Communications*, 1996, **98**, 879–883.
96. Y. G. Chen, M. Ogura, M. Kondo and H. Okushi, *Applied Physics Letters*, 2004, **85**, 2110.
97. M. Boutchich, J. Alvarez, D. Diouf, P. Roca i Cabarrocas, M. Liao, I. Masataka, Y. Koide and J.-P. Kleider, *Journal of Non-Crystalline Solids*, 2012, **358**, 2110–2113.
98. H. Kiyota, H. Okushi, K. Okano, Y. Akiba, T. Kurosu and M. Iida, *Applied Physics Letters*, 1992, **61**, 1808.
99. C. Hu and R. V. Mahajan, Electronic packages, assemblies, and systems with fluid cooling, 2006, United States Patent: 7126822.
100. K. Küsová, O. Cibulka, K. Dohnalová, I. Pelant, J. Valenta, A. Fučíková, K. Žídek, J. Lang, J. Englich, P. Matějka, P. Štěpánek and S. Bakardjieva, *ACS Nano*, 2010, **4**, 4495–4504.
101. V. Petráková, A. Taylor, I. Kratochvílová, F. Fendrych, J. Vacík, J. Kučka, J. Štursa, P. Cígler, M. Ledvina, A. Fišerová, P. Kneppo and M. Nesládek, *Advanced Functional Materials*, 2012, **22**, 812–819.
102. P. Fojtík, K. Dohnalová, T. Mates, J. Stuchlík, I. Gregora, J. Chval, A. Fejfar, J. Kočka and I. Pelant, *Philosophical Magazine B*, 2002, **82**, 1785–1793.
103. E. Verveniotis, J. Čermák, A. Kromka, M. Ledinský, Z. Remeš and B. Rezek, *Physica Status Solidi (a)*, 2010, **207**, 2040–2044.
104. G. Cao, *Nanostructures & Nanomaterials: Synthesis, Properties & Applications*, Imperial College Press, London; Hackensack, NJ, 2004.
105. B. Rezek, E. Šípek, M. Ledinský, P. Krejza, J. Stuchlík, A. Fejfar and J. Kočka, *Journal of Non-Crystalline Solids*, 2008, **354**, 2305–2309.
106. B. Rezek, E. Šípek, M. Ledinský, J. Stuchlík, A. Vetushka and J. Kočka, *Nanotechnology*, 2009, **20**, 045302.

107. E. Verveniotis, B. Rezek, E. Šípek, J. Stuchlík and J. Kočka, *Thin Solid Films*, 2010, **518**, 5965–5970.
108. E. Verveniotis, B. Rezek, E. Šípek, J. Stuchlík, M. Ledinský and J. Kočka, *Nanoscale Research Letters*, 2011, **6**, 145.
109. H. Bach and N. Neuroth, *The Properties of Optical Glass*, Springer-Verlag, Berlin, New York, 1995.
110. Y. Muranaka, H. Yamashita and H. Miyadera, *Diamond and Related Materials*, 1994, **3**, 313–318.
111. M. Füner, C. Wild and P. Koidl, *Applied Physics Letters*, 1998, **72**, 1149–1151.
112. L. Ward, *The Optical Constants of Bulk Materials and Films*, Institute of Physics Pub, Bristol; Philadelphia, 2nd ed., 1994.
113. K. M. McNamara, D. H. Levy, K. K. Gleason and C. J. Robinson, *Applied Physics Letters*, 1992, **60**, 580.
114. W. Jacob and M. Unger, *Applied Physics Letters*, 1996, **68**, 475.
115. S. Bühlmann, E. Blank, R. Haubner and B. Lux, *Diam. Relat. Mater.*, 1999, **8**, 194–201.
116. J. Zemek, J. Houdkova, B. Lesiak, A. Jablonski, J. Potmesil and M. Vanecek, *J. Optoelectron. Adv. Mater.*, 2006, **8**, 2133–2138.
117. H. A. Girard, T. Petit, S. Perruchas, T. Gacoin, C. Gesset, J. C. Arnault and P. Bergonzo, *Physical Chemistry Chemical Physics*, 2011, **13**, 11517.
118. N. J. Harrick, *Internal Reflection Spectroscopy*, John Wiley & Sons Inc, 1967.
119. M. Davydova, A. Kromka, P. Exnar, M. Stuchlik, K. Hruska, M. Vanecek and M. Kalbac, *Physica Status Solidi (a)*, 2009, **206**, 2070–2073.
120. L. Bacakova, L. Grausova, J. Vacik, A. Kromka, H. Biederman, A. Choukourov, and V. Stary, in *Advances in Diverse Industrial Applications of Nanocomposites*, ed. B. Reddy, InTech, Rijeka, Croatia, 2011, pp. 399–436.
121. C. E. Nebel, D. Shin, B. Rezek, N. Tokuda, H. Uetsuka and H. Watanabe, *Journal of The Royal Society Interface*, 2007, **4**, 439–461.
122. A. Kromka, B. Rezek, Z. Remes, M. Michalka, M. Ledinsky, J. Zemek, J. Potmesil and M. Vanecek, *Chem. Vap. Deposition*, 2008, **14**, 181–186.
123. O. Babchenko, Z. Remes, T. Izak, B. Rezek and A. Kromka, *Physica Status Solidi (b)*, 2011, **248**, 2736–2739.
124. Z. Remes, H. Kozak, B. Rezek, E. Ukraintsev, O. Babchenko, A. Kromka, H. A. Girard, J.-C. Arnault and P. Bergonzo, *Applied Surface Science*, 2013, **270**, 411–417.
125. A. H. Lettington and C. Smith, *Diamond and Related Materials*, 1992, **1**, 805–809.
126. C. A. Rego, P. W. May, E. C. Williamson, M. N. R. Ashfold, Q. S. Chia, K. N. Rosser and N. M. Everitt, *Diamond and Related Materials*, 1994, **3**, 939–941.

CHAPTER 14

P-type and N-type Conductivity in Nanodiamond Films

OLIVER A. WILLIAMS

Cardiff School of Physics and Astronomy, Queen's Buildings, The Parade, Cardiff, CF24 3AA, UK
Email: williamso@cf.ac.uk

14.1 Introduction

Nanocrystalline diamond films have generated significant interest in application areas as diverse as heat spreading, micro-electro-mechanical systems (MEMS), electrochemistry, tribology and field emission to name a few.[1] A large proportion of these applications require some level of electrical conductivity, usually *pseudo*-metallic conductivity. Due to the small size of the diamond crystals, free carrier mobilities are rather low (around 1 cm^2 V^{-1} s^{-1}) and, thus, nanocrystalline diamond is generally inappropriate for active electronic devices.[2] However, diamond electrodes are readily fabricated for electrochemistry and MEMS.[3,4]

The conductivity mechanisms of nanocrystalline or nanodiamond films fall generally into two categories: conventional doping with substitutional boron (p-type) and the generation of an enhanced density of states within the bandgap due to sp^2 bonding (n-type).[5,6] These differences are strongly related to the micro/nanostructure of the diamond films and will be discussed in detail in this chapter.

14.2 Film Structure and sp² Content

Conventional diamond films have crystallite sizes that evolve with the distance from the substrate in a classical van der Drift mechanism.[1] After several microns these crystallites are generally around two thirds of the film thickness along their largest axis, but if the film growth is interrupted after a short growth duration, the crystallite size can be closer to the film thickness, for example around 100 nm. These films are termed nanocrystalline diamond (NCD) after their crystallite size and are ultimately conventional thin film diamond grown with very high nucleation densities.[7] An example of such a film is shown in Figure 14.1(a), the film is around 200 nm thick with crystal sizes approximately 50 nm.

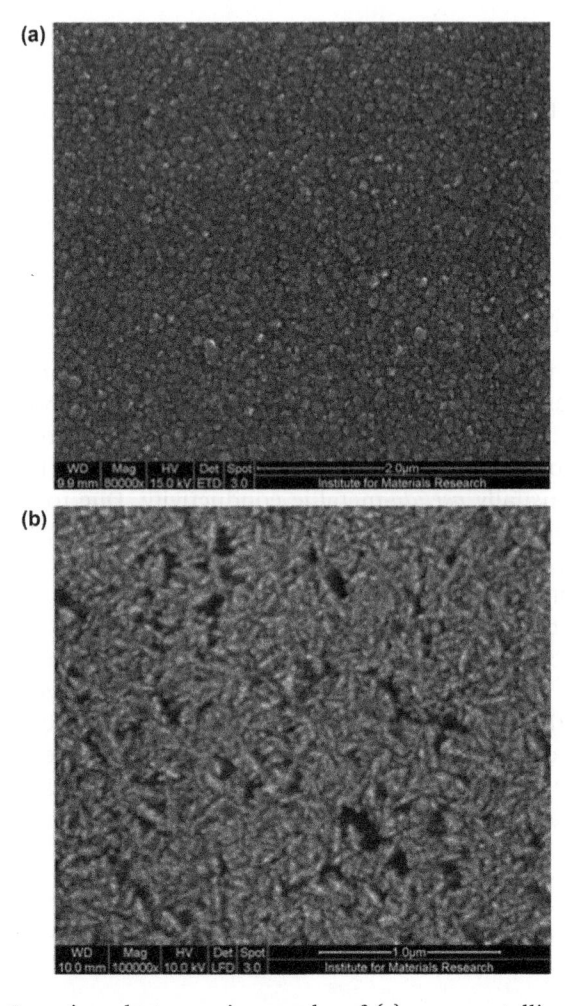

Figure 14.1 Scanning electron micrographs of (a) nanocrystalline diamond and (b) ultra-nanocrystalline diamond.
Reproduced from *Diamond and Related Materials*, 2008, **17**, 1080–1088.[9]

It is also possible to interrupt this evolution of crystallite size by disrupting the layer-by-layer growth mechanism and triggering a new crystal. This process, termed "re-nucleation", is based on the introduction of defects into the evolving crystallite that triggers the nucleation of a new crystal. Films of this type are often called ultra-nanocrystalline diamond (UNCD). A detailed description of this mechanism is available elsewhere, the result being the growth of films with grain sizes of 3–5 nm or 30 nm depending on the process, regardless of film thickness.[1,8] Figure 14.1(b) shows an example of such a film, which is around 1 µm thick.[9] It is clear that there is no clear crystal faceting at this magnification, in fact one requires high resolution transmission electron microscopy (TEM) to resolve the crystallinity in this material.[10]

The profound differences in crystallite size have profound implications on the sp^3/sp^2 ratio of the films. Obviously, the smaller the grain size, the larger the surface-to-volume fraction and thus the larger the overall proportion of grain boundaries in the material. Non-diamond bonded (sp^2) carbon is predominantly located at grain boundaries and, generally, the larger the proportion of grain boundaries, the larger the sp^2 content of the film. This is clearly exhibited in the Raman spectra in Figure 14.2 and the literature.[5,8] The Raman spectra of Figure 14.2 shows both films' growth with and without re-nucleation. The spectra labelled UNCD represents diamond growth with re-nucleation, whereas that labelled SCD is single crystal diamond, *i.e.*, almost no sp^2 material. These two spectra represent the two extremes, *i.e.*, almost no sp^2 bonding to greater than 5% sp^2 bonding. The spectra in between represent nanocrystalline diamond films (*i.e.*, grown without re-nucleation), but with varying concentrations of boron doping. The Raman spectra of the NCD films are comparable with others in the literature. Typically, in addition to the first-order diamond peak (1332 cm^{-1}), D ($\sim$1350 cm^{-1}) and G (1600 cm^{-1}) bands related to sp^2 bonding are

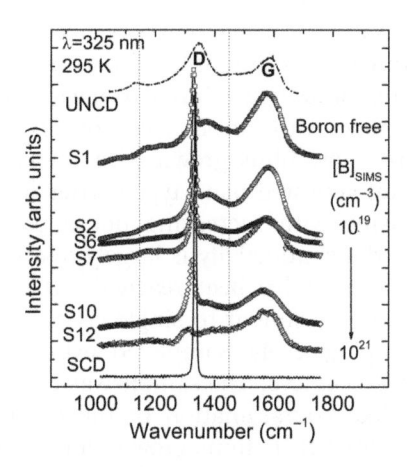

Figure 14.2 Raman spectra of NCD and UNCD films. Reproduced from *Physical Review B*, 2009, **79**, 045206.[5]

observed. This is due to sp² bonding at the grain boundaries and the interface between the silicon wafer and the diamond film. It can vary with the CH_4/H_2 ratio and the microwave power density during the growth process, as well as the nucleation density of the seeding process prior to growth.[1,11] It should be considered that even the highest quality diamond growth process will exhibit some sp² Raman signatures at these film thicknesses (<200 nm).

The position of the first-order diamond peak shifts slightly from that of natural diamond (1332.5 cm^{-1}) due to the small grain sizes and the influence of doping. Boron doping is well known to result in a downshift in the first-order diamond peak due to Fano-interference between the continuum of electronic states introduced by the dopants and the discrete zone-centre phonon.[5] This is most pronounced for films with boron concentrations exceeding 10^{20} cm^{-3}, which corresponds to the onset of the metal–insulator transition for boron-doped diamond films.[12] A smaller shift is seen with lower doping levels, sample S2 ($B_{SIMS} \sim 10^{17}$ cm^{-3}) has a first-order line position of 1328.9 cm^{-1}, whereas sample S12 ($B_{SIMS} \sim 3.3 \times 10^{21}$ cm^{-3}) exhibits a strong shift to 1305.3 cm^{-1}. This Fano-interference is also accompanied by a band at 1210 cm^{-1}, which some authors have attributed to B–C complexes.[12]

The intensities of the G and D bands are considerably less in the NCD films than in UNCD films. This is due to the enhanced sp² content in UNCD films grown with re-nucleation compared with high quality NCD films grown with the suppression of re-nucleation processes.[13] The sp² content has a profound effect on the optical and electronic properties of the resulting films, which is discussed in the following sections.

14.3 Film Conductivity

Both NCD and UNCD films can be made conductive by the addition of either boron- or nitrogen-containing precursors into the gas phase.[5,6] However, the origin of conductivity and the carrier transport mechanism of these two film types is drastically different. Figure 14.3 shows the conductivity of films varies with temperature for (a) NCD films grown with various levels of boron and (b) UNCD films grown with various levels of nitrogen.

Figure 14.3(a) shows how films grown with increasing levels of boron doping (S1–S10) exhibit activated band type carrier transport with the activation energy decreasing with increasing boron concentration. At large boron concentrations the conductivity is largely temperature insensitive as the metal–insulator transition has been reached.[5] At lower doping levels (S2) the activation energy is very similar to single crystal diamond films grown with similar boron doping levels. Sample S1 exhibits a higher activation energy due to the low boron level (the sample was grown with no additional boron, only the residual contamination of the chamber). Figure 14.3(b) shows a similar trend for UNCD films grown with increasing nitrogen.[6] In the case of UNCD, far higher concentrations of nitrogen are required than typical substitutional doping levels, such as with boron-doped NCD.

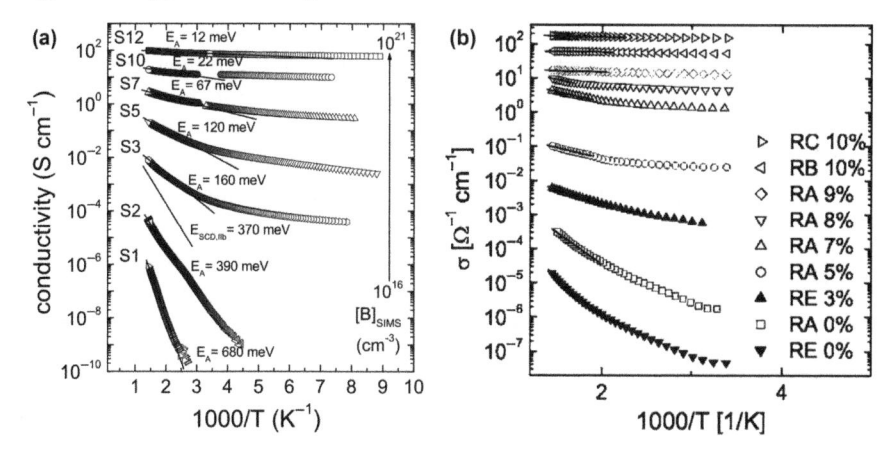

Figure 14.3 Film conductivity vs reciprocal temperature for (a) NCD films and (b) UNCD films.
Reproduced from *Physical Review B*, 2009, **79**, 045206[5] and *Physical Review B*, 2006, **74**, 155429.[6]

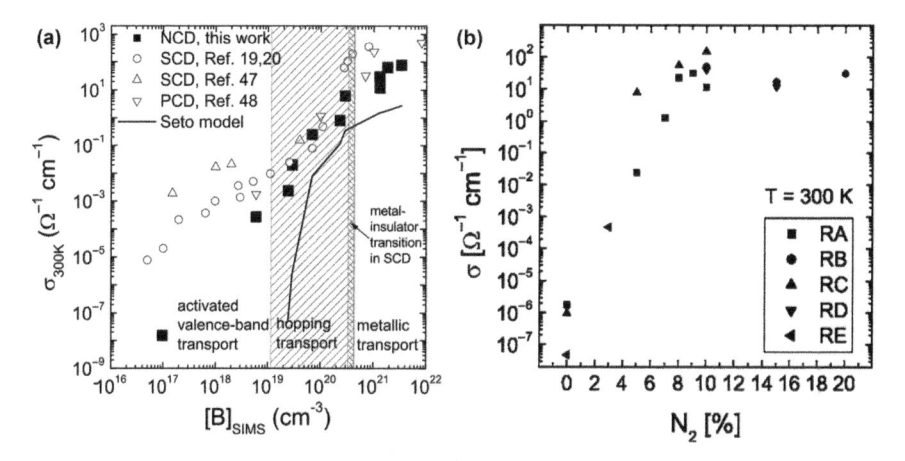

Figure 14.4 Room temperature film conductivity vs impurity concentration for (a) NCD films and (b) UNCD films.
Reproduced from *Physical Review B*, 2009, **79**, 045206[5] and *Physical Review B*, 2006, **74**, 155429.[6] *Note*: the references cited in panel (a) refer to the original publication.

The variation of conductivity with temperature of UNCD films does not exhibit classical band transport like NCD films, and gas-phase nitrogen quantities in excess of 10% are required to reach the metallic transition. Undoped UNCD films are also intrinsically more conductive than undoped NCD films due to their increased grain boundary volume fraction and thus increased sp^2 content.

In Figure 14.4(a) the conductivity of the NCD films is plotted as a function of the boron concentration within the films as determined by secondary ion mass spectroscopy (SIMS).[5] For comparison, values for single crystal diamond (SCD) and microcrystalline diamond (here PCD = polycrystalline diamond) are also plotted on the same chart. It is clear that the data for NCD films conductivity *vs* dopant concentration correlate very well with those of both microcrystalline diamond and single crystal diamond. At lower doping levels, the conductivity of NCD films drops considerably faster than SCD and PCD films due to the lower carrier mobilities in NCD films arising from their small grain sizes and thus enhanced grain boundary scattering. Increasing the boron concentration in all films, whether NCD, PCD or SCD, demonstrates the classical nature of the doping as valence-band transport gives way to hopping transport and, ultimately, metallic transport at levels of boron exceeding 5×10^{20} cm^{-3}.

The case of UNCD is depicted in Figure 14.4(b); however, the nitrogen levels in this figure represent gas-phase concentrations during growth rather than those actually measured in films by secondary ion mass spectrometry (SIMS) after growth. Undoped films are at least two orders of magnitude more conductive than intrinsic NCD films. The addition of nitrogen increases the conductivity until it saturates at levels of 10% in the gas phase. This yields UNCD film conductivity in the range of 10^2 Ω^{-1} cm^{-1}, similar to that of metallically doped NCD. However, these nitrogen concentrations are far in excess of the levels used in conventional doping.

14.4 Carrier Transport in NCD Films

The carrier concentration as determined by Hall effect measurements can be plotted as a function of the concentration of impurity atoms measured in the material. This is shown in Figure 14.5 for both NCD films grown with boron

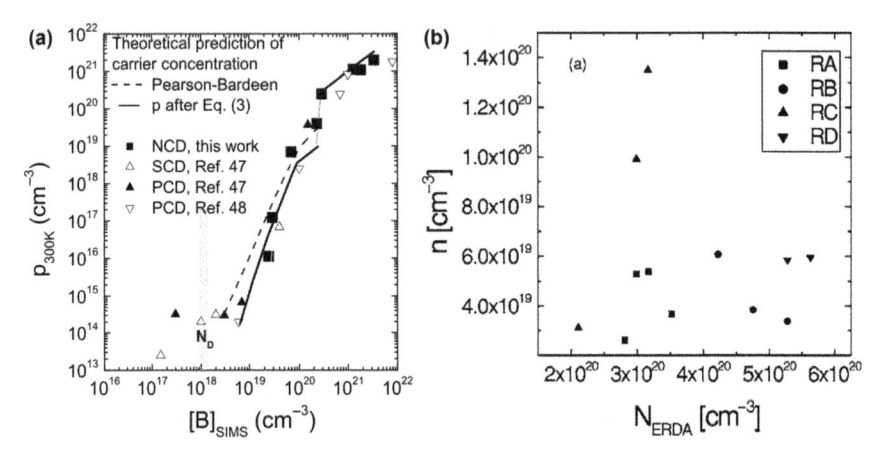

Figure 14.5 Film-free carrier concentrations *vs* impurity concentration for (a) NCD films and (b) UNCD films.
Reproduced from *Physical Review B*, 2009, 79, 045206[5] and *Physical Review B*, 2006, 74, 155429.[6] *Note*: the references cited in panel (a) refer to the original publication.

and UNCD films grown with nitrogen.[5] In Figure 14.5(a) the carrier concentration of boron-doped NCD films at room temperature is plotted against the boron concentration measured in the film as determined by SIMS. There is a clear correlation between incorporated boron and free holes in the films, with NCD, PCD and SCD films exhibiting similar values. Because of the reduced mobility in NCD films, it is difficult to obtain Hall measurements and boron impurity levels below 2×10^{19} cm^{-3}. Clearly, at lower boron concentrations, there is less carrier activation until above the metallic transition ($\sim 5 \times 10^{20}$ cm^{-3}), where the free carrier concentration roughly corresponds to the impurity density, i.e., there is negligible compensation. Thus, boron doping of NCD films is very similar to PCD and SCD films, i.e., classic substitutional doping.

In the case of UNCD the correlation between the free electron concentration and the nitrogen incorporation in the films is very different. Figure 14.5(b) shows the correlation between free electrons as measured by the Hall effect and the actual nitrogen concentration of the UNCD films as determined by elastic recoil detection analysis (ERDA). It is clear that there is no real correlation between the free carrier concentration and the incorporated nitrogen. A similar effect has also been observed using SIMS.[14] In both cases the actual nitrogen content is roughly constant when films are grown with 5% or more nitrogen in the gas phase. The electron spin resonance (ESR) nitrogen spin density has also been shown to be constant as a function of nitrogen incorporation.[6] Thus, there is no evidence that nitrogen acts as a conventional dopant in UNCD films from these measurements.

Figure 14.6(a) shows how the hole mobility at room temperature correlates with the hole concentration in NCD films doped with boron.[5] Again, data

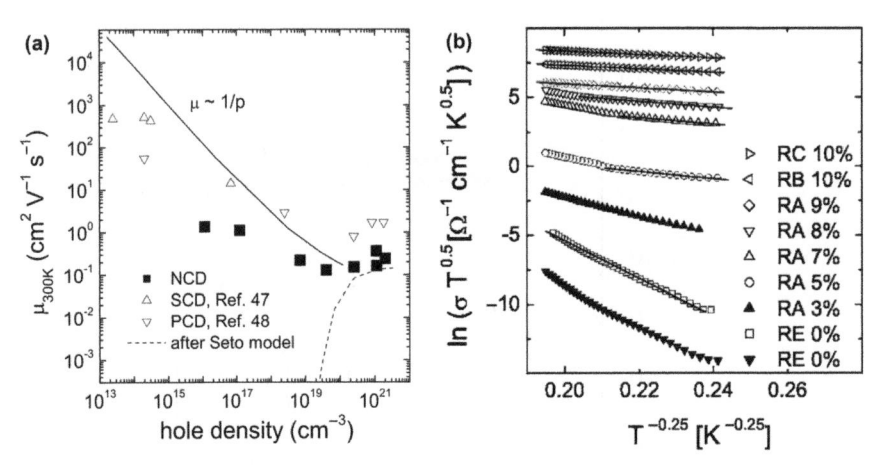

Figure 14.6 (a) Mobility vs free hole concentration for boron-doped NCD films and (b) $\sigma\sqrt{T}$ vs $T^{-0.25}$ for UNCD films.
Reproduced from *Physical Review B*, 2009, **79**, 045206[5] and *Physical Review B*, 2006, **74**, 155429.[6] Note: the references cited in panel (a) refer to the original publication.

from PCD and SCD films are also plotted for comparison. The result is a very similar message to that of Figure 14.5(a), with NCD films correlating well with SCD and PCD films, except at low doping levels where the mobility fails to rise due to grain boundary scattering processes. The solid line represents the expected mobility variation due to ionised impurity scattering, which is a reasonable fit for SCD films. At lower carrier concentrations, PCD films also show lower mobilities as expected. Thus, the mobility behaviour of NCD films can be reasonably well explained by ionised impurity scattering with additional scattering at grain boundaries.[5]

In Figure 14.6(b) the conductivity of UNCD films is plotted against $T^{-0.25}$ in order to investigate three-dimensional (3D) variable-range hopping.[6] The lower conductivity samples correlate quite well with variable-range hopping in three dimensions, whereas the more conductive samples deviate from this behaviour at high temperatures. The mobility of UNCD films is mostly limited to values around 1 cm^2 V^{-1} s^{-1} due to the significant grain boundary scattering mechanism resulting from its reduced crystallite size.[2]

An unusual property of metallically doped NCD films is its superconductivity at low temperatures, as shown in Figure 14.7.[5,15-20] The resistivity of NCD increases with decreasing temperature above the critical temperature (T_c), in contrast with metallically doped single crystal diamond. The T_c is around 2 K in this case; however, higher values have been reported since this work.[21-24] The NCD films doped below the metallic transition exhibit behaviours similar to disordered metals, in that the resistivity rises with reducing temperature.

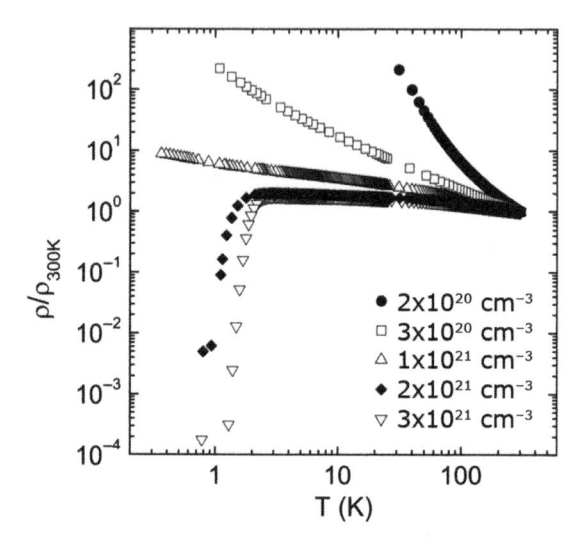

Figure 14.7 Superconductivity of heavily boron-doped NCD films as a function of dopant concentration.
Reproduced from *Physical Review B*, 2009, 79, 045206.[5]

14.5 Optical Properties of NCD Films

The optical properties of NCD films are a complex area with strong convolution between grain size and sp² bonding effects, which has been summarised in recent reviews.[1,9] Generally, the larger the grain size the closer the optical properties correlate with single crystal diamond. This is clearly seen in the optical transparency of the films. Re-nucleating diamond, such as the aforementioned UNCD, is black in appearance, whereas NCDs can have 80% transparency in the visible spectrum when they are a few hundred nanometres thick. This is demonstrated more qualitatively in Figure 14.8.

Figure 14.8(a) shows the absorption coefficient of re-nucleating diamond types as measured by photo-thermal deflection spectroscopy (PDS).[25] The spectra are clearly dominated by mid-gap absorption starting at around 0.8 eV, which is attributed to transitions between π and π* states.[26] The addition of nitrogen during growth (labelled here as 5%) broadens these bands substantially and saturates the PDS. The films labelled R contain grains around 3–5 nm, whereas the grain size of sample AAu is nearer to 20 nm. Thus, the slightly larger grain size shows a marked improvement in transparency. This is far clearer in Figure 14.8(b), where the grain size is around 100 nm. Here, the undoped material displays two orders of magnitude lower optical absorption coefficient compared to Figure 14.8(a), although there is still some evidence of mid-gap absorption due to sp² bonding. The addition of boron results in a substantially increased optical absorption coefficient but is still significantly below that of re-nucleating diamond grown with nitrogen. Thus, the grain size is a critical determining factor in the optical transparency as small grain sizes result in a larger grain boundary volume and thus higher sp² content. Whilst it is true that some grain boundaries are wider and contain more sp² bonding that others, the general rule is that

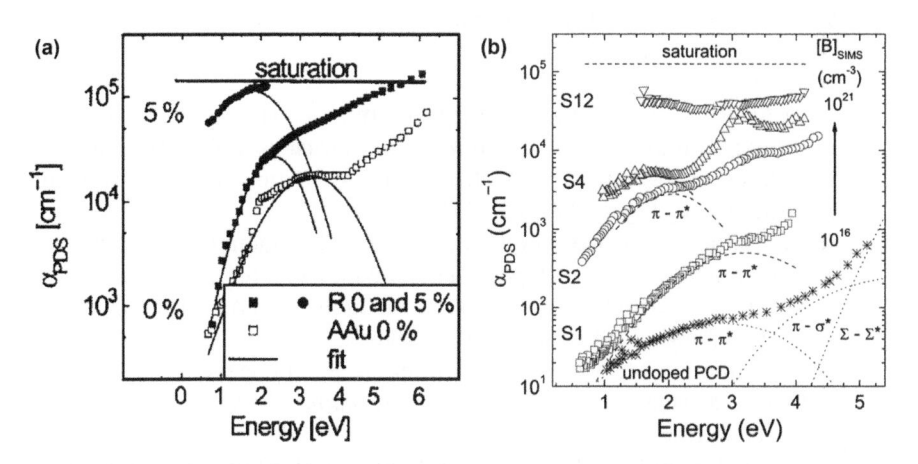

Figure 14.8 Photothermal deflection spectroscopy adsorption coefficient of (a) UNCD films and (b) NCD films.
Reproduced from *Applied Physics Letters*, 2006, **88**, 101908[25] and *Physical Review B*, 2009, **79**, 045206.[5]

smaller grain sizes result in increased optical absorption due to the enhanced grain boundary volume.

14.6 Conclusions

NCD and UNCD films have very different electrical and optical properties due to their different grain boundary volume fractions. NCD films have grain sizes that evolve with film thickness and thus have grain sizes greater than those of UNCD films grown with re-nucleating processes. In general, smaller grain sizes result in larger sp^2 concentrations due to increased grain boundary volume fractions. This non-diamond carbon enhances the density of states within the bandgap and absorbs light, causing the material to darken. Larger grain sizes reduce the grain boundary volume and thus the overall sp^2 content.

UNCD films can exhibit high conductivity levels due to their relatively high sp^2 content and mid-gap states, which are enhanced by growth with added nitrogen. NCD films exhibit higher intrinsic resistivity levels and can be substitutionally doped by the addition of boron into the growth process.

References

1. O. A. Williams, *Diamond and Related Materials*, 2011, **20**(5–6), 621–640.
2. O. A. Williams, S. Curat, J. E. Gerbi, D. M. Gruen and R. B. Jackman, *Applied Physics Letters*, 2004, **85**(10), 1680–1682.
3. J. Hees, R. Hoffmann, A. Kriele, W. Smirnov, H. Obloh, K. Glorer, B. Raynor, R. Driad, N. J. Yang, O. A. Williams and C. E. Nebel, *ACS NANO*, 2011, **5**(4), 3339 3346.
4. W. Smirnov, A. Kriele, R. Hoffmann, E. Sillero, J. Hees, O. A. Williams, N. J. Yang, C. Kranz and C. E. Nebel, *Analytical Chemistry*, 2011, **83**(12), 4936–4941.
5. W. Gajewski, P. Achatz, O. A. Williams, K. Haenen, E. Bustarret, M. Stutzmann and J. A. Garrido, *Physical Review B*, 2009, **79**(4), 045206.
6. P. Achatz, O. A. Williams, P. Bruno, D. M. Gruen, J. A. Garrido and M. Stutzmann, *Physical Review B*, **74**(15), 155429.
7. O. A. Williams, O. Douheret, M. Daenen, K. Haenen, E. Osawa and M. Takahashi, *Chemical Physics Letters*, 2007, **445**(4–6), 255–258.
8. D. Zhou, D. M. Gruen, L. C. Qin, T. G. McCauley and A. R. Krauss, *Journal of Applied Physics*, 1998, **84**(4), 1981–1989.
9. O. A. Williams, M. Nesladek, M. Daenen, S. Michaelson, A. Hoffman, E. Osawa, K. Haenen and R. B. Jackman, *Diamond and Related Materials*, 2008, **17**(7–10), 1080–1088.
10. S. Jiao, A. Sumant, M. A. Kirk, D. M. Gruen, A. R. Krauss and O. Auciello, *Journal of Applied Physics*, 2001, **90**(1), 118–122.
11. O. A. Williams, A. Kriele, J. Hees, M. Wolfer, W. Muller-Sebert and C. E. Nebel, *Chemical Physics Letters*, 2010, **495**(1–3), 84–89.

12. S. Ghodbane and A. Deneuville, *Diamond and Related Materials*, 2006, **15**(4-8), 589-592.
13. P. Achatz, J. A. Garrido, O. A. Williams, P. Brun, D. M. Gruen, A. Kromka, D. Steinmuller and M. Stutzmann, *Physica Status Solidi a-Applications and Materials Science*, 2007, **204**(9), 2874-2880.
14. S. Bhattacharyya, O. Auciello, J. Birrell, J. A. Carlisle, L. A. Curtiss, A. N. Goyette, D. M. Gruen, A. R. Krauss, J. Schlueter, A. Sumant and P. Zapol, *Applied Physics Letters*, 2001, **79**(10), 1441-1443.
15. P. Achatz, E. Bustarret, C. Marcenat, R. Piquerel, T. Dubouchet, C. Chapelier, A. M. Bonnot, O. A. Williams, K. Haenen, W. Gajewski, J. A. Garrido and M. Stutzmann, *Physica Status Solidi a-Applications and Materials Science*, 2009, **206**(9), 1978-1985.
16. P. Achatz, W. Gajewski, E. Bustarret, C. Marcenat, R. Piquerel, C. Chapelier, T. Dubouchet, O. A. Williams, K. Haenen, J. A. Garrido and M. Stutzmann, *Physical Review B*, 2009, **79**(20), 201203.
17. E. Bustarret, P. Achatz, B. Sacepe, C. Chapelier, C. Marcenat, L. Ortega and T. Klein, *Philosophical Transactions of the Royal Society a-Mathematical Physical and Engineering Sciences*, 2008, **366**(1863), 267-279.
18. F. Dahlem, P. Achatz, O. A. Williams, D. Araujo, E. Bustarret and H. Courtois, *Physical Review B*, 2010, **82**(3), 033306.
19. F. Dahlem, P. Achatz, O. A. Williams, D. Araujo, H. Courtois and E. Bustarret, *Physica Status Solidi a-Applications and Materials Science*, 2010, **207**(9), 2064-2068.
20. M. P. Villar, M. P. Alegre, D. Araujo, E. Bustarret, P. Achatz, L. Saminadayar, C. Bauerle and O. A. Williams, *Physica Status Solidi a-Applications and Materials Science*, 2009, **206**(9), 1986-1990.
21. S. Mandal, T. Bautze, O. A. Williams, C. Naud, É. Bustarret, F. Omnès, P. Rodière, T. Meunier, C. Bäuerle and L. Saminadayar, *ACS NANO*, 2011, **5**(9), 7144-7148.
22. S. Mandal, C. Naud, O. A. Williams, E. Bustarret, F. Omnes, P. Rodiere, T. Meunier, L. Saminadayar and C. Bauerle, *Nanotechnology*, 2010, **21**(19), 195303.
23. S. Mandal, C. Naud, O. A. Williams, E. Bustarret, F. Omnes, P. Rodiere, T. Meunier, L. Saminadayar and C. Bauerle, *Physica Status Solidi a-Applications and Materials Science*, 2010, **207**(9), 2017-2022.
24. P. Szirmai, T. Pichler, O. A. Williams, S. Mandal, C. Bäuerle and F. Simon, *Physica Status Solidi (b)*, 2012, **249**(12), 2656-2659.
25. P. Achatz, J. A. Garrido, M. Stutzmann, O. A. Williams, D. M. Gruen, A. Kromka and D. Steinmuller, *Applied Physics Letters*, 2006, **88**(10), 101908-101910.
26. M. Nesladek, K. Meykens, L. M. Stals, M. Vanecek and J. Rosa, *Physical Review B*, 1996, **54**(8), 5552-5561.

CHAPTER 15

Electrochemistry of Nanocrystalline and Microcrystalline Diamond

INGA V. SHPILEVAYA AND JOHN S. FOORD*

Chemistry Research Laboratory, Department of Chemistry, University of Oxford, Mansfield Road, Oxford, OX1 3TA, UK
*Email: john.foord@chem.ox.ac.uk

15.1 Introduction

Electrochemistry studies the chemical phenomena associated with charge transfer across an electrode and electrolyte interface. An electrode is a conductor through which charge is carried by electronic movement; electrolyte is a phase through which charge is carried by the transport of ions. Electrochemical measurements can be employed both in fundamental scientific and industrial applications. One can be interested in obtaining kinetic data about the unstable intermediate in the reaction, or the goal might be the development of a new power source or large-scale chemical processes, such as electroplating. The term 'electrochemical measurement' encompasses not only a broad spectrum of techniques, but also a wide range of electrical devices employed, such as electrochemical sensors (*e.g.*, for blood glucose detection), batteries, fuel cells and electrochromic displays.

In many electrochemical processes the choice of the electrode material is of utmost importance as it contributes to the reaction of interest. Depending on the application, the electrodes can differ significantly. They can be either metallic or semiconducting, in a solid or liquid state. Bulk electrochemical processes employ electrodes that may be square metres in area, whereas an electrochemical biosensor may be of a centimetre scale. The former electrode is expected to work for many months; the latter may be shed after a single use.

It is evident that the optimal choice of an electrode for a particular application depends on the diverse properties of the material. If the electrode is to be used in hostile environments, the robustness, strength and resistance to corrosion are therefore the most important properties to look for. For example, in many electrochemical measurements an electrode can be poisoned by adsorption of electrically insulating compounds on its surface, which can finally lead to the overall system failure. Therefore, an inherent resistance to electrode fouling is a necessary requirement. In electrochemical sensing applications the electrode should provide a measurable and reproducible response to a particular reaction in a certain potential range. In addition, such a response should not be masked by competing electrochemical reactions or background signals arising from undesirable capacitance in the circuit.

Many readers are familiar with diamond as a highly insulating material, or more recently as a semiconductor in modern device technologies. However, if the material is highly doped with boron, it acquires metallic qualities and, in particular, shows an electrical resistance that is comparable to that of many conventional electrodes. As a result, boron-doped diamond has now been established as an electrode with superior performance in many electrochemical applications.

In Figure 15.1 the annual number of articles regarding diamond electrochemistry is compared to that of a well-established analogue electrode, known as glassy carbon. Before the mid-1990s, the diamond electrode was utilised mainly for scientific curiosity. However, from 1995, the growth of interest in diamond electrochemistry is illustrated by a rapid increase in the number of scientific publications. After this stage, the alternation in 'diamond' activity has broadly tracked that of the electrochemical field as a whole, and now has a superior application in this area. This chapter will provide reviewed information on the fundamental aspects of diamond electrochemistry and on recent important achievements in the field.

15.2 Electrochemical Measurements

A simple electrochemical measurement requires at least two electrodes to be connected together by conducting paths both through an ionic conductor (electrolyte) and externally (electric wires), so that the charge can be transferred and monitored. These electrodes comprise an electrochemical cell, which consists of the working electrode—an electrode where the electrochemical reaction of interest is taking place—and a counter electrode. The latter can operate as a cathode, whilst the working electrode can work as an anode or *vice versa*. A typical electrochemical cell can either be employed to

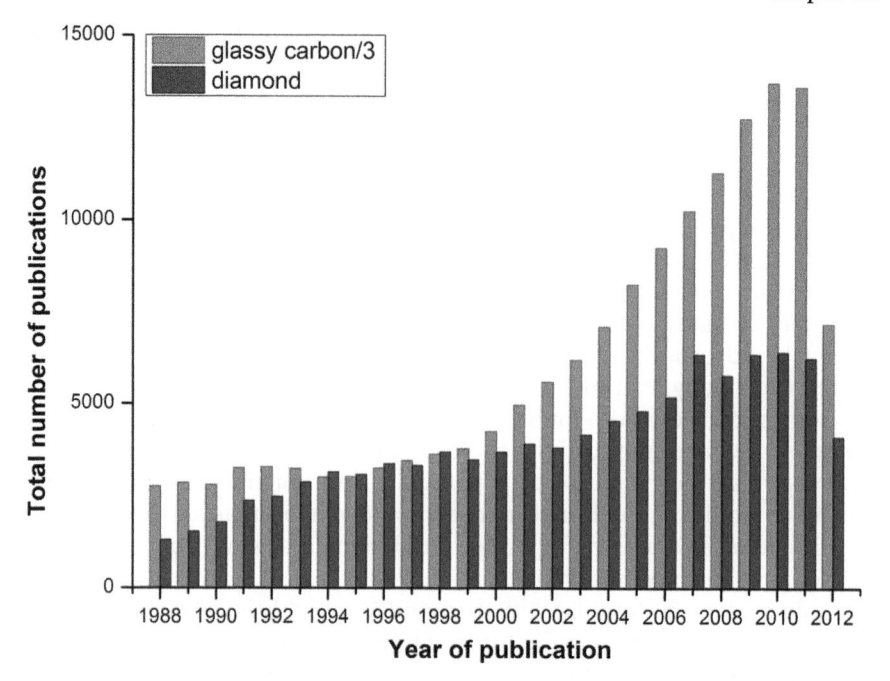

Figure 15.1 Historical data for the period 1988–2012, illustrating the annual number of scientific publications regarding diamond electrochemistry in comparison to similar data for glassy carbon. The columns for glassy carbon were divided by a factor of three.

produce electrical energy as a result of a spontaneous reaction occurring inside it (*i.e.*, a galvanic cell), or it can consume electricity from an external source to drive a specific reaction (*i.e.*, an electrolytic cell).

As the electrodes are immersed in the solution, the potential difference between the two electrodes can be measured whether the cell is passing current or not. In general, electrochemical measurements differ with regard to the type of signal used for the quantification of the charge transfer in the cell and can be principally classified as either potentiometric or potentiostatic.

In potentiometric techniques, if two connected dissimilar electrodes are immersed in the analyte, then the difference between the two interface potentials can be measured under the zero current conditions, where equilibrium exists at each electrode surface. The established potential of the cell depends on the standard electrode potentials of the electrochemical couples and on the solution concentrations of the electroactive species. Thus, if the working electrode is made truly selective to the target species, then an accurate electrochemical sensor can be produced. Various types of sensors have been developed to monitor species, such as protons, fluoride anions, and calcium and potassium cations; a well-known potentiometric sensor is the pH meter glass electrode. A considerable advantage of such potentiometric sensors is their cost-effectiveness; however, problems arise due to the lack of the required selectivity when electroanalytical tasks are more complex.

Potentiostatic or controlled-potential techniques study charge transfer at the electrode/electrolyte interface under dynamic (non-zero current) conditions. In such systems an applied potential to the working electrode drives a charge transfer reaction. As a result, the recorded current can not only provide information on the concentration of the target analyte, but also give an insight into the kinetics and thermodynamics of the electrochemical process.

One of the most versatile approaches in the whole field of potential-controlled measurements is voltammetry. This enables a very sensitive detection (down to the picomolar level) to be carried out; as well as the distinction of several electroactive species to be made simultaneously, providing they react at different potentials. These voltammetric measurements require the use of three electrodes rather than just two as in potentiometry. In this case a potentiostat is employed to control the current flowing through the working electrode as a function of potential applied between it. Additionally, a reference electrode is connected, through which no current flows. It is used to provide a stable, solution properties-independent value of potential to which the potential of the working electrode can be referred. Finally, a third electrode, known as an auxiliary or counter electrode, provides a balancing counter current to that flowing through the working electrode.

Cyclic voltammetry is one of the most widely employed methodologies in the whole area of voltammetry. In such a system, current flowing through the working electrode is recorded as a function of an applied potential. The potential of the working electrode is set up to an E_1 value, then it is swept linearly to a potential, E_2, at which point the direction of the scan is reversed to return to the potential E_1. The resultant current–potential graph, known as a voltammogram, is a plot of the current signal *versus* the excitation potential.

An example of the cyclic voltammetry data obtained from a diamond electrode in $K_3Fe(CN)_6$ solution is illustrated in Figure 15.2.

The ordinate of Figure 15.2 represents the current flowing through the diamond electrode; the abscissa corresponds to the potential applied to the working electrode *versus* the Ag/AgCl reference electrode. At the beginning of the experiment, the potential is swept from 0.5 V to 0.1 V. At this time, electrons are supplied to the $Fe(CN)_6^{3-}$ species, giving rise to the reduction peak at 0.13 V. As the potential is swept back from −0.1 V to 0.5 V, an oxidation process takes place (electrons are removed from the species). It is noteworthy that the redox reaction of interest occurs only in the potential region that makes the electron transfer thermodynamically and kinetically favourable. The electron transfer kinetics are presented by the peak-to-peak separation of cathodic and anodic processes. If peak separation is below the limit of 57 mV, the kinetics are fast. As the kinetics become slower, the peak-to-peak separation increases as a greater electrical driving force, known as the overpotential, is needed. The voltammetric data allows a fundamental

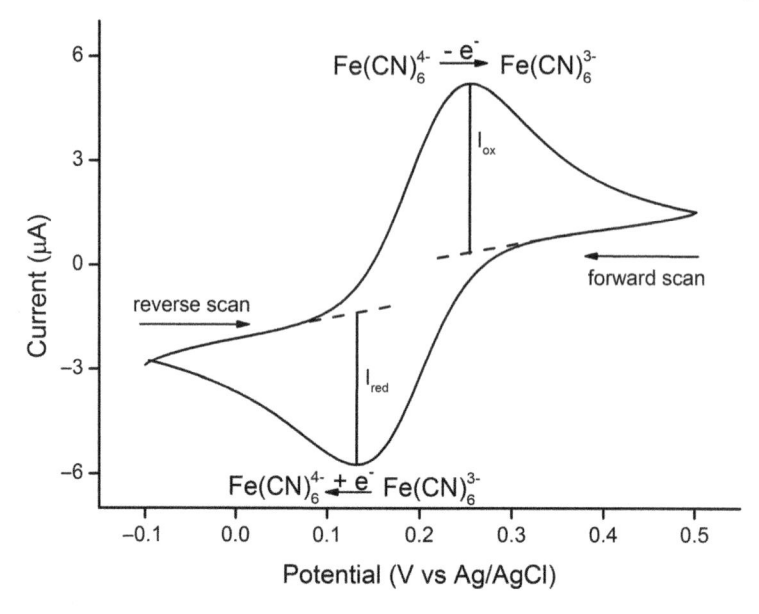

Figure 15.2 Cyclic voltammogram obtained from a diamond electrode in 0.1 mM K$_3$Fe(CN)$_6$, 0.1 M KCl solution at 50 mV s^{-1}.

understanding of the electrochemical system, therefore underpinning the production of the electrochemical system related to the sensor. In practical applications, greater sensitivity can be accomplished if pulsed potential waveforms are employed. This has resulted in the development of several alternative voltammetric methods, closely related to the cyclic voltammetry approach described above. There are several books available as references[1–3] to the reader, which provide in-depth coverage of dynamic electrochemistry.

15.3 Diamond Films in Electrochemistry

There are several advantages to boron-doped diamond electrodes compared to other conventional electrode materials, notably glassy carbon and graphitic electrodes, which have often been employed as amperometric sensors, where a rapid, sensitive, reproducible and easily measured signal response is the primary requirement.[1–3] Firstly, diamond electrodes exhibit a relatively low signal-to-background-current ratio due to their smaller double layer capacitance relative to other materials ($\sim$10 µF cm^{-2}) and a lower density of ionizable or redox active groups on the surface. The minimal presence of these groups on diamond normally results both in the significant decrease of undesirable distinct background peaks and *pseudo-*capacitive background in the electrochemical data. Secondly, the chemically inert nature of the diamond surface provides a wide potential window with the width of 3–4 V in aqueous and 5–7.5 V in non-aqueous media.[4] The potential window here can be defined as a potential range between the onset

Electrochemistry of Nanocrystalline and Microcrystalline Diamond

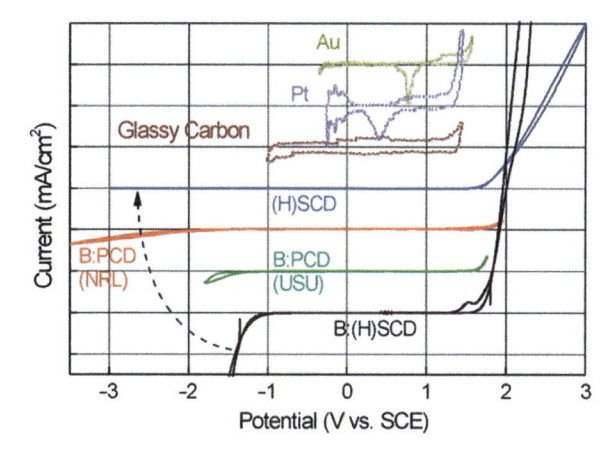

Figure 15.3 Cyclic voltammograms of water electrolysis in 0.5 M H_2SO_4 at two boron-doped polycrystalline films. B: PCD(NRL) with 5×10^{19} [B] cm^{-3} and B: PCD(USU) with 5×10^{20} [B] cm^{-3} are compared to a single crystal boron-doped diamond B: (H)SCD with 3×10^{20} [B] cm^{-3} and with an undoped diamond (H)SCD. Data for Pt, Au and glassy carbon are also presented. The plots were shifted vertically for the comparison.
Reproduced with permission from ref. 5. PCD = polycrystalline diamond; SCD = single crystal diamond.

of hydrogen evolution (rise of current at negative potentials) and oxygen evolution (rise of current at positive potentials) reactions. Figure 15.3 shows cyclic voltammograms obtained during water electrolysis in 0.5 M H_2SO_4 solution (data taken from ref. 5). The plots were shifted vertically for better comparison. It is noteworthy that the electrochemical potential window of the diamond electrode is the greatest, whereas the background current is the smallest in comparison to conventional electrodes. In addition, by changing the boron-doping level it is possible to control the range of the potential window.

Due to the exceptional microstructural stability of the material at high temperatures and high current densities, diamond is also of a great interest in the development of important electroanalytical applications, such as sonoelectroanalysis, where the application of ultrasound in electrochemical systems results in the development of cavitational processes, which enhance mass transport to the electrode surface, therefore improving sensitivity and preventing the electrode from undesirable adsorption.[6]

Electrode fouling is a general problem in electrochemistry. The decreased amount of groups on the diamond surface results in weak adsorption of the analyte to the diamond electrode, thus leading to improved resistance to fouling. Another striking characteristic of this electrode is the high reproducibility of the electrochemical response it provides, also attributed to the reduced amount of surface groups. Finally, the long-term responsiveness and robustness, as well as the optical transparency of diamond in the UV/vis and IR regions of the electromagnetic spectrum, make it attractive for spectroelectrochemical applications.

Although natural diamond is an electrical insulator with a bandgap of 5.47 eV, its boron-doped form is electrically conductive. Boron doping became most prevalent due to chemical vapour deposition (CVD) growth techniques, where the boron incorporation was accomplished by the addition of gaseous boron compounds, such as diborane or trimethyl-borane, to the growth chemistry. It is noteworthy that other types of dopants, such as phosphorus,[7] nitrogen,[8] hydrogen[9] or sulphur,[10] have been explored; however, most of the work in diamond electrochemistry is focussed on the boron-doped type materials.

Although boron forms an acceptor level approximately 0.35 eV above the valence band when present at low concentrations ($\sim 10^{17}$ cm^{-3}), superior conductivity and electrochemical properties are achieved with boron concentrations of 10^{19}–10^{21} cm^{-3} and carrier mobilities of 0.2–0.5 cm^2 V^{-1} s^{-1} as boron centers interact and broaden into an impurity band that moves towards the valence band and mediates conduction when the B concentration increases.

The choice of diamond, particularly for sensing applications, depends on a broad range of factors, from commercial availability and material cost to selection of the appropriate phase purity, microstructure of the crystallites comprising the film, electrical and surface conductivity of the material, and surface chemistry. For the purpose of clarity, it can be useful to describe the material available at the present time in accordance with the classification widely accepted in the literature: by the grain size—microcrystalline, nanocrystalline and ultrananocrystalline diamond films—and by the crystalline orientation, i.e., polycrystalline and monocrystalline films. The following section summarises several important structural properties of diamond films and their influence on electrochemistry.

The majority of research in diamond electrochemistry has been performed on microcrystalline B-doped diamond (MCD). It is available in the form of large area wafers and as-grown films on large-area substrates with a crystal size of different crystallographic orientation, which varies in the range of 5–50 μm. A typical electron micrograph of MCD is shown in Figure 15.4(a). The electrochemical properties of this diamond are mainly influenced by the phase purity, and the boron concentration and its distribution. These characteristics can effectively be controlled by the choice of the growth conditions, in which low methane–hydrogen ratios in the reactor feedstock result in phase-pure materials with large grain sizes, and a higher methane concentration leads to the opposite effect. B-doped MCD with relatively low sp^2 content has found widespread use in the electroanalysis, electrosensing[11] and electrosynthesis of organic[12,13] and inorganic compounds,[14] and in waste water treatment.[15] Detailed information on the electrochemistry of MCD are available to the reader.[6]

Nanocrystalline diamond (NCD) is generally grown in hydrogen-rich carbon-lean CVD growth environments and can be produced in a

Figure 15.4 Scanning electron microscopy (SEM) images of (a) MCD B-doped diamond; (b) NCD B-doped diamond.
Reproduced with permission from ref. 16.

boron-doped form with very low (<1%) to moderate (up to 3–5%) non-diamond content, normally located at grain boundaries or defects.[17] The crystallite sizes of NCD vary in the range of 100–200 nm. A typical scanning electron microscopy (SEM) image of NCD is presented in Figure 15.4(b). Although the intrinsic electrochemical properties of NCD are rather similar to MCD, there are several specific features that degrade for nanocrystalline films. If we compare the width of the potential window as the phase purity of diamond is varied, it is decreased generally from 3–4 V for a phase-pure diamond to around 2.5 V for the nanocrystalline analogue. This clearly illustrates a reduction in the electrochemical selectivity of the electrode. In order to explain this feature, it should be mentioned that phase-pure diamond displays good electrochemical responsiveness for simple electron transfer processes, but chemically complicated steps, such as water decomposition, are hindered—presumably because intermediates cannot be stabilised by bonding to the inert electrode surface. As the phase purity falls, the selectivity reduces and the potential window is decreased. In addition, the surface conductivity of NCD can be deteriorated by the high concentration of grain boundaries also caused by a lower phase-purity of the nanocrystalline material.

The formation of ultrananocrystalline diamond microstructures with characteristic crystallite sizes in the range of 2–5 nm prevails in argon-rich, hydrogen-poor environments. These films consist of up to 5% sp^2 carbon content, which contributes to the enhanced through-film conductivity due to the greater proportion of intercrystallite regions. It is noteworthy that UNCD conductivity can be significantly increased by n-type doping with nitrogen.[18] The electrochemistry of diamond films is greatly influenced by their sp^2/sp^3 carbon ratio, and the differences between NCD and UNCD films that arise are likely to be related to the greater number of active sites located within the sp^2 region, contributing to improved catalytic activity of the latter.

Although some studies of single crystal diamond electrodes have been performed,[19–21] the availability of boron-doped single crystal diamond films is relatively scarce. The use of single-crystal type diamond electrodes is significant in the evaluation of the fundamental properties of electrochemical systems, whereas polycrystalline diamond, containing sp^2 type carbon and a variety of non-uniformly-doped crystal faces, does not fully reflect its intrinsic features.

15.4 Electrochemistry of Nanodiamond Powder Films

Many conventional carbon-based electrodes are successfully employed for the determination of many target compounds, such as H_2O_2, glucose or DNA, although this electrochemical detection is generally confined to the limitations of selectivity. For example, many organic molecules can be detected by the current arising from electrochemical oxidation or reduction; however, various redox species present in the analyte medium may be oxidised or reduced at similar potentials, therefore giving rise to a false signal. In addition, such detection may occur in the same potential region as the oxygen evolution reaction from water, which also masks the signal. Finally, organic compounds in solution or, in particular, the products of their oxidation are likely to adsorb to the electrode surface and consequently block its activity.

In order to overcome these problems, conductive porous layers were recently introduced on the surfaces of conventional electrodes.[22] The utilization of such porous layers can result in the improved detection of target compounds. In addition, by varying both the hydrophobicity/hydrophilicity of the porous layer and the diffusion regime, adsorption on the electrode surface can be significantly decreased, therefore prolonging the operational life-time of the material.[23]

Diamond electrodes are the least susceptible to fouling compared to any other conventional carbon electrode and their wide potential window minimises undesirable masking effects arising from oxygen or hydrogen evolution reactions. However, as aforementioned, some chemically complicated electrochemical reactions have imposed the requirement of improved diamond sensitivity. Thus, the porous layer approach seems to be especially appealing for diamond sensitivity enhancement, which can be reached by creating chemically selective permeable coatings of a high surface area as schematically shown in Figure 15.5.

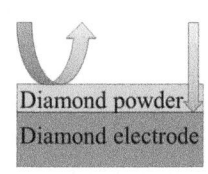

Figure 15.5 Scheme of a porous layer on top of a diamond electrode.

Two particular modifiers, which are finely divided diamond powders, namely detonation nanodiamond powder and boron-doped nanocrystalline diamond powder, can be employed in this 'porous layer' modification strategy in order to provide an enhanced surface area, unique surface chemistry and improved selectivity. These diamond powders differ fundamentally with regard to their conductivity. Whereas the latter is B-doped and conductive, the former is undoped and insulating. A comparative analysis of the essential electrochemistry of powder films immobilised on diamond electrode is presented below.

Detonation nanodiamond (DND) is a carbon-based nanomaterial. It is formed, as the name suggests, by the detonation of carbon containing explosive mixtures, where the application of high temperature and high pressure shock waves thereby result in the formation of small quantities of diamond particulates, approximately 5–10 nm in diameter. The material recently became freely available at low cost and in large quantities, mainly from China and Russia. Nanodiamond particles consist of an sp^3 diamond core surrounded by an amorphous shell with a highly functionalised surface.[24] Raman and X-ray photoelectron spectroscopy (XPS) studies of the DND surface revealed that the sp^2-carbon content is mainly related to the specific surface defects.[25]

The surface chemistry of DND particles strongly depends on production and purification procedures. Commercially obtained DNDs are normally found in a highly oxidised form, containing hydroxyl, carbonyl, carboxyl and lactone surface groups.[26] The existence of these hydrophilic groups and sp^2 carbon residues results in the strong agglomeration of 100–1000 nm particles, of which it is possible to break apart only partially by powerful ultrasound techniques or ball milling.[27]

Recently, the synthesis of metallically B-doped diamond nanopowders (BDDP) comprised of 100–500 nm nanoparticles has been reported.[28] The material can be produced by either milling of B-doped diamond films[29] or CVD growth of the conductive layer over the DND surface.[30] Initial studies have shown that the two physical forms of BDDP and DND are quite different; where the latter, as expected, consists of aggregated nanoparticles, milled BDDP displays a very broad size distribution as shown in Figure 15.6. By carrying out a centrifugation step, uniform BDDP films with particle sizes of approximately 100 nm can be produced. The Raman spectrum of BDDP is found to be similar to that of boron-doped polycrystalline diamond films with asymmetrical broadening of the characteristic diamond peak at 1331 cm^{-1}. This distinctive feature of the peak shape is related to Fano interference, which results from a quantum interference between the discrete zone centre phonon and the continuum of electronic levels at the same energy.

Typical XPS C 1s spectra of boron-doped nanodiamond powder and detonation nanodiamond surfaces are presented in Figure 15.7. The changes in the profiles arise from chemical shifts due to differing amounts of sp^2 carbon and highly oxidised centres on the DND surface, such as carboxylic acid terminations (-COOH) and carbonate ester groups (-C(O)–O–C(O)-).

Figure 15.6 SEM image of as-milled BDDP.

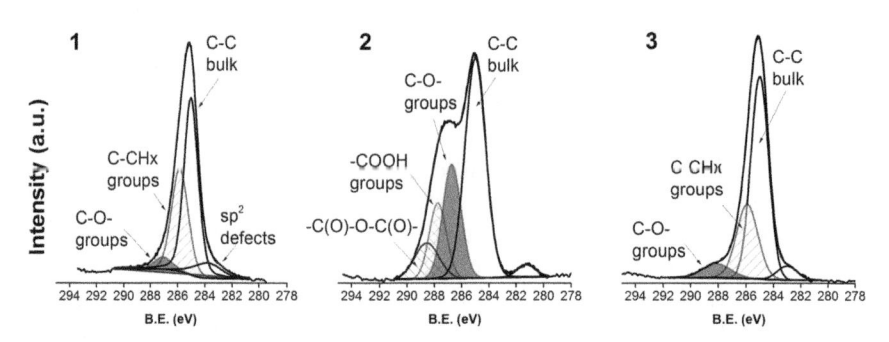

Figure 15.7 XPS spectra of (1) BDDP; (2) DND; and (3) boron-doped diamond electrode (BDDE).

These differences in surface termination affect the observed electrochemistry, since they change not only the electronic properties of the surface, but also the ability of the surface to form bonds with intermediates.

Although DND is undoped and nominally insulating, which would rule out electrochemical application of the material, detonation nanodiamond powders, immobilised onto electrode surfaces, were found to show complex electrochemical behaviour related to the presence of redox-active surface groups on DND.[31] Differential pulse voltammetry studies have also revealed that the DND layer itself can be either oxidised or reduced through the oxidation or reduction of its surface states at particular potentials.[25]

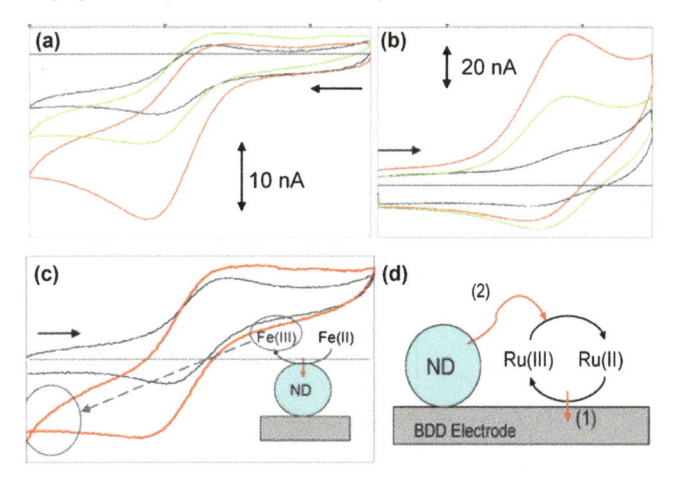

Figure 15.8 (a) Cyclic voltammogram (CV) of 1 μM Fe(CN)$_6^{3-}$ reduction in 0.2 M pH 4 phosphate-buffered saline (PBS) at clean BDD electrode (black) and BDD modified with two monolayers of DND (red, first scan; green, second scan) 10 mV s^{-1}; (b) CV of 1 μM Ru(CN)$_6^{4-}$ oxidation in 0.2 M pH 4 PBS at clean BDD electrode (black) and BDD modified with two monolayers of DND (red, first scan; green, second scan) 10 mV s^{-1}; (c) CV of 1 μM Fe(CN)$_6^{4-}$ oxidation in 0.2 M pH 4 PBS at clean BDD (black) and BDD modified with two monolayers of DND (red) 10 mV s^{-1}; inset shows posited spontaneous oxidation of Fe(CN)$_6^{4-}$ by ND surface state * to form Fe(CN)$_6^{3-}$, red arrow indicates direction of electron transfer (ET); (d) schematic of suggested mechanism for current enhancement of Ru(CN)$_6^{4-}$ oxidation in the presence of ND: (1) Ru(CN)$_6^{4-}$ is oxidised at underlying BDD electrode to form Ru(CN)$_6^{3-}$ and (2) Ru(CN)$_6^{3-}$ is spontaneously reduced by ND surface states to regenerate Ru(CN)$_6^{4-}$; red arrow indicates direction of ET. Reproduced from ref. 31 with permission from the *Phys. Chem. Chem. Phys.* Owner Societies.

The experimental evidence confirmed that the presence of the DND layer on the electrode surface was also responsible for significant enhancement of redox currents of the standard electrochemical couples in diluted solutions. It was found that the reduction current at the DND surface in 1 μM Fe(CN)$_6^{3-}$ solutions (red line, Figure 15.8(b)), as well as the oxidation current of 1 μM Ru(CN)$_6^{4-}$ (red line, Figure 15.8(c)) were several times greater than that of B-doped diamond electrodes BDDE (black line). This gave the suggestion that the 'feedback' mechanism was feasible for the observed electrochemistry. In particular, it was proposed that it is the underlying electrode that reduces the redox-active compound. However, when this reduced form is in contact with DND surface groups, it becomes spontaneously oxidised, thus giving rise to the amplified currents as shown in Figure 15.8(e).[31]

Several voltammetry studies have been carried out either at DND powder film electrodes[32] or BDDP film electrodes;[28] however, the comparative analysis of their electrochemical properties has not been reported. Here, we will present the relative study[33] on two types of these powder films. Figure 15.9 shows

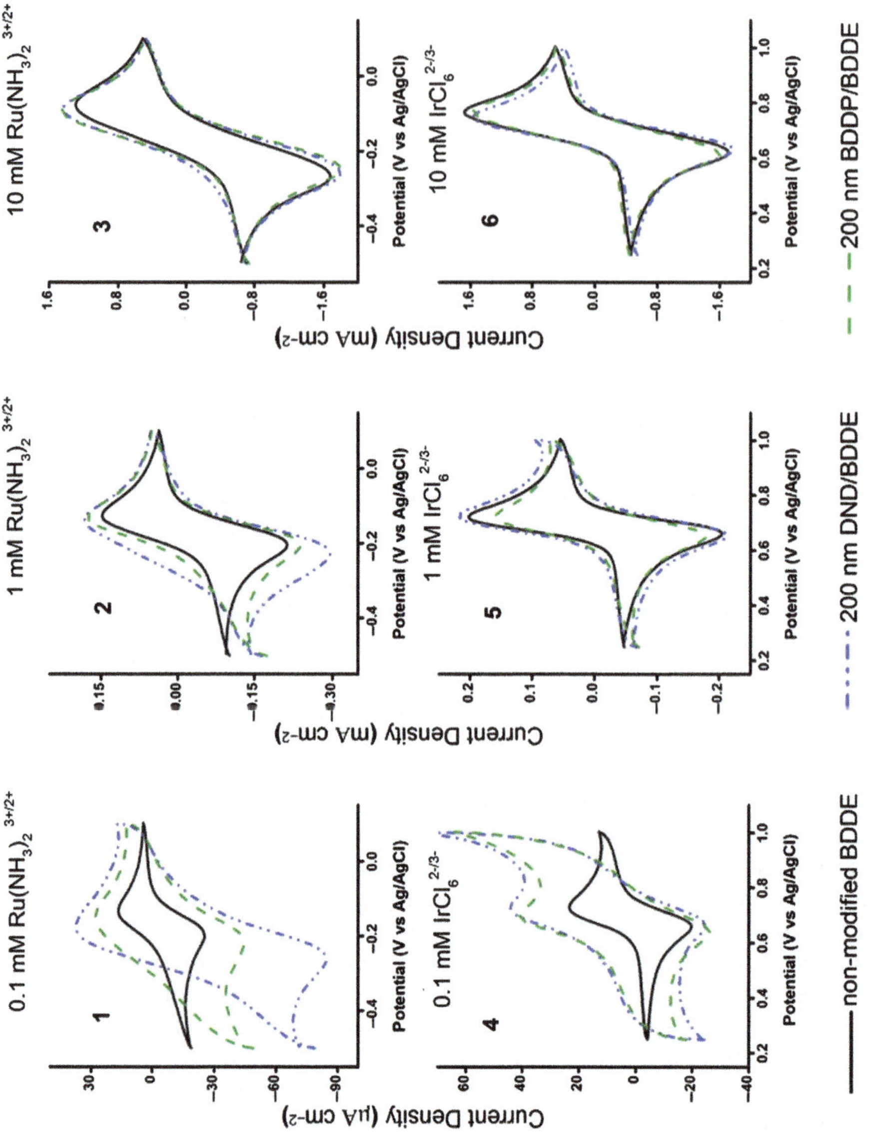

0.1 mM Ru(NH3)2 3+/2+
1 mM Ru(NH3)2 3+/2+
10 mM Ru(NH3)2 3+/2+
0.1 mM IrCl6 2-/3-
1 mM IrCl6 2-/3-
10 mM IrCl6 2-/3-
1
2
3
4
5
6
Current Density (μA cm-2)
Current Density (mA cm-2)
Current Density (mA cm-2)
Potential (V vs Ag/AgCl)
non-modified BDDE
200 nm DND/BDDE
200 nm BDDP/BDDE

Figure 15.9 CV of 0.1, 1 and 10 mM Ru(NH$_3$)$_6$$^{3+/2+}$; IrCl$_6$$^{2-/3-}$; Fe(CN)$_6$$^{3-/4-}$; and 0.1 and 1 mM CH$_3OH^{+/0}$ at bare BDDE, 200 nm DND/BDDE and 200 nm BDDP/BDDE. Reproduced from ref. 33 with permission from the *Phys. Chem. Chem. Phys.* Owner Societies.

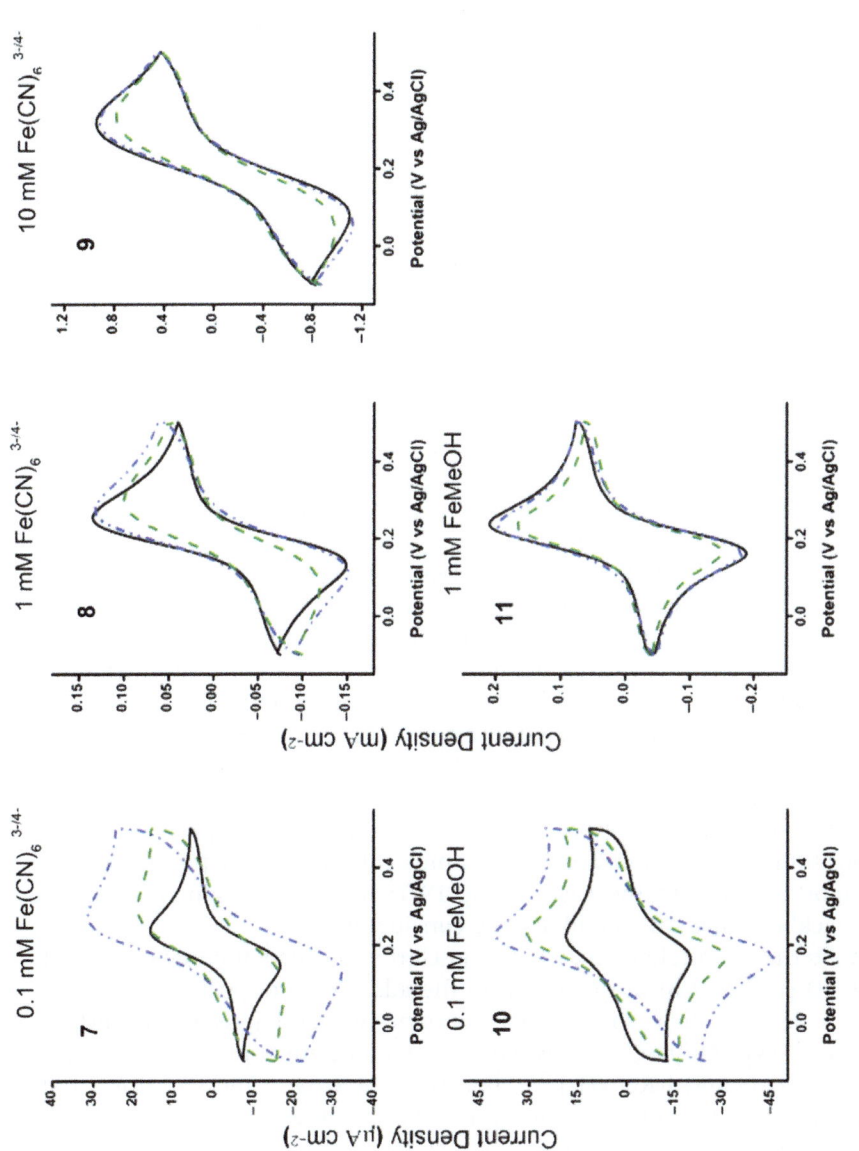

examples of cyclic voltammograms obtained in standard redox solutions at naked diamond electrodes and diamond electrodes modified with DND and BDDP layers correspondently.

At high concentrations of redox species, it is observed that diffusional electrochemistry at films prepared from each of the two powders is relatively similar to that of the planar diamond electrode. However, significant differences arise at low concentrations, where the powder-modified electrodes have shown amplified currents.

The CV response of DND porous films was found to be greater than that of BDDP and naked BDDE at analyte concentrations of 0.1 mM. The redox response obtained in hexaammineruthenium (III) chloride (Figure 15.9(1)) and potassium hexachloroiridate (IV) (Figure 15.9(4)) solutions has shown asymmetry in the peak currents: the maximum increase of the reduction current in $Ru(NH_3)_6^{3+}$ and the maximum increase of the oxidation current in $IrCl_6^{2-}$ solutions were observed. However, the current peaks in potassium ferricyanide and ferrocenemethanol solutions were found to be symmetric, as shown in Figure 15.9(7) and Figure 15.9(10).

Although this current enhancement on the porous film layer can arise from the diffusional regime switch (from purely planar diffusion to 'thin-layer' diffusion), the mathematical 'semi-infinite' diffusion model[1,34] was not sufficient to describe the experimental data recorded at these layers. It is clear that, if the current increase was caused by the diffusional regime switch, then the electrochemical behaviour of two porous films of the same thickness, particle size and pore size should have been similar. However, significant differences, which were recorded between the DND and BDDP films, suggest that this electrochemical behaviour cannot be fully attributed to the switch of the diffusional regime at the porous layer.

In order to explain such dissimilarities, the contributions from three processes can be taken into account: (1) the planar diffusion of redox molecules from the bulk solution; (2) the diffusion of these redox molecules through the solution trapped within the interconnected pockets of the porous film; (3) the adsorption of redox molecules within the powder film. It is important that, in the case of the DND powder, the aforementioned 'feedback' mechanism should also be considered.

An alternative mechanism, which can clarify the current enhancement at the DND surface, can be attributed to the adsorption of positively charged redox molecules, such as $Ru(NH_3)_6^{3+}$, onto negatively charged COO^- surface groups of DND. Although this explains the enhancement of the redox peaks of positively charged $Ru(NH_3)_6^{3+}$ species at the oxygenated and untreated surfaces, such a type of adsorption would not be necessarily feasible for the negatively charged redox couples, such as ferricyanide, ferrocyanide or initially neutral ferrocenemethanol, which are unlikely to be susceptible to adsorption.[35]

In order to explain this behaviour, the redox activity of the DND surface groups was studied *in situ* by infrared (IR) spectroscopy, which was used to monitor nanodiamond surface transformations in the presence of redox

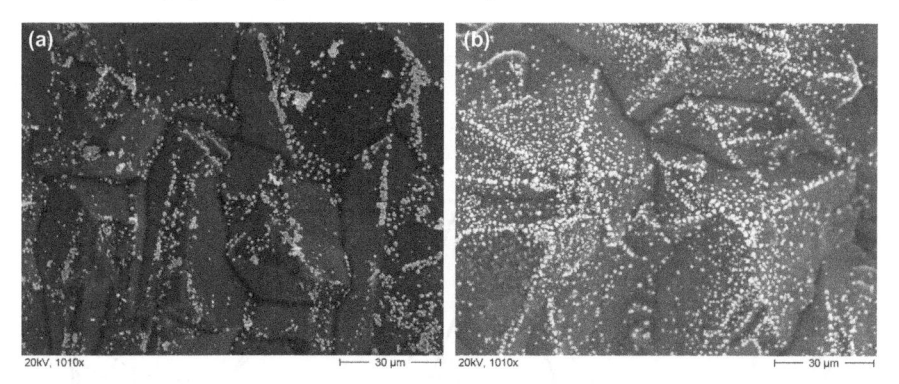

Figure 15.10 SEM images of platinum particles deposited at a potential of −0.2 V on: (a) naked BDDE; (b) 200 nm DND/BDDE.

couples.[36] This study revealed that the electron transfer between the DND surface and the probe redox molecule results in the oxidation of ~8.5% of surface alcohol groups, with the synchronous formation of unsaturated quinone-like groups. The authors concluded that the redox nature of such transformations (oxidation of the alcohol surface groups to carbonyl groups) was responsible for the changes in amplified response rather than simple adsorption of redox compounds within the film.

At higher concentrations, this feature of the DND film becomes invisible due to the higher overall current densities, and therefore the contribution of the porous layer to the electrochemical response becomes insignificant.

Another interesting area, where the properties of the film influence the electrochemistry observed, relates to electrocatalytic applications. Pt nanoparticles immobilised on diamond electrodes have a range of interesting electrocatalytic applications in fuel cells and hydrogen production. However, a problem arises from the lack of nanocatalyst stability on the diamond interface, leading to Pt loss and, as a result, an overall reduction in performance. Therefore, the implementation of porous nanodiamond powder coatings can be explored to improve the electrode performance.[37] Our work has shown that nanodiamond permeable layers provide not only highly uniform nanocatalyst dispersions, as shown in Figure 15.10, but also the highest activity in a model reaction of methanol electrooxidation (Figure 15.11) and the best long-term stability related to Pt bounded in porous film (Figure 15.12). It is noteworthy that a significant increase in the electrocatalytic activity and nanoparticle stability was observed on Pt catalysts immobilised on conductive BDDP powder films.

15.5 Functionalised Diamond Electrodes

A fundamental requirement for an electrochemical sensor is that the device provides a selective and stable response to the target compound without generating a false signal in the presence of co-interfering species. Various

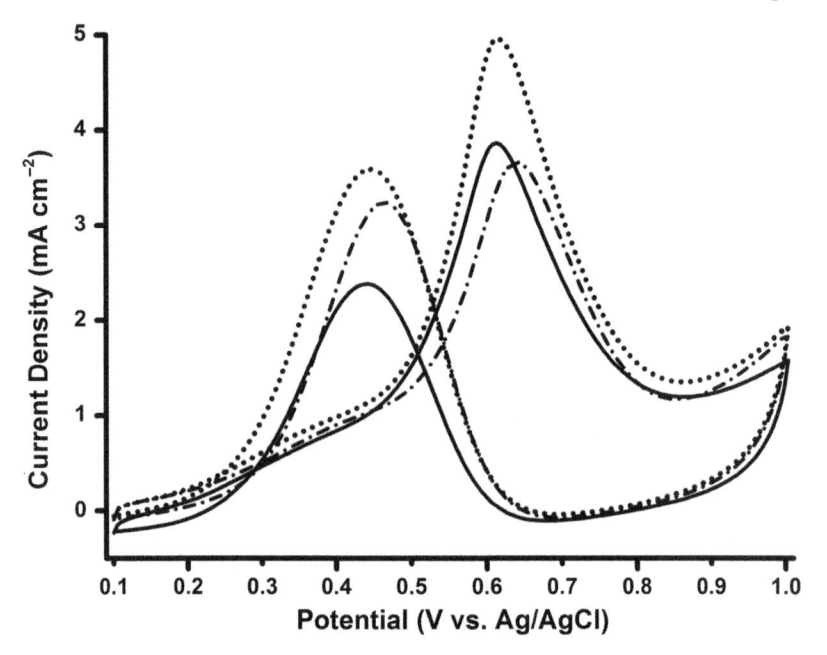

Figure 15.11 CV in 0.1 M CH$_3$OH, 0.5 M H$_2$SO$_4$ at (· · ·) Pt/200 nm BDDP/BDDE; (–) Pt/200 nm DND/BDDE and (–··–)Pt/bare BDDE.
Reproduced from ref. 33 with permission from the *Phys. Chem. Chem. Phys.* Owner Societies.

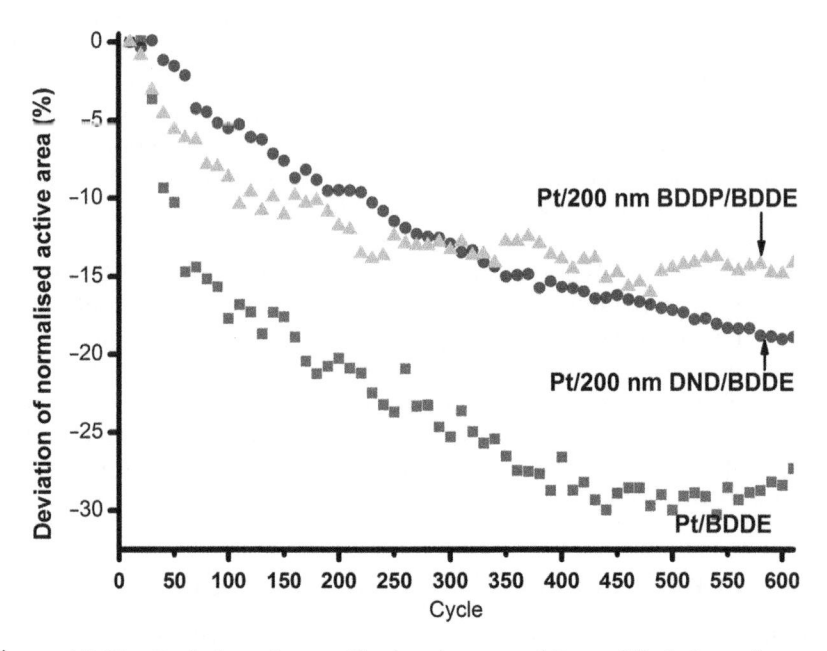

Figure 15.12 Deviation of normalised active area of Pt-modified electrodes.
Reproduced from ref. 33 with permission from the *Phys. Chem. Chem. Phys.* Owner Societies.

electrochemical approaches allow characterisation of differing redox molecules on the basis of half-wave potential of the redox reaction, or rather subtle features such as the rate of diffusion to the electrode surface, which can be complicated by the presence of chemically similar species. This situation can be resolved if external techniques are employed. For instance, chemically similar compounds can be separated by passing through a chromatographic system, or a selective membrane can be introduced, so only the target species are allowed to reach the electrode surface.

A sophisticated alternative was suggested with a 'molecular recognition' strategy, where any electrochemical signal measured in a system corresponds only to the target compound. This is possible to accomplish by chemical functionalisation of an electrode surface, so it only recognises the target compound. As an example, undesirable nonspecific protein adsorption, observed in diamond biosensors, can be minimised by the alteration of the electrode termination. On the other hand, if the specific adsorption is desirable, it can be manipulated by the surface termination. In some cases changes in diamond surface termination can result in improved electrode kinetics. Another way to achieve high electrode specificity is related to the immobilisation of functional biomolecules, such as DNA or proteins, which can be covalently grafted to the diamond surface. It has been shown that since this suggests the formation of strong C–C bonds, such a functionalisation is extremely stable on the diamond surface.[38]

During the past few years versatile methods of surface modification have been developed. Many of them have been further extended to the functionalisation of diamond nanoparticles. The following section discusses the types, methods and electrochemical properties of functionalised diamond electrodes.

15.5.1 Types of Diamond Surface Termination

Properties of hydrogen[39] and oxygen surface terminations of the diamond electrodes have been widely reviewed. NH_2-, F- and Cl-terminations are less common and their electrochemical characteristics are still under investigation.

Carbon atoms at the diamond interface are presented in their reduced form. In order to restore the valence the diamond surface undergoes a reconstruction to form a π-bonded atomic array. These surfaces are very reactive and, if in contact with other phases, tend to chemisorb species. Chemisorbed hydrogen termination of the diamond surface is normally generated in the CVD reactor on account of the growth process, which involves the use of hydrogen plasma; however, some weakly chemisorbed hydrocarbons can be also found.

H-termination is very stable chemically. In order to reproduce the reconstructed surface and remove hydrogen termination it should be heated more than 800 °C in a vacuum or inert atmosphere. The hydrogen-terminated surface was also found to be very stable in air at room temperatures, although slow oxidation does take place if it is exposed to air for long periods.

Hydrogen termination gives rise to a unique phenomena: characteristic surface-induced conductivity, only feasible in the presence of electrolyte.[40] This conductivity can be explained through the 'transfer doping' mechanism, which suggests that the surface conductivity emerges from the missing valence band electrons being tunnelled into the electrolyte.[41,42] For this tunnelling to be possible, the energy level of the valence-band maximum of the diamond surface should be greater than the chemical potential of the electrolyte. As hydrogen termination results in the formation of a C–H surface dipole layer with a negative charge concentrated on the carbon ($C^{\sigma-}$) atom and a positive charge on the hydrogen ($H^{\sigma+}$), this dipole generates electrostatic potential, perpendicular to the surface over a distance of the order of the C–H bond length (1.1 Å), and shifts all energy levels of diamond up to 1.6 eV with regard to the chemical potential of an electrolyte.[43] This results in the 'negative electron affinity', which is caused by the conduction-band states of the diamond being above the vacuum level of the electrolyte. Therefore, it is energetically favourable for the electrons from the diamond valence band to tunnel into the empty electronic states of the electrolyte until a thermodynamic equilibrium between the Fermi level of diamond and electrochemical potential of the electrolyte is established.

A general trend to arise from studies of diamond electrodes is that redox couples, such as $Ru(NH_3)_6^{3+/2+}$ and $IrCl_6^{2-/3-}$, show electron transfer kinetics for simple 'outer sphere' electron transfer processes, which involve minimum changes in bonding configuration of the solvated redox molecule. Although the effect of the hydrogen-terminated surface on the electrode kinetics was not observed for 'outer sphere' reactions, where the chemical interaction between the relevant species and the surface is lowest, it otherwise becomes a significant feature in diamond electrochemistry. For example, $Fe(CN)_6^{3-/4-}$ couples were found to be quite sensitive to the surface chemistry of carbon electrodes and the electron transfer kinetics can change by several orders of magnitude with regard to the electrode termination. Typical results are presented in Figure 15.13 [44] The peak-to-peak separation observed on a chemically oxidised diamond (WCO) was found to be 243 mV, thus indicating a slow electron transfer. On the contrary, the peak-to-peak separation on the hydrogenated surface (HPT and CT) was estimated to be 60 mV, hence indicating that a faster electron transfer was taking place. These differences in the electrode kinetics can be related both to the decreased electron affinity of the oxygenated diamond surface[45] and to the electrostatic repulsion between the negatively charged $Fe(CN)_6^{3-/4-}$ species and negatively charged functionalities on the oxygenated diamond surface.

If the hydrogenated diamond has been exposed to air for prolonged periods, changes in the surface termination occur. The exact nature of the oxygen groups formed at the diamond surface during this exposure is less well known; however, the available evidence suggests that carbonyl and ether groups dominate when diamond is in contact with dry oxygen. Hydroxyl

Electrochemistry of Nanocrystalline and Microcrystalline Diamond 373

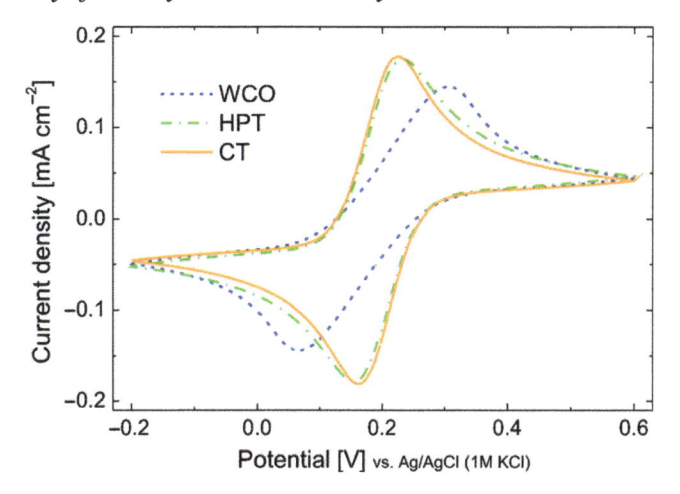

Figure 15.13 CV of 1 mM Fe(CN)$_6^{3-/4-}$ in 0.1 M KCl at 50 mV s^{-1} at wet chemically oxidised (WCO), hydrogen plasma-terminated (HPT) and cathodically treated (CT) diamond electrodes.
Reprinted with permission from ref. 44. Copyright 2010, American Institute of Physics.

groups were found to prevail if the surface was treated in atmospheric moisture.[46]

Diamond oxidisation can be produced by oxygen plasma treatment, anodic electrochemical oxidation, ultrasonic treatment, ozone oxidation or wet chemistry in the presence of mineral acids mixtures, such as aqua regia or piranha solution.

It should be noted that the variation in the termination from hydrogen to oxygen results in the loss of hydrogen-induced conductivity, and charge transfer from the oxygenated surface is inhibited through the positive electron affinity, as in the example with Fe(CN)$_6^{3-/4-}$ kinetics given above. Therefore, the electrochemical activity of the diamond electrode can be controlled by means of surface termination. The difference in electrochemical activity of hydrogen- and oxygen-terminated diamond introduces a way to produce a diamond surface in conductive–nonconductive patterns, which is of interest for biosensing applications.

Fluorine termination can be generated by the exposure of diamond electrode to F radicals, CF$_4$ and C$_4$F$_8$ plasmas,[47] and atomic or molecular fluorine beams,[48] as well as chemical treatment in anhydrous HF. Surface fluorination results in an exceptional extension of the potential window of diamond electrodes up to 5 V before hydrogen and oxygen evolution reactions, therefore allowing the electrochemical detection of various compounds in the widest potential range.[49] Due to the imposed hydrophobicity, fluorinated electrodes were found to exhibit the lowest electrocatalytic activity in reactions that involve adsorption states, thus controllable adsorption of bioactive compounds can be achieved.

Chlorine termination can be obtained either by plasma, or photochemical or thermal decomposition of chlorine-containing gases. Although Cl-terminated diamond is also hydrophobic, the stability of the C–Cl bond was reported to be much weaker; it tends to form C–OH groups in the presence of water or moisture. Chlorine termination is generally used as an intermediate step in more complex diamond functionalisations.[50]

Amination of the diamond surface may be achieved through the exposure of H- or Cl-terminated diamond to NH_3 gas under UV irradiation.[51] NH_2-terminated surfaces allow various bioactive compounds to be bonded to the diamond electrodes, which will be discussed further.

A representative example of differences between the electrochemical activity of aminated, fluorinated and oxidised detonation nanodiamond powder films in 0.05 M $K_3[Fe(CN)_6]$ solution is presented in Figure 15.14 [52]

The electrochemical behaviour of aminated DND (A-ND) electrodes differ significantly from that of the fluorinated surface (F-ND). Positively charged NH_3^+ surface groups attract $Fe(CN)_6^{3-/4-}$ species by electrostatic interaction, therefore resulting in enhanced currents, whereas a highly hydrophobic C–F surface may repel ferri/ferro cyanide anions.

Among the surface terminations described above, detonation nanodiamond was reported to be terminated with hydroxyl (DND-OH), ester (DND-O-R) and peptide surface groups (DND-CO-NHR). In addition, such processes as Suzuki couplings,[53] Diels–Alder cycloadditions[54] and Prato reactions[55] have been successfully performed on DND surfaces. Detailed reviews, such as in ref. 50, are available to the reader.

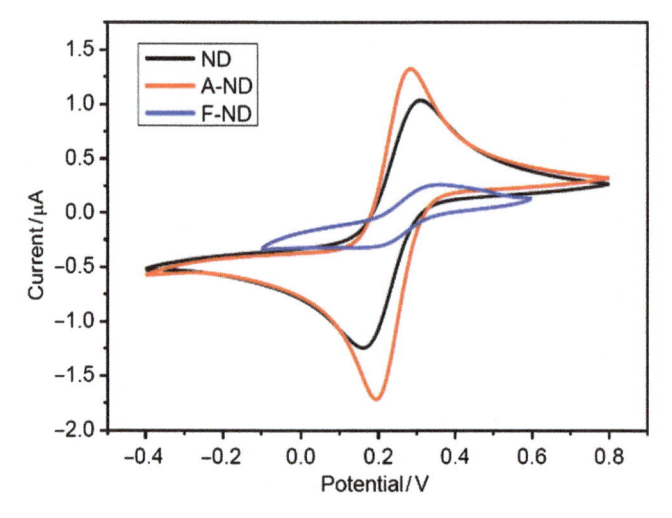

Figure 15.14 Cyclic voltammogram of aminated DND (A-ND), fluorinated DND (F-ND) and as-received DND (ND) electrodes in 0.05 M $K_3[Fe(CN)_6]$ in 0.1 M KCl *vs* Ag/AgCl.
Reprinted with permission from ref. 52.

15.5.2 Photochemical, Electrochemical and Chemical Functionalisation

Photochemical functionalisation of the diamond surface is a very efficient way to implement so-called molecular recognition strategies. It allows not only the modification of the surface groups of the diamond, but also the attachment of various important molecules. The first report on photochemical chlorination, amination and carboxylation with subsequent modification of DNA onto the diamond surface was presented by Takahashi et al.[56] Yang et al showed that alkene molecules can be photochemically attached to the diamond surface.[38] This has opened a new area in diamond chemistry, where biologically important oligonucleotides, proteins and enzymes can be covalently bonded to the surface.

Photochemical functionalisation is based on the activation of C–H bonds at the diamond surface under UV irradiation, followed by covalent C–C coupling of the diamond with external molecules containing vinyl groups (-CH=CH$_2$). This photochemical process has been reviewed thoroughly in several publications.[70] Although the mechanism of photochemical functionalisation of diamond is still not completely explained, it is expected to be similar to those of silicon surfaces. The mechanism can be described in three main steps: (1) ejection of the photoexited electron from the diamond valence band into empty hydrogen-induced states above the vacuum level, as shown in Figure 15.15;

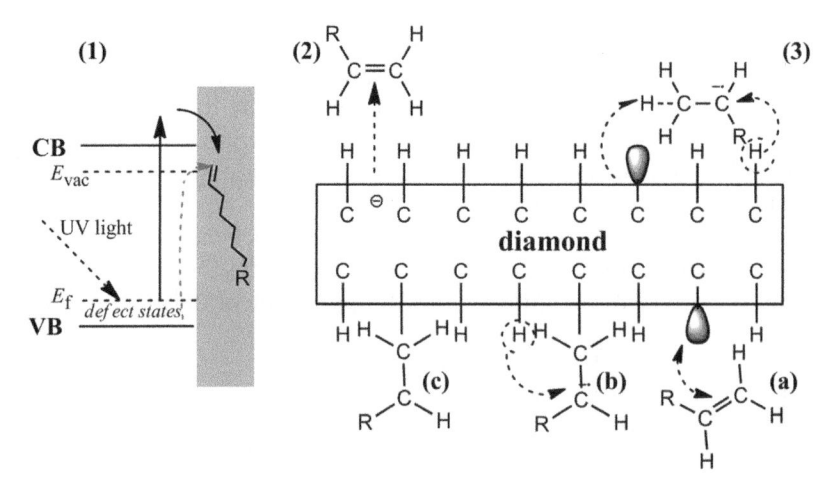

Figure 15.15 Proposed scheme of electron ejection into the liquid phase: (1) excitation from occupied defects and/or surface states to the conduction band followed by diffusion and emission (solid arrow) and direct photoemission from valence band to the vacuum level (dashed arrow); (2) electron transfer from diamond to the olefin; (3) generation of radical–anion with subsequent abstraction of hydrogen atom from diamond, followed by the formation of a dangling bond; (a)–(c) diamond–olefin coupling.
Adapted, with permission, from refs 57 and 58. Copyright 2005 and 2007 American Chemical Society.

the ejection is feasible due to the induced surface conductivity and a lowered ionisation potential of the *H*-terminated diamond; (2) a transfer of the electron from hydrogen-induced states to the lowest unoccupied molecular orbital (LUMO) of the unsaturated group of the reactant, which is then followed by the formation of a radical–anion in the vicinity of the diamond surface; (3) this radical–anion removes hydrogen from the C–H bond of the diamond surface and creates a carbon dangling bond (unsatisfied electron density), which is itself very reactive towards alkenes, where C–C coupling occurs.[57]

Nebel *et al.* have shown that photochemical functionalisation can be utilised for biosensor construction by means of DNA attachment to diamond electrodes.[5] In this study a long chain of ω-unsaturated amine was firstly photoattached on the *H*-terminated diamond. It was then functionalised with a hetero-bifunctional crosslinker, and finally, covalently bonded with a thiol-modified single stranded DNA (ssDNA). This immobilisation allows molecular recognition of probe DNA, which binds to ssDNA in solution if it has the complementary base sequence. Such an approach forms the basis of modern array technology for DNA sequencing, generally carried out by means of optical methods. On the other hand, in electrochemical sensors the binding to the complementary (probe) DNA to the DNA-modified electrode surface can be confirmed directly by rapid electrical readout. In addition, a certain character of the electrochemical response in cyclic voltammetry in the presence of standard redox couples can be used to identify the hybridisation type of DNA.

Although photochemical attachment has been successfully utilised, it is noteworthy that it does not display desirable selectivity in the case of conjugated systems, whereas the presence of two or more double bonds results in non-selective surface modification. As an alternative, electrochemical functionalisation presents a selective immobilisation strategy of differing molecules on diamond electrodes.

Swain and co-workers have introduced the electrochemical grafting of aryl diazonium salts on boron-doped diamond electrodes.[59] Schematically, the process is represented Figure 15.16. An applied potential to the diamond

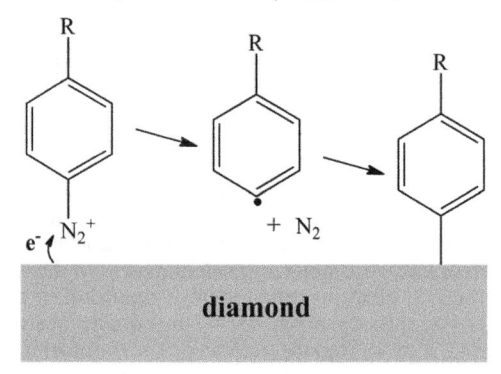

Figure 15.16 Schematic illustration of electrochemical functionalisation by aryl diazonium ions.

electrode causes the electrochemical reduction of the aryldiazonium ion and generates the aryl radical, which then subsequently forms a covalent bond with the electrode surface.

Electrochemical grafting permits a diverse chemistry on the electrode surface not only by varying functional groups (R) of the diazonium compounds, but also by grafting molecules to the pre-modified diamond, therefore resulting in heterogeneous multilayer formation. In addition, the functional group introduced through this process can be reduced or oxidised, therefore yielding suitable anchor sites for the sensing of bio-active compounds. For example, nitrophenyl molecules electrochemically attached to the diamond electrode can be further reduced to an aminophenyl layer, which reacts *in situ* with either the probe DNA or a target protein.[60]

Oxidised diamond surfaces with zero surface conductivity and higher ionisation potentials were found to be inert both to UV photochemical functionalisation and electrochemical grafting. However, the presence of radical initiators, such as ammonium iron sulphate hexahydrate (AISH) in the aryldiazonium solution, can result in successful attachment to the inactive oxygenated diamond.[53]

A somewhat different electrofunctionalisation approach comes from the electrodeposition of metal (metal oxide) nanoparticles on boron-doped diamond electrodes, which can be easily implemented for biosensor production. Electrochemically functionalised diamond electrodes with metal nanoparticles have received increased attention in the past decade and have been widely used to detect target compounds, such as adenine, guanine, DNA, glucose, methanol and H_2O_2. Metal nanoparticles, such as platinum, gold, silver, nickel, cobalt, copper, bismuth and antimony, have been mainly employed in these studies. In electroanalytical applications the use of nanoparticle-modified diamond electrodes has certain advantages compared to the aforementioned systems as they provide enhanced mass transport, high surface area, improved signal-to-noise ratio and long-term stability.[61]

An interesting example of electrochemical sensor manufacturing for DNA detection has been developed on the basis of ZrO_2-modified boron-doped diamond electrodes.[62] ZrO_2 nanoparticles were employed to attach phosphate-terminated DNA oligonucleotides; the scheme of this DNA functionalised electrode is show in Figure 15.17. The system displayed an increased redox response to redox mediator methylene blue, which was quenched in the presence of complementary DNA oligonucleotides, therefore achieving a highly sensitive and selective detection of these molecules in solution.

In addition to photochemical and electrochemical functionalisation through the linker molecule, it is also possible to use a direct modification, where bio-active species can be covalently bonded to the hydrogenated diamond. This approach benefits from the better stability of bioactive compounds on the diamond surface. For example, an enzyme-based amperometric biosensor was produced by direct photochemical functionalisation of the *H*-terminated nanocrystalline diamond with the enzyme catalase from bovine liver for enhanced H_2O_2 detection.[63]

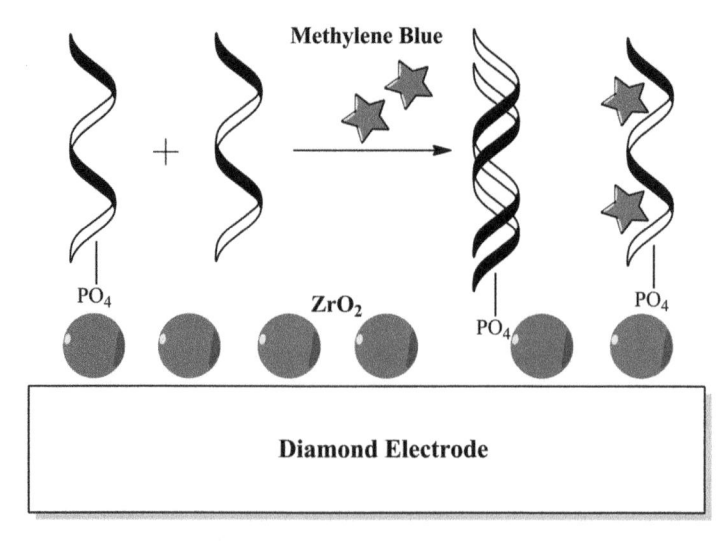

Figure 15.17 Scheme of electrochemical detection of DNA hybridisation by a zirconia-modified diamond electrode.

15.6 Adsorption Properties of Diamond Films

Boron-doped diamond is a well-established electrode material possessing a range of advantageous properties, including chemical inertness. It is generally claimed that diamond displays a significant resistance to electrode fouling, which is of particular advantage in biological media, where adsorption on the electrode is a frequent reason of electrode failure. Although there have been some studies attempting to quantify the extent to which fouling occurs at diamond electrodes, a comprehensive analysis of adsorption on diamond and diamond powder film electrodes has not been reported. Our work presented below discusses both the adsorption electrochemistry of organic molecules and proteins on diamond surfaces. The influence of the nature of the material, as well as the surface termination on the adsorption properties, will be discussed.

The first electrochemical studies of surface-bound reactions on diamond electrodes was reported by Swain et al.[64] Chronocoulometric and cyclic voltammetry investigations of anthraquinone-2,6-disulfonate (2,6-AQDS) reduction was chosen as a model reaction. The chronocoulometry technique measures the charge that has been transferred across the electrode/electrolyte interface when the potential of the electrode is changed from a value where no current is flowing to a value where a faradaic current flows. The charge detected provides quantitative information on the surface coverage of the adsorbed molecules. The authors concluded that both diamond electrodes and glassy carbon (GC) were less susceptible to fouling compared to highly oriented pyrolytic graphite (HOPG) and oxygenated GC electrodes. In the latter case it was suggested that 2,6-AQDS physisorbs on various defect sites of HOPG electrodes and approximately forms a monolayer coverage on the surface, 40% of which can be removed by surface rinsing.

Electrochemistry of Nanocrystalline and Microcrystalline Diamond 379

It is evident that the surface chemistry of a film plays a key role in the adsorption process. In order to understand the level of material fouling, the adsorption properties of DND and BDDP powder films were compared against the naked diamond electrode with regard to the surface termination in the model study with 2,6-AQDS and methyl viologen (MV) solutions. The CVs of 1 mM MV in 0.1 M KCl are presented in Figure 15.18.

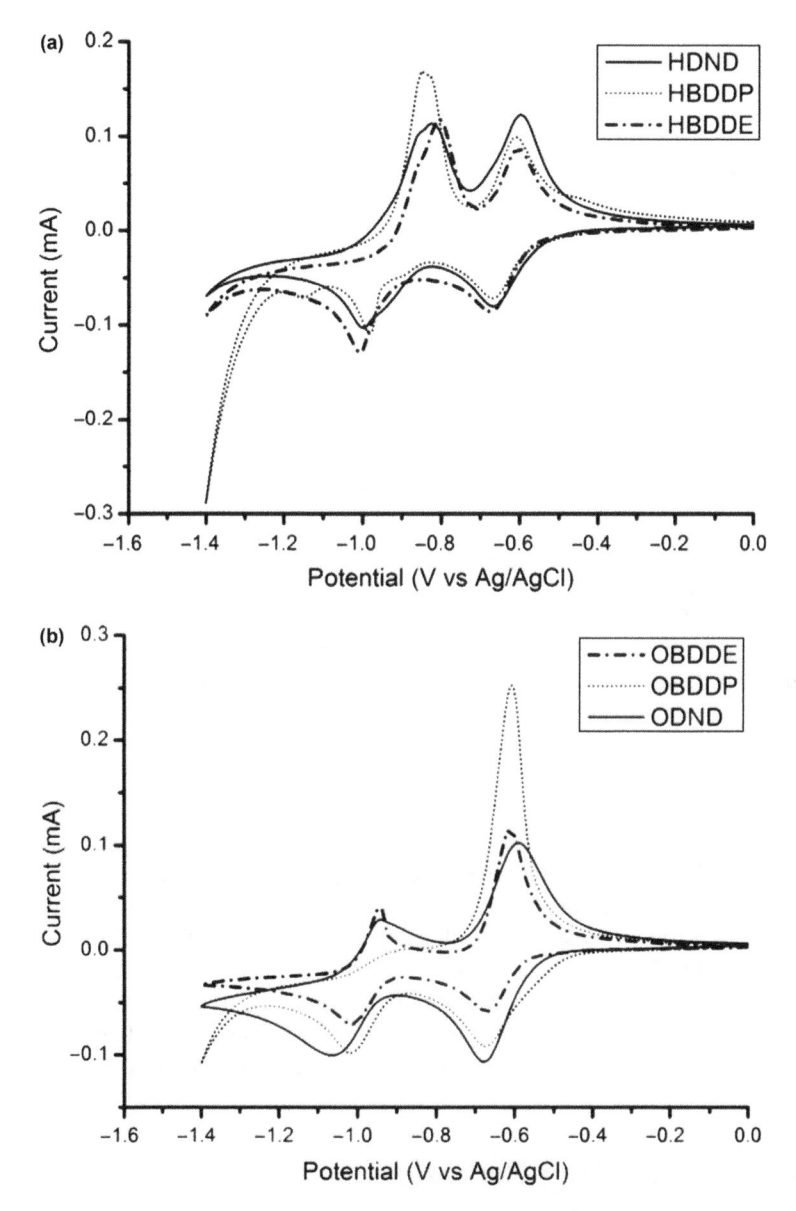

Figure 15.18 CVs of 1 mM MV in 0.1 M KCl, 50 mV s^{-1} at hydrogenated (a) and oxygenated (b) diamond films.

This work showed that, regardless of the surface termination, DND films are subjected to the greatest fouling, whilst BDDP and naked diamond electrodes exhibit a certain chemical inertness. 2,6-AQDS was found to bond preferably to the hydrogenated diamond surface, whereas methyl viologen preferentially bonds to the oxygenated surface. The poorer adsorption of 2,6-AQDS to the oxygen-terminated surface can be attributed to the repulsion of the negatively charged AQDS from the negatively charged oxidised surface; whereas a strong coulombic interaction of $MV^{+/0}$ species with the negatively charged oxygenated surface can explain the dominant trend in MV adsorption.

Another voltammetric study of the quinone species quinizarin and its electrocatalytic reduction of oxygen at boron-doped diamond electrodes revealed that, contrary to the widespread belief that adsorption of organic molecules on BDD is minimal, not only did quinizarin bond to the electrode surface, but this adsorption was also influenced by the pre-exposure of the electrode to organic solvents at low quinizarin concentrations.[65]

Several reports on protein adsorption to diamond surfaces have also been published. It was reported that hydrophilic oxygen-terminated diamond shows a positive affinity for proteins, whilst hydrogen-terminated hydrophobic diamond was found to be inert to the adsorption of proteins, such as fibrinogen.[66]

Nebel et al. have shown that cytochrome C adsorbs both on hydrogenated and oxygenated diamond electrodes; however, it was reported that the adsorption to the hydrophilic O-diamond electrode proceeds without protein denaturation, whereas the adsorption to the hydrophobic H-diamond leads to the protein denaturation.[67] Steinmueller-Nethl et al.[68] investigated the adsorption of the bone morphogenetic protein-2 to oxygen-terminated nanocrystalline diamond and concluded that not only the adsorption of protein prevailed on the diamond surface, but also that the protein was found to be fully active, which made it particularly attractive for medical implants applications.

Another study of bovine serum albumin proteins in the presence of standard ruthenium (III) hexammonium redox complexes has revealed that, although the diamond electrode retains good activity in such solutions, electrode fouling was observed in the long-term due to protein adsorption.[69] Two adsorption processes were distinguished. The first was described as a simple physical adsorption scheme, which can be easily reversed by washing the electrode in water. The second was observed only when negative potentials were applied to the diamond electrode and was attributed to the attraction and interaction of the positively charged protein at the electrode interface.

15.7 Concluding Remarks

The current field of diamond and diamond powder film electrochemistry has been reviewed. It has been summarised that CVD diamond can be synthesised in several forms with differing surface terminations, which have a significant influence on the electrochemical properties of the electrode, and so have to be carefully controlled if predictable results are to be obtained.

Diamond powder films and their superior electrocatalytic applications have been discussed. Various applications of diamond sensors in electrochemical analysis have been shown with regards to the modification strategy, which can be applied in order to provide preferable performance of the device.

References

1. R. G. Compton and C. E. Banks, *Understanding Voltammetry*, Imperial College Press, London, 2011.
2. J. Wang, *Analytical Electrochemistry*, John Wiley & Sons, New York/Chichester, 2000.
3. A. J. Bard and L. R. Faulkner, *Electrochemical Methods: Fundamentals and Applications*, John Wiley & Sons, New York/Chichester, 2001.
4. M. Yoshimura, K. Honda, T. Kondo, R. Uchikado, Y. Einaga, T. N. Rao, D. A. Tryk and A. Fujishima, *Diamond and Related Materials*, 2002, **11**, 67–74.
5. C. E. Nebel, B. Rezek, D. Shin, H. Uetsuka and N. Yang, *Journal of Physics D-Applied Physics*, 2007, **40**, 6443–6466.
6. R. G. Compton, J. S. Foord and F. Marken, *Electroanalysis*, 2003, **15**, 1349–1363.
7. S. Koizumi, M. Kamo, Y. Sato, H. Ozaki and T. Inuzuka, *Applied Physics Letters*, 1997, **71**, 1065–1067.
8. N. Jiang and T. Ito, *Journal of Applied Physics*, 1999, **85**, 8267–8273.
9. O. Gaudin, M. D. Whitfield, J. S. Foord and R. B. Jackman, *Diamond and Related Materials*, 2001, **10**, 610–614.
10. I. Sakaguchi, M. N. Gamo, Y. Kikuchi, E. Yasu, H. Haneda, T. Suzuki and T. Ando, *Physical Review B*, 1999, **60**, R2139–R2141.
11. M. Hupert, A. Muck, R. Wang, J. Stotter, Z. Cvackova, S. Haymond, Y. Show and G. M. Swain, *Diamond and Related Materials*, 2003, **12**, 1940–1949.
12. J. D. Wadhawan, F. J. Del Campo, R. G. Compton, J. S. Foord, F. Marken, S. D. Bull, S. G. Davies, D. J. Walton and S. Ryley, *Journal of Electroanalytical Chemistry*, 2001, **507**, 135–143.
13. U. Griesbach, D. Zollinger, H. Putter and C. Comninellis, *Journal of Applied Electrochemistry*, 2005, **35**, 1265–1270.
14. C. Saez, M. A. Rodrigo and P. Canizares, *Aiche Journal*, 2008, **54**, 1600–1607.
15. M. A. Rodrigo, P. A. Michaud, I. Duo, M. Panizza, G. Cerisola and C. Comninellis, *Journal of the Electrochemical Society*, 2001, **148**, D60–D64.
16. F. Fendrych, A. Taylor, L. Peksa, I. Kratochvilova, J. Vlcek, V. Rezacova, V. Petrak, Z. Kluiber, L. Fekete, M. Liehr and M. Nesladek, *Journal of Physics D: Applied Physics*, 2010, **43**, 374018.

17. D. M. Gruen, *Annual Review of Materials Science*, 1999, **29**, 211–259.
18. J. E. Butler and A. V. Sumant, *Chemical Vapor Deposition*, 2008, **14**, 145–160.
19. H. B. Martin, A. Argoitia, J. C. Angus and U. Landau, *Journal of the Electrochemical Society*, 1999, **146**, 2959–2964.
20. T. Kondo, Y. Einaga, B. V. Sarada, T. N. Rao, D. A. Tryk and A. Fujishima, *Journal of the Electrochemical Society*, 2002, **149**, E179–E184.
21. T. Kondo, K. Honda, D. A. Tryk and A. Fujishimad, *Journal of the Electrochemical Society*, 2005, **152**, E18–E23.
22. M. C. Henstridge, E. J. F. Dickinson, M. Aslanoglu, C. Batchelor-McAuley and R. G. Compton, *Sensors and Actuators B-Chemical*, 2010, **145**, 417–427.
23. L. Xiao, G. G. Wildgoose and R. G. Compton, *New Journal of Chemistry*, 2008, **32**, 1628–1633.
24. V. N. Mochalin, O. Shenderova, D. Ho and Y. Gogotsi, *Nature Nanotechnology*, 2012, 7, 11–23.
25. K. B. Holt, D. J. Caruana and E. J. Millan-Barrios, *Journal of the American Chemical Society*, 2009, **131**, 11272–11273.
26. A. M. Schrand, S. A. C. Hens and O. A. Shenderova, *Critical Reviews in Solid State and Materials Sciences*, 2009, **34**, 18–74.
27. A. Krueger, M. Ozawa, G. Jarre, Y. Liang, J. Stegk and L. Lu, *Physica Status Solidi a-Applications and Materials Science*, 2007, **204**, 2881–2887.
28. A. E. Fischer and G. M. Swain, *Journal of the Electrochemical Society*, 2005, **152**, B369–B375.
29. G. R. Salazar-Banda, K. I. B. Eguiluz and L. A. Avaca, *Electrochemistry Communications*, 2007, **9**, 59–64.
30. A. Ay, V. M. Swope and G. M. Swain, *Journal of the Electrochemical Society*, 2008, **155**, B1013–B1022.
31. K. B. Holt, *Physical Chemistry Chemical Physics*, 2010, **12**, 2048–2058.
32. L. Y. Bian, Y. H. Wang, J. B. Zang, J. K. Yu and H. Huang, *Journal of Electroanalytical Chemistry*, 2010, **644**, 85–88.
33. I. V. Shpilevaya, S. Hirsz and J. S. Foord, *Phys. Chem. Chem. Phys.*, 2012, in preparation.
34. F. G. Chevallier, N. Fietkau, J. del Campo, R. Mas, F. X. Munoz, L. Jiang, T. G. J. Jones and R. G. Compton, *Journal of Electroanalytical Chemistry*, 2006, **596**, 25–32.
35. K. B. Holt, C. Ziegler, D. J. Caruana, J. Zang, E. J. Millan-Barrios, J. Hu and J. S. Foord, *Physical Chemistry Chemical Physics*, 2008, **10**, 303–310.
36. J. Scholz, A. J. McQuillan and K. B. Holt, *Chemical Communications*, 2011, **47**, 12140–12142.
37. J. Hu, X. Lu and J. S. Foord, *Electrochemistry Communications*, 2010, **12**, 676–679.
38. W. S. Yang, O. Auciello, J. E. Butler, W. Cai, J. A. Carlisle, J. Gerbi, D. M. Gruen, T. Knickerbocker, T. L. Lasseter, J. N. Russell, L. M. Smith and R. J. Hamers, *Nature Materials*, 2002, **1**, 253–257.
39. C. E. Nebel, B. Rezek, D. Shin and H. Watanabe, *Physica Status Solidi a-Applications and Materials Science*, 2006, **203**, 3273–3298.
40. M. I. Landstrass and K. V. Ravi, *Applied Physics Letters*, 1989, **55**, 975–977.

41. H. J. Looi, L. Y. S. Pang, A. B. Molloy, F. Jones, J. S. Foord and R. B. Jackman, *Diamond and Related Materials*, 1998, **7**, 550–555.
42. F. Maier, M. Riedel, B. Mantel, J. Ristein and L. Ley, *Physical Review Letters*, 2000, **85**, 3472–3475.
43. S. J. Sque, R. Jones and P. R. Briddon, *Physical Review B*, 2006, **73**, 085313.
44. R. Hoffmann, A. Kriele, H. Obloh, J. Hees, M. Wolfer, W. Smirnov, N. Yang and C. E. Nebel, *Applied Physics Letters*, 2010, **97**.
45. J. B. Cui, J. Ristein and L. Ley, *Physical Review Letters*, 1998, **81**, 429–432.
46. C. H. Goeting, F. Marken, A. Gutierrez-Sosa, R. G. Compton and J. S. Foord, *Diamond and Related Materials*, 2000, **9**, 390–396.
47. V. S. Smentkowski, J. T. Yates, X. J. Chen and W. A. Goddard, *Surface Science*, 1997, **370**, 209–231.
48. A. Freedman and C. D. Stinespring, *Applied Physics Letters*, 1990, **57**, 1194–1196.
49. S. Ferro and A. De Battisti, *Analytical Chemistry*, 2003, **75**, 7040–7042.
50. A. Krueger and D. Lang, *Advanced Functional Materials*, 2012, **22**, 890–906.
51. G.-J. Zhang, K.-S. Song, Y. Nakamura, T. Ueno, T. Funatsu, I. Ohdomari and H. Kawarada, *Langmuir*, 2006, **22**, 3728–3734.
52. Y. Wang, H. Huang, J. Zang, F. Meng, L. Dong and J. Su, *Int. J. Electrochem. Sci.*, 2012, **7**, 6807–6815.
53. W. S. Yeap, S. Chen and K. P. Loh, *Langmuir*, 2009, **25**, 185–191.
54. G. Jarre, Y. Liang, P. Betz, D. Lang and A. Krueger, *Chem. Commun. (Cambridge, U. K.)*, 2011, **47**, 544–546.
55. D. Lang and A. Krueger, *Diamond Relat. Mater.*, 2011, **20**, 101–104.
56. K. Takahashi, M. Tanga, O. Takai and H. Okamura, *Diamond and Related Materials*, 2003, **12**, 572–576.
57. N. Yang, H. Uetsuka, H. Watanabe, T. Nakamura and C. E. Nebel, *Chem. Mater.*, 2007, **19**, 2852–2859.
58. B. M. Nichols, J. E. Butler, J. N. Russell and R. J. Hamers, *The Journal of Physical Chemistry B*, 2005, **109**, 20938–20947.
59. T. C. Kuo, R. L. McCreery and G. M. Swain, *Electrochemical and Solid State Letters*, 1999, **2**, 288–290.
60. S. Q. Lud, M. Steenackers, R. Jordan, P. Bruno, D. M. Gruen, P. Feulner, J. A. Garrido and M. Stutzmann, *J. Am. Chem. Soc.*, 2006, **128**, 16884–16891.
61. F. W. Campbell and R. G. Compton, *Analytical and Bioanalytical Chemistry*, 2010, **396**, 241–259.
62. B. Liu, J. Hu and J. S. Foord, *Electrochemistry Communications*, 2012, **19**, 46–49.
63. A. Hartl, E. Schmich, J. A. Garrido, J. Hernando, S. C. R. Catharino, S. Walter, P. Feulner, A. Kromka, D. Steinmuller and M. Stutzmann, *Nature Materials*, 2004, **3**, 736–742.
64. J. S. Xu, Q. Y. Chen and G. M. Swain, *Analytical Chemistry*, 1998, **70**, 3146–3154.
65. I. B. Dimov, C. Batchelor-McAuley, L. Aldous and R. G. Compton, *Phys. Chem. Chem. Phys.*, 2012, **14**, 2375–2380.

66. J. M. Garguilo, B. A. Davis, M. Buddie, F. A. M. Kock and R. J. Nemanich, *Diamond Relat. Mater.*, 2004, **13**, 595–599.
67. R. Hoffmann, A. Kriele, S. Kopta, W. Smirnov, N. Yang and C. E. Nebel, *Physica Status Solidi a-Applications and Materials Science*, 2010, **207**, 2073–2077.
68. D. Steinmueller-Nethl, F. R. Kloss, M. Najam-U-Haq, M. Rainer, K. Larsson, C. Linsmeier, G. Koehler, C. Fehrer, G. Lepperdinger, X. Liu, N. Memmel, E. Bertel, C. W. Huck, R. Gassner and G. Bonn, *Biomaterials*, 2006, **27**, 4547–4556.
69. J. Foord and D. Opperman, in *Diamond Electronics - Fundamentals to Applications*, eds. P. Bergonzo, R. Gat, R. B. Jackman and C. E. Nebel, Cambridge University Press, Cambridge, 2007, vol. 956, pp. 97–103.
70. S. Szunerits and R. Boukherroub, *J. Solid State Electrochem*, 2008, **12**, 1205–1218.

CHAPTER 16

Superconductivity in Nanostructured Boron-doped Diamond and its Application to Device Fabrication

SOUMEN MANDAL,* TOBIAS BAUTZE AND
CHRISTOPHER BÄUERLE*

Institut Néel, CNRS, Université Joseph Fourier, 38042, Grenoble, France
*Email: soumen.mandal@gmail.com; bauerle@grenoble.cnrs.fr

16.1 Introduction

Carbon, the material that all known life-forms contain, has plenty of allotropes. The most widely known among them are graphite, diamond and amorphous carbon. Diamond in particular has been known to mankind for almost 6000 years. Although the most common use of diamond is in jewellery, some of its extraordinary properties make it suitable for various engineering and scientific usages. It is one of the hardest materials known, which makes it an excellent candidate for cutting and grinding tools. In its pure form it is an insulator but it can be turned into a semiconductor or even a superconductor by sufficient doping. In nature, diamond is formed at high temperature and high pressure inside the earth at a depth of 140–190 km over a period of 1–3 billion years. It is also possible to grow diamond artificially by a high-pressure high-temperature process, which mimics the conditions present inside the earth.[1] Apart from this, chemical vapor

RSC Nanoscience & Nanotechnology No. 31
Nanodiamond
Edited by Oliver Williams
© The Royal Society of Chemistry 2014
Published by the Royal Society of Chemistry, www.rsc.org

deposition (CVD) has also been used to grow diamond artificially.[2] The main advantage of this CVD process is the ability to grow diamond films on large substrates. It opens the possibility to use diamond in a variety of applications, like heat sinks for high heat-producing electronics, coatings to reduce wear and tear in mechanical tools, to name a few.[3] The CVD process also allows control of various other properties of the film. It is possible to grow films of varying grain sizes from a few nanometres to microns. Apart from this, one can add a variety of dopants to drastically change some of the properties of pure diamond. Naturally occurring blue diamond, which contains boron impurities, is semiconducting. The use of CVD has made it possible to alter the boron concentration such that doped diamond can even be turned superconducting. The first reported superconducting diamond was grown in 2004 by Ekimov *et al.*[4] using a high-pressure, high-temperature method. They achieved superconductivity by heavily doping the material with boron above a critical concentration. This discovery led to a flurry of activity in the study of superhard superconducting materials.[5–12] Diamond is now known to be a type-II superconductor with a critical temperature close to 4 K and a critical field in excess of 4 T.[4] However, recently, transition temperatures as high as 11 K have been achieved in homoepitaxial CVD-grown diamond films.[13]

The origin of superconductivity in diamond is still highly debated. At present there are three competing theories to explain its origin but any serious experimental evidence is missing to date. The three competing theories are: i) phonon mediated pairing-driven conventional BCS theory,[14] ii) correlated impurity band theory[15] and iii) weak localization of spin-flip-driven hole pairs close to the Fermi level.[16] Some recent studies on the isotopic shift of the transition temperature upon isotopic substitution of boron and carbon atoms point to conventional BCS theory,[17] but more experimental studies are required before a conclusion on the origin of superconductivity can be drawn.

Although the origin of superconductivity in diamond is an interesting field of study, our main interest lies in the fact that, in addition to its superconductivity, diamond also has a very high Young's modulus. This makes it an excellent candidate for making superconducting nano-electromechanical systems (NEMS) of very high quality. However, the realization of NEMS is only possible if it can be proven that the superconducting,[7–9] as well as the mechanical, properties[10] of the nanocrystalline layers of diamond grown on a non-diamond substrate, like silicon, are preserved even when the material is nanopatterned. This chapter will focus on a comprehensive study of nanopatterned superconducting diamond films grown on silicon dioxide on a silicon substrate. First, we will discuss the details of the nanofabrication process of diamond thin films. Then, we will present our results on low temperature studies of these nanopatterned circuits, followed by studies on diamond superconducting quantum interference devices (SQUIDs). We will conclude this chapter with the very first results on electromechanical systems made from boron-doped diamond.

16.2 Nanofabrication of Polycrystalline Boron-doped Diamond

Figure 16.1 shows the schematic of the complete nanofabrication process. We start with CVD-grown, boron-doped diamond on 500 nm thick silicon dioxide on a silicon substrate. The details of the growth process can be found elsewhere.[18-20] The film is thoroughly cleaned in mild acid solution followed by rinsing with water. Then, the film is dipped in acetone for a few minutes to remove further organic pollutants that may be present on the surface. As a final step, the sample is dipped in isopropyl alcohol, which is then allowed to evaporate to obtain a clean surface. This cleaned sample is spin coated with 4% polymethyl-methacrylate (PMMA) to form a 250 nm thick layer, which is then prebaked at 180 °C for 5 min. This layer of PMMA is exposed to an electron beam with a dose of ~300 µC cm^{-2} with an acceleration voltage of 20 kV. The exposed PMMA layer is then developed for 35 s in a solution of 1 : 3 methyl isobutyl ketone (MIBK) and isopropyl alcohol (IPA). The developed layer is then covered with 65 nm of nickel and evaporated in a standard electron-gun evaporator. This is followed by the lift-off process by dipping the sample in acetone, which results in a nickel pattern as shown in the middle panel of the second line in Figure 16.1. This nickel layer acts as a mask for the plasma etching process. The etching is done using an inductively coupled plasma-based reactive ion etching process. The sample is cooled to 10 °C while being exposed to the oxygen plasma for approximately 8 min. The anisotropic nature of the etching process results in structures with a height-to-width aspect ratio greater than three.

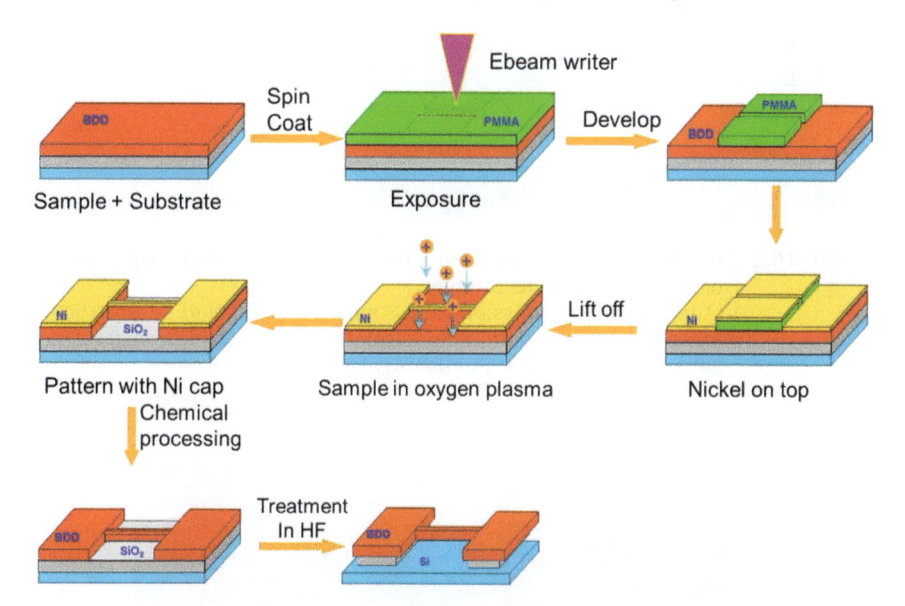

Figure 16.1 Schematic of the nanofabrication process, starting from a diamond thin film to a nanomechanical system.

After this process, the nickel mask is removed by dipping the sample in FeCl$_3$ solution followed by cleaning in dilute acid solution, acetone and finally isopropyl alcohol. At this stage, ohmic contacts are evaporated on the contact pads using photolithography and a standard electron-gun evaporation technique. The ohmic contacts consist of a trilayer of titanium, platinum and gold with thicknesses of 30, 50 and 40 nm, respectively. This trilayer is then annealed in high vacuum at 750 °C for 30 min to form a titanium carbide interface between the metals and the diamond layer. This completes the process for realizing micro- and nano-circuits out of boron-doped diamond.

In Figure 16.2 we have shown scanning electron microscope (SEM) images of a typical circuit used for low-temperature characterization of nano-patterned diamond films. Note the small section in the last panel of the figure showing a line with a thickness of 350 nm, a length of 500 nm and a width of 90 nm. Such a high aspect ratio is possible due to the anisotropic nature of the etching process. In order to fabricate electromechanical systems, an additional step is added to this process, as shown in the last panel of Figure 16.1. The circuit is exposed to highly reactive hydrogen fluoride gas, which etches away the SiO$_2$ layer underneath the diamond, resulting in suspended structures of desired geometry. Diamond itself is inert to this chemical etching process. A typical suspended beam is shown in Figure 16.3.

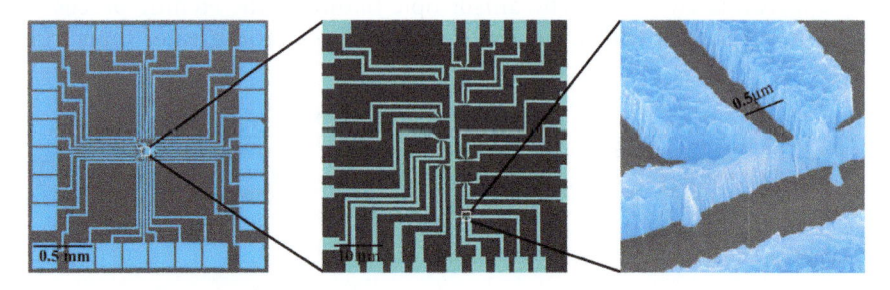

Figure 16.2 A typical circuit made from boron-doped diamond using electron beam lithography techniques.
(Adapted with permission from Soumen Mandal *et al.* Nanostructures made from superconducting boron-doped diamond, *Nanotechnology*, 2010, **21**, 193503, IOP Publishing, UK.)

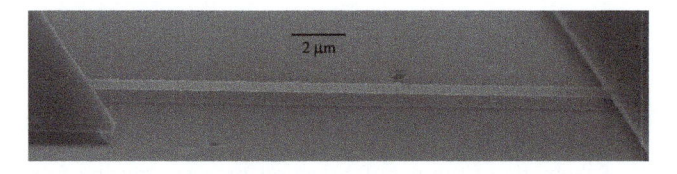

Figure 16.3 A typical suspended beam fabricated out of boron-doped diamond.

16.3 Low-temperature Characterization of Nanofabricated Samples

In this section we will discuss the low temperature results for our nanofabricated circuits. The measurements were done using both standard DC and lock-in based AC measurement SEM techniques in a common four-probe geometry. The measurements were performed in both ^{3}He and ^{3}He/^{4}He dilution refrigerators. For the electrical characterization of the film, silver paste contacts were deposited on the film within a distance of approximately 5 mm. The resistance was measured as a function of temperature with a bias current of 1 μA. The results are displayed in Figure 16.4. The transition temperature of the film was close to 3 K with a transition width of 0.7 K using 10–90% of the onset resistance criterion. The width in the transition temperature can be attributed to a variation of the grain sizes in the sample.

In Figure 16.5 we show the superconducting critical temperature T_C of wires with various widths. For the measurements we used very low currents, typically of the order of 100 nA. The transition temperatures of the wires with widths greater than 100 nm are close to 2 K, which is similar to the bulk value of the wafer from which the circuit was fabricated. Only in the case of the narrowest wire, which is less than 100 nm wide, T_C is close to 1.5 K.

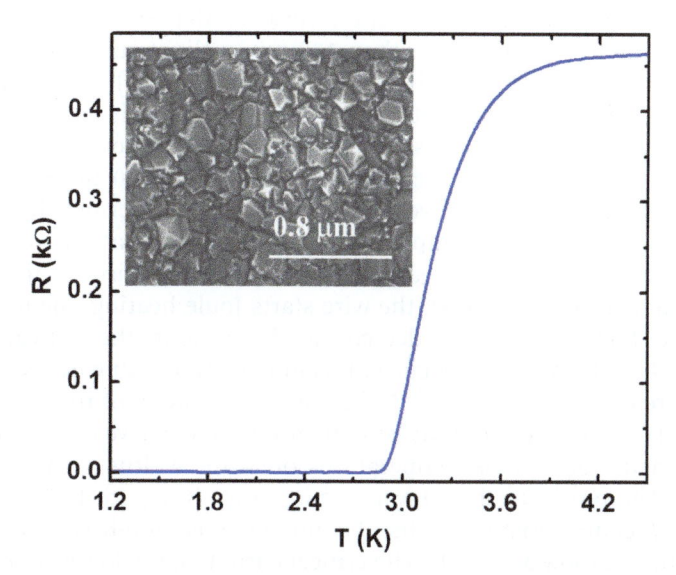

Figure 16.4 Superconducting transition temperature of a boron-doped thin film. The inset shows the SEM image of the surface of an as-grown layer. The surface consists of grains with a typical size of 150 nm for a 250 nm thick film.
(Adapted with permission from Soumen Mandal et al. Nanostructures made from superconducting boron-doped diamond, *Nanotechnology*, 2010, **21**, 193503, IOP Publishing, UK.)

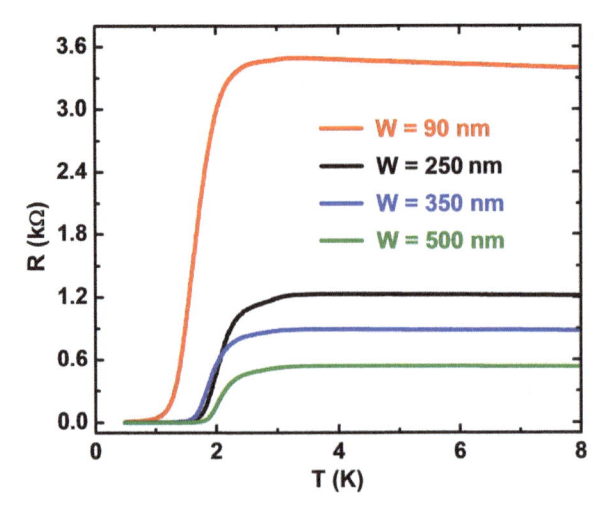

Figure 16.5 Resistance *versus* temperature of four representative diamond wires of similar lengths and height with varying widths.
(Adapted with permission from Soumen Mandal *et al.* Nanostructures made from superconducting boron-doped diamond, *Nanotechnology*, 2010, **21**, 193503, IOP Publishing, UK.)

This observation is confirmed for other wafers, where wires with widths greater than 100 nm have T_C values close to that of the non-patterned sample, while for wires thinner than 100 nm, a slight suppression of T_C is seen. We have also measured the voltage–current (*V*–*I*) characteristics of these wires. Some of the results will be presented in this chapter, whereas detailed descriptions of the results can be found elsewhere.[21,22]

Figure 16.6 shows the *V*–*I* curves measured on a 500 nm wide wire at various temperatures. The curves are hysteretic due to thermal effects. When the current in the wire is ramped up from zero, the voltage across the wire stays zero as long as the bias current is well below the critical current. Once the critical current is reached, the wire starts Joule heating due to its finite resistance. Subsequently, on decreasing the current, the critical current, usually called the re-trapping current, is much lower than what is measured while increasing the current.[23,24] The slight asymmetry of the *V*–*I* curves is attributed to an offset in our measurement set-up. We have also done similar measurements at various temperatures on wires widths of 350, 250 and 90 nm.[22] The result for the 90 nm wire is shown in Figure 16.7.

The *V*–*I* characteristic for the 90 nm wire is non-hysteretic even at temperatures as low as 50 mK. The critical current at the lowest temperature is close to 20 nA. The reason for the non-hysteretic nature of these curves is due to low power dissipation when the sample reaches the critical current. For a sample to show hysteretic behaviour, the power dissipated at the critical current must be higher than the re-trapping power of the sample. Here, the re-trapping power is close to 4.5 pW,[22] while the power dissipated at the critical current is of the order of 0.65 pW. Hence, in this case thermal

Superconductivity in Nanostructured Boron-doped Diamond and its Application 391

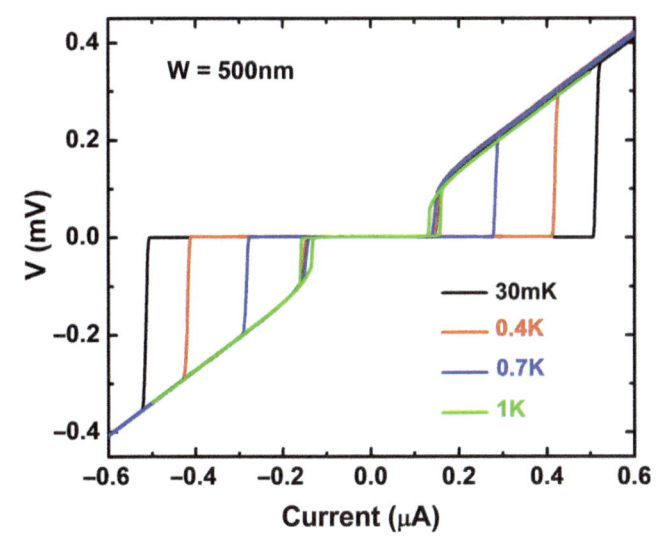

Figure 16.6 Voltage–current (V–I) characteristics of a 500 nm wide wire taken at different temperatures. The characteristics are hysteretic due to Joule heating.
(Adapted with permission from Soumen Mandal *et al.* Nanostructures made from superconducting boron-doped diamond, *Nanotechnology*, 2010, **21**, 193503, IOP Publishing, UK.)

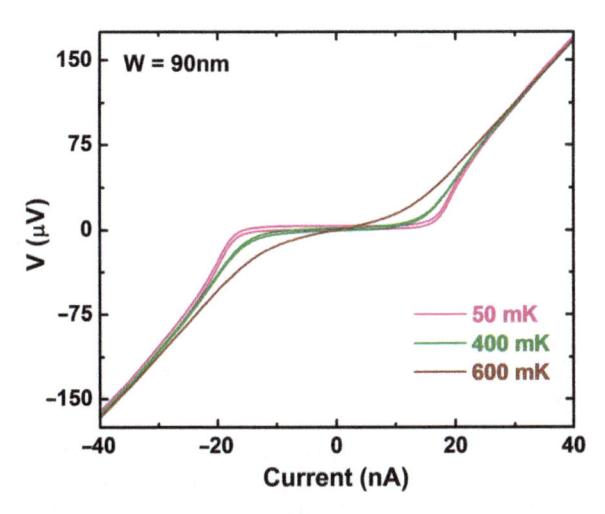

Figure 16.7 Voltage–current characteristics at different temperatures for a wire with dimensions of 500 nm length, 90 nm width and 300 nm height.
(Adapted with permission from Soumen Mandal *et al.* Detailed study of superconductivity in nanostructured nanocrystalline boron-doped diamond thin films, *Phys. Status Solidi A* 2010, **207** (9), 2017–2022, WILEY-VCH Verlag GmbH & Co. KGaA, Weinheim.)

effects do not come into play when the sample transits from the normal state to superconducting state and *vice versa*.

Apart from the temperature dependence of the *V–I* characteristics, we have also measured the isothermal *V–I* characteristics at various applied magnetic fields for in-plane, as well as out-of-plane, configurations. In Figure 16.8 we have presented the out-of-plane configuration results for a 500 nm wide wire. The curves are hysteretic until an applied field of 30 mT is reached.[21] Above 30 mT the curves are no longer hysteretic due to the fact that the critical current of the device is lower than the re-trapping current. The lowering of the critical current is because of the suppression of superconductivity due to magnetic field induced pair breaking effects. For a closer look at the data, we have numerically calculated the differential resistance of the sample, as presented in panel B of Figure 16.8. For fields as high as 500 mT, we have seen a clear zero resistance state close to zero injection current.[21] Above this field, even though the resistance never becomes completely zero, a local minimum at zero current suggests the existence of superconducting pockets, even at fields as high as 3 T. We have also measured the *V–I* characteristics of the wire by putting an in-plane field perpendicular to the flow of current. The results are shown in Figure 16.9.

In this case the curves are hysteretic until the in-plane applied field reaches a value of 130 mT, which is clearly different from what is seen in the case of the out-of-plane applied field. One quick look at the curves in panel A of Figures 16.8 and 16.9 tells us that, for the same applied field, the critical current is higher for the case when the applied field is in the plane of the sample. We believe that the suppression of the critical current in this system

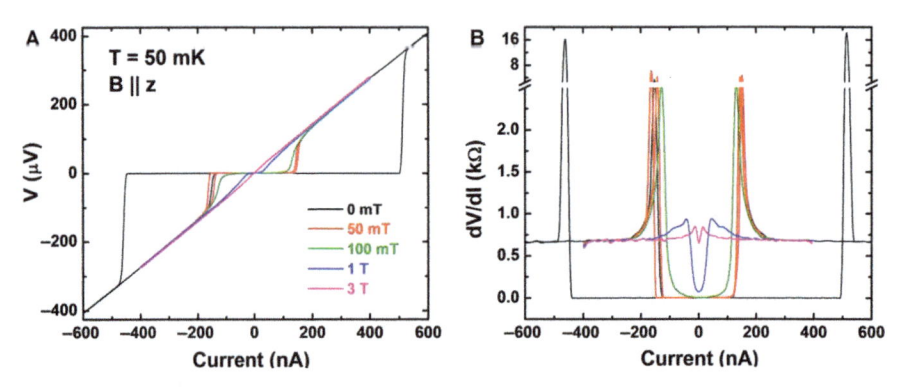

Figure 16.8 Isothermal (*T* = 50 mK) voltage–current characteristics of a 500 nm wide wire taken at various applied magnetic fields. The field is perpendicular to the sample. The *V–I* characteristics are hysteretic below 30 mT. In panel B we have shown the differential resistance extracted numerically from the *V–I* curves. The resistance goes to zero when the wire is in its superconducting state.
(Adapted with permission from Soumen Mandal *et al.* Nanostructures made from superconducting boron-doped diamond, *Nanotechnology*, 2010, **21**, 193503, IOP Publishing, UK.)

Superconductivity in Nanostructured Boron-doped Diamond and its Application 393

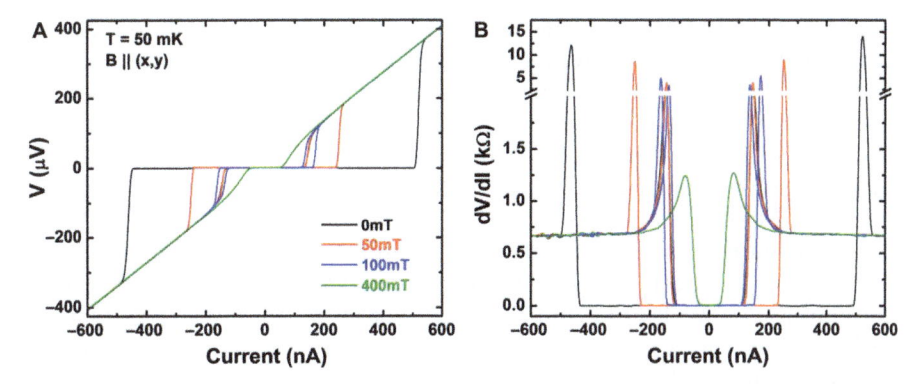

Figure 16.9 *V–I* characteristics of a 500 nm wide wire at 50 mK under different magnetic fields applied in the plane of the sample, perpendicular to the current direction. A hysteretic behavior is seen until the field reaches 130 mT. Panel B shows the differential resistance extracted numerically from the *V–I* curves.

is due to vortex penetration in the devices. Here, the difference in the aspect ratio of the devices when applying the field parallel or perpendicular to the surface may result in a higher vortex density when the field is applied perpendicular to the sample surface. In the case of in-plane fields the measurements were done only until 400 mT and, for the 500 nm wide wire, we see a clear zero resistance state close to zero injection current even at the highest applied field.

In conclusion we have shown that nanofabricated samples made from boron-doped diamond retain their superconducting properties. We have also seen that larger grain sizes in the boron-doped diamond (BDD) film lead to more robust superconductivity with higher critical field and transition temperatures. Based on our findings on nanofabricated samples, we are in a position to fabricate and study quantum devices made from this material with larger grain sizes, which will be presented in the next section.

16.4 Superconducting Quantum Interference Device Made From Superconducting Diamond

As we have seen in the preceding section, superconductivity persists in nanoscale BDD samples and hence opens the possibility to realize superconducting devices with this very interesting material. The main idea is to explore the possibility of harnessing the high critical field of the material for making sensitive quantum devices, which can operate at magnetic fields of a few Teslas. Such a device is the superconducting quantum interference device (SQUID).

In this section we will describe the realization of a micron-scale superconducting quantum interference device (μ-SQUID). A SQUID is a superconducting loop containing one or more Josephson junctions and the phase accumulated across the junctions results in a modulation of the critical current when an external magnetic field is applied. First invented by Jaklevic et al.,[25] it is an extremely sensitive device that can be used to measure extremely small magnetic fields. These devices find applications in various fields of science, such as scanning SQUID microscopy,[26,27] magnetoencephelography (MEG)[28] and magnetic property measurements,[27,29] to name a few.

A schematic of the measurement set-up for SQUID oscillations is presented in Figure 16.10. The SQUID is formed by a superconducting loop containing two weak links and is current-biased by adding a large biasing resistor R_b in series with a voltage source V_s. The value of R_b is chosen in such a way that R_n, the normal state resistance of the device, is much smaller than the biasing resistor. In our case with a normal state resistance of the order of a few kilo-ohms we chose 1 MΩ for the resistor such that the current in the circuit is determined by the biasing resistor at all times. To measure a V–I curve, we swept the source voltage and, therefore, the current in the circuit. We measured the output voltage with a low noise amplifier and recorded the bias current when the sample transited into its normal state. Such a measurement is repeated by varying the magnetic flux ϕ and the critical current shows a periodic behaviour as a function of the magnetic flux with the oscillation period being one flux quantum $\phi_o = h/2e$ in the SQUID loop, as shown schematically in panel B of Figure 16.10. Another way of measuring these oscillations is to bias the SQUID with a constant current close to I_c and to measure the output voltage as a function of applied flux, although this technique is limited to shunted devices and will be discussed below. Now, if one biases the SQUID at the point of steepest slope (black point in panel B of Figure 16.10), a small change in flux through the SQUID loop results in a large change in the measured critical current. In this way this device acts as an extremely sensitive flux detector and, hence, a magnetic field detector capable of detecting fields as low as 10^{-14} T is

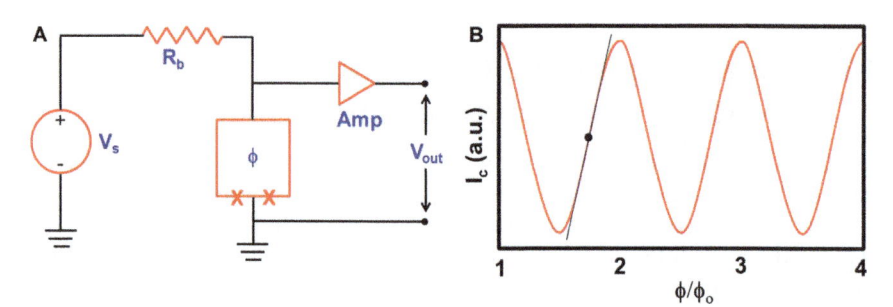

Figure 16.10 Schematic of a SQUID showing a voltage source with a bias resistance R_b, which sets the current in the circuit. The crosses in the superconducting loop represent Josephson junctions. The critical current oscillation as a function of applied flux through the loop is schematically shown in panel B.

possible.[30,31] For comparison, the magnetic field of a heart is around 10^{-10} T and that of the human brain is around 10^{-13} T.

In general this detector has serious limitations in the form that it can only be used in environments where the magnetic field perpendicular to the SQUID plane is well below 1 T or by using a pick-up coil, which is placed within the magnetic field, while the SQUID has to be placed in a magnetic field protected environment.[29] In the literature one can find SQUIDs made from materials like Nb$_3$Sn[32] and Nb$_3$Ge,[33,34] which have high critical fields; however, their operation at high fields has not been proven to date. In the following subsection we will describe the study of µ-SQUIDs made from boron-doped diamond, which can be operated in magnetic fields in excess of 1 T. Such a device can find applications in the fields of ultrasensitive motion detection of nanomechanical systems[35] and high-field SQUID microscopy, to name a few.

16.4.1 Low Temperature Studies on Non-shunted Devices

Our µ-SQUID devices were fabricated out of 300 nm thick boron-doped diamond films using the steps depicted in Figure 16.1. A typical circuit, along with its components, is shown in Figure 16.11. Each circuit consists of six µ-SQUIDs—the difference being in the widths of the weak links, which in this case serve as the Josephson junctions.[36,37] We have fabricated devices with three different widths: 100 nm, 170 nm and 250 nm. The lengths of the links are kept fixed at 250 nm and the arms of the devices are 500 nm wide. The data presented here will be mostly from the µ-SQUID with weak links, a width of 100 nm and a length of 250 nm, unless otherwise stated. The area of the µ-SQUID is 2.5 µm×2.5 µm. The measurements were done in both a ^{3}He close cycle refrigerator with a base temperature of 400 mK and a dilution refrigerator with a base temperature of 40 mK.

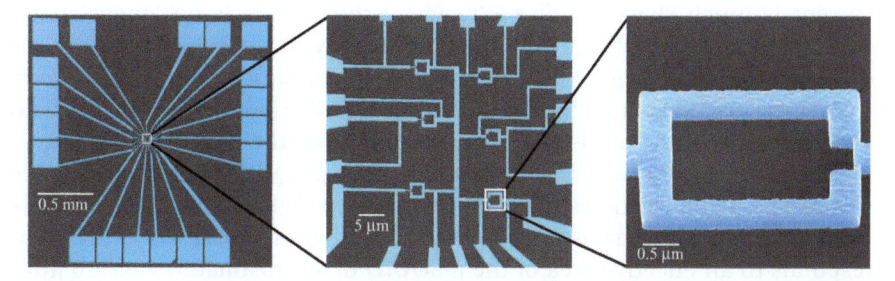

Figure 16.11 Scanning electron micrographs of the µ-SQUID circuit. The left panel shows the complete circuit with several pads for ohmic contacts. The chip is 2 mm×2 mm in size. The middle panel shows the close-up view of a 50 µm×50 µm area in the middle of the sample. The right panel shows a tilted view of one of the µ-SQUIDs.
(Adapted with permission from Soumen Mandal *et al*. The Diamond Superconducting Quantum Interference Device, *ACS Nano*, 2011, 5, 7144–7148, © 2011 American Chemical Society.)

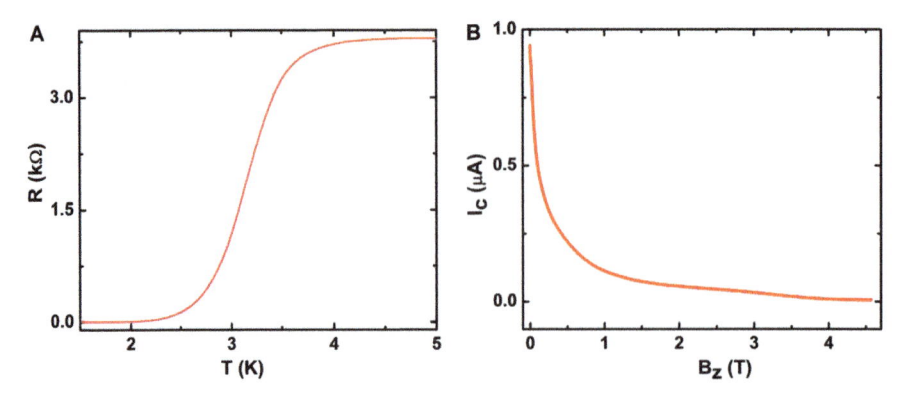

Figure 16.12 Resistance *vs* temperature curve for our μ-SQUID showing a transition temperature of 3 K, which is also the transition temperature of the bulk wafer. Critical current as a function of out-of-plane magnetic field for a device with 100 nm wide weak link is shown in panel B. (Adapted with permission from Soumen Mandal *et al.* The Diamond Superconducting Quantum Interference Device, *ACS Nano*, 2011, 5, 7144–7148, © 2011 American Chemical Society.)

We have performed detailed low-temperature characterizations of our devices.[38] In this subsection we will present the main results of our measurements. The resistance data from the measurements reveals a transition temperature of 3 K, as shown in panel A of Figure 16.12. The *V–I* characteristics of the samples were hysteretic but, on application of a magnetic field, the hysteresis disappeared once the critical current dropped below the re-trapping current due to pair-breaking effects. The hysteretic behavior is due to thermal effects.[24,39,40] The variation of the critical current with magnetic field is shown in panel B of Figure 16.12. The critical current drops exponentially until the field reaches 1 T and then the decay in the critical current is much slower until 4.5 T, where the sample is no longer superconducting. This existence of superconductivity at such high fields gives diamond the advantage of being a material for high field quantum devices.

We have also measured the critical current oscillation by applying the field both in-plane and out-of-plane.[38] A summary of the results is shown in Figure 16.13, where we present results from a μ-SQUID with 100 nm wide weak links. Panel A shows the oscillation at low fields, where the device is hysteretic. We see clear oscillations with a period of ∼0.31 mT, which corresponds to an effective area of the μ-SQUID of approximately 2.6×2.6 μm^2, which is in agreement with our geometrical dimensions. The modulation amplitude is 15% for the device with 100 nm wide weak links, while our measurements for 170 nm wide weak links show a modulation of approximately 5%.[38] These modulation values are comparable to values seen in case of aluminium and niobium μ-SQUIDs.[41]

The sensitivity of our device can be estimated by repetitively measuring the critical current at a point of highest slope on the oscillation curve.

We obtained a sensitivity of 40 µΦ$_o$ Hz$^{-1/2}$, which is comparable to sensitivities of µ-SQUIDs with a similar design made from niobium and aluminium.[42] The sensitivity of the diamond µ-SQUID can be increased by carefully optimizing the SQUID design, as well as the material properties.[43] At present, the sensitivity is limited by our electronic setup, as well as external noise sources.

When a large magnetic field (>150 mT) is applied to the superconducting device, it reduces the critical current below the re-trapping current. This leads to a non-hysteretic nature of the voltage–current characteristic. Since the characteristic is non-hysteretic, an AC-based measurement technique can be used to record the SQUID oscillations. The measurement technique has been detailed in an earlier publication.[38]

We have measured voltage oscillations in our µ-SQUID at various magnetic fields up to 4 T. In panel B of Figure 16.13 we have shown the voltage

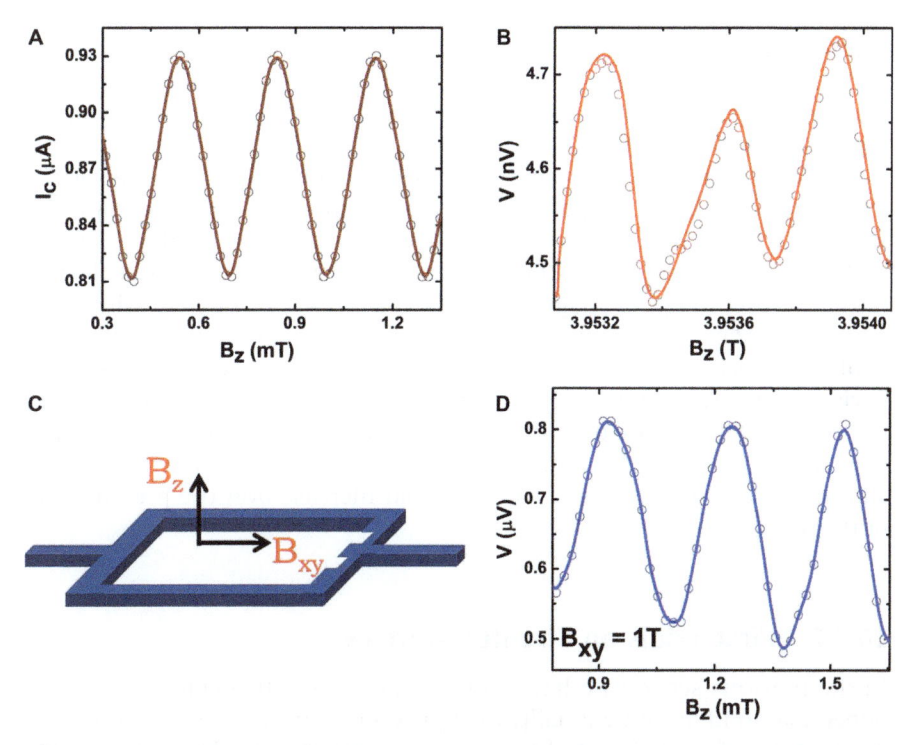

Figure 16.13 Low field oscillations of the critical current for a µ-SQUID with 100 nm wide weak links are shown in panel A. The high field data for perpendicular field orientation ($B=B_z$) for the same device is shown in panel B. Panel D shows voltage oscillations for the parallel configuration ($B=B_{xy}$). The measurement configuration is shown schematically in panel C.
(Adapted with permission from Soumen Mandal *et al.* The Diamond Superconducting Quantum Interference Device, *ACS Nano*, 2011, 5, 7144–7148, © 2011 American Chemical Society.)

oscillations close to 4 T. The periodicities of the oscillations are not perfectly symmetric. This is due to the fact that we are using the same coil with a field perpendicular to the μ-SQUID to apply a steady magnetic field up to 4 T, as well as the small variations of the order of 0.02 mT to probe the SQUID oscillations, which are of the order of 0.3 mT. Because of such small variations in the field, we measured the actual field values by averaging the value over a time scale of approximately 1 min. This limitation in the measurement setup can easily be removed by putting a feed line close to the μ-SQUID for probing the oscillations.[35]

We have also measured the voltage oscillations of our μ-SQUID by applying a constant field in the plane of the sample as schematized in panel C of Figure 16.13. The in-plane applied field, B_{xy}, is perpendicular to both weak links. A small probing field B_z is applied perpendicular to the plane to measure the oscillations. In our measurements we have kept B_{xy} constant at 0.5 T and 1 T, which is the highest possible field for our xy-coil and the result of the measurement with an in-plane field of 1 T is shown in panel D of Figure 16.13. The point worth mentioning here is that the thickness of our μ-SQUID is close to 300 nm, and in standard superconductors such oscillations are not visible under similar conditions when the layer thickness exceeds a few nanometers.[44]

In the literature one can find reports of extremely sensitive μ-SQUIDs being used in a variety of applications, like scanning SQUID microscopy,[26,42] magnetization measurements in mesoscopic systems[23,45] and in isolated molecules, to name a few.[46] However, a severe drawback of all these devices is the narrow range of magnetic fields that they can operate within. In addition, when μ-SQUIDs were operated at fields above 1 T, the field was applied perfectly in the plane of the sample,[44] keeping the thickness of the device close to few nanometers, which severely reduces the critical current of the system. The diamond μ-SQUID detailed in this section does not suffer from any such disability and can operate at fields as high as 4 T, as shown above, which is more than a six-fold increase over the present state of the art.[47]

16.4.2 First Trials on Shunted Devices

In the previous section we have shown that a μ-SQUID made from boron-doped diamond has functionality in applied fields in excess of 1 T; however, we have also found that its low field characteristics are hysteretic, which severely limits its usability with regards to measurement speed and sensitivity. Hence, it is important to see whether this hysteresis can be removed. As pointed out above, the hysteresis is mainly due to thermal effects and one way of removing the hysteresis is to put a heat shunt in contact with the weak links. In this way the heat generated by the device when it transits from the superconducting to the normal state can be evacuated instantaneously and the system will be void of any thermal effects.

Superconductivity in Nanostructured Boron-doped Diamond and its Application 399

For making shunted µ-SQUIDs, two different approaches can be considered. One possibility is to grow the diamond films on a metal surface, while the other is to cover the whole device with metal in the same way as ohmic contacts are made for measuring the devices. In this section we will show preliminary results using the first method. To choose an appropriate metal for the heat shunt, one has to consider the growth conditions. For a good quality diamond film, it is important that the diamond film sticks firmly to the substrate. For metal films, this adhesion is directly dependent on the ease with which the diamond film can form metal carbides. In this regard we chose tungsten, which readily forms carbide and thus gives rise to a good adhesion to the diamond film. To start with, we used two boron-doped diamond films grown on a tungsten film on a silicon wafer with different thicknesses of tungsten. The thickness of the tungsten layer was fixed at 20 nm and 50 nm. The steps for realizing the nano-circuits are the same as detailed in Figure 16.1. Only, in this case, the etching has to be done very carefully since the oxygen plasma also attacks the tungsten layer. We have made two circuits; one in which the tungsten layer has been left untouched and the other in which tungsten has been left only underneath the etched diamond layer, as shown schematically in Figure 16.14.

In this case the ohmic contacts were made by a silver paste instead of the usual titanium, platinum and gold trilayer. This is because the trilayer process requires 30 min baking in a high vacuum environment to generate the ohmic contact between the diamond and the metals. In order to avoid any diffusion of tungsten into the diamond layer this step was omitted. In Figure 16.15A and B we show scanning electron micrographs of the two types of circuit. Although the two sets of images look similar, on careful observation of the right panel in Figure 16.15B, one can identify a thin dark layer underneath the SQUID device, which corresponds to the tungsten layer left over after the etching process.

Here, µ-SQUID designs and dimensions are essentially the same as for our unshunted SQUIDs. The weak links have the same lengths (250 nm), while the width was varied from 100 to 170 and then to 250 nm. The different widths give rise to different critical currents of the system. In this section we will present data from the µ-SQUID with weak links of widths of 100 nm.

In the following we present the experimental results of a device with a shunt metal (50 nm of W) on the complete chip (panel A Figure 16.15), as

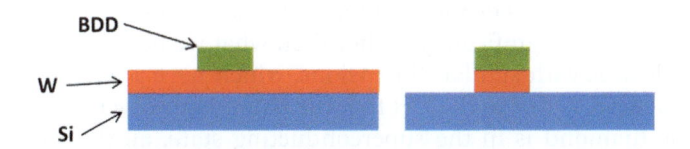

Figure 16.14 Schematic of the two different circuits assessed for shunted SQUIDs. The blue layer is the silicon wafer, the red layer is tungsten (20 nm and 50 nm) and the green layer is boron-doped diamond (BDD).

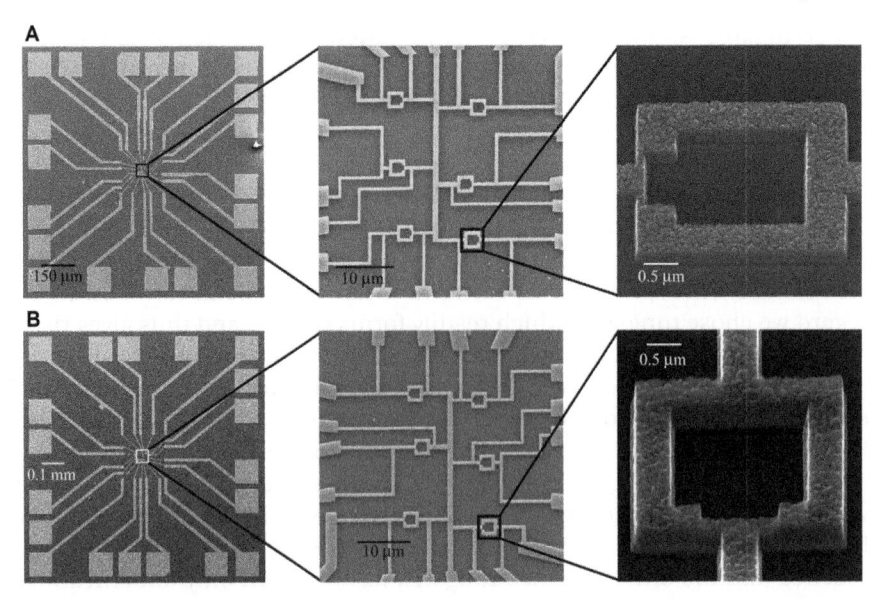

Figure 16.15 A: SEM images of circuits with a full shunt layer underneath the SQUID. B: SEM image of the circuit with metal layer only under the diamond nanostructure. Note the thick dark layer underneath the structure in the right panel. This dark layer corresponds to the metal shunt.

shown in Figure 16.16. The resistance *versus* temperature curve shows two clear transitions. We associate the first transition, which is close to 1.2 K, to the weak links, while the other one is from the entire circuit and close to the bulk transition. We have measured the *V–I* characteristics of the device using the DC measurement technique, as previously described. The DC technique is particularly useful for checking the presence of hysteresis in the circuit. The *V–I* characteristic is S-shaped, which is typical for shunted superconducting devices and similar to what is observed for unshunted SQUIDs at high magnetic fields. The voltage oscillation can hence be measured with the same AC-measurement technique. We fix the bias current to 0.12 mA and record the resulting voltage oscillations as depicted in panel C of Figure 16.16.

To understand the behaviour of the shunt layer it is essential to understand the difference in critical current that we see in the case of non-shunted and shunted SQUIDs. For the shunted SQUID, we observe a critical current of 50 µA, which is significantly higher than what we have observed for non-shunted devices with similar dimensions, which have a value close to 1 µA. This, we believe, is due to the fact that the shunt layer covers the whole chip. When the diamond is in the superconducting state, all the current in the system passes through the diamond layer, but as soon as it reaches the normal state, the current in the system is divided between the shunt layer and the diamond layer. In this kind of geometry, when the diamond layer is

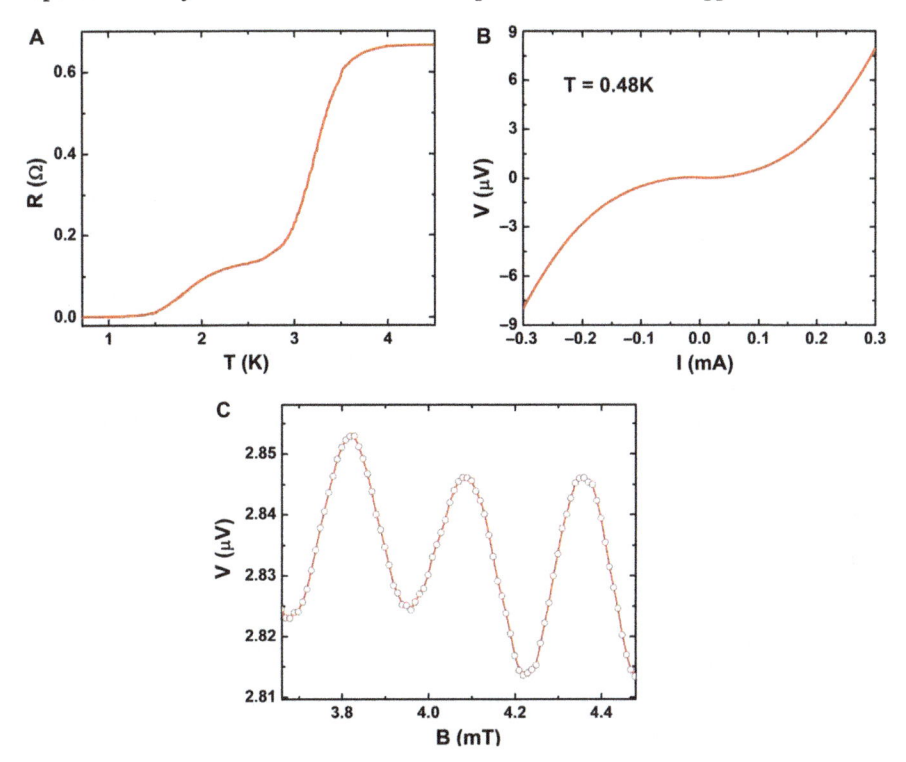

Figure 16.16 Characteristics of a SQUID device with 100 nm wide weak links. The metal layer (50 nm W) under the circuit covers the whole chip area. Panel A shows the resistance *versus* temperature curve for the device. Two transitions are seen: one corresponding to the weak links and the other due to the rest of the circuit. The zero field *V–I* characteristics are shown in panel B. No hysteresis is seen in the curve. Panel C shows the low field voltage oscillation for the device when biased at a current of 0.12 mA.

in its normal state, there are many low resistance paths in the shunt for the current to flow, which are not only confined to the nanostructure. Since the resistance of the diamond layer is of the order of few kΩ and the shunt layer sheet resistance is close to 1 $\Omega\,\text{sq}^{-1}$, a large portion of the current passes through the shunt layer and then the diamond layer transits back to the superconducting state. This leads to the increase in the critical current that we see. This argument is validated further by our results for the second set of samples, where the 50 nm shunt layer is geometrically confined below the nanostructure only. In that case a large increment in critical current ($I_c \sim 0.8\,\mu\text{A}$) is not observed due to the higher resistance of the tungsten layer. Hence, a large current is needed in the case where the metal is spread over the entire chip to transit the diamond film into the normal state so that, even if the current is divided between the two layers, the current through the diamond layer is large enough to retain its normal state. For completeness, let us add that, for films with a 20 nm thick tungsten shunt layer, the μ-SQUIDs were hysteretic for temperatures much below the transition temperature, and hence the shunt layer was not sufficient to evacuate the generated heat in the weak links.

In conclusion we have shown in the first part of this section that boron-doped diamond can be used to make sensitive quantum devices, which can be operated at very high magnetic fields. This opens up a variety of new application, such as their use with nanomechanical oscillators to fabricate sensitive motion detectors. Apart from that, we have also demonstrated that it is possible to build quantum devices void of any thermal effects using tungsten in contact with the nanostructure, improving its usability for low field applications. It is to be noted that further studies on device optimization with respect to an appropriate shunt layer thickness and shunt geometry is needed before final conclusions on the device capabilities can be drawn.

16.5 Nanomechanical Systems

In the preceding section we exploited the truly remarkable electric properties of boron-doped diamond and demonstrated that nanocrystalline boron-doped diamond is a promising material to realize superconducting quantum devices. In this section we would like to exploit the mechanical properties to show that boron-doped diamond can also find its application in the fields of micro- and nano-electromechanical systems.

More generally, nanomechanical resonators offer a wide range of applications in modern technology, as well as in state-of-the art research. On the one hand, they can be used as detectors for single atom masses,[48] single charges[49] and single spins.[50] On the other hand, an emerging class of highly miniaturized timing devices start to rival the electrical performance of the well-established quartz-based oscillators[51] that are used in modern technology. Applications range from real time tracking, frequency up and down conversion in radio frequency (RF) transceivers and the clocking of logical circuits.

The miniaturization of the resonators brings along a variety of new problems to the well-established field of micromechanical resonators.[52] A new set of transducers must be developed, surface properties start to play a significant role at submicron dimensions and it becomes more difficult to maintain a reproducible nanofabrication. Miniaturization, on the other hand, allows exploration of novel regimes, in particular the quantum regime.[53–55]

The challenge we address in the research presented here is the pursuit of high quality factors in nanomechanical systems. In a first approximation the quality factor describes the rate of energy loss compared to the energy stored in the resonator. An ideal dissipation-free resonator only vibrates at an infinitely narrow fundamental frequency with corresponding frequency harmonics, but since real systems are limited by interactions with their surroundings, the oscillator is damped and its energy is dissipated. Electric-LC resonators, for example, exhibit quality factors that rarely surpass 100, while mechanical systems can exhibit very high quality factors up to several millions.[56–58]

Superconducting diamond is the perfect candidate for the realization of nanomechanical oscillators with very high quality factors. Its sp³ hybridization, along with the resulting strong sigma bonds, make it the hardest material on Earth, resulting in a very high Young's modulus of the order of one terapascal.[59] In addition, its exceptional superconducting properties make it a perfectly dissipation-free system with regard to its electric properties. All energy loss is hence of a mechanical nature, which in turn should lead to extremely high quality factors. Furthermore, the high quality factor goes directly with a low energy exchange with the environment. Along with a high resonant frequency, this allows in principle the implementation of these kinds of structures in modern quantum information technology, for example, as tools for quantum metrology or as coupled hybrid quantum systems.

In the following we will present the first realization of BDD resonators. We will then introduce a simple experimental setup to detect and characterize nanomechanical resonators directly and will briefly underline the superb characteristics of superconducting diamond. An outline for future research aspects will be given at the end of this section.

In Figure 16.17 we present two nanomechanical BDD resonators. The fabrication process has already been detailed in Figure 16.1. We chose two different structures for our measurements: the first resonator consists only of boron-doped diamond to show that it is feasible to drive the resonator in its superconducting regime and the second resonator has a thin 50 nm gold layer on the top. This allows study of its behavior even at temperatures above the superconducting critical temperature, although these measurements will not be discussed here.

A well-established technique for measuring nanomechanical resonators is the so-called magneto-motive detection scheme.[60] It is mainly used to determine the characteristic properties of a resonator, such as its resonance frequency, non-linear behavior and its quality factor. In addition, this technique allows one to distinguish a mechanical resonance from an electrical one by simply varying the magnetic field. When an oscillating current

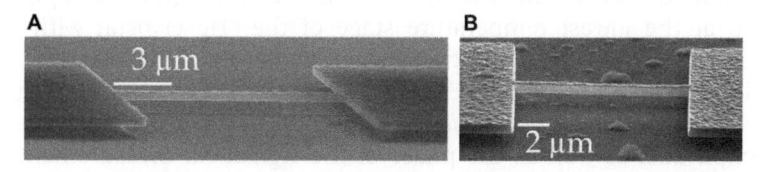

Figure 16.17 BDD nanomechanical resonator. Panel A shows a pure BDD resonator with dimensions 12 µm×300 nm×325 nm (length×width× thickness), whereas panel B shows a BDD resonator with a thin gold layer on the top. Its dimensions are 12 µm×500 nm×(325+50) nm.

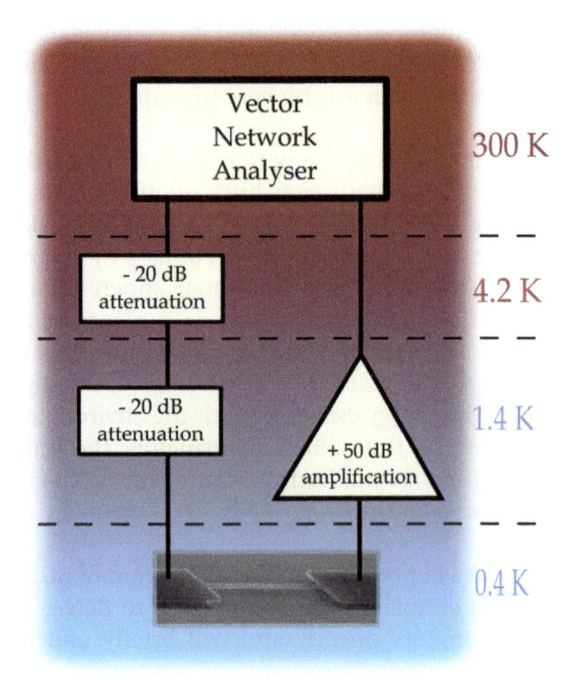

Figure 16.18 Schematic of the experimental cryogenic setup. The sample is actuated with an AC-signal from the vector network analyzer and actuated *via* the Lorentz force in a magnetic field at low temperatures.

is passed through the resonator in a magnetic field perpendicular to the current flow, the resonator is actuated *via* the Lorentz force when the driving frequency matches the resonance frequency of the structure. At the same time, the motion of the oscillator induces a time-varying voltage due to Faraday's law, which manifests itself as an increase of the impedance of the device. In our setup this impedance is measured *via* a transmission measurement, which is carried out using a Rhode and Schwartz vector network analyzer (VNA), as depicted in Figure 16.18.

The signal from the generator port is fed through a 50 Ω adapted coaxial line with two cryogenic attenuation stages. The first thermalization takes place at the 4.2 K stage, the second at the 1.2 K stage with two identical attenuators of 20 dB each. The signal is then fed through the sample that is located at the lowest temperature stage of the ^{3}He cryostat with a base temperature of 400 mK. The transmitted signal is amplified with a commercial low noise amplifier (Caltech CITLF1 SN120) that is thermally anchored at 4.2 K and recorded *via* the input port of the VNA, which measures the amplitude, as well as the phase of the signal. All measurements were carried out using this setup.

The transmission has been measured as a function of input power and external magnetic field in order to identify the superconducting–normal transition of the mechanical resonator. Figure 16.19 shows this behavior for

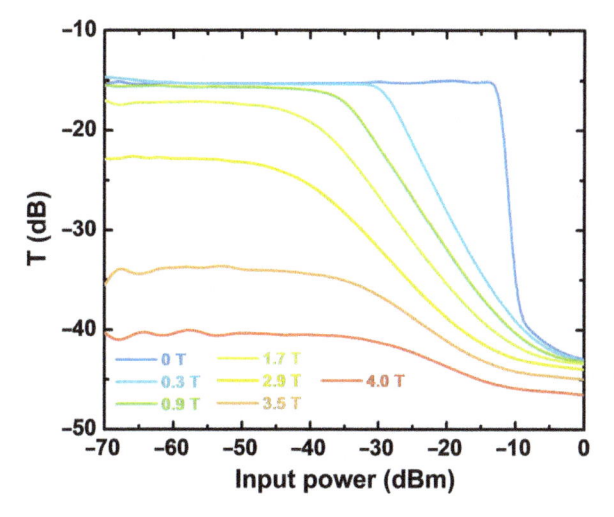

Superconductivity in Nanostructured Boron-doped Diamond and its Application 405

Figure 16.19 Transmission through the sample for different input powers at different magnetic fields. The x-axis shows the input power and the transmission is plotted along the y-axis.

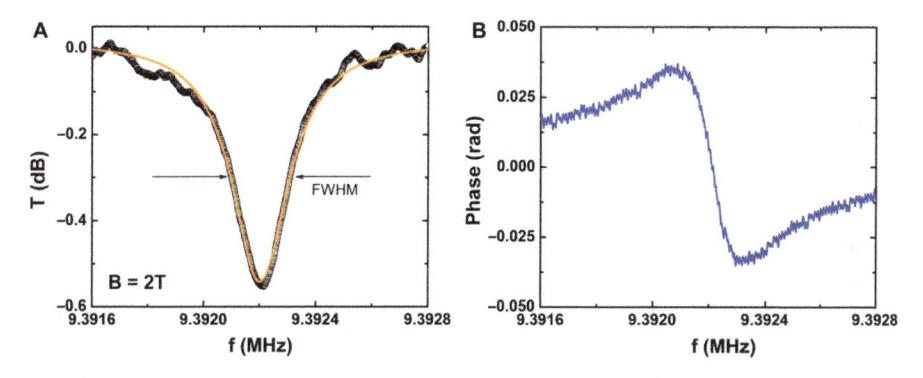

Figure 16.20 A: the transmitted signal at resonance of the pure diamond resonator at a magnetic field of 2 T. B: the phase response. The full-width at half-maximum (FWHM) is indicated for the calculation of the quality factor.

the pure diamond beam from Figure 16.17. At zero magnetic field (blue curve), one can clearly identify the superconducting state: the transmitted signal is constant up to −10 dBm input power.

At this point, the superconducting-to-normal-state transition occurs and the transmission drops about 27 dB. From this value we can calculate the series resistance of the device to be approximately ∼2.1 kΩ, which corresponds to the normal state resistance of the mechanical resonator. The transition occurs at lower input power for higher magnetic field until about a magnetic field of 4 T, above which the sample stays normal.

In Figure 16.20 we present the mechanical properties of the resonator, where we measure the frequency response of our oscillator. Panel A in

Figure 16.20 depicts the transmitted signal at resonance at a magnetic field of 2 T. From a Lorentzian fit we can extract the loaded quality factor of about 40000 and, from the phase response (panel B), we determine the exact resonance frequency to be 9.392110 MHz (zero crossing of the phase). The quality factor in our system is mainly limited by clamping and surface losses due to the strong undercut of the clamping pads by the HF etching procedure, as seen in Figure 16.17. Further measurements are, however, necessary to confirm this statement. The loss mechanism can, in principle, be studied *via* the length dependence of the quality factor in greater detail.[61]

One disadvantage of the magneto-motive detection scheme is the onset of eddy currents inside the structure, which gives rise to an additional magnetic field that is opposed to the applied external magnetic field. This leads to an additional force acting on the resonator that is opposite to its movement. This additional damping adds linearly to the intrinsic mechanical damping, whereas its contribution is quadratic as depicted in Figure 16.21. From the fit we can extract the unloaded or intrinsic quality factor of about 41000.

In order to compare to other nanomechanical resonators made from different materials, it is common to calculate the product of resonance frequency and quality factor for which we obtain 3.37×10^{11}. This value is similar to state-of-the-art mechanical resonators.[62]

The ability to realize superconducting diamond resonators has a variety of possible applications in modern research. Their mechanical and superconducting properties make them an outstanding candidate for the integration into superconducting quantum circuits. It is feasible to incorporate such a high quality resonator into a SQUID loop[35] or to couple it to a superconducting cavity[63] for ultrasensitive displacement detection.

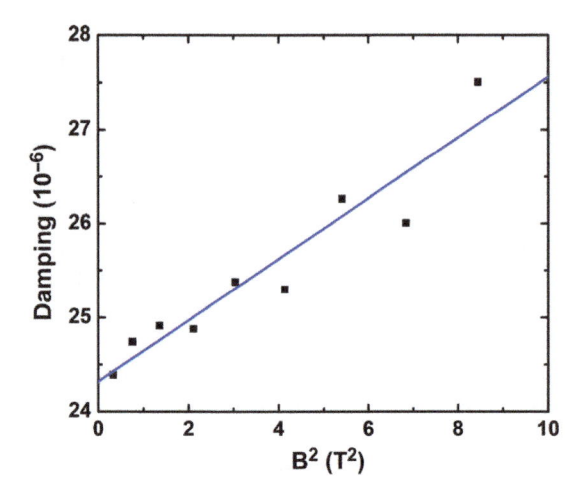

Figure 16.21 Eddy current damping as a function of the square of the magnetic field. The intrinsic mechanical damping can be extracted by extrapolation to zero magnetic field.

16.6 Summary and Conclusions

In conclusion we have presented in this chapter a comprehensive study on nanostructures made from boron-doped diamond. It has been shown that the material retains its remarkable properties even when it is nanofabricated. We have also demonstrated the working of a micron-sized SQUID made from boron-doped diamond. SQUIDs made from boron-doped diamond retain their functionality even at fields as high as 4 T. We have also attempted to make shunted devices out of this material but rigorous efforts are needed before realizing a device that can be usefully employed in various applications. Lastly, we have presented our results from nanomechanical resonators, which clearly show that this material is indeed a good candidate for making monolithic superconducting nanomechanical devices. The next steps in realizing better devices can be divided into two parts: a first approach can be to investigate various design parameters to obtain the best sensitivity and quality factors. A second approach can address the material considerations, which may lead to nanocrystalline samples with higher critical parameters, like critical temperature, critical field, Young's modulus *etc.* In particular, boron-doped diamond grown on substrates to induce compressive stress in the diamond layer would significantly enhance the quality factor.[64–67] It also remains to be seen what happens when these devices are made from monocrystalline diamond films. It is expected that the monocrystalline films may result in better devices due to the absence of inherent defects, like grain boundaries and sp^2 carbon present in the nanocrystalline BDD films; however, the fabrication and measurement techniques may have to be adapted for handling monocrystalline diamond-based devices.

Acknowledgement

We would like to acknowledge technical assistance from the Nanofab team of the Institut Néel, in particular B. Fernandez. We also acknowledge valuable discussions and help from Y. Baines, V. Bouchiat, E. Bustarret, K. Hasselbach, T. Meunier, C. Naud, F. Omnes, L. Saminadayar, W. Wernsdorfer and O. Williams. This work has been supported by the French National Funding Agency (ANR) in the frame of its program in "Nanosciences and Nanotechnologies" (SUPERNEMS Project no. ANR-08-NANO-033).

References

1. H. T. Hall, *Science*, 1958, **128**, 445–449.
2. M. Werner and R. Locher, *Reports on Progress in Physics*, 1998, **61**, 1665.
3. M. C. Costello, D. A. Tossell, D. M. Reece, C. J. Brierley and J. A. Savage, *Diamond and Related Materials*, 1994, **3**, 1137–1141.
4. E. A. Ekimov, V. A. Sidorov, E. D. Bauer, N. N. Mel'nik, N. J. Curro, J. D. Thompson and S. M. Stishov, *Nature*, 2004, **428**, 542–545.

5. X. Blase, E. Bustarret, C. Chapelier, T. Klein and C. Marcenat, *Nat Mater*, 2009, **8**, 375–382.
6. G. Dubitskiy, V. Blank, S. Buga, E. Semenova, V. Kul'bachinskii, A. Krechetov and V. Kytin, *JETP Letters*, 2005, **81**, 260–263.
7. M. Nesladek, D. Tromson, C. Mer, P. Bergonzo, P. Hubik and J. J. Mares, *Applied Physics Letters*, 2006, **88**, 232111.
8. W. Gajewski, P. Achatz, O. A. Williams, K. Haenen, E. Bustarret, M. Stutzmann and J. A. Garrido, *Physical Review B*, 2009, **79**, 045206.
9. P. Achatz, W. Gajewski, E. Bustarret, C. Marcenat, R. Piquerel, C. Chapelier, T. Dubouchet, O. A. Williams, K. Haenen, J. A. Garrido and M. Stutzmann, *Physical Review B*, 2009, **79**, 201203.
10. M. Imboden, P. Mohanty, A. Gaidarzhy, J. Rankin and B. W. Sheldon, *Applied Physics Letters*, 2007, **90**, 173502.
11. A. Gaidarzhy, M. Imboden, P. Mohanty, J. Rankin and B. W. Sheldon, *Applied Physics Letters*, 2007, **91**, 203503–203503.
12. M. Imboden and P. Mohanty, *Physical Review B*, 2009, **79**, 125424.
13. Y. Takano, *Journal of Physics: Condensed Matter*, 2009, **21**, 253201.
14. J. Bardeen, L. N. Cooper and J. R. Schrieffer, *Physical Review*, 1957, **106**, 162–164.
15. G. Baskaran, *Science and Technology of Advanced Materials*, 2008, **9**, 044104.
16. J. J. Mareš, P. Hubik, J. Krištofik and M. Nesládek, *Science and Technology of Advanced Materials*, 2008, **9**, 044101.
17. E. A. Ekimov, V. A. Sidorov, A. V. Zoteev, J. B. Lebed, J. D. Thompson and S. M. Stishov, *Science and Technology of Advanced Materials*, 2008, **9**, 044210.
18. O. A. Williams, O. Douhéret, M. Daenen, K. Haenen, E. Ōsawa and M. Takahashi, *Chemical Physics Letters*, 2007, **445**, 255–258.
19. O. A. Williams, M. Nesladek, M. Daenen, S. Michaelson, A. Hoffman, E. Osawa, K. Haenen and R. B. Jackman, *Diamond and Related Materials*, 2008, **17**, 1080–1088.
20. O. A. Williams, *Diamond and Related Materials*, 2011, **20**, 621–640.
21. S. Mandal, C. Naud, O. A. Williams, É. Bustarret, F. Omnès, P. Rodière, T. Meunier, L. Saminadayar and C. Bäuerle, *Nanotechnology*, 2010, **21**, 195303.
22. S. Mandal, C. Naud, O. A. Williams, É. Bustarret, F. Omnès, P. Rodière, T. Meunier, L. Saminadayar and C. Bäuerle, *Physica Status Solidi (a)*, 2010, **207**, 2017–2022.
23. W. Rabaud, L. Saminadayar, D. Mailly, K. Hasselbach, A. Benoît and B. Etienne, *Physical Review Letters*, 2001, **86**, 3124–3127.
24. H. Courtois, M. Meschke, J. T. Peltonen and J. P. Pekola, *Physical Review Letters*, 2008, **101**, 067002.
25. R. C. Jaklevic, J. Lambe, A. H. Silver and J. E. Mercereau, *Physical Review Letters*, 1964, **12**, 159–160.
26. J. R. Kirtley, C. C. Tsuei, J. Z. Sun, C. C. Chi, L. S. Yu-Jahnes, A. Gupta, M. Rupp and M. B. Ketchen, *Nature*, 1995, **373**, 225–228.

27. R. C. Black and F. C. Wellstood, in *The SQUID Handbook*, eds. J. Clarke and A. I. Braginski, WILEY-VCH Verlag GmbH & Co. KGaA, Weinheim, 2006, vol. 2, pp. 392–436.
28. D. Cohen, *Science*, 1972, **175**, 664–666.
29. R. Kleiner, D. Koelle, F. Ludwig and J. Clarke, *Proceedings of the IEEE*, 2004, **92**, 1534–1548.
30. J. Vrba, J. Nenonen and L. Trahms, in *The SQUID Handbook*, eds. J. Clarke and A. I. Braginski, WILEY-VCH Verlag GmbH & Co. KGaA, Weinheim, 2006, vol. 2, pp. 299–300.
31. M. B. Ketchen, W. M. Goubau, J. Clarke and G. B. Donaldson, *Journal of Applied Physics*, 1978, **49**, 4111–4116.
32. C. T. Wu and C. M. Falco, *Applied Physics Letters*, 1977, **30**, 609–611.
33. H. Rogalla, B. David and J. Ruhl, *Journal of Applied Physics*, 1984, **55**, 3441–3443.
34. M. Dilorio, A. de Lozanne and M. Beasley, *Magnetics, IEEE Transactions on*, 1983, **19**, 308–311.
35. S. Etaki, M. Poot, I. Mahboob, K. Onomitsu, H. Yamaguchi and H. S. J. van der Zant, *Nat Phys*, 2008, **4**, 785–788.
36. P. W. Anderson and A. H. Dayem, *Physical Review Letters*, 1964, **13**, 195–197.
37. K. K. Likharev, *Reviews of Modern Physics*, 1979, **51**, 101–159.
38. S. Mandal, T. Bautze, O. A. Williams, C. Naud, É. Bustarret, F. Omnès, P. Rodière, T. Meunier, C. Bäuerle and L. Saminadayar, *ACS Nano*, 2011, **5**, 7144–7148.
39. W. J. Skocpol, *Physical Review B*, 1976, **14**, 1045–1051.
40. D. Hazra, L. M. A. Pascal, H. Courtois and A. K. Gupta, *Physical Review B*, 2010, **82**, 184530.
41. K. Hasselbach, D. Mailly and J. R. Kirtley, *Journal of Applied Physics*, 2002, **91**, 4432–4437.
42. K. Hasselbach, C. Veauvy and D. Mailly, *Physica C: Superconductivity*, 2000, **332**, 140–147.
43. R. Voss, R. Laibowitz, A. Broers, S. Raider, C. Knoedler and J. Viggiano, *IEEE Transactions on Magnetics*, 1981, **17**, 395–399.
44. L. Chen, W. Wernsdorfer, C. Lampropoulos, G. Christou and I. Chiorescu, *Nanotechnology*, 2010, **21**, 405504.
45. H. Bluhm, N. C. Koshnick, J. A. Bert, M. E. Huber and K. A. Moler, *Physical Review Letters*, 2009, **102**, 136802.
46. W. Wernsdorfer and R. Sessoli, *Science*, 1999, **284**, 133–135.
47. A. Finkler, Y. Segev, Y. Myasoedov, M. L. Rappaport, L. Ne'eman, D. Vasyukov, E. Zeldov, M. E. Huber, J. Martin and A. Yacoby, *Nano Letters*, 2010, **10**, 1046–1049.
48. K. Jensen, K. Kim and A. Zettl, *Nat Nano*, 2008, **3**, 533–537.
49. G. A. Steele, A. K. Hüttel, B. Witkamp, M. Poot, H. B. Meerwaldt, L. P. Kouwenhoven and H. S. J. van der Zant, *Science*, 2009, **325**, 1103–1107.
50. D. Rugar, R. Budakian, H. J. Mamin and B. W. Chui, *Nature*, 2004, **430**, 329–332.

51. J. T. M. van Beek and R. Puers, *Journal of Micromechanics and Microengineering*, 2012, **22**, 013001.
52. K. L. Ekinci and M. L. Roukes, *Review of Scientific Instruments*, 2005, **76**, 061101.
53. J. Chan, T. P. M. Alegre, A. H. Safavi-Naeini, J. T. Hill, A. Krause, S. Groblacher, M. Aspelmeyer and O. Painter, *Nature*, 2011, **478**, 89–92.
54. J. D. Teufel, T. Donner, D. Li, J. W. Harlow, M. S. Allman, K. Cicak, A. J. Sirois, J. D. Whittaker, K. W. Lehnert and R. W. Simmonds, *Nature*, 2011, **475**, 359–363.
55. A. D. O'Connell, M. Hofheinz, M. Ansmann, R. C. Bialczak, M. Lenander, E. Lucero, M. Neeley, D. Sank, H. Wang, M. Weides, J. Wenner, J. M. Martinis and A. N. Cleland, *Nature*, 2010, **464**, 697–703.
56. J. E.-Y. Lee and A. A. Seshia, 5.4-MHz single-crystal silicon wine glass mode disk resonator with quality factor of 2 million, *Sensors and Actuators A: Physical*, 2009, **156**(1), 28–35.
57. A. K. Hüttel, G. A. Steele, B. Witkamp, M. Poot, L. P. Kouwenhoven and H. S. J. van der Zant, *Nano Letters*, 2009, **9**, 2547–2552.
58. E. A. Laird, F. Pei, W. Tang, G. A. Steele and L. P. Kouwenhoven, *Nano Letters*, 2011, **12**, 193–197.
59. O. A. Williams, A. Kriele, J. Hees, M. Wolfer, W. Müller-Sebert and C. E. Nebel, *Chemical Physics Letters*, 2010, **495**, 84–89.
60. M. Poot and H. S. J. van der Zant, *Physics Reports*, 2012, **511**, 273–335.
61. X. M. H. Huang, X. L. Feng, C. A. Zorman, M. Mehregany and M. L. Roukes, *New Journal of Physics*, 2005, **7**, 247.
62. D. R. Schmid, P. L. Stiller, C. Strunk and A. K. Hüttel, *New Journal of Physics*, 2012, **14**, 083024.
63. C. A. Regal, J. D. Teufel and K. W. Lehnert, *Nat Phys*, 2008, **4**, 555–560.
64. S. S. Verbridge, J. M. Parpia, R. B. Reichenbach, L. M. Bellan and H. G. Craighead, *Journal of Applied Physics*, 2006, **99**, 124304–124308.
65. Q. P. Unterreithmeier, T. Faust and J. P. Kotthaus, *Physical Review Letters*, 2010, **105**, 027205.
66. T. Faust, P. Krenn, S. Manus, J. P. Kotthaus and E. M. Weig, *Nat Commun*, 2012, **3**, 728.
67. J. Rieger, T. Faust, M. J. Seitner, J. P. Kotthaus and E. M. Weig, *Applied Physics Letters*, 2012, **101**, 103110–103114.

CHAPTER 17

Diamond Nano-electromechanical Systems

PRITIRAJ MOHANTY*[a] AND MATTHIAS IMBODEN[b]

[a] Boston University Physics Department, 590 Commonwealth Avenue, Boston MA 02215, USA; [b] Boston University Department of Electrical and Computer Engineering, 8 St Mary's Street, Boston MA 02215, USA
*Email: mohanty@bu.edu

17.1 Introduction

Nano-electromechanical systems (NEMS) are currently used in an incredibly wide range of fields, both in applications and fundamental research. Miniaturization allows for unprecedented detection and transduction of tiny signals. This enhancement of the transduction signal is exploited in applications ranging from single molecule detection,[1] quantum mechanics[2,3] and spintronics,[4] to biomolecular recognition of proteins and DNA and chemical sensing. Such devices are typically fabricated out of silicon. But many other materials are also widely used, depending on the specific application. These include silicon carbide, gallium arsenide, aluminium nitride, a host of metals, and even carbon nanotubes and graphene. This chapter outlines the advantages of using diamond as a NEMS material. To start, a brief overview of elasticity theory is given along with general considerations that apply to nanoscale devices. Typical fabrication, as well as drive and detection methods, are presented. This is followed by an overview of dissipation mechanisms present in NEMS. For all topics, examples relevant to diamond are presented. As diamond MEMS (micro-electromechanical systems) devices have already been extensively discussed in the literature,[5–8] the focus here is on NEMS, but

---

RSC Nanoscience & Nanotechnology No. 31
Nanodiamond
Edited by Oliver Williams
© The Royal Society of Chemistry 2014
Published by the Royal Society of Chemistry, www.rsc.org

relevant examples from the larger MEMS are also included. In the final section, a few examples of future diamond NEMS experiments are discussed.

17.1.1 What NEMS Do

As the name suggests, NEMS devices or systems are both mechanical and electrical in nature. Typically, the defining element is a mechanical feature that is coupled in a particular way to an electrical circuit. This circuit will detect and possibly actuate the mechanical degrees of freedom as is discussed in section 17.3.2.

In NEMS and also in many MEMS the mechanical element is coupled to a particular signal of interest. This signal could be mechanical in nature, as is the case for gyroscopes and accelerometers that measure angular momentum and acceleration,[5] a change in mass down to a single atom,[9] a magnetic pulse,[10] a biological signal or even quantum states.[2] What is common to all of these devices is that they are based on monitoring a change in the mechanical response of a structure that is coupled to the degree of freedom being measured. Some examples of devices are depicted in Figure 17.1.

17.1.2 Why They Work

This diverse range of applications results from the favorable scaling effects. In miniaturization, volumetric properties, such as mass, are most strongly affected due to scaling. This implies that, for forces acting on surface areas, the corresponding acceleration will increase, resulting in larger relative signal sizes and shorter response times. High mechanical frequencies and high quality factors essentially decouple from the mechanical noise background, which falls off as $1/f$. In addition, an appropriate electrical readout scheme allows for the detection of minute changes of the mechanical state the device is in. All of this results in the ability to detect signals at short timescales, with small amounts of power. In addition to increased sensitivity, small size, and scalable manufacturing techniques, a large number of devices can be patterned on small areas. The following sections describe basic fabrication methods, as well as electrical coupling methods, which are used in current diamond NEMS technology.

17.1.3 Diamond: The Ideal NEMS Material

The mechanical response, as well as the ability to couple it to an electrical readout element, is critically dependent on the material. Diamond can be used to leverage its unique properties to further enhance NEMS performance. This is true for mechanical, electrical, and chemical properties, and examples are given for all cases.

With advances in chemical vapor deposition (CVD) diamond-growth techniques, especially in improvements of nucleation density,[12] it is now possible to grow the high-quality nanocrystalline diamond thin films needed as the starting point for NEMS fabrication. Table 17.1 summarizes some of the properties that make diamond a very interesting NEMS material.

Diamond Nano-electromechanical Systems 413

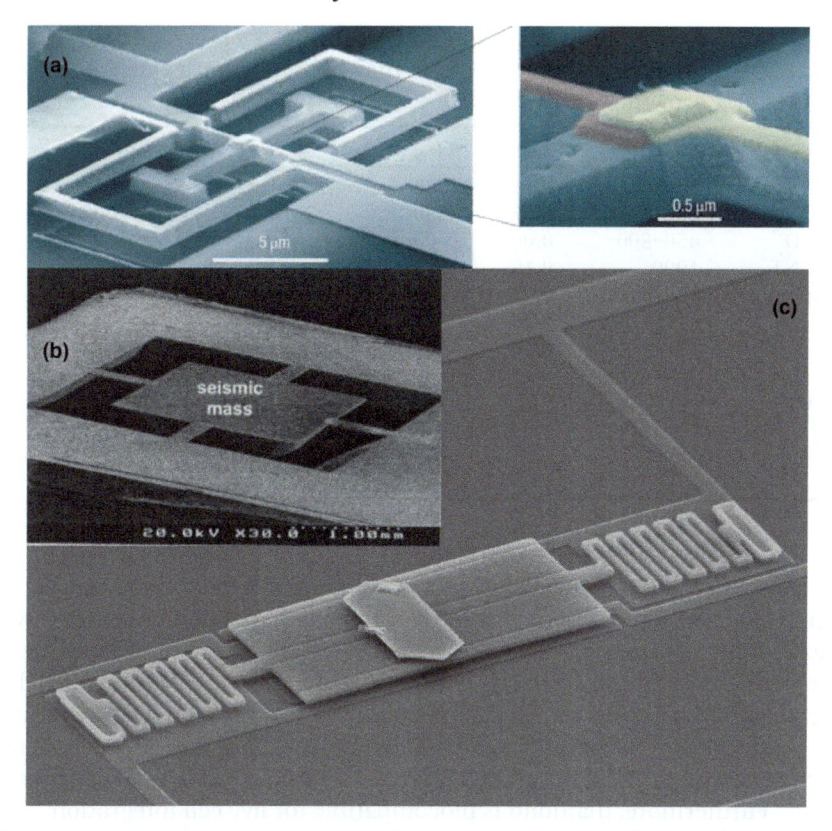

Figure 17.1 a) Silicon-based electron spin-flip resonator.[4] b) Diamond-based acceleration sensor.[6] c) Silicon-based torsion oscillator to measure single flux vortices.[11]
(a) Reprinted by permission from Macmillan Publishers Ltd: Nature Nanotechnology, copyright (2008). (b) Reprinted from ref. 6, Copyright (1999), with permission from Elsevier. (c) Reprinted by permission from Macmillan Publishers Ltd: Nature, copyright (1999).

Although there are exceptions,[13–15] diamond MEMS and NEMS are essentially all made of polycrystalline thin films, where nano- and ultrananocrystalline materials are used in NEMS devices with grain sizes on the order of or less than 200 nm. As discussed later in the section on fabrication, this arises from the need to grow diamond on a sacrificial layer that allows the structure to be suspended, thus providing the three-dimensional release required for most mechanical devices.

The obvious benefits of diamond are its mechanical properties, specifically the high Young's modulus and fracture strength, which is rivalled only by exotic materials, such as carbon nanotubes. This allows for tiny devices to remain mechanically solid, while pushing resonance frequencies beyond the GHz mark.[16,17] For larger structures, such as gears or rotors involving sliding elements, low tribology enhances operation efficiency and lifetime. Beyond

Table 17.1 Common NEMS material properties as reported in the literature; some values may differ greatly depending on growth conditions.

	E (GPa)	ν	ρ (kg m^{-3})	α (10^{-6} K^{-1})	κ (W m^{-1} K^{-1})
SCD	1200	0.20	3520	1.2	2000
NCD	800–1100	0.07	3500	1.0	1000
UNCD	450–800	0.06	3500	1.5	1–12
SWCNT	1000	0.10	1350	1.7	3
Graphene	1000	0.17	2250	−7.4	500–2500
Si	112	0.28	2329	2.49	124
SiC	415	0.16	2160	4.6	114
TiC	450	0.18	4940	7.7	330
GaAs	85.5	0.31	5316	5.4	50
AlN	350	0.24	3260	4.5	180
SiN	160–230	0.27	2800–3300	3.3	30
Au	77.2	0.42	19320	14.4	300

the mechanical properties, diamond is an excellent thermal conductor. It has low electromagnetic absorption over a wide frequency span and can be tuned from an almost perfect electrical insulator to a semiconductor with piezoresistive properties by adding a wide range of dopants, such as boron. Furthermore, it is an excellent field emitter of electrons. With further doping, diamond will even become superconducting at cryogenic temperatures. Its chemical properties make it possible to use diamond NEMS in harsh environments as it has high radiation hardness, especially at high energies.[18] Furthermore, diamond is biocompatible for live cell integration[19] and allows for biological functionalization,[20] enabling cells-on-chips technology and bio-electronics.

All of these properties can be leveraged to make novel NEMS devices with applications in diverse areas of fundamental and technological interest.

17.2 NEMS Dynamics

17.2.1 Elasticity Theory and the Euler–Bernoulli Equations

There are generally two types of mechanical devices: resonant and non-resonant. Larger MEMS structures are typically either resonant or non-resonant, while all NEMS devices are essentially driven on resonance to optimize sensitivity and minimize noise. Examples of non-resonant structures are larger comb actuated shutters,[21] mechanical switches, and pressure sensors.[7] Such devices will respond to external stimuli on transient timescales, which can be extremely fast. This will set the switching speed of an RF switch, as well as the minimum response time to changes in pressure. Alternatively, resonant devices benefit from a high quality factor by minimizing coupling to noise and maximizing response sensitivity. Resonant modes of operation are strongly favored in NEMS where static displacements are often too small to detect.

Diamond Nano-electromechanical Systems

The static Euler–Bernoulli equation describes the deflection of a solid due to an applied load. For a solid object, this is written as:

$$\frac{d^2}{dx^2}\left(EI\frac{d^2a}{dx^2}\right) = q(x) = EI\frac{d^4a}{dx^4}, \qquad (17.1)$$

where E is the Young's modulus, $I = t^3w/12$ is the second moment of inertia of a beam of thickness t and width w, q is the distributive shear force and $a(x)$ is the amplitude of the structure at point x. The second expression is true if the product EI is constant, as is often the case. This expression results from balancing all the forces and moments for each infinitesimal section along the structure. Eqn 17.1 is derived from considering the bending moment M and shear force Q along the beam due to the applied stress σ_x:

$$M = -\iint \sigma_x y\, dA = -EI\frac{d^2a}{dx^2}$$

$$Q = \frac{dM}{dx} = -\frac{d}{dx}\left(EI\frac{d^2a}{dx^2}\right). \qquad (17.2)$$

A complete discussion is given in ref. 22, which contains a detailed discussion on NEMS dynamics. Using this, as well as Hooke's law, one can determine the spring constant for a given geometry. Both the clamping boundary conditions, which are set during fabrication, and the load distribution, which results during actuation and sensing, determine the exact response. Examples of typical setups are summarized in Table 17.2. More complex MEMS devices are a combination of multiple such spring elements. Additional examples can be found in other sources.[22]

17.2.2 Dynamic Solutions

The resonant modes can be derived from the Euler–Bernoulli equation by calculating the action. The result is a differential equation, fourth order in space (for a constant product EI) and second order in time:

$$\frac{d^2}{dx^2}\left(EI\frac{d^2a}{dx^2}\right) = -\mu\frac{d^2a}{dt^2} + f(x,t). \qquad (17.3)$$

$\mu = \rho A$ is the mass per unit length and f the force per unit length. This equation can be solved by separation of variables ($a(x,t) = X(x)T(t)$), and then finding the general solution to the homogeneous equation and adding a particular solution to the inhomogeneous solution. The most general ansatz for the homogeneous equation is:

$$T(t) = A\sin(\omega t + \varphi)$$
$$X(x) = a_1 \cos(\beta x) + a_2 \cosh(\beta x) + a_3 \sin(\beta x) + a_4 \sinh(\beta x).$$

$$\beta_n = \sqrt[4]{\frac{\rho A \omega_n^2}{EI}}, \qquad (17.4)$$

Table 17.2 Beam loading and clamping boundary conditions with corresponding spring constants. Each beam in the figures has one unit of moment of inertia I.

Geometry	Clamping	Figure	Load	Spring constant
Cantilever	Fixed-free		Point	$3\dfrac{EI}{L^3}$
Cantilever	Fixed-free		Distributive	$8\dfrac{EI}{L^3}$
Doubly clamped	Fixed-fixed		Point	$192\dfrac{EI}{L^3}$
Doubly clamped	Fixed-fixed		Distributive	$384\dfrac{EI}{L^3}$
Doubly clamped	Fixed-fixed		Point, off-center	$3\dfrac{EIL^3}{a^3 b^3}$
Simply supported	Pivot-sliding		Point	$48\dfrac{EI}{L^3}$
Axially loaded bar	Free		Both ends	$\dfrac{AE}{L}$
Coil torsional spring	Fixed-free		Point	$\dfrac{EI}{L}$
Shear frame fixed base				$24\dfrac{EI}{L^3}$
Folded-flexure				$24\dfrac{EI}{L^3}$

Diamond Nano-electromechanical Systems

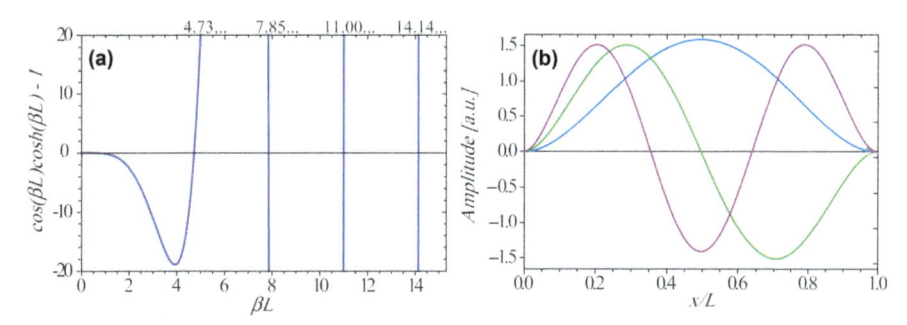

Figure 17.2 a) Doubly clamped beam condition for non-zero solution, first four eigenmodes. b) Normalized amplitude of first three modes along the length of a doubly clamped beam.[31]

Table 17.3 Boundary conditions for the Euler-Bernoulli beam eqn 17.4.

Clapping condition	1st boundary condition	2nd boundary condition	Physical meaning
Clamped	$X=0$	$X'=0$	No displacement, horizontal at attachment
Free	$X''=0$	$X'''=0$	No stress, no shear force
Pivot (simply supported)	$X=0$	$X''=0$	No displacement, no stress
Guided	$X'=0$	$X'''=0$	Horizontal at attachment, no shear force

where β takes on discrete values for non-trivial solutions. These are given by eqn 17.5 and are depicted in Figure 17.2 for the doubly clamped beam boundary conditions. The boundary conditions are given by the particular geometry and clamping setup of a given structure and are summarized in Table 17.3.

Applying these boundary conditions for the doubly clamped beam ($x_{BC}=0,L$) results in the matrix equation whose determinant condition guarantees non-zero solutions:

$$\cos(\beta L)\cosh(\beta L) - 1 = 0 \quad \beta_n L = 0, 4.73, 7.85, 11.00, 14.14, \ldots.$$

$$X(x) = A_n[\sin(\beta_n x) - \sinh(\beta_n x)] + B_n[\cos(\beta_n x) - \cosh(\beta_n x)].$$

$$\frac{A_n}{B_n} = -\frac{\cos(\beta_n L) - \cosh(\beta_n L)}{\sin(\beta_n L) - \sinh(\beta_n L)} \tag{17.5}$$

Solving for the frequency $f = \omega/2\pi$ results in one of the most important expressions used in resonant NEMS devices:

$$f_n = \eta_n \sqrt{\frac{E}{\rho}} \frac{t}{L^2}. \tag{17.6}$$

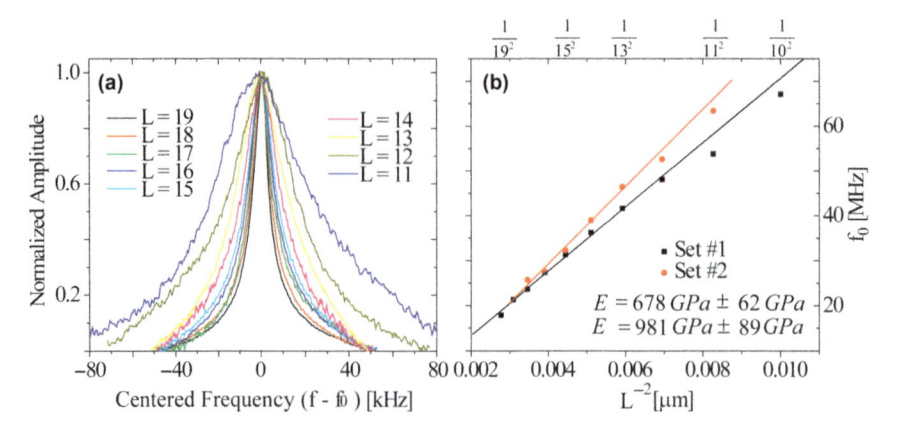

Figure 17.3 a) Resonance response for doubly clamped diamond resonators of varying lengths ($L = 11–19$ μm). b) Young's modulus derived from the slope of plotting f vs $1/L^2$.
Reprinted with permission from ref. 23. Copyright (2007), AIP Publishing LLC.

$\eta_1 = 1.028$ for the fundamental mode of a doubly clamped beam. The first term describes the mode number, the second the material properties, and the third the beam geometry. It becomes clear that a high Young's modulus and a low density (the root of the ratio is the longitudinal sound velocity) results in high resonance frequencies. This property is what allows relatively large diamond NEMS devices to achieve GHz frequency modes.[16,17] The geometric term illustrates how downscaling results in an increase in resonance frequency. Considering uniform thickness wafers, varying the length of a resonator results in a NEMS-based method to measure the dynamic Young's modulus.[23,24] Such results are depicted in Figure 17.3.

So far, only flexural beam modes have been considered. Other modes that are commonly found in NEMS devices are torsional and thickness modes. Torsional modes suffer from lower mechanical noise than flexural modes. Thickness modes are particularly common in piezoelectric materials and are found at higher frequencies. These are known as bulk acoustic wave (BAW) resonators. Such bulk modes operate well in air and even more viscous environments, making them a popular mode choice for biosensing applications. Other modes that are related to BAW resonators are shear-mode, area expansion mode, or surface acoustic wave resonators, which are all types of acoustic modes. These modes have been studied in diamond devices.[25] Surface acoustic wave resonators have been fabricated with NCD;[26] they function as chemical and biological sensors due to their ability to operate at high pressures. Such NCD-based devices typically include a non-diamond piezoelectric material, such as aluminium nitride, which serves as the electrically active component, whereas the diamond dominates the mechanical properties, also referred to as the mechanical load.

17.2.3 Beyond the Standard Euler–Bernoulli Equation

So far, the solutions presented are valid only for uniform thin-beam devices, where $t, w \ll L$. Furthermore, no intrinsic strain, forcing, or even dissipation was considered. Such terms are often significant and must be included in eqn 17.3. Dissipation can be added by making the frequency complex and replacing the ansatz of the differential equation in time with:

$$T(t) = A\sin(\omega_n t + \varphi) \rightarrow e^{-i\omega_n t - \gamma_n t + \varphi}. \qquad (17.7)$$

The complex frequency $\omega_n \rightarrow \omega_n - i\gamma_n$ results in a damping term characterized by $\gamma_n = \omega_n/2Q$. This is equivalent to making the Young's modulus complex $E = E_R + E_I$ in eqn 17.3.

The dissipative effects are a defining property of the performance and sensitivity of NEMS resonators. A further correction of interest to the Euler–Bernoulli equation is the addition of intrinsic tension. For dynamic structures, this can result in an increase of the resonance frequency due to tensile strain, or a decrease in frequency and even buckling for a large compressive strain. The derivation on buckling is extensively covered in the literature. Here, the correction terms and effects on the resonance frequency are presented for tensile strain and compressive strain below the buckling limit:

$$\frac{d^2}{dx^2}\left(EI\frac{d^2 a}{dx^2}\right) - \left(T_0 + \frac{EA}{2L}\int_0^L \left(\frac{\partial z}{\partial x}\right)^2 dx\right) = -\mu\frac{d^2 a}{dt^2} + f(x,t) \qquad (17.8)$$

This results in a resonance frequency of:

$$f_n = \eta_n \sqrt{\frac{E}{\rho}\left(1 + \frac{L^2 T_0}{4\pi^2 EI}\right)} \frac{t}{L^2}. \qquad (17.9)$$

For very high tension, the beam behaves like a string whose dynamic properties are no longer dependent on the resonator's thickness but only on its length and tension. Additional phenomena, such as thermoelastic effects, can be included by adding terms to eqn 17.8.[27]

17.2.4 The Simple Harmonic Oscillator and Nonlinear Terms

Along each point of the resonator, the amplitude follows the simple harmonic oscillator equation. This captures the frequency response and dissipative nature of the device, but not the extended mode shape. The second order differential equation of the damped driven harmonic oscillator is:

$$m\ddot{x} + m\gamma\dot{x} + kx = F, \qquad (17.10)$$

where F is the forcing term, m is the mass, k is the spring constant, and γ is the dissipation. By assuming harmonic forcing $(F = F_0\cos(\omega t))$ and taking the Fourier transformation of eqn 17.10, one can determine the amplitude and phase response in terms of drive frequency. This harmonic response can be measured in a vector network analyzer (or lock-in amplifier). It is used to characterize a resonance.

$$x(t) = A(\omega)e^{-i(\omega_d t + \varphi(\omega))},$$

$$A(\omega) = \frac{F/m}{\sqrt{(\omega^2 - \omega_R^2)^2 + \gamma^2 \omega_R^2}},$$

$$\varphi(\omega) = \tan^{-1}\left(\frac{\omega \gamma}{\omega^2 - \omega_0^2}\right), \quad (17.11)$$

where $\omega_0 = \sqrt{\frac{k}{m}}$ is the intrinsic undamped resonance frequency and $\omega_R = \sqrt{\frac{k}{m} - \frac{\gamma^2}{2}}$ is the true resonance frequency including dissipation. This response is illustrated by the blue trace in Figure 17.4.

The small scale of NEMS devices makes them prone to exceeding the limit for linear drive amplitude, beyond which nonlinear effects must be taken into account. While this may degrade device performances for some applications, there are many scenarios where this effect can be exploited not only to increase sensitivity but also to enable fundamentally new devices. For example, the dynamic two-state solution creates discrete and switchable amplitude states that can serve as bits in a mechanical logic circuit.[28,29] Furthermore, mechanical nonlinearities have been exploited in signal amplification and noise squeezing, with applications ranging from energy harvesting to macroscopic quantum behavior. A detailed theoretical description of a NEMS resonator with nonlinear spring constant and nonlinear damping is given elsewhere.[30] Here, we briefly describe the results. Expanding the harmonic oscillator eqn 17.10 and only including bounded terms, as well as assuming harmonic forcing, one can write:

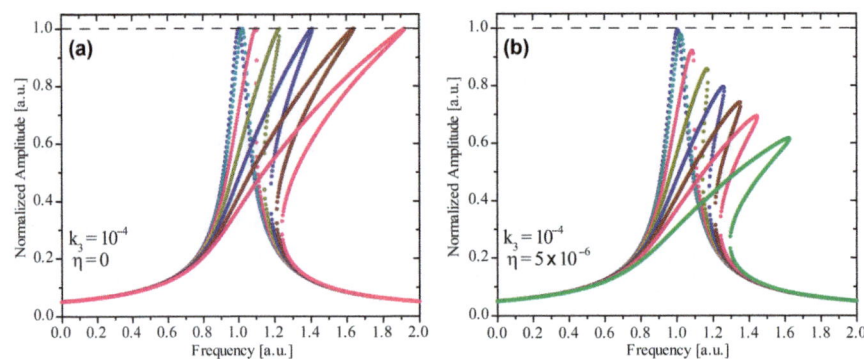

$$m\ddot{x} + m\gamma \dot{x} + kx + k_3 x^3 + \eta x \dot{x}^3 = F\cos(\omega t), \quad (17.12)$$

Figure 17.4 Driven harmonic resonator with a) nonlinear spring constant, b) nonlinear spring constant and nonlinear damping.[31]
Reprinted with permission from ref. 56. Copyright 2013, AIP Publishing LLC.

where k_3 is the nonlinear spring constant and η is the nonlinear damping coefficient. The solution for the amplitude is a transcendental equation, which can only be solved numerically:

$$x^2 = \frac{\left(\frac{F}{2m\omega_0^2}\right)^2}{\left(\frac{\omega - \omega_0}{\omega_0} - \frac{3k_3}{8m\omega_0^2}x^2\right)^2 + \left(Q^{-1} + \frac{\eta}{8m\omega_0}x^2\right)^2}. \quad (17.13)$$

Here, $Q^{-1} = m\gamma/\omega$ is the quality factor. From this equation, one can determine the backbone curve, which relates the maximum amplitude to the frequency at which the maximum amplitude occurs, as well as the critical amplitude, defined as the onset of bifurcation (pink trace in Figure 14.4 a)):

$$\omega_{max} = \omega_0 + \frac{3}{8}\frac{k_3}{m\omega_0^2}x_{max}^2; \quad (17.14)$$

$$x_c^2 = \frac{8}{3}\frac{\gamma\omega_0}{k_3}\frac{1}{\sqrt{3} - \frac{\eta\omega_0}{k_3}}. \quad (17.15)$$

These equations reveal some interesting behaviors that are visible in Figure 17.4. First, adding a nonlinear spring constant does not change the maximum amplitude. Hence, Hooke's law still applies, and there is no overextension beyond the elastic limit. If the forcing is reduced, the linear nature of the oscillator will be recovered. The nonlinear spring constant changes the frequency at which maximum amplitude occurs (eqn 17.14). For positive k_3, this leads to beam hardening and an increase in frequency and for negative k_3, this results in beam softening. Both cases are observed in NEMS devices, and experiments have even demonstrated switching from one to the other by manipulating external fields.

The nonlinear damping, on the other hand, does reduce the maximum amplitude of a resonator for a given drive force. This is clearly visible in Figure 17.4b), where the normalized amplitude is reduced with increasing drive force. Correspondingly, it does not affect the frequency at which the maximum amplitude occurs. In a sense, k_3 only affects the frequency, and η primarily affects the amplitude. They both work together when considering the critical amplitude. It is noteworthy that the nonlinear damping suppresses the onset of bifurcation and, for a damping coefficient $\eta = \frac{\sqrt{3}k_3}{\omega_0}$ and above, the bifurcation is eliminated for all drive amplitudes.

The origins of the nonlinear spring constant k_3 are well understood. They may arise from external fields, as observed with capacitive drive and detection techniques, where $k_3 = \frac{1}{6}V_G^2 C^{IV}|_0 < 0$ (V_G is the gate voltage of the capacitor and C^{IV} the forth derivative of the capacitance), or from geometric nonlinearities $k_3 = \omega_n^2 \beta_n \frac{6m}{\ell^2} > 0$ for a doubly clamped beam (β_n is a mode-dependant numerical constant ranging from 0.2 to 0.5). Typically, the intrinsic material nonlinearities are much weaker and do not affect NEMS dynamics; although these effects may be pronounced at ultra-low temperatures.

The physical interpretation of nonlinear dissipation is less well understood. There have been a number of experiments that indicate the presence of nonlinear damping through the widening of the Lorentzian. However, detailed theoretical explanations have yet to be presented. Furthermore, it is commonly claimed that only the smallest NEMS devices, such as carbon nanotube (CNT) resonators, exhibit nonlinear damping. There is no reason to believe that such nonlinear damping would not also be visible in significantly larger diamond NEMS devices as high quality factors would make nonlinear damping effects more apparent. Further studies should help reveal the physical nature of this phenomenon and the constraints it sets on devices.

17.3 Methods of Fabrication, Actuation, and Detection

In this section standard fabrication steps, and actuation and detection methods for diamond NEMS devices are discussed. These methods are applicable to a range of materials. However, some of these methods are particularly suitable for diamond. Such cases are highlighted.

17.3.1 Fabrication

The structures described here are manufactured using top-down lithographic methods. MEMS are essentially based on optical lithography and NEMS on e-beam lithography, where the spatial resolution is on the order of a few micrometers and tens of nanometers, respectively. As a basic rule, the amount of material deposited is more than required and the unwanted material is subsequently etched away as a part of the process. This is in contrast to the growth techniques typical for carbon nanotubes or other exotic materials. Even in these cases, a base structure can be made using conventional methods, and then additional components, such as the CNT, can be added.[32]

17.3.1.1 Multilayer Wafer

Figure 17.5 depicts typical fabrication steps, as well as the resulting diamond harp structure. In this method one starts with a handle wafer (silicon >100 μm), an insulation layer and pedestal (silicon oxide, silicon 0.1–~10 μm), and finally the device layer (CVD diamond, 0.1–~10 μm). A single or a number of lithography steps (Figure 17.5b) are used to thermally apply metallic masks (Figure 17.5c) that will withstand the anisotropic reactive ion etch (RIE) (Figure 17.5d), which defines the devic's two-dimensional (2D) geometry. 300 W O_2/CF_4 plasmas at pressures of ~25 mTorr and flow rates (50/5 sccm) result in typical etch rates of ~100 nm min^{-1} of diamond.

Diamond Nano-electromechanical Systems 423

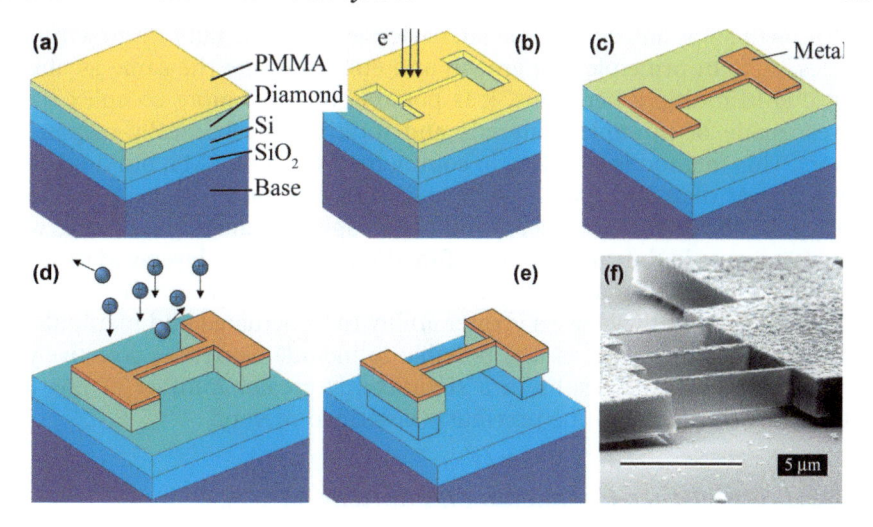

Figure 17.5 a) Wafer, b) e-beam patterning, c) metallization, d) and e) etch steps, f) final product.[31]

The metallic masks can also serve as electrodes, as is the case for the doubly clamped beam structures depicted below.

The final step (Figure 17.5e) is the release, where an isotropic etch removes the sacrificial layer below the mechanical structure, allowing it to move freely. For silicon oxide sacrificial layers, hydrogen fluoride is used as an isotropic wet etch. The structures can be dried using a critical point dryer if needed. It turns out that larger, softer MEMS structures are often more susceptible to damage caused by surface tensions than high frequency NEMS resonators are, even though they are orders of magnitude smaller. This is a result of the high spring stiffness of NEMS devices. Diamond has proven to be a robust material for such structures. It should be noted that using silicon on top of oxide as a sacrificial layer has many benefits. Once the structure is suspended, the remaining oxide provides an excellent insulating base, preventing any unwanted shorts. Furthermore, an SF_6 plasma can be used at relatively high pressures (200 mTorr) to etch away the silicon. This means that no etchants, such as KOH, are required and the structure is never submerged in a liquid; hence, it cannot be damaged by surface tension during release.

Devices manufactured by this method tend to be essentially two dimensional, with the thickness of each material being constant. Also, motion is predominantly in the plane of the wafer with only small flexure or torsion occurring out of the plane. The device behavior is strongly defined by the initial wafer.

17.3.1.2 Multistep Growth

An alternative method is based on depositing thin films over the entire wafer, followed by selective etching. However, instead of starting with all of the growth steps already completed, devices can also be built by growth/etch cycles.[5,17,21] In principle, the growth and etch steps are the same as above, but due to multiple iterations, it is possible to build more complex structures, made of multiple materials and multiple mechanically responsive layers. This method is standard in commercial silicon MEMS, but due to the difficulty of diamond CVD growth, it is less commonly seen in diamond structures. Nevertheless, cheap and rapid optical lithography, along with improvements in the quality of thin-film diamond, can be leveraged to create even more complex diamond MEMS. These devices are also inherently two dimensional. But they have a greater ability to be actuated and manipulated in the third dimension, as it is possible to include electrodes selectively at multiple levels. This results in a significantly greater number of possible structures compared to the previously described method.

17.3.1.3 Further Comments on Fabrication

There have been attempts to selectively grow diamond by controlling the location where seeding occurs. This removes the need for etching the diamond structure after deposition. The results, however, are not competitive with the methods described above. Growing continuous diamond thin films into pre-existing molds has been successful in manufacturing field emission devices and nano-probes,[33] among other devices.

For many sensor applications, surface termination is crucial. This is critically important for structures that require surface functionalization. For example, one may choose hydrogen or hydroxyl termination resulting in hydrophobic or hydrophilic surfaces, respectively. This surface treatment can be achieved using oxygen or hydrogen plasmas, where a simple "water test" reveals the hydro-philic and -phobic properties.

So far, all structures considered were made of polycrystalline diamond. Freestanding, single-crystal diamond structures have been fabricated, as shown in Figure 17.6. This is achieved by growing single-crystal diamond, then adding a sacrificial layer through ion implantation.[13,14] This allows

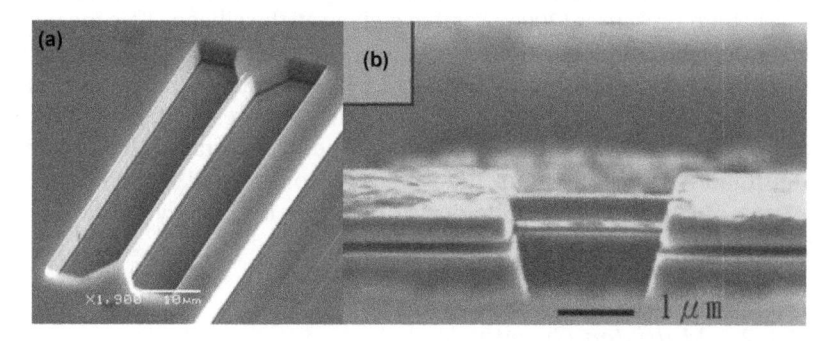

Figure 17.6 Single-crystal diamond freestanding structure; a) Reprinted with permission from ref. 14. Copyright (2005), Wiley; b) Reprinted with permission from ref. 13. Copyright (2007), American Vacuum Society.

selective etching, which is needed for suspending the structure. Highly specialized optical or thermal devices could be manufactured using this method. However, the mechanical properties of nanocrystalline diamond (NCD) are close to those of the single crystal, making NCD the simpler choice. More recently, a wafer bonding technique has been reported. It allows single-crystal resonators to be manufactured on a silicon dioxide sacrificial layer,[15] enabling integration of single-crystal diamond with standard silicon-based NEMS fabrication methods.

17.3.2 Drive and Detection

A critical component of all MEMS/MEMS devices is the drive and detection technique. This determines the method of coupling between the mechanical degree of freedom and the electrical readout circuit. Some devices are passive and require no external drive, such as vibration sensors, whose mechanical energy is transferred from the signal they are designed to sense directly into the device. More often, a device is actuated on resonance by applying mechanical energy through a given transduction method. On resonance, the structure is most sensitive to external stimuli, and hence, more effective as a sensor. In this section the most common actuation and detection methods are presented and compared. For the drive, this corresponds to the forcing term introduced in eqn 17.3. The common drive and detection methods used in MEMS and NEMS are compatible with diamond or hybrid diamond-and-metallic structures. In addition, some specific properties of diamond allow for less conventional actuation/detection methods.

17.3.2.1 Capacitive Technique

Probably the most common drive and detection method for both on- and off-resonance devices is capacitive. In this method selective modes can be targeted. With the use of appropriate geometries, linear drive forces can be applied over a long actuation range. The capacitive method also allows for simultaneous drive and detection capabilities. The voltage applied to a capacitor results in a force. Correspondingly, any resulting change in geometry will change the capacitance of the device.

For two parallel electrodes, the capacitive force is given by:

$$F_c = \frac{1}{2}V^2 \frac{dC}{dx}. \qquad (17.16)$$

For parallel plate capacitors, this results in a nonlinear forcing term, resulting in spring softening. A drive containing a d.c. voltage and an a.c. voltage is applied to the resonator. Considering the first order terms in displacement, the harmonic oscillator equation is given by:

$$m\ddot{x} + m\gamma\dot{x} + (k - \frac{C''V^2}{2})x \approx C'V_G V_{dr}, \qquad (17.17)$$

where $\frac{dC}{dx}$ is expanded into a Tyler series. Typically, $V_G \gg V_{dr}$.

For a constant bias voltage, but varying capacitance due to the mechanical motion, charges will flow on and off the capacitor. This current is proportional to the amplitude of the resonator and can be used to monitor the dynamics:

$$I = \frac{dQ}{dt} = V_G \frac{dC}{dt} = V_G C' \dot{x},$$

$$\tilde{I} = iV_G C' \omega \tilde{x}. \qquad (17.18)$$

The resonator acts as a high-impedance current source, and the signal is typically pushed through a transimpedance amplifier before being detected in a lock-in amplifier or a network analyser, as illustrated in Figure 17.7.

Combining equation 17.17 and 17.18, one finds that in the linear regime the detected current (or voltage) is proportional to the drive voltage and the gate voltage squared. Sweeping these externally controlled parameters, C'^2/m and C''/m can be directly measured. The displacement on resonance is:

$$x = Q \frac{C' V_G V_{dr}}{k_0 - \frac{1}{2} C'' V_G^2}. \qquad (17.19)$$

The capacitive plate method is limited by the $\frac{1}{3}$ pull-in effect. As the drive force is nonlinearly dependent on the gap of the capacitor plates, for any spring and capacitor geometry, it can be shown that after the voltage causes a $\frac{1}{3}$ reduction of the gap between the two plates, a snap-in takes effect as the

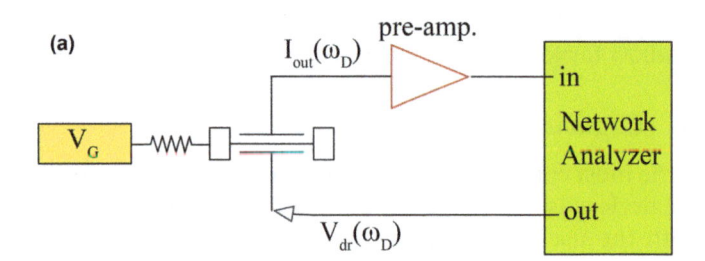

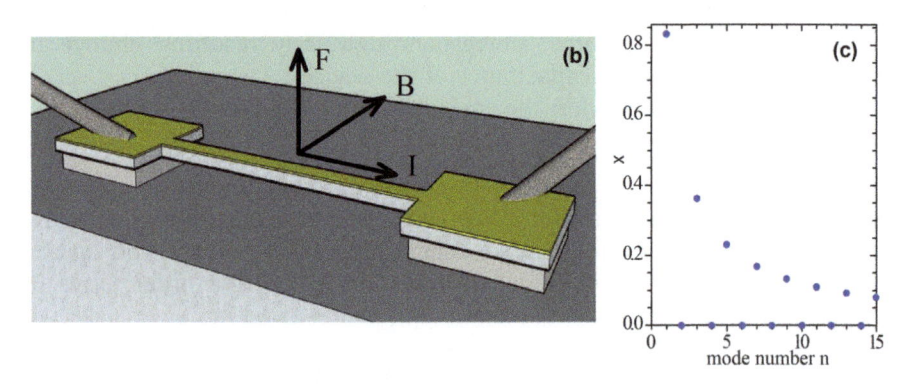

Figure 17.7 Drive and detection schematics a) capacitive, b) magnetomotive methods.[31] c) Mode dependent numerical value of the effective area swept over a resonance cycle (used in eqn 17.24).

Figure 17.8 a) Doubly clamped beam resonator with capacitive drive and detection plates. b) Diamond comb actuator with folded spring supports.
Reprinted with permission.
Reprinted from ref. 21, Copyright (2004), with permission from Elsevier.

electrical force exceeds the mechanical force. This sets a fundamental limit to the useful range of such devices.

For larger MEMS devices, capacitive comb actuators have been developed. In these devices the motion is parallel to the plates, which allows for much longer linear travel distances. In such devices, the drive force becomes:

$$F_c = \frac{n}{2} V^2 \frac{t}{d}, \qquad (17.20)$$

where n is the number of electrodes, V is the applied voltage, t is the comb thickness and d is the gap between electrodes.

Combining the drive and detection terms of the resonator results in the frequency dependent current:

$$\tilde{I}(\omega) = \frac{iC'^2/mV_G^2 V_{dr}}{\omega_R^2 - \omega^2 + \frac{i\omega\omega_R}{Q}}. \qquad (17.21)$$

Due to the two-dimensional nature of thin-film structures, the capacitance is limited by the number of combs and the material thickness. This may be large compared to the doubly clamped beam depicted in Figure 17.8. However, compared to parallel plates, where the capacitive area is not limited by film thickness, the capacitance is typically quite small and on the order of picofarad.

For the capacitive method to work directly on diamond structures, they must be doped considerably so that a current to and from the electrodes can flow. This limits the purity of the diamond for which this method is applicable. An alternative approach would require integration of metals.

17.3.2.2 Piezoelectric Technique

In piezoelectric materials a mechanical strain can be induced by applying an electrical voltage. Correspondingly, a mechanical strain applied though an external force will result in a voltage drop across the material. This means

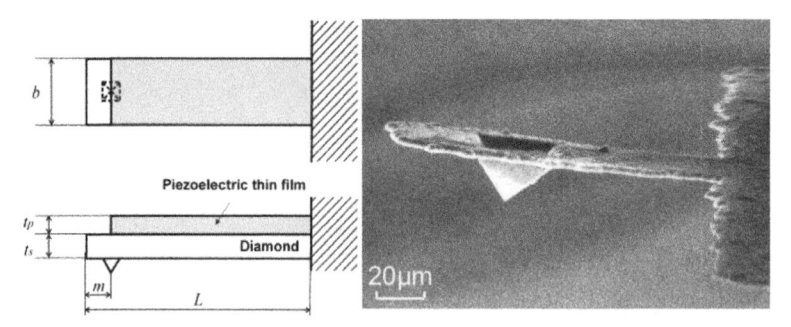

Figure 17.9 476 MHz AFB resonator, ZnO on NCD.
Reprinted from ref. 34, Copyright (2002), with permission from Elsevier.

that electrical signals can be used to drive and detect a mechanical mode. The field-strain relation is given by:

$$\epsilon_i = \frac{\Delta L_i}{L_i} = d_{ji}E_j, \qquad (17.22)$$

where i is the strain axis, L_i is the length, d_{ji} is the contour piezoelectric coefficient, and E_j the electric field along the j axis. Both frequency and electrode placement will define which modes are sensitive to this effect.

As diamond is not a piezoelectric material, this method of actuation and detection can only be used in a hybrid setup. This can be done in two ways. One, diamond can be grown directly on top of (or beneath) a piezoelectric material, such as aluminum nitride. In such experiments the piezoelectric material provides the drive and detection, whereas the diamond layer acts as the mechanically relevant bulk material (Figure 17.9). Such setups are used for thickness modes, as well as surface acoustic wave resonators. Flexural modes are possible, but these are typically less suitable for this actuation method. As the two materials are strongly coupled, the transfer of mechanical strain from the diamond to the piezoelectric material is maximized. A drawback is that the mechanical element is no longer pure and the contribution from the other materials must be taken into account when studying mechanical properties, such as the Young's modulus or dissipation.

By mounting a diamond resonator on top of a macroscopic piezoelectric shaker, mechanical energy can be transferred to the resonator. Hence, the resonator can be made of diamond alone. One drawback, however, is that the coupling is weak and the energy transfer is low, which makes the system inefficient. Also, this method can only be used to actuate; therefore, a second independent method for detection must be implemented.

17.3.2.3 Magnetomotive Technique

A low impedance drive and detection circuit can be set up by the magnetomotive method. This setup is popular for cryogenic NEMS devices that operate at high frequencies and low temperatures. If a current is passed through the device in the presence of a magnetic field, the corresponding Lorentz force will actuate the mechanical mode. A strong magnetic field increases the coupling, which is why a cryogenic setup with a superconducting magnet is preferred. The drive force becomes:

$$\vec{F} = I\vec{L} \times \vec{B}, \quad (17.23)$$

where $\vec{L}$ is the length of the resonator, $I = V/R$ the applied current, and $\vec{B}$ is the external magnetic field. Plugging this forcing term into eqn 17.10 results in a resonator displacement linear in the drive voltage and magnetic field. The readout is given by Faraday's law that describes how the change in magnetic flux induces an electromotive force:

$$V_{\text{emf}} = -\frac{d\phi_B}{dt} \rightarrow \tilde{V}_{\text{emf}} = i\omega\xi_n BL\tilde{x}(\omega), \quad (17.24)$$

where ϕ_B is the magnetic flux, and ξ_n is a mode-dependent geometric number equal to zero for all even values of n and ranges from 0.83 for $n = 1$ to zero as n approaches infinity, as depicted in Figure 17.7. Combining equations 17.10 and 17.24 results in the magnetomotive frequency-dependent response voltage:

$$\tilde{V}_{\text{emf}} = \frac{\frac{i\omega\xi_n B^2 L^2}{m}}{\omega_R^2 - \omega^2 + \frac{i\omega\omega_R}{Q}} \tilde{I}(\omega). \quad (17.25)$$

As this is a low impedance measurement, it is capable of detecting resonances above one GHz.[16] The significant drawback is that a large magnetic field is required, which is in the range of 1–8 T for NEMS devices. Correspondingly, this coupling can be used for magnetic field detection as well. A fundamental drawback is the increase in dissipation due to circuit loading, as discussed in the next section. Another limitation is that only odd modes can be observed. Similar to the capacitive case, in order for a current to flow it is necessary to apply metallic leads over the resonator to ensure a low impedance current-carrying element.

17.3.2.4 Piezoresistive Technique

The piezoresistive effect is analogous to the piezoelectric effect: a change in strain results in a change in resistance of the material characterized by the piezoresistive gauge factor:

$$\gamma_P = \frac{\Delta R}{R}\frac{1}{\epsilon_s} = (1+\nu) + \frac{\delta\rho_e}{\rho_e}\frac{1}{\epsilon_s}. \quad (17.26)$$

The first term on the right-hand side is the geometric factor of order one and is dependent on Poisson's ratio ν. For diamond, γ_P can range from 5–1000 depending on doping and crystal purity. For large γ_P this method can result in a sensitive strain measurement. Unlike the piezoelectric effect, where strain can be induced by applying a voltage, it is not possible to apply or

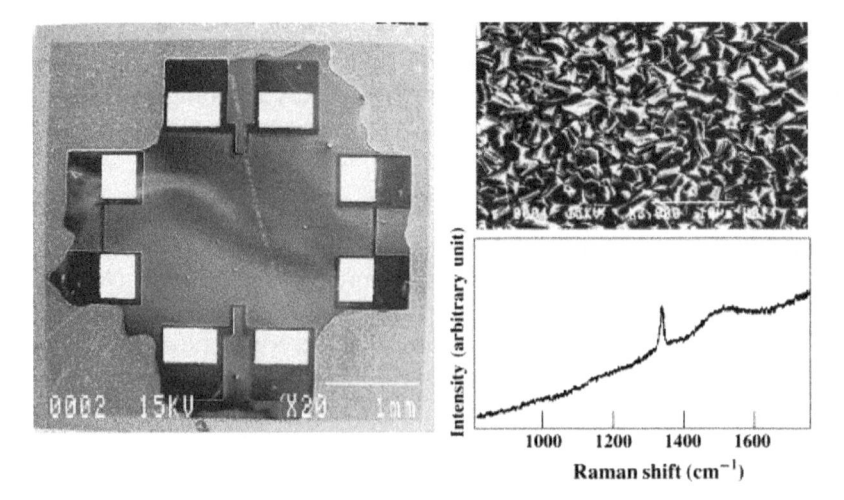

Figure 17.10 Piezoresistive diamond MEMS pressure sensor. Reprinted from ref. 37, Copyright (2005), with permission from Elsevier.

induce a resistance. Hence, the drive-detection symmetry is lost and the piezoresistive method can only be used for detection.

For static measurements diamond piezoresistivity has been exploited for biological[35,36] and pressure-sensing devices,[37] an example is depicted in Figure 17.10. In short, the diamond window will bulge as the pressure on either side of it changes, resulting in a change of strain in the window. Simply by measuring the resistance of the window (or an element on the window), one can determine the mechanical bulge and hence the pressure surrounding the device. Piezoresistivity has also been exploited in resonant NEMS devices where mixing drive and detection signals results in sensitive frequency-dependent detection methods. Dynamic piezoresistive methods have been implemented in diamond MEMS.[31,38]

17.3.2.5 Other Drive and Detection Methods

There are other drive and detection methods that can be applicable to diamond devices. One example is electrothermal drive, where an applied current heats the structure, which couples to the mechanical mode through the thermal expansion coefficient as $\Delta L = \alpha L \Delta T$. The induced thermal strain is given by:

$$\varepsilon^{th} = \int_{T_0}^{T_1} \alpha(T) dT. \qquad (17.27)$$

Diamond switches have been actuated by this method, but there is no reason why it should not also be used for dynamic MHz actuation as this method has been well tested in other materials.[39] The thermal expansion coefficient

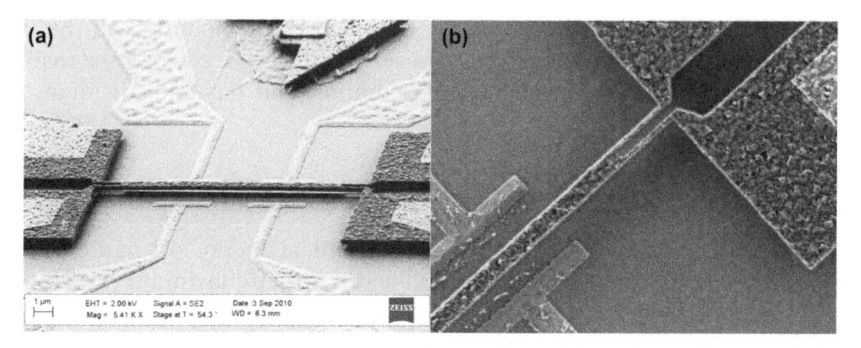

Figure 17.11 a) Diamond doubly clamped beam with gold side electrodes for dielectric force actuation. b) Close-up of piezoresistive end-clamp geometry.[31]

of diamond is not particularly high compared to other materials. However, its thermal stability and high thermal conductance can offset this shortcoming. For materials with high thermal conductivity, the small dimensions of NEMS result in an extremely short thermal relaxation time $\tau_{th} = \frac{t^2 \rho C_P}{\pi^2 \kappa}$, allowing for resonances to be actuated well into the MHz range. The thermal coupling to the mechanical mode can only be used for driving the device as it would be extremely difficult to measure changes in temperature over such small distances and short timescales.

A relatively new method for NEMS actuation and detection based on dielectric force has been proposed.[40] Diamond, with a comparable dielectric constant to these experiments and also high dielectric breakdown, is a good candidate for this actuation method. It is based on the forces due to electric fields acting on a polarized material. It is assumed that a strong d.c. electric field polarizes the dielectric, and the superimposed a.c. field results in forcing, analogous to the capacitive drive. The forcing term takes the form:

$$\vec{f} = (\vec{P}\vec{\nabla})\vec{E} = \varepsilon_0 \chi (\vec{E}\vec{\nabla})\vec{E}, \qquad (17.28)$$

where $\vec{f}$ is the force per unit length, χ is the dielectric constant of the resonator material, $\vec{P}$ is the polarization and $\vec{E}$ is the externally applied electric field. For symmetric electrodes (but not in plane with the resonator), only forces in the z axis can be generated. But, if the symmetry is broken, significant in-plane forces are generated. An example of such a device is depicted in Figure 17.11. This expands the number and type of modes that can be actuated. To increase the sensitivity of this technique in diamond NEMS, the dielectric constant can be increased by adding impurities, such as water, to the growth process.

To conclude, further detection methods that are commonly used in both MEMS and NEMS are optical detection techniques. These include reflecting a laser beam off the moving resonator and recording the angle of deflection, as is common for AFM applications. More sensitive methods, where optical fibers are used to couple the fringe field with specific modes, such as gallery modes, have also proven suitable at high frequencies. Unlike the electrical readout methods, optical methods do not add circuit loading, so the true mechanical response is observed. These methods are covered extensively in the literature. Examples of optical actuation and detection in diamond devices are given here.[41,42]

17.4 Dissipation in Diamond NEMS

Dissipation is a measure of the rate of mechanical energy lost in a resonant device. It can be determined by the full-width at half-maximum (FWHM) of the Lorentzian response in frequency domain described in section 17.2.4 or by the ring-down time in time domain.[43] The quality factor Q is the inverse of the dissipation factor, which is written as:

$$Q^{-1} = \frac{\Delta W}{2\pi W} \cong \frac{\gamma}{\omega_0} \cong \frac{\Delta \omega}{\omega_0}, \quad (17.29)$$

where W is the total mechanical energy in the system and ΔW is the energy lost per cycle of oscillation. ω_0 is the resonance frequency and $\Delta \omega$ is the full-width at half-maximum of the Lorentzian response. The standard linear solid model modifies Hooke's law to take the relaxation timescales of the strain and stress into account. Assuming a time delay between stress and strain, as well as harmonic forcing, one obtains a frequency-dependent complex Young's modulus. The imaginary component is responsible for energy dissipation. In this section common dissipation sources in diamond NEMS are discussed. Some of these are exemplified in Figure 17.12.

These dissipative mechanisms are of great interest for device engineering, as well as for understanding the material itself. For nonresonant structures such as switches, MEMS mirror, or pressure gauge, dissipation sets the timescale at which a structure can respond to an external stimulus. For such

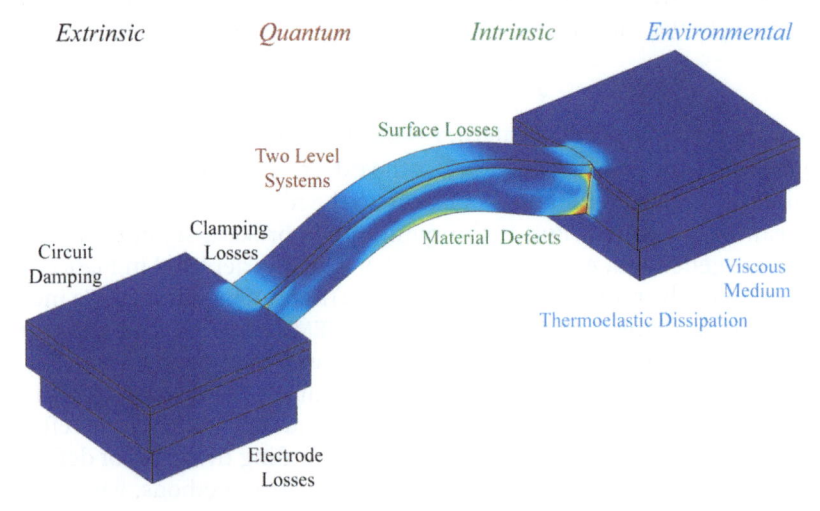

Figure 17.12 Examples of dissipation in a doubly clamped beam resonator; the color depicts strain intensity.[31]

systems, extremely low dissipation may be problematic as the transient time diverges as $Q^{-1} \to 0$. For example, in a pressure sensor the mechanical settling time due to a change in pressure would be long, making it difficult to detect rapid fluctuations. In such cases, critical damping with $Q=1$ is the most efficient condition. Resonant devices typically require high quality factors. It reduces the mechanical and electrical noise that can couple into the devices. A higher quality factor also allows for more accurate frequency measurement. Many sensors rely on frequency shifts as the method of sensing as this is one of the most precise measurements possible. It is common for a frequency to be measured with eight-digit accuracy, a feat not possible for the amplitude or dissipation. Lower dissipation also means lower energy consumption and improved signal-to-noise ratio, which is important for signal processing and in communication applications.

Dissipation sources and mechanisms can be categorized depending on their nature. They may be bulk defect effects, surface effects, or external sources. Some are temperature-sensitive, while others are geometry-dependent. The following subsections offer some insight into what mechanisms degrade diamond NEMS resonators and how this can be remedied. For multiple uncorrelated dissipation mechanisms, the total dissipation is the sum of contributions from each dissipation mechanism:

$$Q^{-1} = \frac{\Delta W}{2\pi W} = \frac{1}{2\pi W} \sum_i \Delta W_i = Q_{CL}^{-1} + Q_{TE}^{-1} + Q_{MD}^{-1} + \cdots \quad (17.30)$$

A resonator may have multiple relevant dissipation sources acting simultaneously.

17.4.1 Clamping Losses (Q_{CL}^{-1})

Clamping losses occur whenever mechanical strain energy is concentrated at the attachment points of the structure. This is the case for the flexural modes of cantilevers and doubly clamped beams. Mechanical strain radiates out of the structure into the base and is not recovered. The firmer the base, the less energy is transferred; hence, maximizing the mechanical impedance mismatch maximizes the reflection of strain energy back into the resonant structure. This is reflected in the scaling behavior of dissipation with regards to device geometry:[44]

$$Q_{\text{in plane}}^{-1} = \alpha \frac{w^3}{L^3}$$

$$Q_{\text{out of plane}}^{-1} = \beta \frac{t}{L}$$

$$Q_{\text{large base}}^{-1} = \gamma \frac{w}{L} \frac{t^4}{L^4}, \quad (17.31)$$

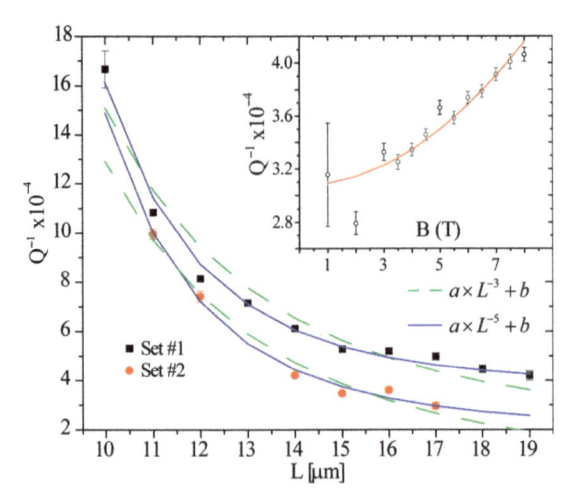

Figure 17.13 Example of circuit loading, clamping losses, and surface losses in diamond MHz doubly clamped beam resonators.
Reprinted with permission from ref. 23. Copyright (2007), AIP Publishing LLC.

where L, w, and t are the length, width, and thickness of a beam resonator. α, β, and γ, are mode- and material-dependent coefficients. Experimentally, it was found that diamond NCD has $\alpha = 10$–11.2.[23,43] An example of dissipation scaling with resonator length is given in Figure 17.13. For this experiment, the beam thickness and width was held constant, while only varying the length. Three dissipation mechanisms are required to explain the observations. Clamping losses account for the length dependency, circuit loading for the field dependency (insert of Figure 17.13), and surface losses for the length-independent tail observable in Figure 17.13.

Clamping losses can be minimized by increasing the aspect ratio. This makes higher frequencies at constant thickness more vulnerable, and hence, typical doubly clamped beams have quality factors below 10^4 or even below 10^3 as the frequency increases from 10^7 to 10^9 Hz. Alternatively, clever attachment points can result in energy trapping modes, where a minimum of elastic energy is radiated into the base. An example of this is the diamond antenna structure depicted in Figure 17.14. Here, it was shown how a large structure can be forced to resonate at high frequencies through an array of smaller devices. As the strain energy is located far from the attachment points, these modes have quality factors beyond 20 000 at 630 MHz and still 8660 at 1.446 GHz, resulting in exceptional fQ factors on the order of 10^{13}.

17.4.2 Thermoelastic Dissipation (Q_{TE}^{-1})

When a solid is rapidly squeezed or stretched, heating or cooling will occur as a result of scattering between he acoustic and thermal phonons. If this temperature gradient has the time to thermalize, then the heat transfer

Diamond Nano-electromechanical Systems 435

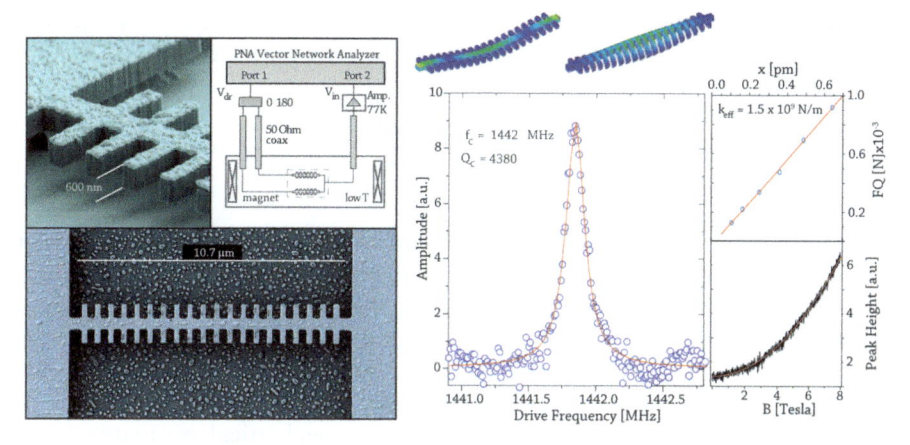

Figure 17.14 High order collective mode maximizes high frequency response and minimizes clamping losses.
Reprinted with permission from ref. 16. Copyright (2007), AIP Publishing LLC.

results in the removal of mechanical energy from the resonant mode, and hence, in dissipation. This is essentially the inverse of a thermal drive. There are two components to this phenomenon: 1) the amplitude of heat generated by the squeezing, and 2) the rate at which this heat can dissipate, which is the inverse of the thermal relaxation time. If the resonance is much slower than the thermal relaxation time, the system is in the isothermal limit, there is no heat flow, and hence, no thermoelastic dissipation. The opposite extreme occurs when the resonator period is much shorter than the thermal relaxation time. In this adiabatic limit there is no heat transfer as the process is too slow to keep pace. The corresponding dissipation takes the form:

$$Q_{TE}^{-1} = \frac{\alpha^2 TE}{\rho C_p} \frac{\omega \tau_{th}}{1 + (\omega \tau_{th})^2}. \quad (17.32)$$

The first fraction is a measure of how much heat is generated (α is the thermal expansion coefficient, T the temperature, E the Young's modulus, ρ the density, and C_p the heat capacity). The second part includes the timescale for thermalization $\tau_{th}^{-1} = \frac{\pi^2 \kappa}{l^2 \rho C_p}$. This dissipation source results in a fundamental limit for all resonators, given by their material properties, independent of defects or clamping geometries. Fortunately, it is also easy to avoid this source of dissipation by engineering the structure away from the resonant timescale as thermoelastic dissipation is suppressed at both low and high frequencies.

17.4.3 Circuit Loading (Q_{CL}^{-1}) and Multiple Materials (Q_{MM}^{-1})

A resonator is often constructed out of more than one material. In a typical resonator one may combine diamond with piezoelectric materials or metals to enable actuation and detection. For material-dependent dissipation mechanisms, each layer can be taken individually into account:

$$Q_{MM}^{-1} = \sum_{j,k\cdots \neq i}^{i} \left(1 + \frac{t_j E_j + t_k E_k + \cdots}{t_i E_i}\right)^{-1} Q_i^{-1}, \quad (17.33)$$

where i, j, k are material-type indices. It can be seen from the first coefficient that, for large thickness differences in materials, as is the case for typical diamond–gold magnetomotive structures, the correction to the total dissipation is minimal and is strongly dominated by the diamond layer. This is often not the case for hybrid piezoelectric structures that require the piezoelectric material to be reasonably thick.

Circuit loading is a result of coupling the mechanical resonator to an electrical readout or drive circuit. Hence, this dissipation mechanism is not present in optical setups. Magnetomotive, capacitive, and piezoelectric drive and detection schemes, however, can contribute or even dominate the dissipation in the device. One way to model the setup is a RLC circuit, where the mechanical resonator can be mapped to an equivalent RLC element. By identifying the electrical equivalents, one can determine the resulting dissipation. For example, in devices driven by the magnetomotive method, the voltage generated by the moving resonator across the beam will try to equilibrate. In practice this system behaves like a parallel shunted RLC circuit with the following equivalent electrical components:

$$R_m = \frac{\xi L^2 B^2}{m\omega_0} Q_M^{-1},$$

$$L_m = \frac{\xi L^2 B^2}{m\omega_0^2},$$

$$C_m = \frac{m}{\xi L^2 B^2}, \quad (17.34)$$

where Q_M^{-1} is the intrinsic dissipation. Following ref. 45, this results in an additional circuit dissipation term for magnetomotive drive and detection:

$$Q_{CL\ \text{Magnetomotive}}^{-1} = \frac{\xi L^2 B^2}{m\omega_0} \frac{R}{|Z|^2}, \quad (17.35)$$

where $Z = R + iX$ is the shunt impedance.

As illustrated in Figure 17.13, this contribution is dependent on the square of the magnetic field. Depending on intrinsic dissipation, this contribution can become significant at fields over 2–3 T. These effects can be calculated or measured experimentally and subtracted out to obtain the true mechanical dissipation.

Analogously, circuit dissipation for capacitive and piezoelectric setups for doubly clamped beam resonators becomes:

$$Q^{-1}_{\text{CL Capacitive}} = \frac{V_G^2 C'^2}{2\pi m \omega_0} R,$$

$$Q^{-1}_{\text{CL Piezo}} = \frac{6.03 \ twd_{31}^2 V/t}{\rho L^3 \omega_0} R. \quad (17.36)$$

17.4.4 Mechanical Defects, Surfaces (Q^{-1}_{SL}) and Bulk (Q^{-1}_{MD})

Defects can cause additional dissipation in imperfect crystals. For any crystalline system, defects are concentrated at the surface as the crystal symmetry must be broken there. This results in dislocations, surface contamination, and dangling bonds that can all contribute to dissipation. For macroscopic objects, this surface effect can be neglected. Through scaling down to the sub-micron realm, the surface-to-volume ratio grows, making surface effects increasingly important. To include these surface-related dissipation mechanisms, a complex surface Young's modulus E^S is introduced. Assuming that the real part deviates only slightly from the bulk but the imaginary part is still large, one can write an expression for the surface contribution to dissipation. Calculating the mechanical energy and energy loss by integrating over the strain field, an expression can be obtained for a cantilever or doubly clamped beam:

$$Q^{-1}_{\text{SL}} = \frac{2\delta(3w+t)}{wt} \frac{E_I^S}{E}, \quad (17.37)$$

where δ is the depth affected by the surface Young's modulus. The product δE_I^S is determined experimentally, as in Figure 17.13, by the length-independent contribution of the dissipation.

Dissipation due to mechanical defects can also arise in the bulk of the resonator. These are typically thermally activated and will show up as a Debye peak in the temperature-*versus*-dissipation plot. This has been demonstrated along with the Arrhenius relation in diamond NEMS resonators.[45] The defect is characterized by an activation energy and attempt rate, resulting in frequency- and temperature-dependent dissipation, analogous to thermoelastic dissipation:

$$Q^{-1}_{\text{MD}} = \sigma \frac{\omega \tau_{\text{MD}}}{1 + (\omega \tau_{\text{MD}})^2}, \quad (17.38)$$

where σ is a unitless constant dependent on defect density and the time constant is given by $\tau_{\text{MD}} = \tau_{\text{MD0}} \exp(\frac{E_A}{k_B T})$, where E_A is the activation energy and τ_{MD0} is the attempt period. The activation energy is on the scale of the self-diffusion energy, typically a few electron volts or below, for NCD $E_A \approx 0.02$ eV has been reported.[43] For systems with a few well-defined defects, the Debye peaks are discrete. If the defects become smeared over a large temperature range, as would be the case in a more amorphous material, the resulting peaks add up to a linear temperature dependence.

Growing high-quality NCD will result in a Young's modulus approaching that of single-crystal diamond. Where the increase in the bulk value for the Young's modulus from 900 GPa to 1200 GPa only has a moderate impact on the resonance frequency (15% effect), the improvement in the quality factor is much more significant. Once clamping and circuit losses are eliminated, dissipation due to mechanical defects can dominate. The improvement is amplified for surface effects as illustrated in Figure 17.13. For data set 2, a much higher Young's modulus was reported and a considerably lower surface dissipation was measured, compared to data set 1. The resonators were manufactured from two separate CVD NCD wafers.

17.4.5 Viscous Damping (Q_{VD}^{-1})

Many NEMS experiments are operated in a vacuum. Decoupling the system from the environment as much as possible improves device sensitivity and performance. This is true for oscillators, as well as mechanical switches. However, NEMS and MEMS are often used as sensors for gas or biological agents; hence, they must operate in gaseous environments and viscous fluids. Such devices must be engineered carefully to minimize viscous damping resulting from the friction that occurs as a solid moves through a fluid. As diamond NEMS are biocompatible and can be functionalized, this dissipation constraint will become relevant for biosensing devices.

At low pressures, the molecular regime applies and it is assumed that the molecules do not interact with each other over the length scales relevant to the resonator. This is valid if the mean free path is longer than the resonator's width or thickness. Dissipation is a result of momentum transfer between gas molecules and the resonator. For a doubly clamped beam resonator, this dissipation becomes:[46]

$$Q_{VDLP}^{-1} = 0.002 \frac{P}{\sqrt{E\rho}} \left(\frac{L}{t}\right)^2, \qquad (17.39)$$

where P is the pressure. Shortening the length of the resonator enables higher resonator frequency and lower dissipation. Typical resonators are not affected by air damping at pressures below 1 Torr. High-frequency resonators will deviate from the linear pressure dependency where a $\sqrt{P}$ relation has been observed experimentally.[45]

In the viscous limit (strongly interacting molecules), this dissipation is dependent on material and geometric properties, as well as the viscosity μ of the surrounding medium, but is not dependent not on pressure. The model assumes that the resonator is dragging the surrounding medium with it, which results in mass loading and energy exchange. Dissipation for a doubly clamped beam mode is given by:

$$Q_{\text{VD}}^{-1} = \frac{3.8\mu}{\sqrt{E\rho}w}\left(\frac{L}{t}\right)^2. \tag{17.40}$$

Here, the response time of the medium is not included. Analogous to thermoelastic dissipation, the rate at which the viscous fluid reacts to the resonator will determine the rate at which energy is transferred.

For large area structures in close proximity to static solids, as is the case in many capacitive-based MEMS, squeeze-film damping occurs. Essentially, the term $\left(\frac{w}{g}\right)^3$ is multiplied to eqn 17.40, where g is the gap size for $w \gg g$. Adding holes will mitigate this for small $\frac{w}{g}$ geometries. Such holes are often already present as a result of the release process described in the section 17.3.1.

Flexure modes are most susceptible to air damping and should be avoided unless particular geometries can be implemented. Generally, breathing modes show superior performance in viscous environments due to their small mechanical amplitudes and higher frequencies.

17.4.6 Quantum Dissipation at Ultra-low Temperatures

Great efforts in low-temperature NEMS research have been made towards the discovery of macroscopic quantum states in mechanical resonators. At low frequencies and high temperatures, thermal energies are larger than the quantum of energy in a given mode. At low temperatures and high frequencies, the quantum nature of the oscillator will eventually dominate. This crossover occurs as $\omega\hbar > k_{\text{B}}T$. Experimentally, this corresponds to approximately 1 GHz at 50 mK. Such frequencies are attainable in various types of NEMS at dilution cryostat temperatures of a few mK. To this end, exotic detection schemes have been developed. These include superconducting single electron transistors and superconducting quantum interference devices (SQUIDs). There are a range of materials used in these devices: metals, dielectrics and single-walled carbon nanotubes.[3] In 2011 an aluminum-based drum resonator coupled to a quantum dot was used to detect and controllably switch between quantum states.[2]

In this section, a dissipative mechanism based on quantum tunneling is presented. It has been observed in a range of materials, including ultrananocrystalline diamond UNCD. This mechanism dominates in some NEMS structures at subkelvin temperatures, and hence is of greatest importance in this emerging field of research.

17.4.6.1 The Standard Tunneling Model of Two-level Systems

To explain the low-temperature thermal properties of glasses, the standard tunneling model of two-level systems was presented by Anderson *et al.*[47]

The basic idea is that, at low temperatures, the lattice remains in an almost degenerate ground state. The degeneracy is caused by defects separated by an energy barrier too high to allow for thermal hopping. Phonon-assisted tunneling does, however, allow for transitions between these states. A material is glassy when the energy distribution of the asymmetry is wide, a result of the many defects and allowed states. Initially, acoustic propagation experiments showed the presence of such tunneling through acoustic attenuation, and it is not surprising that mechanical resonators exhibit this form of dissipation as well. There are a number of models that describe the exact mechanisms; each split the temperature range into two regions. At low temperature, resonant absorption occurs, where the tunneling of the defects is fast compared to the thermal energy. At higher temperatures, relaxation absorption occurs, where the thermodynamic equilibrium is no longer established. Specifically for NEMS, this is described by Seonez et al.,[48] where the periodic mechanical strain modulates the energy splitting of the two-level systems and hence influences the tunneling rate. This mechanism prevents the thermal equilibrium from being established and pumps mechanical energy out of the mode into the phonon bath in an irreversible manner, resulting in dissipation due to phonon-assisted tunneling of two-level systems. The transition temperature splitting of the resonant and absorption regimes is frequency-dependent and follows a $T \sim f^{1/3}$ power law. This holds true over an astonishingly wide frequency space, as is depicted in Figure 17.15.

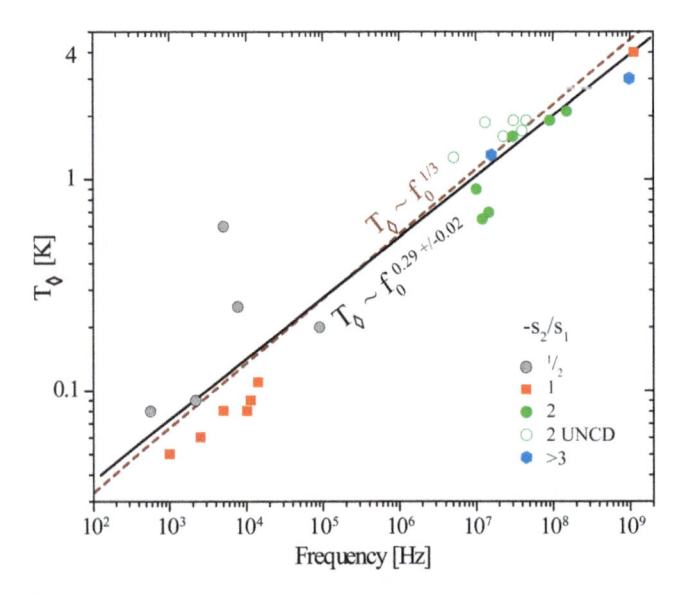

Figure 17.15 Crossover temperature as a function of frequency.
Reprinted from ref. 57, Copyright (2013), with permission from Elsevier.

Diamond Nano-electromechanical Systems 441

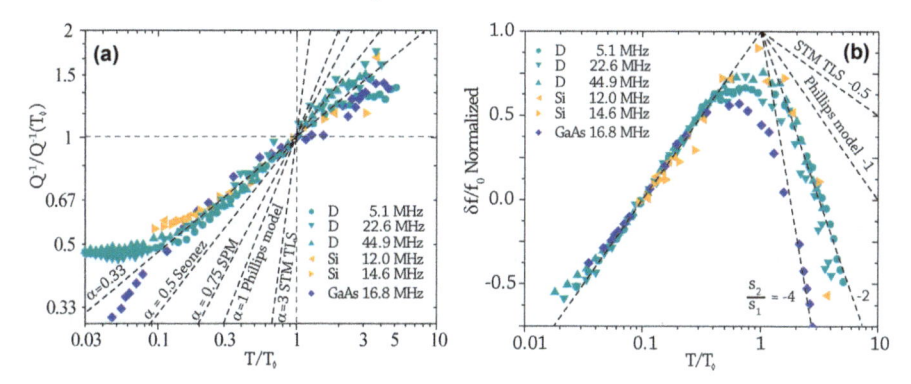

Figure 17.16 Quantum friction: scaled results for three NEMS materials, including UNCD of dissipation and frequency shift.[49]
Copyright (2009) by The American Physical Society.

17.4.6.2 Quantum Dissipation in NEMS Resonators

These two-level systems cause a frequency shift, which results in a characteristic logarithmic temperature dependency. While models predict a linear or square root temperature dependency on dissipation, NEMS in silicon, gallium arsenide, and UNCD experiments have all shown a $Q^{-1} \sim T^{1/3}$ dependency as depicted in Figure 17.16.

It is interesting that these inherently crystalline materials exhibit glassy dissipation mechanisms in NEMS. It can be shown that only a small number of participating defects ($<10^4$) are needed for the TLS to dominate dissipation at subkelvin temperatures.

More experimental and theoretical work is still needed to fully explain and quantify this low-temperature phenomenology, including the saturation of the measured dissipation at the lowest temperatures, the $\frac{1}{3}$ power law, and the increase in slope of frequency shift in the relaxation regime of TLS. Now that macroscopic quantum states have been observed, the need to fully understand dissipation at ultra-low temperatures is greater than ever.

17.5 Novel Diamond NEMS Devices and Future Capabilities

MEMS have already entrenched themselves in commercial markets in the form of gyroscopes, RF switches and filters, oscillators, pressure sensors, and more. NEMS have proven to be extraordinarily sensitive, resulting in single-atom detection, spin-flip detectors, and macroscopic quantum states, amongst others. Diamond has a special role to play due to its extraordinary mechanical, thermal, electrical, and chemical properties. Due to the complex growth compared to more established materials, such as silicon, it may be a while before diamond MEMS and NEMS become widespread in commercial markets. For niche applications, however, where the properties of diamond allow new enabling technologies, one would expect diamond to flourish. In this final section some exploratory ideas are presented as the many established uses of diamond MEMS and NEMS are described in detail elsewhere.[7]

17.5.1 Field Emission in Diamond

Diamond is known to be a good source of cold field emitted electrons,[50] and it has been proposed to use this property for cathode sources in displays, as well as RF amplifiers. MEMS-based field emitter devices can be operated at low powers and high frequencies. This has resulted in the renaissance of the almost-forgotten technology of cathode tubes.

The emitted current is described by the Fowler Nordheim equation:

$$I_{FN} = \Sigma_2 \beta^2 E^2 e^{-\frac{\Sigma_1}{\beta E}}, \tag{17.41}$$

where Σ_1 and Σ_2 are geometric- and material-dependent constants, β is a geometric form factor, and $E = \frac{V}{g}$ is the electric field between the anode and cathode. In a teeth geometry configuration the geometric enhancement factor β has been reported to exceed 200, which greatly increases the emission current. Examples of such devices and the performance is depicted in Figure 17.17. The critical component here is the exponential dependency of the current on distance. This is what makes it possible to use field emission as a displacement sensor in NEMS devices. Using e-beam lithography, sharp tips (cathodes) can be manufactured and placed in close proximity to resonant structures (the anode). Expanding eqn 17.41 in terms of $\frac{x}{g}$ results in the static and harmonic currents resulting from the motion of the resonator:

$$I_{FN} = \frac{\Sigma_2 \beta^2 E^2 e^{-\frac{g\Sigma_1}{\beta V}}}{g^2}\left[1 - \left(2 + \frac{g\Sigma_1}{\beta V}\right)\frac{x}{g} + \cdots\right] \tag{17.42}$$

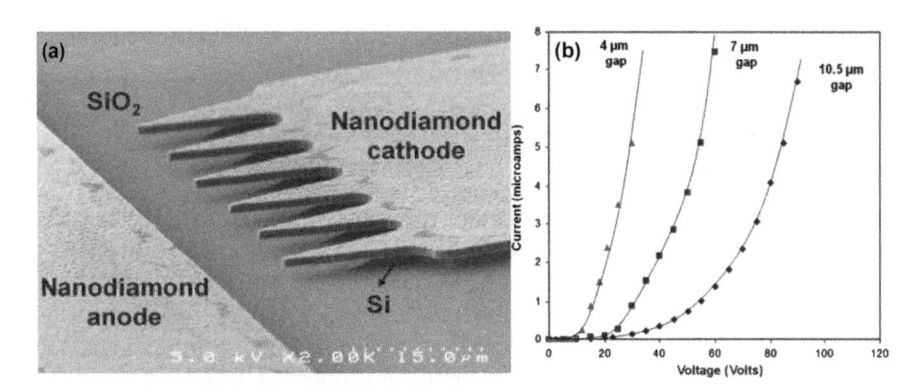

Figure 17.17 Nanocrystalline diamond lateral field emitter.
Reprinted from ref. 50, Copyright (2011), with permission from Elsevier.

where x is the displacement from equilibrium. The signal size scales as $\frac{1}{g^3}$, making it extremely sensitive as the gap size is reduced to the 100–200 nm range.

Field emission could be used for ultrasensitive displacement detection, on-chip GHz frequency amplification, and signal manipulation.

17.5.2 Superconductivity and Diamond NEMS

Boron-doped NCD is a type II superconductor with a crossover temperature around 2–6 K and a critical magnetic field on the order of 4 T. The superconductivity is maintained in structures machined smaller than the micron scale, and such NCD SQUIDs have been manufactured.[51] One can consider the possibility of coupling the mechanical degrees of freedom of a resonator with its superconducting state. A NEMS embedded in a SQUID loop has already been used for ultraprecise displacement sensing.[52] In that experiment the mechanical and electrical structures were different materials. It should be possible integrate both the mechanics and superconductivity in a single device as the one depicted in Figure 17.18.

Instead of measuring the displacement through the change in flux, it may be possible to measure the change in superconductivity due to the mechanical strain induced in the resonator. This would result in measurement analogous to the piezoresistive detection method described above. It is known that applying strain to a superconductor, including diamond superconductors,[53] results in a shift in the transition temperature. Sitting on the transition point and modulating the strain would result in a large piezoresistive response.

Beyond strain detection, such experiments could help shed light on how superconductivity is affected by strain, and hence the cooper-pair phonon coupling. Diamond is an ideal material to study NEMS interacting with superconductivity and can be used as a platform to explore new scientific and technological opportunities.

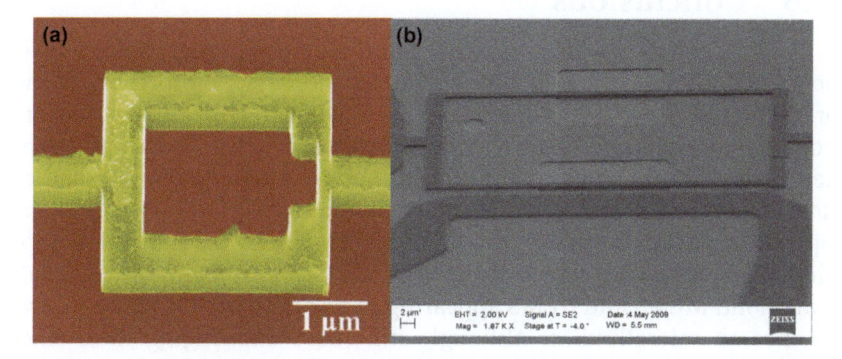

Figure 17.18 a) A diamond SQUID loop with weak link junctions, b) Diamond SQUID loop with embedded resonator.
Reprinted with permission from ref. 51. Copyright (2011) American Chemical Society.

17.5.3 The Nitrogen Vacancy Defect in Diamond

Nitrogen vacancy centers (NV) are a special type of defect that form in diamond made of a substitutional nitrogen impurity adjacent to a missing carbon site on the lattice. The strain induced by the nitrogen holds the vacancy in place. This point defect is the realization of a room-temperature quantum state that can be manipulated using electric and magnetic fields. The NV forms a triplet electron spin state with long enough lifetimes to be probed with spectroscopic studies. For these reasons, it is often considered as a possible building block for quantum entanglement, spintronics, quantum communication, and even quantum computing.[54]

In the presence of a magnetic field, the electron energy levels are split. This results in a number of sharp observable transitions in the microwave region. These transitions are sensitive to various forms of external radiation and to the local strain field in the crystal. It is known that the intrinsic strain field in a diamond lattice splits the degeneracy of the excited electron states, resulting in unique energy levels for each defect. The presence of strain is analogous to small electric fields,[55] and the spin Hamiltonian must be extended to include the strain component.

Considering the high precision at which the electron splitting can be measured, as well as their sensitivity to the local strain fields, it is not difficult to consider tuning these transitions by applying a static strain field using MEMS devices. By actuating a given actuation mode, the strain field could be manipulated in space and time. Some of the transitions observed in NV defects occur at frequencies similar to diamond NEMS resonators. Such coupling would enable new types of photonic NEMS devices that are coupled to quantum dots. A prerequisite for such photonic applications is the further development in single-crystal diamond NEMS devices, but the recent techniques described above prove this is a real possibility in the near future.

17.6 Conclusions

In this chapter an overview of MEMS and NEMS was presented. Elasticity theory and fabrication, as well as actuation and detection methods relevant for diamond, have been discussed. It is shown how the extraordinary properties of diamond can be leveraged to build novel devices and experiments. In addition to its high Young's modulus, diamond's thermal, electrical, and biological properties make it interesting for a wide range of applications. The ability to grow high-quality thin films and the compatibility of diamond with standard optical and e-beam lithography has resulted in diamond MEMS and NEMS appearing in technologies that range from RF switches to microfluidics. Although still more challenging than silicon, single-crystal diamond, NCD, and UNCD are becoming more prevalent as devices and the material of choice for cutting-edge experiments. It can be assumed that this trend will continue as there is still much to be learned and gained.

References

1. H.-Y. Chiu, P. Hung, H. W. Ch Postma and M. Bockrath, *Nano Lett.*, 2008, **8**, 4342.
2. A. D. O'Connell, M. Hofheinz, M. Ansmann, R. C. Bialczak, M. Lenander, E. Lucero and A. N. Cleland, *Nat.*, 2010, **464**, 697.
3. M. Poot and H. S. J. van der Zant, *Phys. Rep.*, 2012, **511**, 243.
4. G. Zolfagharkhani, A. Gaidarzhy, P. Degiovanni, S. Kettemann, P. Fulde and P. Mohanty, *Nat. Nanotechnol.*, 2008, **3**, 720.
5. Z. Cao and D. Aslam, *Diamond Relat. Mater.*, 2010, **19**, 1263.
6. E. Kohn, P. Gluche and M. Adamschik, *Diamond Relat. Mater.*, 1999, **8**, 934.
7. R. S. Sussmann, *CVD Diamond for Electronic Devices and Sensors*. Wiley & Sons Ltd, 2009, vol. 26.
8. T. P. Knowles and M. J. Buehler, *Nat. Nanotechnol.*, 2011, **211**, 469.
9. V. Aksyuk, F. F. Balakirev, G. S. Boebinger, P. L. Gammel, R. C. Haddon and D. J. Bishop, *Science*, 1998, **280**, 720.
10. T. Braun, M. K. Ghatkesar, N. Backmann, W. Grange, P. Boulanger, L. Letellier, H.-P. Lang, A. Bietsch, C. Gerber and M. Hegner, *Nat. Nanotechnol.*, 2009, **4**, 179.
11. C. A. Bolle, V. Aksyuk, F. Pardo, P. L. Gammel, E. Zeldov, E. Bucher, R. Boie, D. J. Bishop and D. R. Nelson, *Nat.*, 1999, **399**, 43.
12. O. A. Williams, O. Douhéret, M. Daenen, K. Haenen, E. Ōsawa and M. Takahashi, *Chem. Phys. Lett.*, 2007, **445**, 255.
13. C. F. Wang, E. L. Hu, J. Yang and J. E. Butler, *J. Vac. Sci. Technol., B: Microelectron. Nanometer Struct.—Process., Meas., Phenom.*, 2007, **25**, 730.
14. P. Olivero, S. Rubanov, P. Reichart, B. C. Gibson, S. T. Huntington, J. Rabeau, A. D. Greentree, J. Salzman, D. Moore, D. N. Jamieson and S. Prawer, *Adv. Mater.*, 2005, **17**, 2427.
15. P. L. Ovartchaiyapong, M. A. Pascal, B. A. Myers, P. Lauria and A. C. Jayich, *Appl. Phys. Lett.*, 2012, **101**, 163505.
16. A. Gaidarzhy, M. Imboden, P. Mohanty, J. Rankin and B. W. Sheldon, *Appl. Phys. Lett.*, 2007, **91**, 203503.
17. J. Wang, J. E. Butler, T. Feygelson, and C. C. Nguyen, *C. C. Proc. – IEEE Annu. Int. Conf. Micro Electro Mech. Syst.*, 17th, 2007, 641.
18. W. De Boer, J. Bol, A. Furgeri, S. Müller, C. Sander, E. Berdermann and M. Huhtinen, *Phys. Status Solidi A*, 2007, **204**, 3004.
19. G. Christian, Specht, Oliver A. Williams, Richard B. Jackman and Ralf Schoepfer, *Biomaterials*, 2004, **25**, 4073.
20. R. J. Hamers, J. E. Butler, T. Lasseter, B. M. Nichols, J. N. Russell, K. Y. Tse and W. S. Yang, *Diam. Rel. Mater.*, 2005, **14**, 661.
21. N. Sepúlveda-Alancastro and Dean M. Aslam, *Microelectron. Eng.*, 2004, **73**, 435.
22. A. N. Cleland, *Foundations of Nanomechanics: From Solid-state Theory to Device Applications*, Springer-Verlag, Berlin, 2003.

23. M. Imboden, P. Mohanty, A. Gaidarzhy, J. Rankin and B. W. Sheldon, *Appl. Phys. Lett.*, 2007, **90**, 173502.
24. V. P. Adiga, A. V. Sumant, S. Suresh, C. Gudeman, O. Auciello, J. A. Carlisle and R. W. Carpick, *Phys. Rev. B*, 2009, **79**, 245403.
25. M. Benetti, D. Cannata, F. Di Pietrantonio, E. Verona, S. Almaviva, G. Prestopino, C. Verona and G. Verona-Rinati, *IEEE Int. Ultrason. Symp*, 2008, 1924.
26. V. Mortet, O. A. Williams and K. Haenen, *Phys. Status Solidi A*, 2008, **205**, 1009.
27. R. Lifshitz and M. L. Roukes, *Phys. Rew. B*, 2000, **61**, 5600.
28. D. N. Guerra, M. Imboden and P. Mohanty, *Appl. Phys. Lett.*, 2008, **93**, 033515.
29. D. N. Guerra, A. R. Bulsara, W. L. Ditto, S. Sinha, K. Murali and P. Mohanty, *Nano Lett.*, 2012, **10**, 1168.
30. R. Lifshitz and M. C. Cross, *Reviews of Nonlinear Dynamics and Complexity*, 2008, **1**, 1.
31. M. Imboden, *Thesis, Diamond Nanoelectromechanical Resonators: Dissipation and Superconductivity*, Boston University, Boston, 2012.
32. M. Muoth, T. Helbling, L. Durrer, S.-W. Lee, C. Roman and C. Hierold, *Nat. Nanotechnol.*, 2010, **5**, 589.
33. K. Okano, K. Hoshina, S. Koizumi and K. Nishimura, *Diamond Relat. Mater.*, 1996, **5**, 19.
34. T. Shibata, K. Unno, E. Makino, Y. Ito and S. Shimada, *Sens. Actuators, A*, 2002, **102**, 106.
35. M. W. Varney, D. M. Aslam, A. Janoudi, H.-Y. Chan and D. H. Wang, *Biosensors*, 2011, **1**, 118.
36. Z. Cao, and D. Aslam, *IEEE Nanotechnol. Mater. Devices Conf.*, 2009. NMDC'09, 190.
37. A. Yamamoto, N. Nawachi, T. Tsutsumoto and A. Terayama, *Diamond Relat. Mater.*, 2005, **14**, 657.
38. J. Lu, Z. Cao, D. M. Aslam, N. Sepulveda and J. P. Sullivan, 3rd, *IEEE International Conference on Nano/Micro Engineered and Molecular Systems*, 2008, 873.
39. I. Bargatin, I. Kozinsky and M. L. Roukes, *Appl. Phys. Let.*, 2007, **90**, 9093116.
40. Q. P Unterreithmeier, E. M. Weig and J. P. Kotthaus, *Nat.*, 2009, **458**, 1001.
41. J. W. Baldwin, M. K. Zalalutdinov, T. Feygelson, B. B. Pate, J. E. Butler and B. H. Houston, *Diamond Relat. Mater.*, 2006, **15**, 2061.
42. N. Sepúlveda-Alancastro, *Polycrystalline Diamond RF MEMS Resonator Technology and Characterization*, Thesis, Michigan State University, 2005.
43. A. B. Hutchinson, P. A. Truitt, K. C. Schwab, L. Sekaric, J. M. Parpia, H. G. Craighead and J. E. Butler, *Appl. Phys. Lett.*, 2004, **84**, 972.
44. M. W. Varney, D. M. Aslam, A. Janoudi, H.-Y. Chan and D. H. Wang, *Biosensors*, 2011, **1**, 118.

45. P. Mohanty, D. A. Harrington, K. L. Ekinci, Y. T. Yang, M. J. Murphy and M. L. Roukes, *Phys. Rev. B*, 2002, **66**, 085416.
46. W. E. Newell, *Science*, 1968, **161**, 1320.
47. P. W. Anderson, B. I. Halperin and C. M. Varma, *Philos. Mag.*, 1972, **25**, 1.
48. C. Seoánez, F. Guinea and A. H. C. Neto, *Phys. Rev. B*, 2008, **77**, 125107.
49. M. Imboden and P. Mohanty, *Phys. Rev. B*, 2009, **79**, 125424.
50. K. Subramanian, W. P. Kang, J. L. Davidson, N. Ghosh and K. F. Galloway, *Microelectron. Eng.*, 2011, **88**, 2924.
51. S. Mandal, T. Bautze, O. A. Williams, C. Naud, E. Bustarret, F. Omnes, P. Rodière, T. Meunier, C. Bäuerle and L. Saminadayar, *ACS Nano*, 2011, **5**, 7144.
52. S. Etaki, M. Poot, I. Mahboob, K. Onomitsu, H. Yamaguchi and H. S. J. Van der Zant, *Nat. Phys*, 2008, **4**, 785.
53. E. A. Ekimov, V. A. Sidorov, E. D. Bauer, N. N. Mel'Nik, N. J. Curro, J. D. Thompson and S. M. Stishov, *Nat*, 2004, **428**, 542.
54. F. Jelezko and J. Wrachtrup, *Phys. Status Solidi A*, 2006, **203**, 3207.
55. Ph. Tamarat, N. B. Manson, J. P. Harrison, R. L. McMurtrie, A. Nizovtsev, C. Santori, R. G. Beausoleil *et al.*, *New J. Phys.*, 2008, **10**, 045004.
56. M. Imboden, O. Williams and P. Mohanty, *Applied Physics Letters*, 2013, **102**(10), 103502.
57. M. Imboden and P. Mohanty, *Physics Reports*, 2014, **534**, 89–146.

CHAPTER 18

Diamond-based Resonators for Chemical Detection

EMMANUEL SCORSONE* AND ADELINE TROUVÉ

CEA-LIST, Diamond Sensors Laboratory, Gif-sur-Yvette 91191, France
*Email: emmanuel.scorsone@cea.fr

18.1 Introduction

Chemical/biochemical sensors are devices that transform chemical or biological information into an analytically useful signal. Generally speaking, they are the result of coupling a selective layer to a physical part known as the transducer. The sensors may be classified according to the operating principle of the transducer. In brief, the main transduction phenomena are optical, electrochemical, electrical and gravimetric.[1] A wide variety of innovative chemical and biochemical sensor technologies are being reported in the literature every year, often showing highly promising performances. However, when it comes to commercial sensors, one has to admit that the market is very conservative. A possible explanation for this is the lack of robustness and reliability of the newly developed sensors and, therefore, their difficulty to maintain performance specifications under adverse operating conditions.[2]

The robustness of the sensors is linked, on the one hand, to their ability to withstand mechanical shocks, stresses, or vibrations. On the other hand, it is related the chemical stability of the selective layer in the operating environment. The variety of exceptional physicochemical properties of diamond materials are generally extremely resilient to chemical degradation. They also feature outstanding mechanical properties, such as a high Young's

---

RSC Nanoscience & Nanotechnology No. 31
Nanodiamond
Edited by Oliver Williams
© The Royal Society of Chemistry 2014
Published by the Royal Society of Chemistry, www.rsc.org

modulus and a strong resistance to fracture. Furthermore, diamond is mainly composed of carbon atoms, which can be made available on the material surface for covalent grafting of chemical receptors *via* strong carbon–carbon bonds. Hence, it becomes obvious that diamond can play an important role toward improving the stability and reliability of chemical or biochemical sensors. Additionally, when looking more closely at the range of remarkable physical and chemical properties of diamond, one can also predict that it is of high interest for improving the transducing characteristics of sensors, such as their sensitivity levels. Let us consider boron-doped diamond electrodes for instance: their wide potential window in aqueous media allows them to address target analytes that could not be measured with standard platinum or other carbon-based electrodes;[3] also, the low background current, resulting from low electrical double layer capacitance of the electrode, generally gives rise to better signal-to-noise ratios when compared with other electrode materials.[4] Thus, there has been an increasing interest in diamond materials for the development of chemical sensors, as reflected by the increasing number of scientific publications in this area over the last decade.

Many interesting chemical/biochemical diamond-based sensor architectures have been developed recently that will be summarised in section 18.2. Nevertheless, we have chosen in this chapter to focus on two particular types of transducer technologies: cantilevers and surface acoustic wave (SAW) sensors, which offer a number of significant advantages over other types of transducers. For example, today, biosensors are mainly based on electrochemical and optical transduction methodologies. Electrochemical transduction answers the main requirements associated to real time *in situ* portable sensors. These capabilities, as well as material costs, which allow the design of one-shot sensors, constitute the basis of their success, for example, towards glucose detection in point-of-care diagnostics. However, even though electrochemistry is well suited to enzymatic biochemical sensors in which redox enzymes are intimately associated to the electrode transducer, it suffers from a lack of accuracy, sensitivity and selectivity in the case of non-metabolistic detection that uses biological receptors, such as antibodies, proteins or DNA. For these latter applications, the sensing principles are mainly based on optical instrumentations (fluorescence microscopy, surface plasmon resonance or colorimetry), which are usually cumbersome and unfit for handheld applications. In contrast, since SAW sensors and cantilevers measure mass (sometimes amongst other parameters), and since all molecules have a mass, they are capable, in principle, of detecting any type of analyte independently, for instance, of their optical or electrochemical properties. These sensors generally offer also the possibility to achieve higher sensitivity performances than with other types of transducers. Finally, they can be fabricated according to the advances of micro- and nano-technologies. As a result, they are compatible with low cost manufacturing and miniaturised systems. Hence they appear highly promising for a wide range of "low cost" portable sensing applications.

In this context the goal of this chapter is to give some insight into the potential of diamond for the development of high performance chemical sensors. The first section summarises a few examples of sensors and the various exceptional properties of diamond that can be exploited in chemical sensing. The next two sections describe in more detail, for the reasons given above and also because they illustrate well the benefit of diamond for such applications, cantilever and SAW transducer technologies.

18.2 Diamond Materials: Some Remarkable Properties for the Development of High Performance Chemical Sensors

18.2.1 Physical Properties

Over the last decades, silicon has been used extensively for microelectromechanical systems (MEMS) fabrication, taking advantage of the available silicon-micromachining techniques developed for integrated circuits technology. For example, silicon is used for the fabrication of accelerometers, pressure sensors, resonators and chemical/biochemical sensors. Nevertheless, diamond exhibits superior mechanical properties over silicon, including, for instance, a lower friction coefficient, a higher resistance to fracture and a much higher Young's modulus[5–7] (Table 18.1). These outstanding mechanical properties of diamond materials are well known. They have been exploited for many years for cutting, polishing or drilling tools. Actually, those assets are also very promising for the development of MEMS, and such devices may be useful for chemical detection.

Table 18.1 Outstanding linear mechanical properties of single-crystal diamond (calculated or measured) and CVD diamond (measured). (Reprinted with permission from ref. 7; copyright (2012) America Institute of Physics.)

	Single-crystal diamond	CVD diamond
Bulk modulus	433 GPa	443 GPa
Shear modulus	502 GPa	507 GPa
Young's modulus, anisotropy	1050–1210 GPa	
(Random) crystallites	1143 GPa	500–1200 GPa
Poisson ratio, anisotropy	0.00786–0.115	
(Random) crystallites	0.0691	0.075
Sound velocity, long. (111)	19039 m s^{-1}	18784 m s^{-1}
Sound velocity, long. (100)	18038 m s^{-1}	
Sound velocity, long. (110)		18182 m s^{-1}
Rayleigh velocity, (110) texture	10753 m s^{-1}	10326 m s^{-1}
Rayleigh velocity, polycrystal.	10930 m s^{-1}	10850 m s^{-1}
Vibrational frequency (Raman)	1332.2 cm^{-1}	1332 cm^{-1}
Expansion coefficient		0.8×10^{-6} K^{-1}
		0.9×10^{-6} K^{-1}

Various diamond-based MEMS have been investigated and reported in the literature, such as RF switches.[8] Diamond cantilevers with high Q-factors were also characterised (Figure 18.1).[9] Some were recently used for chemical detection (section 18.3). Other systems, like film bulk acoustic resonators (FBAR) operating at 3.5 GHz, have been reported.[10] Such devices hold promise too for the development of gravimetric chemical sensors of extreme sensitivity.

Diamond also exhibits some remarkable electronic properties. Doped diamond is a wide gap semiconductor that may reach intrinsic electron and hole mobilities of, typically, ≥ 3100 cm^2 V^{-1} s^{-1} and ≥ 3200 cm^2 V^{-1} s^{-1} at 300 K, respectively. In practice, only p-doping with boron is well controlled and enables such performances. Boron doping in polycrystalline diamond may also be achieved and, although its electronic properties are not as good as that of single crystals, such a material remains very attractive in, for example, electrochemical applications. In contrast with other semiconductors like silicon, boron-doped diamond does not contain a surface oxide layer. Hence, the semiconducting material is available directly to interact with the external medium, which is clearly very advantageous for many chemical sensing applications.

Several types of solid state chemical sensors exploiting the semiconducting properties of diamond have been investigated. For instance, Gurbuz and coworkers demonstrated the performance of catalyst/adsorptive-oxide/insulator/semiconductor (CAIS) devices based on doped diamond/intrinsic diamond/metal catalyst for the detection of hydrogen, oxygen, carbon monoxide (Figure 18.2) or various hydrocarbons.[11] The influence of gases used in the semiconductor industry at temperatures typically above

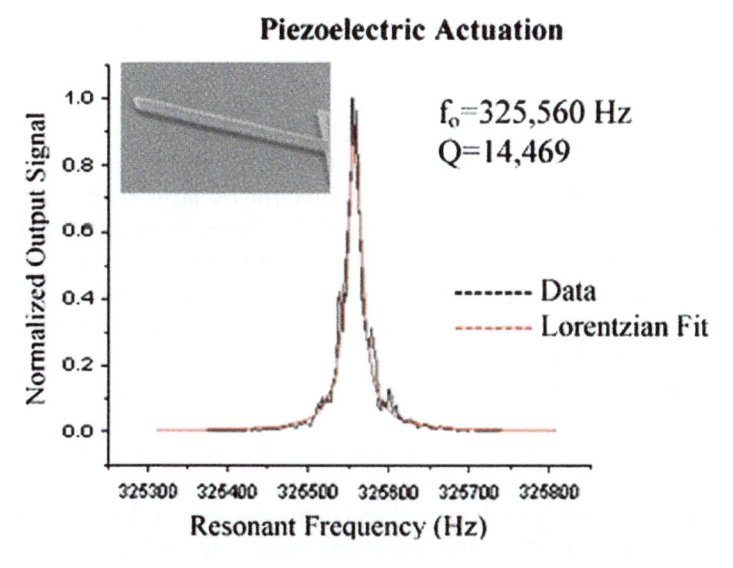

Figure 18.1 Typical response of piezoelectric actuated poly-C cantilever resonator. (Adapted from ref. 9 with permission from Elsevier.)

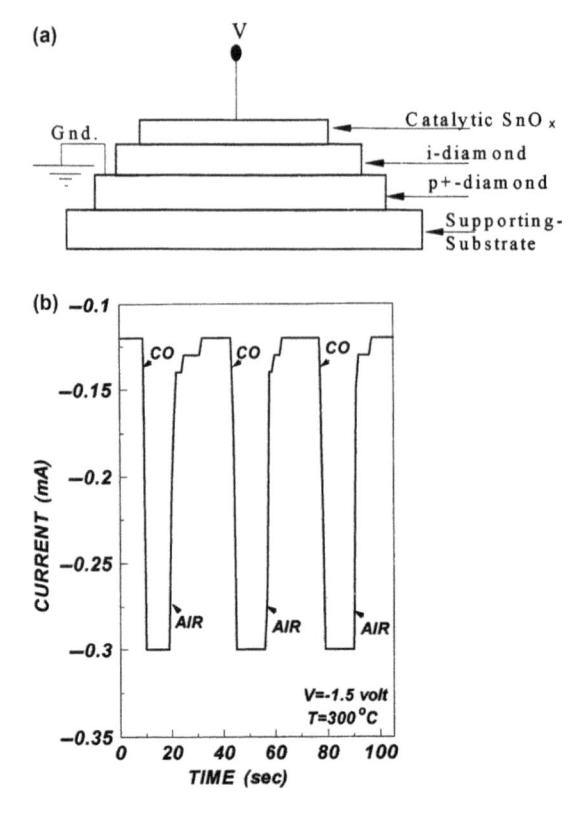

Figure 18.2 a) Diamond-based CAIS sensor, b) typical response of CAIS sensor to 9.4 Torr CO in 0.4 Torr oxygen atmosphere.
(Reprinted from ref. 22 with permission from Elsevier.)

≥100 °C,[12,13] as well as that of CO gas at room temperature,[14] on the resistivity of doped diamond layers was also investigated.

Furthermore, hydrogen-terminated diamond exhibits a p-type surface conductivity due to the presence of an electrolytic layer resulting from adsorption of water molecules and other adsorbates from the environment onto the surface.[15] This surface conductivity is highly dependent on the pH in aqueous media when the surface is slightly oxidised, which has led to the development of pH sensors based on those conductivity phenomena.[16] Flat[17,18] or nano-structured[19,20] hydrogenated diamond surfaces were also studied for gas sensing, where surface conductivity variations are measured upon gas exposure.

The thermal conductivity of diamond single crystal lies in the order of 2000 W m^{-1} K^{-1} at 300 K, which is approximately five times higher than copper. This extremely high value has been exploited in the development of high power components or even for silicon-on-diamond (SOD) substrates for heat dissipation applications in microelectronics.[21] This property of diamond may also be beneficial for designing catalytic gas sensors in which fast temperature cycles may be useful, either for energy saving management or because the selectivity of the sensor may be tuned by changing its operating temperature.[22] In addition, diamond, being a wide bandgap semiconductor, exhibits electronic properties that remain more stable than silicon at high temperatures. Therefore, it may be seen as a good candidate for the design of stable gas sensors operating at temperatures up to 400 °C. Finally, it is interesting to mention that a thermocouple containing a diamond/palladium junction has been assessed for the detection of hydrogen gas.[23]

Diamond's wide bandgap offers extreme optical transparency from ultraviolet to infrared, together with a high refractive index, in the order of 2.5 at 600 nm. These optical properties, when combined with the chemical properties, are promising for the development of robust and sensitive chemical/biochemical transducers. For instance, label-free photonic crystal biosensors have recently been shown to be a highly sensitive method for performing a wide variety of biochemical assays. These devices are generally made of silicon. They operate in the infrared spectral region and are somehow limited in the chemistry that can be used to immobilise bio-receptors onto their surface. In contrast, diamond photonic crystals are particularly interesting here because they can potentially operate in the visible region with an extremely high quality factor and small volume cavities, and offer again highly stable carbon chemistry. Recently, nanocrystalline diamond photonic crystals were reported with a Q-factor as high as 2800 and it is hoped in the future to reach Q-factors up to 10 000.[24,25] Therefore, diamond offers a bright future in the area of optical biosensor chips.

18.2.2 Chemical Properties

Diamond materials are obviously also very attractive for their chemical properties. First of all, they are very inert and chemically resilient to, for instance, corrosion, at temperatures up to around 400 °C, where they may start to deteriorate when in the presence of oxygen. These properties make it highly stable even in very harsh environments. This is one reason, for instance, why radiation diamond detectors were developed in the past because they could withstand the high temperature, high pressure and highly acidic operating conditions in nuclear power plants.[26] This high stability of diamond may open new fields of applications for chemical sensing in harsh environments, but it also contributes to the increasing robustness and reliability of chemical sensors in more conventional operating conditions.

Diamond also offers a carbon-terminated surface, which may be made available for grafting a wide range of organic receptors through strong covalent carbon–carbon bonds. Several chemical[27,28] and electrochemical routes[29] for surface grafting on diamond have been reported. Such modified surfaces have been used for the development of a wide range of chemical or biochemical sensors.[30,31]

Furthermore, boron-doped diamond features remarkable electrochemical properties. These include a wide potential window that can exceed 3 V in

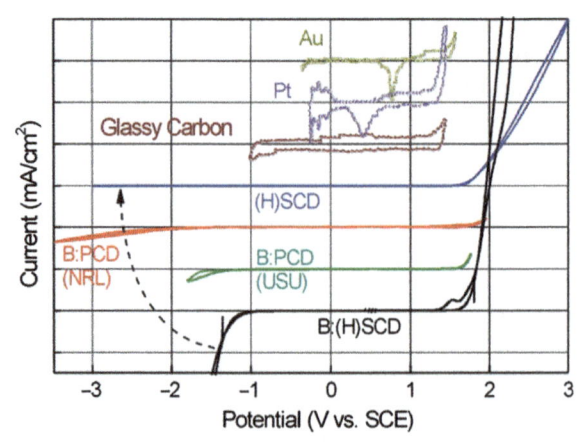

Figure 18.3 Voltammograms for water electrolysis on various electrodes. The supporting electrolyte is 0.5 M H_2SO_4. The graphs are shifted vertically for comparison. Two polycrystalline films, B: PCD (provided by the Naval Research Laboratory, NRL) with 5×10^{19} B cm^{-3} and B: PCD (provided by Utah State University, USU) with 5×10^{20} B cm^{-3}, are compared with a single crystalline boron-doped diamond, B: (H)SCD (single crystalline diamond) with 3×10^{20} B cm^{-3}, and with an undoped diamond (H)SCD. Also shown are data for Pt, Au and glassy carbon. Oxidation reactions, e.g., oxygen evolution, have positive currents and emerge around 1.8 V for all diamond samples. Reduction reactions, e.g., hydrogen evolution, have negative currents and show very different properties. Note that the background current within the regime between hydrogen and oxygen evolution for diamond is very low and the electrochemical potential window is large compared with glassy carbon, Pt and Au.
(Reprinted from ref. 29 with permission from IOP.)

aqueous media (Figure 18.3) and a low background current, as well as a high reactivity comparable to that of platinum after electrochemical activation.[32] Moreover, because of the high atomic density of diamond, diffusion of chemical or biological species into the material is impossible; hence, fouling of the electrodes is only limited to physico-chemical adsorption of high molecular weight species at the surface. Thus, the electrodes are generally less prone to fouling. Moreover, it was demonstrated that such bare diamond electrodes may be reactivated after fouling in biological mediums and, therefore, potentially used for *in situ* continuous monitoring (Figure 18.4).[33] Thus, BDD electrodes are promising for chemical detection and several applications have been investigated over the years. The wide potential window of diamond enabled the detection of nitroaromatic pollutants in seawater where platinum failed because the reduction potential of the target chemicals falls outside the solvent reduction front of platinum electrodes.[3] Diamond electrodes have also proven to be efficient for the detection of glucose,[34] lectins,[35] DNA,[36] *etc.* in the liquid phase, as well as arsine in the gas phase,[37] to name but a few.

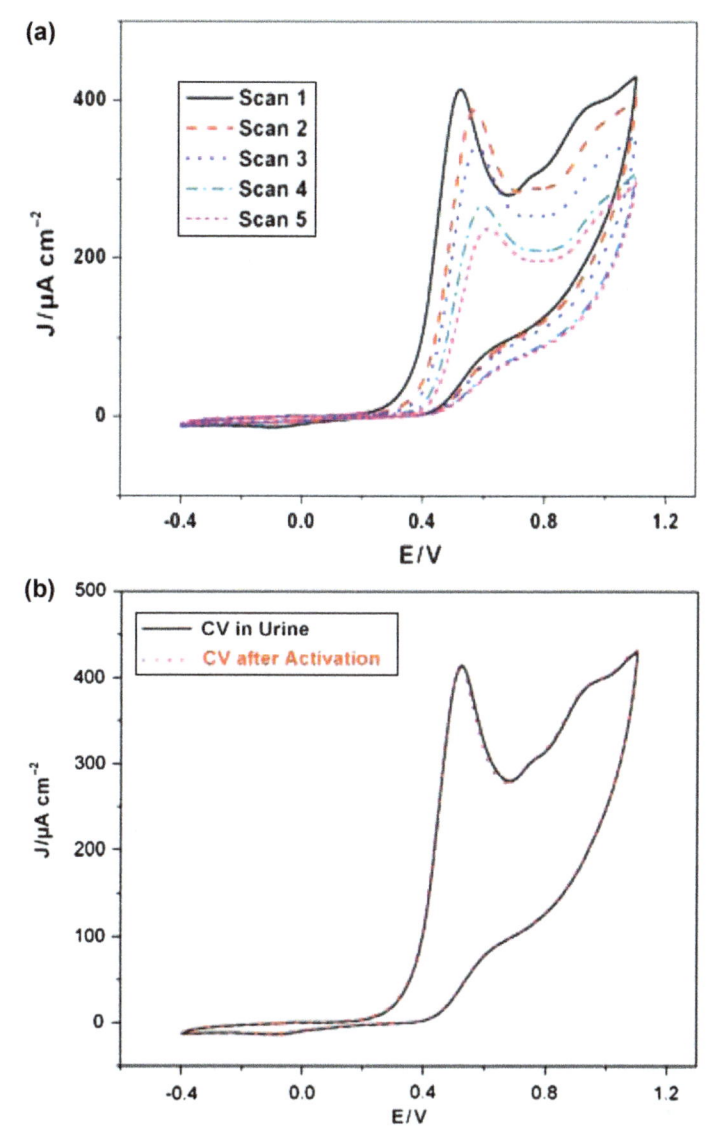

Diamond-based Resonators for Chemical Detection 455

Figure 18.4 a) Cyclic voltammogram in human urine from -0.4 V to 1.1 V vs. Ag/AgCl at 100 mV s^{-1} measured with BDD electrode, where J is the current density in µA cm^{-2} and E is the applied voltage in V. The electrode was cleaned thoroughly in deionised water prior to each scan and, hence, the attenuation of the peak is due to fouling and not because the solution surrounding the electrode is depleted of electro-active species; b) comparison of the cyclic voltammogram of "as-grown" electrode (solid line) and the same electrode after activation (dotted line), where J is the current density in µA cm^{-2} and E is the applied voltage in V. The electrolyte is human urine and the scan rate is 100 mV s^{-1}.
(Reproduced with permission of the Electrochemical Society from ref. 33.)

Sometimes, electrochemical measurements may be coupled to simultaneous gravimetric or viscosity monitoring using typically quartz crystal microbalances (QCM) and, quite logically, the fabrication of diamond-QCM has been attempted. Here, the difficulty is related to the lack of compatibility between diamond growth temperature conditions (typically >600 °C) and the phase transition temperature for α-quartz (513 °C). Approaches to overcome this issue consist of attaching a previously grown freestanding diamond film onto the QCM[38] or to use a high-temperature-stable piezoelectric material, such as langasite[39] (Figure 18.5).

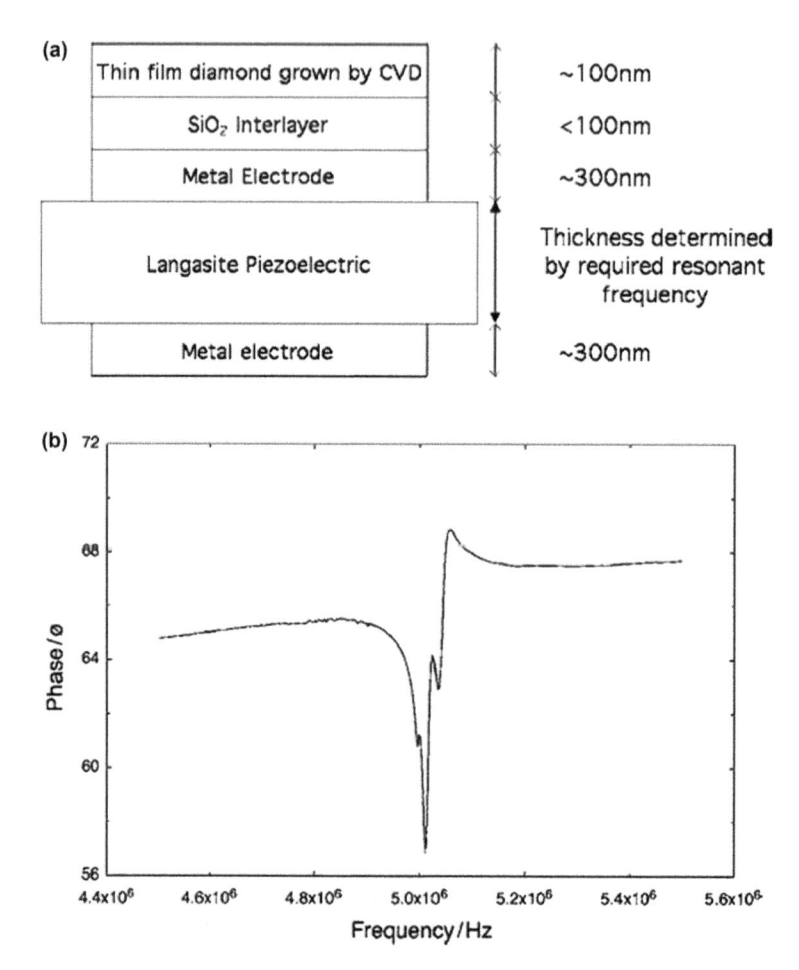

Figure 18.5 a) Schematic of the nanocrystalline diamond-coated langasite thickness shear mode resonator; b) phase shift plotted against frequency, showing the resonance of the nanocrystalline diamond-coated langasite thickness shear mode resonator.
(Reprinted with permission from ref. 39, American Institute of Physics.)

18.3 Diamond Cantilevers

Cantilevers were originally developed in the 1980s as tips for scanning probe microscopy. Since then, extensive research efforts has been dedicated to the fabrication and characterisation of microscale cantilevers useful for atomic force microscopy (AFM) probes. As a result, more advanced probes have become available; but when combined with research work from other areas of science, these efforts have also led to the development of new families of mechanical sensors, including, for instance, accelerometers and also chemical and biochemical sensors. Thus, the feasibility to detect specific gases, enzymes, proteins or DNA sequences using accurate mass-sensitive micro-cantilevers was demonstrated.[40] In contrast with most other gravimetric transducers, cantilevers are not made of, nor do they depend on, piezoelectric materials; although, in some instances, piezoelectric gauges have been integrated to the transducer as part of the resonance frequency readout system. While MEMS can take many different shapes, devices with very simple rectangular beam configurations appear highly suitable for transducing physical, chemical or biological stimuli into useful electrical signals. AFM probes are generally fabricated from a highly inert material, such as silicon nitride but, for chemical sensing applications, a wide range of materials have been tested. Amongst those, diamond offers probably the best performance in terms of mechanical properties and chemical inertness. Micro-electro-mechanical systems (MEMS) and, more precisely, cantilevers are relatively recent technologies when compared to more conventional transducer technologies, like optical or electrochemical methods. Nevertheless, they are recognised as promising sensing devices for chemical and biological analysis because they can offer low instrumental detection limits along with high miniaturisation and integration capabilities.

18.3.1 Transduction Principles

Cantilevers may operate either in the static mode or in the dynamic mode. In the static mode the surface energy changes due to chemical interactions occuring at one surface of the device according to Stoney's law.[2] This results in a bending of the cantilever that is often detected optically. In this case the transduction mechanism is based on asymmetrical changes in the surface stress and there is no mass measurement involved. The sensitivity of the device increases with decreasing stiffness of the cantilever materials. Therefore, diamond is clearly not suitable for this mode of transducer operation and polymer-based cantilevers are preferred.

In the resonant regime the transduction mechanism is rather complex and is still subject to debate. Rectangular cantilevers operating in vacuum or in gases can be considered as weakly damped mechanical oscillators,[40] for which the spring constant k can be approximated as:

$$k = \frac{Et^3 w}{4l^3}, \qquad (18.1)$$

where E is the Young's modulus of the material composing the cantilever, and w, t and l are the width, thickness and length of the cantilever, respectively. In this regime the cantilever resonance frequency extracted from Euler–Bernoulli beam theory is given by:

$$f_0 = \frac{1}{2\pi}\sqrt{\frac{k}{m_0}}, \qquad (18.2)$$

where m_0 is the effective mass of the cantilever. In such low damping media the cantilever response may be affected both by gravimetric loading on the cantilever and spring constant variation as a result of chemical and physical interactions occurring at the surface of the transducer. Thus, the relative cantilever resonance frequency change due to a variation of mass (Δm) and stiffness (Δk) can be expressed by:[41]

$$\frac{\Delta f}{f_0} \cong \left(\sqrt{\left(1+\frac{4l^3\Delta k}{Ewt^3}\right)} \times \sqrt{\left(1-\frac{\Delta m}{m}\right)}\right) - 1. \qquad (18.3)$$

Therefore, in low damping media, the cantilever is a true mass-sensitive device only if the spring constant k does not change during the measurement. This is a condition that may not be satisfied, for instance, when the elasticity of the selective layer is affected by chemical exposure. According to eqn 18.2, the higher the Young's modulus, the higher the resonance frequency of the cantilever. This equation implies, for instance, that, for equivalent geometries, a rectangular diamond cantilever will typically resonate at twice the resonant frequency of the same silicon cantilever in a vacuum. Let us assume now that our two isometric cantilevers, resonating for instance at 10 kHz for the silicon device and at 20 kHz for the diamond device, are subject to a mass uptake of 10 ng without affecting the spring constant of both resonators. In this case eqn 18.3 would show that the resonance frequency variation of the diamond cantilever is approximately 400 Hz, while that of the silicon device is around 290 Hz. Therefore, in low damping media, it is reasonable to accept that the gravimetric sensitivity of the diamond transducer is significantly higher than its silicon counterpart.

In fluids, mechanical deformations of the cantilevers involve appreciable dissipation of mechanical energy into thermal energy. This energy dissipation can be quantified using the quality factor (or Q-factor). This depends on several parameters, such as cantilever material and shape and clamping losses, as well as the viscosity of the medium. When the cantilevers are long enough, clamping loss may be neglected and the Q-factor can be calculated using the vibrating sphere model in viscous fluids.[42] Here, the Q-factor is given by:

$$Q = \frac{\chi^2 \sqrt{\rho E}}{12\pi\sqrt{3}} \frac{wt^2}{\mu l R\left(1+\frac{R}{\delta}\right)} \qquad (18.4)$$

$$R = \sqrt{\frac{wl}{\pi}} \qquad (18.5)$$

$$\delta = \sqrt{\frac{\mu}{\pi \rho f_0}}, \qquad (18.6)$$

where χ is a constant relative to the first vibration mode ($\chi = 1.875$), w, l and t are the width, length and thickness of the cantilever, respectively, E is Young's modulus, μ is the viscosity of the medium, R is the radius of the equivalent sphere, which can be approximated by eqn 18.5 and δ is the boundary layer thickness calculated from eqn 18.6, where ρ is the density of the medium. In a given medium and for a given geometry, these equations imply that the Q-factor of a diamond cantilever (for $\rho = 3.50\,\mathrm{g\,cm^{-3}}$ and $E = 1000\,\mathrm{GPa}$) is typically three times higher than a silicon counterpart (for $\rho = 2.33\,\mathrm{g\,cm^{-3}}$ and $E = 160\,\mathrm{GPa}$). This was verified experimentally by Bongrain et al. (Figure 18.6).[43] In high damping media one has to take this dissipation of mechanical energy into account when calculating the resonance frequency of the cantilevers. The later can thus be expressed as:

$$f_{0,Q} = \frac{1}{2^{\frac{3}{2}}\pi} \sqrt{\frac{k}{m_0}} \frac{\sqrt{2Q-1}}{Q}. \qquad (18.7)$$

Theoretically, this expression shows that the higher the Q-factor, the higher the resonance frequency. Hence, the resonance frequency of microscale diamond cantilevers tends also to be higher than cantilevers made of other materials in viscous media. This is a very important observation because, today, damping is a serious limitation for measurements in the resonant regime in liquids. As a result, most cantilever-based chemical or biological sensors reported in the literature are operating in the bending mode. Here, diamond holds significant potential for improving the transduction

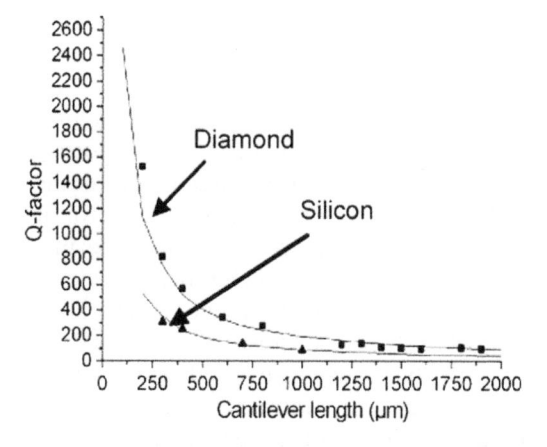

Figure 18.6 Measured and calculated Q-factor values of fabricated 4.6 μm thick bare diamond cantilevers and comparison with the calculated values of identical silicon structure Q-factors; values found in the literature for equivalent silicon cantilevers.
(Reprinted from ref. 43 with permission from Elsevier.)

performance of cantilevers in liquid environments, a prerequisite for biosensor applications.

Nevertheless, there is some strong evidence showing that chemical and physical interactions between the cantilever surface and its environment do affect the cantilever stiffness, thus making the gravimetric analysis much less straightforward. Several models were proposed to explain the cantilevers sensitivity to surface stress induced by molecular interactions based on strain-independent contributions, but they were questioned because they do not take into account stress relaxation possibilities. Moreover, the classical one-dimensional beam theory also predicts that cantilever stiffness is independent of strain-independent surface stress. Today, this still has not been elucidated as other groups have shown that three-dimensional models indicate that strain-independent surface stress indeed influences cantilever stiffness. In contrast others groups have proposed that cantilever resonance frequency surface stress is influenced by strain-dependent surface stress. The influence of this contribution upon the cantilever flexural rigidity can be described from one-dimensional beam theory by:

$$(EI)^b = (EI)^0 \left(1 + 3\frac{b_{\text{top}} + b_{\text{bottom}}}{Et}\right), \quad (18.8)$$

where $(EI)^b$ and $(EI)^0$ are the cantilever flexural rigidity with and without strain-dependent surface stress, respectively, and b_{top} and b_{bottom} are the strain-dependent surface stress on the top and the bottom surface of the cantilever, respectively.[41] The relation between a rectangular shape cantilever stiffness coefficient k and EI is given by:

$$k = \frac{3EI}{l^3}. \quad (18.9)$$

Eqns 18.1, 18.2 and 18.8 predict a weak sensitivity of cantilevers upon surface stress variations in the order of a few hundred µHz for typical surface stress variations of several hundred mN m^{-1} for silicon cantilevers. According to eqn 18.8, the sensitivity would be even lower for a diamond cantilever of the same size, since it is inversely proportional to the Young's modulus. Hence, according to the strain-dependent model, the advantage of using diamond structures for their superior Q-factor in liquids may be negated by a lower sensitivity. Nevertheless, the experimental values of the frequency shifts induced by biological/chemical species immobilisation on cantilever surfaces are reported in the range of several tens of Hz for, typically, hundreds of mN m^{-1} surface stress variations. These later values are in closer agreement with the values calculated using strain-independent surface stress models, which are controversial in the literature. These considerations illustrate evidence of a serious mismatch existing between theoretical predictions and experimental observations. Until this is resolved, the real advantage of using diamond cantilevers in the resonant regime in liquids for their exceptional mechanical properties is still to be demonstrated.

18.3.2 Fabrication Methods

Single-crystal and polycrystalline diamond cantilevers have been reported in the literature. Nevertheless, most of them fall into the second category since the advantages of single-crystal cantilevers, typically a slightly higher Q-factor,[44] does not add much value to chemical sensors, as opposed to the main drawback associated with low availability of single crystals and poor compatibility with mass production. Therefore, in the following we will concentrate only on polycrystalline diamond cantilevers.

The patterning of diamond for MEMS fabrication poses serious challenges because the chemical resilience of the material prevents the use of wet etching protocols. Despite this limitation, there is still a real interest for the use of patterning processes that remain compatible with conventional silicon and microsystem fabrication techniques. Taking this into account, two main approaches have been considered. Perhaps the most obvious one consists of selectively etching a continuous film of diamond by dry-etching techniques (Figure 18.7a). Thus, four primary diamond dry-etching methods that are compatible with classical clean room lithographic techniques have been reported over the years.[45] They include electron cyclotron etching (ECR), ion bean etching (IBE), reactive ion etching (RIE) and inductively coupled plasma etching (ICP). The most frequently used gases for etching diamond using these techniques are O_2, CF_4 and SF_6. These dry-etching techniques may be used either to etch monocrystalline or polycrystalline diamond.

In the case of polycrystalline diamond, another way of patterning diamond consists of selectively seeding the substrate in order to grow only in areas of interest. Several bottom-up methods of this kind have been reported that include the use of a photoresist loaded with diamond powder, the use of masks, such as SiO_2 or Ti/Pt, or even by direct spray writing.[45] Seeding may be performed by the deposition of diamond nanoparticles over the substrate or by bias enhanced nucleation.[46] However, these methods are limited in the resolution that can be achieved and, hence, to large size cantilevers, typically of several tens of microns in length and width. Moreover, unwanted nucleation generally occurs during growth in undesired areas, which contribute also to the loss of resolution of the patterns. In order to improve the resolution of patterns, variations of the technique were developed using SiO_2[47] or silicon molds[43,46] (Figure 18.7b). Here, silicon molds are preferred since SiO_2 molds may be etched by the hydrogen plasma, which contribute to a loss of resolution.

18.3.3 Diamond-based Resonant Cantilever Chemical Sensors

Chemical or biochemical sensing with resonant cantilevers is a fairly recent approach, although several proofs of concept have already been reported in the literature in the last twenty years or so. The materials used to fabricate

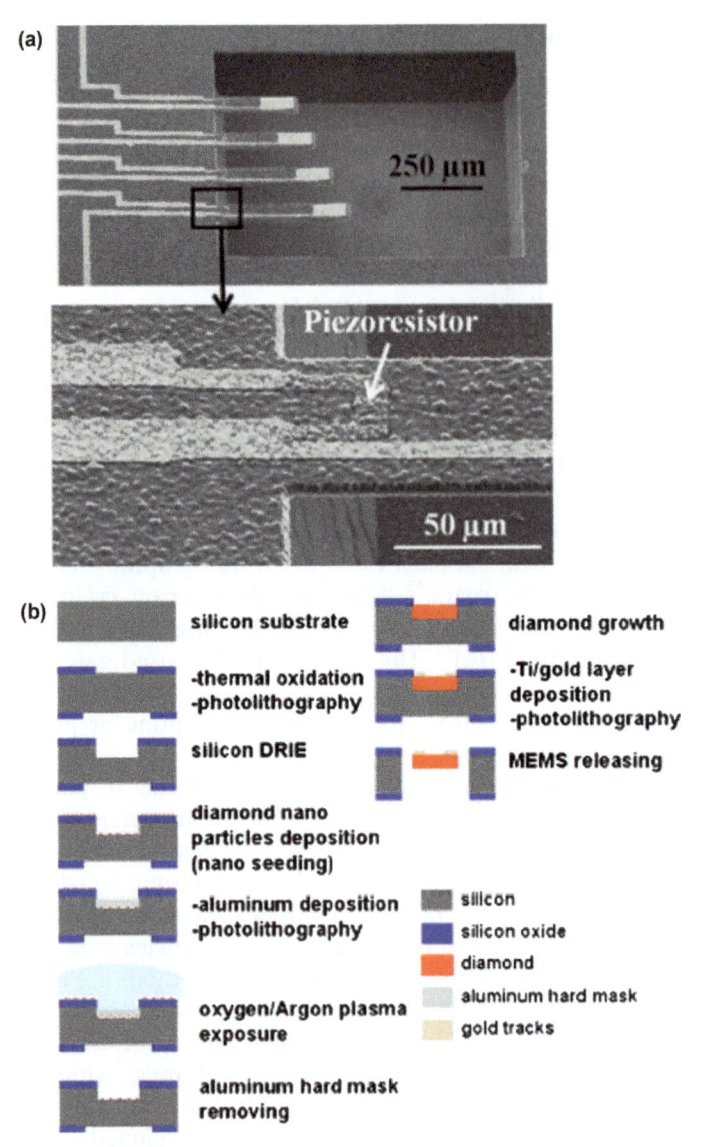

Figure 18.7 a) SEM image of the polycrystalline diamond cantilever structure used to characterise the piezoresistive effect (reprinted from ref. 45 with permission from Elsevier); b) fabrication process of diamond MEMS devices.
(Reprinted from ref. 43 with permission from Elsevier).

cantilever transducers are mostly silicon, silicon oxide or silicon nitride. Then, a selective coating is added onto one surface of the device, generally in the form of a polymer coating or chemically grafted receptors. Since the chemistry of silicon-based materials is somehow limited, a layer of gold is often deposited onto the transducer surface in order to allow immobilisation of receptors *via* thiol attachment.[48] Most cantilever chemical sensors to date operate in the bending mode in liquid because the Q-factors of resonant structures in liquids are too low to perform high S/N measurements. The use of diamond holds promise for working in the dynamic mode in liquids since its Young's modulus is sufficiently high with respect to more conventional materials, thus significantly increasing the Q-factor and enabling measurements in liquids. Moreover, the carbon surface of diamond can be used directly to attach, covalently, chemical or bio-receptors onto the transducer without the addition of a gold layer. It was shown, for instance, that the addition of a gold layer onto a diamond cantilever surface affects significantly its Q-factor.[43] Diamond may also be doped with boron in order to integrate either BDD electrodes onto the structure, which can be used for localised electro-grafting on the surface,[49] or piezoresistive gauges as a frequency readout strategy.[45] Therefore, in principle, it is possible to fabricate single-material diamond cantilevers that can include all the necessary functions for making chemical sensors without significantly affecting the mechanical properties of the cantilever. Despite all these assets, there are very few reports of diamond cantilevers for chemical sensing, probably because of the poor accessibility of diamond materials in the chemical sensor community.

Amongst the few reports of chemical sensing with diamond cantilevers, one can mention the study by Bongrain *et al.* on the response of a diamond cantilever to protonation/deprotonation events of an acid group immobilised on the surface of a bulk diamond cantilever (Figure 18.8).[41] This experiment was set up to investigate the effects of surface stress variations, while avoiding significant mass loading on the cantilever. Frequency

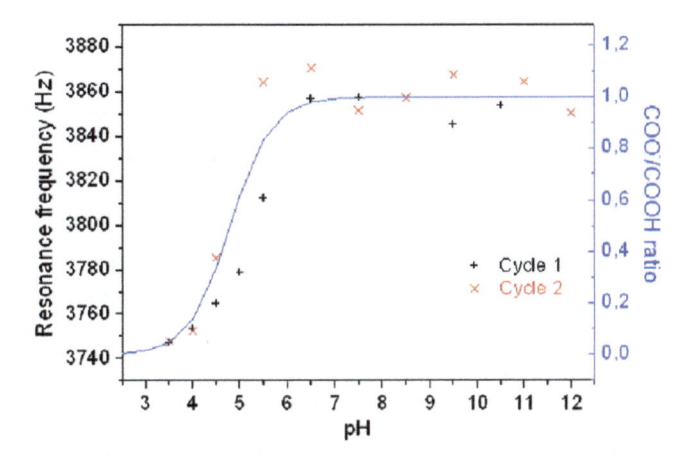

Figure 18.8 Resonance frequency variations of a diamond cantilever functionalised with carboxyl groups up pH cycling.
(Reprinted with permission from ref. 41. Copyright (2011) American Chemical Society.)

variations over 100 Hz could be observed when cycling the pH from three to twelve, with a main contribution to the response arising from stiffness variations of the cantilevers due to electrostatic interactions at the surface of the sensor. Interestingly, in this experiment the carboxylic function was also immobilised covalently on the diamond surface by a one-step protocol[27] comparable, in that respect, to the processes used on gold layers. This experiment confirmed that cantilevers can be extremely sensitive to chemical events occurring at their surfaces, even when abstracting the mass parameter. This effect was exploited for the detection of DNA hybridisation/denaturation events at the surface of a diamond cantilever (Figure 18.9).[49] Indeed, DNA, being itself a highly charged molecule, is liable to charge interactions with

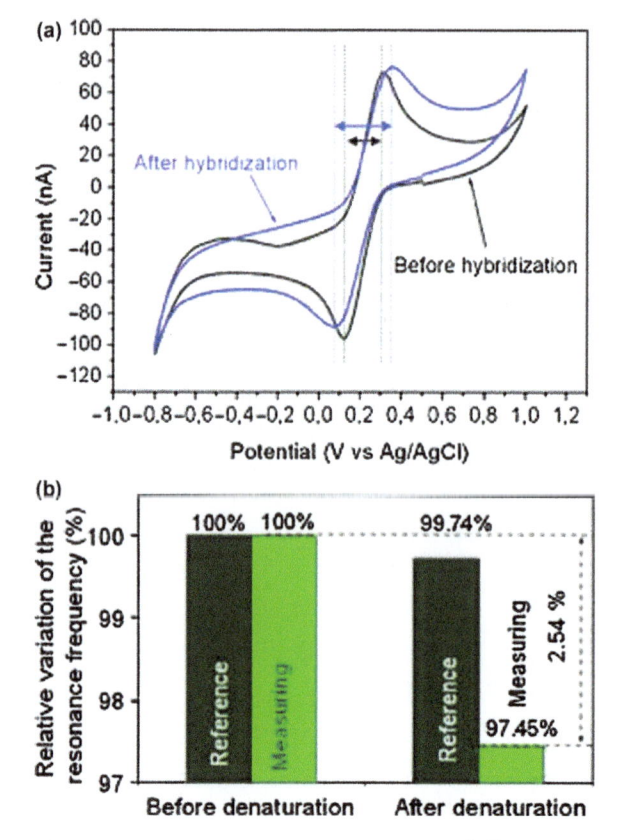

Figure 18.9 a) Cyclic voltamograms in 0.5 mM Fe(CN$_6$)$^{3-/4-}$, 100 mM KCl, 100 mM KNO$_3$ before and after DNA hybridisation; b) resonance frequency of both reference (dark) and measuring (light) cantilevers before and after DNA denaturation in phosphate-buffered saline (PBS) solution. For the reference cantilever: $f = 3121$ Hz before, and $f = 3113$ Hz after denaturation; for the measuring cantilever: $f = 2948$ Hz before, and $f = 2873$ Hz after denaturation.
(Reprinted from ref. 49 with permission, copyright (2012) John Wiley and Sons.)

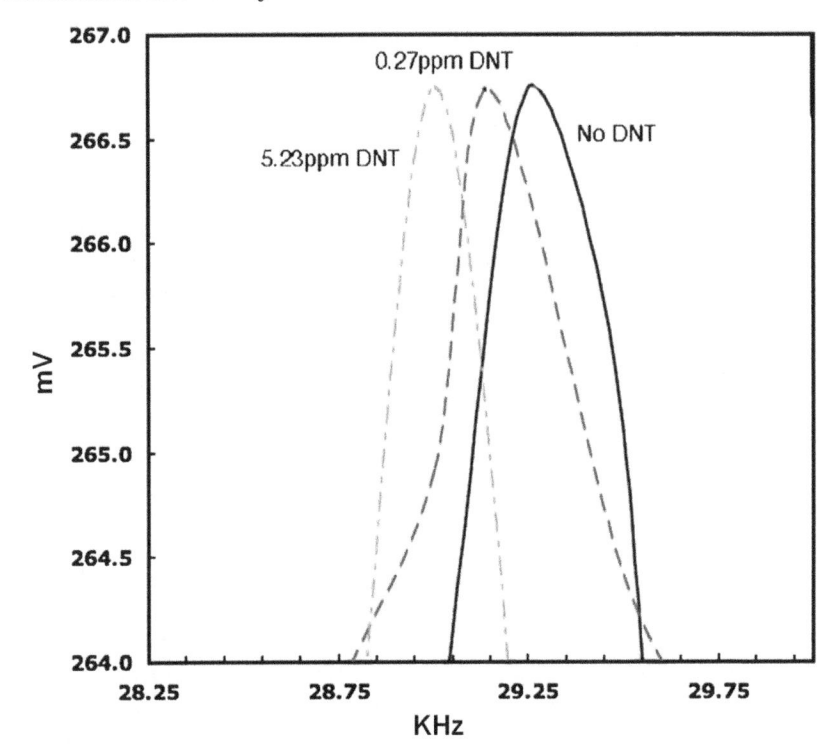

Figure 18.10 Dynamic response (expressed in mV) of a nanodiamond-coated cantilever before exposure to DNT and following exposure to 0.27 and 5.23 ppm DNT vapour.
(Reprinted with permission from ref. 50, American Institute of Physics.)

neighbouring DNA molecules or ions from the surrounding medium. Probe DNA was immobilised *via* an electrochemical process on the surface of a diamond cantilever onto which a BDD electrode was overgrown. The possibility to detect complementary 32 base pair DNA was demonstrated.

Finally, diamond may also be used as a chemically sensitive coating deposited onto a silicon cantilever. This approach allows the use of well-established silicon cantilever technology, while benefiting from the properties of diamond surfaces for sensing applications. 2,4-dinitrotoluene (DNT), a safe analog for the detection of the explosive trinitrotoluene, was used in order to demonstrate this proof of concept.[50] A respectable sensitivity of 0.77 Hz ppb^{-1} was achieved by such a sensor (Figure 18.10).

18.4 Surface Acoustic Wave Resonators

18.4.1 Generality

Surface acoustic wave (SAW) chemical sensors constitute another family of so-called "mass sensitive sensors", although the sensing mechanisms are actually fairly complex and include some visco-elastic and electro-acoustic components along with the gravimetric sensitivity.[51,52] There are many types of SAW sensor devices but they all rely on the same fundamental principle: the perturbation upon chemical stimuli of an acoustic wave piezoelectrically generated and travelling across the surface of the transducer. In this chapter we will focus on Rayleigh SAW (R-SAW) transducers since there are very few reports, if any, of diamond-based SAW chemical transducers based on other propagation modes. R-SAW transducers are particularly interesting for sensing in the gas phase. The most basic R-SAW structure is a two-terminal delay line transducer, as illustrated in Figure 18.11. Here, the acoustic wave is generated at the transmitter electrode. It travels along the surface of the substrate and is then transformed back into an electrical signal by the reverse piezoelectric effect at the receiving electrode. A selective layer is generally deposited on the delay line. Interactions of this layer with target chemicals induce amplitude attenuation, phase difference or frequency shift.[2] Other SAW sensors employ a resonator arrangement, which allows better sensor performances, particularly because of higher Q-factors. In general, since the wave energy is confined to within one wavelength from the surface, this characteristic yields sensors that are very sensitive to chemical interactions occurring at the surface.[51] As a result, the sensitivity of SAW transducers is typically one order of magnitude higher than that of QCM.[2]

The delay line transducer operates most efficiently when the SAW wavelength λ matches the transducer periodicity d. This occurs when the

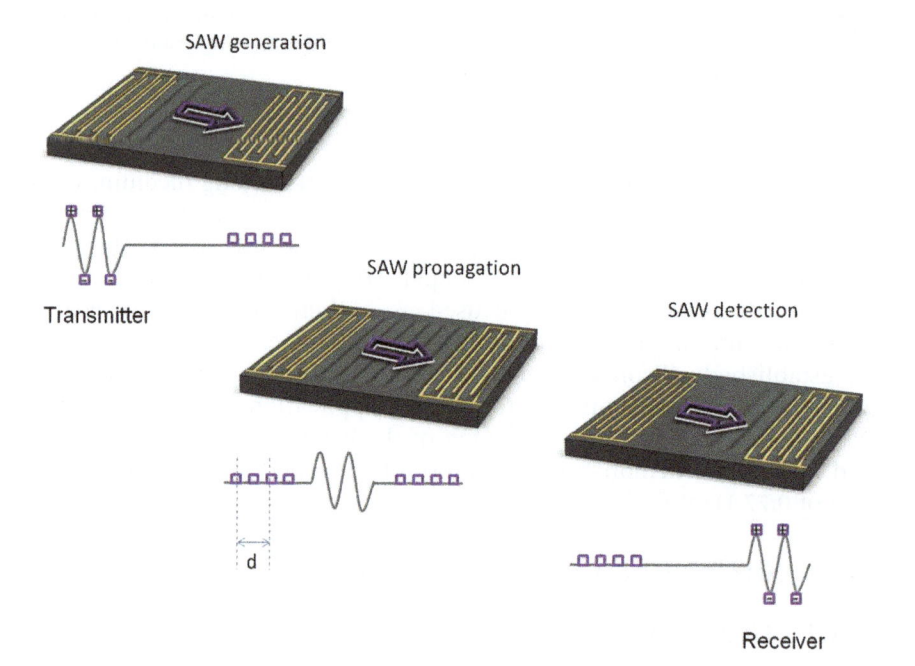

Figure 18.11 Generation of an acoustic wave on a delay line resonator (schematic by B. Bazin).

transducer is excited at the synchronous frequency f_0, which satisfies eqn 18.10, where v_O is the SAW propagation velocity.

$$f_0 = \frac{v_O}{d} \qquad (18.10)$$

The acoustic velocity in diamond is highest (typically 20 000 m s^{-1} in the single crystal); therefore, SAWs travel faster in diamond than in conventional piezoelectric materials. Besides, according to eqn 18.10, the higher v_O, the higher the frequency. This has been the basis for the development of high frequency diamond-based resonators operating in the GHz range for electronic applications. Since diamond is not a piezoelectric material, typically ZnO or AlN layers are deposited onto the diamond to piezoelectrically generate the acoustic wave.[53,54]

In the case of R-SAW transducers, the effect of mass loading on the wave propagation may be expressed by:

$$\frac{\Delta v}{v_0} = -C_m f_0 \rho_s, \qquad (18.11)$$

where C_m is a mass sensitivity factor and ρ_s the crystal density. The mass sensitivity S_m is given by:

$$S_m = \frac{1}{\Delta m}\left(\frac{\Delta v}{v_0}\right). \qquad (18.12)$$

Thus, the sensitivity of a Y-cut quartz R-SAW delay line transducer operating at 97 MHz is approximately 12 200 Hz cm^2 µg^{-1}.[51] Nevertheless, eqn 18.11 suggests that, for R-SAWs, the sensitivity increases with the frequency of the device. Therefore, one way to increase the mass sensitivity of a SAW transducer intuitively is to work at high frequencies. Hence, diamond SAW resonators should be good candidates for designing high sensitivity chemical sensors. In practice this is not entirely true because for high frequencies typically in the GHz range, the acoustic wave propagation becomes highly affected by the presence of the selective coating deposited onto the transducer surface, which degrades also the Q-factor significantly. The sensitivity to other environmental parameters is also potentially higher and so is the noise level. Hence, a compromise has to be found and, generally, SAW sensors operate in the range 100–500 MHz. At those frequencies, there is no need to use diamond as a waveguide. However, diamond has been used on such devices as a suitable sensing interface layer since it offers a highly stable versatile surface for the design of selective coatings.

18.4.2 Diamond Nanoparticles-coated SAW Chemical Sensors

R-SAW sensors are promising for the detection of chemicals in the gas phase due to their extremely high sensitivity. The selective coating here is generally the limiting element in terms of sensor-to-sensor repeatability, reliability and long-term stability. In most cases the sensitive coatings are based on polymers that are generally difficult to deposit homogeneously onto the transducer surface.[55–57] Other types of coatings have also been considered, including carbon nanotubes[58] or, more recently, graphene sheets,[59] but they also suffer from coating difficulties. Indeed, a strict control of the thickness uniformity, viscosity and film adherence of the coating is necessary in order to obtain reliable performances as any defects present on the SAW propagation path is known to degrade the performances of the sensors.[57]

In this context nanoparticles of diamond have been considered as an alternative sensitive coating that could solve some of the issues encountered with other known selective coatings. Recent developments in conformal coating with diamond nanoparticles using layer-by-layer deposition methods have enabled the deposition of diamond nanoparticle thin films onto SAW transducers[60] (Figure 18.12). For example, negatively charged diamond particles were immobilised onto R-SAW resonators using a cationic polymer.[61,62] Diamond nanoparticles from detonation can be found in nanometer sizes and can thus be deposited using this approach as single or multiple layers on sensors.

In addition to the high stability of the particles due to their carbon sp^3 nature, thin films of diamond nanoparticles offer several advantages for gas sensing applications using R-SAW transducers. First of all, the high density of diamond prevents absorption of any molecules onto the bulk of the coating. Hence, chemical interactions occur only at the surface of the particles, thus avoiding potential poisoning of the selective layer by trapping of molecules into the material. Moreover, in contrast with polymers for instance, the lack of diffusion into the bulk of the layer may improve the sensor response times since it relies mainly on adsorption kinetics at the diamond surface. For example, Chevallier *et al.* reported the detection of 240 ppb 2,4-dinitrotoluene using a selective coating based on a porphyrin

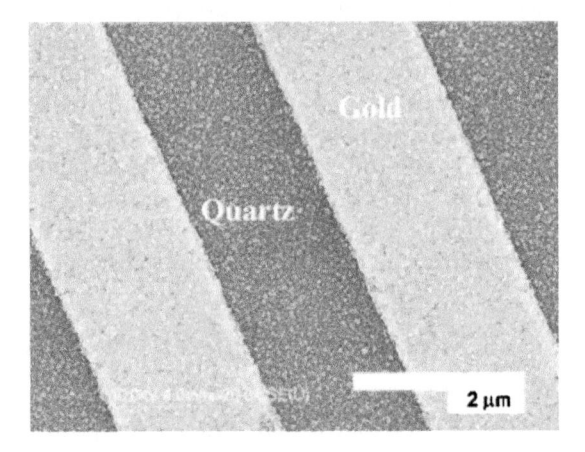

Figure 18.12 SEM image of a SAW sensor with a slight nanodiamond coating. (Reprinted from ref. 61 with permission from Elsevier.)

Figure 18.13 a) Transient response of SAW sensor coated with ZnTMPy-nanodiamond at room temperature to different concentrations of DNT; b) calibration curve obtained from four separate SAW sensors prepared with three layers of ZnTMPy-nanodiamond (red curve), three layers of nanodiamond (gray curve) and three layers of ZnTMpy (blue curve) to DNT gas at room temperature.
(Reprinted from ref. 62 with permission from Elsevier.)

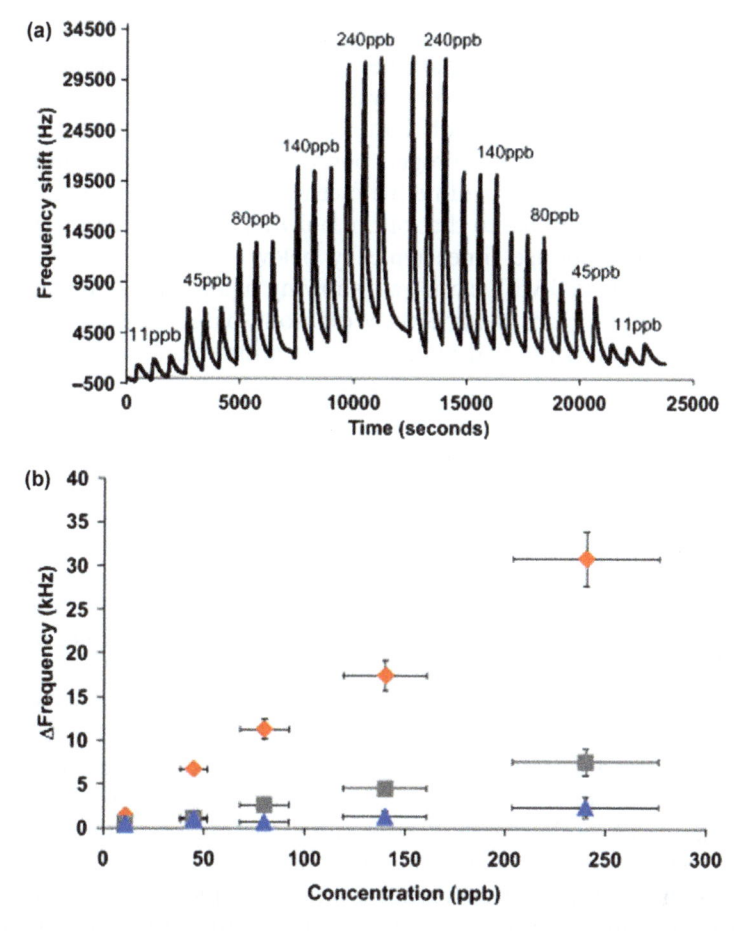

complex immobilised on diamond nanoparticles with a response time $T_{90\%}$ of typically 250 s.[62] Here, the porphyrin complex was chosen for its strong affinity with nitroaromatic compounds.[64,65] The same team showed that the use of diamond nanoparticles in this case could improve the sensitivity of the sensor with respect to the same transducer coated with porphyrin alone by, typically, a factor ten (Figure 18.13), most probably because of the higher sensing surface area created by the nanoparticle coating.

As already discussed in the earlier sections, the carbon nature of diamond offers many opportunities for attachment of a wide range of chemical or biological receptors *via* C–C covalent bonding. Tard and coworkers showed, for example, that fluoroalcohols groups (-(CF$_3$)$_2$-OH) could be immobilised covalently on diamond nanoparticles for the detection of organophosphorous compounds.[63] This work was inspired from previous work on polymers containing fluoroalcohol moieties that exhibit strong hydrogen bond acidic affinities with hydrogen bond basic organophosphorus nerve agents.[66] The sensitivity to dimethyl-methylphosphonate, a simulant for sarin gas, was comparable to that of the polymer coatings (typically 750 Hz ppm v^{-1}), with response times typically below 1 min. The same R-SAW sensor was also tested with sarin gas, showing again similar results. Other examples of applications using R-SAW diamond-based sensors include the detection of mustard gas and phosgene,[63] or ammonia.[61]

18.4.3 Toward Artificial Olfaction Using Diamond-based SAW Sensors

In many applications of chemical or biochemical sensing, selectivity is a crucial parameter. However, the only approach to make 100% selective coatings is probably through the use of bio-receptors, such as antibodies, enzymes or DNA. Those receptors are generally suitable to detect biological targets; hence, they are particularly useful for the development of sensors for medical diagnostic. For instance, the enzyme glucose oxidase is highly suitable for the specific detection of glucose. Unfortunately, the use of bio-receptors for the detection of non-metabolic species is very limited. Amongst the few examples Kalaji and co-workers developed specific proteins for the identification of explosives.[67] Furthermore, bio-receptors have a limited lifetime, especially if not kept at an adequate temperature, and in suitable pH or buffer conditions. Hence, they are not suitable selective coatings for measurements in the gas phase. For these applications, more stable selective coatings must be developed, but chemistry generally fails to provide 100 % selective coatings. One approach has been to fabricate imprints of target species using molecularly imprinted polymers, although this approach has been unconvincing so far. The other approach is to use sensor arrays in which each individual sensor features a broad selectivity to chemicals or family of chemicals. Upon exposure to vapours, the responses of different sensors in the array are analysed using a statistical multi-parametric approach, such as an artificial neural network, in order to identify the fingerprint of the target vapour. Such an approach has been known for many years as an "electronic nose".[68]

For such artificial olfaction applications, R-SAW sensors are highly promising because they offer a high sensitivity in the ppb range along with the possibility to address, as discussed before, any molecule, since they are mass transducers. Diamond nanoparticle coatings, as described in the previous section, appear as highly promising for such applications since they may be used as a robust generic sensing interfaces of which the chemical affinity may be tuned by various surface chemical terminations.

Diamond-based Resonators for Chemical Detection

This principle was demonstrated by Chevallier and co-workers by comparing the sensitivity of two diamond nanoparticle-coated R-SAWs with different surface terminations to various chemical vapours.[61] In one sensor the surface of diamond was hydrogen terminated and in the other sensor the diamond surface was highly oxidized. Figure 18.14a shows the typical response of the two sensors to 30 ppm volume ammonia vapours. Here, both sensors are giving a response; hence, neither of them is very specific. Nevertheless, the "oxidized diamond" sensor is typically ten times more sensitive than the "hydrogen-terminated diamond" sensor due to the polar affinity of the oxide groups with ammonia vapours. These two sensors were also exposed with other vapours and, again, the affinities of the two sensors to the different chemicals appeared to be clearly different (Figure 18.14b).

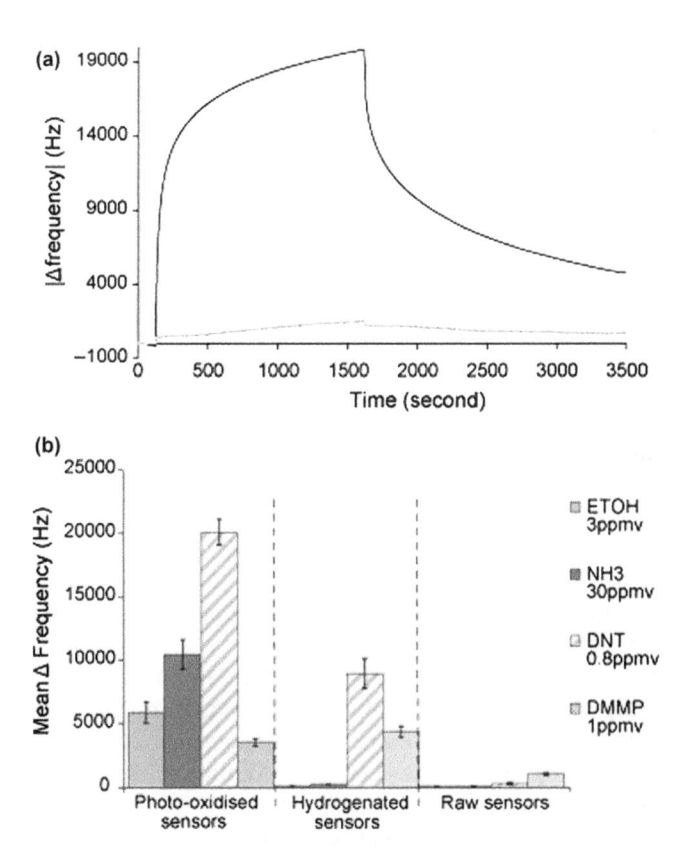

Figure 18.14 a) Response of hydrogenated nanodiamond-coated transducer (dark gray curve) and a photo-oxidised nanodiamond-coated transducer (black curve) to a 30 ppmv NH_3 exposure; b) mean response of four hydrogenated and four photo-oxidised nanodiamond-coated transducers, respectively, to successive 60 s exposures to ethanol, NH_3, DNT and DMMP.
(Reprinted from ref. 61 with permission from Elsevier.)

This somewhat basic experiment demonstrated that simple surface modifications can have a drastic influence on the selectivity of the diamond based R-SAW sensors. Many wet or plasma processes have been described in the literature to modify the chemical termination of diamond surfaces, thus highlighting the opportunities of development for such sensor arrays. Research in this field is very recent and considerable further work will be necessary in order to build specific reliable sensor arrays to comply with the needs of potential users.

18.5 Conclusions

In summary, diamond features exceptional chemical and physical properties that can be beneficial to improve not only the robustness and reliability of chemical sensors in the field but also their sensing performances. Thus, a number of diamond-based chemical or biochemical sensors have been reported in the last twenty years, often showing highly promising performances. They include, for instance, solid-state sensors, field effect transistors and electrochemical sensors. Amongst them, diamond-based gravimetric sensors, such as SAW sensors and cantilevers, offer the possibility of high sensitivity detection of a wide range of possible analytes at potentially low costs and with a high level of miniaturisation. Research in this field is still at a fairly early stage and diamond material in its various forms, from nanoparticles to bulk single crystals, has still a lot to offer toward highly innovative new generations of sensors.

References

1. A. Hulanicki, S. Geab and F. Ingman, *Pure Appl. Chem.*, 1991, **63**, 1247.
2. J. Janata, *Principles of Chemical Sensors*, Springer, Verlag New York Inc., 2nd edition, 2009, p. 374.
3. J. de Sanoit, E. Vanhove, P. Mailley and P. Bergonzo, *Electrochimica Acta*, 2009, **544**, 5688.
4. B. V. Sarada, Tata N. Rao, D. A. Tryk and A. Fujishima, *Anal. Chem.*, 2000, **72**, 1632.
5. O. Auciello, S. Pacheco, A. V. Sumant, C. Gudeman, S. Sampath, A. Datta, R. W. Carpick, V. P. Adiga, P. Zurcher, Z. Ma, H. C. Yuan, J. A. Carlisle, B. Kabius and J. Hiller, *IEEE Microwave Magazine*, 2008, **8**, 61.
6. J. K. Luo, Y. Q. Fu, H. R. Le, J. A. Williams, S. M. Spearing and W. I. Milne, *J. Micromech. Microeng.*, 2007, **17**, S147.
7. P. Hess, *J. Appl. Phys.*, 2012, **111**, 051101.
8. S. Balachandran, T. Weller, A. Kumar, S. Jeedigunta, H. Gomez, J. Kusterer, E. Kohn, *Nanocrystalline Diamond for RF-MEMS Applications, Emerging Nanotechnologies for Manufacturing*, ed. W. Ahmed and M. J. Jackson, Elsevier, Oxford, UK, 2010, pp. 277–300.
9. N. Sepulveda, D. Aslam and J. P. Sullivan, *Diamond Relat. Mater.*, 2006, **15**, 398.

10. S. Shikata, S. Fujii and T. Sharda, *Diamond Relat. Mater.*, 2009, **18**, 253.
11. Y. Gurbuz, W. P. Kang, J. L. Davidson and D. V. Kerns, *Sens. Actuators B*, 2004, **99**, 207.
12. T. Takada, T. Fukunaga, K. Hayashi, Y. Yokota, T. Tachibana, K. Miyata and K. Kobashi, *Sens. Actuator A-Phys.*, 2000, **82**, 97.
13. Y. Isamu, K. Nobuyuki, N. Masayuki, S. Hiroshi, T. Katsunobo, *Semiconductor gaz sensor*, Brevet EP 0 488 352 A2, 1991.
14. R. K. Joshi, J. E. Weber, Q. Hu, B. Johnson, J. W. Zimmer and A. Kumar, *Sens. Actuators B*, 2010, **145**, 527.
15. G. Swain, A. Anderson and J. Angus, *MRS Bulletin*, 1998, **23**, 56.
16. J. A. Garrido, A. Härtl, M. Kuch, M. Stutzmann, O. A. Williams and R. B. Jackmann, *Appl. Phys. Lett.*, 2005, **86**, 073504.
17. A. Helwig, G. Müller, J. A. Garrido and M. Eickhoff, *Sens. Actuators B*, 2008, **133**, 156.
18. A. Kromka, M. Davydova, B. Rezek, M. Vanecek, M. Stuchlik, P. Exnar and M. Kalbac, *Diamond Relat. Mater.*, 2010, **19**, 196.
19. Q. Wang, S. L. Qu, S. Y. Fu, W. J. Liu, J. J. Li and C. Z. Gu, *J. Appl. Phys.*, 2007, **102**, 103714.
20. M. Davydova, A. Kromka, B. Rezek, O. Babchenko, M. Stuchlik and K. Hruska, *App. Surf. Sci.*, 2010, **256**, 5602.
21. J.-P. Mazellier, M. Mermoux, F. Andrieu, J. Widiez, J. Dechamp, S. Saada, M. Lions, M. Hasegawa, K. Tsugawa, P. Bergonzo and O. Faynot, *J. Appl. Phys.*, 2011, **110**, 084901.
22. Y. Gurbuz, W. P. Kang, J. L. Davidson and D. V. Kerns, *Sens. Actuators B*, 1998, **49**, 115.
23. A. Balducci, A. D'Amico, C. Di Natale, M. Marinelli, E. Milani, M. E. Morgada, G. Pucella, G. Rodriguez, A. Tucciarone and G. Verona-Rinati, *Sens. Actuators B*, 2005, **102**, 111.
24. X. Checoury, D. Neel, P. Boucaud, C. Gesset, H. Girard, S. Saada and P. Bergonzo, *Appl. Phys. Lett.*, 2012, **101**, 171115.
25. M. J. Burek, N. de Leon, B. J. Shields, B. J. M. Hausmann, Y. Chu, Q. Quan, A. S. Zibov, H. Park, M. D. Lukin and M. Loncar, *Nano Lett.*, 2012, **12**, 6084.
26. P. Bergonzo, R. B. Jackman, in *Thin Film Diamond*, Part II, eds C. E. Nebel, J. Ristein, vol. 77, 2003, Academic Press, San Diego, USA, p. 197.
27. C. Agnès, S. Ruffinatto, E. Delbarre, A. Roget, J.-C. Arnault, F. Omnès and P. Mailley, *IOP Conf. Series: Materials Science and Engineering*, 2010, **16**, 012001.
28. S. Q. Lud, M. Steenackers, R. Jordan, P. Bruno, D. M. Gruen, P. Feulner, J. A. Garrido and M. Stutzmann, *J. Am. Chem. Soc.*, 2006, **128**, 16884.
29. C. E. Nebel, B. Rezek, D. Shin, H. Uetsuka and N. Yang, *J. Phys. D: Appl. Phys.*, 2007, **40**, 6443.
30. J.-H. Yang, M. Degawa, K.-S. Song, C. Wang and H. Kawarada, *Mater. Lett.*, 2010, **64**, 2321.

31. K.-S. Song, T. Hiraki, H. Umezawa and H. Kawarada, *Appl. Phys. Lett.*, 2007, **90**, 063901.
32. E. Vanhove, J. de Sanoit, P. Mailley, M.-A. Pinault, F. Jomard and P. Bergonzo, *Phys. Status Solidi A*, 2007, **206**, 2063.
33. R. Kiran, E. Scorsone, J. de Sanoit, J.-C. Arnault, P. Mailley and P. Bergonzo, *J. Electrochem. Soc.*, 2013, **160**, H67.
34. J. Lee and S.-M. Park, *Anal. Chim. Acta*, 2005, **545**, 27.
35. S. Szunerits, J. Niedziołka-Jönsson, R. Boukherroub, P. Woisel, J.-S. Baumann and A. Siriwardena, *Anal. Chem.*, 2010, **82**, 8203.
36. J. Weng, J. Zhang, H. Li, L. Sun, C. Lin and Q. Zhang, *Anal. Chem.*, 2008, **80**, 7075.
37. T. A. Ivandini, D. Yamada, T. Watanabe, H. Matsuura, N. Nakano, A. Fujishima and Y. Einaga, *J. Electroanal. Chem.*, 2010, **645**, 58.
38. Y. Zhang, S. Asahina, S. Yoshihara and T. Shirakashi, *J. Electrochem. Soc.*, 2002, **149**, H179.
39. O. A. Williams, V. Mortet, M. Daenen and K. Haenen, *Appl. Phys. Lett.*, 2007, **90**, 063514.
40. N. V. Lavrik, M. J. Sepaniak and P. G. Datskos, *Rev. Sci. Instrum.*, 2004, **75**, 2229.
41. A. Bongrain, C. Agnès, L. Rousseau, E. Scorsone, J-C Arnault, S. Ruffinatto, F. Omnès, P. Mailley, G. Lissorgues and P. Bergonzo, *Langmuir*, 2011, **27**, 12226.
42. K. Naeli and O. Brand, *J. Appl. Phys.*, 2009, **105**, 014908.
43. A. Bongrain, E. Scorsone, L. Rousseau, G. Lissorgues and P. Bergonzo, *Sens. Actuators B*, 2011, **154**, 142.
44. M. Liao, C. Li, S. Hishita and Y. Koide, *J. Micromech. Microeng.*, 2010, **20**, 085002.
45. Z. Cao and D. Aslam, *Diamond Relat. Mater.*, 2010, **19**, 1263.
46. A. Bongrain, E. Scorsone, L. Rousseau, G. Lissorgues, C Gesset, S Saada and P Bergonzo, *J. Micromech. Microeng.*, 2009, **19**, 074015.
47. G. M. R. Sirineni, H. A. Naseem, A. P. Malshe and W. D. Brown, *Diamond Relat. Mater.*, 1997, **6**, 952.
48. M. Álvarez, L. G. Carrascosa, M. Moreno, A. Calle, A. Zaballos, L. M. Lechuga, C. Martínez and J. Tamayo, *Langmuir*, 2004, **20**, 9663.
49. A. Bongrain, H. Uetsuka, L. Rousseau, L. Valbin, S. Saada, C. Gesset, E. Scorsone, G. Lissorgues and P. Bergonzo, *Phys. Status Solidi A*, 2010, **207**, 2078.
50. R. K. Ahmad, A. C. Parada, S. Hudziak, A. Chaudhary and R. B. Jackman, *Appl. Phys. Lett.*, 2010, **97**, 093103.
51. D. S. Ballantine, R. M. White, S. J. Martin, A. J. Ricco, E. T. Zellers, C. G. C. Frye and H. Wohltjen, *Acoustic Waves Sensors, Theory, Design, and Physico-chemical Applications*, Academic Press, 1st edition, San Diego, USA, 1996, 436.
52. E. A. Ash, G. W. Farnell, H. M. Gerard, A. A. Oliner, A. J. Slobodnik, JR. and H. I. Smith, *Acoustic surface waves*, Springer, Verlag Berlin and Heidelberg GmbH & Co. K, 1978, p. 331.

53. J. G. Rodriguez-Madrid, G. F. Iriarte, D. Araujo, M. P. Villar, O. A. Williams, W. Müller-Sebert and F. Calle, *Mater. Lett.*, 2012, **66**, 339.
54. W.-C. Shih and R.-C. Huang, *Vacuum*, 2008, **83**, 675.
55. G. Harsanyi, *Polymer Films in Sensor Applications: Technology, Materials, Devices and Their Characteristics*, Technomic Publishing Co., Lancaster, Basel, 1995, p. 435.
56. J. W. Grate, *Chem. Rev.*, 2000, **100**, 2627.
57. J. W. Grate and R. A. McGill, *Anal. Chem.*, 1995, **67**, 4015.
58. M. Penza, P. Aversa, G. Cassano, W. Wlodarski and K. Kalantar-Zadeh, *Sens. Actuators B*, 2007, **127**, 168.
59. R. Arsat, M. Breedon, M. Shafiei, P. G. Spizziri, S. Gilje, R. B. Kaner, K. Kalantarzadeh and W. Wlodarski, *Chem. Phys. Lett.*, 2009, **467**, 344.
60. H. A. Girard, E. Scorsone, S. Saada, C. Gesset, J. C. Arnault, S. Perruchas, L. Rousseau, S. David, V. Pichot, D. Spitzer and P. Bergonzo, *Diamond Relat. Mater.*, 2012, **23**, 83.
61. E. Chevallier, E. Scorsone and P. Bergonzo, *Sens. Actuators B*, 2011, **154**, 238.
62. E. Chevallier, E. Scorsone, H. A. Girard, V. pichot, D. Spitzer and P. Bergonzo, *Sens. Actuators B*, 2010, **151**, 191.
63. B. Tard, A. Trouvé, E. Scorsone, A. Voigt, M. Rapp and P. Bergonzo, *The 14th International Meeting on Chemical Sensors*, 2012, 481.
64. S. Tao, G. Li and H. Zhu, *J. Mater. Chem.*, 2006, **16**, 4521.
65. S. Tao, J. Yin and G. Li, *J. Mater. Chem.*, 2008, **18**, 4872.
66. Q. Zheng, Y.-C. Fu and J.-Q. Xu, *Procedia Eng.*, 2010, **7**, 179.
67. C. D. Gwenin, M. Kalaji, P. A. Williams and R. M. Jones, *Biosens. Bioelectron.*, 2007, **22**, 2869.
68. J. Gardner, P. N. Bartlett, *Sensors and Sensory Systems for an Electronic Nose*, Springer, Dordrecht, Netherlands, 1992, p. 340.

CHAPTER 19

All-diamond Electrochemical Devices: Fabrication, Properties, and Applications

NIANJUN YANG,* WALDEMAR SMIRNOV AND JAKOB HEES

Fraunhofer Institute for Applied Solid State Physics (IAF), Freiburg 79108, Germany
*Email: nianjun.yang@iaf.fraunhofer.de

19.1 Introduction

Since the introduction of diamond films as electrodes by Iwaki *et al.* in 1983,[1] and later by Pleskov *et al.* in 1987,[2] diamond electrochemistry has been widely investigated.[3–8] Heavily boron-doped diamond has been recognized as one of the best materials for electrochemical[7–11] and bio-electrochemical sensing applications[12–14] due to its unique physical and chemical properties.[7,8,14–18] It is metallically conductive and shows the widest electrochemical potential window. If the window is defined with an absolution value of current density less than 1.0 mA cm^{-2}, the typical potential window of heavily boron-doped diamond is about 3.2 V in aqueous solutions, 4.6 V in organic solutions, and 4.9 V in room temperature ionic liquid. Its capacitance current is ultra-low, 10 times lower than that of Au and 400 times lower than that of glassy carbon. It is chemically inert, does not swell in electrolyte solutions, and does not show surface fouling, especially when hydrogen-terminated. Its surface can be terminated[18] with hydrogen, oxygen or mixtures of both, which allow convenient optimization of electronic properties of the solid/electrolyte interfaces. It is biocompatible and can be biofunctionalized *via* carbon chemistry.[12–14,19,20] Furthermore, its surface can be structured as designed from nanotextures with dimensions of typical a few nanometers[21–23] to nanowires with lengths of a few micrometers.[24]

In most studies reported up to now for electrochemical and biochemical sensing applications, planar macroscopic diamond electrodes have been frequently applied. To obtain better sensing performance in chemical and biochemical sensing applications, well-designed and market-available all-diamond devices are critical. Low dimension electrodes in micrometer or even nanometer ranges with special designs (*e.g.*, ensembles/arrays, integrated electrochemical sensors) are especially promising because they offer various benefits over planar macroscopic electrodes,[25–30] such as reduced Ohmic resistance, enhanced mass transport, decreased charging currents, decreased deleterious effects of solution resistance, and increased potential for fast voltammetric measurements. Defined by the sizes/diameters of electrodes, small dimensional electrodes are classified as microelectrode (25 µm < electrode diameter < 100 µm), ultramicroelectrode (0.1 µm < electrode diameter < 25 µm), or nanoelectrode arrays (electrode diameter < 100 nm).[25,26]

As for the fabrication of diamond-based microelectrodes and ultramicroelectrodes, Cooper and his co-workers reported for the first time in 1988 the fabrication of a boron-doped diamond microelectrode.[31] One frequently applied approach to produce diamond-based microelectrodes and ultramicroelectrodes is to coat a sharpened metal wire with a boron-doped diamond film. The metals used are mainly from tungsten[31–36] and platinum.[37,38] One alternative approach developed by Martin and coworkers[39,40] is to coat a tungsten wire that is sealed in a quartz glass capillary. These microelectrodes and ultramicroelectrodes have been characterized[31–40] with scanning electron microscopy (SEM), optical microscopy, and various electrochemical techniques (mainly voltammetry). They have been applied for electrochemical sensing applications in non-aqueous[31] and aqueous electrolytes.[32–40] For example, they have been used in biological media to detect dopamine in mouse brains,[33] to monitor norepinephrine release in a mesenteric artery,[38] to investigate the role of adenosine in the modulation of breathing within animal tissues,[39] and to inspect serotonin as a neuromodulator.[40] As expected, these microelectrodes and ultramicroelectrodes have shown lower detection limits towards analytes in solution and increased signal-to-noise ratios than those obtained on planar macroscopic diamond electrodes. In addition, they have exhibited the potential to conduct spatially and temporally resolved electrochemical measurements when they are adopted as the tip for scanning electrochemical microscopy (SECM)[36,41] and as the sensor for *in vivo* detection of dopamine by fast scan voltammetry.[33]

However, a single microelectrode or ultramicroelectrode only generates a small current that is relatively difficult to detect with conventional electrochemical setups. This has been circumvented by fabricating arrays or ensembles of these electrodes that operate in parallel. By applying an array or ensemble, the signals are amplified but the characteristics of the individual microelectrodes or nanoelectrodes are not lost. Regarding the fabrication, characterization, and applications of diamond-based microelectrode arrays (MEAs), ultramicroelectrode arrays (UMEAs), and nanoelectrode arrays/ensembles (NEAs/NEEs), several works have been conducted.[42–56]

In this chapter we first summarize the technologies developed for the fabrication of all-diamond devices, including MEAs, UMEAs, and NEAs/NEEs. In the second part the properties of these devices are shown and characterized with various techniques. A diamond-based atomic force microscope–scanning electrochemical microscope (AFM-SECM) tip, as a special all-diamond device, will also be demonstrated. The applications of these devices are shown in the third part. Some devices fabricated in our laboratory are shown in detail. The last part is a summary and outlook.

19.2 Fabrication

19.2.1 Microelectrode Arrays and Ultramicroelectrode Arrays

In 2002 Fujishima and coworkers used structured silicon substrates to fabricate diamond microelectrode arrays (MEAs).[42] The array consisted of 200 micro-disks with diameters between 25 and 30 µm and electrode spacings of 250 µm. Rychen and coworkers[43] produced diamond UMEAs by depositing a boron-doped film onto patterned silicon nitride (5 µm in thickness). The diameter of the microelectrodes was 5 µm, the distance between microelectrodes 150 µm, and the number of electrodes was 106. Kang, Swain and coworkers[44–48] realized diamond UMEAs with different shapes, spacing, and numbers of electrodes. They utilized the "as-grown" diamond surface with randomly micro-structured topology as a planar diamond electrode. They also used a micro-patterning technique to produce a well-defined pyramidal tips array with controlled uniformity. Compton and coworkers[49–51] realized for the first time all-diamond UMEAs in 2005. The diameters of the electrodes were between 10 and 25 µm with a separation of 100 to 250 µm. Bergonzo et al.[52] and Carabelli et al.[53] utilized nanocrystalline diamond films to generate electrode arrays.

We recently demonstrated the batch production of integrated UMEAs using polycrystalline diamond films.[55,56] Insulting polycrystalline diamond films were grown on a two inch silicon wafer in an ellipsoidal-shaped microwave plasma enhanced chemical vapor deposition system (CVD).[56] The temperature was in the range of 750–900 °C, the microwave power was 3 kW, the gas mixture was 3% methane in hydrogen, and the pressure was 60 mbar. A boron-doped polycrystalline diamond film was grown by adding 7000 ppm trimethylboron (TMB) to the gas phase. Prior to fabrication of electrodes, diamond films were cleaned wet-chemically in a mixture of concentrated sulfuric acid (98%) and concentrated nitric acid (65%) ($v:v = 3:1$) at 200 °C for 1.5 h.

Figure 19.1 shows schematically the fabrication process of integrated UMEAs.[55,56] It includes three photolithography steps, two etching steps, one

All-diamond Electrochemical Devices: Fabrication, Properties, and Applications   479

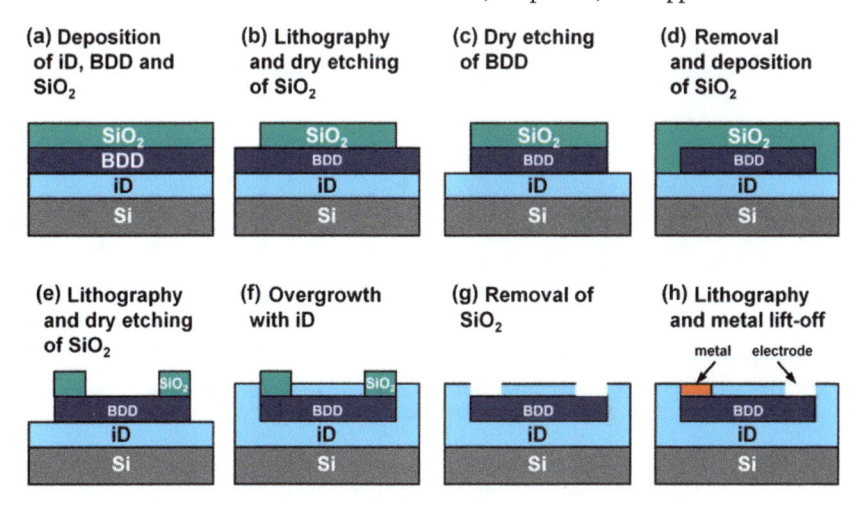

Figure 19.1 Schematic plots of the fabrication of integrated all-diamond ultramicroelectrode arrays. iD: insulating diamond, PR: photo-resist, BDD: boron-doped diamond, and RIE: reactive ion etching.
(Reprinted from ref. 55.)

over-growth step, one metal-deposition step, and one lift-off step. Figure 19.1a) shows the first step where a boron-doped polycrystalline diamond (B-PCD) film (with a thickness 200–500 nm) is deposited on a polished insulating diamond (iD) film (with a thickness of 8–10 μm). The second step (Figure 19.1b) is the application of the first photolithography treatment. After spin-coating of the wafer with photo-resist (PR), a 350 nm SiO_2 layer is deposited. SiO_2-based patterns are generated by etching with SF_6 gas (Figure 19.1c). To produce boron-doped diamond-based structures, reactive ion etching (RIE) of the wafer (Figure 19.1d) is conducted in a gas mixture of oxygen and hydrogen.[58] Please note that these structures are protected. The conductivity of unprotected areas is checked frequently to make sure boron-doped diamond (BDD) is etched away. For the overgrowth of insulating diamond, the second photolithography process is applied (Figure 19.1e). The overgrowth (Figure 19.1f) coats all areas with insulating diamond except the parts of the counter, reference, and ultramicroelectrodes, which are protected by SiO_2. The third photolithography step is finally applied (Figure 19.1g) and the Ti/Pt/Au (20/60/200 nm) metal layers are deposited for electrical contacts. The last step is the application of the lift-off technique as shown in Figure 19.1h).

Figure 19.2 shows the fabricated diamond UMEA chip.[56] It has a size of 5 × 5 mm^2 (Figure 19.2a). The counter and quasi-reference electrodes are oxidized boron-doped diamond films, which are integrated on the chip (Figure 19.2b). The diameter of the ultramicroelectrodes is 10 μm. The vertical and horizontal centre-to-centre spacing in between electrodes is 60 μm (Figure 19.2c). The total number of ultramicroelectrodes on one chip is 45. For one 3-inch wafer, more than 40 integrated diamond UMEA chips are produced.

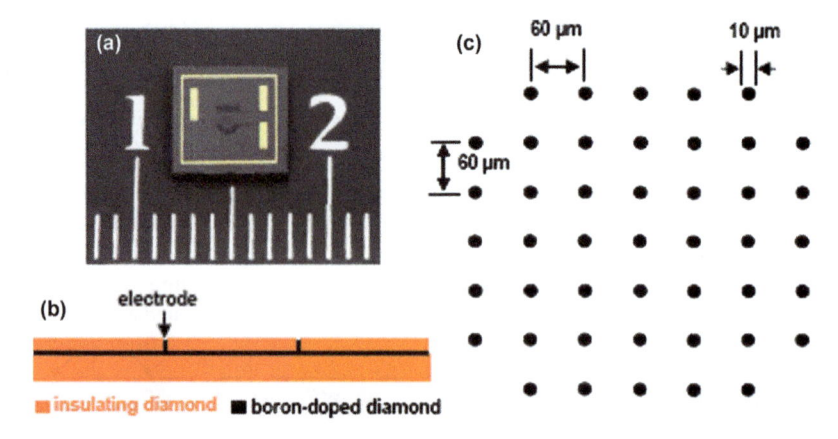

Figure 19.2 (a) Image of one integrated all-diamond ultramicroelectrode array. The yellow parts are metal contacts. The dark part with a semi-circle is the counter electrode, the dark rectangle is the reference electrode and the center electrode is the working electrode. (b) Schematic plot of the structure of all-diamond electrode. (c) The arrangement of the microelectrode in a 500 μm (diameter) circle. (Reprinted from ref. 55.)

19.2.2 Nanoelectrode Ensembles and Arrays

In the literature, e-beam lithography,[59] focused ion beam milling[60,61] nanoimprint,[59] and nanosphere lithography[62] have been applied to fabricate nanoelectrodes and their arrays (NEAs). For nanoelectrode ensembles (NEEs), different approaches have been developed, such as the deposition of metals into pores of polycarbonate nanoporous membranes,[63–65] nanosphere lithography,[66] or block copolymer self-assembly.[67–69] In addition, spatially separated carbon nanofibers and diamond nanograss, as well as porous diamond film, have been utilized as NEEs for electrochemical applications.[70–79]

For the first time, we fabricated all-diamond NEAs and NEEs with nanocrystalline diamond (NCD) films.[57] The use of NCD films has some advantages over polycrystalline diamond films. For example, the grain sizes of NCD films vary from only a few tens of nanometers up to hundreds of nanometers (*e.g.*, 300 nm), resulting in better electrochemical activity than that obtained on polycrystalline diamond films, which have grain sizes in the range of micrometers.[80,81] NCD films were grown on 3-inch silicon substrates in an ellipsoid reactor using microwave-assisted chemical vapor deposition.[56] Seeding was conducted by immersing silicon wafers in a nanodiamond suspension with an average particle size of 5 nm.[82] The densities of diamond seeds was more than 10^{11} cm^{-2}. The growth of insulating NCD was performed using H_2/CH_4 plasma with a methane admixture of 1 or 2%. Boron doping of NCD films was achieved by adding trimethylboron (TMB) to the gas phase with B/C ratios of 6000 ppm. The boron concentrations of doped NCD films were measured by secondary ion mass spectroscopy and found to be in the range of 1×10^{21} to 4×10^{21} cm^{-3}.[57]

Figure 19.3 (left) shows schematically the approach we used to fabricate NEAs.[57] E-beam lithography was applied. On a 200 nm thick boron-doped NCD film, a 200 nm thick layer of SiO$_2$ is deposited. This oxide layer is structured using e-beam lithography with subsequent nickel deposition and SF$_6$ etching of SiO$_2$. In the next step metal contacts are deposited using photolithography to allow electrical contact for electrochemical characterization. In the crucial step a 140 nm thin, insulating NCD film is grown on the part of the boron-doped NCD layer that is exposed to the CVD plasma and not protected by SiO$_2$ islands. With the removal of SiO$_2$ in hydrofluoric acid, arrays of recessed boron-doped NCD electrodes surrounded by insulating diamond are obtained.

Figure 19.3 (right) shows schematically the method we developed to fabricate NEEs.[57] Nanosphere lithography was utilized. Initially, a photolithography step is used to deposit metal contacts. Thereafter, samples are immersed in a solution of SiO$_2$ spheres with a radius of 500 nm. In an ultrasonic bath an equilibrium between spheres sticking to and leaving the sample surface occurs. Thus, the concentration of the solution is directly correlated to the density of spheres on the sample surface, as well as to the average distance of neighboring spheres. The density of electrodes obtained from a SiO$_2$ sphere solution of a given concentration is derived from a large area scan of 50 μm × 50 μm. To obtain sigmoidal voltammograms, we chose a concentration of 9.55×10^8 cm^{-3}, corresponding to a surface density of 9.7×10^5 cm^{-2} and an average distance from neighboring spheres of ∼10 μm. The next step involves the growth of insulating diamond around the abovementioned spheres. Insulating diamond selectively grows on the area exposed to the plasma. After removal of the SiO$_2$ spheres in hydrofluoric

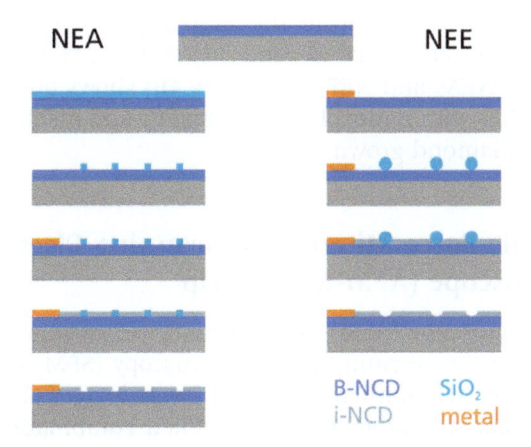

Figure 19.3 Process steps for the fabrication of all-diamond NEA (left) and NEE (right).
(Reprinted from ref. 57.)

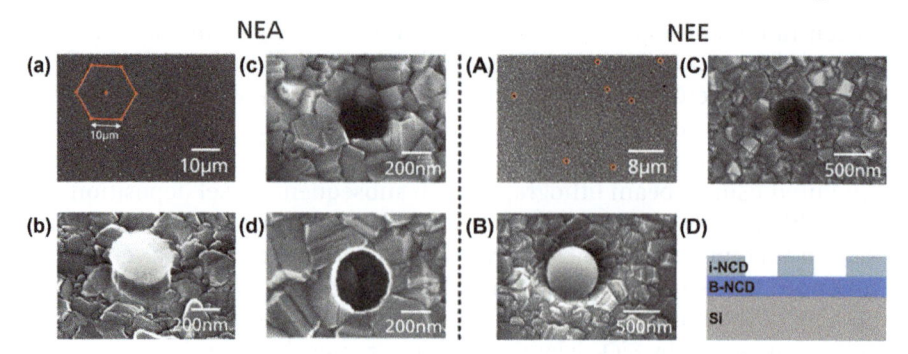

Figure 19.4 SEM images of a NEA and a NEE at different stages of the fabrication process: (a) overview of the design with distances of 10 μm of neighboring electrodes with hexagonal order (indicated in red), (b) structured SiO$_2$/Ni islands on boron-doped NCD layer, (c) insulating diamond grown around SiO$_2$, and (d) the final recessed diamond electrode. (A) Overview of statistically distributed electrodes on boron-doped NCD substrate indicated by red circles, (B) SiO$_2$ sphere after growth of insulating diamond, (C) final boron-doped NCD electrode after removal of SiO$_2$, and (D) schematics of the cross section of fabricated electrodes.
(Reprinted from ref. 57.)

acid, we obtain electrodes of a concave shape. Electrode radii of 175 nm are achievable with spheres of 500 nm radius.

Figure 19.4 shows several SEM images of fabricated NEAs and NEEs. Figure 19.4D) sketches the cross section of the fabricated NEAs and NEEs, composed of a structured insulating NCD layer on top of a conductive boron-doped NCD film. SEM images show a hexagonal order for NEAs nanoelectrodes (red lines in Figure 19.4a), while, for NEEs, nanoelectrodes are distributed randomly (red dots in Figure 19.4A). Figure 19.4b) and c) illustrates two different steps during the fabrication of a nanoelectrode by e-beam lithography. Figure 19.4d) and 4C) show SEM images of a single nanoelectrode in NEAs and NEEs. Figure 19.4B) shows one step during the fabrication of NEEs with nanosphere lithography and displays a SiO$_2$ sphere with insulating diamond grown around it.

19.2.3 Atomic Force Microscope–Scanning Electrochemical Microscope (AFM–SECM) Tip

Atomic force microscopy (AFM) and scanning electrochemical microscopy (SECM) belong to the scanning probe microscopy (SPM) family of techniques, which is a branch of microscopy that forms images of surfaces using a physical probe that scans the specimen.[25] Their combination (AFM–SECM) provides the possibility to obtain topographical and electrochemical properties of a specimen simultaneously.[83–85] For these experiments, AFM–SECM cantilevers/probes have electroactive areas at the apex of the AFM tip[86–90] or with an electrode recessed from the tip apex.[83,91,92] This electrode area or recessed electrode is as small as an ultramicroelectrode and can be the tip for SECM measurements.

Several fabrication schemes have been proposed in the literature for the fabrication of such AFM–SECM probes, such as the mounting[93,94] or growth[95,96] of sharp objects, such as carbon nanotubes on commercial AFM tips. However, most manufactured tips are either sharp but not wear-resistant, or mechanically strong but blunt. To overcome the wear resistance of AFM cantilevers, diamond has been used to coat different AFM tips[98–104] because diamond is exceptionally hard, stable, and wear resistant owing to its high Young's modulus of 1220 GPa in comparison to much lower values for silicon (185 GPa) or silicon nitride (250 GPa).[97] Most commercially available diamond AFM probes are fabricated by overgrowth of silicon tips using plasma-enhanced chemical vapor deposition (CVD),[98] hot filament-assisted CVD,[99] or by growth of diamond into a silicon mold.[98,100–104] The wear resistance of ultra-nanocrystalline diamond (UNCD)-coated AFM tips is at least one order of magnitude greater than that of silicon nitride AFM tips. However, tips coated with very small diamond crystals may suffer from abrupt fracture and detachment of clusters.[104–106] Large diamond crystals and higher surface roughness decrease the image resolution once diamond-coated AFM cantilevers are applied.[107]

Recently, we demonstrated a method allowing the formation of sharp diamond nanowhiskers at the apex of a diamond-coated AFM probe, serving as ultra-sharp diamond AFM tip.[108] Boron-doped diamond electrodes are integrated into AFM probes for combined AFM–SECM measurements.[108] Our approach is based on CVD diamond coating of silicon cantilevers and diamond nanowire technology.[58] Figure 19.5A) shows a schematic of the fabrication steps of sharpened diamond AFM cantilevers. The commercial silicon tips are first seeded with nanodiamond particles[82,109,110] and then coated with a NCD film.[56,57] After overgrowth, the tip becomes rough and its radii is enlarged. Figure 19.5B) reveals one example where a commercial silicon tip is coated with a 350 nm thick diamond film. The tip radius is enlarged to about 450 nm. In the second stage the tip is sharpened by use of self-organized metal nano-masks in a plasma etching process. To generate these masks, a 20 nm thin gold layer is sputtered on the diamond-coated silicon tips followed by annealing at 1000 °C. The annealing is performed on an inductively heated graphite plate in a vacuum for 10 min in a home-build induction furnace. Figure 19.5C) shows the SEM image of gold nano-masks, where small and bright dots are seen with the size of 40 ± 10 nm. Etching is finally carried out using a Sentech ICP SI500 at 5×10^{-2} mbar, 1000 W and 50 sccm oxygen gas flow for 30 s. During this process the Au masks are continuously removed leaving conically formed diamond nanowires. Figure 19.5D) shows one typical plasma-etched (sharpened) diamond-coated silicon tip. Nanowires appear on the tip of the cantilever. One nanowire at the tip apex with a radius of 5 nm is magnified and shown in Figure 19.5E).

AFM–SECM tips are obtained by coating commercial silicon cantilevers with a heavily boron-doped diamond layer. In a consecutive step the coated

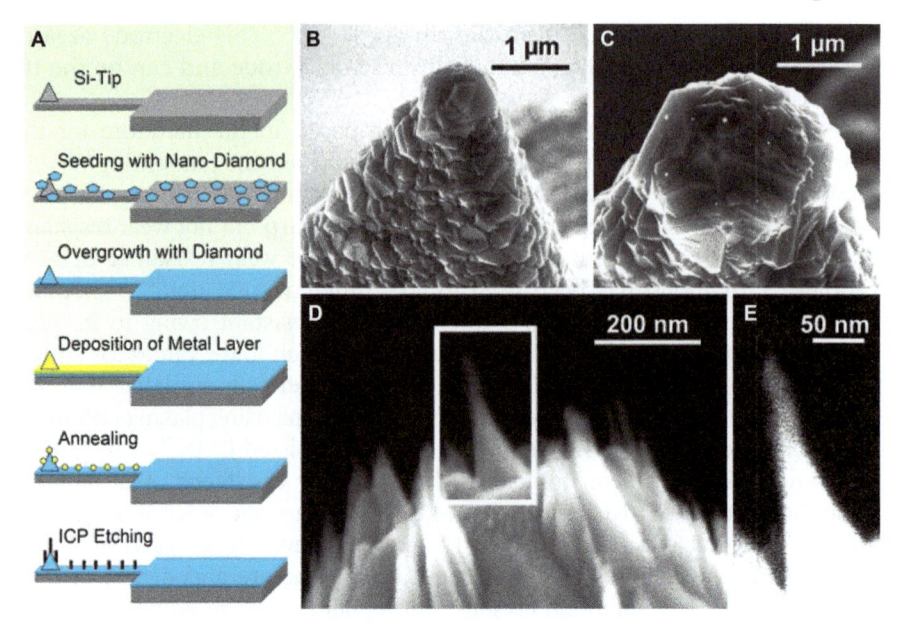

Figure 19.5 A) A schematic of the fabrication process: from diamond deposition to sharpening of the tip. B) SEM image of a diamond-coated silicon AFM tip. C) SEM image of self-organized gold nano-mask at the tip apex. D) SEM image of etched diamond nanowires at the tip apex. E) Magnified SEM image of one single nanowire.
(Reprinted from ref. 108.)

probes are insulated with intrinsic diamond. Please note that a small part of the AFM chip is covered during the growth of insulating diamond on the boron-doped diamond, leaving the possibility for an electric contact to the tip. Figure 19.6A) shows a schematic of a hybrid silicon/diamond cantilever with multiple diamond layers. A SEM image is shown in Figure 19.6B). Exposing the electroactive boron-doped diamond electrode was achieved by focused ion beam (FIB) milling.[83] This results in a sharp insulating diamond AFM tip (Figure 19.6C) and a boron-doped diamond electrode recessed from the tip apex (Figure 19.6D). The height of the diamond tip and the ring diameter of the exposed ring electrode depend on the deposited film thickness of the boron-doped diamond and the FIB milling depth. The typical height of the diamond tip is about 1 to 2 µm and the electroactive area of the boron-doped diamond ring is about 0.5 to 1 µm^2.

19.3 Properties

19.3.1 Microelectrode Arrays and Ultramicroelectrode Arrays

The properties of these devices have been investigated using different techniques, such as electron and optical microscopy, Raman spectroscopy,

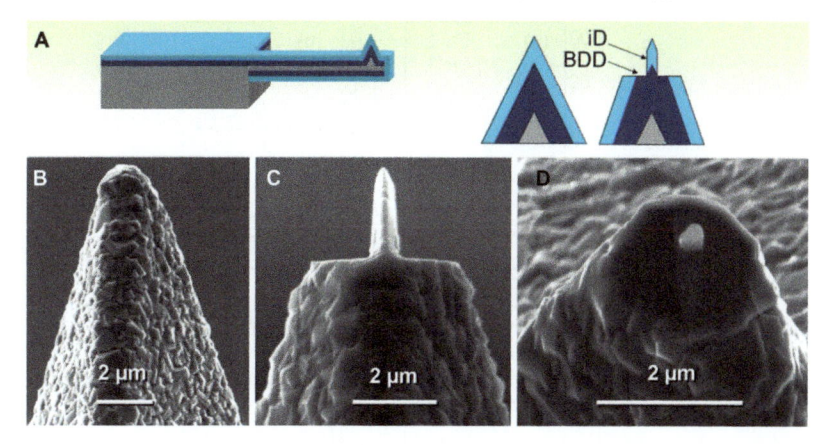

Figure 19.6 A) Schematic cross sections of the combined AFM–BDD–SECM probe with conducting and isolating diamond layers. B) SEM image of Si AFM tip coated with conducting and insulating diamond. C) and D) SEM images of the same tip after FIB milling, exposing an electrode (images shown from different angles).
(Reprinted from ref. 108.)

and electrochemical techniques. We applied voltammetry to investigate the effect of surface termination of diamond electrodes and boron doping levels on the Faradaic currents on diamond UMEAs.[55,56] One redox couple of Fe(CN)$_6^{3-/4-}$ was adopted as the probe. Hydrogen-terminated diamond UMEAs with a boron concentration of 4.2($\pm$2) × 10^{20} cm^{-3} shows the highest Faradaic current, indicating the fastest electron transfer process. As expected for UMEAs, the variation of the supporting electrolyte does not change the capacitive current much but alters the Faradaic current dramatically.[25–30]

Figure 19.7 shows the effect of the scan rate on Faradaic currents (solid lines) for hydrogen-terminated diamond UMEA in 1.0 mM Fe(CN)$_6^{3-/4-}$/KCl (0.1 M) solution. The magnitude of the Faradaic currents increases and the shape of the voltammograms vary with increasing scan rate. At low scan rates ranging from 0.02 (a) to 0.2 (b) V s^{-1} and at fast scan rates of 20 V s^{-1} (d), peak-shaped voltammograms were obtained, indicating the linear diffusion-limited transport of analytes. A sigmoidal-shaped voltammogram (c) was detected at a scan rate of 2 V s^{-1}, which is consistent with hemispherical diffusion to the ultramicroelectrodes on the array. To calculate the diffusion layer thickness δ, we applied the equation[25] $\delta = (2D\Delta E/v)^{1/2}$ (where v is the scan rate, $D = 7.6 \times 10^{-6}$ cm s^{-1} is the diffusion coefficient of analytes, ΔE is the potential range over which electrolysis occurs). For example, to estimate the size of the diffusion layer thickness at $E = 0.4$ V, we used the value $\Delta E = 0.8$ V, since significant electrolysis current started at $E = -0.4$ V. The calculated values of δ at $E = 0.4$ V for the scan rates of 0.02, 0.2, 2, and 20 V s^{-1} were 250, 78, 25, and 7.8 μm, respectively. The center-to-center separation and the diameter of the microelectrodes are 60 μm and 10 μm,

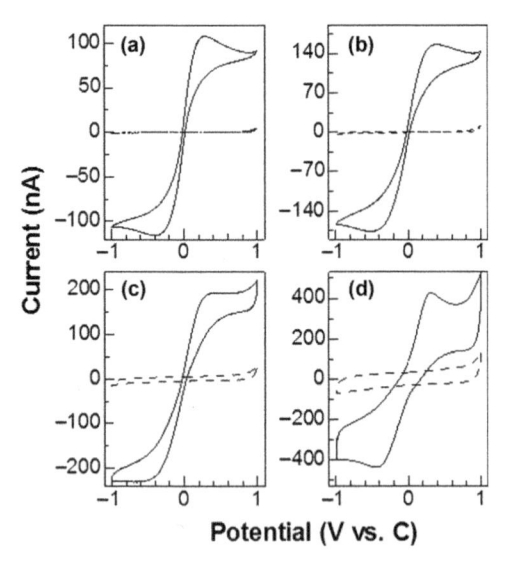

Figure 19.7 Scan rate dependency: cyclic voltammograms in 0.1 M KCl solution with (solid lines) and without (dashed lines) 1.0 mM Fe(CN)$_6^{3-/4-}$ at a scan rate of (a) 0.02, (b) 0.2, (c) 2, and (d) 20 V s^{-1}. (Reprinted from ref. 55.)

respectively. Namely, the separation between electrodes is 50 µm. At a scan rate of 0.02 V s^{-1}, $\delta = 250$ µm, which is much larger than the spacing between electrodes, indicating a complete overlap of redox molecule diffusion of individual electrodes and, subsequently, a linear diffusion profile. At a higher scan rate of 0.2 V s^{-1}, $\delta = 78$ µm, which is only slightly greater than the separation of electrodes, an overlap of adjacent diffusion profiles is still dominating. When δ becomes larger than the diameter of a microelectrode but is still smaller than the separation between electrodes, the voltammetric response is the response of an individual microelectrode (sigmoidal curve) multiplied by the total number of electrodes in the array. This can be detected, for example, at a scan rate of 2 V s^{-1}. However, further increases of the scan rate (*e.g.*, to 20 V s^{-1}) leads to even smaller values of δ (*e.g.*, to 7.8 µm) than the size of microelectrodes (10 µm). In this case the linear diffusion dominates the mass transport, resulting in peak-shaped voltammograms.

On the other hand, the capacitive currents (dashed lines in Figure 19.7) recorded in 0.1 M KCl increase linearly with scan rate. The S/B ratios (Faradaic current to capacitive current) for scan rates of 0.02, 0.2, 2, and 20 V s^{-1} were estimated to be 1817 ± 40, 215 ± 18, 28 ± 6, and 10 ± 1, respectively. The highest ratio was achieved at the slowest scan rate. In this case the diffusion profiles at neighboring microelectrodes overlapped and, thus, peak-shaped voltammograms were detected. The magnitude of the Faradaic current is, thus, proportional to the geometric area, which is comprised of all microelectrodes and the insulating parts. The contribution of the capacitive current to the total current is small since the capacitive

All-diamond Electrochemical Devices: Fabrication, Properties, and Applications 487

current is proportional to the scan rate and to the electrochemical active area, which is only the area of all the microelectrodes. This gives rise to the enhanced ratios of S/B, leading to increased sensitivity for analytes. Since the geometric area of an UMEA chip is always 50–1000 times larger than the electrochemical active area, a 50–1000 times better sensitivity is expected for such electrode arrays.[55,56]

19.3.2 Nanoelectrode Ensembles and Arrays

Our NEEs were characterized using conductive atomic force microscopy (C-AFM).[57] The density determined by AFM was 8.5×10^5 cm^{-2}, which is in good agreement with the expected values from the sphere concentration in the solution. The size of electrodes was 175 nm in radius. The cross sections of the topography for two neighboring electrodes clarify that these conductive nanoelectrodes are surrounded by 140 nm thick insulating diamond.

Our NEA and NEE were characterized with voltammetry in 1.0 mM Fe(CN)$_6^{3-/4-}$ in 0.1 M KCl. The electrode in the NEA had a radius of 250 nm and a distance of 10 μm from adjacent electrodes. The electrode in the NEE had a radius of 175 nm and the density of electrodes is 8.5×10^5 cm^{-2}. The scan rate was varied from a few mV s^{-1} up to 10 V s^{-1}. At small scan rates (e.g., 20 mV s^{-1} for the NEA and 1 mV s^{-1} for the NEE), the voltammograms have mixed shapes, indicating partially overlapping diffusion hemispheres. Increasing the scan rate leads to typical steady-state sigmoidal voltammograms on both electrodes, indicating spherical diffusion.[111–115]

In addition to cyclic voltammetry, electrochemical impedance spectroscopy (EIS) was performed in order to investigate the characteristic properties of the fabricated NEE and NEA. As shown in Figure 19.8, the impedance spectra for the NEA (a) and the NEE (b) show similar characteristic features. Both graphs display a large semicircle in the high-frequency regime. At low frequencies, a transition to linear diffusion with unity slope occurs, particularly observable for the NEA in Figure 19.8(a). A semicircle at

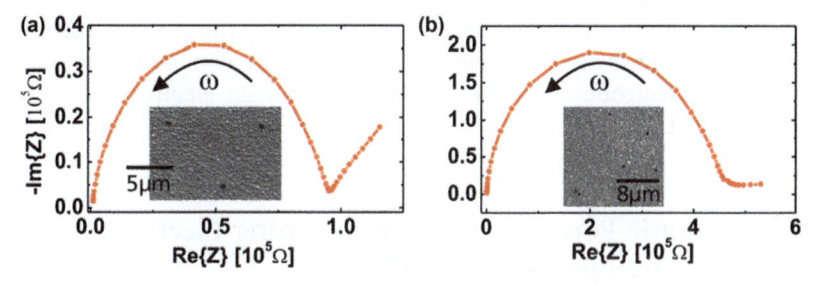

Figure 19.8 Impedance spectra at an open circuit potential of 1 mM Fe(CN)$_6^{3-/4-}$ in 0.1 M KCl for (a) the nanoelectrode array with 250 nm electrode radius and (b) the nanoelectrode ensemble with radii of 175 nm, with insets of the distribution and dimensions of the array and the ensemble. (Reprinted from ref. 57.)

high frequency regime is due to a three-dimensional hemispherical diffusion on diamond NEA and NEE.[85,111–113] The transition at low frequency represents the regime of overlapping diffusion hemispheres. The change from typical three-dimensional diffusion to overlapping diffusion hemispheres is very distinct for the NEA (Figure 19.8a). Thus, impedance spectroscopy was further utilized to determine average distances between adjacent electrodes, since the impedance semicircle decreases with the number of electrodes and inversely scales with the square of the radius of a single electrode.[111,113] For all arrays, the distance between electrodes is 10 µm. The density of electrodes for one NEA is 11×10^5 cm^{-2} and the number of electrodes is 18 000.

Moreover, the voltammetric response of Fe(CN)$_6^{3-/4-}$, Ru(NH$_3$)$_6^{2+/3+}$, and IrCl$_6^{2-/3-}$ show the dependence of surface termination on the charge of the analytes.[57] Please note that, on planar macroscopic diamond electrodes, both analytes show no dependence of electron transfer rate constant on the surface termination of diamond electrodes.[57] On the diamond NEA, the voltammogram of the anion IrCl$_6^{2-/3-}$ shows a fast electron transfer on the hydrogen-terminated surface, whilst at the oxygen-terminated surface, the steady-state current, as well as the slope of the transition from reduction to oxidation, decreases, which is indicative of a slower electron transfer. This tendency is similar to that seen on another negatively charged redox couple of Fe(CN)$_6^{3-/4-}$. For the positively charged redox molecule, Ru(NH$_3$)$_6^{2+/3+}$, the opposite effect is observed. On an oxygen-terminated diamond surface, the electron transfer rate for Ru(NH$_3$)$_6^{2+/3+}$ is faster than that on a hydrogen-terminated surface. This effect for IrCl$_6^{2-/3-}$ and Ru(NH$_3$)$_6^{2+/3+}$ is small compared to that for Fe(CN)$_6^{3-/4-}$. It is known that the hydrogen-terminated diamond surface has a positive surface dipole layer ("positive" refers to the interface of diamond to the liquid) and the oxygen-terminated surface results in a negative surface dipole layer. A macroscopic diamond electrode shows a higher degree of inhomogeneity with respect to the boron-doping level and termination effects, due to its macroscopic dimensions. For the investigated NEA, one would expect a homogenized behavior due to the small grains of the NCD films, as well as a more effective termination of the small electrochemically active area. Therefore, the possible effects responsible for the decrease of electron transfer rate constant upon oxygenation of the electrode surface of NEAs are either an electrostatic or a site blocking effect.

19.3.3 Atomic Force Microscope–Scanning Electrochemical Microscope (AFM–SECM) Tip

The resonance properties of our overgrown cantilevers were analyzed using a vibrometer and a standard AFM setup with a piezo-actuator and a laser.[108] The resonance frequencies of these cantilevers are also simulated. The simulation was performed using the Ansys™ software for a silicon cantilever with the dimensions of 225 μm (length) ×35 μm (width) ×7 μm (height). The thickness of the diamond coating was simulated in the range of 0 μm to 1 μm. The Young's moduli of silicon and NCD grown at 3% of methane are 130 GPa and 950 GPa, respectively.[116] The simulation shows that the resonance frequencies of overgrown cantilevers shift to higher values, depending on the thickness of the NCD films. When the thickness of the NCD film is zero, the simulation gives the same resonance frequency as for bare silicon cantilevers. The resonant frequencies obtained from experiments agree with the simulated data. This suggests that a defined resonant frequency can be adjusted by choosing a defined diamond coating thickness. Moreover, peak-shape analysis of the resonant frequencies before and after overgrowth shows an enhancement in the quality factor, which is defined by $Q = f/w$, where f denotes the resonant frequency and w the full-width at half-maximum (FWHM) of the resonance peak. For example, after coating a silicon cantilever with a 850 nm diamond film, the resonant frequency increases from 130 kHz to 370 kHz and the quality factor also increases from 120 to 710. The increased ratios for the resonance frequency and quality factor are 185% and 490%, respectively. The enhancement in Q-factor is higher than expected. This is attributed to the uniform diamond coating, which is hermetically wrapped around the silicon cantilever and results in a strong and rigid structure. It thus confirms that the interface between silicon and diamond is of high quality since no dissipative effects are occurring.

The overgrown and sharpened nanowire tips were tested with respect to their imaging quality in tapping and contact modes. The achieved resolutions, as well as the wear properties, were investigated and compared to commercially available silicon (Figure 19.9A) and commercially available diamond AFM probes (Figure 19.9B). As for resolution measurements, single-stranded λ-phage DNA molecules on mica were imaged in a tapping mode. A sharp silicon tip (Figure 19.9A1, A2) and our sharpened AFM tip (Figure 19.9C1, C2) generate high quality images with high resolution. By using a commercially available diamond-coated AFM tip (Figure 19.9B), DNA structures cannot be resolved. Analyzing cross sections of the DNA images (Figure 19.9A2, C2) yields the tip radius. The real width of the DNA is assumed to be of the order of 1 nm. The measured width of the DNA, however, is a convolution of the scanning tip geometry and the actual DNA width (Figure 19.9D). Taking this into consideration, similar tip radii of 7 to 9 nm for both the standard silicon tip and our diamond nanowire probes were obtained.

Measurements regarding the wear characteristics were carried out by imaging an ultra-hard and rough substrate (in this case a NCD surface) in contact mode. For direct comparison, the scanned area was selected to be the same spot of the imaged substrate for different probes. The resolution of the standard silicon tip (Figure 19.9A3) is comparable to our sharpened tip (Figure 19.9C3), while the commercial diamond-coated tip (Figure 19.9B3) was inferior in performance. The imaging quality obtained with the silicon tip deteriorated within the first recorded scan (Figure 19.9A4) due to abrasion of the tip. The commercial diamond tip suffered from fracture during

490 Chapter 19

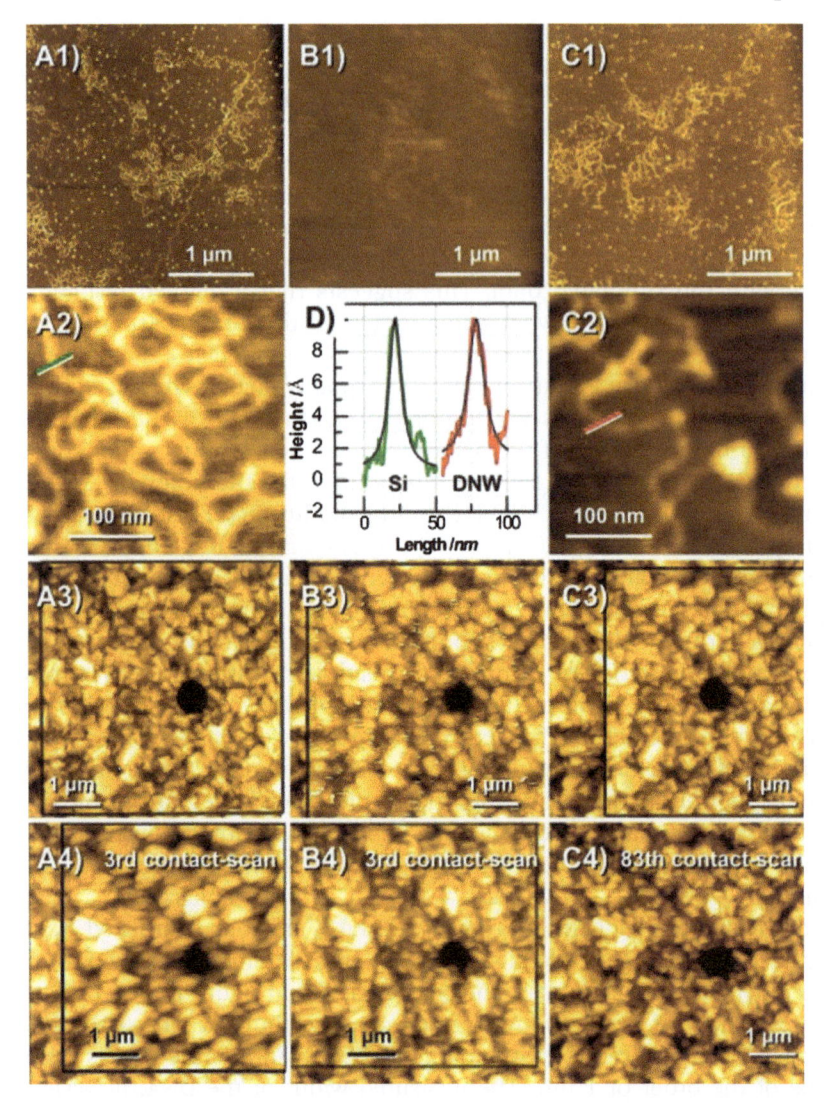

Figure 19.9 AFM images showing wear and resolution characteristics of a standard silicon probe (A), of a commercial diamond tip (B) and of our sharpened diamond tip (C). Shown are tapping mode scans of λ-phage DNA on mica (2) and a higher magnification image (3). Tapping mode and contact mode AFM scans on the same spot of a nanocrystalline diamond sample are shown in (3) and (4), respectively. D) Line scan and Lorentz fits of the Si tip and diamond nanowire tip (DNW) are denoted in A2 and C2 as green and red lines, respectively. Images were recorded in air. (Reprinted from ref. 108.)

the first three scans in contact mode (Figure 19.9B4). Our sharpened tip sustained for more than 80 scans with only slight degradation in imaging quality (Figure 19.9C4). The imaging experiment using our nanowire tip was stopped after 83 scans and a total scanned length of 320 000 µm, which were recorded within 18 h in contact mode operation.

A proof of principle for measurements of the diamond AFM–SECM tip was conducted in our laboratory.[108] In this test, a model sample was used. It was a boron-doped NCD film that was partially overgrown by insulating diamond. The AFM–SECM tip remains at a fixed and constant distance from the imaged surface during the measurement. The distance is given by the height of the sharp diamond tip. AFM images were measured in a contact mode. SECM images were recorded with a feedback mode in a $Fe(CN)_6^{3-}$ solution. The potential applied at the AFM–SECM tip was 0.15 V (vs. Ag/AgCl) and the potential at the substrate (model sample) was 0.35 V (vs. Ag/AgCl). The images show clearly the topography and the faradaic current on the insulating and conductive areas, indicative of the successful fabrication of an all-diamond AFM–SECM tip.

19.4 Applications

These all-diamond devices have been applied in many different sensing applications. Diamond MEAs, UMEAs, and NEEs/NEAs have been used for the detection of environmental analytes (e.g., nitrate, 4-nitrophenol,[117–119] Cr(VI)ions, Ag(I)ions,[120] sulfate, peroxodisulfate,[43] hydrogen peroxide),[55] for bio-detections (e.g., detection of dopamine,[44–48] neuronal activity measurements,[121,122] quantal catecholamine secretion from chromaffine cells[123]), and for the generation and detection of peroxidisulfate with the aid of scanning electrochemical microscopy (SECM).[124] We have shown that dopamine can be sensitively and selectively detected on diamond UMEAs in the presence of ascorbic acid.[55] Compared with the results shown on other diamond electrodes (including macro-sized electrodes,[125–130] MEA,[131] UMEA,[47,48] and diamond nanograss[132]), on diamond UMEAs we achieved the lowest detection limit (1.0 nM) for dopamine detection, which is 50–100 times lower than previously reported.[55] Diamond UMEAs are thus promising for the detection of low dopamine concentrations (0.01–1 µM) in biological samples individually or in the presence of other similar compounds, such as ascorbic acid.

Besides these electrochemical and biochemical sensing applications, all-diamond NEAs and NEEs are promising for the investigation of the adsorption behavior of species on surfaces, to study fast electron charge-transfer processes on different surface terminations, and to realize sensitive, fast detections. As for the applications of all-diamond AFM–SECM tips, they can be used to image/detect biomolecules, like single-strand DNA and proteins, at high resolution and with a long lifetime, and to study the reactivity of technical surfaces with features in the nanometer range in complex media, etc.

19.5 Summary and Outlook

Batch production of all-diamond devices is possible using different technologies, such as photolithography, e-beam lithography, nanosphere lithographay, and etching processes. In particular, the NEEs fabricated by nanosphere lithography offer a cheap and simple alternative to e-beam lithography to obtain electrodes of a few hundreds of nanometers or less. These small-dimensional electrode arrays show advantages over planar macro-sized diamond electrodes. By optimization of Faradaic and capacitive currents in diamond UMEAs, a 50–1000 times better sensitivity than planar macro-sized electrodes is achieved; In combination with the efficiency and suitability of the selective electrochemical surface termination of NEAs or NEEs, it offers a new versatile system for electrochemical sensing with high selectivity. Sharpened diamond AFM tips show better image resolution and have long lifetimes with respect to wear resistance. All-diamond AFM–SECM tips have been demonstrated as powerful and efficient tools for recording topography and electrochemical signals of samples.

In future work more attention will be paid to *in vivo* applications of all-diamond devices (*e.g.*, fast and sensitive detection of biomolecules released from cells, *in situ* monitoring of cell/bacterial growth, *etc.*), surface cleaning/refreshing of biomolecule-functionalized/adsorbed diamond surfaces, the effects of surface chemistry (terminations and linker chemistry), and surface structures (nanotextures, nanowires, and porosity) on their sensing performance. Besides these, future work needs to concentrate on simplifying fabrication processes and marketing them at reasonable prices, as well as developing their industrial applications.

Acknowledgements

This work was supported by the Baden-Württemberg Stiftung under the contract research "Einzelmolekulare Strukturanalyse und zelluläre Transportmechanismen an Poren-formenden Peptiden" (P-LS-SPII/23). The authors thank Dr Christine Kranz from University of Ulm, Germany for FIB experiments, Rene Hoffmann, Armin Kriele, Harald Obloh, Wolfgang Müller-Sebert, Oliver A. Williams, Karlheinz Glorer, Brian Raynor, Rachid Driad, and Christoph E. Nebel from the Fraunhofer IAF for technical assistance and fruitful discussions.

References

1. M. Iwaki, S. Sato, K. Takahashi and H. Sakairi, *Nucl. Instrum. Methods in Phys. Res.*, 1983, **209-210**, 1129.
2. Y. V. Pleskov, A. Y. Sakharova, M. D. Krotova, L. L. Bouilov and B. V. Spitsyn, *J. Electroanal. Chem.*, 1987, **228**, 19.
3. R. Tenne and C. Levy-Clement, *Israel J. Chem*, 1998, **38**, 57.
4. G. M. Swain, A. B. Andreson and J. C. Angus, *MRS Bulletin*, 1998, **23**, 56.
5. Y. V. Pleskov, *Russ. Chem. Rev.*, 1999, **68**, 381.
6. C. E. Nebel and J. Ristein, *Thin Film Diamond II: Semiconductors and Semimetals*, vol. 77, Elsevier Academic Press, Amsterdam, 2004.

7. A. Fujishima and Y. Einaga, T. N. Rao and D. A. Tryk, *Diamond Electrochemistry*, Elsevier Academic Press, Tokyo, 2005.
8. E. Brillas, C. A. Martinez-Huitle, *Synthetic Diamond Films: Preparation, Electrochemistry, Characterization, and Applications*, Wiley, New Jersey, 2011.
9. R. L. McCreery, *Chem. Rev.*, 2008, **108**, 2646.
10. O. Chailapakul, W. Siangproh and D. A. Tryk, *Sens. Lett.*, 2006, **4**, 99.
11. Y. L. Zhou and J. F. Zhi, *Talanta*, 2009, **79**, 1189.
12. C. E. Nebel, B. Rezek, D. Shin, H. Uetsuka and N. Yang, *J. Phys. D: Appl. Phys.*, 2007, **40**, 6443.
13. R. Linares, P. Doering and B. Linares, *Stud. Health Technol. Infor*, 2009, **149**, 284.
14. V. Vermeeren, S. Wenmackers, P. Wagner and L. Michiels, *Sensor*, 2009, **9**, 5600.
15. A. Argoitia, H. B. Martin, E. J. Rozak, U. Landau and J. C. Angus, *Mater. Res. Soc. Proc*, 1996, **416**, 349.
16. G. M. Swain and R. Ramesham, *Anal. Chem.*, 1993, **65**, 345.
17. G. M. Swain, *Adv. Mater.*, 1994, **6**, 388.
18. R. Hoffmann, A. Kriele, H. Obloh, J. Hees, M. Wolfer, W. Smirnov, N. Yang and C. E. Nebel, *Appl. Phys. Lett.*, 2010, **97**, 052103.
19. W. Yang, O. Auciello, J. E. Butler, W. Cai, J. A. Carlisle, J. E. Gerbi, D. M. Gruen, T. Knickerbocker, T. L. Lasseter, J. N. Russell Jr, J. M. Smith and R. J. Hamers, *Nat. Mater.*, 2002, **1**, 253–257.
20. A. Hartl, E. Schmich, J. A. Garrido, J. Hernando, S. C. R. Catharino, S. Walter, P. Feulner, A. Kromka, D. Steinmuller and M. Stutzmann, *Nat. Mater.*, 2004, **3**, 736.
21. Y. S. Zou, Y. T. Yang, W. J. Zhang, Y. M. Chong, B. He, I. Bello and S. T. Lee, *Appl. Phys. Lett.*, 2008, **92**, 053105.
22. N. Yang, H. Uetsuka, E. Osawa and C. E. Nebel, *Nano. Lett.*, 2008, **11**, 3572.
23. C. E. Nebel, N. Yang, H. Uetsuka, E. Osawa, N. Tokuda and O. William, *Diam. Rel. Mat.*, 2009, **18**, 910.
24. W. Smirnov, A. Kriele, N. Yang and C. E. Nebel, *Diam. Rel. Mat.*, 2010, **19**, 186.
25. A. J. Bard, and L. R. Faulkner, *Electrochemical Methods, Fundamentals and Applications*, 2nd Edition, Wiley-VCH, MA, 2001.
26. J. Wang, *Analytical Electrochemistry*, 2nd Edition, Wiley-VCH, MA, 2000.
27. X. J. Huang, A. M. O'Mahony and R. G. Compton, *Small*, 2009, **5**, 776.
28. D. W. M. Arrigan, *Analyst*, 2004, **129**, 1157.
29. R. G. Compton, G. G. Wildgoose, N. V. Rees, I. Streeter and R. Baron, *Chem. Phys. Lett.*, 2008, **459**, 1.
30. O. Ordeig, J. del Campo, F. X. Munoz, C. E. Banks and R. G. Compton, *Electroanalysis*, 2007, **19**, 1973.
31. J. B. Cooper, S. Pang, S. Albin, J. Zheng and R. M. Johnson, *Anal. Chem.*, 1998, **70**, 464.

32. B. V. Sarada, T. N. Rao, D. A. Tryk and A. Fujishima, *J. Electrochem. Soc.*, 1999, **146**, 1469.
33. A. Suzuki, T. A. Invadini, K. Yoshimi, A. Fujishima, G. Oyama, T. Nakazato, N. hattori, S. Kitazawa and Y. Einaga, *Anal. Chem.*, 2007, **79**, 8608.
34. K. B. Holt, J. P. Hu and J. S. Foord, *Anal. Chem.*, 2007, **79**, 2556.
35. J. P. Hu, J. S. Foord and K. B. Holt, *Phys. Chem. Chem. Phys.*, 2007, **9**, 5469.
36. J. P. Hu, K. B. Holt and J. S. Foord, *Anal. Chem.*, 2009, **81**, 5663.
37. J. Cvacka, V. Quaiserova, J. W. Park, Y. Show, A. Muck and G. M. Swain, *Anal. Chem.*, 2003, **75**, 2678.
38. J. W. Park, Y. Show, V. Quaiserova, J. J. Galligan, G. D. Fink and G. M. Swain, *J. Electrochem. Soc.*, 2005, **583**, 56.
39. J. M. Halpern, S. Xie, G. P. Sutton, B. T. Higashikubo, C. A. Chestek, H. Lu, H. J. Chiel and H. B. Martin, *Diam. Rel. Mat.*, 2006, **15**, 183.
40. S. Xie, G. Shafer, C. G. Wilson and H. B. Martin, *Diam. Rel. Mat.*, 2006, **15**, 225.
41. A. L. Colley, C. G. Williams, U. D'Haenens Johnsson, M. E. Newton, P. R. Uniwin, N. R. Wilson and J. V. Macpherson, *Anal. Chem.*, 2006, **78**, 2539.
42. K. Tsunozaki, Y. Einaga, T. N. Rao and A. Fujishima, *Chem. Lett.*, 2002, 502.
43. C. Provent, W. Haenni, E. Santoli and P. Rychen, *Electrochim Acta*, 2004, **49**, 3737.
44. K. L. Soh, W. P. Kang, J. L. Davidson, Y. M Wong, A. Wisisoraat, G. Swain and D. E. Cliffel, Sens, *Actuators B*, 2003, **91**, 39.
45. K. L. Soh, W. P. Kang, J. L. Davidson, S. Basu, Y. M. Wong, D. E. Cliffel, A. B. Bonds, G. Swain and G. Diam, *Rel. Mat.*, 2004, **13**, 2009.
46. K. L. Soh, W. P. Kang, J. L. Davidson, Y. M. Wong, D. E. Cliffel and G. Swain, *Diam. Rel. Mat.*, 2008, **17**, 240.
47. K. L. Soh, W. P. Kang, J. L. Davidson, Y. M. Wong, D. E. Cliffel and G. Swain, *Diam. Rel. Mat.*, 2008, **17**, 900.
48. S. Raina, W. P. Kang and J. L. Davidson, *Diam. Rel. Mat.*, 2010, **19**, 256.
49. M. Pagels, C. E. Hall, N. S. Lawrence, A. Meredith, T. G. L. Jones, H. P. Godfried, C. S. J. Pickles, J. Wilman, C. E. Banks, R. G. Compton and L. Jiang, *Anal. Chem.*, 2005, **77**, 3705.
50. A. O. Simm, C. E. Banks, S. Ward-Jones, T. J. Davies, N. S. Lawrence, T. G. J. Jones, L. Jiang and R. G. Compton, *Analyst*, 2005, **130**, 1303.
51. N. S. Lawrence, M. Pagels, A. Meredith, T. G. J. Jones, C. E. Hall, C. S. Pickles, H. P. Godfried, C. E. Banks, R. G. Compton and L. Jiang, *Talanta*, 2006, **69**, 829.
52. M. Bonnauron, S. Saada, L. Rousseau, G. Lissorgues, C. Mer and P. Bergonzo, *Diam. Rel. Mat.*, 2008, **17**, 1399.
53. M. Bonnauron, S. Saada, C. Mer, C. Gesset, O. A. Williams, L. Rousseau, E. Scorsone and P. Mailley, *Phys. Stat. Sol. (a)*, 2008, **205**, 2126.

54. V. Carabelli, S. Gosso, A. Marcantoni, Y. Xu, E. Colombo, Z. Gao, E. Vittone, E. Kohn, A. Pasquarelli and E. Carbone, *E. Biosens. Bioelectron.*, 2010, **26**, 92.
55. W. Smirnov, N. Yang, R. Hoffmann, J. Hees, H. Obloh, W. Muller-Sebert and C. E. Nebel, *C. E. Anal. Chem.*, 2011, **83**, 7438.
56. N. Yang, W. Smirnov, J. Hees, R. Hoffmann, A. Kriele, H. Obloh, W. Müller-Sebert and C. E. Nebel, *Phys. Stat. Sol. (a)*, 2011, **208**, 2087.
57. J. Hees, R. Hoffmann, A. Kriele, W. Smirnov, H. Obloh, K. Glorer, B. Raynor, R. Driad, N. Yang, O. A. Williams and C. E. Nebel, *ACS Nano*, 2011, **5**, 3339.
58. W. Smirnov, A. Kriele, N. Yang and C. E. Nebel, *Diam. Rel. Mat*, 2010, **19**, 186.
59. M. E. Sandison and J. M. Cooper, *Lab Chip*, 2006, **6**, 1020.
60. Y. H. Lanyon and D. W. M. Arrigan, *Sens. Actuators B*, 2007, **121**, 341.
61. Y. H. Lanyon, G. De Marzi, Y. E. Watson, A. J. Quinn, J. P. Gleeson, G. Redmond and D. W. M. Arrigan, *Anal. Chem.*, 2007, **79**, 3048.
62. H. Li and N. Wu, *Nanotechnology*, 2008, **19**, 275301.
63. R. M. Penner and C. R. Martin, *Anal. Chem.*, 1987, **59**, 2625.
64. V. P. Menon and C. R. Martin, *Anal. Chem.*, 1995, **67**, 1920.
65. M. Yang, F. Qu, Y. Lu, Y. He, G. Shen and R. Yu, *Biomaterials*, 2006, **27**, 5944.
66. T. Lohmuller, U. Muller, S. Breisch, W. Nisch, R. Rudorf, W. Schuhmann, S. Neugebauer, M. Kaczor, S. Linke, S. Lechner, J. Spatz and M. Stelzle, *J. Micromech. Microeng.*, 2008, **18**, 115011.
67. E. Jeoung, T. H. Galow, J. Schotter, M. Bal, A. Ursache, M. T. Tuominen, C. M. Stafford, T. P. Russell and V. M. Rotello, *Langmuir*, 2001, **17**, 6396.
68. C. Wang, X. Shao, Q. Liu, Y. Mao, G. Yang, H. Xue and X. Hu, *Electrochim. Acta*, 2006, **52**, 704.
69. C. Wang, Q. Liu, X. Shao, G. Yang, H. Xue and X. Hu, *Talanta*, 2007, **71**, 178.
70. J. Li, J. Koehne, A. Cassell, H. Chen, H. T. Ng, Q. Ye, W. Fan, J. Han, M. Meyyappan, *Inlaid Multi-Walled Carbon Nanotube Nanoelectrode Arrays for Electroanalysis*, Wiley-VCH, Verlag, Germany, 2005, vol. 17, pp. 15–27.
71. J. Koehne, J. Li, A. M. Cassell, H. Chen, Q. Ye, H. T. Ng, J. Han and M. Meyyappan, *J. Mater. Chem.*, 2004, **14**, 676–684.
72. Y. Tu, Y. Lin, W. Yantasee, Z. Ren, *Carbon Nanotubes Based Nanoelectrode Arrays: Fabrication, Evaluation, and Application in Voltammetric Analysis*, Wiley-VCH, Verlag, Germany, 2005, vol. 17, pp. 79-84.
73. S. Siddiqui, P. U. Arumugam, H. Chen, J. Li and M. Meyyappan, *ACS Nano*, 2010, **4**, 955.
74. M. Wei, C. Terashima, M. Lv, A. Fujishima and Z.-Z. Gu, *Chem. Commun.*, 2009, **45**, 3624.
75. D. Luo, L. Wu and J. Zhi, *ACS Nano*, 2009, **8**, 2121.
76. M. Lv, M. wei, F. Rong, C. Terashima, A. Fujishima and Z.-Z. Gu, *Electroanalysis*, 2010, **22**, 199.

77. W. Wu, L. Bai, X. Lin, Z. Tang and Z. Gu, *Electrochem. Commun.*, 2011, **13**, 872.
78. D. Luo and J. Zhi, *Electrochem. Commun.*, 2009, **11**, 1093.
79. Y. Yang, J.-W. Oh, Y.-R. Kim, C. Terashima, A. Fujishima, J. S. Kim and H. Kim, *Chem. Commun.*, 2010, **46**, 5793.
80. M. C. Granger and G. M. Swain, *J. Electrochem. Soc.*, 1999, **146**, 4551.
81. W. Gajewski, P. Achatz, O. A. Williams, K. Haenen, E. Bustarret, M. Stutzmann and J. A. Garrido, *Phys. Rev. B*, 2009, **79**, 045206.
82. O. A. Williams, O. Douheret, M. Daenen, K. Haenen, E. Osawa and M. Takahashi, *Chem. Phys. Lett.*, 2007, **445**, 255.
83. C. Kranz, G. Friedbacher and B. Mizaikoff, *Anal. Chem.*, 2001, **73**, 2491.
84. C. Kranz, A. Kueng, A. Lugstein, E. Bertagnolli and B. Mizaikoff, *Ultramicroscopy*, 2004, **100**, 127.
85. C. Kranz and J. Wiedemair, *Anal. Bioanal. Chem.*, 2008, **390**, 239.
86. J. V. Macpherson and P. R. Unwin, *Anal. Chem.*, 2000, **72**, 276.
87. J. V. Macpherson and P. R. Unwin, *Anal. Chem.*, 2001, **73**, 550.
88. P. S. Dobson, J. M. R. Weaver, M. N. Holder, P. R. Unwin and J. V. Macpherson, *Anal. Chem.*, 2004, **77**, 424.
89. M. R. Gullo, P. L. T. M. Frederix, T. Akiyama, A. Engel, N. F. deRooij and U. Staufer, *Anal. Chem.*, 2006, **78**, 5436.
90. A. Avdic, A. Lugstein, M. Wu, B. Gollas, I. Pobelov, T. Wandlowski, K. Leonhardt, G. Denault and E. Bertagnolli, *Nanotechnology*, 2011, **22**, 145306.
91. A. Kueng, C. Kranz, A. Lugstein, E. Bertagnolli and B. Mizaikoff, *Angewandte Chemie International Edition*, 2003, **42**, 3238.
92. H. Shin, P. J. Hesketh, B. Mizaikoff and C. Kranz, *Anal. Chem.*, 2007, **79**, 4769.
93. H. Kado, K. Yokoyama and T. Tohda, *Rev. Sci. Instrum*, 1992, **63**, 3330.
94. H. J. Dai, J. H. Hafner, A. G. Rinzler, D. T. Colbert and R. E. Smalley, *Nature*, 1996, **384**, 147.
95. G. Janchen, P. Hoffmann, A. Kriele, H. Lorenz, A. J. Kulik and G. Dietler, *Appl. Phys. Lett.*, 2002, **80**, 4623.
96. C. L. Cheung, J. H. Hafner, T. W. Odom, K. Kim and C. M. Lieber, *Appl. Phys. Lett.*, 2000, **76**, 3136.
97. K. E. Spear, J. P. Dismukes, *Synthetic Diamond: Emerging CVD Science and Technology*. Wiley, New York, 1994.
98. P. Niedermann, W. Hanni, N. Blanc, R. Christoph and J. Burger, *J. Vac. Sci. Technol. A-Vac. Surf. Films*, 1996, **14**, 1233.
99. G. Tanasa, O. Kurnosikov, C. F. J. Flipse, J. G. Buijnsters and W. J. P. van Enckevort, *J. Appl. Phys.*, 2003, **94**, 1699.
100. A. Malave, E. Oesterschulze, W. Kulisch, T. Trenkler, T. Hantschel and W. Vandervorst, *Diam. Relat. Mat.*, 1999, **8**, 283.
101. T. Shibata, Y. Kitamoto, K. Unno and E. Makino, *J. Microelectromech. S*, 2000, **9**, 47.
102. K. Unno, T. Shibata and E. Makino, *Sens. Actuators, A*, 2001, **88**, 247.

103. K. Unno, Y. Kitamoto, T. Shibata and E. Makino, *Smart. Mater. Struct.*, 2001, **10**, 730.
104. K. H. Kim, N. Moldovan, C. H. Ke, H. D. Espinosa, X. C. Xiao, J. A. Carlisle and O. Auciello, *Small*, 2005, **1**, 866.
105. R. Agrawal, N. Moldovan and H. D. Espinosa, *J. Appl. Phys.*, 2009, **106**, 064311.
106. J. Liu, D. S. Grierson, N. Moldovan, J. Notbohm, S. Li, P. Jaroenapibal, S. D. O'Connor, A. V. Sumant, N. Neelakantan, J. A. Carlisle, K. T. Turner and R. W. Carpick, *Small*, 2010, **6**, 1140.
107. K.-H. Chung and D.-E. Kim, Wear characteristics of diamond-coated atomic force microscope probe, *Ultramicroscopy*, 2007, **108**, 1.
108. W. Smirnov, A. Kriele, R. Hoffmann, E. Sillero, J. Hees, O. A. Williams, N. Yang, C. Kranz and C. E. Nebel, *Anal. Chem.*, 2011, **83**, 4936–4941.
109. O. A. Williams, J. Hees, C. Dieker, W. Jager, L. Kirste and C. E. Nebel, *ACS Nano*, 2010, **4**, 4824.
110. R. K. Ahmad, A. C. Parada, S. Hudziak, A. Chaudhary and R. B. Jackman, *Appl. Phys. Lett.*, 2010, **97**, 093103.
111. J. Guo and E. Lindner, *Anal. Chem.*, 2009, **81**, 130.
112. M. Fleischmann, S. Pons and J. Daschbach, *J. Electroanal. Chem.*, 1991, **17**, 1.
113. M. Fleischmann and S. Pons, *J. Electroanal. Chem.*, 1988, **250**, 277.
114. L. M. Abrantes, M. Fleischmann, L. M. Peter, S. Pons and B. R. Scharifker, *J. Electroanal. Chem.*, 1988, **256**, 229.
115. O. Koster, W. Schuhmann, H. Vogt and W. Mokwa, *Sens. Actuators B*, 2001, **76**, 573.
116. A. Kriele, O. A. Williams, M. Wolfer, D. Brink, W. Muller-Sebert and C. E. Nebel, *Appl. Phys. Lett.*, 2009, **95**, 3.
117. M. Pagels, C. E. Hall, N. S. Lawrence, A. Meredith, T. G. J. Jones, H. P. Godfried, C. S. J. Pickles, J. Wilman, C. E. Banks, R. G. Compton and L. Jiang, *Anal. Chem.*, 2005, **77**, 3705.
118. A. O. Simm, C. E. Banks, S. Ward–Jones, T. J. Davies, N. S. Lawrence, T. G. J. Jones, L. Jiang and R. G. Compton, *Analyst*, 2005, **130**, 1303.
119. N. S. Lawrence, M. Pagels, A. Meredith, T. G. J. Jones, C. E. Hall, C. S. Pickles, H. P. Godfried, C. E. Banks, R. G. Compton and L. Jiang, *Talanta*, 2006, **69**, 829.
120. C. Madore, A. Duret, W. Haenni, A. Perret, in *Proceedings of the Symposium on Microfabricated Systems and MEMS V*, Electrochemical Society Proceeding Series, Pennington, USA, 2000, p. 159.
121. M. Bonnauron, S. Saada, L. Rousseau, G. Lissorgues, C. Mer and P. Bergonzo, *Diam. Rel. Mat.*, 2008, **17**, 1399.
122. M. Bonnauron, S. Saada, C. Mer, C. Gesset, O. A. Williams, L. Rousseau, E. Scorsone and P. Mailley, *Phys. Stat. Sol. (a)*, 2008, **205**, 2126.
123. V. Carabelli, S. Gosso, A. Marcantoni, Y. Xu, E. Colombo, Z. Gao, E. Vittone, E. Kohn, A. Pasquarelli and E. Carbone, *Biosens. Bioelectron*, 2010, **26**, 92.

124. D. Khamis, E. Mahe, F. Dardoize and D. Devilliers, *J. Appl. Electrochem.*, 2010, **40**, 1829.
125. E. Popa, H. Notsu, T. Miwa, D. A. Tryk and A. Fujishima, *Electrochem. Solid-State Lett.*, 1999, **2**, 49.
126. A. Fujishima, T. N. Rao, E. Popa, B. V. Sarada, I. Yagi and D. A. Tryk, *J. Electroanal. Chem.*, 1999, **473**, 179.
127. D. Sopchak, B. Miller, R. Kalish, Y. Avyigal and X. Shi, *Electroanalysis*, 2002, **14**, 473.
128. W. C. Poh, K. P. Loh, W. D. Zhang, S. Triparthy, J.-S. Ye and F.-S. Sheu, *Langmuir*, 2004, **20**, 5484.
129. P. S. Siew, K. P. Loh, W. C. Poh and H. Zhang, *Diam. Rel. Mat.*, 2005, **14**, 426.
130. G.-H. Zhao, M.-F. Li and M.-L. Li, *Central European J. Chemistry*, 2007, **5**, 1114.
131. A. Suzuki, T. A. Ivandini, K. Yoshimi, A. Fujishima, G. Oyama, T. Nakazato, N. Hattori, S. Kitazawa and Y. Einaga, *Anal. Chem.*, 2007, **79**, 8608.
132. M. Wei, G. Terashima, M. Lv, A. Fijishima and Z.-Z. Gu, *Chem. Comm.*, 2009, 3624.

CHAPTER 20

Electron Field Emission from Diamond

TRAVIS C. WADE

Applied Diamond Inc., 3825 Lancaster Pike, Wilmington, DE 19807, USA
Email: travis.c.wade@gmail.com

20.1 Mechanisms of Electron Emission

Electron emission is generally stimulated by thermal excitation, application of an electric field, or a combination of the two. In thermionic emission, electrons gain kinetic energy as the temperature increases. If the temperature is sufficiently high (typically >1500 °C), some distribution of electrons have energy exceeding the vacuum level. These electrons are spontaneously emitted into vacuum with no applied electric potential.

Thermally assisted field emission, in which emission is achieved by a combination of thermal energy and reducing the electric barrier potential, is called thermionic-field or Schottky emission. At moderate temperatures (700–1500 °C), electrons have total energy above the Fermi level (E_F) but below the vacuum level (E_{vac}). In order for these electrons to emit, a moderate electric field must be applied to reduce the width of the potential barrier.

Field emission is characterized by a strong electric field applied to dramatically reduce the width of the potential barrier, thereby allowing electrons to quantum-mechanically tunnel into vacuum in the absence of elevated temperatures.

20.2 Field Emission

In order to eject electrons from a solid surface, energy must be applied so that electrons can overcome or tunnel through the potential barrier. The potential barrier to emission is related to the electron affinity (χ) and the work function (Φ). The electron affinity (χ) embodies the energy difference between the conduction band and the vacuum, whereas the work function is the energy difference between the Fermi level and the vacuum. Discussion of surface treatments that can result in a negative electron affinity is found in section 20.5.3.

20.3 Fowler–Nordheim Field Emission Theory

Electron field emission from metals has been verified theoretically and experimentally to follow the Fowler–Nordheim equation[1]:

$$J = K_1 \left(E^2/\Phi\right) exp\left(-K_2 \Phi^{3/2}/E\right), \qquad (20.1)$$

where J is the emission current density (A cm^{-2}); K_1 and K_2 are constants: $K_1 = 1.54 \times 10^{-6}$ A eV V^{-2}, $K_2 = 6.83 \times 10^7$ V cm^{-1} eV$^{-3/2}$; Φ is the work function (eV) of the emitting surface and E is the macroscopic electric field (V cm^{-1}) across the parallel plates, which is given by $E = V/D$. V is the applied voltage and D is the anode–cathode spacing.

Eqn 20.1 can be expressed in Fowler–Nordheim (F-N) form to demonstrate the exponential relationship of the equation:

$$Ln\left(I/E^2\right) = Ln\left(AK_1\beta^2/\phi\right) - \lfloor\left(K_2\phi^{1.5}/\beta\right)\left(1/E\right)\rfloor \qquad (20.2)$$

In eqn 20.2 the image effect has been ignored since it is considered to have minor effects on emission current in all but the most extreme current densities. Space charge is negligible below current densities of the order of 10^7 A cm^{-2}.[2] As the cathode–anode spacing becomes smaller, higher interanode pressures can be tolerated so long as the mean free path of electrons is sufficiently long to allow efficient electron transfer. To facilitate emission at low electric fields, the cathode is normally made of a material with a low barrier to field emission.

Apparent in the equation, the emission current strongly depends on the work function of the cathode. A low work function contributes to a higher emission current for a given applied electric field. In the second form, a plot of Ln(I/E^2) versus $1/E$ should be linear with slope equal to $-K_2\Phi^{1.5}/\beta$ and the y-intercept equal to Ln($AK_1\beta^2/\Phi$). This plot is generally referred to as a Fowler–Nordheim (F-N) plot.

20.4 Improvements upon the F-N Form

The Fowler–Nordheim equation is derived for a planar cathode with a uniform electric field in the vacuum gap. These assumptions are not valid for sharp non-metallic structures. The precise calculation of potential distribution, electric field, and emission current for a sharp microstructure involves numerical calculation of the three-dimensional Poisson equation and Schrodinger equation for electron emission. The difficulty of solving these equations numerically has led to the development of a number of simplifications and empirical factors.

There are three experimentally unknown device parameters involved in the slope and intercept of an F-N plot: the emitting area (A), the work function (Φ), and the field enhancement factor (β). These three parameters cannot empirically be determined because only two equations can be obtained from the slope and intercept of the F-N plot. Additional methods must be used to determine one of these unknown parameters.

The emitting area (A) is easy to estimate but there is no presently known experimental technique to directly measure the emitting area of diamond field emission cathodes. Furthermore, it is expected that the area is field-dependent and geometry specific. Spindt et al.,[3] provide a formalism for the calculation of emitting area based on measurable dimensions and known quantities such as applied field. Gotoh et al., have attempted to measure emission area but with ambiguous results.[4]

The work function (Φ) of diamond is difficult to determine because diamond is a wide-band-gap material. Work by Gomer et al.,[5] describes a linear dependence of work function on field. In addition, the work function of polycrystalline diamond depends on composition and surface structure. Measurements of the work function of polycrystalline diamond films have been performed by many research groups worldwide.[6–12]

The field enhancement factor (β) depends on the geometry of the cathode, and it may be estimated by physical measurement of the diamond tip geometry by SEM. Consider the sharp cone structure as illustrated in Figure 20.1. The sharp cone structure is generally referred to as a "Spindt cathode" in reference to Dr Charles Spindt's work on the fabrication of

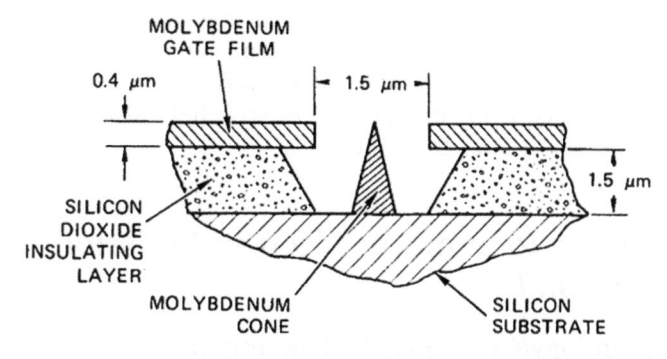

Figure 20.1 Spindt field emission cathode made of a sharp molybdenum cone. Reprinted with permission from American Institute of Physics, copyright 1976.[3]

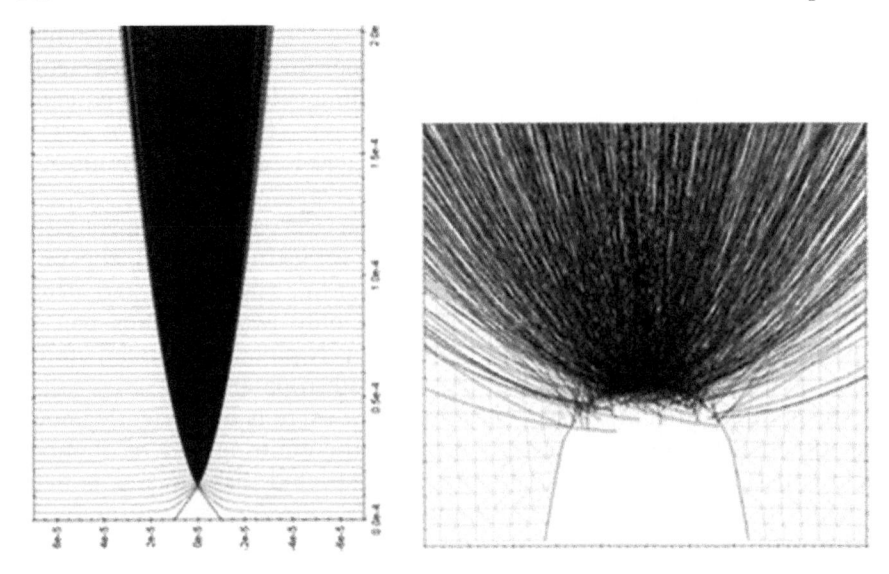

Figure 20.2 Result of modeling efforts showing field lines over a diamond tip showing the local enhancement of field at sharp features. Modeling performed by Dr Brau's laboratory at Vanderbilt University.
Reproduced with permission from *"Development of high-brightness electron sources for free-electron lasers"* Ph.D. Thesis of Jonathan Jarvis (2009), Vanderbilt University.

conical metal field emitters.[3] The sharp cone structure results in a non-uniform electric field at the emitter surface as illustrated in Figure 20.2.

An empirical approximation of emission current for a sharp microstructure can be obtained with a simple modification of Fowler–Nordheim equation for a planar metal cathode by replacing the parallel electric field with electric field at the apex of the sharp microstructure that is $E = \beta V/d$. β is the factor by which electric field is increased due the sharp microstructure relative to the planar structure. This approximation implies that the emission current for a sharp microstructure is equivalent to the emission current of a planar cathode of the same vacuum gap but the effective electric field is increased by a factor of β. This approximation agrees well with experimental results, likely because the electric field of a sharp tip decays rapidly for the region away from the apex.

20.5 Factors Relevant to Electron Emission from Diamond

20.5.1 Topology/Geometric Enhancement

The virtues of geometric field enhancement have been understood since Dr Charles Spindt[3] developed sharp metal features for field emission

Electron Field Emission from Diamond

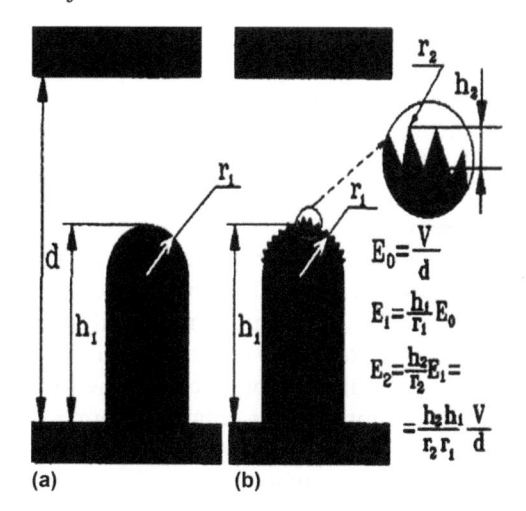

Figure 20.3 Geometric field enhancement model.
Reprinted from *Applied Surface Science*, 1995, 87–88, with permission from Elsevier.[13]

applications. The field enhancement factor of the sharpened diamond tip can be estimated using a two-step field emission enhancement model. In this model[13] the sharpened diamond tip is modeled as a large pillar with a semicircular top (similar to a Norman window) with tip height of h_1 and tip radius curvature of r_1. The semicircular top is further decorated with small pyramids of height h_2 and tip radius of curvature r_2, as illustrated in Figure 20.3. The electric field at the sharpened tip apex arises from the two-tip cascaded structure. In the first step the electric field at the apex of large conical tip is enhanced by the factor of h_1/r_1 from the planar base. In the second step the electric field at the apex of the sharp tiny conical tip is enhanced by the factor of h_2/r_2 from the apex of large conical tip. Thus, the total geometrical field enhancement factor of the sharpened tip is the product of field enhancement factor of two cascaded tip structures.

$$\beta_{\text{sharpened}} = (h_1/r_1)(h_2/r_2) \tag{20.3}$$

It can be assumed that the emission current arises entirely from electron tunneling within the vicinity of this highest electric field region. Geometric field enhancement factors calculated by this method agree reasonably well with results derived from experimental F–N emission analysis.[14]

After the Fowler–Nordheim theoretical work in 1928, a notable advance in geometric enhancement came with the development in 1937 by Erwin W. Müeller of the field electron microscope (FEM). Müeller used a sharply pointed tungsten wire as a field emitter tip in a vacuum enclosure opposite a phosphor screen. As a voltage is applied between the tip and screen, an image forms which reflects the current-density distribution across the emitter apex. Smaller emitting surfaces and longer drift distances between the cathode and anode increase the effective magnification, which is commonly $\sim 10^6\times$.

Calculations of emission dependence on β performed by Dyke et al.,[2] indicate that an increase in β of only 20% would decrease the emission current by more than a factor of ten. A tip radius of 5 nm differs by 20% from a tip radius of 6 nm. A high degree of control is therefore required for predictable field emission.

The geometrically enhanced emitter concept has been extended to silicon cathodes. The silicon cathode is usually heavily doped ($n+$) to achieve a low work function for silicon ($\Phi \approx \chi = 4.12$ eV) and a good ohmic contact with the metal back plate. Silicon emitters have shown some improvement over metal cathodes.[15] Since the work function of silicon is the same order of magnitude as most metals, the improvement obtained from a silicon emitter is in increasing the geometric field enhancement factor due to the availability of mature integrated circuit manufacturing technology, which facilitates mass production of the emitters. However, silicon emitters have limited applications because the operating voltage of a silicon cathode is still high compared to that of a solid-state device. In addition, silicon emitters have a serious surface adsorption problem, which leads to instability and lifetime issues.

20.5.2 Temperature

Due to the large band gap, diamond field emitters are largely temperature independent. At high temperatures (>700 °C), diamond field emitters have been found to emit by thermionic mechanisms consistent with the Richardson equation.[12,16,17]

20.5.3 Surface States and Electron Affinity

Diamond has a low and, under some conditions, negative electron affinity (NEA). NEA indicates that any electron that manages to occupy the conduction band will be spontaneously emitted from the surface. This combination of low surface barrier to electron emission in an otherwise robust material has attracted attention to diamond's promise as a high performance cold cathode material.

Electrons in the conduction band are generally prevented from escaping into the vacuum by the electron affinity barrier (χ). However, the diamond (100), (110), and (111) surfaces are known to exhibit negative electron affinity when terminated with hydrogen.[10,18–21] Hydrogen exists as an ionic species forming an affinity lowering surface dipole sufficient to result in a negative electron affinity surface. Material systems in which the electron affinity is negative are rare. Himpsel et al. have observed negative electron affinity from unterminated, unreconstructed diamond (111) surfaces.[18]

Diederich et al.,[10] have investigated band bending, electron affinity, and work function of differently terminated, doped, and oriented diamond surfaces by X-ray and ultraviolet photoelectron spectroscopy (XPS and UPS, respectively). The hydrogen-terminated diamond surfaces have negative electron affinity (NEA), whereas the hydrogen-free surfaces present positive electron affinity (PEA). The NEA peak is only observed for the boron-doped diamond (100)-(2×1):H surface, whereas it is not visible for the nitrogen-doped diamond (100)-(2×1):H surface due to strong upward band bending. Electron emission from energy levels below the conduction band minimum (CBM) up to the vacuum level E_{vac} allowed the electron affinity to be measured quantitatively for PEA, as well as for NEA. The lightly boron-doped diamond (100)-(2×1):H surface presents a high-intensity NEA peak. Its cut-off is situated at a kinetic energy of 4.9 eV, whereas the upper limit of the vacuum level is situated at 3.9 eV, resulting in a NEA of at least −1.0 eV and a maximum work function of 3.9 eV. The highly boron-doped diamond (100) surface behaves similarly, showing that the NEA peak is present due to the downward band bending independent of the boron concentration. The nitrogen-doped (100)-(2×1):H surface shows a low NEA of −0.2 eV but no NEA peak due to the strong upward band bending. E_{vac} is situated at 4.2 eV or below, resulting in a NEA of at least −0.9 eV and a maximum work function of 4.2 eV. The high-intensity NEA peak of boron-doped diamond seems to be due to the downward band bending together with the reduced work function because of hydrogen termination. Although unaffected by anneals up to 600 °C, hydrogen desorption at higher annealing temperatures (1100 °C) increases the work function and NEA disappears. For the nitrogen-doped diamond (100) surface, the work function behaves similarly, but the observation of a NEA peak is absent because of the surface barrier formed by the high upward band bending.

For hydrogen-free diamond (100) surfaces, the electron affinity is small and positive, as shown in Figure 20.4. Partially hydrogenated diamond surfaces display negative electron affinity (NEA), as illustrated in Figure 20.5. The diamond (100)-(2×1):H hydrogenated surface is believed to be a true NEA surface as illustrated in Figure 20.4. The small electron affinity of diamond is believed to be critical to the observed emission from diamond at very low fields.

Diamond surfaces coated with a thin layer of certain metals, including zirconium (Zr),[6] Co,[22] Ni,[19] TiO,[23] and Cs[24] have also been observed to exhibit NEA.

Hydrogen is observed to impart a negative electron affinity to the (100) surface of diamond up to a temperature of 1100 °C. Deuterium imparts the same negative electron affinity characteristics to the diamond surface as hydrogen but with a higher binding energy. Ultraviolet photoemission spectroscopy studies of diamond surface termination with deuterium performed by Baumann and Nemanich[21] confirm the predicted higher binding energy of deuterium. Deuterium imparts the same affinity lowering effect but is observed to be stable up to 1250 °C. Similarly, annealing hydrogen-terminated diamond (110) surfaces to 800 °C was sufficient to remove the negative electron affinity, but temperatures of 900 °C were

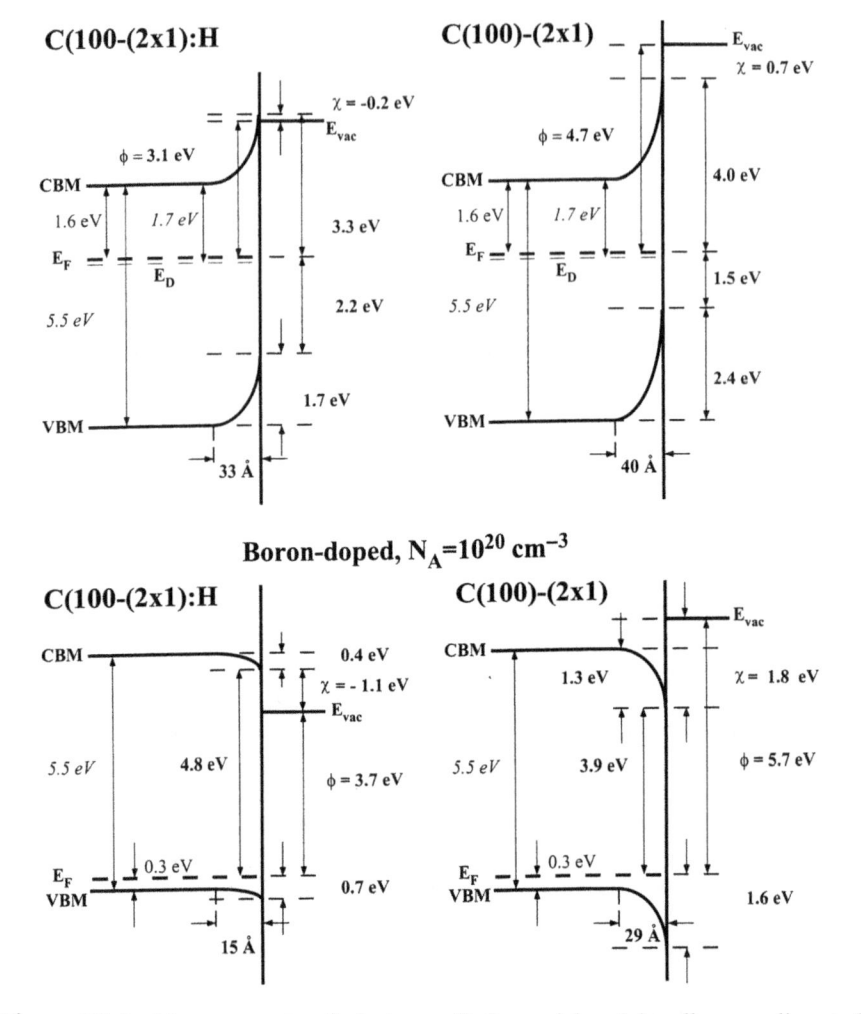

Figure 20.4 Measurements of electron affinity and band bending on diamond surfaces with boron and nitrogen doping. Hydrogen-terminated surfaces (two plots on left) show negative electron affinity. Boron doping leads to downward band bending, while nitrogen causes upward band bending.
Reprinted from Surface Science, 1998, **418**(1), 219–239, with permission from Elsevier.[10]

required to remove the negative electron affinity from the deuterium-terminated surface. Deuterium has been found to preferentially abstract and replace hydrogen on polycrystalline diamond surfaces (Figure 20.6) by Koleske *et al.*[25] Desorbed atoms/molecules were analyzed by mass analysis of time-of-flight experiments. Hydrogen was found to be abstracted 3× faster than deuterium on diamond surfaces.

Electron Field Emission from Diamond 507

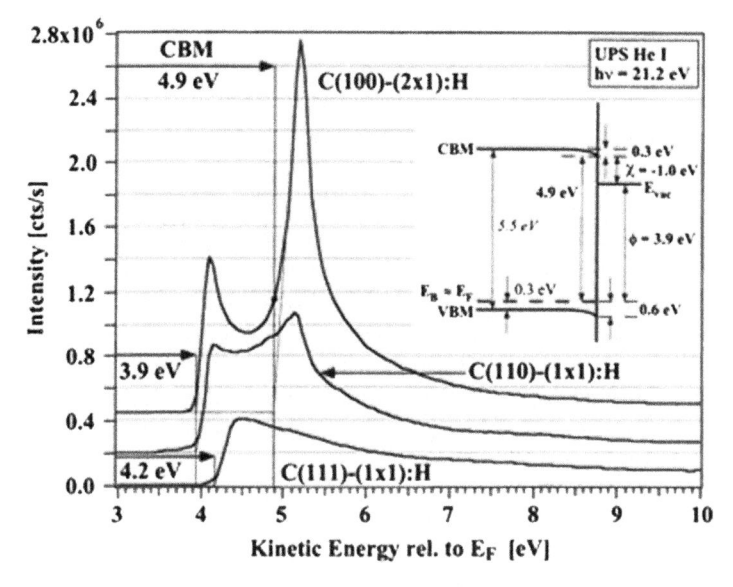

Figure 20.5 Ultraviolet photoelectron spectroscopy (UPS) results of three hydrogen-terminated diamond surfaces, all displaying negative electron affinity. Reprinted from *Surface Science*, 1999, **424**(2–3), L314–L320, with permission from Elsevier.[20]

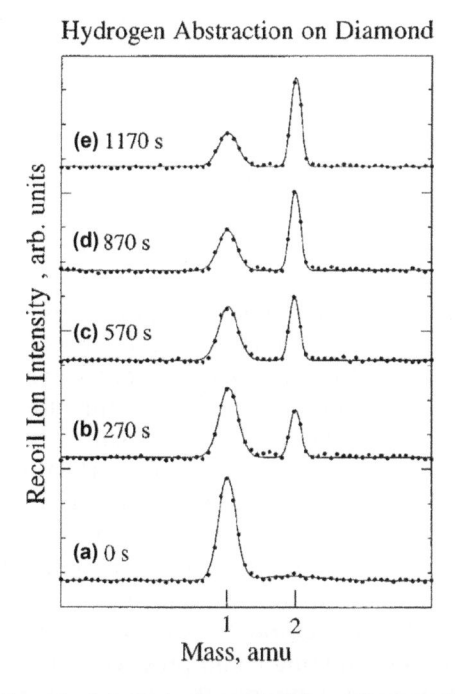

Figure 20.6 Over time, hydrogen is abstracted from diamond by a deuterium flow. Reprinted with permission from *J. Chem. Phys.*, 1995, **102**(2), 992–1002. Copyright 1994, American Institute of Physics.[25]

20.5.4 Carrier Transport

While the low (or negative) electron affinity of diamond surfaces may be important and can make diamond an efficient emission material, it is not adequate, by simply invoking this property, to explain diamond's potential as a field emission material. To sustain field emission, there must be a continuous supply of electrons and a sustainable transport mechanism for the electrons to reach the surface. Emission related to a small electron affinity requires population of states in the conduction band which is limited due to the wide bandgap of diamond. A low electron affinity surface is an effective contributor to electron emission only when the energy levels of occupied states, possibly including surface states, are positioned sufficiently close to the conduction band minimum in diamond. Unless n-type doping is performed, electrons must be injected into the conduction band of diamond in order to populate these states.

Bobrov et al.[26] explored single crystal diamond electronic structure *via* scanning tunneling microscopy (STM). In high purity single crystal diamond they observed one-dimensional, fully delocalized states and very long diffusion lengths for conduction band electrons. The insulating nature of diamond makes it difficult to probe electronic states, and indeed the STM tip was crushed when trying to establish a tunneling current. By annealing *in situ*, they were able to remove adsorbed hydrogen and realize a 5 MΩ resistance through 200 μm of diamond. Only at high voltages (> 5.9 V) was imaging possible, indicating that the STM is operating in the near-field emission regime.

In the *"Physics and Chemistry of Color"* by K. Nassau, it is pointed out that nitrogen, aluminum, beryllium, boron, and lithium are the only known materials to readily enter the diamond lattice.[27]

20.5.4.1 N-type Doping

Diamond has many superior material properties, yet its application in electronic devices has been limited due to the difficulty of producing n-type thin films of sufficiently high conductivity. Previous efforts to synthesize diamond or diamond-like carbon thin films with high n-type conductivity have been largely unsuccessful.[28-31] Phosphorous has recently been shown to be a n-type dopant in CVD diamond.[32] However, phosphorous is a significantly larger atom than carbon, nitrogen, or boron (covalent radius of P is 1.57 times larger than N and 1.25 times larger than B) and this places limitations on the amount of phosphorous, which can be incorporated into the diamond and limits its potential electrical performance in a device. Although part of the prior work to date[31] demonstrated that n-type doping could produce shallow donor levels close to the conduction band of diamond, the room-temperature conductivities are still too low for the application of these materials in conventional electronic devices.

The donor state of nitrogen lies in the bandgap of diamond, 2.2 eV below the conduction band. Since nitrogen is a deep donor impurity that is not ionized at room temperature, field emission enhancement is not expected to arise from the direct emission from the donor dopant but could instead arise from dopant related defect centers. Known nitrogen-related defects identified by photoluminescence and cathodoluminescence include vacancies trapped adjacent to the substitutional nitrogen atom (1.94 and 2.15 eV above the valence band maximum) at A centers (2.3 and 2.46 eV), and at B centers (2.49 eV). However, electrons in these defect states would seem to require too much energy to couple to the vacuum or conduction band.

N-doped nanocrystalline films have been grown in work by Bhattacharyya et al.[33] Nitrogen-doped ultrananocrystalline diamond (UNCD) thin films were synthesized using chemical vapor deposition (CVD) with a $CH_4/Ar/N_2$ gas mixture. The morphology and transport properties of the films were observed to be greatly affected by the presence and amount of nitrogen in the plasma. The transmission electron microscopy (TEM) data indicate that the grain size and grain boundary width increase with the addition of N_2 in the plasma. Transport measurements indicate that these films have n-type electrical conductivity as shown in the Arrhenius plot (Figure 20.7) for a series of films synthesized using different nitrogen concentrations in the plasma.

In recent years, work function values of less than 2 eV for nitrogen-incorporated diamond films have been reported. In work by Koeck et al.,[34] doped diamond films with a negative electron affinity (NEA) surface showed a reduced effective work function for electron emission of ∼1.5 eV for

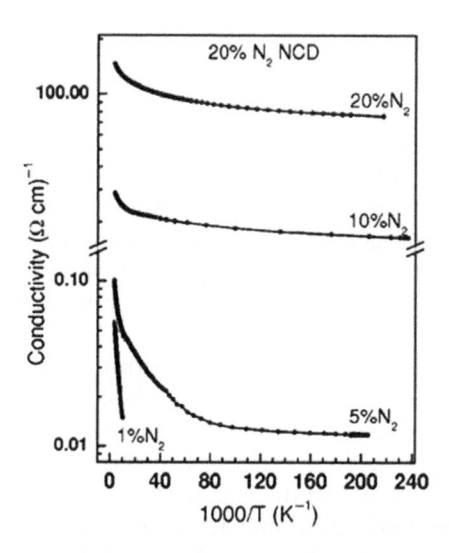

Figure 20.7 Arrhenius plot for a series of nanocrystalline films. Reprinted with permission from *Applied Physics Letters*, 2001, 79(10), 1441–1443. Copyright 2001, American Institute of Physics.[33]

nitrogen-doped diamond films prepared by microwave plasma CVD. The film exhibited a resistance that decreased with temperature suggesting the role of the dopant. An analysis based on the Richardson–Dushman equation indicated an emission barrier of <1.3 eV. In similar systems, work by Suzuki et al.,[35] derived a work function of 1.99 eV in nitrogen-incorporated diamond.

20.5.4.2 P-type Doping

P-type doping lowers the Fermi level of the bulk and traditionally increases the work function. Boron is the most prevalent and well understood dopant in diamond. The boron acceptor state lies in the diamond band gap 0.4eV above the valence band. However, in boron-doped (p-type) diamond films, the turn-on electric field decreases with p-type doping. This is due to surface band bending and is explained by Diederich et al. in Figure 20.4, which demonstrates the effect of boron doping on the electron affinity and work function of diamond surfaces.

20.5.4.3 Defect States

Defect states or bands are very probable origins of electrons for field emission enhancement. Vacancy defects in diamond thin films can be substantial.[36,37] Defects in the form of graphite and multiply twinned quintuplet wedges[38] have also been observed. Calculations indicate that defect states may exist in the bulk bandgap. The effects of negative electron affinity, band bending, image interaction, and surface states have been examined by Huang et al.[39] Conventional theory of electron field emission applied to crystalline diamond does not explain the measured high-current emission at low fields. They postulate two sub-bands in the intrinsic bandgap, which may be generated by defects or impurities. With reasonable band parameters, the I–V characteristics calculated by Huang et al., agree with experimental data. This unfortunately relies on the assumption that emission from defect states will have an infinite supply of electrons like a valence band, when in fact carrier replenishment is a major barrier to field emission from defect states. More theoretical study is needed to determine how these defect states couple to each other to form a defect-induced conducting band.

A detailed theoretical examination of electron transport mechanisms in diamond by Huang et al. did not reveal any viable process to populate these tunneling states.[40] The density of states for a lattice with a single vacancy has been calculated and it was found that defect states exist only within a narrow energy range ($\sim$1–2 eV) above the valence band maximum. However, since this study was performed, additional defect states have been recently located 2.0 eV below the conduction band minimum of diamond by photoelectron yield spectroscopy.[41] Defect states such as these may be viable candidates for electron emission from diamond.

XPS studies were performed in 2008 on diamond films by Yamaguchi et al.[42] in order to shed light on the origin of emitted electrons. In this work the authors theorized that, since the emitted electrons were observed to be mono-energetic, they must originate in the same energy state. Since emission currents are large, the energy state from which the electrons are emitting must have a high density of states. Ergo, emission must be occurring directly from the valence band as it is the only known energy level with a sufficiently high density of states.

Zhu et al., have fabricated diamond samples with varying defect densities by CVD[43,44] and found that the potential barrier and turn-on threshold of undoped diamond films declines rapidly with increasing structural defect density. Similar correlations between the field emission and the structural defect density were also observed for p-type (boron) diamond films. It was also found that emission from p-type diamond films was more stable than undoped diamond films. These results demonstrate that the creation of defects and, most probably, defect-induced energy band(s) enhances field emission. If these bands are wide enough or closely spaced, electron hopping within the band(s) or excitation from the valence band could provide the necessary conduction path for electrons to reach the emitter surface to sustain stable emission of electrons into the vacuum. Electrons may tunnel directly from these bands or hop to surface states for emission. The position(s) of these defect-induced energy bands could not be determined by Zhu et al. because of uncertainty in the local field enhancement factor and emission area. These results are in agreement with other work on diamond-coated emitters and planar diamond films.[44–47]

Defects created by ion implantation have also been studied for electron field emission.[43,48] The emission threshold field was observed to decline rapidly as the implantation dose increases. In addition, implantation of silicon ions resulted in more dramatic electron emission enhancement than carbon implantation. This was attributed to the larger mass of silicon ions producing more defects and damage. The emission characteristics of implanted diamond films were found to be insensitive to atmospheric exposure, suggesting that the modified surface produced by the implantation process is stable and chemically inert. Defects introduced in the surface regions by ion implantation were observed to increase the conductivity and reduce the work function of the diamond. The effect of carbon, hydrogen, argon, and xenon ion irradiation on a pure graphite carbon fiber has also been studied.[49] Experimental results indicate that the field emission threshold has potential for optimization with careful choice of dose, beyond which the emission threshold was degraded by further implantation.

References

1. R. H. Fowler and L. Nordheim, Electron Emission in Intense Electric Fields. Proceedings of the Royal Society of London, *Series A, Containing Papers of a Mathematical and Physical Character*, 1928, **119**(781), 173–181.

2. W. P. Dyke and J. K. Trolan, Field Emission: Large Current Densities, Space Charge, and the Vacuum Arc, *Physical Review*, 1953, **89**(4), 799.
3. C. A. Spindt, I. Brodie, L. Humphrey and E. R. Westerberg, Physical properties of thin-film field emission cathodes with molybdenum cones, *Journal of Applied Physics*, 1976, **47**(12), 5248–5263.
4. Y. Gotoh, T. Kondo, M. Nagao, H. Tsuji, J. Ishikawa, K. Hayashi and K. Kobashi, Estimation of emission field and emission site of boron-doped diamond thin-film field emitters, *J. Vac. Sci. Tech. B*, 2000, **18**(2), 1018–1023.
5. R. Gomer, *Field Emission and Field Ionization*. 1961, Cambridge, MA: Harvard University Press.
6. P. K. Baumann and R. J. Nemanich, Electron affinity and schottky barrier height of metal-diamond (100), (111), (110) interfaces, *Journal of Applied Physics*, 1997, **82**, 5148.
7. F. A. M. Koeck, R. J. Nemanich, A. Lazea and K. Haenen, Thermionic electron emission from low work-function phosphorus doped diamond films, *Diamond and Related Materials*, 2009, **18**(5–8), 789–791.
8. J. I. B. Wilson, J. S. Walton and G. Beamson, Analysis of chemical vapour deposited diamond films by X-ray photoelectron spectroscopy, *Journal of Electron Spectroscopy and Related Phenomena*, 2001, **121**(1–3), 183–201.
9. P. Abbott, E. D. Sosa and D. E. Golden, Effect of average grain size on the work function of diamond films, *Applied Physics Letters*, 2001, **79**(17), 2835.
10. L. Diederich, O. M. Küttel, P. Aebi and L. Schlapbach, Electron affinity and work function of differently oriented and doped diamond surfaces determined by photoelectron spectroscopy, *Surface Science*, 1998, **418**(1), 219–239.
11. J. Robertson, Electron affinity of carbon systems, *Diamond and Related Materials*, 1996, **5**(6–8), 797–801.
12. W. F. Paxton, T. C. Wade, M. Howell, N. H. Tolk, J. L. Davidson and W. P. Kang, Thermionic emission characterization of boron-doped microcrystalline diamond films at elevated temperatures, *Physica Status Solidi A*, 2011, **209**(10), 1993–1995.
13. E. I. Givargizov, V. V. Zhirnov, A. N. Stepanova, E. V. Rakova, A. N. Kiselev and P. S. Plekhanov, Microstructure and field emission of diamond particles on silicon tips, *Applied Surface Science*, 1995, **24**, 87–88.
14. A. Wisitsora-at, W. P. Kang, J. L. Davidson, Q. Li, J. F. Xu and D. V. Kerns, Efficient electron emitter utilizing boron-doped diamond tips with sp2 content, *Applied Surface Science*, 1999, **146**(1–4), 280–286.
15. M. A. R. Alves, P. H. L. de Faria and E. S. Braga, Current-voltage characterization and temporal stability of the emission current of silicon tip arrays, *Microelectronic Engineering*, 2004, **75**(4), 383–388.
16. O. W. Richardson, The Electrical Conductivity Imparted to a Vacuum by Hot Conductors, *Phil. Trans. R. Soc. Lond. A*, 1903, **201**, 497–549.
17. A. Modinos, *Field, Thermionic, and Secondary Electron Emission Spectroscopy*. 1st edn, 1984, New York: Plenum Press, p. 375.

18. F. J. Himpsel, J. A. Knapp, J. A. VanVechten and D. E. Eastman, Quantum photoyield of diamond(111)& A stable negative-affinity emitter, *Physical Review B*, 1979, **20**(2), 624.
19. J. Van Der Weide, Z. Zhang, P. K. Baumann, M. G. Wensell, J. Bernholc and R. J. Nemanich, Negative-electron-affinity effects on the diamond (100) surface, *Physical Review B*, 1994, **50**(8), 5803.
20. L. Diederich, P. Aebi, O. M. Kuttel and L. Schlapbach, NEA peak of the differently terminated and oriented diamond surfaces, *Surface Science*, 1999, **424**(2-3), L314-L320.
21. P. K. Baumann and R. J. Nemanich, Surface cleaning, electronic states and electron affinity of diamond (100), (111), and (110) surfaces, *Surface Science*, 1998, **409**, 320-335.
22. P. K. Baumann and R. J. Nemanich, Negative electron affinity effects and Schottky barrier height measurements of cobalt on diamond (100) surfaces, *Proceedings of the Applied Diamond Conference*, 1995, **1**, 41.
23. C. Bandis, D. Haggerty and B. B. Pate, Electron emission properties of the negative electron affinity (111) 2×1 diamond TiO interface, *Proc. Material Research Society*, 1994, **339**, 75.
24. W. E. Pickett, Negative electron affinity and low work function surface cesium on oxygenated diamond (100), *Physical Review Letters*, 1994, **73**, 1664.
25. D. D. Koleske, S. M. Gates, B. D. Thoms, J. N. Russell and J. E. Butler, Hydrogen on polycrystalline diamond films: Studies of isothermal desorption and atomic deuterium abstraction, *J. Chem. Phys.*, 1995, **102**(2), 992-1002.
26. K. Bobrov, A. J. Mayne and G. Dujardin, Atomic-scale imaging of insulating diamond through resonant electron injection, *Nature*, 2001, **413**(6856), 616-620.
27. K Nassau., *The Physics and Chemistry of Color*, 1983, New York, NY: Wiley.
28. T. Saito, M. Kameta, K. Kusakabe, S. Morooka, H. Maeda, Y. Hayashi, T. Asano and A. Kawahara, Morphology and semiconducting properties of homoepitaxially grown phosphorus-doped (1 0 0) and (1 1 1) diamond films by microwave plasma-assisted chemical vapor deposition using triethylphosphine as a dopant source, *Journal of Crystal Growth*, 1998, **191**(4), 723-733.
29. K. Okano, S. Koizumi, S. R. P. Silva and G. A. J. Amaratunga, Low-threshold cold cathodes made of nitrogen-doped chemical-vapour-deposited diamond, *Nature*, 1996, **381**(6578), 140-141.
30. J. Robertson, J. Gerber, S. Sattel, M. Weiler, K. Jung and H. Ehrhardt, *Mechanism of Bias-enhanced Nucleation of Diamond on Si, Applied Physics Letters*, vol. 66, 1995, AIP, pp. 3287-3289.
31. J. F. Prins, n-type semiconducting diamond by means of oxygen-ion implantation, *Physical Review B*, 2000, **61**(11), 7191.
32. S. Koizumi, M. Kamo, Y. Sato, S. Mita, A. Sawabe, A. Reznik, C. Uzan-Saguy and R. Kalish, Growth and characterization of phosphorus doped n-type diamond thin films, *Diamond and Related Materials*, 1998, 7(2-5), 540-544.

33. S. Bhattacharyya, O. Auciello, J. Birrell, J. A. Carlisle, L. A. Curtiss, A. N. Goyette, D. M. Gruen, A. R. Krauss, J. Schlueter, A. Sumant and P. Zapol, Synthesis and characterization of highly-conducting nitrogen-doped ultrananocrystalline diamond films, *Applied Physics Letters*, 2001, 79(10), 1441–1443.
34. F. A. M. Koeck and R. J. Nemanich, Low temperature onset for thermionic emitters based on nitrogen incorporated UNCD films, *Diamond and Related Materials*, 2009, **18**(2-3), 232–234.
35. M. Suzuki, T. Ono, N. Sakuma and T. Sakai, Low-temperature thermionic emission from nitrogen-doped nanocrystalline diamond films on n-type Si grown by MPCVD, *Diamond and Related Materials*, 2009, **18**(10), 1274–1277.
36. G. V. Saparin, Microcharacterization of CVD diamond films by scanning electron microscopy: morphology structure and microdefects, *Diamond and Related Materials*, 1994, **3**, 1337.
37. W. S. Lee, Y.-J. Baik, K. Y. Eun and D. Y. Yoon, Metallographic etching of polycrystalline diamond films by reaction with metal, *Diamond and Related Materials*, 1995, **4**, 989.
38. J. C. Lin, K. H. Chen, H. C. Chang, C. S. Tsai, C. E. Lin and J. K. Wang, The vibrational dephasing and relaxation of CH and CD stretches on diamond surfaces: an anomaly, *J. Chem. Phys.*, 1996, **105**(10), 3975.
39. Z. H. Huang, P. H. Cutler, N. M. Miskovsky and T. E. Sullivan, Theoretical study of field emission from diamond, *Applied Physics Letters*, 1994, **65**(20), 2563.
40. Z. H. Huang, P. H. Cutler, N. M. Miskovsky and T. E. Sullivan, A Band-to-band Tunneling Injection Mechanism For Charge Carriers in Composite Wide Bandgap Field Emission Sources, *J. Vac. Sci. Tech. B*, 1995, **15**, 337.
41. J. Reistein, W. Stein and L. Ley, Defect spectroscopy and determination of the electron diffusion length in single crystal diamond by total photoelectron yield spectroscopy, *Physical Review Letters*, 1997, **78**(9), 1803.
42. H. Yamaguchi, I. Saito, Y. Kudo, T. Masuzawa, T. Yamada, M. Kudo, Y. Takakuwa and K. Okano, Electron emission mechanism of hydrogenated natural type IIb diamond (111), *Diamond and Related Materials*, 2008, **17**(2), 162–166.
43. W. Zhu, G. P. Kochanski, S. Jin and L. Seibles, Defect-enhanced electron field emission from chemical vapor deposited diamond, *Journal of Applied Physics*, 1995, **78**(4), 2707–2711.
44. K. H. Park, S. Lee, K. Song, J. I. Park, K. J. Park, S. Han, s. J. Na, N. Lee and K. H. Koh, Field emission characteristics of defective diamond films, *Journal of Vacuum Science and Technology B*, 1998, **16**(2), 724.
45. C. Wang, A. Garcia, D. C. Ingram, M. Lake and M. E. Kordesch, Cold field emission from CVD diamond films observed in emission electron microscopy, *Electronics Letters*, 1991, **27**(16), 1459–1461.
46. N. S. Xu, Y. Tzeng and R. V. Latham, Similarities in the cold electron-emission characteristics of diamond coated molybdenum electrodes

and polished bulk graphite surfaces, *Journal of Physics D-Applied Physics*, 1993, **26**(10), 1776–1780.
47. V. L. Humphreys and J. Khachan, Spatial correlation of electron field emission sites with non-diamond carbon content in CVD diamond, *Electron Device Letters, IEEE*, 1995, **31**(12), 1018.
48. T. Habermann, Modifying chemical vapor deposited films for field emission displays, *Journal of Vacuum Science & Technology B*, 1998, **16**(2), 693.
49. K. C. Walter, H. H. Kung and C. J. Maggiore, Improved field emission of electrons from ion irradiated carbon, *Applied Physics Letters*, 1997, **71**(10), 1310.

Subject Index

abrasion techniques, 233–234
absorption coefficient, 351–352
activation energy, 309–312
adhesion transfer seeding
 technique, 240–242
adsorption properties, diamond
 films, 378–380
AFM. *See* atomic force microscopy
agglomeration, 51–53
alkylation, surface modification
 linkers, 68–69
all-diamond devices
 AFM-SECM tip
  applications, 491
  fabrication, 482–484
  properties, 488–491
 microelectrode arrays
  applications, 491
  fabrication, 478–479
  properties, 484–487
 nanoelectrode arrays
  applications, 491
  fabrication, 480–482
  properties, 487–488
 nanoelectrode ensembles
  applications, 491
  fabrication, 480–482
  properties, 487–488
 ultramicroelectrode
  arrays
   applications, 491
   fabrication, 478–480
   properties, 484–487
amides, surface modification
 linkers, 71–72

amorphous silicon
 nucleation and growth, 317–320
 selective growth, diamond
  nanocrystals, 320–322
annealing, 63–65
Arrhenius equation, 310
Arrhenius factor, 310
artificial olfaction, diamond-based
 SAW sensors, 470–472
arylation
 diazonium salts, 73
 Diels–Alder reactions, 74–75
atomic force microscopy (AFM),
 199–201
atomic force microscopy-scanning
 electrochemical microscopy
 (AFM–SECM) tip
  applications, 491
  fabrication, 482–484
  properties, 488–491
attenuated total reflectance-Fourier
 transmittance infrared
 (ATR–FTIR) spectroscopy
  definition, 328
  silicon prism, diamond
   growth, 331–333

ballas
 definition, 254
 diamond growth mechanism,
  260–261
BDDP. *See* B-doped diamond
 nanopowders
B-doped diamond nanopowders
 (BDDP), 363

Subject Index

BEN. *See* bias enhanced nucleation
bias enhanced nucleation (BEN)
    heteroepitaxy, 232–233
    induced mechanisms, 230–232
    principle, 229
    relevant parameters, 229–230
    vs. seeding technique, 245
Bingel–Hirsch reaction, 75–76
biochemical sensor, 448
bio-imaging applications, fluorescent nanodiamonds, 157–158
biomolecules, surface functionalizations/conjugations, 171–173
biophysical interaction
    carbon-based nanomaterials
        interactions with bacteria, 178–184
        interactions with protozoa, 184–187
    characterization studies, 173–177
    surface functionalizations/conjugations, 171–173
biosensors, electrochemical behaviour, 143–144
boron-doped diamond, 378
    nanomechanical systems, 402–406
    polycrystalline, 387–388
    superconductivity
        non-shunted devices, 395–398
        shunted devices, 398–402
        transition temperature, 389
        voltage–current characteristics, 390–393
boron-doped nanodiamond, 146–147
bulk acoustic wave (BAW) resonators, 418

cantilevers
    fabrication methods, 461
    resonant chemical sensors, 461–465
    transduction principle, 457–460
capacitive technique, 425–427
carbonado, 254
carbon-based nanomaterials
    interactions with bacteria, 178–184
    interactions with protozoa, 184–187
carbon surface termination, 371
carboxylated nanodiamond, 59–63
carrier transport
    electron field emission
        defect states, 510–511
        n-type doping, 508–510
        p-type doping, 510
    NCD films, 348–350
chemical reactivity, 294–296
chemical sensors
    artificial olfaction, 470–472
    definition, 448
    diamond-based resonant cantilever, 461–465
    diamond nanoparticles-coated, 467–470
    high performance
        chemical properties, 453–456
        physical properties, 450–453
chlorine surface termination, 374
chronocoulometry technique, 378
circuit loading, 435–437
clamping losses, 433–434
coefficient of thermal expansion (CTE), 293–294
composites, electrochemical behaviour, 146
conductivity, NCD films, 346–348
contact measurement methods, 307–308
controlled light hydro-dynamic pulse, 123–124
controlled nanodiamond synthesis, 123–124

correlation coefficient analysis, 214
covalent grafting, 171
(100) crystal facet growth, 260
(111) crystal facet growth, 260
CTE. *See* coefficient of thermal expansion
CVD deposition systems, low temperature diamond growth
    DC plasma-assisted CVD, 300
    electron cyclotron resonance plasma, 300
    magnetic field-assisted HF plasma, 300
    magnetic field-assisted MW plasma, 300
    other techniques, 301
    process pressure, 299
    pulsed MW plasma, 299
    surface wave MW plasma, 300–301
cyclic voltammetry, 357

DC plasma-assisted CVD, 300
deagglomeration, 51–53
decoherence, 177
defect-free diamond nanoparticles, 3–7
defect states, 510–511
detection technique, nanodiamond-based, 159–160
detonation carbon
    isolation of nanodiamonds, 30–33
    synthesis of, 28–30
detonation nanodiamonds (DNDs)
    applications
        biological applications, 40–41
        composites, 36–39
        heat-transfer media, 39
        intrinsic catalytic activity, 39
        liquid chromatography, 40
        thermoelectric energy converters, 41

detonation carbon
    isolation of nanodiamonds, 30–33
    synthesis of, 28–30
    structure of soot particles, 33–35
    surface chemistry, 363
    suspensions, 35–36
    unsolved problems
        disaggregation mechanism, 42–43
        electrostatic interaction model, 42
detonation soot. *See* detonation carbon
detonation synthesis *vs.* light hydrodynamic pulse synthesis, 124
diagnostic studies, bacterial detection, 159–160
diamond
    electrochemistry, 129
    superconductivity
        non-shunted devices, 395–398
        shunted devices, 398–402
        transition temperature, 389
        voltage–current characteristics, 390–393
    surface-wave plasma, 270–273
    work function, 501
diamond-based resonators
    cantilevers
        fabrication methods, 461
        resonant chemical sensors, 461–465
        transduction principle, 457–460
    high performance chemical sensors
        chemical properties, 453–456
        physical properties, 450–453

Subject Index

surface acoustic wave
  resonators
    diamond nanoparticles-
      coated chemical
      sensors, 467–470
    generality, 465–467
diamond-based SAW sensors,
  artificial olfaction, 470–472
diamond films
  adsorption properties, 378–380
  electrochemistry, 358–362
diamond morphology
  correlation of growth
    conditions
      gas activation, 261–262
      gas-phase composition,
        262
      impurities, 262–263
  growth mechanism
    ballas diamond growth,
      260–261
    (100) crystal facet growth,
      260
    (111) crystal facet growth,
      260
    graphitic carbon
      morphology, 261
    nanocrystalline diamond,
      261
    single crystal growth, 259
    ultra-nanocrystalline
      diamond, 261
  Raman spectroscopy, 255–258
diamond nanoparticles
  defect-free, 3–7
  electrochemical behaviour
    applications, 143–146
    boron-doped
      nanodiamond, 146–147
    electrode preparation
      methods, 130–131
    high temperature high
      pressure type 1b
      diamond, 147–148
    impedance spectroscopy,
      142–143

    100 nm diamond
      particles, 147
    in situ spectroscopy
      studies, 141–142
    solution pH effects,
      134–135
    solution redox species,
      136–141
    surface modification,
      131–134
  modeling defects
    kinetic stability, 9–10
    mechanical stability, 8–9
    probability of
      observation, 9–10
    thermodynamic stability,
      8–9
  point defects
    comparative and
      concentration studies,
      20–22
    H3 centres, 17–18
    incidental impurities,
      13–14
    intrinsic defects, 10–12
    N3 centres, 18–19
    photoactive
      nitrogen–vacancy (N–V)
      centres, 14–17
    V–N–V defect, 19–20
diamond nanoparticles-coated
  chemical sensors, 467–470
diamond NEMS
  devices and capabilities
    field emission, 442–443
    nitrogen vacancy defects,
      444
    superconductivity, 443
  dissipation
    circuit loading, 435–437
    clamping losses, 433–434
    intrinsic, 436
    mechanical defects, 437
    multiple materials,
      435–437
    quantum, 439–441

diamond NEMS (*continued*)
    thermoelastic, 434–435
    viscous damping, 438–439
diamond nucleation
    heterogeneous
        experimental observations, 224–225
        models and theory, 223–224
        structural defects, 226–227
        surface mechanisms, 225–226
    homogeneous, 223
diamond thin films synthesis
    chemical vapor deposition
        chemical reactivity, 294–296
        coefficient of thermal expansion, 293–294
        diffusion coefficient, 294–296
        low temperature melting, 293
        softening substrates, 293
    nucleation/seeding process, 292
diazonium salts, arylation, 73
Diels–Alder reactions, 74–75
direct surface termination
    annealing, 63–65
    carboxylated nanodiamond, 59–63
    hydrogenated nanodiamond, 55–57
    hydroxylated nanodiamond, 58–59
    other terminations, 65–67
dissipation
    circuit loading, 435–437
    clamping losses, 433–434
    intrinsic, 436
    mechanical defects, 437
    multiple materials, 435–437
    quantum
        NEMS resonators, 441
        standard tunneling model, 439–441
    thermoelastic, 434–435
    viscous damping, 438–439
DNDs. *See* detonation nanodiamonds
Dox. *See* doxorubicin
doxorubicin (Dox), 152–153
drug resistance, 152
dynamic friction behavior, NCD films, 283–284

EELS. *See* electron energy-loss spectroscopy
elasticity theory, 414–415
electrochemical behaviour
    applications
        biosensors, 143–144
        composites, 146
        electrochemical sensors, 143
        electrode coatings, 146
        fuel cell catalyst electrode, 144–145
        photocatalysis, 146
        supercapacitor electrode material, 145–146
    boron-doped nanodiamond, 146–147
    diamond films, 358–362
    electrode preparation methods, 130–131
    high temperature high pressure type 1b diamond, 147–148
    impedance spectroscopy, 142–143
    nanodiamond powder films, 362–369
    100 nm diamond particles, 147
    in situ spectroscopy studies, 141–142
    solution pH effects, 134–135
    solution redox species, 136–141
    surface modification, 131–134
electrochemical grafting, 376–377

Subject Index

electrochemical measurements, 355–358
electrochemical network, nanodiamond monolayer coatings, 205
electrochemical sensors, 143
electrode
    coatings, 146
    definition, 354
electrode-immobilised nanodiamond particles
    electrode preparation methods, 130–131
    solution pH effects, 134–135
    surface modification, 131–134
electrolyte, definition, 354
electron affinity, 504–507
electron cyclotron resonance (ECR) plasma, 300
electron emission, 499
electron energy-loss spectroscopy (EELS), 120–121
electron field emission
    carrier transport
        defect states, 510–511
        n-type doping, 508–510
        p-type doping, 510
    characterization, 499
    electron affinity, 504–507
    Fowler–Nordheim theory, 500–502
    geometric enhancement, 502–504
    potential barrier, 500
    surface states, 504–507
    temperature, 504
    topology, 502–504
ellipsometry, 309
emitting area, 501
enhanced diamond nucleation
    bias enhanced nucleation
        heteroepitaxy, 232–233
        induced mechanisms, 230–232
        principle, 229

        relevant parameters, 229–230
    techniques, 228–229
esters, surface modification linker, 71–72
Euler–Bernoulli equation
    standard, 419
    static, 414–415
extracellular matrix (ECM) proteins, 196

FE-MISPC. *See* field-enhanced, metal-induced solid-phase crystallization
field emission, 442–443
field-enhanced, metal-induced solid-phase crystallization (FE-MISPC), 320–321
field enhancement factor, 501–502
fluorescent nanodiamonds (FNDs), 157–158
fluorine surface termination, 373
FNDs. *See* fluorescent nanodiamonds
focused MW plasma, 323–326
Fowler–Nordheim field emission theory, 500–502
fuel cell catalyst electrode
    electrochemical behaviour, 144–145
functional defects, 1–2
functionalised diamond electrodes
    electrochemical grafting, 376–377
    photochemical functionalisation, 375–376
    surface termination types
        carbon, 371
        chlorine, 374
        fluorine, 373
        hydrogen, 371–373
functional neuronal networks
    formation, 201–202
    patterned nanodiamond coatings, 214–216

gas activation, diamond growth conditions, 261-262
gas chemistry, low temperature diamond growth
   adding halogens, 302-304
   adding oxygen, 304-306
   other gas mixtures, 306-307
gas-phase composition, diamond growth conditions, 262
gas-phase purification, 99-100
gene delivery, nanodiamond-based, 161-163
general radiation (GR1) defect, 2
Ge substrates, diamond growth conditions, 333-335
glass
   focused MW plasma, 323-326
   material properties, 323
   pulsed linear antenna microwave plasma, 326-327
glial scarring, 195
gliosis, 195
grafting of complex moieties, pre-functionalized nanodiamond
   click chemistry, 78-80
   miscellaneous reactions, 80-81
   peptide coupling, 77-78
graphitic carbon morphology, 261
grazing angle reflectance-Fourier transmittance infrared (GAR-FTIR) spectroscopy)
   definition, 328
   metallic optical mirrors, diamond growth, 329-331

heteroepitaxial diamond nucleation, 232-233
heterogeneous diamond nucleation
   experimental observations, 224-225
   models and theory, 223-224
   structural defects, 226-227
   surface mechanisms, 225-226
heterogeneous nucleation, 223

high performance chemical sensors
   chemical properties, 453-456
   physical properties, 450-453
high pressure, high temperature nanodiamonds, 156-157
high temperature high pressure (HTHP) type 1b diamond, 147-148
homogeneous diamond nucleation, 223
homogeneous nucleation, 222
hot filament chemical vapor deposition (HFCVD), 297-298
H3 point defects, 17-18
hydrogenated nanodiamond, 55-57
hydrogen surface termination, 371-373
hydroxylated nanodiamond, 58-59

ICP-MS. *See* inductively coupled plasma mass spectrometry
imaging studies
   bio-imaging applications, fluorescent nanodiamonds, 157-158
   magnetic resonance imaging, 158
   safety and biocompatibility studies
      pre-clinical evaluation, 159
      in vitro validation, 159-160
   impedance spectroscopy, electrochemical behaviour, 142-143
implantable devices, nanodiamond-based
   nanodiamond-chemotherapeutic tablets, 155-156
   nanodiamond-PLLA complexes, tissue engineering, 156

impurities
    diamond growth conditions, 262–263
    incidental, 13–14
induced thermal strain, 430
inductively coupled plasma mass spectrometry (ICP-MS), 122
industrial detonation synthesis, 29
infrared (IR) pyrometers, 308–309
in situ spectroscopy studies, electrochemical behaviour, 141–142
interferometric thermometry, 309
intrinsic defects, 10–12
intrinsic dissipation, 436
intrinsic electric excitability, nanodiamond-grown neurons, 204–205
in vitro validation, nanodiamond safety, 159–160
ion flux, 229–230

kinetic stability, diamond nanoparticles, 9–10

laser nanodiamond synthesis
    advanced applications, 125
    characterization
        inductively coupled plasma mass spectrometry, 122
        Raman spectroscopy, 122
        scanning electron microscopy, 119–120
        transmission electron microscopy, 120–121
        X-ray diffraction analysis, 117–119
    controlled light hydro-dynamic pulse, 123–124
    light hydro-dynamic pulse
        physical mechanism, 117
        technological process, 116–117
    pulsed laser ablation in liquid, 114–115

LHDP. *See* light hydro-dynamic pulse
light hydro-dynamic pulse (LHDP)
    controlled, 123–124
    vs. detonation synthesis, 124
    physical mechanism, 117
    technological process, 116–117
liquid chromatography, 40
liquid phase pulse laser ablation (LP-PLA). *See* pulsed laser ablation in liquid (PLAL)
liquid-phase purification, 97–99
low-pressure diamond
    growth mechanism, diamond morphology
        ballas diamond growth, 260–261
        (100) crystal facet growth, 260
        (111) crystal facet growth, 260
        graphitic carbon morphology, 261
        nanocrystalline diamond, 261
        single crystal growth, 259
        ultra-nanocrystalline diamond, 261
    Raman spectroscopy, diamond morphology, 255–258
low temperature diamond growth (LTDG)
    amorphous silicon
        nucleation and growth, 317–320
        selective growth, diamond nanocrystals, 320–322
    CVD deposition systems
        DC plasma-assisted CVD, 300
        electron cyclotron resonance plasma, 300
        magnetic field-assisted HF plasma, 300

low temperature diamond growth
(LTDG) (*continued*)
    magnetic field-assisted
        MW plasma, 300
    other techniques, 301
    process pressure, 299
    pulsed MW plasma, 299
    surface wave MW plasma,
        300–301
  gas chemistry
    adding halogens,
        302–304
    adding oxygen, 304–306
    other gas mixtures,
        306–307
  glass
    focused MW plasma,
        323–326
    material properties, 323
    pulsed linear antenna
        microwave plasma,
        326–327
  hot filament chemical vapor
    deposition, 297–298
  infrared spectroscopy
    ATR-FTIR applications,
        331–333
    GAR-FTIR substrates,
        329–331
    Ge substrates, 333–335
  microwave plasma chemical
    vapor deposition, 298
  pulsed linear antenna
    microwave plasma, 312–316
  substrate temperature
    ellipsometry, 309
    infrared pyrometers,
        308–309
    interferometric
        thermometry, 309
    thermocouple, 307–308
low-temperature growth, NCD films
  advantages, 268–269
  definition, 268
  growth mechanism, 277–282
  plasma conditions, 269–270
  surface-wave plasma, 273–276
  tribological properties, 282–287
LP-PLA. *See* liquid phase pulse laser
  ablation
LTDG. *See* low temperature diamond
  growth

magnetic field-assisted HF plasma,
  300
magnetic field-assisted MW plasma,
  300
magnetic resonance imaging, 158
magnetomotive technique, 428–429
mass sensitivity factor, 467
MEAs. *See* microelectrode arrays
measurement methods
  contact, 307–308
  non-contact
    ellipsometry, 309
    infrared pyrometers,
        308–309
    interferometric
        thermometry, 309
mechanical defects, 437
mechanical stability, 8–9
mEPSCs. *See* miniature excitatory
  postsynaptic currents
microelectrode arrays (MEAs)
  applications, 491
  fabrication, 478–479
  properties, 484–487
micron-scale superconducting
  quantum interference device
  (μ-SQUID)
    non-shunted devices, 395–398
    shunted devices, 398–402
microwave plasma chemical vapor
  deposition (MWCVD), 298
miniature excitatory postsynaptic
  currents (mEPSCs), 204–205
modeling and simulation
  techniques, 161–163
modeling defects, diamond
  nanoparticles
    kinetic stability, 9–10
    mechanical stability, 8–9

Subject Index

probability of observation, 9–10
thermodynamic stability, 8–9
molecular recognition strategy, 371
multilayer wafer fabrication, 422–423
multistep growth fabrication, 423–424
myelosuppression, 152–153

nanocrystalline diamond (NCD) films
    advantages, 268–269
    carrier transport, 348–350
    conductivity, 346–348
    definition, 268
    growth mechanism, 277–282
    optical properties, 351–352
    plasma conditions, 269–270
    structure and sp^2 content, 344–346
    surface-wave plasma, 273–276
    tribological properties, 282–287
    growth mechanism, 261
nanodiamond–chemotherapeutic tablets, 155–156
nanodiamond coatings
    characterisation, 208–210
    neuronal adhesion mechanism, 213–214
    neuronal cell attachment, 211–213
    other growth platforms, 207
    patterned, 214–216
    simplicity and universality, 206
    surface properties, 213–214
nanodiamond–doxorubicin complexes, 152–153
nanodiamond-grown neurons, intrinsic electric excitability, 204–205

nanodiamond layering
    neuronal cell attachment, 198–199
    neurons direct attachment, 207
    surface roughness, 199–201
nanodiamond monolayer coatings
    electrochemical network, 205
    functional neuronal networks formation, 201–202
    patterned nanodiamond coatings, 214–216
    neuronal attachment, 201
    synaptic connectivity, 204–205
nanodiamond particles
    direct surface termination
        annealing, 63–65
        carboxylated nanodiamond, 59–63
        hydrogenated nanodiamond, 55–57
        hydroxylated nanodiamond, 58–59
        other terminations, 65–67
    linkers, surface modification
        alkylation, 68–69
        amides, 71–72
        arylation, diazonium salts, 73
        arylation, Diels–Alder reactions, 74–75
        Bingel–Hirsch reaction, 75–76
        esters, 71–72
        Prato reaction, 75
        silanization, 69–71
    surface structure
        agglomeration, 51–53
        deagglomeration, 51–53
        initial surface stucture, 49–51
nanodiamond–PLLA complexes, tissue engineering, 156
nanodiamond powder films, 362–369

nanodiamond powders
   applications, 113–114
   definition, 112–113
   problems and issues, 113
   purification
      assessment, 92–94
      oxidation behavior, 94–96
      structure and composition, 90–92
      thermal stability, 94–96
nanodiamond surface, electronic structure, 135–136
nanoelectrode arrays (NEAs)
   applications, 491
   fabrication, 480–482
   properties, 487–488
nanoelectrode ensembles (NEEs)
   applications, 491
   fabrication, 480–482
   properties, 487–488
nano-electromechanical systems (NEMS)
   defining element, 412
   devices and capabilities in diamond
      field emission, 442–443
      nitrogen vacancy defects, 444
      superconductivity, 443
   dissipation in diamond
      circuit loading, 435–437
      clamping losses, 433–434
      intrinsic, 436
      mechanical defects, 437
      multiple materials, 435–437
      quantum, 439–441
      thermoelastic, 434–435
      viscous damping, 438–439
   drive and detection
      capacitive technique, 425–427
      magnetomotive technique, 428–429
      other methods, 430–432
      piezoelectric technique, 427–428
      piezoresistive technique, 429–430
   dynamic solutions, 415–418
   elasticity theory, 414–415
   Euler-Bernoulli equation, 414–415, 419
   fabrication
      further comments, 424–425
      multilayer wafer, 422–423
      multistep growth, 423–424
   material properties, 412–414
   nonlinear dissipation, 419–422
   resonators, quantum dissipation, 441
   simple harmonic oscillator, 419–422
nanofabrication, polycrystalline boron-doped diamond, 387–388
nanomechanical systems, boron-doped diamond, 402–406
NEA. See negative electron affinity
NEAs. See nanoelectrode arrays
NEEs. See nanoelectrode ensembles
negative electron affinity (NEA), 504–507
NEMS. See nano-electromechanical systems
neuronal adhesion mechanism, 213–214
neuronal attachment, 201
neuronal biomaterials, 195–197
neuronal cell attachment, 211–213
nitrogen vacancy defects, 444
100 nm diamond particles, 147
non-contact measurement methods
   ellipsometry, 309
   infrared pyrometers, 308–309
   interferometric thermometry, 309
nonlinear dissipation, 419–422
nonlinear spring constant, 421
non-shunted devices, 395–398

Subject Index

N3 point defects, 18–19
n-type doping, 508–510
nucleation
    amorphous silicon, 317–320
    definition, 222
    heterogeneous, 223
    homogeneous, 222
nucleus, definition, 222

optically clear diamond, 254
optical properties, NCD films, 351–352
oxidation behavior
    metal impurities, 100–102
    nanodiamond powders, 94–96
oxygen-containing gas chemistry, 304–306

patterned nanodiamond coatings, 214–216
patterned seeding technique, 242–244
PCD. See polycrystalline diamond
photoactive nitrogen–vacancy (N-V) centres, 14–17
photocatalysis, electrochemical behaviour, 146
photochemical functionalisation, 375–376
photoluminescence, 174–175
piezoelectric technique, 427–428
piezoresistive technique, 429–430
PLAL. See pulsed laser ablation in liquid
PLAMWP. See pulsed linear antenna microwave plasma
plastics, 293
PLIIR. See pulsed-laser-induced liquid–solid interfacial reaction
PLLA. See poly(L-lactic acid)
point defects, diamond nanoparticles
    comparative and concentration studies, 20–22
    H3 centres, 17–18
    incidental impurities, 13–14
    intrinsic defects, 10–12
    N3 centres, 18–19
    photoactive nitrogen–vacancy (N-V) centres, 14–17
    V–N–V defect, 19–20
poly(L-lactic acid) (PLLA), 156
polycrystalline boron-doped diamond, 387–388
polycrystalline diamond (PCD), 254–255
polyvinyl alcohol (PVA), 238, 240
Prato reaction, 75
pre-functionalized nanodiamond
    click chemistry, 78–80
    miscellaneous reactions, 80–81
    peptide coupling, 77–78
probability of observation, 9–10
p-type doping, 510
pulsed laser ablation in liquid (PLAL), 114–115
pulsed-laser-induced liquid–solid interfacial reaction (PLIIR). See pulsed laser ablation in liquid (PLAL)
pulsed linear antenna microwave plasma (PLAMWP)
    glass, 326–327
    low temperature diamond growth, 312–316
pulsed MW plasma, 299
purified nanodiamond
    assessment, 92–94
    electrical properties, 105–107
    gas-phase purification, 99–100
    liquid-phase purification, 97–99
    metal impurities, oxidation behavior, 100–102
    optical properties, 105–107
    oxidation behavior, 94–96
    structure and composition, 90–92, 102–105
    surface chemistry, 102–105
    thermal stability, 94–96
PVA. See polyvinyl alcohol

quantum dissipation
   NEMS resonators, 441
   standard tunneling model, 439–441
quantum yield, 2

Raman spectroscopy
   laser synthesis
      characterization, 122
      low-pressure diamond, 255–258
Rayleigh surface acoustic wave (R-SAW) transducers, 466
reaction path, 309
reactive oxygen species (ROS), 179
relative defect energy, 11
re-nucleation, 345
resonant cantilever chemical sensors, 461–465
ROS. *See* reactive oxygen species

safety and biocompatibility studies
   pre-clinical evaluation, 159
   in vitro validation, 159–160
scanning electrochemical microscopy (SECM), 139
scanning electron microscopy (SEM), 119–120
SECM. *See* scanning electrochemical microscopy
Seebeck effect, 307
seeding technique
   abrasion techniques, 233–234
   adhesion transfer, 240–242
   vs. bias enhanced nucleation, 245
   chemical/thermal stability, 237
   patterned, 242–244
   suspensions, nanodiamonds
      evaporation, 237–240
      solvent, 236–237
      surface charge, 236
      synthesis, 235–236
SEM, scanning electron microscopy
shunted devices, 398–402
silanization, 69–71
simple harmonic oscillator, 419–422

softening substrates, 293
SQUID. *See* superconducting quantum interference device
μ-SQUID. *See* micron-scale superconducting quantum interference device
standard Euler–Bernoulli equation, 419–422
static Euler–Bernoulli equation, 414–415
substrates
   CVD diamond films
      chemical reactivity, 294–296
      coefficient of thermal expansion, 293–294
      diffusion coefficient, 294–296
      low temperature melting, 293
      softening substrates, 293
   infrared spectroscopy
      GAR-FTIR, 329–331
      germanium, 333–335
substrate temperature
   ellipsometry, 309
   infrared pyrometers, 308–309
   interferometric thermometry, 309
   thermocouple, 307–308
supercapacitor electrode material, 145–146
superconducting diamond
   non-shunted devices, 395–398
   shunted devices, 398–402
   transition temperature, 389
   voltage–current characteristics, 390–393
superconducting quantum interference device (SQUID)
   non-shunted devices, 395–398
   shunted devices, 398–402
superconductivity
   diamond
      non-shunted devices, 395–398

Subject Index

shunted devices, 398–402
transition temperature, 389
voltage–current characteristics, 390–393
diamond NEMS, 443
surface acoustic wave (SAW) resonators
  diamond nanoparticles-coated chemical sensors, 467–470
  generality, 465–467
surface bombardment, 230
surface chemistry, nanodiamond, 129–130
surface functionalizations/conjugations, biomolecules, 171–173
surface modification linkers
  alkylation, 68–69
  amides, 71–72
  arylation
    diazonium salts, 73
    Diels–Alder reactions, 74–75
  Bingel–Hirsch reaction, 75–76
  esters, 71–72
  Prato reaction, 75
  silanization, 69–71
surface roughness, 231–232
  nanodiamond layering, 199–201
  NCD films, 284–286
surface termination, diamond electrodes
  carbon, 371
  chlorine, 374
  fluorine, 373
  hydrogen, 371–373
surface wave MW plasma, 300–301
surface-wave plasma
  diamond, 270–273
  features, 270–271
  NCD films, 273–276
suspensions
  detonation nanodiamonds, 35–36

seeding technique
  evaporation, 237–240
  solvent, 236–237
  surface charge, 236
  synthesis, 235–236
systemic drug delivery
  nanodiamond–doxorubicin complexes, 152–153
  targeted nanodiamond drug delivery, 153–154

targeted nanodiamond drug delivery, 153–154
TEM, transmission electron microscopy
thermally assisted field emission, 499
thermal stability, nanodiamond powders, 94–96
thermionic emission, 499
thermodynamic stability, diamond nanoparticles, 8–9
thermoelastic dissipation, 434–435
tissue engineering, nanodiamond–PLLA complexes, 156
total probability of escape, 10
transducer, 448
transduction principle, 457–460
transmission electron microscopy (TEM), 120–121
tribological properties, NCD films, 282–287

ultra-low temperatures, quantum dissipation
  NEMS resonators, 441
  standard tunneling model, 439–441
ultramicroelectrode arrays (UMEAs)
  applications, 491
  fabrication, 478–480
  properties, 484–487
ultra-nanocrystalline diamond (UNCD)

UMEAs. *See* ultramicroelectrode arrays
unicellular microorganisms, carbon-based nanomaterials
    interactions with bacteria, 178–184
    interactions with protozoa, 184–187

viscous damping, 438–439
V–N–V point defects, 19–20

work function, 501

X-ray diffraction (XRD) analysis, 117–119